Further Reading

Hashmi, A. S. K., Toste, D. F. (Eds.)

Modern Gold Catalyzed Synthesis

2009

ISBN: 978-3-527-31952-7

Mohr, F. (Ed.)

Gold Chemistry

Highlights and Future Directions

2009

ISBN: 978-3-527-32086-8

Oro, L. A., Claver, C. (Eds.)

Iridium Complexes in Organic Synthesis

2009

ISBN: 978-3-527-31996-1

Van Leeuwen, P. W. N. M. (Ed.)

Supramolecular Catalysis

2008

ISBN: 978-3-527-32191-9

Diederich, F., Stang, P. J., Tykwinski, R. R, (Eds.)

Modern Supramolecular Chemistry

Strategies for Macrocycle Synthesis

2008

ISBN: 978-3-527-31826-1

Modern Supramolecular Gold Chemistry

Gold-Metal Interactions and Applications

Edited by
Antonio Laguna

WILEY-VCH Verlag GmbH & Co. KGaA

The Editor

Prof. Dr. Antonio Laguna
Universidad de Zaragoza
Departamento de Quimica Inorganica
50009 Zaragoza
Spain

All books published by **Wiley-VCH** are carefully produced. Nevertheless, authors, editors, and publisher do not warrant the information contained in these books, including this book, to be free of errors. Readers are advised to keep in mind that statements, data, illustrations, procedural details or other items may inadvertently be inaccurate.

Library of Congress Card No.: applied for

British Library Cataloguing-in-Publication Data
A catalogue record for this book is available from the British Library.

Bibliographic information published by the Deutsche Nationalbibliothek
The Deutsche Nationalbibliothek lists this publication in the Deutsche Nationalbibliografie; detailed bibliographic data are available on the Internet at http://dnb.d-nb.de.

© 2008 WILEY-VCH Verlag GmbH & Co. KGaA, Weinheim

All rights reserved (including those of translation into other languages). No part of this book may be reproduced in any form – by photoprinting, microfilm, or any other means – nor transmitted or translated into a machine language without written permission from the publishers. Registered names, trademarks, etc. used in this book, even when not specifically marked as such, are not to be considered unprotected by law.

Typesetting Thomson Digital, Noida, India
Printing Strauss Gmbh, Mörlenbach
Binding Litges & Dopf GmbH, Heppenheim

Printed in the Federal Republic of Germany
Printed on acid-free paper

ISBN: 978-3-527-32029-5

Contents

Preface *XI*
Abbreviation *XV*
List of Contributors *XIX*

1 **The Chemistry of Gold** *1*
M. Concepción Gimeno
1.1 Introduction *1*
1.1.1 History *1*
1.1.2 Chemical and Physical Properties *2*
1.1.3 Theoretical Considerations *2*
1.2 Chemistry *4*
1.2.1 Gold(-I) Compounds *5*
1.2.2 Gold(0) and Gold Clusters *6*
1.2.3 Gold Nanoparticles and Nanoclusters *8*
1.2.4 Gold(I) Complexes *10*
1.2.4.1 Mononuclear [AuXL], [AuL$_2$]$^+$ or [AuX$_2$]$^-$ Complexes *10*
1.2.4.2 Gold(I) Complexes with Polydentate Ligands *14*
1.2.4.3 Three and Four-Coordinate Gold(I) Complexes *20*
1.2.4.4 Oligomeric Gold(I) Complexes *23*
1.2.4.5 Single Heteroatom Bridged Gold(I) Complexes *26*
1.2.4.6 Organometallic Gold(I) Complexes *30*
1.2.5 Gold(II) Complexes *35*
1.2.5.1 Mononuclear Gold(II) Complexes *35*
1.2.5.2 Polynuclear Gold(II) Complexes *38*
1.2.6 Gold(III) Complexes *41*
1.2.6.1 Organometallic Gold(III) Complexes *41*
1.2.6.2 Gold(III) Complexes with Polydentate Ligands *44*
1.2.6.3 Gold(III) Complexes with Chalcogen Ligands *46*
1.2.7 Gold in Higher Oxidation States *49*
References *50*

Modern Supramolecular Gold Chemistry: Gold-Metal Interactions and Applications.
Edited by Antonio Laguna
Copyright © 2008 WILEY-VCH Verlag GmbH & Co. KGaA, Weinheim
ISBN: 978-3-527-32029-5

2	**Gold–Gold Interactions** 65
	Olga Crespo
2.1	Introduction 65
2.2	Gold(I)–Gold(I) Interactions 66
2.2.1	Aggregation of Mononuclear Dicoordinated Gold Units 67
2.2.1.1	Dinuclear Aggregates 68
2.2.1.2	Tri- and Tetranuclear Associations 73
2.2.1.3	Formation of Infinite Chains and Supramolecular Arrays 76
2.2.2	Clustering of Gold Atoms at One Center 78
2.2.3	Complexes with Bridging Monodentate Ligands 86
2.2.4	Complexes with Bridging Bidentate Ligands 90
2.2.5	Complexes with Polydentate Ligands 110
2.3	Gold(I)-Gold(III) Interactions 113
2.4	Concluding Remarks 117
	References 117
3	**Gold Nanomaterials** 131
	Eduardo J. Fernández and Miguel Monge
3.1	Introduction 131
3.2	Molecular Gold Clusters 131
3.2.1	Synthesis and Structural Characterization of Phosphine-Stabilized Gold Clusters 132
3.2.2	Gold Clusters with Other Ligand Stabilizers 136
3.3	Large-Size Gold Clusters. The Au_{55} Case 139
3.4	Gold Nanoparticles 144
3.4.1	Synthesis of Gold Nanoparticles 144
3.4.1.1	Monolayer Protected Clusters 144
3.4.1.2	Other Ligands 146
3.4.1.3	Polymers 149
3.4.1.4	Dendrimers 157
3.4.1.5	Seeded Growth 160
3.4.1.6	Nanoparticle–Biomolecule Hybrids 163
3.4.1.7	Nanoparticle Assembly 165
3.5	Properties and Applications of Gold Nanoparticles 169
3.6	Concluding Remarks 172
	References 173
4	**Gold–Heterometal Interactions and Bonds** 181
	Cristian Silvestru
4.1	Introduction 181
4.2	Main Group Metal–Gold Compounds 183
4.2.1	Gold–Group 1 Metal Compounds 183
4.2.2	Gold–Mercury Compounds 186
4.2.3	Gold–Group 13 Metal Compounds 192
4.2.3.1	Gold–Gallium Compounds 192

4.2.3.2	Gold–Indium Compounds	*194*
4.2.3.3	Gold–Thallium Compounds	*196*
4.2.4	Gold–Group 14 Metal Compounds	*217*
4.2.4.1	Gold–Silicon Compounds	*221*
4.2.4.2	Gold–Germanium Compounds	*222*
4.2.4.3	Gold–Tin Compounds	*227*
4.2.4.4	Gold–Lead Compounds	*231*
4.2.5	Gold–Group 15 Metal Compounds	*231*
4.2.5.1	Gold–Antimony Compounds	*232*
4.2.5.2	Gold–Bismuth Compounds	*235*
4.3	Transition Metal–Gold Compounds	*235*
4.3.1	Gold–Titanium Compounds	*237*
4.3.2	Gold–Group 5 Metal Compounds	*237*
4.3.2.1	Gold–Vanadium Compounds	*238*
4.3.2.2	Gold–Niobium Compounds	*238*
4.3.3	Gold–Group 6 Metal Compounds	*238*
4.3.3.1	Gold–Chromium Compounds	*238*
4.3.3.2	Gold–Molybdenum and Gold–Tungsten Compounds	*239*
4.3.4	Gold–Group 7 Metal Compounds	*241*
4.3.4.1	Gold–Manganese and Gold–Rhenium Compounds	*241*
4.3.4.2	Gold–Technetium Compounds	*243*
4.3.5	Gold–Group 8 Metal Compounds	*243*
4.3.6	Gold–Group 9 Metal Compounds	*251*
4.3.7	Gold–Group 10 Metal Compounds	*255*
4.3.7.1	Gold–Nickel Compounds	*255*
4.3.7.2	Gold–Palladium Compounds	*255*
4.3.7.3	Gold–Platinum Compounds	*256*
4.3.8	Gold–Group 11 Metal Compounds	*260*
4.3.8.1	Gold–Copper Compounds	*260*
4.3.8.2	Gold–Silver Compounds	*263*
4.4	Concluding Remarks	*272*
	References	*272*

5 Supramolecular Architecture by Secondary Bonds *295*
María Elena Olmos

5.1	Introduction	*295*
5.2	Supramolecular Gold Entities Built with Gold–Non-Metal Secondary Bonds	*296*
5.2.1	Secondary Bonds: Au···NM Interactions	*296*
5.2.2	Hydrogen Bonds: Au···H–NM Interactions	*314*
5.2.3	Au···π Interactions	*318*
5.3	Supramolecular Gold Entities Built with Non-Metal–Non-Metal Secondary Bonds	*320*
5.3.1	Secondary Bonds: NM···NM Interactions	*320*
5.3.2	Hydrogen Bonds: NM···H–NM Interactions	*326*

5.3.3	C–H···π Interactions and π···π Stacking 337
5.4	Concluding Remarks 341
	References 342

6	**Luminescence of Supramolecular Gold-Containing Materials** 347
	José María López-de-Luzuriaga
6.1	Introduction. Conditions for Luminescence in Gold Complexes 347
6.2	Luminescent Supramolecular Gold Entities 351
6.2.1	Networks from Mononuclear Units 352
6.2.2	Networks from Dinuclear Units 359
6.2.3	Networks from Trinuclear Units 367
6.2.4	Networks from Higher Nuclearity Systems 375
6.3	Luminescent Supramolecular Gold–Heterometal Entities 376
6.3.1	Supramolecular Gold–Group 11 Metal Complexes 377
6.3.2	Other Supramolecular Gold–Heterometal Complexes 385
6.4	Concluding Remarks 398
	References 398

7	**Liquid Crystals** 403
	Manuel Bardají
7.1	Introduction 403
7.1.1	What Are Liquid Crystals? 403
7.1.2	What Are Metallomesogens? 406
7.1.3	Liquid Crystals Characterization 407
7.1.4	Liquid Crystals Applications 407
7.1.5	The Advantages of Gold 408
7.2	Gold Mesogens 408
7.2.1	Alkynyl Ligands 409
7.2.2	Azobenzene Ligands 410
7.2.3	Carbene Ligands 410
7.2.4	Dithiobenzoate Ligands 413
7.2.5	Isocyanide Ligands 413
7.2.5.1	Combined with an Alkynyl Ligand 414
7.2.5.2	Combined with Halides or Pseudohalides 414
7.2.5.3	Combined with a Perhaloaryl Ligand 417
7.2.5.4	Cationic bis(Isocyanide) Compounds 419
7.2.5.5	Chiral Isocyanide 420
7.2.5.6	Binary Mixtures of Isocyanide Derivatives 421
7.2.6	Perhaloaryl Ligands 422
7.2.7	Pyrazolate Ligands 423
7.2.8	Stilbazol Ligands 426
7.3	Concluding Remarks 426
	References 426

8	**Catalysis** *429*	
	M. Carmen Blanco Ortiz	
8.1	Introduction *429*	
8.1.1	Transition Metal Catalysis *430*	
8.1.2	Gold Catalysis *430*	
8.2	Homogeneous Catalysis *432*	
8.2.1	Nucleophilic Additions to C–C Multiple Bonds *432*	
8.2.1.1	Allenes as Substrates *433*	
8.2.1.2	Alkenes as Substrates *439*	
8.2.1.3	Alkynes as Substrates *446*	
8.2.1.4	Enynes as Substrates *466*	
8.2.2	Activation of Carbonyl Groups and Alcohols *472*	
8.2.3	Hydrogenation Reactions *474*	
8.3	Heterogeneous Catalysis *476*	
8.3.1	Hydrogenation Reactions *476*	
8.3.1.1	Alkenes Hydrogenation *476*	
8.3.1.2	Hydrogenation of α,β-Unsaturated Aldehydes *478*	
8.3.2	Oxidation Reactions *478*	
8.3.2.1	C–H Bond Activation *478*	
8.3.2.2	Oxidation of Alcohols and Aldehydes *479*	
8.3.2.3	Epoxidation *479*	
8.3.2.4	Direct Synthesis of Hydrogen Peroxide *480*	
8.3.3	Reactions Involving Carbon Monoxide *481*	
8.3.3.1	Carbon Monoxide Oxidation *481*	
8.3.3.2	Water Gas Shift Reaction *481*	
8.4	Concluding Remarks *482*	
	References *482*	
9	**Concluding Remarks** *491*	
	M. Concepción Gimeno and Antonio Laguna	
	References *494*	
	Index *495*	

Preface

Gold is not only the first metal utilized by humans but it is also a very special element, known and cherished by everybody. It has been present in Man's life since the earliest civilizations and has occupied an important place in the history of mankind for over 7000 years, as shown by the excellent goldsmith works found, for instance, in Minoan, Egyptian or American tombs.

Why is gold so attractive? It is not a very rare element on earth, and other metals, for example, platinum, rhodium, osmium or rhenium, are less abundant and more expensive. It is probably coveted most for its noble character; gold coins and jewellery do not tarnish, even after long exposure to extremely aggressive conditions, and the metal is resistant to enzymatic attack in biological systems. Its relevance since the dawn of civilization is reflected by various applications, for example, as a general means of exchange, in the production of coins, in jewelry or in decoration of china ceramic and glass. More recently other applications have emerged, particularly in the fields of microelectronics, medicine, material sciences and catalysis.

However, the chemistry of gold has played a minor role in history. It has been dominated by the metallic state and initially it only focused on different methods for converting gold metal into all possible forms for exchange or decoration, or for anticorrosive or electrical applications. In the last three decades its chemistry has been in continuous expansion with the opening of a new and advanced branch of science and applications. The spectacular development of gold chemistry reconfirms the unique character of this element. Its chemistry is also different to that of other metals and the main differences are the consequence of the relativistic effects, which are especially important for gold. There is a significant contraction/stabilization of s and p orbitals and an expansion of the d orbitals [1]. Some of the unusual facts about gold are as follows:

1. The value of its electrochemical potential is the lowest of any metal, which coincides with the relevance of the metallic state, and is a consequence of the significant contraction and stabilization of the 6s orbital.

2. Gold is the most electronegative of all the metals, once again confirming its noble character.

Modern Supramolecular Gold Chemistry: Gold-Metal Interactions and Applications.
Edited by Antonio Laguna
Copyright © 2008 WILEY-VCH Verlag GmbH & Co. KGaA, Weinheim
ISBN: 978-3-527-32029-5

3. Also related to the stabilization of the 6s orbital is the strikingly high electron affinity of this element, being closest to that of iodine. The reaction of equimolar amounts of gold and cesium readily produces a compound CsAu, which is not an alloy but an ionic compound with the gold center as the anion and a salt structure like cesium iodide. Clear evidence of the formation of a monoatomic ion is given by the electrolytic dissociation of CsAu in liquid ammonia or in the formation of other ionic compounds with alkali cations or $[NMe_4]^+$.

4. Other similarities with halogens are the presence of diatomic molecules (Au_2) in gold vapor, with dissociation energies comparable to those of halogens, or the spontaneous disproportionation of gold metal into Au^- and Au^+, in a melt reaction in the presence of cesium.

5. The small difference in energy between the s, p and d states leads to the efficient formation of s/d or s/p hybridizations, which are important for explaining the pronounced tendency of gold(I) to form linear two-coordinate complexes. This tendency for two coordination is much greater than for other isoelectronic centers, such as platinum(0), silver(I), or mercury(II), which normally yield compounds with higher coordination numbers.

6. The closed shell $5d^{10}$ in gold(I) is no longer chemically inert and can interact with other gold atoms. It also becomes possible to rationalize bonding between different gold(I) centers, which is a very difficult fact to explain in terms of classical bonding, confirming the great affinity between gold atoms. They try to be as close as possible to each other and Au–Au distances are normally even shorter than those found in metallic gold. This effect is known as aurophillic attraction or aurophilicity [2]. According to Pyykkö, the simplest picture to describe aurophilicity is that "it is just another van der Waals (dispersion) interaction, but an unusually strong one in an initially unexpected place" [1d].

7. Surprisingly, the noble gas xenon (a very weak coordinating agent) is bonded to gold in the oxidation states I, II and III. The first species were prepared by reaction of AuF_3 with elemental xenon, which produces a spectacular new class of gold(II) compounds. Again relativity plays an important role in stabilizing these unusual compounds as well as other Au–Xe species predicted by theoretical calculations.

8. As a consequence of the destabilization of the 5d orbitals in gold, we can explain the formation of oxidation states as high as III and V, which are almost absent or unknown, respectively, in the case of silver.

To date, several books have been published on the chemistry of gold [3, 4] or more specifically on its organometallic chemistry [5]. Other recent reviews cover the coordination [6] or the organometallic chemistry of this element [7]. This book offers a general overview of the supramolecular chemistry of gold up to the end of 2007, in order to highlight the usefulness of this metal. Thus, we have chosen to describe three different aspects of this chemistry: (i) the synthesis and properties of supramolecular aggregates due to the formation of gold–gold, gold–metal interactions or other secondary bonds; (ii) the preparation of nanomaterials, such as clusters or nanoparticles; and (iii) new applications for VOC (vapor organic compound) sensors, LEDs (light emitting devices), solvoluminescent and electroluminescent materials, liquid crystals and catalysis. We have therefore included a first general chapter describing the current situation of gold chemistry, its special characteristics, oxidation states and the main type of complexes.

This book could be of the interest for advanced chemistry students. They will find an interesting overview of gold chemistry, with representative situations in different oxidation states, examples of aurophillic interactions, interesting optical properties, and actual and potential applications of gold complexes. It will also be useful not only for researchers interested in the chemistry of gold and its applications, but also for those involved in metal–metal interactions, heteronuclear chemistry or in the optical properties of coordination compounds. This book is intended to be of considerable help for newly formed research groups in this fascinating field of gold compounds, offering basic knowledge of the chemistry of gold compounds and some of their particular and uncommon properties.

The layout of the book is as follows, Chapter 1 is an introductory chapter presenting an overview of the current status of gold chemistry (oxidation states, relativistic effects and different types of compounds). It emphasizes gold's many special characteristics and the most frequent and stable oxidation states. The main type of complexes and applications will be commented. Chapter 2 focuses on aurophillic interactions, which are similar in energy to hydrogen bonds and are responsible for the formation of interesting aggregates of different nuclearity. Many examples are provided of the distorted geometries observed around the gold atoms and the possibility of synthesizing supramolecular compounds. The formation of aggregates with gold–gold bonds and nanoparticles is an active area of research and is discussed in Chapter 3. This analyzes the synthesis of gold nanomaterials using different methods and their properties and applications. The synthesis and properties of other supramolecular derivatives, formed by interactions of gold centers with other metals or with ligands or substituents of adjacent molecules, are discussed in Chapters 4 and 5. The last three chapters are devoted to the study of specific properties and applications which have emerged strongly in recent decades, such as luminescence (with potential applications, such as sensors or light-emitting devices), liquid crystals (calamitic or discotic thermotropic mesogens based on gold) and homogeneous and heterogeneous catalysis.

Most of the authors belong to the Material Science Institute of Aragon or its Associated Unit at the University of Zaragoza or the University of La Rioja, respectively, and some of them have been working on the chemistry of gold since before 1990. Others such as Dr. M. Bardají, belonged to the same group, which is actually at the University of Valladolid, and Prof. C. Silvestru, from "Babes-Bolyai" University in Cluj Napoca, has been collaborating with our group during the last decade. We would like to thank the others members of the group and particularly our students for their patience and tolerance during the preparation of the different chapters.

Finally, we are grateful to the editors and staff of VCH (especially Dr. M. Koehl and Dr. R. Muenz) for their help and guidance throughout the preparation of this book.

Zaragoza, July 2008 *Antonio Laguna*

References

1 Pyykkö, P. (1997) Chemical Reviews, **97**, 597; Pyykkö, P. (2004) Angewandte Chemie-International Edition, **43**, 4412; Pyykkö, P. (2005) Inorganica Chimica Acta, **358**, 4113; Pyykkö, P. Chemical Society Reviews, DOI 1039/b708613j.
2 Schmidbaur, H. (1990) Gold Bulletin, **23**, 11; Schmidbaur, H. (1992) Interdisciplinary Science Reviews, **17**, 213.
3 Puddephatt, R.J. (1978) The Chemistry of Gold, Elsevier, Amsterdam.
4 Schmidbaur, H. (ed.) (1999) Gold, Progress in Chemistry, Biochemistry and Technology, John Wiley & Sons.
5 Patai, S. and Rappoport, Z. (eds) (1999) The Chemistry of Organic Derivatives of Gold and Silver, John Wiley & Sons, Chichester.
6 Gimeno, M.C. and Laguna, A. (2003) Comprehensive Coordination Chemistry II, **5** (eds J.A. McCleverty and T.J. Meyer), Elsevier, New York, p. 911.
7 Schmidbaur, H. and Schier, A. (2007) Comprehensive Organometallic Chemistry III, **2** (eds R. Crabtree, M. Mingos and K. Meyer), Elsevier, New York, p. 251.

Abbreviations

APS	aminopropyltriethoxylsilane
BDAB	benzyldimethylammonium bromide
BIB	2-bromoisobutyryl bromide
2,2′-bipy	2,2′-bipyridine
4,4′-bipy	4,4′-bipyridine
t-Bu-fy	2,7-di-tert-butylfluoren-9-ylidene
bzim	1-benzyl-2-imidazolate
carb	-C(OEt) = $NC_6H_4CH_3$-p
$CH_3im(CH_3py)$	1-methyl-3-[2-(3-methyl)pyridinyl]imidazol-2-ylidene
$CH_3im(CH_2py)$	1-methyl-3-(2-pyridinylmethyl)imidazol-2-ylidene
CH_3impy	1-methyl-3-(2-pyridinyl)imidazol-2-ylidene
Cp^*	pentamethylcyclopentadienyl
[2.2.2]crypt	4,7,13,16,21,24-hexaoxa-1,10-diazabicyclo[8.8.8]hexacosane
CTAB	cetyltrimethylammonium bromide
dbbpy	4,4′-di-tert-butyl-2,2′-bipyridyl
dbfphos	4,6-bis(diphenylphosphino)dibenzofuran
dcpm	bis(dicyclohexylphosphino)methane
DDP	2-[(2,6-diisopropylphenyl)amino]-4-[(2,6-diisopropylphenyl)imino)]-2-pentene
DDT	dodecanethiol
DENs	dendrimer encapsulated nanoparticles
dimen	1,8-diisocyano-p-menthane
DIPCDI	di-iso-propylcarbodiimide
DMF	N,N-dimethylformamide
dmpi	2,6-dimethylphenyl isonitrile
DMSA	dimercaptosuccinic acid
DMSO	dimethylsulfoxide
DNCs	dendrimer–nanoparticle composites
dpen	1,2-bis(diphenylphosphino)ethylene
dpma	bis(diphenylphosphinomethyl)phenylarsine
dpmp	bis(diphenylphosphinomethyl)phenylphosphine

Modern Supramolecular Gold Chemistry: Gold-Metal Interactions and Applications.
Edited by Antonio Laguna
Copyright © 2008 WILEY-VCH Verlag GmbH & Co. KGaA, Weinheim
ISBN: 978-3-527-32029-5

dppb	1,2-bis(diphenylphosphino)butane
dppe	1,2-bis(diphenylphosphino)ethane
dppf	1,1′-bis(diphenylphosphino)ferrocene
dppm	bis(diphenylphosphino)methane
dppp	bis(diphenylphosphino)propane
DSNs	dendrimer-stabilized nanoparticles
EGDMA	glycol dimethacrylate
Fc	ferrocenyl
GMA	poly(glycidyl methacrylate)
Hacac	acetylacetone
HAD	hexadecylamine
LAM	laurylamine
mes	mesityl
MMA	methyl methacrylate
MPCs	monolayer protected clusters
MUD	11-mercaptoundecano
nbd	norbornadiene
NCDs	nanoparticle-cored dendrimers
OA	octylamine
ODA	octadecylamine
OLA	oleylamine
P_2phen	2,9-bis(diphenylphosphino)-1,10-phenanthroline
P_2napy	2,7-bis(diphenylphosphino)-1,8-naphthyridine
P2VP	poly(2-vinyl pyridine)
P4VP	poly(4-vinylpyridine)
PAA	polyacrylic acid
PAMAM	poly(amidoamine)
PAnP	9,10-bis(diphenylphosphino)anthracene
1,10-phen	1,10-phenanthroline
3,5-Ph_2pz	3,5-diphenylpyrazolate
PDDA	poly(diallyl dimethylammonium)chloride
PDMS	poly(dimethylsiloxane)
PEG	poly(ethylene glycol)
PEI	poly(ethyleneimine)
D-pen	D-penicillaminate
PEO-SH	thiol-terminated poly(ethylene oxide)
PMMA	poly(methyl methacrylate)
PNIPAM	poly(*N*-isopropylacrylamide)
PNP	2-[bis(diphenylphosphino)methyl]pyridine
PPh_2py	diphenyl-2-pyridylphosphine
PPN	bis(triphenylphosphine)iminium
PS-SH	thiol-terminated polystyrene
PVA	poly(vinyl alcohol)
PVME	poly(vinyl methyl ether)
PVP	poly(*N*-vinylpyrrolidone)

py	pyridine
(py)$_2$im	1,3-bis(2-pyridinyl)imidazol-2-ylidene
(pyCH$_2$)$_2$im	1,3-bis(2-pyridinylmethyl)imidazol-2-ylidene
S^{t}BuDED	1,1-dicarbo-*tert*-butoxyethylene-2,2-thioperthiolate
SH-PEG	thiolated polyethyleneglycol
Spy	2-pyridinethiolate
Tab	4-(trimethylammonio)benzenethiolate
tdt	toluene-3,4-dithiolate
terpy	2,2′:6′,2″-terpyridine
tfbb	tetrafluorobenzobarrelene
tfepma	bis[bis(trifluoroethoxy)phosphino]methylamine
THF	tetrahydrofuran
TOAB	tetraoctylammonium bromide
TOPO	tri-n-octylphosphine oxide
trip	2,4,6-triisopropylphenyl
triphos	1,1,1-tris(diphenylphosphinomethyl)ethane
tht	tetrahydrothiophene

List of Contributors

Manuel Bardají
Química Inorgánica
Facultad de Ciencias
Universidad de Valladolid
E-47011 Valladolid, Spain
e-mail: bardaji@qi.uva.es

María del Carmen Blanco
Química Inorgánica
Instituto de Ciencia de Materiales de Aragón.
Universidad de Zaragoza – C.S.I.C.
E-50009 Zaragoza, Spain
e-mail: mablanor@unizar.es

Olga Crespo
Química Inorgánica
Facultad de Ciencias – I.C.M.A.
Universidad de Zaragoza – C.S.I.C.
E-50009 Zaragoza, Spain
e-mail: ocrespo@unizar.es

Eduardo J. Fernández
Departamento de Química
Universidad de La Rioja, U.A.– C.S.I.C.
Complejo Científico-Tecnológico
E-26004-Logroño, Spain
e-mail: eduardo.fernandez@unirioja.es

M. Concepción Gimeno
Química Inorgánica
Instituto de Ciencia de Materiales de Aragón.
Universidad de Zaragoza – C.S.I.C.
E-50009 Zaragoza, Spain
e-mail: gimeno@unizar.es

Antonio Laguna
Química Inorgánica
Facultad de Ciencias – I.C.M.A.
Universidad de Zaragoza – C.S.I.C.
E-50009 Zaragoza, Spain
e-mail: alaguna@unizar.es

José M. López-de-Luzuriaga
Departamento de Química
Universidad de La Rioja, U.A.– C.S.I.C.
Complejo Científico-Tecnológico
E-26004-Logroño, Spain
e-mail: josemaria.lopez@unirioja.es

Miguel Monge
Departamento de Química
Universidad de La Rioja, U.A.– C.S.I.C.
Complejo Científico-Tecnológico
E-26004-Logroño, Spain
e-mail: miguel.monge@unirioja.es

Modern Supramolecular Gold Chemistry: Gold-Metal Interactions and Applications.
Edited by Antonio Laguna
Copyright © 2008 WILEY-VCH Verlag GmbH & Co. KGaA, Weinheim
ISBN: 978-3-527-32029-5

Elena Olmos
Departamento de Química
Universidad de La Rioja, U.A.–
C.S.I.C.
Complejo Científico-Tecnológico
E-26004-Logroño, Spain
e-mail: m-elena.olmos@unirioja.es

Cristian Silvestru
Catedra de Chimie Anorganica
Facultatea de Chimie si Inginerie
Chimica
Universitatea Babes-Bolyai,
RO-400028 Cluj-Napoca, Romania
e-mail: cristi@chem.ubbcluj.ro

1
The Chemistry of Gold
M. Concepción Gimeno

1.1
Introduction

1.1.1
History

Gold was discovered as shining yellow nuggets and is undoubtedly the first metal known to early civilizations. The symbol derives from the Latin word *aurum*, which is related to the goddess of dawn, Aurora. Early civilizations equated gold with gods and rulers, and gold was sought in their name and dedicated to their glorification. Humans almost intuitively attribute a high value to gold, associating it with power, beauty, and the cultural elite. And since gold is widely distributed all over the globe, it has been perceived in the same way throughout ancient and modern civilizations everywhere.

Archeological digs suggest gold was first used in the Middle East where the first known civilizations developed. Experts in the study of fossils have observed that pieces of natural gold were found in Spanish caves used by Paleolithic Man in about 40 000 BC. The oldest pieces of gold jewellery were discovered in the tombs of Queen Zer of Egypt and Queen Pu-abi of Ur in Sumeria and date from the third millennium BC. Most Egyptian tombs were raided over the centuries, but the tomb of Tutankhamun was discovered undisturbed by modern archeologists. The largest collection of gold and jewellery in the world included a gold coffin whose quality showed the advanced state of Egyptian craftsmanship and goldworking (second millennium BC). The Persian Empire, in what is now Iran, made frequent use of gold in artwork as part of the religion of Zoroastrianism. Persian goldwork is most famous for its animal art, which was modified after the Arabs conquered the area in the seventh century AD. Gold was first used as money in around 700 BC, when Lydian merchants (western Turkey) produced the first coins. These were simply stamped lumps of a 63% gold and 27% silver mixture known as "electrum."

Modern Supramolecular Gold Chemistry: Gold-Metal Interactions and Applications.
Edited by Antonio Laguna
Copyright © 2008 WILEY-VCH Verlag GmbH & Co. KGaA, Weinheim
ISBN: 978-3-527-32029-5

Nevertheless, gold as a metal has been omnipresent since the dawn of civilisation and its chemistry has played a minor role in history. Initially, all that was known of its chemistry was its concentration, recovery, and purification. Moreover, compounds, such as tetrachloroauric acid or salts of the anions $[Au(CN)_2]^-$ or $[Au(CN)_4]^-$, were very important because they were intermediates in the recovery of the metal. The chemistry of gold was merely regarded as an art to recover and convert gold metal into all possible forms of ornamental, monetary, anticorrosive or electrical usage. It is therefore no surprise that the chemistry of gold, which is so clearly dominated by the metallic state, remained undeveloped for so long. The associated development of an old handicraft, with all its secrets, into a field of research based on technology is a good example of how gold chemistry has finally matured into an advanced branch of science with a significant bearing on applications in many fields [1].

1.1.2
Chemical and Physical Properties

What is it about gold that makes it so attractive and so useful? Gold is not very hard; a knife can easily scratch pure gold and it is very heavy or even dense for a metallic mineral. Some of the other characteristics of gold are ductility, malleability and sectility, meaning it can be stretched into a wire, pounded into other shapes, and cut into slices. Gold is the most ductile and malleable element on our planet. It is a great metal for jewellery because it never tarnishes.

The color and luster of gold are what make this metal so attractive. Gold is found as the free metal and in tellurides. It is widely distributed and almost always associated with pyrite or quartz. It is found in veins and in alluvial deposits. Gold also occurs in seawater in concentrations of 0.1–$2\,\text{mg ton}^{-1}$, depending on the location of the sample. In the mass, gold is a yellow-colored metal, although it may be black, ruby, or purple when finely divided. One ounce of gold can be beaten out to $300\,\text{ft}^2$. Gold is a good conductor of electricity and heat. It is not affected by exposure to air or to most reagents. It is inert and a good reflector of infrared radiation. Gold is usually alloyed to increase its strength. Pure gold is measured in troy weight, but when gold is alloyed with other metals the term *karat* is used to express the amount of gold present.

Gold has an electrochemical potential which is the lowest of any metal. This means that gold in any cationic form will accept electrons from virtually any reducing agent to form metallic gold. It is the most electronegative of all metals, which once again confirms its noble character. Moreover, although the electron affinity for metals is not usually included in textbooks for gold; the process from gold(0) to gold(–I) can be easily accomplished, and it has, in fact, been known since 1930.

1.1.3
Theoretical Considerations

The electronic configuration of gold(0) is $5d^{10}6s^1$; for gold(I) it is $5d^{10}6s^0$ and for the gold(–I) anion it is $5d^{10}6s^2$. These configurations would justify the relative stability of gold(I) compounds, with 10 electrons in a closed set of 5d orbitals, or even, to some

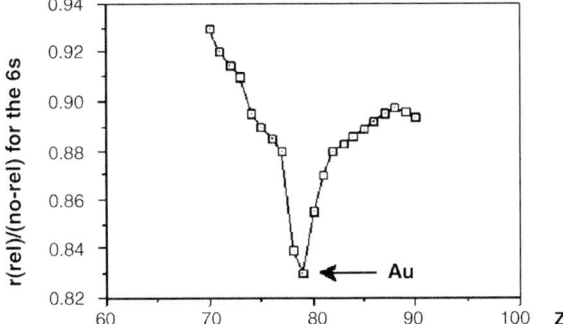

Figure 1.1 Ratio of r(rel)/r(non-rel) versus atomic number for the 6s electrons.

extent, the formation of the aurate anion, but they do not allow us to understand the predominance of the metallic form.

Post-lanthanide elements contain a large number of protons in their atomic nuclei; therefore, the electrons move in a field of very high nuclear charge, which leads to velocities approaching that of light and, consequently, they have to be treated according to Einstein's theories of relativity. This is particularly true for electrons that are in s orbitals, which have wavefunctions that correspond to a finite electron density at the atomic nucleus, but it is less important for electrons in p or d orbitals. Electrons moving at a speed close to the speed of light cannot be treated in terms of classical physics, but they are assigned a relativistic mass that is larger than the mass of the electron at rest. The effect on the 6s electrons, in the post-lanthanide elements, is that the orbital radius is contracted and the distance of the electron from the nucleus is reduced. Figure 1.1 shows a plot where the ratio of the relativistic radius of the valence electrons to their non-relativistic radius is presented as a function of atomic number. It is clear that this ratio strongly deviates from unity as Z increases, and it reaches a pronounced local minimum for gold. Thus, without making any further special assumptions, this theoretical approach leads to the conclusion that gold actually occupies a unique position among the elements.

There are several consequences of this effect in gold chemistry:

1. The color of gold. Gold has an absorption beginning at 2.4 eV, attributed to a transition from the filled 5d band to the Fermi level (essentially the 6s band). It therefore reflects red and yellow light and strongly absorbs blue and violet. The analogous absorption for silver, however, lies in the ultraviolet, at around 3.7 eV.

2. A marked reduction in the lengths of covalent bonds involving gold atoms is often found and the covalent radius of gold is smaller than that of silver. Schmidbaur has recently confirmed this in the isomorphous di-coordinated complexes [M(PMes$_3$)$_2$]BF$_4$ (M = Ag, Au; Mes = mesithyl) [2]. The estimated covalent radii for silver and gold are 1.33 and 1.25 Å, respectively, so di-coordinated gold is smaller than silver by about 0.09(1) Å.

3. In gold, more than in silver, both states are now available for bonding. The closed shell $5d^{10}$ is no longer chemically inert and can interact with other elements, that is, with other gold atoms in molecules or clusters. Bonding between two gold(I) centers with equal charge and a closed shell $5d^{10}$ configuration can also be rationalized, and this is a very difficult fact to explain in terms of classical bonding. The metal atoms approach each other to an equilibrium distance of between 2.7 and 3.3 Å. This range includes the distance between gold atoms in gold metal and approaches, or even overlaps with, the range of distances established for the few authentic Au–Au single bonds. Schmidbaur has called this effect *aurophillic attraction* or *aurophilicity* [3].

4. The small difference in energy between the s, p and d orbitals leads to the efficient formation of linear two-coordinate complexes in gold(I). However, silver(I) prefers the formation of three- and four coordinate derivatives.

5. The destabilization of the 5d orbitals allows the easy formation of the oxidation state III in gold to be explained; this is almost absent in silver; and the stabilization of the 6s orbitals explains the formation of the gold(–I), oxidation state, which is unknown in silver.

Theoretical calculations have played a key role in understanding the origin of these differences and also in the development of gold and silver chemistry. Bonding between closed-shell atoms was successfully traced in several early theoretical investigations by extended Hückel quantum chemical calculations [4–7]. Based on the hybridization concept, the nature of the bonding interaction could be qualitatively rationalized in the language of chemists. The introduction of relativistic effects in more advanced calculations has shown that bonding between closed-shell metal atoms or ions may be strongly enhanced by these effects [8–18]. Since relativistic effects have been known to reach a local maximum for gold in particular, aurophilicity was accepted as a logical consequence of these contributions. In fact, aurophilic bonding is considered as an effect based largely on the electron correlation of closed-shell components, somewhat similar to van der Waals interactions but unusually strong [15, 16]. All these studies have consistently shown that calculations will only reproduce the attractive forces between the gold atoms very well if relativistic affects are included.

1.2
Chemistry

Since the early 1980s, the chemistry of gold has undergone continuous expansion; not only well-established areas of research have developed, but also new innovative approaches have enabled great diversification of the fields of research. The metal and its complexes also have special characteristics that make them suitable for several uses. Gold possesses special characteristic features that make it unique, such as high chemical and thermal stability, mechanical softness, high electrical conductivity, and beautiful appearance. All these attributes gave rise to many relevant applications. For example, gold is an essential element for nanoscale electronic devices because it is resistant to oxidation and mechanically robust. The well-known elegant red color in Venetian crystal glass stems from surface plasmon absorption of blue light by gold nanoparticles a few tens of nm in size. Gold compounds have been used successfully for the treatment of rheumatoid arthritis [19–21]. Gold is also an outstanding element for use as a heterogeneous catalyst operating at ambient temperature because it is catalytically active at low temperature (200–350 K compared with Pd and Pt at 400–800 K) [22]; in the last few years several uses in homogeneous catalysis have been reported [23–26].

1.2.1
Gold(-I) Compounds

Many of the peculiarities of the chemical and physical properties of gold are due to its relativistic effects, including the high electron affinity which explains the high propensity of gold to adopt the negative oxidation state −I. The binary alkali metal aurides RbAu and CsAu have been known for about 50 years [27, 28], but the ternary auride oxides M_3AuO (M = K, Rb, Cs) and the aurideaurates Rb_5Au_3O and M_7Au_5O (M = Rh, Cs) have been discovered only recently [29–32].

The anionic character of gold is also emphasized by its similarities to the heavier halides Br^- and I^-. Using a macroreticular ion exchange resin with a high affinity toward cesium ions in liquid ammonia, it has been possible to exchange cesium for thetramethylammonium and the first compound of negatively-charged gold with a non-metal cation was thus isolated, $NMe_4^+Au^-$ [33]. An estimation has been made of the auride radius as 1.9 Å, which is very similar to the bromide anion; in fact, the compound crystallizes isotypically to $(NMe_4)Br$. CsAu dissolves in liquid ammonia forming a yellow solution and, when ammonia is removed, CsAu is not recovered directly. Instead, a new, intense blue solid crystallizes which has the composition $CsAu \cdot NH_3$ [34]. The crystal structure exhibits features characteristic of low-dimensional systems. Slabs of overall composition CsAu are separated by single NH_3 layers. The gold atoms are shifted towards one another forming zig-zag chains with $Au \cdots Au$ separation of 3.02 Å (4.36 Å in CsAu). This auridophilic attraction takes place between $d^{10}s^2$ anions but surprisingly yields a similar distance to the aurophilic attraction in d^{10} cations. The theoretical nature of this effect is not clear. Both the dispersion effects and the net bonding, resulting from partial oxidation from the top of the 6s band, could play a role. The observed Mössbauer isomer shift at gold (CsAu 7.00 mm S^{-1}, RbAu 6.70 mm S^{-1}, $CsAu \cdot NH_3$ 5.96 mm S^{-1}, Au −1.23 mm S^{-1}) does not show a full −1 auride charge and means the latter is still a possibility [35]. Following the concept of similarity between gold(–I) and the halides, since the latter are good acceptors in hydrogen bonding systems, complexes with hydrogen bonds involving Au^- ions as acceptors should be possible. The first complex in which neutral ammonia molecules act as donors and auride anions as acceptors has been found in the complex $[Rb([18]crown-6)(NH_3)_3]Au \cdot NH_3$ [36].

1.2.2
Gold(0) and Gold Clusters

The synthesis of silicalix[n]phosphinines, a new class of macrocycles incorporating sp^2-hybridized phosphorus atoms, has enabled the preparation of gold(0) macrocycles [37]. These ligands have an adequate balance between σ-donating and π-accepting properties and can then act as a macrocyclic equivalent of carbonyl groups. The reaction of these macrocycles with [AuCl(SMe$_2$)] in the presence of GaCl$_3$ yields the complexes [AuL][GaCl$_4$]. The electrochemical reduction of one of these derivatives enables the synthesis of the gold(0) compound (Equation 1.1).

$$ \text{(structures)} \qquad (1.1) $$

However, there is an important class of gold compounds in which the formal oxidation state is intermediate between 0 and +1. They can be homo- or heteronuclear compounds and the nuclearity can range from 4 to 39 [38–41]. These compounds are available through several synthetic routes, which are mainly:

1. Reaction of a suitable gold(I) precursor such as [AuX(PR$_3$)] (X = halide, SCN, NO$_3$) or HAuCl$_4$ with a reducing agent such as NaBH$_4$, CO, Ti(η-C$_6$H$_5$Me)$_2$, and so on, a representative example is given in Equation 1.2 [42]:

$$[Au(NO_3)(PPh_3)] + NaBH_4 \rightarrow [Au_9(PPh_3)_8](NO_3)_3 \qquad (1.2)$$

2. Treatment of gold vapor with phosphine ligands (Equation 1.3) [43]:

$$Au(g) + dppm \rightarrow [Au_5(dppm\text{-}H)(dppm)_3](NO_3)_2 \qquad (1.3)$$

3. Reactions of gold clusters with other ligands or metal complexes. These types of reactions can progress with or without change in the metal skeleton and in the latter with an increase or decrease in nuclearity (Equations 1.4 and 1.5) [44, 45]:

$$[Au_9(PPh_3)_8]^{3+} + Bu_4NI \rightarrow [Au_8I(PPh_3)_6]^+ \qquad (1.4)$$

$$[Au_9(PPh_3)_8]^{3+} + 3[Au(C_6F_5)_2]^- \rightarrow [Au_{10}(C_6F_5)_4(PPh_3)_5] \qquad (1.5)$$

4. Nucleophilic addition, elimination, and substitution reactions in which the cluster core remains intact (Equation 1.6) [46, 47]:

$$[Au_8(PPh_3)_6]^{2+} + 2PPh_3 \rightarrow [Au_8(PPh_3)_8]^{2+} \qquad (1.6)$$

The bonding in gold cluster molecules has been interpreted using free electron models based on Stone's tensor surface harmonic theory [48, 49]. High similarity has

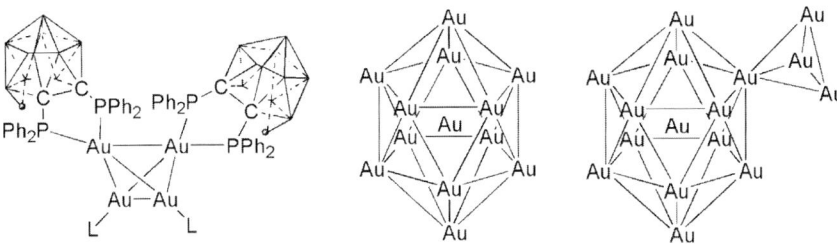

Figure 1.2 Gold clusters representative of Au$_4$, Au$_{13}$ and Au$_{16}$ cores.

been observed between alkali metals and gold in the spectra of molecular orbitals when they form clusters, and the primary bonding and antibonding interactions result from the overlap of the s valence orbitals. For gold, the 5d orbitals are core-like and only hybridize slightly with the 6s, where the 6p valence orbitals are too high-lying to contribute significantly to the bonding. The geometries have been classified as (i) spherical polyhedral clusters characterized by a total of $12n + 8$ valence electrons (n is the number of peripheral gold atoms); and (ii) toroidal or elliptical polyhedral clusters characterized by a total of $12n + 6$ valence electrons [50].

All the nuclearities from 4 to 13 gold atoms are known in homonuclear clusters, the latest is a regular icosahedron with a central gold atom [51, 52]. This icosahedron may add in one of the corners three more gold atoms, forming a pendant tetrahedron and thus resulting in a 16-gold-atom cluster [53]. Some representative examples are shown in Figure 1.2.

The largest structurally-characterized gold cluster is [Au$_{39}$Cl$_6$(PPh$_3$)$_{14}$]Cl$_2$ [54], which has a structure related to a hexagonal packed geometry with $1:9:9:1:9:9:1$ individual layers of gold atoms (Figure 1.3).

Several heteronuclear clusters are known and with early transition metals based on carbonyl clusters [55]. With the platinum group metals, the clusters [M(AuPPh$_3$)$_8$]$^{2+}$ (M = Pd, Pt), synthesized by the reduction of mixtures of [M(PPh$_3$)$_4$] and [Au(NO$_3$)(PPh$_3$)], have been thoroughly studied [56–58]. They undergo several types of reactions such as nucleophilic additions ranging from 16-electron to 18-electron clusters, which induce a change in the geometry from toroidal to spheroidal, as shown in Figure 1.4 [59, 60].

The most outstanding examples of heteronuclear gold clusters are the series of Au–Ag supraclusters whose metal frameworks are based on vertex-sharing. In the structure of these compounds, the basic building block is the 13-metal atom (Au$_7$Ag$_6$) icosahedra. These high nuclearity clusters have been termed "clusters of clusters" and they follow a well-defined growth sequence by successive additions of

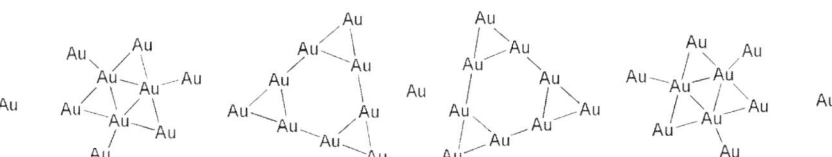

Figure 1.3 Layers of the cluster [Au$_{39}$Cl$_6$(PPh$_3$)$_{14}$]Cl$_2$.

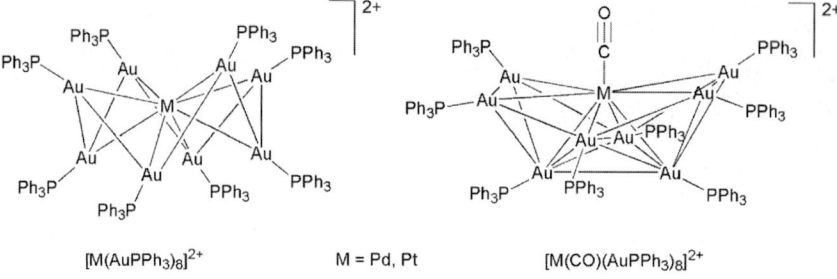

Figure 1.4 Structure of the 16- and 18-electron clusters.

icosahedral units via vertex sharing. The first member of this vertex-sharing poly-icosahedral cluster series is the 25-metal-atom cluster, $[(Ph_3P)_{19}Au_{13}Ag_{12}Br_8]^+$ [61, 62], whose metal core can be considered as two icosahedra sharing a common vertex (Figure 1.5). They have been obtained by reducing a mixture of $[AuX(PR_3)]$ and $[AgX(PR_3)]$ with $NaBH_4$. The growth sequence by successive additions of icosahedral units via vertex-sharing has been used to obtain the triicosahedral $[(p\text{-}TolP)_{12}Au_{18}Ag_{20}Cl_{14}]$ [63], and the tetraicosahedral $[(p\text{-}TolP)_{12}Au_{22}Ag_{24}Cl_{10}]$ [64]. The structure of the triicosahedral clusters is three M_{13} atom (Au_7Ag_6) Au-centered icosahedra sharing three Au vertices in a cyclic manner (Figure 1.6).

1.2.3
Gold Nanoparticles and Nanoclusters

One area related to gold clusters is the field of nanoparticles and nanoclusters. These materials have actually been known since ancient times when they were used for their esthetic appeal and also for their therapeutic properties in the form of colloidal gold. In recent decades, the field of nanoparticle research has emerged and in-depth research has dealt with the properties and potential applications of these systems.

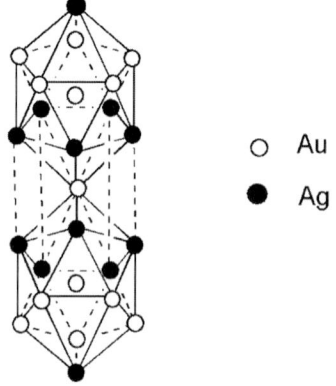

Figure 1.5 Core of $[(Ph_3P)_{19}Au_{13}Ag_{12}Br_8]^+$.

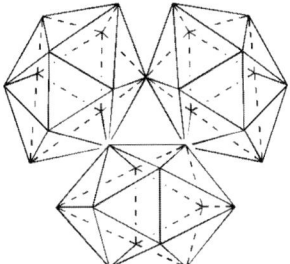

Figure 1.6 Core of the triicosahedral clusters.

Gold nanoparticles have dimensions in the range 1 nm to 1 μm, although there is often no clear demarcation between a nanoparticle and a non-nanoparticle system. For example, the 39-atom gold cluster has dimensions in this range and could therefore be considered to be a nanoparticle. However, a well-defined large-atom aggregate (X-ray characterization) is usually classified as a cluster. The compound $[Au_{55}Cl_6(PPh_3)_{12}]$, synthesized from $[AuCl(PPh_3)]$ by reduction with B_2H_6 [64, 65], is usually referred to as a cluster, although the crystal structure is not available and it can be classified as nanoparticles with dimensions of 1.4 ± 0.4 nm. Many attempts have been made to characterize this gold cluster and its physical properties and reactivity have been widely studied [66].

The synthetic methods used to produce gold nanoparticles range from the reduction of $HAuCl_4$ with citrate in water, reported by Turkkevitch in 1951 [67], to the reduction of Au(III) complexes and stabilization with different ligands from which the thiolates of different chain lengths [68] are the most popular (Equation 1.7), although others such as xanthates, disulfides, phosphines, isocyanide, and so on, have been used. Other techniques such as microemulsion, copolymer micelles, seeding growth, and so on, have also been utilized to synthesize gold nanoparticles [69]. With the discovery of the self-assembled monolayers (SAMs) absorbed onto metal colloids, the surface composition of gold nanoparticles can be modified to contain a variety of functional groups and even mixtures of functional groups.

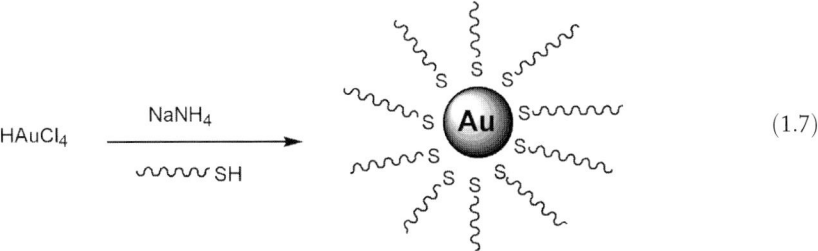

(1.7)

These Au nanoparticles, protected by a monolayer of thiolated ligands, display interesting properties, such as single-electron charging and molecule-like HOMO–LUMO energy gaps, and can be used in optical and chemical sensing [70, 71]. Their physicochemical properties are closely related to their size and size distribution. Therefore, the ability to synthesize nanoparticles in a size-controlled manner has been one of the goals of materials science. Consequently, new procedures have been developed and one of the most successful synthetic routes consists in the reductive decomposition of polymeric Au(I)-SR complexes [72, 73]. Many clusters of different nuclearities have been obtained by this or other methods, such as Au_{25}(glutathione)$_{18}$ [74], $Au_{38}(SPhX)_{24}$ [75], $Au_{75}(SC_6H_{13})_{40}$ [76] obtained by reduction of $[Au_{55}Cl_6(PPh_3)_{12}]$, or even clusters of higher nuclearity such as the Au_{140} monolayer protected cluster [77].

Much research has also been carried out into determining whether highly ordered clusters with cage-like structures are possible for gold. The closest match to a cage-like cluster is the bimetallic icosahedron W@Au_{12}, which was first predicted by Pyykkö and Runeberg [78] and later synthesized by Li et al. [79]. However, the pure icosahedral form of Au_{12} is unstable and must be stabilized by the endohedral W atom. Theoretical studies of gold fullerenes are also scarce [80] but a highly stable icosahedron Au_{32} fullerene was predicted based on DFT calculations [81] and recently an alternative icosahedral Au_{42} fullerene cage has been shown to be competitive energetically [82]. Gold nanotubes are a synthetic reality [83] and platonic gold nanocrystals have also been prepared through careful growth-rate regulation in different crystallographic directions [84]. These nanocrystals have the perfect symmetry for 2D and 3D packing and therefore could enable the rational tuning of their optical, electrical, and catalytic properties.

1.2.4
Gold(I) Complexes

The chemistry of gold(I) species is by far the most developed. Many monoclear, dinuclear or, in general, polynuclear derivatives with several types of ligands have been described. It is difficult to say which type of complexes are more stable and important for gold, those of phosphine or polyphosphine ligands have been studied in depth, as have organometallic gold complexes or species with chalcogenolate or chalcogenide ligands. In this chapter on the chemistry of gold, the most important types of gold(I) complexes for synthetic applications, structural patterns, properties, and so on, will be discussed.

1.2.4.1 Mononuclear [AuXL], [AuL$_2$]$^+$ or [AuX$_2$]$^-$ Complexes

The chemistry of gold(I) complexes is dominated by linear two-coordinate complexes of the form [AuXL], which, despite being known for a long time, have returned to prominence because they display short intermolecular gold–gold contacts that associate the monomeric units into dimers, oligomers or even uni- and multidimensional polymers. These aurophilic interactions, together with other secondary bonds such as hydrogen bonds, play a role in determining the solid state arrangement of Au(I) complexes [85]. The tendency of molecules containing an Au···Au interaction to crystallize with more than one molecule in the asymmetric unit ($Z' > 1$) has been calculated [86]. This behavior is believed to be related to the size of the differential of the two substituents. If the two ligands have a disparity in size, the $Z' > 1$ is favored, whereas the ligands of equivalent size have a tendency to form structures with $Z' > 1$.

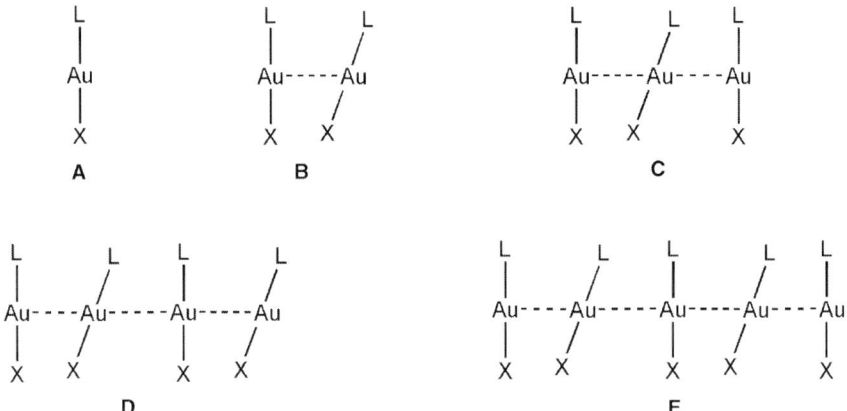

Figure 1.7 Association of [AuXL] complexes in dimers (B), trimers (C), tetramers (D) or chain polymers (E) through short gold–gold contacts.

This fact can be easily seen in neutral complexes of the type [AuXL], which have been prepared for a great variety of donor ligands. The neutral ligand is mainly a phosphine, arsine, isocyanide, carbine, ylide, amine, and so on, and the anionic moiety can be an halide, alkyl or aryl, chalcogenolate, and so on. The complexes with the smallest ligands form the highest-dimensional aggregates and the complexes [AuCl(CO)] [87] or [Au(CN)(CNMe)] [88] form two-dimensional layers, but [AuCl(PMe$_3$)] [89] and [Au(CN)(PMe$_3$)] [90] are chains, and [Au(OCOCF$_3$)(PMe$_3$)] is a trimer [91], and many [AuX(PR$_3$)] complexes with medium-sized ligands are dimers [41]. Figure 1.7 shows the association of these molecules.

A particular case is the gold chloride isocyanide complexes of the type [AuCl(CNR)] (R = C$_n$H$_{2n+1}$, n = 1–11) [92]. The molecules behave like flexible hydrocarbon chains with a rod-like endgroup containing the gold atom. In the solid state the molecules arrange in antiparallel chains formed by aurophilic bonding. This structure is similar to those seen for the (1 – n)-alcohols (with hydrogen bonds) and also shows temperature-dependence polymorphism consistent with the formation of rotator phases. This is further proof of the concept that hydrogen bonding and aurophilic bonding are similar in their binding energies and directionality.

In some cases, the complexes [AuXL] may undergo ligand redistribution in solution and crystallize as homoleptic isomeric forms [AuL$_2$]$^+$[AuX$_2$]$^-$. These derivatives also aggregate in the solid state through aurophilic interactions in the form +−+−+−, such as for example in [AuIL] (L = tetrahydrothiophene, tetrahydroselenophene) [93]. The sequence of ions in the aggregates may vary against all intuition by placing ions of like charge next to each other, for example, in patterns −++− as in the complex [Au(py)$_2$][AuX$_2$] (X = Cl, Br, I) [94] or +−−+ in [Au(PPhMe$_2$)$_2$][Au(GeCl$_3$)$_2$] [95] instead of the expected +−+−. A novel pattern [+ neutral −] has recently been found in the trimeric compound [Au(2-NH$_2$py)$_2$]$^+$[AuCl(2-NH$_2$py)][AuCl$_2$]$^-$ [96] (Figure 1.8).

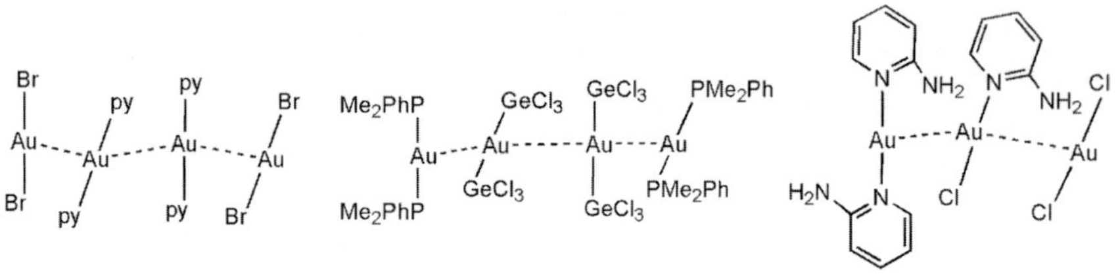

Figure 1.8 Association −++−, +−−+ and + neutral − in [AuXL] complexes.

In other complexes and depending on the ligands present, the formation of aggregates is based on other secondary bonds such as hydrogen bonds or Au···S interactions. The complexes [AuCl(4-PPh$_2$C$_6$H$_4$CO$_2$H)] [97] or [Au(SC$_4$H$_3$N$_2$O$_2$)(PPh$_3$)] [98] associate in pairs through hydrogen bonding and [Au(4-SC$_6$H$_4$CO$_2$H)(PPh$_3$)] forms a supramolecular structure through hydrogen bonding and aurophilic interactions [99]. Some thiolate complexes, such as [Au{SSi(O^{i}Pr)$_3$}(PPh$_3$)], form dimers through Au···S interactions [100]. Supramolecular structures are also achieved in mononuclear gold(I) complexes such as [Au(O$_2$CCF$_3$)(4-PPh$_2$C$_6$H$_4$CO$_2$H)], which crystallizes as infinite polymer chains of "[Au(4-PPh$_2$C$_6$H$_4$CO$_2$H)]$_n$," with a degree of helicity in the chains arising from the propeller-like arrangement of the phenyl groups [97]. The complexes with chiral secondary phosphines [AuX(PHMePh)] (X = Cl, Br, I) show chain-like supramolecular aggregates through aurophilic interactions, the chain contains both enantiomers following the sequence ···RSRSRS···[101].

An interesting physical property exhibited by gold(I) compounds with short Au···Au interactions is the visible luminescence observed under UV excitation in the solid state. In the complexes [AuCl(TPA)] (Figure 1.9) and [AuCl(TPA·HCl)] (TPA = 1,3,5-triaza-7-phosphaadamantane), which have a phosphine ligand with a small cone angle that allows short aurophilic interactions, their luminescence spectra are substantially different and are correlated with the change in the Au···Au interaction [102]. At 78 K [AuCl(TPA)] luminesces intensely red (674 nm), while the protonated [AuCl(TPA·HCl)] luminesces yellow (596 nm). The emission bands in both complexes blue-shift as the temperature is increased. The interpretation of the phenomena is supported by extended Hückel MO calculations.

Although the [AuXL] complexes are well represented for many donor L ligands and for many X anionic ligands and have long been used as precursors in gold chemistry, curiously, analogous fluoride complexes [AuFL] have only been observed recently. AuF has long been considered a "non-existent compound"; its existence in the gas phase was theoretically predicted [103] and was later identified and characterized in the gas phase [104]. A series of neutral gold complexes [AuXNg] (Ng = Ar, Kr, Xe; X = F, Cl, Br)

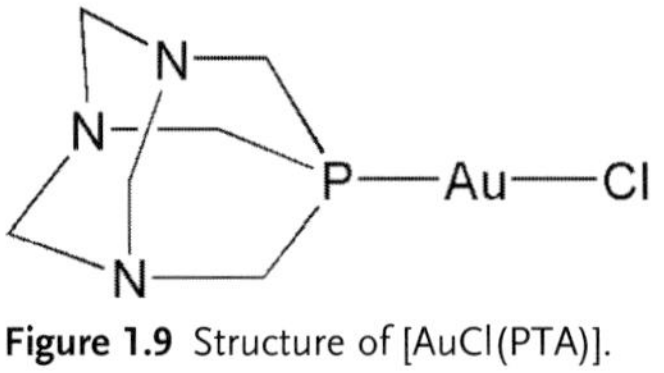

Figure 1.9 Structure of [AuCl(PTA)].

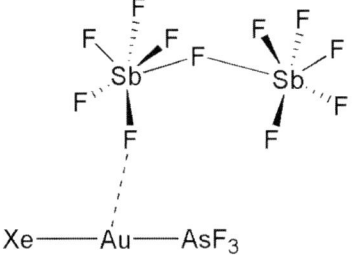

Figure 1.10 Structure of [AuXe(AsF$_3$)][Sb$_2$F$_{11}$].

have been generated by laser ablation of the metal in the presence of a suitable precursor, and stabilized in a supersonic jet of Ar [105]. The complex [AuFXe] has been detected and characterized in the gas phase using microwave rotational spectroscopy. As expected, it is the noble gas–noble metal halide complex more strongly bonded to a very short Au–Xe distance of 2.54 Å [106]. All evidence is consistent with an AuXe covalent bonding in [AuFXe]. The first gold(I) complex with an XeAu bond is the cationic [AuXe(AsF$_3$)][Sb$_2$F$_{11}$] (Figure 1.10), it was prepared by reaction of [Au(AsF$_3$)]$^+$[SbF$_6$]$^-$ with xenon in SbF$_5$-rich HF/SbF$_5$ solutions. The cation interacts only weakly with the anion and has an Au–Xe distance of 2.607 Å [107].

Another type of neutral gold(I) complex is [AuX(PR$_3$)] where X is an anionic oxygen or nitrogen donor ligand such as [Au{N(SO$_2$CF$_3$)$_2$}(PR$_3$)] or [Au(OR)(PR$_3$)] or even [Au(OSO$_2$CF$_3$)(PR$_3$)], which are relevant as catalytically-active species or catalytic precursors [26, 108].

Gold thiolates of the form [Au(SR)(PR$_3$)] are important complexes that are known for a great variety of thiolate ligands and also serve as building blocks to obtain polynuclear species [109]. Several applications have been found, for example, in medicine with the commercialization of the antiarthritic drug Auranofin [Au(SR)(PEt$_3$)] (Figure 1.11). Many other examples with this stoichiometry have been reported, which also have antiarthritic, antitumoral or antimicrobial activity. The structure of another antiarthritic drug, gold thiomalate (myocrysine), has been reported and crystallized as a mixed sodium/cesium salt Na$_2$Cs[Au$_2$L(LH)] (Figure 1.12), which is a polymer that consists of two interpenetrating spirals, with approximately fourfold symmetry [110].

Anionic [AuX$_2$]$^-$ with halide or pseudohalide ligands are well known and have been widely used as a starting materials. The salts [Au(SCN)$_2$]$^-$ have been structurally characterized showing, for alkaline metals, infinite linear chains with alternating

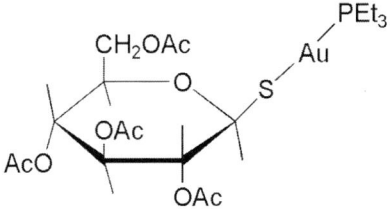

Figure 1.11 Structure of Auranofin.

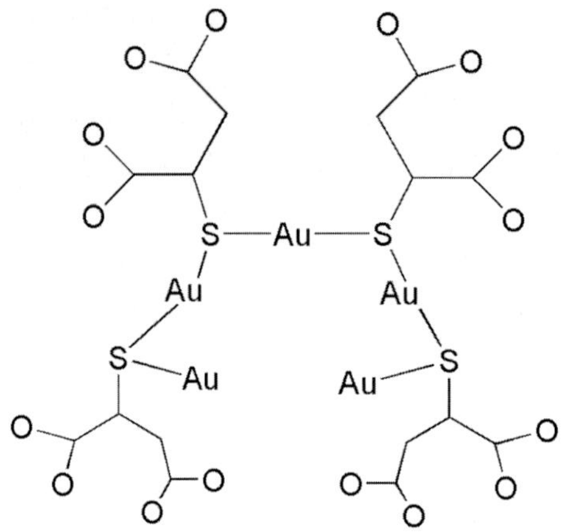

Figure 1.12 Structure of gold thiomalate.

short and long gold–gold distances along the chain, the $[NMe_4]^+$ salt forms a kinked chain of trimers joined at a shared gold atom as the kink, and the $[NBu_4]^+$ salts contain isolated dimers with a short aurophilic interaction [111]. All these complexes with Au–Au interactions are emissive and, as suggested by Fackler and Schmidbaur [102], the emission correlates inversely with distance. The anionic gold(I) thiolates, $Bu_4N[Au(SC_6H_4R)_2]$ show luminescence in the solid state, the emission maxima range from 438 nm (blue) to 529 nm (green), depending on the substituent R [112]. The anionic complexes with the hydrosulfide ligand, $[(PPh_3)_2N][Au(SH)_2]$ and $[AuR(SH)]^-$ are the only gold(I) complexes described with this ligand [113].

1.2.4.2 Gold(I) Complexes with Polydentate Ligands
Polydentate ligands have been widely used as bridging or quelating ligands in gold(I) chemistry. One of the most important types are diphosphines, which give very stable gold(I) complexes with different structural patterns. Dinuclear (chloro)gold(I) complexes with the phosphorus atoms bridged by one to eight carbon atoms, of the type $[Au_2Cl_2\{\mu\text{-}(PPh_2)_2(CH_2)_n\}]$ ($n = 1$–8), have been prepared from $[AuCl(CO)]$ or $[AuCl(SR_2)]$ with the corresponding bidentate phosphine. They can adopt different structural patterns, as a result of the formation of Au–Au bonds (see Figure 1.13). The crystal lattice of the 1,4-bis(diphenylphosphino)butane ($n = 4$) or hexane ($n = 6$) derivatives contain independent molecules (type A) [114, 115], which show no intra- or inter-molecular Au–Au interactions, in contrast with the structures of the related gold complexes with shorter or longer chain diphosphines, where intra- ($n = 1$, type B) [116] and intermolecular metal–metal interactions with the formation of discrete dimers (type C) [117, 118] or polymeric chains (type D) are found [114, 115, 119]. The completely different packing of the monomeric molecules of $[Au_2Cl_2\{\mu\text{-}(PPh_2)_2(CH_2)_4\}]$ and the two polymorphic forms of $[Au_2Cl_2\{\mu\text{-}(PPh_2)_2(CH_2)_2\}]$, where the short Au–Au contacts give rise to dimers or to a polymeric chain structure, are particularly remarkable, since the conformations of the free phosphines show a close relationship. It has been suggested that packing effects, with or without solvent

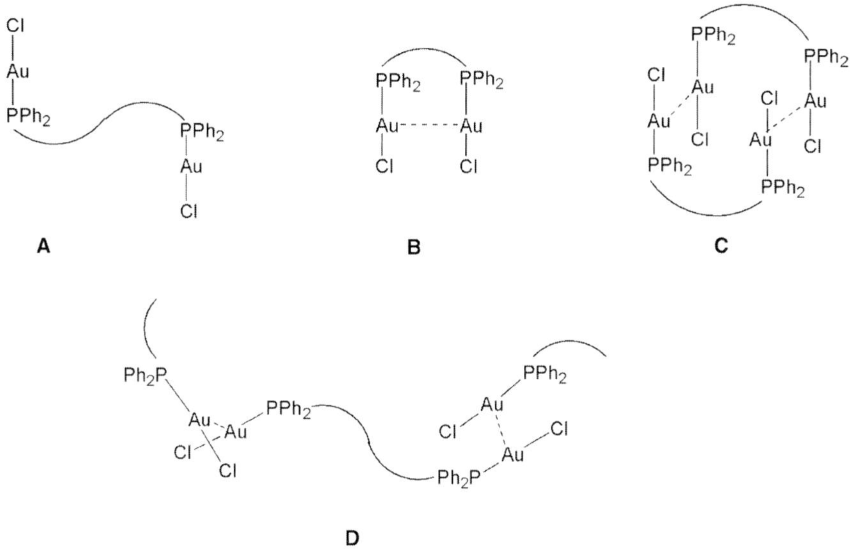

Figure 1.13 Different structural patterns of [Au$_2$Cl$_2${μ-PPh$_2$(CH$_2$)$_n$}].

molecules, determine the conformation of these molecules; in fact, for [Au$_2$Cl$_2${μ-(PPh$_2$)$_2$(CH$_2$)$_2$}] the presence of dichloromethane in the crystal leads to a transformation of the dimeric units present in the solvent-free crystal modification to a polymeric chain. For the $n = 3$ and larger chains, the complexes show monomeric dinuclear molecules connected through intermolecular Au–Au contacts of about 3.30 Å to give polymeric chains.

Analogous structures occur in other diphosphine or diarsine complexes of the type [AuCl$_2${μ-(PPh$_2$)$_2$X}] [120–124] where X can be a great variety of bridging moieties; the structure varies from monomeric molecules without Au–Au interactions to dimeric complexes with intramolecular metal–metal contacts or polymeric compounds with intermolecular Au–Au interactions. In the complexes with the diphosphine PPh$_2$C(=PMe$_3$)PPh$_2$, the ligands change their ground state syn/anti orientation to a symmetrical syn/syn conformation upon coordination to gold(I). From temperature-dependent NMR studies, the energy of the Au–Au interaction was estimated to be of the order of 29–33 kJ mol^{-1} [120, 121]. Figure 1.14 shows some examples of this type of complex.

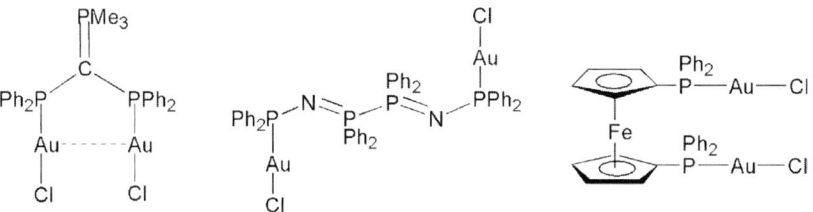

Figure 1.14 Some examples of [AuCl$_2${μ-(PPh$_2$)$_2$X}] complexes.

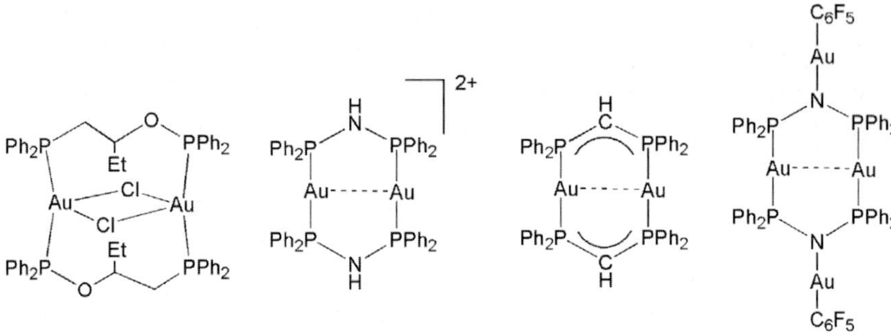

Figure 1.15 Dinuclear diphosphine complexes.

Other types of well-represented dinuclear derivatives with diphosphines have the stoichiometry [Au$_2$(μ-PP)$_2$]$^{2+}$ and are known for a great variety of diphosphines R$_2$PXPR$_2$ (Figure 1.15) where X can vary from the simple methylene CH$_2$ [125–127], bis(diphenylphosphino)methane (dppm), for which many studies have been carried out, to another hydrocarbon chain [128], to NH [129], and so on. Most of these derivatives are three-coordinate by bonding to the anionic ligand, which may be Cl, Br, I, S$_2$CNEt$_2$, or BH$_3$CN and also by formation of Au–Au interactions with molecules such as [AuCl(GeCl$_3$)]$^-$. Some of these complexes are luminescent, such as [Au$_2${(PR$_2$)$_2$CH$_2$}]$^{2+}$ (R = Me, Ph, Cy) for which several studies have been carried out on aurophilic attraction and luminescence [130]. The high luminescent complex [Au$_2${(PPh$_2$)$_2$CH$_2$}](OTf)$_2$ with a triplet excited state has been postulated to be used in light-emitting diodes [131]. The dinuclear derivatives [Au$_2${(PR$_2$)$_2$X}]$^{2+}$ (X = CH$_2$, NH) can be easily deprotonated to give the neutral complexes [Au$_2${(PR$_2$)$_2$Y}] (Y = CH, N) [132, 133]. Further coordination of the C or N atoms to other metal complexes gives tetra or hexanuclear derivatives [133, 134].

Most of the dinuclear gold(I) complexes are homobridged diauracycles with the same bridging ligand on each side, but some examples of heterobridged derivatives have been reported. These contain a diphosphine in addition to other bidentate ligands such as bis(ylide) [135, 136], dithiolate [137], dithiocarbamate [136, 138], xantate [139], phosphoniodithioformate [140], dithiophosphinate [141], pyridine-2-thiolate [136], and so on. They can be obtained by reaction of the [Au$_2$X$_2$(μ-PP)] complexes with the bidentate ligand or by ligand exchange reactions between two different homobridged dinuclear compounds. Examples of these complexes are shown in Figure 1.16.

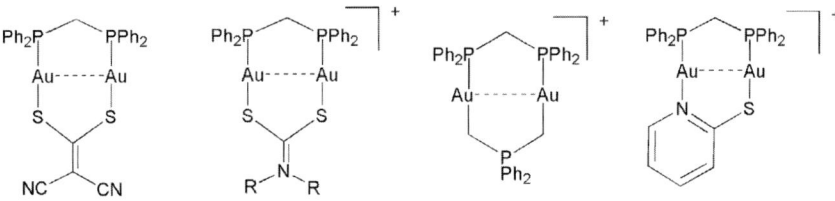

Figure 1.16 Dinuclear heterobridged diphosphine gold(I) complexes.

Figure 1.17 Gold(I) complexes with tritertiary phosphine or phosphine-arsine ligands.

Triphosphines or mixed phosphine-arsine ligands have also been used as ligands to coordinate gold(I) centers and the structures of the complexes can vary, depending on the structural requirements of the phosphine ligands. Then complexes with stoichiometry $[Au_3Cl_3(\mu\text{-LLL})]$ [142–144], $[Au_3(\mu\text{-LLL})_2]^{3+}$ [145, 146], $[Au_2\{(PPh_2)_2CHPPh_2\}_2]^{2+}$ [143] or $[Au_4Cl_2(\mu\text{-LLL})_2]^{2+}$ [147] have been prepared but sometimes with different structural frameworks such as $[Au_3Cl_3\{(PPh_2)_3CH\}_2]$, which forms a triangle of gold atoms, and $[Au_3(PPh_2CH_2PPhCH_2PPh_2)_2]^{3+}$, for which the gold atoms form a linear chain (Figure 1.17).

With tetraphosphines, the usual stoichiometry is $[Au_4Cl_4(\mu\text{-}P_4)]$, such as those with tetrakis(diphenylphosphine)methane [148], or tetrakis(diphenylphosphine)tetrathiafulvalene, $(PPh_2)_4TTF$, [149] (Figure 1.18), and gold(I) complexes with dentritic polymers containing phosphorus atoms as terminal groups have also been described [150]. Some of these polyphosphine complexes have been used to synthesize heteronuclear complexes, such as bis(diphenylphosphino)methane (dppm) [151], $PPh_2CH_2AsPhCH_2PPh_2$ [152], and so on, which lead to several derivatives with the polydentate ligands bridging two or more metal atoms (Figure 1.19).

Another important class of bidentate ligands in gold chemistry are those composed of sulfur donor ligands such as dithiocarbamates, dithiolates, dithiocarboxylates, dithiophosphinates, dithiophosphates, and so on. The complexes prepared with these ligands are generally of the type $[Au_2(\mu\text{-SS})(PR_3)_2]^{n+}$ or $[Au_2(\mu\text{-SS})(\mu\text{-PP})]$ [153–155], $[Au_2(\mu\text{-SS})_2]^{n-}$ [156, 157] or $[Au_3(\mu\text{-SS})_2(PPh_3)_2]^-$ [158]. All these

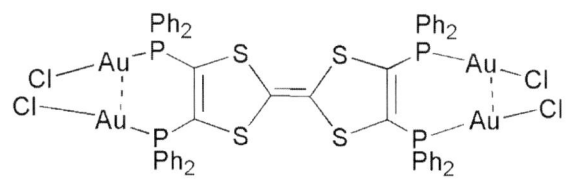

Figure 1.18 Structure of $[Au_4Cl_4\{(PPh_2)_4TTF\}]$.

Figure 1.19 Heteronuclear phosphine gold(I) complexes.

Figure 1.20 Gold(I) complexes with bidentate sulfur ligands.

complexes have intramolecular Au–Au interactions and some of them also present intermolecular contacts. The usual structure for [Au$_2$(μ-SS)(PR$_3$)$_2$] complexes is dinuclear but some of them form a supramolecular structure through Au–Au and Au–S interactions such as in [Au$_2$\{μ-S$_2$P(OMe$_2$)\}(PPh$_3$)$_2$]$^+$. The species [Au$_2$(μ-SS)$_2$]$^{n-}$ can be discrete dinuclear units with intramolecular gold–gold contacts or linear chains through intermolecular aurophilic interactions. As a consequence of these aurophilic interactions, many of these complexes present luminescence properties. The compound [Au$_4$\{μ-S$_2$C$_2$(CN)$_2$\}$_2$(PTA)$_2$] has a dinuclear structure with the "AuPTA" fragment bonded to the gold atom with an unsupported Au–Au interaction and is highly luminescent [159]. A chiral luminescent Au$_{16}$ ring has been reported by reaction of [Au$_2$Cl$_2$(μ-dppm)] with K$_2$(pippzdc) (pippzdc = piperazine-1,4-dicarbodithiolate) which gives the tetramer [Au$_4$(pippzdc)(dppm)$_2$]$_4$ [160]. Figure 1.20 shows some of these complexes.

Homoleptic dithiocarboxilates can be tetrameric as [Au$_4$(S$_2$CMe)$_4$] [161] or hexameric as in [Au$_6$(S$_2$C$_6$H$_4$-Me-2)$_6$] [162] (see Figure 1.21). Analogous complexes with selenium ligands are far less developed although some examples have been reported, such as [Au$_2$\{μ-Se$_2$C$_2$(CN)$_2$\}]$^{2-}$ [163] (Figure 1.22).

Substituted phenylene-dithiolate ligands have been thoroughly studied and di-, tri- and tetranuclear complexes of the type [Au$_2$(μ-S$_2$C$_6$H$_3$R)(PPh$_3$)$_2$], [Au$_3$(μ-S$_2$C$_6$H$_3$R)(PPh$_3$)$_3$]$^+$ or [Au$_4$(S$_2$C$_6$H$_3$R)$_2$L$_2$] have been reported with the *ortho* isomer [137, 153, 164–166]. The *meta* and *para* isomers present different stoichiometries and structures such as [Au$_3$(1,3-S$_2$C$_6$H$_4$)(PPh$_3$)$_3$]$^+$, which shows a one-dimensional aggregate through head-to-tail aurophilic interactions or the

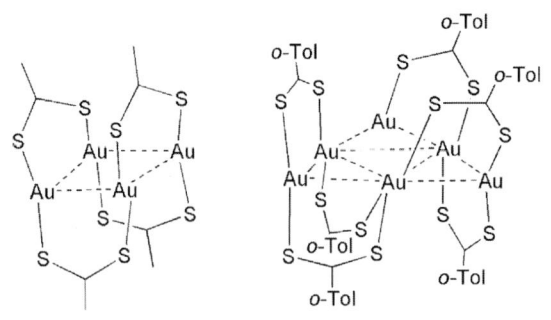

Figure 1.21 Dithiocarboxilate gold(I) complexes.

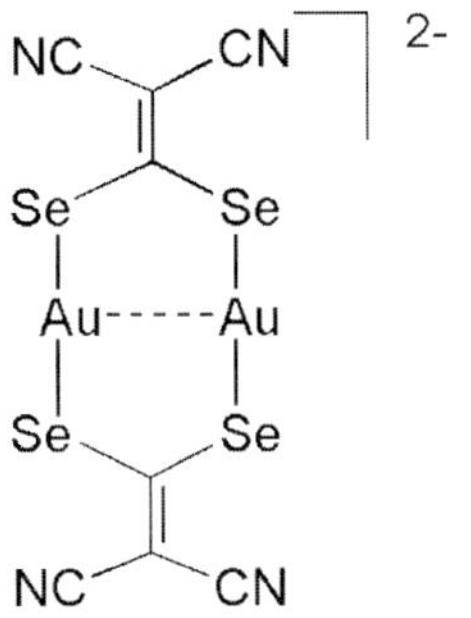

Figure 1.22 Structure of [Au$_2${μ-Se$_2$C$_2$(CN)$_2$}]$^{2-}$.

tetranuclear [Au$_4$(1,4-S$_2$C$_6$H$_4$)(PPh$_3$)$_4$]$^{2+}$ [167]. Benzenehexatiol reacts with [AuCl(PPh$_3$)] to give the hexanuclear golden wheel [Au$_6$(S$_6$C$_6$)(PPh$_3$)$_6$] [168]. Figure 1.23 shows some of these complexes.

Other complexes with these polydentate ligands include species with diphosphine ligands such as the complex with the 2-thioxo-1,3-dithiole-4,5-dithiolate (dmit) ligand, [Au$_4$(dmit)$_2$(dppm)$_2$] [169] (Figure 1.24a), the tetracoordinated compound with the 1,2-dithiolate-o-carborane [Au$_4$(S$_2$C$_2$B$_{10}$H$_{10}$)$_2${(PPh$_2$)$_2$C$_2$B$_{10}$H$_{10}$}$_2$] [170] (Figure 1.24b), or the gold complex [Au$_2$Cl$_2${Fc(S$_2$CNEt)$_2$}] [171] with the bis(dithiocarbamate)ferrocene ligand in which there are η^2 interactions between the gold(I) atoms and the cyclopentadienyl ligands (Figure 1.24c).

Other types of polydentate ligands are those with different donor atoms that can be of the type P,C or N,C or S,S,C,C or P,N, and so on, [172–175]. Many gold(I) complexes with these heterofunctional ligands have been prepared. Figure 1.25 shows some examples.

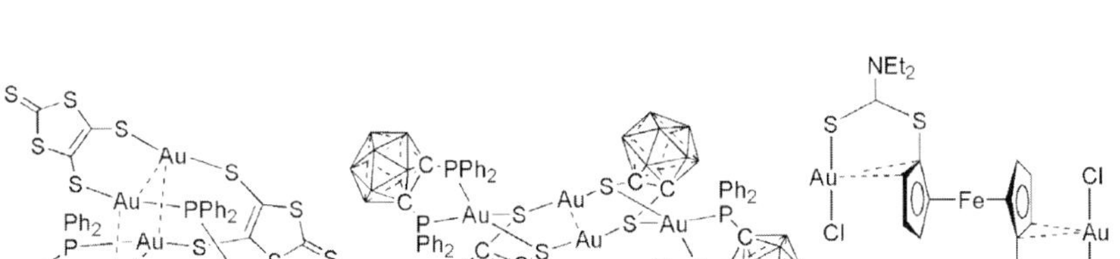

Figure 1.23 Di- and hexa-thiolate gold(I) complexes.

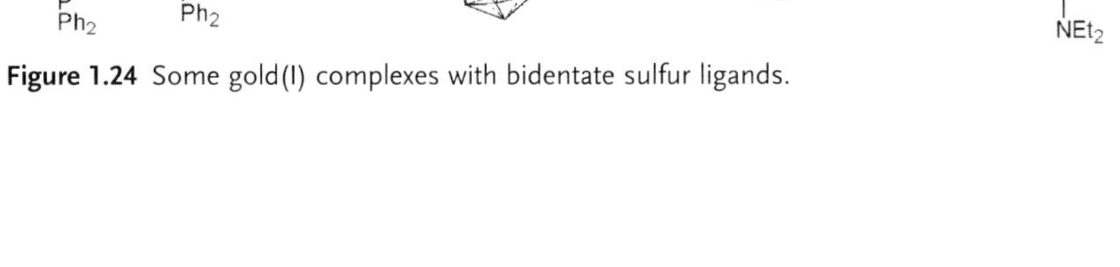

Figure 1.24 Some gold(I) complexes with bidentate sulfur ligands.

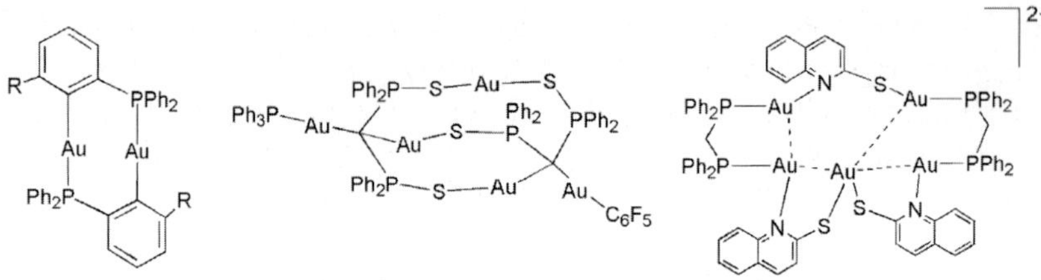

Figure 1.25 Gold(I) complexes with mixed donor ligands.

Gold(I) complexes with polydentate nitrogen ligands are also known, such as with the ferrocenyl-terpyridine ligand (Figure 1.26a) in which the ligand acts as tridentate [176], or the tetranuclear derivative with 2,2′-bibenzimidazolate (Figure 1.26b) [177], or the complexes obtained in the reaction of trans-1,2-bis (4-pyridyl)ethylene with [Au$_2$(O$_2$CCF$_3$)$_2${μ-(PPh$_2$)$_2$(CH$_2$)$_n$}] which gives with $n=2$ the cyclic compound and with $n=3$ or 4 one-dimensional linear or U-shaped polymers (Figure 1.26c) [178]. In the same reaction with 4,4′-bipyridine in solution the complexes exist as an equilibrium mixture of linear oligomers and cyclic compounds; when $n=1$, 3 or 5 they exist as 26-, 30- and 34 membered macrocyclic rings, respectively, and only when $n=1$ are there significant intramolecular Au–Au contacts. Some of these compounds are strongly emissive at room temperature and in the solid state [179].

1.2.4.3 Three and Four-Coordinate Gold(I) Complexes

High coordinated gold(I) complexes with monodentate phosphines, arsines or stibines have been reported. It was first demonstrated by ^{31}P{^{1}H} NMR studies that bis-, tris, and even tetrakis(phosphine)gold(I) complexes exist in solution. Owing to rapid ligand exchange on the NMR time scale the individual complexes can only be observed at low temperature. The linear complexes [AuX(PR$_3$)] interact with an excess of phosphine to give a series of species including primarily [AuX(PR$_3$)$_2$], [Au(PR$_3$)$_3$]X or [Au(PR$_3$)$_4$]X as components of the equilibria, as shown in Equation 1.8. The

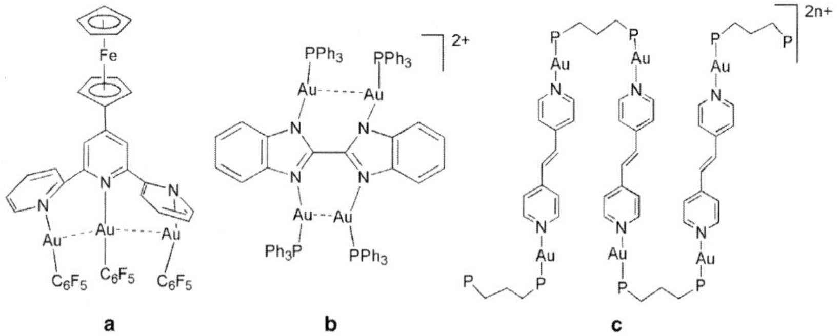

Figure 1.26 Gold(I) derivatives with polydentate nitrogen ligands.

maximum coordination number attainable depends on the particular ligand used. For bulky phosphines, such as $PR_3 = PCy_3$, only the two coordinated cation $[Au(PCy_3)_2]^+$ is accessible, but with $PR_3 = PBu_3$, $P(4\text{-Tol})_3$ or $PPh_2\{CH_2CH_2(2\text{-py})\}$ both $[Au(PR_3)_2]^+$ and $[Au(PR_3)_3]^+$ are detected; for $PR_3 = PEt_3$, PMe_2Ph, $P(OEt)_3$ or $P(OCH_2)_3CEt$ finally the two- three- and four-coordinated species are observed [180–182].

$$[AuX(PR_3)] \overset{PR_3}{\rightleftharpoons} [AuX(PR_3)_2] \rightleftharpoons [Au(PR_3)_2]X \overset{PR_3}{\rightleftharpoons} [Au(PR_3)_3]X \overset{PR_3}{\rightleftharpoons} [Au(PR_3)_4]X \tag{1.8}$$

Three-coordinate complexes of the type $[AuX(PR_3)_2]$ are known and they show Au–P distances longer than those in two-coordinate complexes and P–Au–P angles somewhat wider than 120° [183–185]. Some of these three-coordinate bis(phosphine) gold(I) complexes luminesce in the solid state, as well as in solution. The emission is attributed to the metal-centered pz $\to$ (d_{x2-y2}, d_{xy}) transition [184, 186]. The most regular three-coordination is observed in compounds where all the three ligands are identical, as in $[Au(PPh_3)_3]^+$ salts for which several crystal structures with different anions have been reported [187]. The cation of $[Au(PCy_2Ph)_3]ClO_4$, obtained by reaction of $[AuCl(PCy_2Ph)]$ with an excess of PCy_2Ph in the presence of $(NH_4)ClO_4$, has an almost ideal trigonal planar geometry [P–Au–P angles of 119.3(3)°] [188]. The complexes $[Au(TPA)_3]Cl$ and $[Au(TPPTS)_3]^{8-}$ [TPPTS = tris[(3,3′,3″-phosphinidyne-tris(benzenesulfonate)] show luminescence in aqueous solution [189].

Four-coordinate complexes with monodentate ligands $[Au(L)_4]^+$ have been described for PPh_3, PPh_2Me, $AsPh_3$ and $SbPh_3$ [190–193]. For triphenylphosphine the structures of three modifications of the compound have been determined, none of which shows the expected simple tetrahedral geometry, however, the cation of $[Au(PMePh_2)_4]PF_6$, or those with the arsine or stibine ligands, show a nearly regular tetrahedral geometry. Other tetra-coordinate complexes are of the form $[AuX(PR_3)_3]$ [194] and show the presence of a four-coordinate gold atom in a distorted tetrahedral geometry, with rather long Au–P and Au–X distances. The water soluble and luminescent gold(I) complex $[AuI(MeTPA)_3]I_3$ [(MeTPA)I = 1-methyl-1-azonia-3,5-diaza-7-phosphaadamantane iodide] has been obtained by reaction of $[AuCl(SMe_2)]$ with three equivalents of (MeTPA)I [195]. The coordination environment of the gold center is approximately trigonal planar and the iodine is weakly coordinated to the gold atom perpendicular to the AuP_3 plane [Au–I distance 2.936(1) Å]. The solid shows a yellow emission (598 nm) at 77 K and an orange emission (686 nm) at 140 K. $[AuI(MeTPA)_3]I_3$ undergoes an unusual phenyl-transfer reaction in aqueous solution with $NaBPh_4$ to form $[AuPh(MeTPA)_3]BPh_4$.

Three-coordinate gold(I) complexes with functionalized ligands have been used to prepare heteronuclear derivatives. Thus the reaction of 7-diphenylphosphino-2,4-dimethyl-1,8-naphthyridine (dpnapy) with [AuCl(tht)] affords the tricoordinate compound $[Au(dpnapy)_3]^+$ (Figure 1.27a), the latter shows strong affinity towards Cu(I) and Cd(II) (Figure 1.27b) ions [196]. The 2-(diphenylphosphino)-1-methylimidazole (dpim) also forms a three-coordinate Au(I)–Ag(I) dimer, $[AuAg(dpim)_3]^{2+}$ (Figure 1.27c) with a short metal–metal distance, which shows an intense luminescence [197].

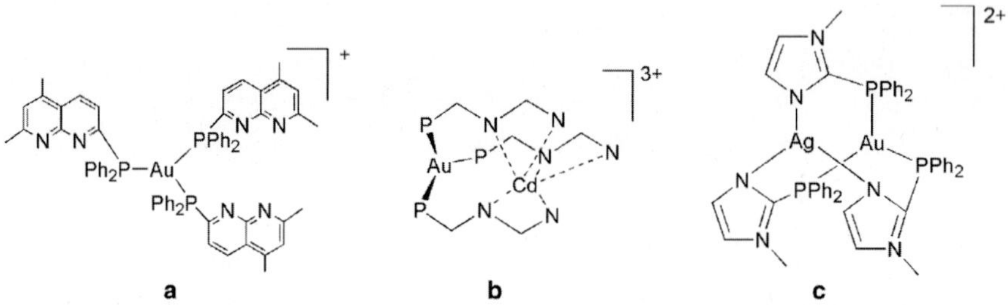

Figure 1.27 Three-coordinate gold(I) complexes with heterofunctional ligands.

Neutral or cationic three-coordinate gold(I) complexes with chelating diphosphines have been synthesized and are of the type [AuX(P–P)], [AuL(P–P)]$^+$ or [Au$_2$(μ-P–P)(P–P)]$^{2+}$, which has chelating and bridging diphosphine ligands [198–201]. Many of these complexes are strongly luminescent. Other three-coordinate complexes are those composed only of bridging diphosphines such as [Au$_2$(μ-P-P)$_3$]$^{2+}$, which have been described for Me$_2$PCH$_2$PMe$_2$, PPh$_2$pyPPh$_2$, PPh$_2$pzPPh$_2$, and so on, [202–204]. Better represented are species of the form [Au$_2$(μ-L–L)(μ-P–P)$_2$]$^{n+}$, where L-L are anionic ligands that can be halogens or pseudohalogens, dithiocarbamate, dithiophosphonate, and so on [205, 206]. Many of these complexes are strongly luminescent. Some examples of these complexes can be seen in Figure 1.28.

Luminescent trigonal gold(I) metallocryptates have been obtained by reaction of 2,9-bis(diphenylphosphino)-1,8-naphthyridine or 2,9-bis(diphenylphosphino)-1,10-phananthroline PPh$_2$(phen)PPh$_2$ with [AuCl(tht)] (molar ratio 3:2) and in the presence of a cation Na$^+$, Tl$^+$ or Hg^0, which is encapsulated in the cavity making Au–M interactions [207]. Similar mixed metal metallocryptands with Pd(0) or Pt(0) have been reported (Figure 1.29) [208].

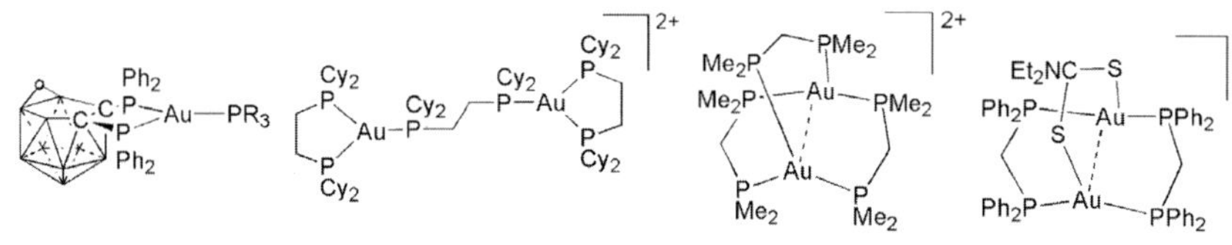

Figure 1.28 Three-coordinate gold(I) complexes with diphosphine ligands.

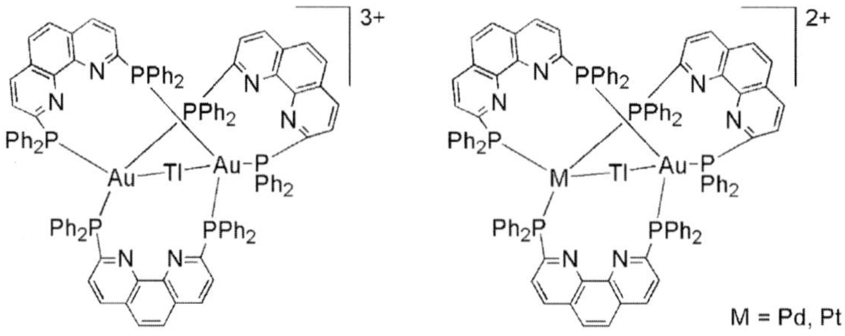

Figure 1.29 Gold(I) and mixed-metal metallocryptands.

Figure 1.30 Tetra-coordinate gold(I) complexes.

Four-coordinate homoleptic species of thetype $[Au(L-L)_2]^+$, where L–L can be a diphosphine or diarsine, have been obtained for several ligands such as $R_2PCH_2PR_2$ [209], cis-$Ph_2PCH=CHPPh_2$ [209], $(PPh_2)_2C_2B_{10}H_{10}$ [210], bis(diphenylphosphino)ferrocene [124], and so on. Some of these tetrahedral diphosphine complexes are cytotoxic and present a broad spectrum of antitumor activity [209]. Dinuclear tetra-coordinate derivatives are achieved with tetradentate phosphines, including tetraphos or tris(2-(diphenylphosphino)ethyl)phosphine [211]. Mixed four-coordinate gold(I) complexes have been prepared with the 1,2-bis(diphenylphosphino)-o-carborane ligands and are of the type $[Au\{(PPh_2)_2C_2B_{10}H_{10}\}(P\text{-}P)]^+$ [212] or with one *nido* and one *closo* diphosphine such as $[Au\{(PPh_2)_2C_2B_9H_{10}\}\{(PPh_2)_2C_2B_{10}H_{10}\}]$ [51]. Figure 1.30 shows some of these complexes.

1.2.4.4 Oligomeric Gold(I) Complexes

These complexes have the general formula $[AuL]_n$ or $[Au_2(L-L)]_n$ and are obtained with different ligands. The dinuclear complexes with bidentate ligands have been commented previously. Organometallic complexes of the type [AuR] are known, for example for mesityl, the structure of which is a pentamer $[Au_5(Mes)_5]$ (Figure 1.31a). This is a useful starting material to prepare other organometallic derivatives. It also displays unusual reactivity, for example with naked Ag^+ ions to give a complex, $[Au_6Ag(Mes)_6]$, with unsupported Au(I)–Ag(I) interactions (Figure 1.31b) [213]. Alkynyl complexes of the type $[Au(C\equiv CR)]_n$ have been formulated as polymeric but the structure of $[Au(C\equiv C^tBu)]_n$ has been shown to be a "catena" species with two interconnected cyclic hexamers [214].

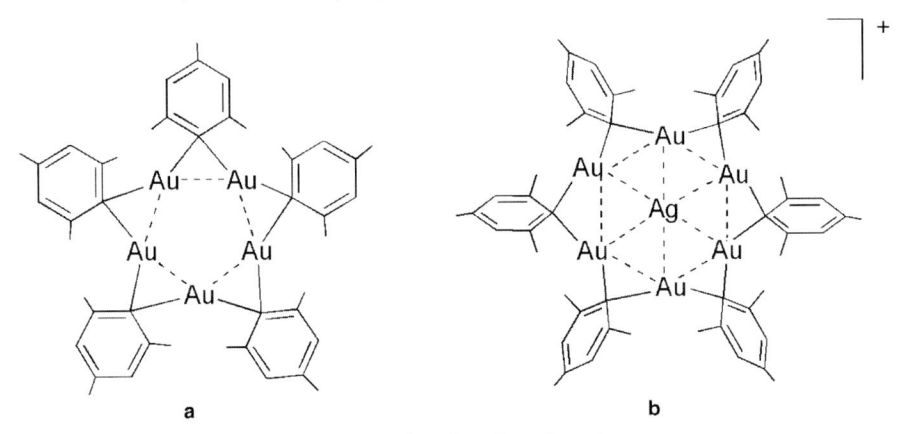

Figure 1.31 Oligomeric gold(I) complexes with carbon donor ligands.

Figure 1.32 Oligomeric gold(I) complexes with phosphide or phosphine ligands.

For phosphorus donor ligands the structures of [Au(PR$_2$)$_2$]$_n$ are oligomeric rings of different sizes ($n = 3,4,6$) depending on the phosphide substituents [215]. With diphosphines, the most common stoichiometry for complexes [Au(P–P)]$^+$ are dimers, as commented above, but other studies of gold(I) cations with empirical formula [Au{PPh$_2$(CH$_2$)$_n$PPh$_2$}]$^+$ show that the complexes crystallize as rings for $n = 3$ or 5 but as a polymer when $n = 4$ (as fused Au$_6$(µ-P–P)$_6$ rings) [216]. With the diphosphine 9,10-bis(diphenylphosphino)anthracene, the 1:1 reaction with [AuCl(SMe$_2$)] gives the trinuclear complex [Au$_3$(µ-PP)$_3$]$^{3+}$ [217]. This luminescent gold ring is shown to be an inorganic analog of cyclohexane in terms of structure and solution dynamics. These complexes are presented in Figure 1.32.

Thiolates, selenolates or tellurollates of general formula [Au(ER)]$_n$ are usually insoluble and only a limited number have been characterized structurally as oligomers. Tetramers have been reported, for example, for [Au$_4${E(SiMe$_3$)}$_4$] (E = S, Te) [218], hexamers for [Au$_6${S(2,4,6-C$_6$H$_2^i$Pr$_3$)}$_6$] [219], or the catenanes [Au$_{10}$(SC$_6$H$_4$-p-CMe$_3$)$_{10}$] and the precursor of the gold(I) drug Auranofin [Au$_{11}$(SR)$_{11}$] R = 2,3,4,6-tetra-O-acetyl-β-1-D-thioglucopyranosato [220] (Figure 1.33).

Triangular trigold(I) complexes of the type [Au$_3$L$_3$] are formed with pyrazolate, orthometallated pyridine, benzimidazole, and so on; they have been known for a long time but their remarkable chemical reactivity and physical properties have only recently been discovered. The gold trimer with carbeniate ligands [Au$_3$(MeN=COMe)$_3$] displays the new phenomenon termed solvo-luminescence. After irradiation with near-UV light, crystals of this compound show a long-lived photoluminescence that can be detected by the human eye for tens of seconds after cessation of irradiation. Addition of dichloromethane or chloroform to these previously-irradiated crystals produces a bright burst of light. For this phenomenon the solid state

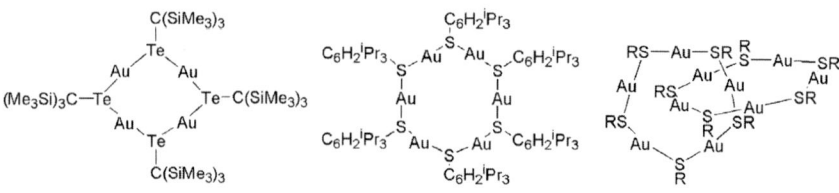

Figure 1.33 Oligomeric gold(I) derivatives with chalcogen donor ligands.

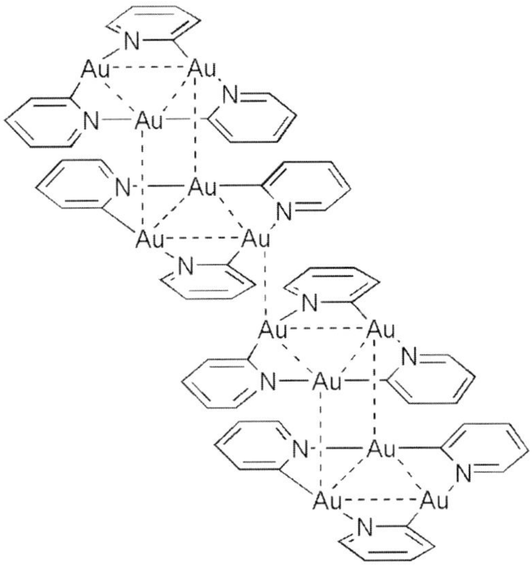

Figure 1.34 Structure of [Au$_3$(NC$_5$H$_4$)$_3$].

structure is crucial and consists of individual molecules of Au$_3$L$_3$ which aggregate to form columnar stacks through Au–Au interactions. It is believed that energy storage involves charge separation within the solid, and this charge separation is facilitated by conduction of electrons along the columnar structure [221]. A similar complex [Au$_3$(PhCH$_2$N=COMe)$_3$] does not associate into a trigonal prismatic array but packs in a stair-step fashion with eight discrete molecules in the asymmetric unit. This compound does not display solvoluminiscence [222]. The structure of the trimer [Au$_3$(NC$_5$H$_4$)$_3$] has been studied and shows that individual molecules self-associate through aurophilic interactions into two distinct structural motifs that involve both extended chains of molecules connected by pairwise and individual Au–Au contacts, and discrete dimers linked by pairwise Au–Au interactions (Figure 1.34) [223].

The trinuclear derivatives [Au$_3$(p-MeC$_6$H$_4$N=OEt)$_3$] and [Au$_3$(Bzim)$_3$] (Bzim = 1-benzylimidazolate) are colorless complexes but can produce brightly colored materials by sandwiching Ag$^+$ or Tl$^+$ ions and forming linear-chain complexes (Figure 1.35) with interesting luminescence properties such as thermochromism [224]. They also

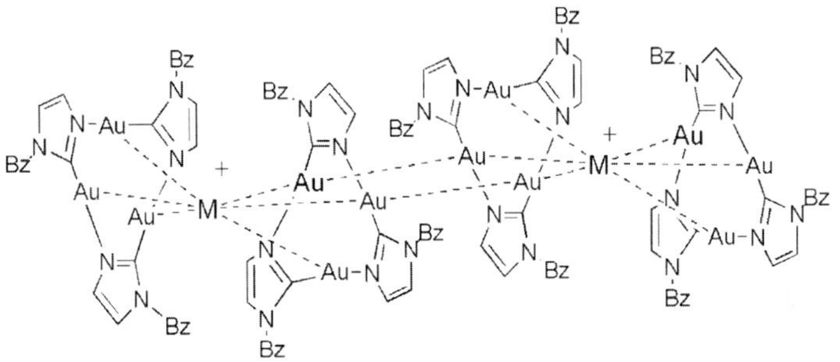

Figure 1.35 Structure of [Au$_3$M(benzimidazolate)$_3$]$^+$.

react with the π molecular acid trinuclear Hg^{II} complex $[Hg(\mu\text{-}C_6H_4)_3]$ to give compounds with acid–base stacking among the planar molecules [225].

Other oligomeric complexes are obtained with nitrogen donor ligands including amides, such as $[Au_4\{NSiMe_3)_3\}_4]$ [226], or formamidates such as $[Au_4\{(NPh)_2CH\}_4]$ [227], triazenides, such as $[Au_4\{(NPh)_2N\}_4]$ [228], or pyrazolates, such as the trinuclear species commented above or the hexanuclear $[Au_6(2\text{-}5\text{-}Ph_2pz)_6]$ [229].

1.2.4.5 Single Heteroatom Bridged Gold(I) Complexes

One of the most fascinating areas of gold(I) chemistry has been the discovery that phosphine–gold fragments coordinate around a central heteroatom. The species formed are exciting, not only from an experimental and structural viewpoint but also theoretically and are the clearest example of the "aurophilic power of gold." Thus, interesting hypercoordinated species of clustering of gold(phosphine) fragments around a central heteroatom such as carbon, phosphorus, nitrogen, sulfur, and so on, have been described, in addition to other complexes with more conventional stoichiometry, and all have in common the presence of short gold–gold interactions of about 3 Å. Usually, the chemistry of the first row elements of the p-block is known to follow classical rules of bonding, and only in cases of extreme electron-deficiency has the traditional electron count to be reconsidered in order to account for special types of molecular or solid state structures. Many of the heteroatom-centered complexes are electron deficient and the gold–gold interactions provide a significant contribution to their stability. The type of structure adopted is greatly influenced by the existence of gold–gold interactions.

Polyauration starts from the carbon atom for which the species with four, five and six gold atoms have been prepared. These are available from the reaction of polyborylmethanes with $[AuCl(PR_3)]$ or trimethylsilyl diazomethanes with $[O(AuPR_3)_3]^+$. The tetranuclear derivatives are formed with bulky phosphines and less sterically demanding phosphines enable the synthesis of the hypervalent species [230]. The structures of these complexes are tetrahedral, trigonal bypiramidal and octahedral, respectively (Figure 1.36). Many complexes of the type $[RC(AuPR_3)_4]^+$ [231] have also been synthesized.

Similar complexes have been prepared with nitrogen as a central hereroatom but to date only the tetranuclear and pentanuclear derivatives have been achieved from the reaction of ammonia with the oxonium complexes (Figure 1.37), with a tetrahedral or trigonal bipyramidal geometry, respectively [232]. Auration of polyamines or related

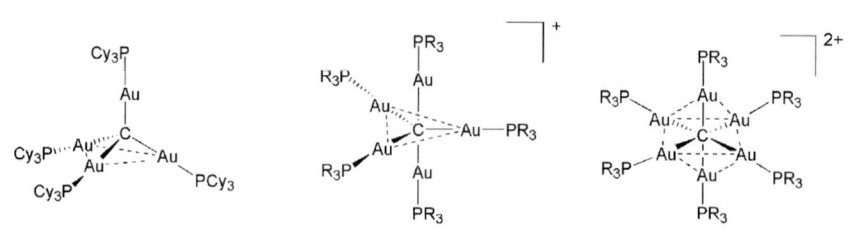

Figure 1.36 Carbon-centered gold(I) complexes.

Figure 1.37 Nitrogen-centered gold(I) complexes.

compounds has also been developed and many examples, such as the auration of hydrazine (Figure 1.37), have been reported [233].

Phosphorus and arsenic atoms also act as bridging ligands for several $AuPR_3^+$ fragments. The $[P(AuPR_3)_4]^+$ [234] and the $[As(AuPR_3)_4]^+$ [235] have different geometry because the phosphonium adopts the expected tetrahedral structure, whereas arsonium has a square-pyramidal geometry with the arsenic in the apical position, probably because the radius of the arsenic is too large to allow for metal–metal bonding in a tetrahedral structure (see Figure 1.38). If a lone pair of electrons is allocated to the As apex, then the $AsAu_4$ core has to be described as electron deficient. The $[P(AuPPh_3)_5]^{2+}$ has a square pyramidal arrangement, in contrast to the trigonal bipyramidal structure expected by classical bonding arguments [236]. For the tricationic species $[P(AuPR_3)_6]^{3+}$ an octahedral structure has been proposed [237].

The field of chalcogen-centered gold(I) complexes is well developed. Among the trinuclear derivatives $[E(AuPR_3)_3]^+$ which exist for all the chalcogens, usually as dimers in the solid state, the oxonium cations have displayed the most synthetic applicability, not only in the preparation of other heteroatom centered species but in many organometallic gold(I) complexes. The tetra(oxonium) compounds have been obtained by reaction of $[O(AuPR_3)_3]^+$ with $[Au(BF_4)(PR_3)]$ and have the expected tetrahedral geometry [238]. Heteronuclear complexes with a triply or quadruply oxo ligand have been reported, such as $[Rh_2(dien)_2\{O(AuPPh_3)_2\}_2](BF_4)_2$ obtained from $[O(AuPPh_3)_3]^+$ and $[RhCl(dien)]_2$ [239], or $[O(AuPPh_2py)_3M](BF_4)_2$ (M = Ag, Cu), prepared by reaction of $[O(AuPPh_2py)_3]^+$ with M^+ [240] (Figure 1.39). The latter has also been obtained for the sulfur or selenium atoms. A distinctive feature of these complexes is that they present an extremely bright luminescence. These species exhibit a large variation λ_{em}^{max} as a function of the μ_3-E capping ligand; The energy of

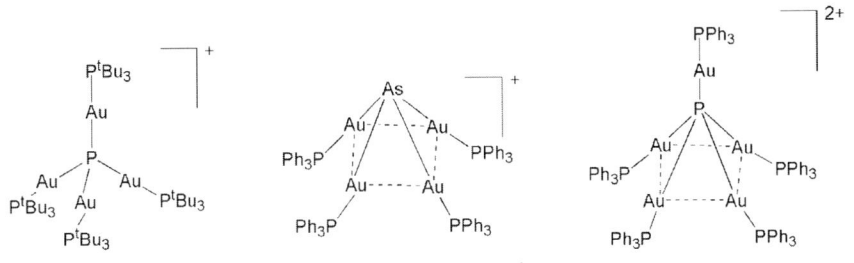

Figure 1.38 Phosphorus and arsenic centered gold(I) complexes.

Figure 1.39 Oxo-centered gold(I) complexes.

the emissions decreases on changing from oxygen, significantly at higher energy, to sulfur and to selenium. The large change in λ_{em}^{max} between these complexes clearly indicates involvement of the Group 16 capping atom in the excited state. Since the lone pair orbitals lie lowest in O, next in S, and highest in Se, an assignment consistent with the emission results in a ligand-to-metal–metal charge transfer (^{3}LMMCT).

The tetranuclear sulfur and selenium derivatives [E(AuPPh$_3$)$_4$]$^{2+}$ show a square pyramidal geometry with the chalcogen atom in the apical position [241]. These cations dimerize in the solid state through Au···E interactions (Figure 1.40). Further auration of sulfur and selenium to give the penta- and hexa-nuclear species has been achieved and trigonal bipyramidal and octahedral structures have been proposed [242].

Diphosphines have also been used as auxiliary ligands in chalcogen-centered gold chemistry and several complexes with a great variety of structural frameworks have been reported, such as the polynuclear derivatives [Au$_{12}$S$_4$(μ-dppm)$_6$]X$_4$, [Au$_{10}$Se$_4$(μ-dppm)$_4$]$^{2+}$ or [Au$_{10}$Se$_4$(μ-dppf)$_4$]$^{2+}$ shown in Figure 1.41 [242, 243]. Some of these

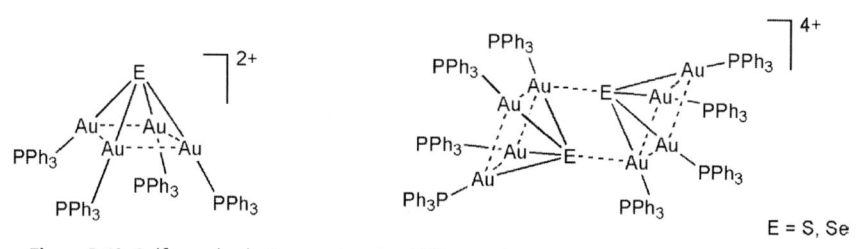

Figure 1.40 Sulfur and selenium-centered gold(I) complexes.

Figure 1.41 Polynuclear gold(I) complexes with bridging chalcogenide ligands.

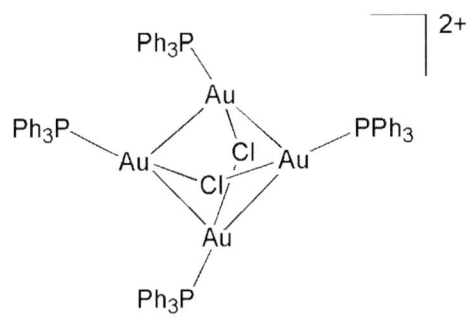

Figure 1.42 Structure of [{Cl(AuPPh$_3$)$_2$}$_2$]$^{2+}$.

compounds display rich luminescence properties with intense green and orange emissions.

Halide anions can also act as single bridges between two gold centers in complexes of the type [X(AuPPh$_3$)$_2$]$^+$ (X = Cl, Br), which are usually monomers but in SbF$_6^-$ salt the chloronium cations undergo an intimate aggregation; dimerization occurs against electrostatic repulsion, which appears to be offset by the gain in the number of aurophilic interactions (Figure 1.42) [244].

Other examples of polyauration around a single bridging heteroatom are those arising at borides, phosphide or chalcogenolate ligands. Therefore the reaction of the phosphineborane (Cy$_3$P)(SiMe$_3$)BH$_2$ with the oxonium salt, [O(AuPPh$_3$)$_3$]BF$_4$, gives the tetra-aurated compound [(Cy$_3$P)B(AuPPh$_3$)$_4$]BF$_4$ (Figure 1.43) [245].

Phosphide ligands of the type PR^{2-} or PR$_2^-$ also yield polynuclear derivatives such as [RP(AuPPh$_3$)]BF$_4$, obtained by reaction of RPH$_2$ with [O(AuPPh$_3$)]BF$_4$; this subsequently reacts with [Au(BF$_4$)(PPh$_3$)] to afford [RP(AuPPh$_3$)$_4$](BF$_4$)$_2$ [246]. Gold clusters with arsenide (AsR^{2-}) or phosphide (P^{3-} and PR$_2^-$) have been synthesized from the reaction of [Au$_2$Cl$_2$(μ-P–P)] with RAs(SiMe$_3$)$_2$, or with RP(SiMe$_3$)$_2$ and P(SiMe$_3$)$_3$, respectively [247]. They have a varied stoichiometry, such as [Au$_{19}$(As^{n}Pr)$_8$(dppe)$_6$]Cl$_3$, [Au$_{10}$(AsPh)$_4$(dppe)$_4$]Cl$_2$ or [Au$_{18}$P$_2$(PPh)$_4$(PHPh)(dppm)$_6$]Cl$_3$ (Figure 1.44).

Chalcogenolate gold(I) complexes of the type [Au(ER)(PR'$_3$)] (E = S, Se, Te) can further aggregate more gold atoms giving di- or trinuclear complexes of the form [Au$_2$(μ–ER)(PR'$_3$)$_2$]$^+$ or [Au$_3$(μ–ER)(PR'$_3$)$_3$]$^{2+}$ [41]. Similar complexes are

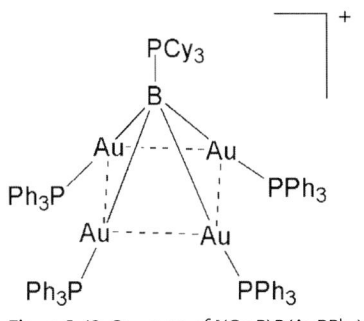

Figure 1.43 Structure of [(Cy$_3$P)B(AuPPh$_3$)$_4$]BF$_4$.

Figure 1.44 Phosphide and arsenide gold(I) complexes.

obtained with diphosphine ligands, for example the oxidation of [Au$_2$(SC$_6$H$_4$-4-Me)$_2$(μ-dppm)] with the ferrocenium cation [FeCp$_2$]PF$_6$ gives the luminescent cluster [Au$_9$(μ-SR)$_6$(μ-dppm)]$^{3+}$ [248]. A dinuclear diphosphine gold(I) thiolate has been functionalized with a macrocycle designed for specific coordination of K$^+$, the macrocycle encapsulates the ion in a sandwich fashion, bridging the two gold atoms in close proximity which triggers luminescence because of the formation of an Au–Au interaction [249]. The doubly-bridging chalcogenolate complexes are very numerous for sulfur, less for selenium and are not well represented for tellurium; one example is the cluster compounds [Au$_8$(TeR)$_8$(PR′$_3$)$_4$] prepared by reaction of TeR$^-$ with [AuCl(PR′$_3$)] in contrast to the expected mononuclear complexes obtained for sulfur and selenium [250]. The triply-bridging species are very scarcely represented and only the sulfur complexes [MeS(AuPMe$_3$)$_3$]$^{2+}$ or [^{t}BuS(AuCl)$_3$]$^-$ have been structurally characterized [251]. Figure 1.45 shows some examples of these chalcogenolate complexes.

1.2.4.6 Organometallic Gold(I) Complexes

Organometallic gold(I) complexes are an important class of complexes in gold chemistry. The Au–C bond is largely covalent and its stability depends largely on the type of ligand. Carbonyl gold(I) complexes are very scarce and unstable, alkyl complexes are not as stable as aryl, and the latter are in turn less stable than ylide or methanide ligands. Other important types of organometallic ligands in gold chemistry are alkynyl and carbene ligands, whose chemistry has recently been developed.

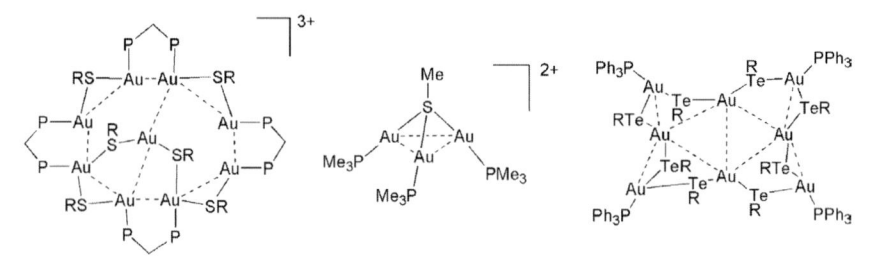

Figure 1.45 Chalcogenolate gold(I) complexes.

Only a few carbonyl complexes have been reported: the long known [AuX(CO)] (C = Cl, Br), [Au(OSO$_2$F)(CO)], [Au(CO)ClAuCl$_3$] and the tris(pyrazolyl)borate [Au{HB(2,5-(CF$_3$)$_2$pz)}$_3$(CO)] [252]. All the compounds exhibit υ(CO) infrared frequencies at higher wavenumbers than free CO, indicating insignificant π-back donation from gold to CO. Alkyl or aryl gold(I) complexes are usually synthesized by reaction of an alkyllithium or a Grignard reagent with a gold(I) compound. Complexes of the type [AuRL] have been reported for a great variety of organic ligands [253], some of them, such as [Au(C$_6$F$_5$)(tht)], with the pentafluorophenyl unit that confers great stability to these complexes, have been widely used as starting material in order to coordinate the "Au(C$_6$F$_5$)" fragment to almost any ligand type [41]. The anionic complexes [AuR$_2$]$^-$, where R is a perhaloalkyl group, have special electronic characteristics that allow them to react as Lewis bases towards many Lewis acid metals such as Ag$^+$, Tl$^+$, Cu$^+$, and so on. With silver, complexes of the type [Au$_2$Ag(C$_6$F$_5$)$_4$(OCMe$_2$)$_2$]$_n$ are obtained; this is a polymeric chain with the tetranuclear units bonded through aurophilic interactions [254]. However, with TlPF$_6$ a chain with unsupported Au–Tl–Au–Tl bonds is formed with [Au(C$_6$Cl$_5$)$_2$]NBu$_4$ [255]. Similar behavior is displayed by the dicyanide salt [Au(CN)$_2$]$^-$, which acts as a building block to synthesize supramolecular coordination polymers [256]. All these complexes show intense luminescence and in some cases reversible vapochromic behavior with several organic solvents. Figure 1.46 shows some of these complexes.

Ylide complexes of gold(I) in which a carbanionic center is σ-bonded to the gold center, are very stable. The ligands are usually phosphonium or sulfonium ylides with tetra-coordinated, positively charged P or S atoms and, consequently, the negative charge should be on the metal center. This is the reason for the excellent donor properties of these ligands. In order to illustrate these donor properties, the complex [Au(CH$_2$PPh$_3$)$_2$]ClO$_4$, in spite of its cationic nature, behaves as a donor towards Ag$^+$ salts giving the tetranuclear derivatives [Au$_2$Ag$_2$(OClO$_3$)$_4$(CH$_2$PPh$_3$)$_4$] (Figure 1.47a) with Au–Ag interactions [257]. Complexes with aurated ylides of the type [Me$_3$PC(AuPPh$_3$)$_3$]$^+$ [258] (Figure 1.47b) or multiaurated [(Me$_2$SO)C(AuPPh$_3$)$_4$]$^{2+}$ [259] (Figure 1.47c) have been reported.

Phosphonium salts of the type Me$_4$P$^+$ or Ph$_2$Me$_2$P$^+$, which has two carbon centers, can, after deprotonation, give the bis(ylide) species and its gold compounds

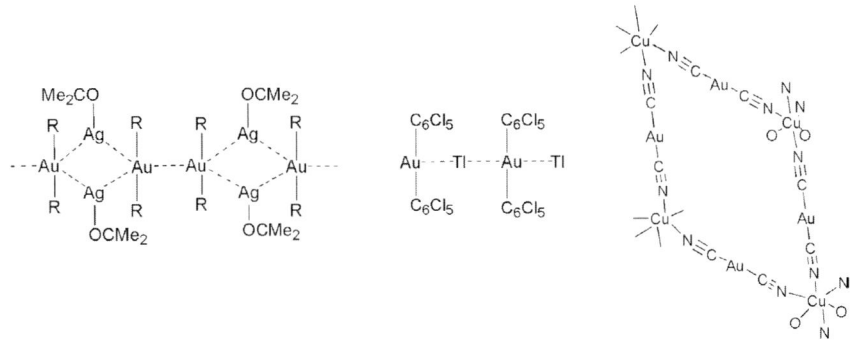

Figure 1.46 Donor–acceptor supramolecular gold(I) complexes.

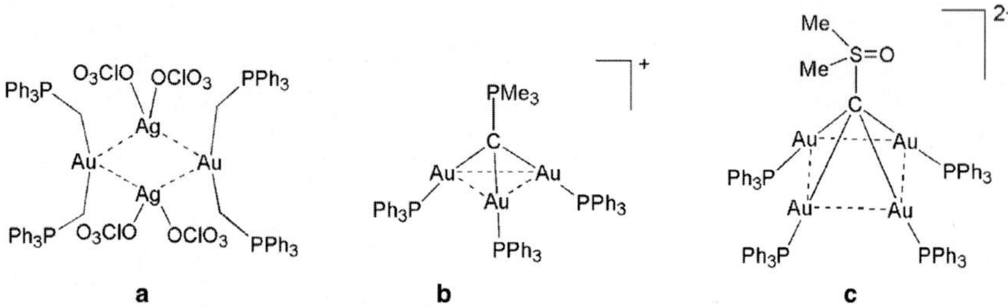

Figure 1.47 Ylide gold(I) complexes.

$[Au_2\{(CH_2)_2PR_2\}_2]$ have remarkable stability and have served as a gateway to a rich chemistry in dinuclear gold(II) ylide complexes [260]. Similar dinuclear complexes have been prepared with the $CH_2PPh_2S^-$ ylide ligand (Figure 1.48) [261]. Ligands related to the ylides are methanides. These are produced after deprotonation of a methylene ligand, usually between two phosphorus atoms. The best example is the dppm ligand which, after deprotonation, gives monoanionic tridentate or dianionic tetradentate ligands as in $[Au_4(CH_2SONMe_2)\{(PPh_2)_2CH\}(dppm)]$ [262] or $[(PPh_3Au)_2C\{(PPh_2)_2Au(PPh_2)_2\}C(AuPPh_3)_2]^{2+}$ [134], respectively, (Figure 1.49).

Carbenes are species with a divalent carbon atom with various substituents and a lone pair of electrons. Classic carbene gold complexes were synthesized in the coordination sphere of the gold atom, addition of amines or alcohols to the coordinated isocyanide ligands. N-Heterocyclic carbenes (Arduengo's carbenes)

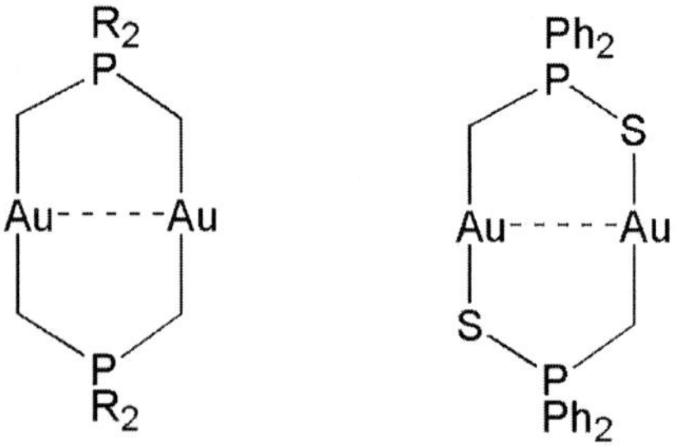

Figure 1.48 Dinuclear gold ylides.

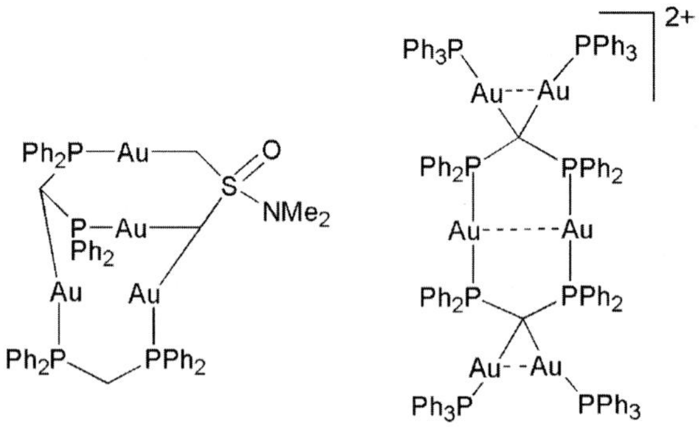

Figure 1.49 Methanide gold(I) complexes.

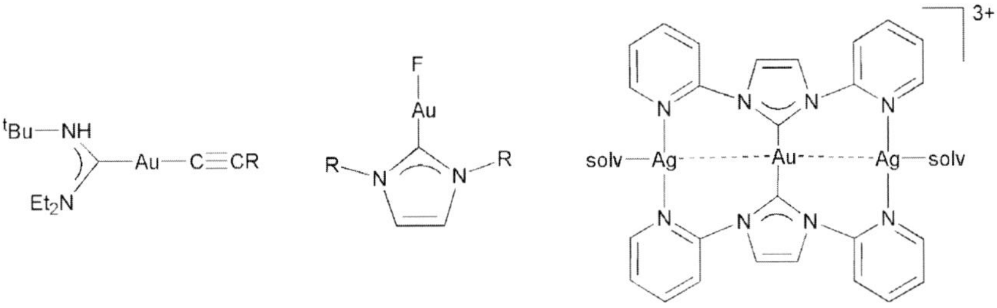

Figure 1.50 Carbene gold(I) complexes.

based on imidazol or benzimidazol have been developed more recently and have been extensively used in the catalysis of organic reactions by gold [26]; these complexes have also been postulated as intermediates in many gold catalyzed reactions (see Chapter 8). Addition of amines to gold(I) isocyanide complexes, including alkynes as auxiliary ligands in gold, gives the carbene derivatives (Figure 1.50a) [263]. Cationic bis(carbene) complexes have been known for a long time but their interesting luminescence properties have only been discovered recently. The complexes [Au{C(NHMe)$_2$}$_2$]PF$_6$ show structures in which the cations are stacked through aurophilic interactions and hydrogen bonds and are emissive. Frozen solutions are also luminescent with different colors, depending on the solvent, but the solutions are nonluminescent. Since the complex [Au{C(NHMe)(NMe$_2$)}$_2$]PF$_6$ is a monomer in the solid state, luminescence clearly occurs in the presence of the metallophillic interactions [264]. Many complexes with N-heterocyclic carbene ligands of different stoichiometries have been reported; they are of the type [AuX(carbene)], [AuL(carbene)]$^+$ or [Au(carbene)$_2$]$^+$ [265]. They derive mainly from imidazol, benzimidazol and benzothiazol salts that, after deprotonation, give the free carbene that can coordinate the gold center. Another route is reaction with the easily available silver carbene complexes. The complexes [AuCl(carbene)] have been successfully used as a catalyst in several organic reactions. A remarkable case is the obtainment of [AuF(carbene)] (Figure 1.50b), which represents the first example of an isolable gold(I) fluoride complex [266]. Theoretical calculations show significant $\pi p/d\pi$ interactions between fluoride and gold(I) and indicate a substantial negative charge on fluorine. Heterofunctional carbene ligands also give heteronuclear species with Au–Ag interactions and luminescence properties (Figure 1.50c) [267].

The chemistry of alkynyl gold(I) complexes has grown in recent years. The preference of gold(I) for a linear coordination, together with the linearity of the C≡C bond of the alkyne unit and its π-unsaturated nature, have made the alkynyl gold complexes attractive building blocks for molecular wires and organometallic oligomeric and polymeric materials, which may have unique properties such as optical nonlinearity, liquid crystallinity or luminescence. Alkynyl gold complexes have been postulated as intermediates in many organic reactions catalyzed by gold complexes (see Chapter 8). Complexes of the type [Au(C≡CR)L] are well represented and are readily obtained through the reaction of the polymeric [Au(C≡CR)]$_n$ species with L or by methathesis of [AuClL] complexes with the

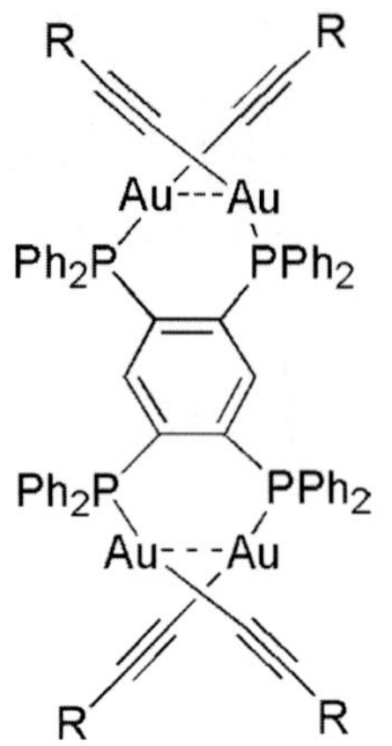

Figure 1.51 Structure of $[Au_4(C{\equiv}CR)_4\{(PPh_2)_4C_6H_2\}]$.

deprotonated alkynyls. The ligand L is usually a tertiary phosphine and bidentate or polydentate phosphines (Figure 1.51) have been used as auxiliary ligands, giving polynuclear complexes not always of the expected nuclearity [41, 268]. These reactions have led to the discovery of organometallic catenanes by self-assembly of an oligomeric digold(I) diacetylide $[\{(AuC{\equiv}CCH_2OC_6H_4)_2X\}_n]$ and a diphosphine ligand $Ph_2P(CH_2)_nPPh_2$. Systematic research of this unusual reaction has revealed that the number of methylene spacer groups n in the diphosphine ligand and the nature of the hinge group X are key factors for determining whether self-assembly will give a simple ring by $1+1$ assembly, a [2]catenane by $2+2$ assembly, or a double braided [2]catenane by $4+4$ asembly (Scheme 1.1) [269].

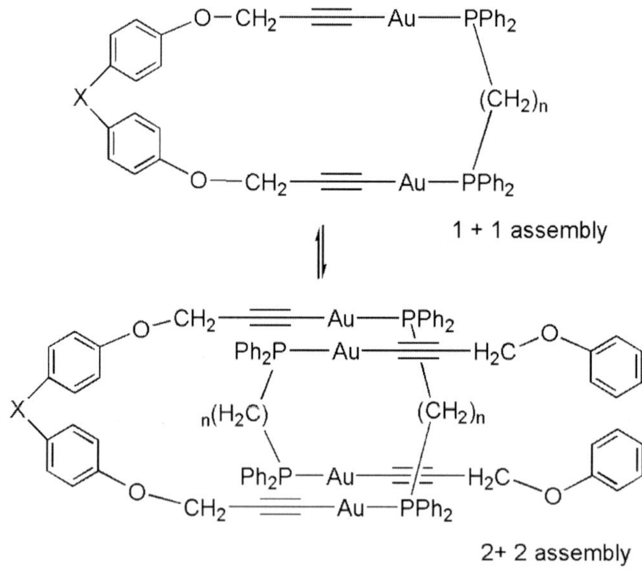

Scheme 1.1 Catenane gold alkynyl complexes.

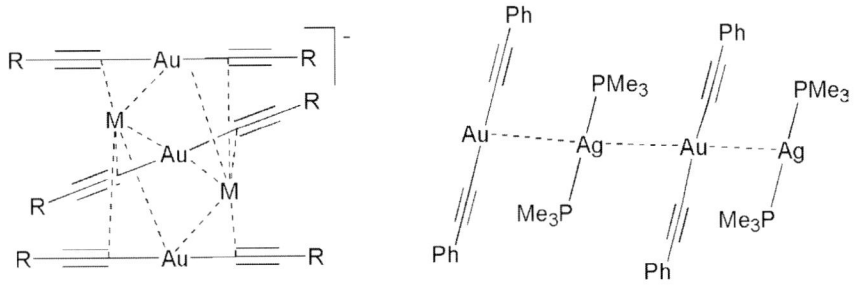

Figure 1.52 Heteronuclear alkynyl gold(I) complexes.

Anionic gold(I) alkynyl complexes of the type [Au(C≡CR)$_2$]$^-$ are also known and reactions with metals such as Ag$^+$, Cu$^+$ give heteropolynuclear complexes [Au$_3$M$_2$(C≡CR)$_6$]$^-$ (M = Cu, Ag). The complex [Ag(PMe$_3$)$_2$][Au(C≡CPh)$_2$] forms a chain of alternating cations and anions through Au–Ag interaction [270]. Figure 1.52 collects some of these alkynyl gold(I) complexes.

1.2.5
Gold(II) Complexes

The number of complexes with gold in a formal oxidation state of two have increased considerably and nowadays this oxidation state can almost be considered as a common oxidation state in gold chemistry; however, the number of gold(II) complexes is very scarce if compared with the more common gold(I) or gold(III) derivatives. The energy required to reach Au^{2+} from atomic gold is not very far from that required to form either Cu^{2+} or Ag^{2+} and to attain M^{3+} less energy is required for Au than for Cu and Ag. Therefore, this argument is not sufficient to justify the lack of stability for the oxidation state 2+ in gold. There is a strong tendency for disproportionation to give Au$^+$ and Au^{3+} because the odd electron in d^9 metal complexes is in a d$_{x2-y2}$ orbital (octahedral tetragonally distorted or square planar arrangement) which has a much higher energy than copper and can be easily ionized. The formation of a gold–gold bond gives more stable compounds and the Au$_2^{4+}$ core species are the most stable and abundant types of gold(II) complexes. Mononuclear gold(II) derivatives are not very abundant and, when reported, further study confirmed that they were mixed gold(I)–gold(III) complexes such as the halides CsAuX$_3$ (=Cs$_2$Au^{I}Au^{III}X$_6$, X = Cl, Br, I) [271].

1.2.5.1 Mononuclear Gold(II) Complexes
Mononuclear gold(II) complexes, consistent with a d^9 configuration, must be paramagnetic (μ_{eff} = 1.79 MB) and show a hyperfine four-line EPR signal, in accordance with the nuclear spin of ^{197}Au (I = 3/2). These two properties are evidence of a real gold(II) complex in addition to their stoichiometry.

In 1992 Herring *et al.* gave clear EPR and magnetic evidence for Au^{2+}, as a species present in partially-reduced Au(SO$_3$F)$_3$ and as a solvated ion in the strong protonic acid HSO$_3$F [272]. Recently, Bartlet et al. prepared and structurally characterized Au(II) fluoro complexes [273]. Gold dissolves at around 20 °C with F$_2$ in anhydrous HF acidified with SbF$_5$, to give a red solution from which orange crystals of Au(SbF$_6$)$_2$ crystallize. Exhaustive fluorination results in total conversion of the gold to an insoluble crystalline red solid, which is Au^{II}(SbF$_6$)$_2$Au^{II}(Au^{III}F$_4$)$_2$. Solvolysis of Au(SbF$_6$)$_2$ in anhydrous HF results in disproportionation to gold and the mixed-valence fluoride Au^{II}Au^{III}$_2$F$_8$. In Au(SbF$_6$)$_2$ the gold(II) atom is at the center of an elongated octahedron of F ligands, the fluor atoms of the approximately square AuF$_4$ unit are at 2.09(2) and 2.15(2) Å × 2, each F provided by a different SbF$_6$ species. The two long Au–F interatomic distances are at 2.64(2) Å.

Most of the compounds already described were prepared with unsaturated S-donor ligands such as dithiocarbamates, dithiolates, dithiolenes or, in general, ligands able to delocalize the unpaired electron, and can be better described as ligand-radical species [274]. The mononuclear gold(II) complex with the thioether macrocycle 1,4,7-trithiacyclononane ([9]aneS$_3$), [Au([9]aneS$_3$)$_2$](BF$_4$)$_2$, (Figure 1.53) obtained by reduction of HAuCl$_4$·3H$_2$O with two equivalents of [9]aneS$_3$ in refluxing HBF$_4$ (40%)/MeOH [275] was described as a truly gold(II) complex with an octahedral arrangement of six sulfur atoms around the gold center, with a Jahn–Teller distortion. The EPR spectrum showed a hyperfine four-line EPR signal at $g = 2.010$ and the lone pair of electrons belonged predominantly to the gold center, with limited delocalization through the ligand. However, DFT calculations and EPR simulations coincided and showed that the singly-occupied molecular orbital (SOMO) has about 27–30% Au 5d$_{xy}$ character and 62–63% equatorial S 3p character, confirming the non-innocence of thioether ligands in this system [276].

Stable gold(II) complexes with nitrogen- and oxygen-containing ligands such as CO$_2$, Me$_2$CO, thf, pyridine, and so on, have been prepared in the gas phase. The successful ligands are characterized by being good σ donor–π acceptor molecules and the most stable are those with large dipole moment and a high ionization energy [277].

Recently, the first metal–xenon compound was obtained by reducing AuF$_3$ with elemental xenon. In attempts to synthesize the elusive AuF, gold(III) reduction was performed with xenon, which is a very mild reducing and very weakly coordinating agent. Surprisingly, the reaction stopped at the Au^{2+} state and resulted in a completely unexpected complex, the cation AuXe$_4$$^{2+}$, the crystals of which can be

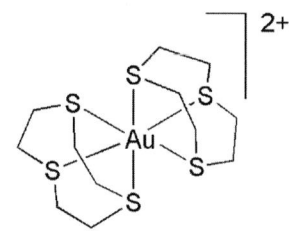

Figure 1.53 Structure of [Au([9]aneS$_3$)$_2$](BF$_4$)$_2$.

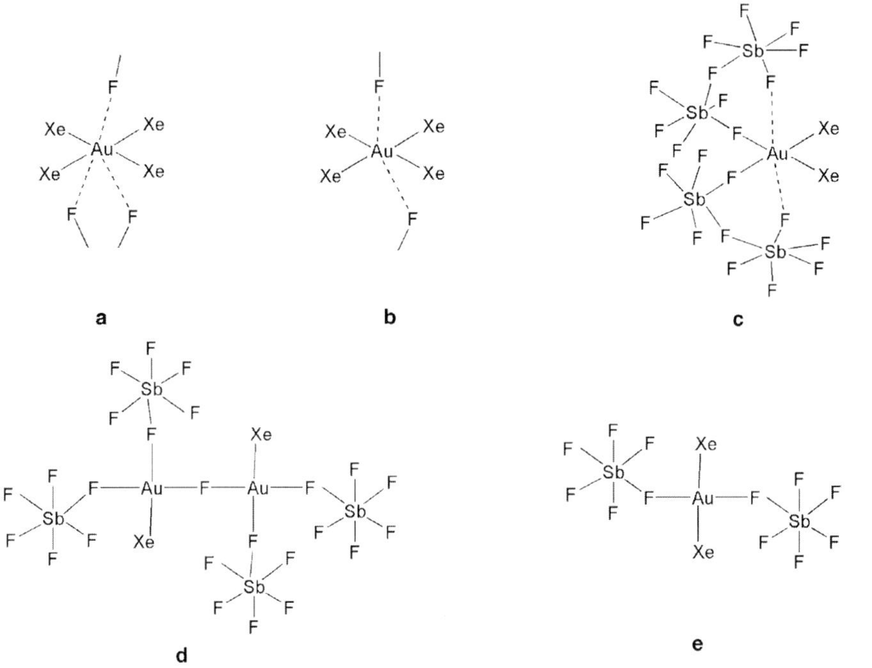

Figure 1.54 (a) triclinic $[AuXe_4]^{2+}$; (b) monoclinic $[AuXe_4]^{2+}$; (c) $[AuXe_2]^{2+}[Sb_2F_{11}]^{2-}$; (d) $[Au_2Xe_2F]^+[SbF_6]^-$; (e) trans-$[AuXe_2]^+[SbF_6]^-$.

grown at −78 °C. Removal of gaseous xenon under vacuum results in the crystallization of $Au(SbF_6)_2$ [278]. The crystal structure of $(Sb_2F_{11})_2[AuXe_4]$ shows a regular square with Au–Xe bond lengths ranging from 2.728(1) to 2.750(1) Å. Three weak contacts between the cation and the anion complete the coordination sphere around the gold atom with Au–F distances of 2.671 and 3.153 Å (Figure 1.54).

In this complex, xenon functions as a σ-donor toward Au^{2+}. This is reflected in the calculated charge distribution within the cation, where the main part of the positive charge resides in the xenon atoms. Relativity plays a large role in stabilizing this and other predicted Au–Xe compounds; about half of the Au–Xe bonding energy comes from relativistic effects [279].

A further study of this reaction [280] has shown that if the concentration of SbF_5 is fairly high, complex $[AuXe_4][Sb_2F_{11}]_2$ forms as the sole product in two crystallographically-different modifications, the triclinic commented previously and another monoclinic, which has been seldom observed. Both modifications differ only in the cation–anion interactions, with a long Au–F contact of 2.928(7) in the monoclinic form. Through a variation in xenon pressure and the acid strength of the HF/SbF_5 other species have been isolated such as cis-$[AuXe_2][Sb_2F_{11}]_2$, trans-$[AuXe_2][SbF_6]_2$, $[Au_2Xe_2F][SbF_6]_3$ and the first Au(III)-Xe complex $[AuXe_2F][SbF_6][Sb_2F_{11}]$. Their structures are shown in Figure 1.54.

1.2.5.2 Polynuclear Gold(II) Complexes

As mentioned previously, one of the reasons for the poor stability of gold(II) complexes is the unfavorable energy of the odd electron. The formation of a metal–metal bond in dinuclear gold(II) complexes giving diamagnetic species provides extra stability. Thus, the number of gold(II) complexes containing the Au_2^{4+} core has increased in recent decades and different stoichiometries are known. Most of the dinuclear gold(II) compounds are synthesized from the corresponding gold(I) precursors by oxidative addition of halogen. This method works properly with a great variety of dinuclear gold(I) complexes. These include symmetric and asymmetric doubly bridged compounds, monobridged dinuclear derivatives or compounds without any bridging ligand. The bridging ligands vary from diphosphines, dithiocarbamates and related species, ylide ligands and mixed-donor ligands of the types C,S or C,P. Experimentally it has been confirmed that the oxidation of a dinuclear gold(I) complex to a gold(II) derivative gives more stable complexes when the gold is bonded to carbon donor ligands. This is true in the case of bis-ylide ligands. Although the most commonly used oxidants are halogens, others have been used such as haloalkyl RX, $[Hg(CN)_2]$, nitroalkanes, N_2O_4, $[Tl(C_6F_5)_2Cl]$, tetraethylthiuram disulfide, $[Ag(OClO_3)(PPh_3)]$ or $[Ag(NCMe)_4]PF_6$.

Dinuclear homoleptic gold(II) complexes with a homo-bridged ligand have been described with many ligands such as sulfate [281], ylide [260] dithiocarbamate [282], dithiolate [283], amidinate [284] or with heterofunctional ligands such as $SPPh_2CH_2^-$ [261], $PR_2C_6H_4^-$ [285] (Figure 1.55). The sulfate salt $AuSO_4$ has long been referred to as a mixed-valent compound in many chemistry textbooks, in spite of the lack of information regarding its crystal structure. Recently, the crystal structure has been determined and shows that $AuSO_4$ is the first simple inorganic compound known containing the cation Au_2^{4+}. This cation is coordinated by two chelating sulfate groups and two monodentate SO_4^{2-} ions. The gold–gold distance is 2.49 Å, the shortest known for dinuclear gold(II) complexes.

The majority and the most stable compounds have been prepared with bis-ylide ligands. Complexes of the type $[Au_2\{\mu\text{-}(CH_2)_2PR_2\}_2]$ have been useful starting materials for the development of gold(II) chemistry. Their oxidation with several reagents such as halogens, alkyl halides, E_2R_2, $[Hg(CN)_2]$ affords the corresponding gold(II) derivatives (see Scheme 1.2). The addition of haloalkanes to dinuclear gold(I) bis-ylide complexes also affords the gold(II) species. Considerable effort has been expended to elucidate the oxidative addition reaction of haloalkanes, and experimental evidence supports the notion that the order of reactivity of such

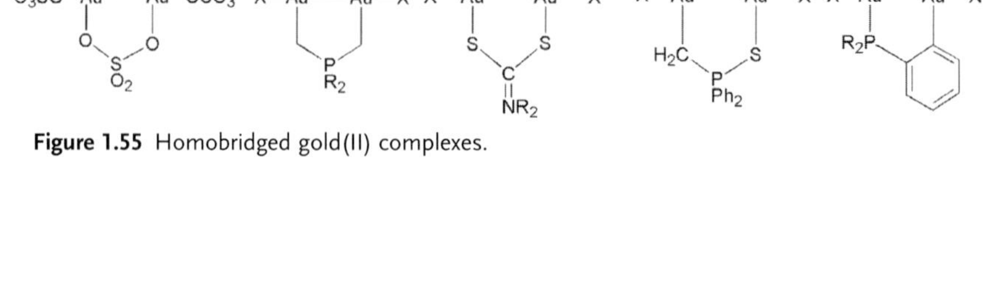

Figure 1.55 Homobridged gold(II) complexes.

Scheme 1.2 Synthesis of bis(ylide) gold(II) complexes.

substrates is inversely proportional to the carbon–halogen bond dissociation energies [286].

Dinuclear gold(II) derivatives with two different bridging ligands have also been synthesized; all of them have the ylide $(CH_2)_2PPh_2^-$ ligand and another bridging ligand that can be diphosphines such as $Ph_2PCH_2PPh_2$ or $Ph_2PNHPPh_2$, dithiocarbamates, xanthate, 2-pyridinethiolate, phosphoniodithioformate, or dithiophosphinate [287] (Figure 1.56).

Another general procedure for preparing gold(II) complexes consists of substitution reactions on gold(II) derivatives. Halide ligands can be substituted by neutral donor ligands such as tetrahydrothiophene, py, ylide, phosphine, and so on, to give the corresponding cationic complexes, or with other anionic ligands such as pseudohalide, carboxylate, dithiocarbamate, thiourea derivatives, and so on, [288]. An interesting class of substitution reactions in these gold(II) derivatives are those where the new

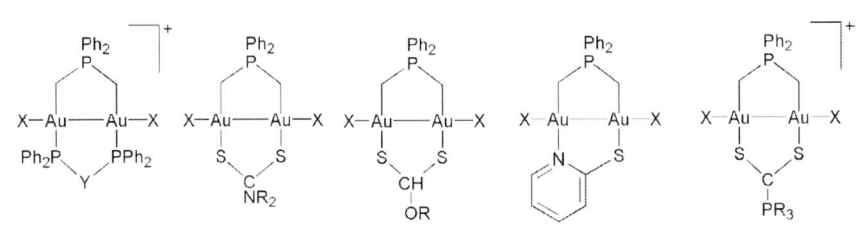

Figure 1.56 Heterobridged gold(II) complexes.

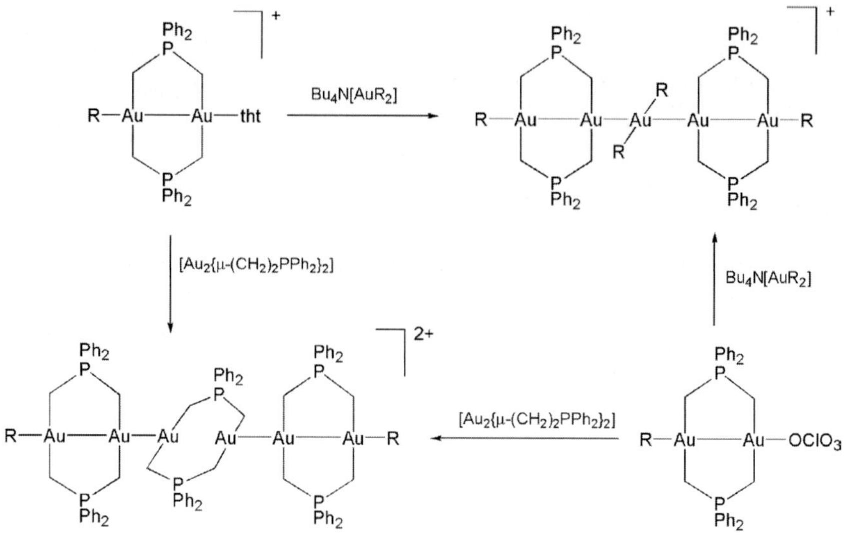

Scheme 1.3 Synthesis of polynuclear gold(II) ylide complexes.

ligands are metal complexes. The reaction of the gold(II) [Au$_2$(C$_6$F$_5$)$_2${μ-(CH$_2$)$_2$PR$_2$}$_2$] with the gold(III) species [Au(C$_6$F$_5$)$_3$(OEt)$_2$] gives the pentanuclear complex [{Au$_2$R{μ-(CH$_2$)$_2$PR$_2$}$_2$}$_2$(AuR$_2$)][AuR$_4$] (R = C$_6$F$_5$) [289]. Its backbone is a linear chain of five gold atoms, all of which have square planar geometry. The Au–Au distances of 2.755(1) and 2.640(1) Å are characteristic of metal–metal bonds, the former corresponding to the unsupported gold–gold bond. The unit [{Au$_2$R{μ-(CH$_2$)$_2$PR$_2$}$_2$}$_2$]$^+$ is readily accessible from many complexes such as [Au$_2$R(tht){μ-(CH$_2$)$_2$PR$_2$}$_2$]$^+$ or [Au$_2$R(OClO$_3$){μ-(CH$_2$)$_2$PR$_2$}$_2$] (R = C$_6$F$_5$, 2,4,6-C$_6$F$_3$H$_2$). The reaction of these gold(II) complexes with either organoaurate Bu$_4$N[AuR$_2$] or the gold(I) bis-ylide [Au$_2${μ-(CH$_2$)$_2$PR$_2$}$_2$], both highly nucleophilic compounds, enables the synthesis of gold chain complexes (Scheme 1.3) [290].

Another interesting polynuclear gold(II) complex has been reported with the ylide ligand SPPh$_2$CH$_2$$^-$ by oxidation with halogens of the trinuclear species [Au$_2$Pt(μ-CH$_2$PPh$_2$S)$_2$]. A Cl–Au–Pt–Au–Cl (Figure 1.57a) is obtained and although an Au^{II}-

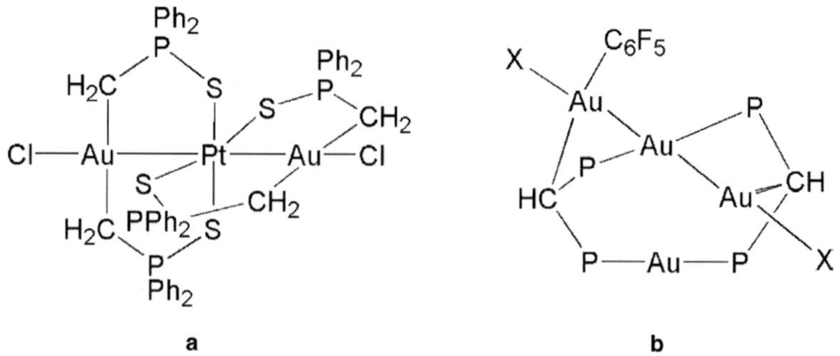

Figure 1.57 Polynuclear gold(II) complexes.

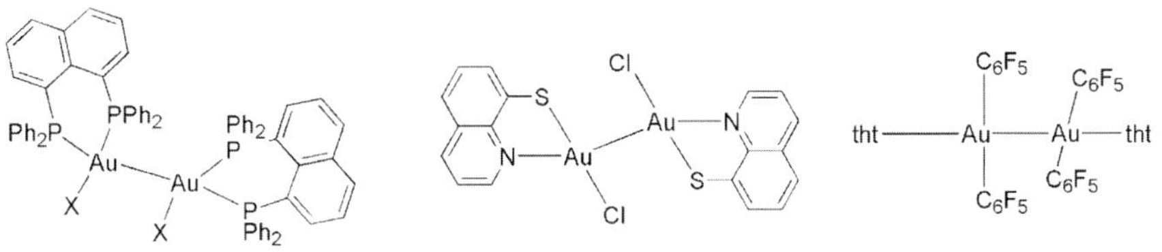

Figure 1.58 Unsupported gold(II) complexes.

Pt^{II}-Au^{II} assignment of the oxidation states can be made, a more correct description, based on Fenske–Hall calculations, is that an [Au-Pt-Au]$^{2+}$ moiety is formed [291]. The reaction of the tetranuclear complex [Au$_2${μ-(PPh$_2$)$_2$CHAu(C$_6$F$_5$)}$_2$] with chlorine or bromine does not oxidize the two gold centers of the diauracycle to give the usual gold(II) derivatives with an X–Au–Au–X backbone. Instead, a new type of reaction occurs, probably because of the presence of other gold(I) centers in close proximity (Figure 1.57b). The proposed assignment of the oxidation states in the linear chain is Au(II)–Au(I)–Au(II) [292].

Dinuclear gold(II) complexes with unsupported Au^{II}–Au^{II} bonds have been reported for chelating ligands such as 1,8-bis(diphenylphosphino)naphthalene (dppn), obtained by oxidation of [Au$_2$X$_2$(dppn)] with [Ag(NCMe)$_4$]PF$_6$ [293], or bis (quinolin-8-ylthio) [294]. The only one described with no chelating ligands is [Au$_2$(C$_6$F$_5$)$_4$(tht)$_2$], which has been prepared by reaction of equimolar amounts of [Au(C$_6$F$_5$)(tht)] and [Au(C$_6$F$_5$)$_3$(tht)] [295]. Figure 1.58 shows these unsupported gold (II) complexes.

1.2.6
Gold(III) Complexes

The chemistry of gold(III) complexes is far less developed than the corresponding gold(I) complexes. Gold(III) gives stable complexes with C, N, P, S, or even O-donor ligands. Organometallic chemistry is very important because many compounds have been prepared starting from organometallic gold(III) precursors.

1.2.6.1 Organometallic Gold(III) Complexes
The majority of gold(III) complexes with carbon donor ligands are with alkyl and mainly aryl ligands and several methods exist for their synthesis:

1. Oxidative addition of halogen to the respective gold(I) complex
2. Electrophilic substitution by gold(III) of one aromatic ring (metallation)
3. Transmetallation reaction with HgR$_2$
4. Substitution reactions on gold(III) derivatives.

Alkyl gold(III) complexes have the general formula [AuR$_4$]$^-$, [AuR$_3$L], [AuR$_2$X$_2$]$^-$ or [AuR$_2$L$_2$]$^+$ where L can be a great variety of donor ligands and R can also be replaced by another anionic ligand such as halogen, giving complexes of the type [AuR$_n$X$_{3-n}$L] or with mixed organometallic ligands [AuRnR$'_{3-n}$L]. Some examples are with methyl or trifluoromethyl ligands such as [Au(CF$_3$)X$_2$(PR$_3$)] or [AuMe$_2$(P,N)]$^+$ [296]

Figure 1.59 Alkyl or aryl gold(III) complexes.

(Figure 1.59a). Aryl gold(III) complexes are more numerous and complexes with perfluoroaryls, such as [Au(C$_6$F$_5$)$_3$(tht)] (Figure 1.59b) obtained by oxidative addition of [Au(C$_6$F$_5$)(tht)] with [Tl(C$_6$F$_5$)$_2$Cl]$_2$, have been widely used as a starting material because of their high stability [297]. The compound [Au(C$_6$F$_5$)$_3$(OEt$_2$)] is a very good acceptor that reacts with the electron-rich bis(ylide) to give the donor–acceptor complex [Au{(CH$_2$)$_2$PPh$_2$}$_2$Au(C$_6$F$_5$)$_3$] (Figure 1.59c) [298]. The complexes [Au(C$_6$F$_5$)$_2$X$_2$]$^-$ and [Au(C$_6$F$_5$)$_2$Cl]$_2$ are also excellent starting materials for obtaining complexes with the "Au(C$_6$F$_5$)$_2$" unit [297]. Arylgold(III) complexes are good catalysts for the addition of nucleophiles to alkynes (see Chapter 8).

The auration of hydrocarbons such as benzene, toluene, and so on, with anhydrous AuCl$_3$ in an inert solvent gives AuRCl$_2$, which, in the presence of ligand L, affords [AuRCl$_2$L] compounds [299]. The reaction of substituted pyridine ligands (HL) with [AuCl$_4$]$^-$ gives the complexes [AuCl$_3$(HL)], which upon heating give the cyclometallated species [Au(C,N)Cl$_2$] or [Au(C,N,N)Cl]$^+$ [300]. Several gold(III) cyclometallated complexes have shown cytotoxic activity to various cancer cells and sometimes with significant antiproliferative effects and promotion of apoptosis to a greater extent than platinum drugs [301]. Sometimes cyclometallation is difficult to achieve and the use of organomercurials as arylating agents is an appropriate route to [Au(N,C)Cl$_2$] complexes [302], which has been used with azobenzene, N,N'-dimethylbenzylamine, and so on. Sequential arylations enable the synthesis of complexes with two different ligands of the type [Au(C,N)ArCl]. Upon treatment with phosphine or chloride, some of the N,C quelated diarylgold(III) chloride complexes undergo reductive elimination of the aryl ligands to give biaryls [303]. Figure 1.60 presents some examples of these complexes.

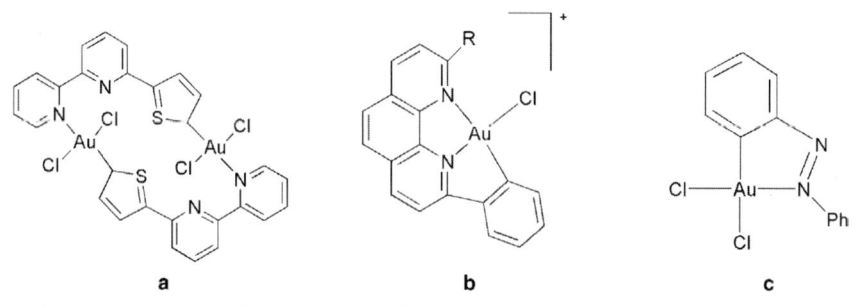

Figure 1.60 Cyclometallated gold(III) complexes.

Figure 1.61 Ylide gold(III) complexes.

Ylide gold(III) complexes have been prepared by oxidation of the corresponding gold(I) complexes with CH_2X_2 or with halogens in the bis(ylide) gold(I) complexes $[Au_2\{(PPh_2)_2PR_2\}_2]$ [304], even by disproportionation of the gold(II) derivatives as in $[Au_2\{(PPh_2)_2PR_2\}_2(C\equiv CPh)_2]$ [305], and also by substitution reactions in gold(III) complexes of the type $[AuMe_3(PPh_3)]$, or by ylide transfer reactions from ylidegold(I) species [306]. Figure 1.61 shows some of these examples.

With methanide ligands a great amount of work has been carried out starting with complexes of the type $[Au(C_6F_5)_2(\eta^2\text{-L-L})]ClO_4$, $[Au(C_6F_5)_3(\eta^1\text{-L-L})]$, in which L–L represents a ligand such as $PPh_2CH_2PPh_2$, $PPh_2CH_2PPh_2R^+$ or $SPPh_2CH_2PPh_2$, and so on. Substitution by deprotonation with a base or with $[Au(acac)(PPh_3)]$ from one to three protons in one or two methylene groups affords a variety of polynuclear complexes with different coordination modes, as shown in Figure 1.62 [307].

Carbene gold(III) complexes have been developed to a much lesser extent than the corresponding gold(I). The reaction of the gold(III) isocyanide complexes $[Au(C_6F_5)_2(CNR)_2]ClO_4$ with hydrazines gives the double carbene species [308]. Related species have been prepared from $[AuCl_3(tht)]$ and lithiated thiazoline [309] or from the tetra(azide)gold(III) complex with isocyanide [310]. N-Heterocyclic carbenes have been prepared only recently by oxidation of the gold(I) derivatives [311]. Figure 1.63 shows these types of complexes.

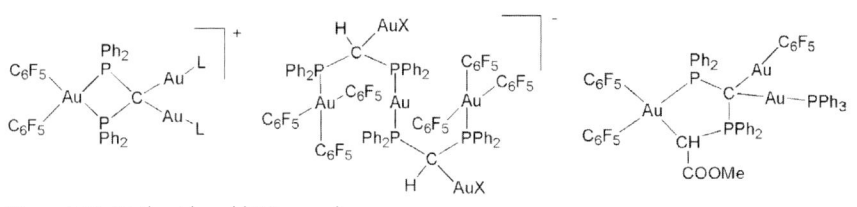

Figure 1.62 Methanide gold(III) complexes.

Figure 1.63 Carbene gold(III) complexes.

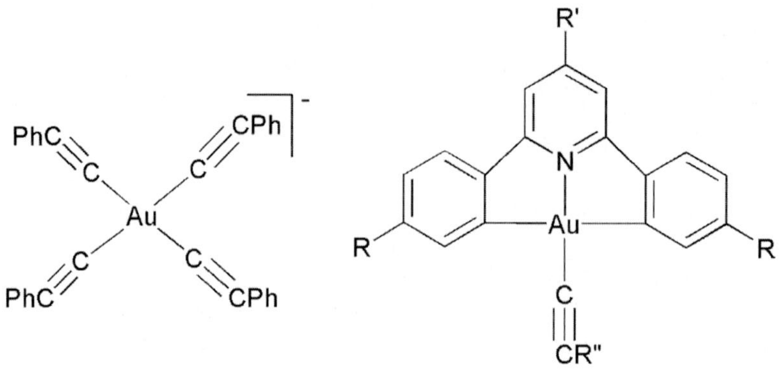

Figure 1.64 Alkynylgold(III) complexes.

Alkynyl gold(III) derivatives are not very numerous and the homoleptic species [Au(C≡CR)$_4$]$^-$ or the neutral [Au(C≡CR)$_3$L] have been prepared, but rapid reduction to dialkynyls or [Au(C≡CR)L] complexes occurs [312]. Several alkynylgold(III) complexes, which contain cyclometallated ligands, have been prepared and display rich luminescence. These are of the type [Au(C≡CR)(C,N,C)] and the origin of the emission is assigned to a π–π* intraligand transition of the cyclometallated moiety [313]. Figure 1.64 shows the alkynyl gold complexes. The tetra(cyano)gold(III) derivative [Au(CN)$_4$]$^-$ also functions as a building block in the synthesis of coordination polymers. It is a weaker ligand than the dicyanoaurate(I) and does not form gold–gold interactions but still yields polymers via M–NC coordinate bonds and also weak Au–NC intermolecular contacts [314] (Figure 1.65).

1.2.6.2 Gold(III) Complexes with Polydentate Ligands

Gold(III) complexes with nitrogen donor ligands are more stable than the corresponding gold(I) species. Several examples with polydentate amine ligands of different types have been reported. Polypyridines such as phenanthroline, terpyridine, pyrazolate ligands, and so on, form square planar gold(III) complexes such as [AuCl(terpy)]$^{2+}$ [315], [AuMe$_2$(py$_2$CHpy)]$^+$ [316] or [AuCl$_2$(μ-pz)]$_2$ or [Au{N,N'-(pz)$_3$BH}$_2$]ClO$_4$ [317]. The rigid bidentate ligand phen enables the synthesis of pseudo-pentacoordinate gold(III) derivatives as in [Au(C$_6$H$_4$CH$_2$NMe$_2$)(phen)(PPh$_3$)]$^{2+}$ [318]. Complexes with polyamines and triethylenetetramine are known [319]. Figure 1.66 shows several of these complexes with polydentate amine

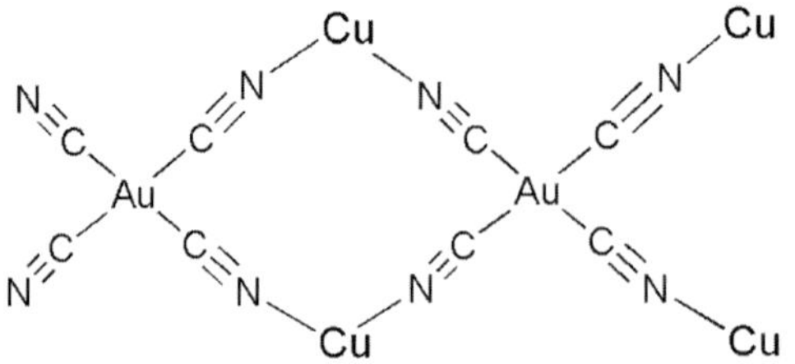

Figure 1.65 1-D chain structure of the cation [Cu(bipy)(H$_2$O)$_2$(Au(CN)$_4$)$_{0.5}$][Au(CN)$_4$]$_{1.5}$.

Figure 1.66 Gold(III) complexes with polydentate amine complexes.

complexes. The cytotoxic activity of some of these complexes with polydentate nitrogen ligands has been tested. The complex [AuCl$_2$(phen)]Cl has shown to be highly cytotoxic toward the A2780 tumor cell line either sensitive or resistant to cisplatin; solution chemistry studies show that the fragment [Au^{III}phen]$^{3+}$ is stable in solution for several hours, even under physiological conditions. The predominant species existing in solution under physiological conditions is likely to be [Au(OH$_2$)$_2$(phen)]$^+$ [320].

Gold(III) porphyrins have been used as acceptors in porphyrin diads and triads due to their ability to be easily reduced, either chemically or photochemically. A new method for incorporating gold(III) into porphyrins (Figure 1.67a) has been described and consists of the disproportionation of [Au(tht)$_2$]BF$_4$ in its reaction with the porphyrin in mild conditions [321]. The metallation of [16]-hexaphyrin with NaAuCl$_4$ yielded the aromatic gold(III) complexes (Figure 1.67b) and the two-electron reduction of the aromatic complexes provided the antiaromatic species [322].

Complexes with diphosphines are numerous and are prepared mainly from substitution reactions in [Au(C$_6$F$_5$)$_3$(tht)] or [Au(C$_6$F$_5$)$_2$(OEt$_2$)$_2$]$^+$ and are of the form [Au(C$_6$F$_5$)$_3$(L-L)], [Au$_2$(C$_6$F$_5$)$_6$(μ-L-L)], [Au(C$_6$F$_5$)$_2$Cl(L-L)] or [Au(C$_6$F$_5$)$_2$(μ-L-L)]$^+$ [323]. The complex [Au(C$_6$F$_5$)$_2$Cl{PPh$_2$C(=CH$_2$)PPh$_2$}] undergoes Michael-type additions with several nucleophiles, with addition to the terminal carbon of the double bond, giving methanide-type complexes [324]. Gold(III) complexes with triphosphines such as 1,1,1'-tris(diphenylphosphinomethyl)ethane, bis(diphenylphosphinomethyl)phenylphosphine have also been obtained [325]. The 2-(diphenylphosphino)aniline reacts with the [Au(C$_6$F$_5$)$_2$(acac)] to give the depronated [Au(C$_6$F$_5$)$_2$(PPh$_2$C$_6$H$_4$NH)],

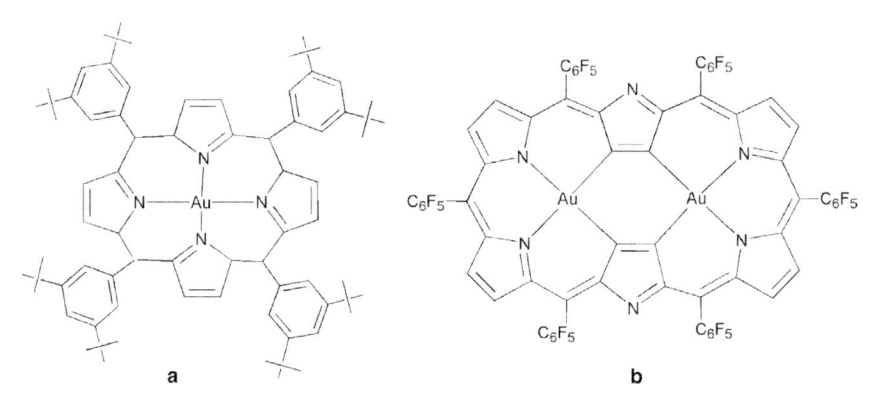

Figure 1.67 Gold(III) porphyrins.

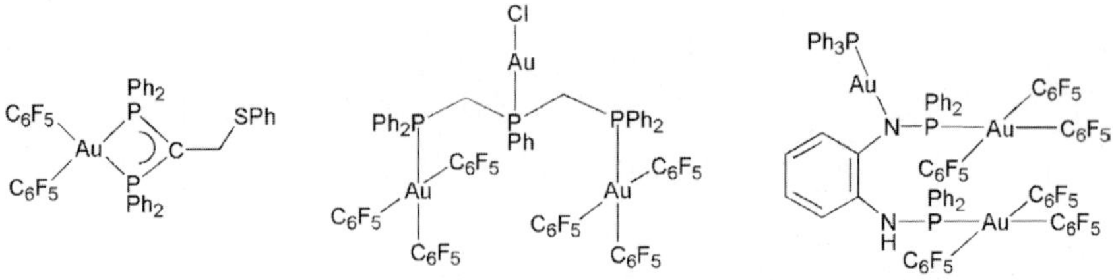

Figure 1.68 Gold(III) complexes with phosphine ligands.

which can be further deprotonated with [Au(acac)(PPh$_3$)] to give the complex [Au(C$_6$F$_5$)$_2$\{PPh$_2$C$_6$H$_4$N(AuPPh$_3$)$_2$\}] [326]. Figure 1.68 shows some of these examples.

The reaction of [Au(C$_6$F$_5$)$_3$(tht)] or [Au(C$_6$F$_5$)$_2$(OEt$_2$)$_2$]ClO$_4$ with diphenylphosphine leads to complexes [Au(C$_6$F$_5$)$_3$(PPh$_2$H)] or [Au(C$_6$F$_5$)$_2$(PPh$_2$H)$_2$]ClO$_4$, which are useful starting materials for preparing polynuclear derivatives with bridging phosphide ligands by reaction with acetylacetonate gold compounds. The treatment of these derivatives with [N(PPh$_3$)$_2$][Au(acac)$_2$] or with salts of Ag$^+$ or Cu$^+$ in the presence of NBu$_4$(acac) gives the complexes shown in Figure 1.69 [327]. When the phosphine is the phenylphosphine PH$_2$Ph, the reaction with [Au(C$_6$F$_5$)$_2$Cl]$_2$ gives the cyclic trinuclear derivative [Au(C$_6$F$_5$)$_2$(μ-PHPh)]$_3$ (Figure 1.69) [328].

1.2.6.3 Gold(III) Complexes with Chalcogen Ligands

Gold(III) forms very stable complexes with chalcogen donor ligands, mainly with sulfur donor compounds but stable species with Au^{III}–O bonds are known, in contrast to those found in Au(I) where these complexes, with some exceptions, are unstable. Hydroxo and oxo complexes have been reported with bypyridine, terpyridine or cyclometallated ligands. The hydroxo complexes are generally synthesized from the corresponding gold(III) precursor with NaOH or KOH in aqueous solution. The oxo species are mostly obtained, by deprotonation or by condensation of the hydroxo species [329]. Oxo complexes have shown antiproliferative effects and DNA and protein binding properties [330]. Other interesting gold(III) derivatives have been prepared by reaction of AuCl$_3$ with metal oxide polytungstate ligands [331]. These complexes, K$_{15}$H$_2$[Au(O)(OH$_2$)P$_2$W$_{18}$O$_{68}$]·25H$_2$O and K$_7$H$_2$[Au(O)(OH$_2$)P$_2$W$_{20}$O$_{70}$(OH$_2$)$_2$]·27H$_2$O are unique because they are the first examples of terminal Au-oxo complexes that exhibit multiple bonding between gold and oxygen. The gold atom is bonded to six oxygen atoms in an octahedral geometry, the equatorial positions are bridging oxide ligands, distances 1.877(11) Å, and the axial positions are occupied by a terminal Au=O bond, 1.763(17) Å, and an aqua molecular, Au–O 2.29(4) Å (see Figure 1.70).

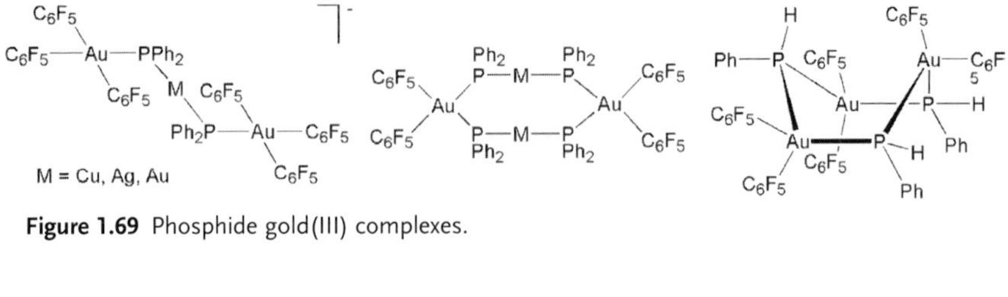

Figure 1.69 Phosphide gold(III) complexes.

Figure 1.70 Gold(III) complexes with μ-oxo ligands and environment of gold(III) in K$_7$H$_2$[Au(O)(OH$_2$)P$_2$W$_{20}$O$_{70}$(OH$_2$)$_2$]·27H$_2$O.

Gold(III) complexes with sulfur or selenium donor ligands such as thiolates or selenolates, dithiolates, dithiocarbamates or bridging sulfide or selenide ligands are more numerous. They are obtained by oxidation of a gold(I) precursor, such as [AuCl(CO)] with PhCH$_2$SSCH$_2$Ph to give the gold(III) species [Au$_2$Cl$_4$(μ-SPh)$_2$] [332] or by substitution reactions such as the diethylether ligand in [Au(C$_6$F$_5$)$_3$(OEt$_2$)] for the thiolate or selenolate metalloligand [Au$_2$(μ-dppf)(ER)$_2$] [333]. The homoleptic complex [Au(SR)$_4$]$^-$ has been obtained for the 1-methyl-1,2,3,4-tetrazole-5-thiolate ligand [334]. The gold(III) hydrosulfide complex NBu$_4$[Au(C$_6$F$_5$)$_3$(SH)] has been prepared by reaction of NBu$_4$[Au(C$_6$F$_5$)$_3$Br] with Na(SH) [335]. Figure 1.71 shows some of these complexes.

Homoleptic gold(III) derivatives with dithiolate, ligands of the type [Au(S–S)$_2$]$^-$ are well known and are usually prepared from [AuCl$_4$]$^-$ with the dithiol; some examples are with 1,2-benzene dithiolate, maleonitriledithiolate, dmit, and so on, [164, 336]. Similar complexes have been reported for bidentate sulfur ligands such as dithiocarbamates, dithiophosphates, and so on, [41]. Other derivatives as the trinuclear species [Au(C$_6$F$_5$)(S$_2$C$_6$H$_4$)]$_3$ [337] or the complex with one *nido* and one *closo*-carborane dithiolate are known [338]. Figure 1.72 collects some of these complexes.

Gold(III) complexes have also been prepared with mixed-donor ligands such as the phosphino thiolates PhP(C$_6$H$_4$SH)$_2$ or P(C$_6$H$_4$SH)$_3$, which give the dinuclear

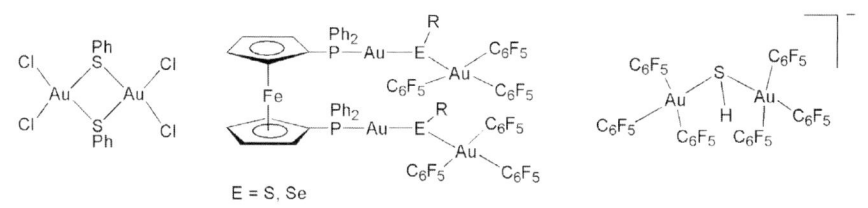

Figure 1.71 Gold(III) complexes with thiolate or selenolate ligands.

Figure 1.72 Dithiolate gold(III) complexes.

Figure 1.73 Gold(III) complexes with phosphino-thiolate ligands.

[Au$_2${PPh(C$_6$H$_4$S)$_2$}$_2$] with gold(I) and gold(III) atoms, or [Au$_2${P(C$_6$H$_4$S)$_3$}$_2$] with two gold(III) centers (Figure 1.73a) [339]. The reaction of tetrachloroaurate(III) with two equivalents of 2-(diphenylphosphino)benzenethiol gives a cation that can be precipitated from solution as tetraphenylborate salt, [Au(2-PPh$_2$C$_6$H$_4$S)$_2$]BPh$_4$ (Figure 1.73b). Surprisingly, the electrochemical behavior corresponds to a reversible single-electron transfer process and indicates that this gold(III) can be reduced to gold(II) and that the gold(II) state has reasonable stability on the electrochemical time scale [340]. The methanide carbon of [Au(C$_6$F$_5$)(PPh$_2$CHPPh$_2$Me)] reacts with carbon disulfide affording the gold(III) derivatives [Au{PPh$_2$C(PPh$_2$Me)C(S)S}$_2$]$^+$ through a carbon–carbon coupling reaction. The reaction of this complex with 2 equivalents of [Au(C$_6$F$_5$)(tht)] gives [Au{PPh$_2$C(PPh$_2$Me)C(SAuC$_6$F$_5$)S}$_2$]$^+$ (Figure 1.73c), which is a vapochromic material and can be used for the detection of some volatile organic compounds [341]. This material changes color from black to orange in the presence of organic vapors.

Gold(III) complexes with a central sulfur or selenium ligand have been reported. The substitution of the proton in NBu$_4$[Au(C$_6$F$_5$)$_3$(SH)] by the isolobal fragments MPPh$_3$$^+$ (M = Au, Ag) affords the compounds NBu$_4$[Au(C$_6$F$_5$)$_3$(SMPPh$_3$)] (Figure 1.74a) [335]. The reaction of [E(AuPPh$_3$)$_2$] or [E(Au$_2$dppf)$_2$] with one or two equivalents of [Au(C$_6$F$_5$)$_3$(OEt$_2$)] gives the mixed valence complexes [E(AuPPh$_3$)$_2${Au(C$_6$F$_5$)$_3$}] and [E(Au$_2$dppf){Au(C$_6$F$_5$)$_3$}], which have a trigonal pyramidal structure, or [E(AuPPh$_3$)$_2${Au(C$_6$F$_5$)$_3$}$_2$] (Figure 1.74b) and [E(Au$_2$dppf){Au(C$_6$F$_5$)$_3$}$_2$], which are tetrahedral in contrast to the quadruply bridging gold(I) complexes, which are square pyramidal [342]. The reaction with [Au(C$_6$F$_5$)$_2$(OEt$_2$)]$^+$ also gives mixed valence complexes of the type [{S(Au$_2$dppf)}$_2$Au(C$_6$F$_5$)$_2$]$^+$ (Figure 1.74c) or [Se(AuPPh$_3$)Au(C$_6$F$_5$)$_2$]$_2$. These complexes show the presence of Au(I)–Au(III) interactions but weaker than the corresponding Au(I)–Au(I) interactions [342].

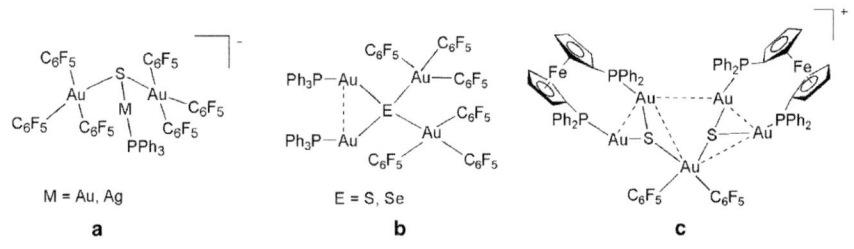

Figure 1.74 Gold(I)–Gold(III) complexes with bridging chalcogen ligands.

1.2.7
Gold in Higher Oxidation States

Complexes with gold in higher oxidation states are known with very electronegative and small ligands such as fluorine. The highest oxidation state that is known beyond doubt is Au^V. It was claimed that the yellow crystalline AuF_7 was prepared by disproportionation of AuF_6, obtained by reaction of AuF_5 with atomic fluorine, and isolated at $-196\,°C$ it decomposes into AuF_5 and F_2 at $100\,°C$ [343]. However, so far these results have never been reproduced or confirmed and recent quantum chemical studies show that this is highly improbable. The strongly exothermic elimination of F_2 with a low activation barrier is not consistent with the reported stability of AuF_7 at room temperature, even the homoleptic dissociation of one equatorial Au–F bond is exothermic and has a barrier only from structural rearrangement. Also, and given the extremely high electron affinity of AuF_6, this species is unlikely to exist in most experimental conditions [344]. The pentafluoride AuF_5 is formed by vacuum pyrolysis of either $[KrF][AuF_6]$ (at 60–65 °C) or $[O_2][AuF_6]$ (at 160–200 °C) [345]. An electron diffraction study of AuF_5 has indicated that the vapor phase consists of di- and trimeric molecules with the gold octahedrally coordinated [346]. A recent single-crystal X-ray diffraction study has shown that AuF_5 exists as a dimer in the solid state [347] (Figure 1.75). Salts of $[AuF_6]^-$ with different cations, including Ag^+, are also known [345, 348]. The compound $[Xe_2F_{11}][AuF_6]$ was first prepared by fluorination of AuF_3 with XeF_2 in the presence of XeF_6. In the solid state, gold is octahedrally surrounded by the fluorine atoms and bonded to the dimeric xenon cation by two of the fluorine atoms (Figure 1.75) [349].

Evidence of the existence of oxidation state +IV is not yet convincing. Some complexes with dithiolene or dithiolate systems have been reported, such as [Au(5,6-dihydro-1,4-dithiin-2,3-dithiolate)$_2$] (Figure 1.76) [350], or [Au(2,3-dithiophenedithiolate)$_2$] [351]. They were prepared by chemical or electrochemical oxidation of

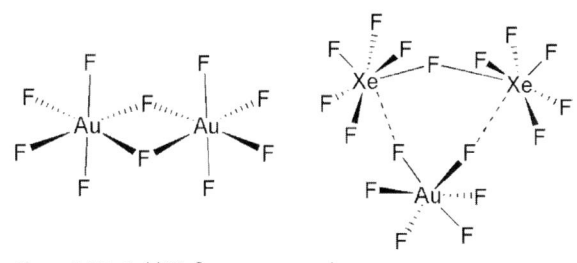

Figure 1.75 Gold(V) fluorocompounds.

Figure 1.76 Gold(IV) dithiolate species.

the corresponding gold(III) derivatives. The molecular orbital calculations suggest that the metal ions of the molecules are best described as d^8 ions, as expected for a square planar gold(III) compound, and therefore the unpaired electron in each molecule resides largely on the ligands.

References

1 Schmidbaur, H.(ed.) (1999) *Gold, Progress in Chemistry, Biochemistry and Technology*, John Wiley & Sons, Chichester, England.
2 Bayler, A., Schier, A., Bowmaker, G.A. and Schmidbaur, H. (1996) *Journal of the American Chemical Society*, **118**, 7006.
3 Scherbaum, F., Grohmann, A., Huber, B., Krüger, C. and Schmidbaur, H. (1988) *Angewandte Chemie-International Edition*, **27**, 1544.
4 Mehrotra, P.K. and Hoffmann, R. (1978) *Inorganic Chemistry*, **17**, 2182.
5 Dedieu, A. and Hoffmann, R. (1978) *Journal of the American Chemical Society*, **100**, 2074.
6 Janiak, C. and Hoffmann, R. (1990) *Journal of the American Chemical Society*, **112**, 5924.
7 Burdett, J.K., Eisenstein, O. and Schweizer, W.B. (1994) *Inorganic Chemistry*, **33**, 3261.
8 Mingos, D.M.P., Slee, T. and Shenyang, L. (1990) *Chemical Reviews*, **90**, 383.
9 Häberlen, O.D., Schmidbaur, H. and Rösch, N. (1994) *Journal of the American Chemical Society*, **116**, 8241.
10 Rösch, N., Görling, A., Ellis, D.E. and Schmidbaur, H. (1991) *Inorganic Chemistry*, **30**, 3986.
11 Rösch, N., Görling, A., Ellis, D.E. and Schmidbaur, H. (1989) *Angewandte Chemie-International Edition*, **28**, 1357.
12 Pyykkö, P., Angermaier, K., Assmann, B. and Schmidbaur, H. *Journal of the Chemical Society. Chemical Communications*, 1995. 1889.
13 Pyykkö, P., Runeberg, N. and Mendizabal, F. (1997) *Chemistry – A European Journal*, **3**, 1451.

14 Pyykkö, P. and Mendizabal, F. (1997) *Chemistry – A European Journal*, **3**, 1458.
15 Pyykkö, P. (1997) *Chemical Reviews*, **97**, 597.
16 Pyykkö, P. (2004) *Angewandte Chemie-International Edition*, **43**, 4412; (2005) *Inorganica Chimica Acta*, **358**, 4113.
17 Schwerdtfeger, P., Bruce, A.E. and Bruce, M.R.M. (1998) *Journal of the American Chemical Society*, **120**, 6587.
18 Kaltsoyannis, N. (1997) *Journal of the Chemical Society-Dalton Transactions*, 1.
19 Shaw, C.F., III (1999) *Chemical Reviews*, **99**, 2589.
20 Sadler, P.J. and Guo, Z. (1998) *Pure and Applied Chemistry*, **70**, 863.
21 *Metallotherapeutic Drugs & Metal-based Diagnostic Agents. The Use of Metals in Medicine.* Marcen Gielen, E.R.T. Tiekink, Wiley, (2005).
22 Catalysis by Gold, G.C. Bond, C. Louis, D.T. Thompson, Catalytic Science Series vol. 6.
23 Gorin, D.J. and Toste, F.D. (2007) *Nature*, **446**, 395.
24 Hashmi, A.S.K. (2004) *Gold Bulletin*, **37**, 51.
25 Hashmi, A.S.K. and Hutchings, G.J. (2006) *Angewandte Chemie-International Edition*, **45**, 7896.
26 Hashmi, A.S.K. (2007) *Chemical Reviews*, **107**, 3180.
27 Biltz, W., Weibke, F., Ehrhorn, H.J. and Wedemeyer, R. (1938) *Zeitschrift für Anorganische und Allgemeine Chemie*, **236**, 12.
28 Sommer, A. (1943) *Nature*, **152**, 215.
29 Feldman, C. and Jansen, M. (1993) *Angewandte Chemie-International Edition*, **32**, 1049.

30 Feldman, C. and Jansen, M. (1995) *Zeitschrift fur Anorganische und Allgemeine Chemie*, **621**, 1907.
31 Mudring, A.V. and Jansen, M. (2000) *Angewandte Chemie-International Edition*, **39**, 3066.
32 Mudring, A.V., Nuss, J., Wedig, U. and Jansen, M. (2000) *Journal of Solid State Chemistry*, **155**, 29.
33 Dietzel, P.D.C. and Jansen, M. (2001) *Journal of the Chemical Society. Chemical Communications*, 2208.
34 Mudring, A.V., Jansen, M., Daniels, J., Krämer, S., Mehring, M., Prates Ramalho, J.P., Romero, A.H. and Parrinelo, M. (2002) *Angewandte Chemie-International Edition*, **41**, 120.
35 Pyykkö, P. (2002) *Angewandte Chemie-International Edition*, **41**, 3573.
36 Nuss, H. and Jansen, M. (2006) *Angewandte Chemie-International Edition*, **45**, 4369.
37 Mézailles, N., Avarvari, N., Maigrot, N., Ricard, L., Mathey, F., Le Floch, P., Cataldo, L., Berclaz, T. and Geoffroy, M. (1999) *Angewandte Chemie-International Edition*, **38**, 3194.
38 Mingos, D.M.P. (1984) *Gold Bulletin*, **17**, 5.
39 Pignolet, L.H. and Krogstad, D.A. (1999) in *Gold Progress in Chemistry, Biochemistry and Technology* (ed. H. Schmidbaur), John Wiley & Sons, Chichester, England, p. 429.
40 Dyson, P.J. and Mingos, M.P. (1999) in *Gold, Progress in Chemistry, Biochemistry and Technology* (ed. H. Schmidbaur), John Wiley & Sons, Chichester, England, p. 511.
41 Gimeno, M.C. and Laguna, A. (2003) in *Comprehensive Coordination Chemistry II*, **5** (eds J.A. McCleverty and T.J. Meyer), Elsevier, New York, p. 911.
42 Bellon, P.L., Cariati, F., Manassero, M., Naldini, L. and Sansoni, M. (1971) *Journal of the Chemical Society. Chemical Communications*, 1423.
43 Van der Velden, J.W.A., Bour, J.J., Wollenbroek, F.A., Beurskens, P.T. and Smits, J.M.M. (1979) *Journal of the Chemical Society. Chemical Communications*, 1162.
44 Van der Velden, J.W.A., Beurskens, P.T., Bour, J.J., Bosman, W.P., Noordik, J.H., Kolenbrander, M. and Buskes, J.A.K.M. (1984) *Inorganic Chemistry*, **23**, 146.
45 Laguna, A., Laguna, M., Gimeno, M.C. and Jones, P.G. (1992) *Organometallics*, **11**, 2759.
46 Yang, Y. and Sharp, P.R. (1994) *Journal of the American Chemical Society*, **116**, 6983.
47 Van der Velden, J.W.A., Bour, J.J., Pet, R., Bosman, W.P. and Noordik, J.H. (1983) *Inorganic Chemistry*, **22**, 3112–3115.
48 Mingos, D.M.P. (1982) *Proceedings of the Royal Society, London Ser. A*, **308**, 75.
49 Evans, D.G. and Mingos, D.M.P. (1982) *Journal of Organometallic Chemistry*, **232**, 171.
50 Hall, K.P. and Mingos, D.M.P. (1984) *Progress in Inorganic Chemistry*, **32**, 237.
51 Crespo, O., Gimeno, M.C., Jones, P.G., Laguna, A. and Villacampa, M.D. (1997) *Angewandte Chemie-International Edition*, **36**, 993.
52 Briant, C.E., Theobald, B.R.C., White, J.W., Bell, L.K. and Mingos, D.M.P. (1981) *Journal of the Chemical Society. Chemical Communications*, 201.
53 Richter, M. and Strähle, J. (2001) *Zeitschrift fur Anorganische und Allgemeine Chemie*, **627**, 918.
54 Teo, B.K., Shi, X. and Zhang, H. (1991) *Journal of the American Chemical Society*, **113**, 4329.
55 Strähle, J., (1999) in *Metal Clusters in Chemistry*, **Vol. 1** (eds P. Braunstein, L.A. Oro and R.R. Raithby), Wiley-VCH, Weinheim, p. 535.
56 Steggerda, J.J. (1990) *Comments on Inorganic Chemistry*, **11**, 113.
57 Bour, J.J., Kanters, R.P.F., Schlebos, P.P.J. and Steggerda, J. (1988) *Recueil Des Travaux Chimiques Des Pays-Bas-Journal of the Royal Netherlands Chemical Society*, **107**, 211.

58 Ito, L.N., Johnson, B.J., Mueting, A.M. and Pignolet, L.H. (1989) *Inorganic Chemistry*, **28**, 2026.
59 Ito, L.N., Felicissimo, A.M.P. and Pignolet, L.H. (1991) *Inorganic Chemistry*, **30**, 387.
60 Kanters, R.P.F., Schlebos, P.P.J., Bour, J.J., Bosman, W.P., Behm, H.J. and Steggerda, J.J. (1988) *Inorganic Chemistry*, **27**, 4034.
61 Teo, B.K. and Keating, K. (1984) *Journal of the American Chemical Society*, **106**, 2224.
62 Teo, B.K. and Zhang, H. (1992) *Angewandte Chemie-International Edition*, **31**, 445.
63 Teo, B.K., Zhang, H. and Shi, X. (1990) *Journal of the American Chemical Society*, **112**, 8552.
64 Teo, B.K., Shi, X. and Zhang, H. (1989) cited in *Chemical & Engineering News*, **67**, 6.
65 Schmid, G., Pfeil, R., Boese, R., Bandermann, F., Meyer, S., Calis, G.H.M. and Van der Velden, J.W.A., (1981) *Chemische Berichte*, **114**, 3634; Schmid, G. (1990) *Inorganic Syntheses*, Vol 32, **27**, 214.
66 Schmid, G. (1992) *Chemical Reviews*, **92**, 1709.
67 Turkevitch, J., Stevenson, P.C. and Hillier, J. (1951) *Discussions of the Faraday Society*, **11**, 55.
68 Giersig, M. and Mulvaney, P. (1993) *Langmuir*, **9**, 3408.
69 Daniel, M.C. and Astruc, D. (2004) *Chemical Reviews*, **104**, 293.
70 Storhoff, J.J. and Mirkin, C.A. (1999) *Chemical Reviews*, **99**, 1849.
71 Nam, J.M., Thaxton, C.S. and Mirkin, C.A. (2003) *Science*, **301**, 1884.
72 Schaaff, T.G., Shafigullin, M.N., Khoury, J.T., Vezmar, I. and Wettern, R.L. (2001) *Journal of Physical Chemistry B*, **105**, 8785.
73 Brust, M., Fink, J., Bethell, D., Schiffrin, D.J. and Kiely, C. (1995) *Journal of the Chemical Society. Chemical Communications*, 1655.
74 Shichibu, Y., Negishi, Y., Tsuuda, T. and Teranishi, T. (2005) *Journal of the American Chemical Society*, **127**, 13464.
75 Donkers, R.L., Lee, D. and Murray, R.W. (2004) *Langmuir*, **20**, 1945.
76 Balasubramanian, R., Guo, R., Mills, A.J. and Murray, R.W. (2005) *Journal of the American Chemical Society*, **127**, 8126.
77 Hicks, J.F., Miles, D.T. and Murray, R.W. (2002) *Journal of the American Chemical Society*, **124**, 13322.
78 Pyykkö, P. and Runeberg, N. (2002) *Angewandte Chemie-International Edition*, **41**, 2174.
79 Li, X., Kiran, B., Li, J., Zhai, H.J. and Wang, L.S. (2002) *Angewandte Chemie-International Edition*, **41**, 4786.
80 Li, J., Li, X., Zhai, H.J. and Wang, L.S. (2003) *Science*, **299**, 864.
81 Johansson, M.P., Sundholm, D. and Vaara, J. (2004) *Angewandte Chemie-International Edition*, **43**, 2678.
82 Gao, Y. and Zeng, X.C. (2005) *Journal of the American Chemical Society*, **127**, 3698.
83 Kondo, Y. and Takayanagi, K. (2000) *Science*, **289**, 606.
84 Kim, F., Connor, S., Song, H., Kuykendall, T. and Yang, P. (2004) *Angewandte Chemie-International Edition*, **43**, 3673.
85 Schmidbaur, H. (2000) *Gold Bulletin*, **33**, 3.
86 Anderson, K.M., Goeta, A.E. and Steed, J.W. (2007) *Inorganic Chemistry*, **46**, 6444.
87 Jones, P.G. (1982) *Zeitschrift fur Naturforschung. Teil B*, **37**, 823.
88 Esperas, S. (1976) *Acta Chemica Scandinavica Series A-Physical and Inorganic Chemistry*, **30**, 527.
89 Angermaier, L., Zeller, E. and Schmidbaur, H. (1994) *Journal of Organometallic Chemistry*, **472**, 371.
90 Ahrland, S., Aurivillius, B., Dreisch, K., Noren, B. and Oskarsson, A. (1992) *Acta Chemica Scandinavica*, **46**, 262.
91 Preinsenberger, M., Schier, A. and Schmidbaur, H. (1999) *Journal of The Chemical Society-Dalton Transactions*, 1645.
92 Bachman, R.E., Fioritto, M.S., Fetics, S.K. and Cocker, T.M. (2001) *Journal of the American Chemical Society*, **123**, 5376.

93. Ahrland, S., Noren, B. and Oskarsson, A. (1985) *Inorganic Chemistry*, **24**, 1330.
94. Adams, H.N., Hiller, W. and Strähle, J. (1982) *Zeitschrift für Anorganische und Allgemeine Chemie*, **485**, 81; Conzelmann, W., Hiller, W., Strähle, J. and Sheldrick, G.M. (1984) *Zeitschrift für Anorganische und Allgemeine Chemie*, **512**, 169; Jones, P.G. and Freytag, M. (2000) *Chemical Communications*, 277.
95. Bauer, A. and Schmidbaur, H. (1996) *Journal of the American Chemical Society*, **118**, 5324.
96. Yip, J.H.K., Feng, R. and Vittal, J.J. (1999) *Inorganic Chemistry*, **38**, 3586.
97. Mohr, F., Jennings, M.J. and Puddephatt, R.J. (2004) *Angewandte Chemie-International Edition*, **43**, 969.
98. Hunks, W.J., Jennings, M.C. and Puddephatt, R.J. (2002) *Inorganic Chemistry*, **41**, 4590.
99. Wilton-Ely, J.D.E.T., Schier, A., Mitzel, N.W. and Schmidbaur, H. (2001) *Journal of The Chemical Society-Dalton Transactions*, 1058.
100. Chojnacki, J., Becker, B., Konitz, A. and Wojnowski, W. (2000) *Zeitschrift für Anorganische und Allgemeine Chemie*, **626**, 2173.
101. Angermaier, K., Sladek, A. and Schmidbaur, H. (1996) *Zeitschrift für Naturforschung. Teil B*, **51**, 1671.
102. Assefa, Z., McBurnett, B.G., Staples, R.J., Fackler, J.P., Jr, Assmann, B., Angermaier, K. and Schmidbaur, H. (1995) *Inorganic Chemistry*, **34**, 75.
103. Schwerdtfeger, P. (1989) *Journal of the American Chemical Society*, **111**, 7261; Söhnel, T., Hermann, H. and Schwerdtfeger, P. (2001) *Angewandte Chemie-International Edition*, **40**, 4382.
104. Schröder, D., Hrusák, J., Tornieporth-Oetting, C., Klapötke, T.M. and Schwarzt, H. (1994) *Angewandte Chemie-International Edition*, **33**, 212.Evans, C.J. and Geary, M.C.L. (2000) *Journal of the American Chemical Society*, **122**, 1560.
105. Evans, C.J., Lesarri, A. and Gerry, M.C.L. (2000) *Journal of the American Chemical Society*, **122**, 6100.Thomas, J.M., Walker, N.R., Cooke, S.A. and Gerry, M.C.L. (2004) *Journal of the American Chemical Society*, **126**, 1235.
106. Cooke, S.A. and Gerry, M.C.L. (2004) *Journal of the American Chemical Society*, **126**, 17000.
107. Hwang, I.C., Seidel, S. and Seppelt, K. (2003) *Angewandte Chemie-International Edition*, **42**, 4392; Seppelt, K. (2003) *Zeitschrift für Anorganische und Allgemeine Chemie*, **629**, 2427.
108. Istrate, F.M. and Gagosz, F.L. (2008) *Journal of Organic Chemistry*, **73**, 730; Komiya, S., Sone, T., Usui, Y., Hirano, M. and Fukuoka, A. (1996) *Gold Bulletin*, **29**, 131.
109. Gimeno, M.C. (2007) "New Perspectives in Sulfur, Selenium and Tellurium" (ed. F.A. Devillanova) *The Royal Society of Chemistry*, 33.
110. Bau, R. (1998) *Journal of the American Chemical Society*, **120**, 9380.
111. Coker, N.L., Bauer, J.A.K. and Elder, R.C. (2004) *Journal of the American Chemical Society*, **126**, 12.
112. Watase, S., Nakamoto, M., Kitamura, T., Kanehisa, N., Kai, Y. and Yanagida, S. (2000) *Journal of The Chemical Society-Dalton Transactions*, 3585.
113. Vicente, J., Chicote, M.T., Gónzalez Herrero, P., Jones, P.G. and Ahrens, B. (1994) *Angewandte Chemie-International Edition*, **33**, 1852; Vicente, J., Chicote, M.T., Gónzalez Herrero, P., Grünwald, C. and Jones, P.G. (1997) *Organometallics*, **16**, 3381.
114. Schmidbaur, H., Bissinger, P., Lachmann, J. and Steigelmann, O. (1992) *Zeitschrift für Naturforschung. Teil B*, **47**, 1711.
115. Van Calcar, P.M., Olmstead, M.M. and Balch, A.L. (1995) *Journal of the Chemical Society. Chemical Communications*, 1773.
116. Schmidbaur, H., Wohlleben, A., Wagner, F., Orama, O. and Huttner, G. (1977) *Chemische Berichte*, **119**, 1748.
117. Bates, P.A. and Waters, J.M. (1985) *Inorganica Chimica Acta*, **98**, 125.

118 Eggleston, D.S., Chodosh, D.F., Girard, G.R. and Hill, D.T. (1985) *Inorganica Chimica Acta*, **108**, 221.
119 Van Calcar, P.M., Olmstead, M.M. and Balch, A.L. (1997) *Inorganic Chemistry*, **36**, 5231.
120 Schmidbaur, H., Graf, W. and Müller, G. (1988) *Angewandte Chemie-International Edition*, **27**, 417.
121 Schmidbaur, H., Graf, W. and Müller, G. (1986) *Helvetica Chimica Acta*, **69**, 1748.
122 Eggleston, D.S., McArdle, J.V. and Zuber, G.E. (1987) *Journal of The Chemical Society-Dalton Transactions*, 677.
123 Schmidbaur, H., Herr, R., Müller, G. and Riede, J. (1985) *Organometallics*, **4**, 1208.
124 Hill, D.T., Girard, G.R., McCabe, F.L., Johnson, R.K., Stupik, P.R., Zhang, J.H., Reiff, W.M., and Eggleston, D.S. (1989) *Inorganic Chemistry*, **28**, 3529; Gimeno, M.C., Laguna, A., Sarroca, C. and Jones, P.G. (1993) *Inorganic Chemistry*, **32**, 5926.
125 Schmidbaur, H., Wohlleben, A., Wagner, F., Orama, O. and Huttner, G. (1977) *Chemische Berichte*, **119**, 1748; Bauer, A. and Schmidbaur, H. (1997) *Journal of The Chemical Society-Dalton Transactions*, 1115.
126 Khan, M.N.I., King, C., Heinrich, D.D., Fackler, J.P. and Porter, L.C. (1989) *Inorganic Chemistry*, **28**, 2150.
127 Puddephatt, R.J. (1983) *Chemical Society Reviews*, **12**, 99.
128 Bayler, A., Schier, A. and Scmidbaur, H. (1998) *Inorganic Chemistry*, **37**, 4353.
129 Usón, R., Laguna, A., Laguna, M., Fraile, M.N., Jones, P.G. and Sheldrick, G.M. (1986) *Journal of The Chemical Society-Dalton Transactions*, 291.
130 King, C., Wang, J.C., Khan, M.N.I. and Fackler, J.P., Jr (1989) *Inorganic Chemistry*, **28**, 2145; Leung, K.H., Phillips, D.L., Tse, M.C., Che, C.M. and Miskowski, V.M. (1999) *Journal of the American Chemical Society*, **121**, 4799; Fu, W.F., Chan, K.C., Miskowski, V.M. and Che, C.M. (1999) *Angewandte Chemie-International Edition*, **38**, 2783.
131 Ma, Y., Che, C.M., Chao, H.Y., Zhou, X., Chan, W.H. and Shen, J. (1999) *Advanced Materials*, **11**, 852.
132 Schmidbaur, H., Mandl, J.R., Bassett, J.M., Blaschke, G. and Zimmer-Gasser, B. (1981) *Chemische Berichte*, **114**, 433.
133 Usón, R., Laguna, A., Laguna, M., Gimeno, M.C., Jones, P.G., Fittschen, C. and Sheldrick, G.M. (1986) *Journal of the Chemical Society. Chemical Communications*, 509.
134 Fernández, E.J., Gimeno, M.C., Jones, P.G., Laguna, A., Laguna, M. and López de Luzuriaga, J.M. (1994) *Angewandte Chemie-International Edition*, **33**, 87.
135 Feng, D.F., Tang, S.S., Liu, C.W., Lin, I.J.B., Wen, Y.S. and Liu, L.K. (1997) *Organometallics*, **16**, 901.
136 Bardají, M., Connelly, N.G., Gimeno, M.C., Jiménez, J., Jones, P.G., Laguna, A. and Laguna, M. (1994) *Journal of The Chemical Society-Dalton Transactions*, 1163.
137 Dávila, R.M., Elduque, A., Grant, T., Staples, R.J. and Fackler, J.P., Jr (1993) *Inorganic Chemistry*, **32**. 1749.
138 Bardají, M., Laguna, A. and Laguna, M. (1995) *Journal of The Chemical Society-Dalton Transactions*, 1255.
139 Bardají, M., Jones, P.G., Laguna, A. and Laguna, M. (1995) *Organometallics*, **14**, 1310.
140 Bardají, M., Laguna, A. and Laguna, M. (1995) *Journal of Organometallic Chemistry*, **496**, 245.
141 141. Zyl, W.E., López de Luzuriaga, J.M., Fackler, J.P., Jr and Staples, R.J. (2001) *Canadian Journal of Chemistry*, **79**, 896.
142 Stüzer, A., Bissinger, P. and Schmidbaur, H. (1992) *Chemische Berichte*, **125**, 367.
143 Fernández, E.J., Gimeno, M.C., Jones, P.G., Laguna, A., Laguna, M. and López de Luzuriaga, J.M. (1993) *Journal of The Chemical Society-Dalton Transactions*, 3401.
144 Zank, J., Schier, A. and Schmidbaur, H. (1998) *Journal of The Chemical Society-Dalton Transactions*, 323.

145 Che, C.M., Yip, H.K., Yam, V.W.W., Cheung, P.Y., Lai, T.F., Shieh, S.J. and Peng, S.M. (1992) *Journal of The Chemical Society-Dalton Transactions*, 427.

146 Bardají, M., Laguna, A., Orera, V.M. and Villacampa, M.D. (1998) *Inorganic Chemistry*, **37**, 5125.

147 Balch, A.L., Fung, E.Y. and Olmstead, M.M. (1990) *Journal of the American Chemical Society*, **112**, 5181.

148 Schmidbaur, H., Stützer, A. and Bissinger, P. (1992) *Zeitschrift fur Naturforschung. Teil B*, **47**, 640.

149 Cerrada, E., Diaz, C., Diaz, M.C., Hursthouse, M.B., Laguna, M. and Light, M.E. (2002) *Journal of The Chemical Society-Dalton Transactions*, 1104.

150 Slany, M., Bardají, M., Casanove, M.J., Caminade, A.M., Majoral, J.P. and Chaudret, B. (1995) *Journal of the American Chemical Society*, **117**, 9764.

151 Balch, A.J., Catalano, V.J. and Olmstead, M.N. (1990) *Inorganic Chemistry*, **29**, 585.

152 Balch, A.J., Catalano, V.J. and Olmstead, M.N. (1990) *Journal of the American Chemical Society*, **112**, 2010.

153 Dávila, R.M., Staples, R.J., Elduque, A., Harlass, M.M., Kyle, L. and Fackler, J.P., Jr (1994) *Inorganic Chemistry*, **33**, 5940.

154 Preisenberger, M., Bauer, A., Schier, A. and Schmidbaur, H. (1997) *Journal of The Chemical Society-Dalton Transactions*, 4753.

155 Crespo, O., Gimeno, M.C., Jones, P.G., Ahrens, B. and Laguna, A. (1997) *Inorganic Chemistry*, **36**, 495.

156 Khan, M.N.I., Fackler, J.P., Jr, King, J.C. and Wang, S. (1988) *Inorganic Chemistry*, **27**, 1672; Khan, M.N.I., Wang, S. and Fackler, J.P., Jr (1989) *Inorganic Chemistry*, **28**, 3579.

157 Vicente, J., Chicote, M.T., González-Herrero, P. and Jones, P.G. (1997) *Chemical Communications*, 2047.

158 Hanna, S.D., Khan, S.I. and Zink, J.I. (1996) *Inorganic Chemistry*, **35**, 5813.

159 Fackler, J.P., Jr, Staples, R.J. and Assefa, Z. (1994) *Journal of the Chemical Society. Chemical Communications*, 431.

160 Yu, S.Y., Zhang, Z.X., Cheng, E.C.C., Li, Y.Z., Yam, V.W.W., Huang, H.P. and Zhang, R. (2005) *Journal of the American Chemical Society*, **127**, 17994.

161 Piovesana, O. and Zanazzi, P.F. (1980) *Angewandte Chemie-International Edition*, **19**, 561.

162 Schuerman, J.A., Fronczek, F.R. and Selbin, J. (1986) *Journal of the American Chemical Society*, **108**, 336.

163 Dietzsch, W., Franke, A., Hoyer, E., Gruss, D., Hummel, H.U. and Otto, P. (1992) *Zeitschrift fur Anorganische und Allgemeine Chemie*, **611**, 81.

164 Gimeno, M.C., Jones, P.G., Laguna, A., Laguna, M. and Terroba, R. (1994) *Inorganic Chemistry*, **33**, 3932.

165 Nakamoto, M., Schier, A. and Schmidbaur, H. (1993) *Journal of The Chemical Society-Dalton Transactions*, 1347.

166 Sladek, A. and Schmidbaur, H. (1996) *Inorganic Chemistry*, **35**, 3268.

167 Ehlich, H., Schier, A. and Schmidbaur, H. (2002) *Inorganic Chemistry*, **41**, 3721.

168 Yip, H.K., Schier, A., Riede, J. and Schmidbaur, H. (1994) *Journal of The Chemical Society-Dalton Transactions*, 2333.

169 Cerrada, E., Laguna, A., Laguna, M. and Jones, P.G. (1994) *Journal of The Chemical Society-Dalton Transactions*, 1325.

170 Crespo, O., Gimeno, M.C., Jones, P.G. and Laguna, A. (1993) *Journal of the Chemical Society. Chemical Communications*, 1696.

171 Gimeno, M.C., Jones, P.G., Laguna, A., Sarroca, C., Calhorda, M.J. and Veiros, L.F. (1998) *Chemistry – A European Journal*, **4**, 2308.

172 Bennett, M.A., Bhargava, S.K., Griffiths, K.D., Robertson, G.B. and Wickramasinghe, W.A. (1987) *Angewandte Chemie-International Edition*, **26**, 258; Bennett, M.A., Bhargava, S.K., Hockless, D.C.R., Welling, L.L. and Willis, A.C. (1996) *Journal of the American Chemical Society*, **118**, 10469; Bhargava, S.K., Mohr, F., Bennett, M.A., Welling,

L.L. and Willis, A.C. (2000) *Organometallics*, **19**, 5628.
173 Gimeno, M.C., Laguna, A., Laguna, M., Sanmartín, F. and Jones, P.G. (1993) *Organometallics*, **12**, 3984.
174 Tzeng, B.C., Che, C.M. and Peng, S.M. (1996) *Journal of The Chemical Society-Dalton Transactions*, 1769.
175 Canales, F., Gimeno, M.C., Jones, P.G., Laguna, A. and Sarroca, C. (1997) *Inorganic Chemistry*, **36**, 5206; Barranco, E.M., Gimeno, M.C., Laguna, A., Villacampa, M.D. and Jones, P.G. (1999) *Inorganic Chemistry*, **38**, 702.
176 Aguado, J.E., Gimeno, M.C. and Laguna, A. (2005) *Chemical Communications*, 3355.
177 Tzeng, B.C., Li, D., Peng, S.M. and Che, C.M. (1993) *Journal of The Chemical Society-Dalton Transactions*, 2365.
178 Irwin, M.J., Vittal, J.J., Yap, G.P.A. and Puddephatt, R.J. (1996) *Journal of the American Chemical Society*, **118**, 13101.
179 Brandys, M.C., Jennings, M.C. and Puddephatt, R.J. (2000) *Journal of The Chemical Society-Dalton Transactions*, 4601.
180 Colburn, C.B., Hill, W.E., McAuliffe, C.A. and Parish, R.V. (1979) *Journal of the Chemical Society. Chemical Communications*, 218.
181 Parish, R.V., Parry, O. and McAuliffe, C.A. (1981) *Journal of The Chemical Society-Dalton Transactions*, 2098.
182 DelZotto, A., Nardin, G. and Rigo, P. (1995) *Journal of The Chemical Society-Dalton Transactions*, 3343.
183 Bowmaker, G.A., Dyason, J.C., Healy, P.C., Engelhardt, L.M., Pakawatchai, C. and White, A.H. (1987) *Journal of The Chemical Society-Dalton Transactions*, 1089.
184 Assefa, Z., Stapless, R.J. and Fackler, J.P., Jr (1994) *Inorganic Chemistry*, **33**, 2790.
185 Barranco, E.M., Crespo, O., Gimeno, M.C., Laguna, A., Jones, P.G. and Ahrens, B. (2000) *Inorganic Chemistry*, **39**, 680.
186 King, C., Khan, M.N.I., Staples, R.J. and Fackler, J.P., Jr (1992) *Inorganic Chemistry*, **31**, 3236.

187 See for example: Guggenberger, L.J. (1974) *Journal of Organometallic Chemistry*, **81**, 271; Clegg, W. (1976) *Acta Crystallographica. Section C, Crystal Structure Communications*, **32**, 2712; Davidson, J.L., Lindsell, W.E., McCullough, K.J. and McIntosh, C.H. (1995) *Organometallics*, **14**, 3497.
188 Muir, J.A., Cuadrado, S.I. and Muir, M.M. (1992) *Acta Crystallographica. Section C, Crystal Structure Communications*, **48**, 915.
189 Forward, J.M., Assefa, Z. and Fackler, J.P., Jr (1995) *Journal of the American Chemical Society*, **117**, 9103.
190 Jones, P.G. (1980) *Journal of the Chemical Society. Chemical Communications*, 1031.
191 Elder, R.C., Zeiher, E.H.K., Onadi, M. and Whittle, R.W. (1981) *Journal of the Chemical Society. Chemical Communications*, 900.
192 Tripathi, U.M., Bauer, A. and Schmidbaur, H. (1997) *Journal of The Chemical Society-Dalton Transactions*, 2865.
193 Usón, R., Laguna, A., Vicente, J., García, J., Jones, P.G. and Sheldrick, G.M. (1981) *Journal of The Chemical Society-Dalton Transactions*, 655.
194 Hitchcock, P.B. and Pye, P.L. (1977) *Journal of The Chemical Society-Dalton Transactions*, 1457; Bauer, A., Schier, A. and Schmidbaur, H. (1995) *Journal of The Chemical Society-Dalton Transactions*, 2919; Jones, P.G., Sheldrick, G.M., Muir, J.A., Muir, M.M. and Pulgar, L.B. (1982) *Journal of The Chemical Society-Dalton Transactions*, 2123.
195 Forward, J.M., Fackler, J.P., Jr and Staples, R.J. (1995) *Organometallics*, **14**, 4194.
196 Chan, W.H., Cheung, K.K., Mak, T.C.W. and Che, C.M. (1998) *Journal of The Chemical Society-Dalton Transactions*, 873.
197 Catalano, V.J. and Horner, S.J. (2003) *Inorganic Chemistry*, **42**, 8430.
198 Gimeno, M.C. and Laguna, A. (1997) *Chemical Reviews*, **97**, 511.
199 Houlton, A., Mingos, D.M.P., Murphy, D.M., Williams, D.J., Phang, L.T. and Hor,

T.S.A. (1993) *Journal of The Chemical Society-Dalton Transactions*, 3629.
200 Viotte, M., Gautheron, B., Kubicki, M.M., Mugnier, Y. and Parish, R.V. (1995) *Inorganic Chemistry*, **34**, 3465.
201 Crespo, O., Gimeno, M.C., Jones, P.G. and Laguna, A. (1996) *Inorganic Chemistry*, **35**, 1361.
202 Bensch, W., Prelati, M. and Ludwig, W. (1986) *Journal of the Chemical Society. Chemical Communications*, 1762; Jaw, H.R.C., Savas, M.M. and Mason, W.R. (1989) *Inorganic Chemistry*, **28**, 4366; Leung, K.H., Phillips, D.L., Mao, Z., Che, C.M., Miskowski, V.M. and Chan, C.K. (2002) *Inorganic Chemistry*, **41**, 2054.
203 Shieh, S.J., Li, D., Peng, S.M. and Che, C.M. (1993) *Journal of The Chemical Society-Dalton Transactions*, 195.
204 Catalano, V.J., Maltwitz, M.A., Horner, S.J. and Vasquez, J. (2003) *Inorganic Chemistry*, **42**, 2141.
205 Khan, M.N.I., King, C., Heinrich, D.D., Fackler, J.P., Jr and Porter, L.C. (1989) *Inorganic Chemistry*, **28**, 2150.
206 Maspero, A., Kani, I., Mohamed, A.A., Omary, M.A., Staples, R.J. and Fackler, J.P., Jr (2003) *Inorganic Chemistry*, **42**, 5311.
207 Catalano, V.J., Bennett, B.L. and Kar, H.M. (1999) *Journal of the American Chemical Society*, **121**, 10235; Catalano, V.J., Kar, H.M. and Bennett, B.L. (2000) *Inorganic Chemistry*, **39**, 121; Catalano, V.J., Malwitz, M.A. and Noll, B.C. (2001) *Chemical Communications*, 581.
208 Catalano, V.J. and Malwitz, M.A. (2004) *Journal of the American Chemical Society*, **126**, 6560.
209 Berners-Price, S. and Sadler, P.J. (1986) *Inorganic Chemistry*, **25**, 3822; Berners-Price, S.J., Jarrett, P.S. and Sadler, P.J. (1987) *Inorganic Chemistry*, **26**, 3074; Berners-Price, S.J., Colquhoun, L.A., Healy, P.C., Byriel, K.A. and Hanna, J.V. (1992) *Journal of The Chemical Society-Dalton Transactions*, 3357.
210 Crespo, O., Gimeno, M.C., Laguna, A. and Jones, P.G. (1992) *Journal of The Chemical Society-Dalton Transactions*, 1601–1605.
211 Delfs, C.D., Kitto, H.J., Stranger, R., Swiegers, G.F., Wild, S.B., Willis, A.C. and Wilson, G.J. (2003) *Inorganic Chemistry*, **42**, 4469; Balch, A.L. and Fung, E.Y. (1990) *Inorganic Chemistry*, **29**, 4764.
212 Crespo, O., Gimeno, M.C., Jones, P.G. and Laguna, A. (1994) *Inorganic Chemistry*, **33**, 6128.
213 Gambarota, S., Floriani, C., Chiesi-Villa, A. and Guastini, C. (1983) *Journal of the Chemical Society. Chemical Communications*, 1304; Cerrada, E., Contel, M., Valencia, A.D., Laguna, M., Gelbrich, T. and Hursthouse, M.B. (2000) *Angewandte Chemie-International Edition*, **39**, 2353.
214 Mingos, D.M.P., Yau, J., Menzer, S. and Willians, D.J. (1995) *Angewandte Chemie-International Edition*, **34**, 1894.
215 Stefanescu, D.M., Yuen, H.F., Glueck, D.S., Golen, J.A. and Rheingold, A.L. (2003) *Angewandte Chemie-International Edition*, **42**, 1046.
216 Brandys, M.C. and Puddephatt, R.J. (2001) *Chemical Communications*, 1280; (2001) *Journal of the American Chemical Society*, **123**, 4839.
217 Yip, J.H.K. and Prabhavathy, J. (2001) *Angewandte Chemie-International Edition*, **40**, 2159.
218 Bonasia, P.J., Gindleberger, D.E. and Arnold, J. (1993) *Inorganic Chemistry*, **32**, 5126.
219 Schroter, I. and Strähle, J. (1991) *Chemische Berichte*, **124**, 2161.
220 Wiseman, M.R., Marsh, P.A., Bishop, P.T., Brisdon, B.J. and Mahon, M.F. (2000) *Journal of the American Chemical Society*, **122**, 12598; Chui, S.S.Y., Chen, R. and Che, C.M. (2006) *Angewandte Chemie-International Edition*, **45**, 1621.
221 Vickery, J.C., Olmstead, M.M., Fung, E.Y. and Balch, A.L. (1997) *Angewandte Chemie-International Edition*, **36**, 1179.
222 Balch, A.L., Olmstead, M.M. and Vickery, J.C. (1999) *Inorganic Chemistry*, **38**, 3494.

223 Hayashi, A., Olmstead, M.M., Attar, S. and Balch, A.L. (2002) *Journal of the American Chemical Society*, **124**, 5791.

224 Burini, A., Fackler, J.P., Jr, Galassi, R., Pietroni, B.R. and Staples, R.J. (1998) *Chemical Communications*, 95.

225 Burini, A., Fackler, J.P., Jr, Galassi, R., Grant, T.A., Omary, M.A., Rawashdeh-Omary, M.A., Pietroni, B.R. and Staples, R.J. (2000) *Journal of the American Chemical Society*, **122**, 11264.

226 Bunge, S.D., Just, O. and Rees, W.S., Jr (2000) *Angewandte Chemie-International Edition*, **39**, 3082.

227 Hartmann, E. and Strähle, J. (1989) *Zeitschrift fur Naturforschung. Teil B*, **44**, 1.

228 Beck, J. and Strähle, J. (1986) *Angewandte Chemie-International Edition*, **25**, 95.

229 Raptis, R.G., Murray, H.H., III and Fackler, J.P., Jr (1987) *Journal of the Chemical Society. Chemical Communications*, 737.

230 Scherbaum, F., Grohmann, A., Huber, B., Krüger, C. and Schmidbaur, H. (1988) *Angewandte Chemie-International Edition*, **27**, 1544; Scherbaum, F., Grohmann, A., Müller, G. and Schmidbaur, H. (1989) *Angewandte Chemie-International Edition*, **28**, 463; Schmidbaur, H. and Steigelmann, O. (1992) *Zeitschrift fur Naturforschung. Teil B*, **47**, 1721; Schmidbaur, H., Brachthäuser, B. and Steigelmann, O. (1991) *Angewandte Chemie-International Edition*, **30**, 1488.

231 Schmidbaur, H., Scherbaum, F., Huber, B. and Müller, G. (1988) *Angewandte Chemie-International Edition*, **27**, 419.

232 Zeller, E., Beruda, H., Kolb, A., Bissinger, P., Riede, J. and Schmidbaur, H. (1991) *Nature*, **352**, 141; Grohmann, A., Riede, J. and Schmidbaur, H. (1990) *Nature*, **345**, 140.

233 Ramamoorthy, V., Wu, Z., Yi, Y. and Sharp, P.R. (1992) *Journal of the American Chemical Society*, **114**, 1526; Shan, H., Yang, Y., James, A.J. and Sharp, P.R. (1997) *Science*, **275**, 1460.

234 Zeller, E., Beruda, H. and Schmidbaur, H. (1993) *Chemische Berichte*, **126**, 2033.

235 Zeller, E., Beruda, H., Kolb, A., Bissinger, P., Riede, J. and Schmidbaur, H. (1991) *Nature*, **352**, 141.

236 Schmidbaur, H., Weidenhiller, G. and Steigelmann, O. (1991) *Angewandte Chemie-International Edition*, **30**, 433.

237 Zeller, E. and Schmidbaur, H. (1993) *Journal of the Chemical Society. Chemical Communications*, 69.

238 Schmidbaur, H., Hofreiter, S. and Paul, M. (1995) *Nature*, **377**, 503.

239 Shan, H. and Sharp, P.R. (1996) *Angewandte Chemie-International Edition*, **35**, 635.

240 Wang, Q.M., Lee, Y.A., Crespo, O., Deaton, J., Tang, C., Gysling, H.J., Gimeno, M.C., Larraz, C., Villacampa, M.D., Laguna, A. and Eisenberg, R. (2004) *Journal of the American Chemical Society*, **126**, 9488; Crespo, O., Gimeno, M.C., Laguna, A., Larraz, C. and Villacampa, M.D. (2007) *Chemistry – A European Journal*, **13**, 235.

241 Canales, F., Gimeno, M.C., Jones, P.G. and Laguna, A. (1994) *Angewandte Chemie-International Edition*, **33**, 769; Canales, S., Crespo, O., Gimeno, M.C., Jones, P.G. and Laguna, A. (1999) *Chemical Communications*, 679.

242 Canales, S., Crespo, O., Gimeno, M.C., Jones, P.G., Laguna, A. and Mendizábal, F. (2000) *Organometallics*, **19**, 4985; Canales, F., Gimeno, M.C., Laguna, A. and Villacampa, M.D. (1996) *Inorganica Chimica Acta*, **244**, 95.

243 Yam, V.W.W., Cheng, E.C.C. and Cheung, K.K. (1999) *Angewandte Chemie-International Edition*, **38**, 197; Fenske, D., Langetepe, T., Kappes, M.M., Hampe, O. and Weis, P. (2000) *Angewandte Chemie-International Edition*, **39**, 1857.

244 Hamel, A., Mitzel, N.W. and Schmidbaur, H. (2001) *Journal of the American Chemical Society*, **123**, 5106.

245 Blumenthal, A., Beruda, H. and Schmidbaur, H. (1993) *Journal of the Chemical Society. Chemical Communications*, 1005.

246 Schmidbaur, H., Zeller, E., Weindenhiller, G., Steigelmann, O. and Beruda, H. (1992) *Inorganic Chemistry*, **31**, 2370.

247 Sevillano, P., Fuhr, O., Kattannek, M., Nava, P., Hampe, O., Lebedkin, S., Ahlrichs, R., Fenske, D. and Cappes, M.M. (2006) *Angewandte Chemie-International Edition*, **45**, 3702; Sevillano, P., Fuhr, O., Hampe, O., Lebedkin, S., Matern, E., Fenske, D. and Cappes, M.M. (2007) *Inorganic Chemistry*, **46**, 7294.

248 Chen, J.H., Mohamed, A.A., Abdou, H.E., Bauer, J.A.K., Fackler, J.P., Jr, Bruce, A.E. and Bruce, M.R.M. (2005) *Chemical Communications*, 1575.

249 Yam, V.W.W., Chan, C.L. and Li, C.K. (1998) *Angewandte Chemie-International Edition*, **37**, 2857.

250 Bumbu, O., Ceamanos, C., Crespo, O., Gimeno, M.C., Laguna, A., Silvestre, C. and Villacampa, M.D. (2007) *Inorganic Chemistry*, **46**, 11457.

251 Sladek, A., Angermaier, K. and Schmidbaur, H. (1996) *Chemical Communications*, 1959; Dell'Amico, D.B., Calderazzo, F., Pasqualetti, N., Hubener, R., Maichle-Mossmer, C. and Strahle, J. (1995) *Journal of The Chemical Society-Dalton Transactions*, 3917.

252 Kharash, M.S. and Isbell, H.S. (1931) *Journal of the American Chemical Society*, **52**, 2919; Willner, H. and Aubke, F. (1990) *Inorganic Chemistry*, **29**, 2195; Willner, H., Schaebs, J., Hwang, G., Mistry, F., Jones, R., Trotter, J. and Aubke, F. (1992) *Journal of the American Chemical Society*, **114**, 8972; Dias, H.V.R. and Jin, W.C. (1996) *Inorganic Chemistry*, **35**, 3687.

253 Schmidbaur, H. and Schier, A. (2007) in *Comprehensive Organometallic Chemistry III*, 2 (eds R. Crabtree, M. Mingos and K. Meyer), Elsevier, New York, p. 251.

254 Fernández, E.J., Gimeno, M.C., Laguna, A., López de Luzuriaga, J.M., Monge, M., Pyykkö, P. and Sundholm, D. (2000) *Journal of the American Chemical Society*, **122**, 7287.

255 Fernández, E.J., López de Luzuriaga, J.M., Monge, M., Olmos, M.E., Pérez, J., Laguna, A., Mohamed, A.A. and Fackler, J.P., Jr (2003) *Journal of the American Chemical Society*, **125**, 2022.

256 Lefebvre, J., Batchelor, R.J. and Leznoff, D.B. (2004) *Journal of the American Chemical Society*, **126**, 16117.

257 Usón, R., Laguna, A., Laguna, M. and Usón, A. (1987) *Organometallics*, **6**, 1778.

258 Schmidbaur, H., Scherbaum, F., Huber, B. and Muller, G. (1988) *Angewandte Chemie-International Edition*, **27**, 419.

259 Vicente, J., Chicote, M.T., Guerrero, R. and Jones, P.G. (1996) *Journal of the American Chemical Society*, **118**, 699.

260 Fackler, J.P., Jr (2002) *Inorganic Chemistry*, **41**, 1.

261 Mazany, A.M. and Fackler, J.P., Jr (1984) *Journal of the American Chemical Society*, **106**, 801.

262 Feng, D.F., Tang, S.S., Liu, C.W., Lin, I.J.B., Wen, Y.S. and Liu, L.K. (1997) *Organometallics*, **16**, 901.

263 Vicente, J., Chicote, M.T., Abrisqueta, M.D. and Jones, P.G. (1997) *Organometallics*, **16**, 5628.

264 White-Morris, R.L., Olmstead, M.M., Jiang, F., Tinti, D.S. and Balch, A.L. (2002) *Journal of the American Chemical Society*, **124**, 2327.

265 Ricard, L. and Gagosz, F. (2007) *Organometallics*, **26**, 4704; de Fremont, P., Stevens, E.D., Eelman, L.D., Fogg, D.E. and Nolan, S.P. (2006) *Organometallics*, **25**, 5824; de Fremont, P., Stevens, E.D., Fructos, M.R., Diaz Requejo, M.M., Pérez, P.J. and Nolan, S. (2006) *Chemical Communications*, 2045.

266 Lactar, D.S., Müller, P., Gray, T.G. and Sadighi, J.P. (2005) *Organometallics*, **24**, 4503.

267 Catalano, V.J., Malwitz, M.A. and Etogo, A.O. (2004) *Inorganic Chemistry*, **43**, 5714.

268 Lu, W., Zhu, N. and Che, C.M. (2003) *Journal of the American Chemical Society*, **125**, 16081; de la Riva, H., Nieuwhuyzen, M., Fierro, C.M., Raithby, P.R., Male, L. and Lagunas, M.C. (2006) *Inorganic Chemistry*, **45**, 1418; Yam, V.W.W., Choi,

S.W.K. and Cheung, K.K. (1996) *Organometalllics*, **15**, 1734.

269 Irwin, M.J., Jennings, M.C. and Puddephatt, R.J. (1999) *Angewandte Chemie-International Edition*, **38**, 3376; McArdle, C.P., Vittal, J.J. and Puddephatt, R.J. (2000) *Angewandte Chemie-International Edition*, **39**, 3819; McArdle, C.P., Van, S., Jennings, M.C. and Puddephatt, R.J. (2002) *Journal of the American Chemical Society*, **124**, 3959.

270 Abu-Salah, O.M., Al-Ohaly, A.R.A. and Knobler, C.B. (1985) *Chemical Communications*, 1502; Schuster, O., Monkowius, U., Schmidbaur, H., Ray, R.S., Kruger, S. and Rosch, N. (2006) *Organometalllics*, **25**, 1004.

271 Kitagawa, H., Kojima, N. and Nakajima, T. (1991) *Journal of The Chemical Society-Dalton Transactions*, 3121.

272 Herring, F.G., Hwang, G., Lee, K.C., Mistry, F., Phillips, P.S., Willner, H. and Aubke, F. (1992) *Journal of the American Chemical Society*, **114**, 1271.

273 Elder, S.H., Lucier, G.M., Hollander, F.J. and Bartlett, N. (1997) *Journal of the American Chemical Society*, **119**, 1020.

274 Ihlo, L., Olk, R.M. and Kirmse, R. (2001) *Inorganic Chemistry Communications*, **4**, 626.

275 Blake, A.J., Gould, R.O., Greig, J.A., Holder, A.J., Hyde, T.I. and Schröder, M. (1989) *Journal of the Chemical Society, Chemical Communications*, 876; Blake, A.J., Greig, J.A., Holder, A.L., Hyde, T.I., Taylor, A. and Schröder, M. (1990) *Angewandte Chemie-International Edition*, **29**, 197.

276 Shaw, J.L., Wolowska, J., Collison, D., Howard, J.A.K., McInnes, E.J.L., McMaster, J., Blake, A.J., Wilson, C. and Schröder, M. (2006) *Journal of the American Chemical Society*, **128**, 13827.

277 Walker, N.R., Wright, R.R., Barran, P.E. and Stace, A.J. (1999) *Organometallics*, **18**, 3569; Walker, N.R., Wright, R.R., Barran, P.E., Murrell, J.N. and Stace, A.J. (2001) *Journal of the American Chemical Society*, **123**, 4223.

278 Siedel, S. and Seppelt, K. (2000) *Science*, **290**, 117.

279 Pyykkö, P. (2000) *Science*, **290**, 64.

280 Drews, T., Siedel, S. and Seppelt, K. (2002) *Angewandte Chemie-International Edition*, **41**, 454.

281 Wickleder, M.S. (2001) *Zeitschrift fur Anorganische und Allgemeine Chemie*, **627**, 2112.

282 Calabro, D.C., Harrison, B.A., Palmer, G.T., Moguel, M.K., Rebbert, R.L. and Burmeister, J.L. (1981) *Inorganic Chemistry*, **20**, 4311.

283 Khan, M.N.I., Fackler, J.P., Jr, King, C., Wang, J.C. and Wang, S. (1988) *Inorganic Chemistry*, **27**, 1672; Khan, M.N.I., Wang, S. and Fackler, J.P., Jr (1989) *Inorganic Chemistry*, **28**, 3579.

284 Abdou, H.E., Mohamed, A.A. and Fackler, J.P., Jr (2007) *Inorganic Chemistry*, **46**, 9692.

285 Bennett, M.A., Bhargava, S.K., Griffiths, K.D. and Robertson, G.B. (1987) *Angewandte Chemie International Edition*, **26**, 260.

286 Murray, H.H. and Fackler, J.P., Jr (1986) *Inorganica Chimica Acta*, **115**, 207.

287 Bardají, M., Gimeno, M.C., Jones, P.G., Laguna, A. and Laguna, M. (1994) *Organometallics*, **13**, 3415; Bardají, M., Jones, P.G., Laguna, A. and Laguna, M. (1995) *Organometallics*, **14**, 1310.

288 Usón, R., Laguna, A., Laguna, M., Jiménez, J. and Jones, P.G. (1991) *Journal of The Chemical Society-Dalton Transactions*, 1361; Fackler, J.P., Jr and Porter, L.C. (1986) *Journal of the American Chemical Society*, **108**, 2750.

289 Usón, R., Laguna, A., Laguna, M., Jiménez, J. and Jones, P.G. (1991) *Angewandte Chemie-International Edition*, **30**, 198.

290 Laguna, A., Laguna, M., Jiménez, J., Lahoz, F.J. and Olmos, E. (1994) *Organometallics*, **13**, 253.

291 Murray, H.H., Briggs, D.A., Garzon, G., Raptis, R.G., Porter, L.C. and Fackler, J.P., Jr (1987) *Organometallics*, **6**, 1992.

292 Gimeno, M.C., Jiménez, J., Jones, P.G., Laguna, A. and Laguna, M. (1994) *Organometallics*, **13**, 2508.

293 Yam, V.W.W., Choi, S.W.K. and Cheung, K.K. (1996) *Journal of the Chemical Society. Chemical Communications*, 1173; Yam, V.W.W., Li, C.K., Chan, C.L. and Cheung, K.K. (2001) *Inorganic Chemistry*, **40**, 7054.

294 Yurin, S.A., Lemenovskii, D.A., Grandberg, K.I., Il'ina, I.G. and Kuz'mina, L.G. (2003) *Russian Chemical Bulletin-International Edition*, **52**, 2752.

295 Coetzee, J., Gabrielli, W.F., Coetzee, K., Schuster, O., Nogai, S.D., Cronje, S. and Raubenheimer, H.G. (2007) *Angewandte Chemie-International Edition*, **46**, 2497.

296 Sanner, R.D., Satcher, J.H., Jr and Droege, M.W. (1989) *Organometallics*, **8**, 1498; Assman, B., Angermaier, K., Paul, M. and Riede, J. (1995) *Chemische Berichte*, **128**, 891.

297 Fernández, E.J., Laguna, A. and Olmos, M.E. (2008) *Coordination Chemistry Reviews*, **252**, 1630.

298 Usón, R., Laguna, A., Laguna, M., Tartón, M.T. and Jones, P.G. (1988) *Journal of the Chemical Society. Chemical Communications*, 740.

299 Fuchita, Y., Utsunomija, Y. and Yasutake, M. (2001) *Journal of The Chemical Society-Dalton Transactions*, 2330.

300 Cinellu, M.A., Zucca, A., Stoccoro, S., Minghetti, G., Manassero, M. and Sansón, M. (1995) *Journal of The Chemical Society-Dalton Transactions*, 2865; Fuchita, Y., Ieda, H., Tsunemune, Y., Kinoshita-Kawashima, J. and Kawano, H. (1998) *Journal of The Chemical Society-Dalton Transactions*, 791; Constable, E.C., Henney, R.P.G., Raithby, P.R. and Sousa, L.R. (1991) *Angewandte Chemie-International Edition*, **30**, 1363; Chan, C.W., Wong, W.T. and Che, C.M. (1994) *Inorganic Chemistry*, **33**, 1266.

301 Coronnello, M., Mini, E., Caciagli, B., Cinellu, M.A., Bindoli, A., Gabbiani, C. and Messori, L. (2005) *Journal of Medicinal Chemistry*, **48**, 6761.

302 Vicente, J., Chicote, M.T. and Bermúdez, M.D. (1982) *Inorganica Chimica Acta*, **63**, 35; Vicente, J., Chicote, M.T. and Bermúdez, M.D. (1984) *Journal of Organometallic Chemistry*, **268**, 191.

303 Vicente, J., Bermúdez, M.D. and Escribano, J. (1991) *Organometallics*, **10**, 3380.

304 Dudis, D.S. and Fackler, J.P., Jr (1985) *Inorganic Chemistry*, **24**, 3758; Jandik, P., Schubert, U. and Schmidbaur, H. (1982) *Angewandte Chemie-International Edition*, **21**, 73.

305 Méndez, L.A., Jiménez, J., Cerrada, E., Mohr, F. and Laguna, M. (2005) *Journal of the American Chemical Society*, **127**, 852.

306 Stein, J., Fackler, J.P., Jr, Paparizos, C. and Chen, H.W. (1981) *Journal of the American Chemical Society*, **103**, 2192; Usón, R., Laguna, A., Laguna, M., Usón, A. and Gimeno, M.C. (1987) *Organometallics*, **6**, 682.

307 Usón, R., Laguna, A., Laguna, M., Manzano, B.R., Jones, P.G. and Sheldrick, G.M. (1984) *Journal of The Chemical Society-Dalton Transactions*, 839; Fernández, E.J., Gimeno, M.C., Jones, P.G., Laguna, A., Laguna, M. and López de Luzuriaga, J.M. (1992) *Journal of The Chemical Society-Dalton Transactions*, 3365; Fernández, E.J., Gimeno, M.C., Jones, P.G., Laguna, A., Laguna, M. and López de Luzuriaga, J.M. (1995) *Organometallics*, **14**, 2918; Usón, R., Laguna, A., Laguna, M., Lázaro, I. and Jones, P.G. (1987) *Organometallics*, **6**, 2326; Gimeno, M.C., Jones, P.G., Laguna, A., Laguna, M. and Lázaro, I. (1993) *Journal of The Chemical Society-Dalton Transactions*, 2223.

308 Usón, R., Laguna, A., Villacampa, M.D., Jones, P.G. and Sheldrick, G.M. (1984) *Journal of The Chemical Society-Dalton Transactions*, 2035.

309 Raubenheimer, H.G., Olivier, P.J., Lindeque, L., Desmet, M., Hrusak, J. and Kruger, G.J. (1997) *Journal of Organometallic Chemistry*, **544**, 91.

310 Wehlan, M., Thiel, R., Fuchs, J., Beck, W. and Fehlhammer, W.P. (2000) *Journal of Organometallic Chemistry*, **613**, 159.
311 311. Frémont, P., Singh, R., Stevens, E.D., Petersen, J.L. and Nolan, S.P. (2007) *Organometallics*, **26**, 1376.
312 Schuster, O. and Schmidbaur, H. (2005) *Organometallics*, **24**, 2289.
313 Yam, V.W.W., Wong, K.M.C., Hung, L.L. and Zhu, N. (2005) *Angewandte Chemie-International Edition*, **44**, 3107; Wong, K.M.C., Hung, L.L., Lam, W.H., Zhu, N. and Yam, V.W.W. (2007) *Journal of the American Chemical Society*, **129**, 11662.
314 Shorrock, C.J., Jong, H., Batchelor, R.J. and Leznoff, D.B. (2003) *Inorganic Chemistry*, **42**, 3917.
315 Hollis, L.S. and Lippard, S.J. (1983) *Journal of the American Chemical Society*, **105**, 4293.
316 Canty, A.J., Minchin, N.G., Healy, P.C. and White, A.H. (1982) *Journal of The Chemical Society-Dalton Transactions*, 1795.
317 Bandini, A.L., Banditelli, G., Bonatti, F., Minghetti, G. and Pinillos, M.T. (1985) *Inorganica Chimica Acta*, **99**, 165; Vicente, J., Chicote, M.T., Guerrero, R. and Herber, U. (2002) *Inorganic Chemistry*, **41**, 1870.
318 Vicente, J., Chicote, M.T., Bermúdez, M.D., Jones, P.G., Fittschen, C. and Sheldrick, G.M. (1986) *Journal of The Chemical Society-Dalton Transactions*, 2361.
319 Messori, L., Abate, F., Orioli, P., Tempi, C. and Marcon, G. (2002) *Chemical Communications*, 612.
320 Messori, L., Abbate, F., Marcon, G., Orioli, P., Fontani, M., Mini, E., Mazzei, T., Carotti, S., O'Connell, T. and Zanello, P. (2000) *Journal of Medicinal Chemistry*, **43**, 3541.
321 Brun, A.M., Harriman, A., Heitz, V. and Sauvage, J.P. (1991) *Journal of the American Chemical Society*, **113**, 8657; Kilså, K., Kajanus, J., Macpherson, A.N., Mårtensson, J. and Albinsson, B. (2001) *Journal of the American Chemical Society*, **123**, 3069; Chambron, J.C., Heitz, V. and Sauvage, J.P. (1997) *New Journal of Chemistry*, **21**, 237.
322 Mori, S. and Osuka, A. (2005) *Journal of the American Chemical Society*, **127**, 8030.
323 Usón, R., Laguna, A., Laguna, M., Fernández, E., Jones, P.G. and Sheldrick, G.M. (1982) *Journal of The Chemical Society-Dalton Transactions*, 1971; Usón, R., Laguna, A., Laguna, M., Fraile, M.N., Jones, P.G. and Sheldrick, G.M. (1986) *Journal of The Chemical Society-Dalton Transactions*, 291.
324 Fernández, E.J., Gimeno, M.C., Jones, P.G., Laguna, A. and Olmos, E. (1997) *Organometallics*, **16**, 1130.
325 Bardají, M., Laguna, A., Orera, V.M. and Villacampa, M.D. (1998) *Inorganic Chemistry*, **37**, 5125; Bardají, M., Laguna, A., Vicente, J. and Jones, P.G. (2001) *Inorganic Chemistry*, **40**, 2675.
326 Bella, P.A., Crespo, O., Fernández, E.J., Fischer, A.K., Jones, P.G., Laguna, A., López de Luzuriaga, J.M. and Monge, M. (1999) *Journal of The Chemical Society-Dalton Transactions*, 4009.
327 Blanco, M.C., Fernández, E.J., Jones, P.G., Laguna, A., López de Luzuriaga, J.M. and Olmos, M.E. (1998) *Angewandte Chemie-International Edition*, **37**, 3042; Blanco, M.C., Fernández, E.J., López de Luzuriaga, J.M., Olmos, M.E., Crespo, O., Gimeno, M.C., Laguna, A. and Jones, P.G. (2000) *Chemistry – A European Journal*, **6**, 4116.
328 Blanco, M.C., Fernández, E.J., Olmos, M.E., Pérez, J., Crespo, O., Laguna, A. and Jones, P.G. (2004) *Organometallics*, **23**, 4373.
329 Cinellu, M.A., Minghetti, G., Pinna, M.V., Stoccoro, S., Zucca, A., Manassero, M. and Sansoni, M. (1998) *Journal of The Chemical Society-Dalton Transactions*, 1735; Cinellu, M.A., Minghetti, G., Pinna, M.V., Stoccoro, S., Zucca, A. and Manassero, M. (1998) *Chemical Communications*, 2397.
330 Casini, A., Cinellu, M.A., Minghetti, G., Gabbiani, C., Coronnello, M., Mini, E. and Messori, L. (2006) *Journal of Medicinal Chemistry*, **49**, 5524.

331 Cao, R., Anderson, T.M., Piccoli, P.M.B., Schultz, A.J., Koetzle, T.F., Geletti, Y.V., Slonkina, E., Hedman, B., Hodgson, K.O., Hardcastle, K.I., Fang, X., Kirk, M.L., Knottenbelt, S., Kögerler, P., Musaev, D.G., Morokuma, K., Takahashi, M. and Hill, C.L. (2007) *Journal of the American Chemical Society*, **129**, 11118.

332 Wang, S. and Fackler, J.P., Jr (1990) *Inorganic Chemistry*, **29**, 4404.

333 Crespo, O., Canales, F., Gimeno, M.C., Jones, P.G. and Laguna, A. (1999) *Organometallics*, **18**, 3142; Canales, S., Crespo, O., Gimeno, M.C., Jones, P.G. and Laguna, A. (2004) *Inorganic Chemistry*, **43**, 7234.

334 Abram, U., Mack, J., Ortner, K. and Müller, M. (1998) *Journal of The Chemical Society-Dalton Transactions*, 1011.

335 Canales, F., Canales, S., Crespo, O., Gimeno, M.C., Jones, P.G. and Laguna, A. (1998) *Organometallics*, **17**, 1617.

336 Kirmse, R., Stach, J., Dietzsch, W., Steimecke, G., and Hoyer, E. (1980) *Inorganic Chemistry*, **19**, 2679; Imai, H., Oksuka, T., Naito, T., Awaga, K. and Inabe, T. (1999) *Journal of the American Chemical Society*, **121**, 8098.

337 Cerrada, E., Fernández, E.J., Jones, P.G., Laguna, A., Laguna, M. and Terroba, R. (1995) *Organometallics*, **14**, 5537.

338 Crespo, O., Gimeno, M.C., Jones, P.G. and Laguna, A. (1997) *Journal of The Chemical Society-Dalton Transactions*, 1099.

339 Ortner, K., Hilditch, L., Zheng, Y., Dilworth, J.R. and Abram, U. (2000) *Inorganic Chemistry*, **39**, 2801.

340 Dilworth, J.R., Hutson, A.J., Zubieta, J. and Chen, Q. (1994) *Transition Metal Chemistry*, **19**, 61.

341 Bardají, M., Gimeno, M.C., Jones, P.G., Laguna, A., Laguna, M., Merchán, F. and Romeo, I. (1997) *Organometallics*, **16**, 1083.Bariain, C., Matias, I.R., Romeo, I., Garrido, J. and Laguna, M. (2001) *Sensors and Actuators B-Chemical*, **76**, 25.

342 Canales, F., Gimeno, M.C., Laguna, A. and Jones, P.G. (1996) *Organometallics*, **15**, 3412; Calhorda, M.J., Canales, F., Gimeno, M.C., Jiménez, J., Jones, P.G., Laguna, A. and Veiros, L.F. (1997) *Organometallics*, **16**, 3837; Canales, S., Crespo, O., Gimeno, M.C., Jones, P.G., Laguna, A. and Mendizabal, F. (2001) *Organometallics*, **20**, 4812.

343 Timakov, A.A., Prusakov, V.N. and Drobyshevskii, Y.V. (1986) *Doklady Akademii Nauk SSSR*, **291**, 125.

344 Riedel, S. and Kaupp, M. (2006) *Inorganic Chemistry*, **45**, 1228.

345 Holloway, J.H. and Schrobilgen, G.J. (1975) *Journal of the Chemical Society. Chemical Communications*, 623; Sokolov, V.B., Prusakov, V.N., Ryzhkov, A.V., Drobyshevskii, Y.V. and Khoroshev, S.S. (1976) *Doklady Akademii Nauk SSSR*, **229**, 884; Vasile, M.J., Richardson, T.J., Stevie, F.A. and Falconer, W.E. (1976) *Journal of The Chemical Society-Dalton Transactions*, 351.

346 Brunvoll, J., Ischenko, A.A., Ivanov, A.A., Romanov, G.V., Sokolov, V.B., Spiridonov, V.P. and Strand, T.G. (1982) *Acta Chemica Scandinavica Series A-Physical and Inorganic Chemistry*, **36**, 705.

347 Hwang, I.C. and Seppelt, K. (2001) *Angewandte Chemie-International Edition*, **40**, 3690.

348 Müller, B.G. (1987) *Angewandte Chemie-International Edition*, **26**, 1081.

349 Leary, K. and Bartlett, N. (1972) *Journal of the Chemical Society. Chemical Communications*, 903; Leary, K., Zalkin, A. and Bartlett, N. (1973) *Journal of the Chemical Society. Chemical Communications*, 131; Leary, K., Zalkin, A. and Bartlett, N. (1974) *Inorganic Chemistry*, **13**, 775.

350 Schultz, A.J., Wang, H.H., Soderholm, L.C., Sifter, T.L., Williams, J.M., Bechgaard, K. and Whangbo, M.H. (1987) *Inorganic Chemistry*, **26**, 3757.

351 Belo, D., Alves, H., Lopes, E.B., Duarte, M.T., Gama, V., Henriques, R.T., Almeida, M., Pérez-Benítez, A., Rovira, C. and Veciana, J. (2001) *Journal of Chemistry – A European Journal*, **7**, 511.

2
Gold–Gold Interactions
Olga Crespo

2.1
Introduction

This chapter focuses mainly on closed shell gold(I) atoms which are found to *feel* attraction among themselves. This attraction is able to modify expected geometries or support one-, two- or three-dimensional arrays and also leads to interesting properties. Closed shell metal cations are expected to repel each other. Nevertheless, an important number of examples containing [d^8–d^{10}–s^2] interactions are known. Particularly strong evidence has been accumulated on the attraction between two or more d^{10} shells. In the case of gold(I) such attraction between two or more monovalent gold(I) ions in compounds has been observed structurally for a long time. In these situations the gold centers approach each other to a distance of between 2.7 and 3.3 Å. This range includes the distance between gold atoms in gold metal and approaches, or even overlaps with the range of distances found in the few *real* Au–Au single bonds. Schmidbaur coined the name *aurophilic attraction* in 1990 and defined it as *"the unprecedented affinity between gold atoms even with "closed-shell" electronic configurations and equivalent electrical charges"* [1]. Since that moment, aurophilicity has become one of the hot topics in gold chemistry, at first with the aim of understanding such interaction, more recently because of the observation that it may play an important role in interesting properties, thus understanding and modulating aurophilic interactions could help to govern these properties, which include the optical behavior. Although different origins are possible for the luminescent emissions of coordination complexes, such as intraligand transitions (ILT), ligand to metal or metal to ligand charge transfer transitions (LMCT, MLCT) and metal centered (MCT) transitions, in many luminescent gold complexes the emissions have been related to the presence of aurophilic interactions.

Information about aurophilicity and chemical situations in which it is observed may be found in different works that may be classified as those which analyse closed shell interactions in general [2], those focused on aurophilicity [1, 3], and others which analyze the role of these interactions in coordination [4] and organometallic gold

Modern Supramolecular Gold Chemistry: Gold-Metal Interactions and Applications.
Edited by Antonio Laguna
Copyright © 2008 WILEY-VCH Verlag GmbH & Co. KGaA, Weinheim
ISBN: 978-3-527-32029-5

chemistry [5]. The strengths of some gold(I)–gold(I) interactions per pair of gold atoms have been theoretically estimated and measured by temperature-dependent NMR spectroscopy, and are in the range 29–46 kJ mol^{-1}, comparable with a *good* hydrogen bond [2]. Experimental and theoretical studies indicate that hydrogen bonding and aurophilic bonding are similar in their binding energies and directionality and thus can be used, alone or combined, in the building of supramolecular structures as will be described below. From theoretical calculations the nature of the aurophilic attraction seems to be understood [2, 6, 7] as based on a dispersion effect with some virtual charge transfer contribution. Although not inherently relativistic, aurophilicity is considerably strengthened by relativistic effects, found in heavy atoms and showing a maximum in gold (see Figure 1.1 in Chapter 1). Thus, relativistic effects have a deep influence on gold chemistry and its applications [8, 9]. A more visual explanation of this phenomenon has been provided by Pyykkö who describes aurophilicity as *"an unusually strong van der Walls interaction."*

Within the scope of this chapter are situations in which the presence of gold–gold contacts has been confirmed by X-ray studies, with minor representation of those species which are *expected to show* such interactions. It is possible to find a wider vision of interesting gold species which are expected to display gold–gold contacts in Chapter 1 of this book. Commentaries are centered on homometallic gold derivatives. Outside the scope of this chapter are those complexes that, in addition to gold–gold interactions display gold–heterometal contacts, which are only commented in particular cases. These complexes are analyzed in Chapter 4. Although the luminescent properties of some representative species are mentioned, such optical behavior is more deeply analyzed in Chapter 6. Other properties are briefly presented. As commented above, finding relations between the presence of aurophilic interactions in gold complexes and their properties is an attractive objective in gold chemistry. For this reason, although this chapter is not meant to be comprehensive, the commentaries have been selected with the aim of giving a wide vision of the sort of ligands involved and the structural types of complexes that present gold–gold contacts.

2.2
Gold(I)–Gold(I) Interactions

The complexes presented here have been classified with regard to different structural situations. Section 2.2.1 shows examples of the association of mononuclear dicoordinated gold units through intermolecular aurophilic interactions leading to different aggregates. Section 2.2.2 is dedicated to polynuclear derivatives in which one center bridges two or more gold atoms. These compounds exhibit intramolecular gold–gold contacts which may lead to unexpected geometries, but some of them also exhibit intermolecular Au–Au interactions, leading to more complicated patterns. The following three sections show examples of compounds in which two or more gold atoms are bridged by monodentate (Section 2.2.3), bidentate (Section 2.2.4) or polydentate ligands (Section 2.2.5) which may show intra and/or intermolecular interations, depending on the ligand.

2.2.1
Aggregation of Mononuclear Dicoordinated Gold Units

One of the first observations, which led to the term *aurophilicity*, was the tendency of dicoordinated mononuclear gold complexes to associate in dimers, oligomers, polymeric chains, 2D or 3D structures. These facts illustrate that gold centers tend to get closer one to another. The principle *"the less the steric hindrance of the ligand the highest the dimensionality of the aggregation"* normally works as a general rule in the formation of the final species. Figure 2.1 shows some situations which are discussed further. The typical linear coordination at the gold centers is frequently distorted by the presence of the aurophilic interactions.

The relevance of this aggregation phenomenon in gold chemistry may be extended to the extraction of gold cyanide from leaching brines. The association of the $[Au(CN)_2]^-$ anion when adsorbed on the surface of activated carbon enhances the efficiency of the process [10]. Another interesting example of the consequences of the presence of these interactions in the properties of gold complexes is the induction of *"rotator phases"* solely built from gold–gold contacts [11] which have been reported in complexes [AuCl(CNR)] ($R = C_nH_{2n+1}$; $n = 2-12$).

Based on the idea that "the molecules containing groups with a preference for assembling via strong intermolecular interactions show preference to form structures with more than one molecule in the asymmetric unit (i.e., $Z' > 1$)" Anderson and coworkers [12] have studied molecules [AuXY]. As expected, these complexes have a tendency to form structures with $Z' > 1$. From molecular volume calculations the authors propose that such behavior can be related to the size differential of the two substituents, such that ligands with similar size show a tendency to form structures with $Z' > 1$. The structures with $Z' > 1$ show preference for discrete motifs (which include di- and polynuclear associations) over infinite chains, whereas those with $Z' = 1$ are evenly distributed between both arrangements. These trends must also consider the electronic characteristics of the ligands for a complete structural design.

The following section shows specific examples of association of dicoordinated mononuclear gold units which have been classified according to the nuclearity of the aggregate. Despite this general classification, the different polymorphic forms of a specific compound are discussed together, as well as the structures of complexes with small ligand variations.

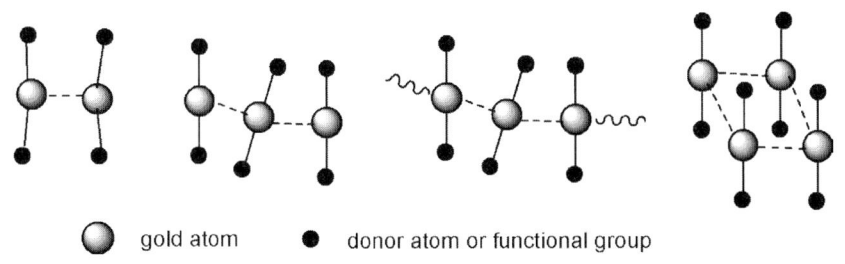

Figure 2.1 Different aggregates of monomeric dinuclear gold units.

2.2.1.1 Dinuclear Aggregates

Formation of dimers is widely observed in the crystal structures of dicoordinated gold (I) mononuclear species. The majority of them are built from [AuXL] units, where X is an anionic ligand and L a neutral one. One type of such species is the well known [AuX(PR$_3$)] (where PR$_3$ is a tertiary phosphine and X a halogen atom). These species can be mainly synthesized: (i) from tetrachloroauric acid (Equation 2.1), or by displacement of a weakly coordinating ligand (Y), such as alkylsulfides or CO, from [AuClY].

$$[AuCl_4]^- + 2PR_3 \rightarrow [AuCl(PR_3)] + PR_3Cl_2 + Cl^- \tag{2.1}$$

Antitumor activity, catalytic properties [13] and luminescent behavior have been described for [AuX(PR$_3$)] (X = halogen) complexes. For instance, the compound [AuIL] (L = (2-aminophenyl)methylphenylphosphine), which crystallizes as a racemic mixture of dimers [14], represents one example of a biologically active species. In the analogous species [AuIL] (L = (2-aminophenyl)diphenylphosphine) gold–gold interactions are absent. Hydrophilic tertiary phosphines have been studied to afford species with unconventional solubilities and reactivity patterns, but also with the aim of exploring potential biological properties. One of such phosphines is TPA (1,3,5-triaza-phospha-adamantane). Crystallographic and temperature dependent photoluminescent studies have been carried out on [AuX(TPA)] (X = Cl, Br, I) complexes. [AuX(TPA)] (X = Cl, Br) crystallize as dimers, as well as the chloride [AuCl(TPA.HCl)] and iodide [AuI(TPA.H)][AuI$_2$] of the protonated ligand. All exhibit crossed dimers built by the association of neutral units, as shown in Figure 2.2a, except for [AuI(TPA.H)][AuI$_2$] in which both ions [AuI(TPA.H)]$^+$ and [AuI$_2$]$^-$ interact in a very nearly perpendicular geometry (Figure 2.2b) [15]. The complexes exhibit interesting luminescent behavior with one emission in the region 580–670 nm (depending on the complex) whose origin has been related to the presence of aurophilic contacts. The unprotonated bromide and iodide also exhibit a high energy emission at about 450 nm, which has been associated with a singlet ligand (halide) to metal charge transfer (LMCT) excited state.

[AuClL] where L is the phosphorus ligand [PH(Ph)(Mes)] (Mes = 2,4,6-(t-Bu)$_3$C$_6$H$_2$ [16], 1-phenyldibenzophosphole [17] (Figure 2.3), PPh$_2$(CF = CF$_2$) [18] or PPh$_2$(CH$_2$)$_2$P(S)Ph$_2$ also form dimers. In the latter the bidentate ligand coordinates to gold only through the phosphorus atom [19]. Among complexes of general formula [AuCl(PR$_3$)] compound [AuCl(PVi$_3$)] (Vi = vinyl) represents one of the scarce examples of structurally characterized coordination of gold to trialkenylphosphines [20]. Dimers are formed with molecular axes crossing approximately at right angles.

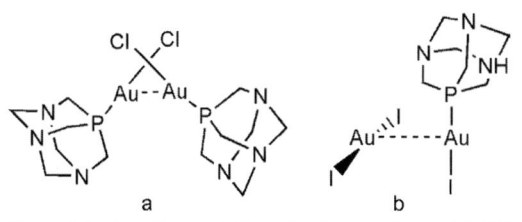

Figure 2.2 Dimeric association of gold complexes with TPA (a) and protonated TPA (b).

Figure 2.3 Dimeric association of [AuCl(1-phenyldibenzophosphole)].

Some studies have been carried out on [AuX(PR$_3$)] (X = halogen) species. The effects of substituents on aurophilic and π–π interactions in [AuX{P(C$_6$H$_4$R)$_3$}] and [AuX{P(C$_6$H$_3$(CF$_3$)$_2$)$_3$}] (X = halogen, SPh, Spy) have been analyzed [21]. Different studies by Pyykkö et al. suggest that the Au–Au interaction in [AuX(PH$_3$)]$_2$ compounds increases with the softness of the X ligand [22] and the Au–Au separation in the dimers decreases in the series Cl > Br > I. This study fits with the results found for [AuX(PPhMe$_2$)] (X = halogen) [23]. Among these complexes, the chloride compound may crystallize as a dimer or trimer, whereas the other halogenides show dinuclear associations. Spectroscopic data indicate the breakdown of the aurophilic associations in solution, nevertheless the iodine complex displays a degree of association in solution, which would be consistent with the theoretical calculations that indicate a stronger gold–gold interaction when the halogen atom is iodide. Emission, photoexcitation absorption and zero-filled optically detected magnetic resonance (ODMR), as well as the kinetic parameters of the emissive excited states have been reported for dimeric [AuX(PPhMe$_2$)]$_2$ (X = Cl, Br, I) and trimeric [AuCl(PPhMe$_2$)]$_3$ complexes and their emissions compared with those of the monomeric [AuX(EPh$_3$)] (E = P, As; X = Cl, Br) [24]. Aurophilic interactions lead to dimers in [Au(OCOMe)(PR$_3$)] (PR$_3$ tris(2-thienyl)phosphine) and are absent in the chloride complexes [AuCl(ER$_3$)] (ER$_3$ = tris(2-thienyl)phosphine, tris(2-thienyl)arsine, tri-(2-fuyl)phosphine) [25]. Other anionic ligands in complexes [AuX(PPh$_3$)] which crystallize as dimers include chalcogenolates [21, 26–29], an example is shown in Figure 2.4, or diphenyldithiophosphinate ligands [30]. Some phosphine thiolate complexes of stoichiometry [Au(SR)L] (SR = phenylthiolate; phenylthiolate OMe or Cl substituted; L = PPh$_3$, TPA) have been studied with the aim of rationalizing the relation between the gold–gold distance and the energies of the luminescent emissions [31].

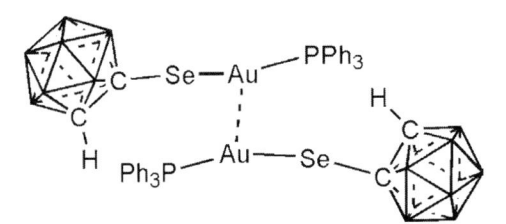

Figure 2.4 Dimeric association in a monoselenotale compound.

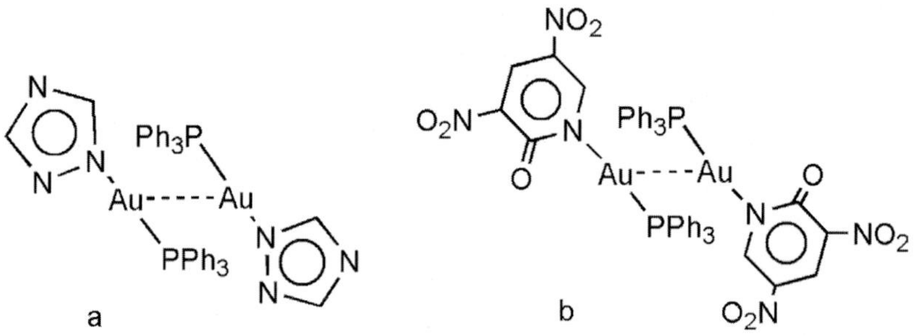

Figure 2.5 Dimeric association of complexes with N-donor ligands.

Triazole, [32] pyridine [33] or pyrazolate [34, 35] ligands may act as anionic N-donors (X) in [AuX(PPh$_3$)] which form dimers in the solid state. Complexes [Au(3,5-R$_2$pz)L] (pz = pirazolate, R = Ph, L = PPh$_3$, TPA; R = C$_4$H$_9$OC$_6$H$_4$, L = PPh$_3$) exhibit emissions in the 450 nm region. Based on the vibronic structure observed in these emissions have been assigned to ligand to metal charge transfer transitions (LMCT) in complexes [Au(3,5-Ph$_2$pz)L] (L = PPh$_3$, TPA). The structure of compound [Au(1,2,4-L)(PPh$_3$)] (L = triazole) is dimeric (Figure 2.5a), whereas that of [Au(1,2,3-L)(PPh$_3$)] is monomeric. Both contrast with the helical polymeric structure obtained for silver under the same reaction conditions in which the triazole ligands bridge two silver atoms. The reaction of the 3,5-dinitro-*ortho*-hydroxypyridine with [Au$_3$(μ-O)(PPh$_3$)$_3$]$^+$ affords the complex [Au(NC$_5$H$_2$(NO$_2$)$_2$O)(PPh$_3$)]$_2$ (Figure 2.5b), whereas in similar conditions the *para* isomer yields the three-nuclear oxonium derivative [Au$_3$(μ-O)(PPh$_3$)$_3$][NC$_5$H$_2$(NO$_2$)$_2$O].

Bis-(3,5-di-*tert*-butyl-1,2,4-triphospholyl)-bis(triethylphosphine)-digold(I) [36] contains an anionic phosphorus donor ligand. In the crystal structure the cation [bis(triethylphosphine)gold(I)]$^+$ and the anion [bis(3,5-*tert*-butyl-1,2,4-triphospholyl)aurate(I)]$^-$ associate in pairs through gold–gold interactions.

Some complexes contain different forms (anionic and neutral) of the same ligand. One example is the formation of dimers between the anion [bis(4-methyl-1,3-thiazol-2-yl)aurate(I)]$^-$ and the cation [bis(4-methyl-1,3-thiazol-2-ylidene)gold(I)]$^+$ [37]. Another example is the compound with the diphenylphosphinite and diphenilphosphinous acid ligands shown in Figure 2.6a [38]. The crystal structure of the latter is an example of the cooperative action of hydrogen bonding and aurophilicity. In the monomer one phosphorus atom coordinates to one oxo ligand, the other to a hydroxyl fragment. This situation allows the formation in the dimer of two hydrogen bonding clamps in the periphery of the molecule. The gold–gold distance is 3.031(1) Å. Hydrogen bonds are also present in the dimeric structure of compounds [AuCl(PPh$_2$OH)] [39] and [AuCl(PPh$_2$(2-OH–C$_6$H$_4$)]. In the latter no aurophilic interactions are present and C–H···Cl hydrogen bonds lead to the formation of dimers (Figure 2.6b) [40]. [Au(Spy)(PPh$_2$py)] crystallizes as the monomer. Upon protonation, complex [Au(SpyH)(PPh$_2$py)]SbF$_6$ is afforded, which crystallizes as dimers connected by gold–gold contacts and hydrogen bonds (Figure 2.6c). The unprotonated compound is weakly emissive, whereas the protonated species displays stronger

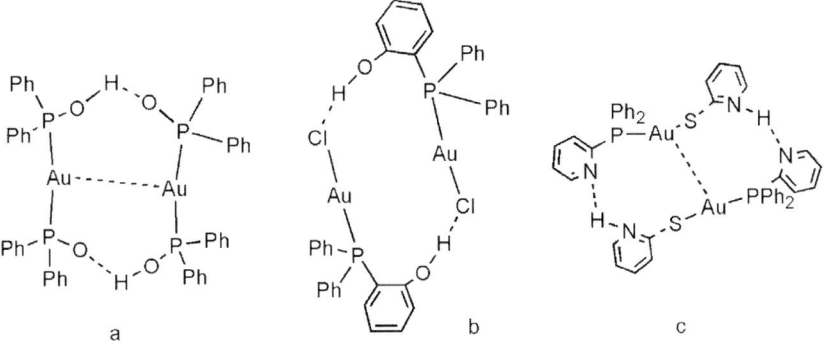

Figure 2.6 Examples of dimers which contain hydrogen bonds.

photoluminescence. The excited state has been tentatively attributed to a S → Au ligand to metal charge transfer transition [41].

Phosphinegold(I) organosulfonates and organosulfinates are attractive because of their potential catalytic activity. Among them the crystal structures of [Au{OS(O$_2$)-p-Tol}(PPh$_3$)] and [Au{S(O)$_2$-p-Tol}(PPh$_3$)] have been analyzed. The former consist of dimers of "Au{OS(O$_2$)-p-Tol}(PPh$_3$)" units which associate through aurophilic interactions. In the dimer each gold atom is bonded to one oxygen of a sulfinate ligand. In the structure of the latter one cation [Au(PPh$_3$)$_2$]$^+$ and one anion [Au{S(O)$_2$-p-Tol}$_2$]$^-$ associate through gold–gold interactions. Two weak oxygen–gold contacts are also present [42]. A different compound is that in which the fragment P{(CH$_2$)$_2$PPh$_2$AuCl}$_3$ acts as a *ligand* in [Au$_4$Cl$_4${P{(CH$_2$)$_2$PPh$_2$}$_3$]. In the crystal structure dimers are formed through gold–gold contacts (Figure 2.7) [43].

Examples of complexes that crystallize as dinuclear aggregates which do not contain tertiary phosphines include the carbene compound [Au{C(NHMe)NEt$_2$}(SCN)] [44]. The change of the substituent in complexes [AuCl(1,3-R$_2$-bimy)]

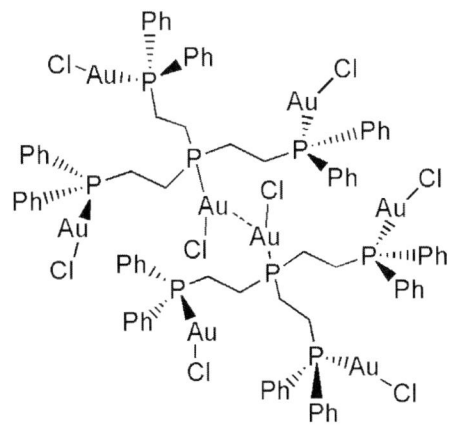

Figure 2.7 Dimeric association of [Au$_4$Cl$_4${P{(CH$_2$)$_2$PPh$_2$}$_3$].

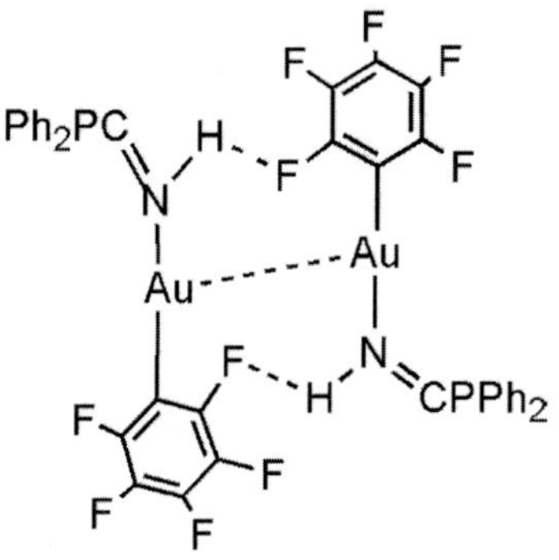

Figure 2.8 Dimeric association of [Au(C$_6$F$_5$){N(H)=CPh$_2$}].

(bimy = benzimidazol-2-ylidene) directs the formation of aurophilic interactions. When the substituent R is methyl both aurophilic and intermolecular π–π contacts are present, whereas if R = Ph only intermolecular π–π interactions exist [45]. Both complexes exhibit high energy emissions attributed to spin-forbidden intraligand transitions. [AuCl(Me$_2$-bimy)] also shows an emission at 620 nm which has been attributed to a spin-forbidden metal centered (^{3}CM) transition. Ab initio calculations have been used to analyze competition between metallophilic and hydrogen bonding in [M(C$_6$F$_5$){N(H)=CPh$_2$}] (M = Au, Ag) systems. Whereas the silver compound shows a ladder-type structure in which [Ag(C$_6$F$_5$){N(H)=CPh$_2$}] units display silver–silver interactions and four hydrogen bonds, the gold complex consists of pairs of molecules associated in antiparallel disposition (Figure 2.8) through aurophilic and C−H···F interactions. The theoretical calculations support that aurophilicity should be comparable in strength with hydrogen bonding and both combine to afford the diauracycles, whereas in the analogous silver compound the hydrogen bonds govern the final arrangement [46].

In the olefin complex [AuCl(C$_{10}$H$_{12}$)] a head to tail association of the two molecules is observed with the direction of the association perpendicular to the [AuCl(C$_{10}$H$_{12}$)] moiety. A very similar geometry has been observed in [AuCl(2,4,6-^{t}Bu$_3$C$_6$H$_2$PH$_2$)], different from the frequently observed "crossed torch" pairs [47]. Reaction of [AuCl(Ph$_2$Me)] with [Au(SiPh$_3$)(Ph$_2$Me)] affords [Au$_2$Cl(SiPh$_3$)(Ph$_2$Me)$_2$]. In its crystal structure one cation [Au(PPh$_2$Me)$_2$]$^+$ and one anion [Au(SiPh$_3$)Cl]$^-$ are connected through a short gold–gold contact of 2.9807(4) Å (Figure 2.9) [48].

These reorganization processes depend on small changes in the ligands. Reaction of [AuCN]$_n$ and imidazolidine-2-thione (etu) leads to [Au(etu)(CN)] which crystallizes

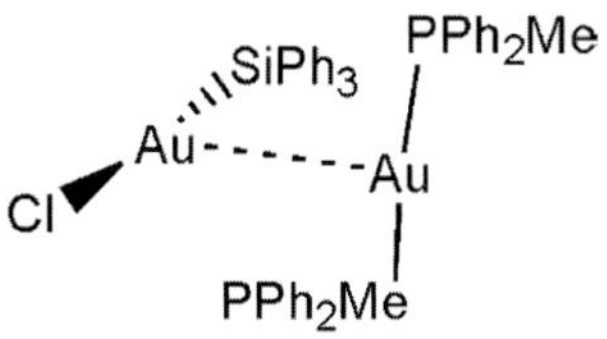

Figure 2.9 [Au(PPh$_2$Me)$_2$][AuCl(SiPh$_3$)].

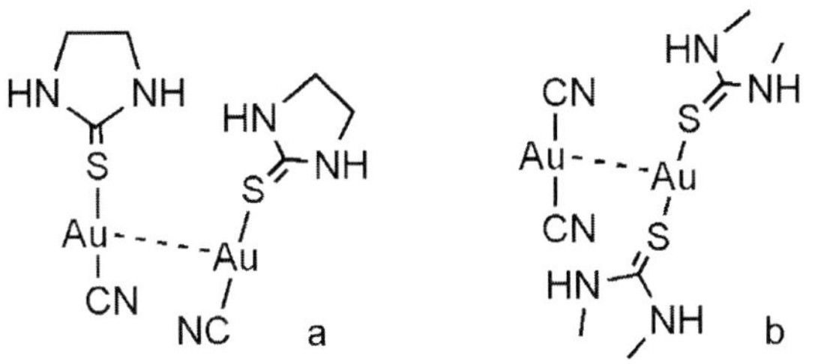

Figure 2.10 [Au(etu)(CN)]$_2$ (a) and [Au(dmtu)$_2$][Au(CN)$_2$] (b).

as dimers (Figure 2.10a), whereas in the crystal structure of the compound obtained by the same reaction with dimethylthiourea (dmtu) [Au(dmtu)$_2$]$^+$ and [Au(CN)$_2$]$^-$ ions are connected through gold–gold contacts (Figure 2.10b) [49]. Aurophilic interaction between oppositely charged ions [Au(CEP)$_2$]$^+$ [Au(CN)$_2$]$^-$ is also observed in the crystal structure of [Au$_2$(CEP)$_2$(CN)$_2$] (CEP = tris(2-cyanoethyl)phosphine) [50].

The majority of dimers are based on the association of neutral units or different charged ions ($\pm$). Although association of ions of like charges is not expected it has been observed, indicating that aurophilic interactions may overrule coulombic repulsions. Aggregation of anions is observed in [Au(SCN)$_2$]$^-$ [51, 52]. A series of salts of the anion [Au(SCN)$_2$]$^-$ exhibits different structural features depending on the cation. The tetrabutylamonium salt is dimeric, the phosphonium Ph$_4$P$^+$ salt is monomeric, the isostructural (Na$^+$, K$^+$, Cs$^+$, NH$_4^+$) salts consist of infinite linear chains with alternating short and long gold–gold contacts and the tetramethylammonium salt forms a kinked chain of trimers. The authors describe an inverse correlation between the gold–gold distance and the emission energies observed (between 506 and 670 nm) for all the species, with the exception of the non-emissive monomeric phosphonium salt. Dimers of cations [AuL$_2$]$^+$ have also been described in the compound with the ligand 4-(trimethylamonio)benzenethiol (Tab) [Au(SC$_6$H$_4$NMe$_3$)$_2$]$^+$ [53], [Au(PMe$_3$)(SeU)]$^+$ (SeU = selenourea) [54] or in the carbene complex [Au{C(NHMe)$_2$}$_2$]$^+$ [55]. In the latter the formation of dimers or chains through aurophilic interactions, as well as the luminescent properties (which have been related to the presence of gold–gold contacts) depend on the anion.

2.2.1.2 Tri- and Tetranuclear Associations

Association of [Au(py)$_2$]$^+$ and [AuX$_2$]$^-$ (X = Cl, Br, I) [56, 57] ions following the pattern $- + + -$ (Figure 2.11a) is another example of aurophilic attraction versus coulombic repulsions. The luminescent complex [Au(Tab)$_2$][Au(CN)$_2$] crystallizes as a tetranuclear species (Figure 2.11b) [53] in which the ions are ordered in the sequence $+ - - +$, whereas the corresponding [Au(Tab)$_2$]X$_2$ (X = PF$_6$ or I) crystallize as dimers, as commented above. Hydrogen bonds complicate the final structures. These hydrogen bonds involve the oxygen atom of a water molecule and hydrogen atoms of phenyl rings (X = I) or the fluorine atoms of the anion and hydrogen atoms of methyl or phenyl groups (X = PF$_6$) to afford a 3D or 2D structure, respectively. In the tetranuclear compound {[Au(Tab)$_2$][Au(CN)$_2$]}$_2$ C–H$\cdots$Au and hydrogen interactions between the cyanide nitrogen or the Tab sulfur with hydrogen atoms of

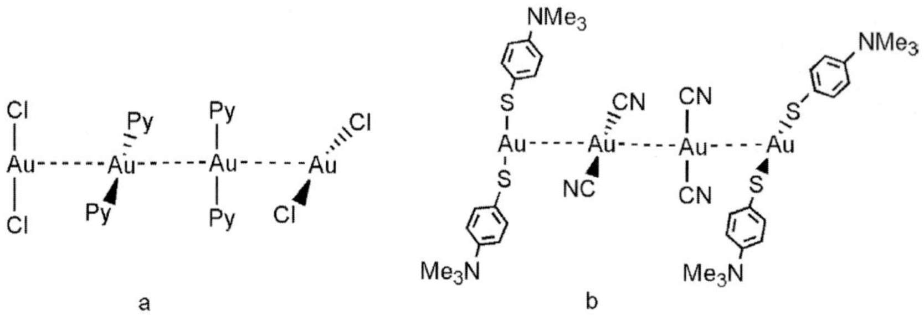

Figure 2.11 Association of like charge ions in tetranuclear aggregates.

methyl or phenyl rings of different chains build a three-dimensional structure. The gold silver complex {[Ag(Tab)$_2$][Au(CN)$_2$]}$_2$ has also been synthesized. It exhibits a + − − + pattern with aurofilic interactions between both [Au(CN)$_2$]$^-$ ions and C−H···Au hydrogen bonds build a 3D structure. By reaction of [Hg(Tab)$_2$](PF$_6$)$_2$ with K[Au(CN)$_2$] compound {[Hg(Tab)$_2$][Au(CN)$_2$]$_2$} is afforded. The crystal structure displays pairs of [Au(CN)$_2$]$^-$ anions which associate through gold–gold contacts and bridge [Hg(Tab)$_2$]$^{2+}$ units as shown in Figure 2.12. The presence of hydrogen bonds leads to a three-dimensional network. All these complexes are luminescent in the solid state [Au(Tab)$_2$]X$_2$ (X = PF$_6$ or I) and {[Au(Tab)$_2$][Au(CN)$_2$]}$_2$ display one emission near 450 nm assigned to a CM excited state and another at about 534 or 670 nm assigned to ligand to a metal–metal charge transfer processes (LMMCT). Complexes {[Ag(Tab)$_2$][Au(CN)$_2$]}$_2$ and {[Hg(Tab)$_2$][Au(CN)$_2$]$_2$} display only one emission at about 520 nm assigned to an MC excited state. Solutions of these species are not luminescent.

Insertion of the fragment GeCl$_2$ (from the 1,4-dioxane complex) in the [AuCl(PPh$_3$)] compound leads to [Au(GeCl$_3$)(PPh$_3$)] [58], which shows a dimeric structure (Figure 2.13a). In the tetranuclear compound {[Au(PPh$_2$Me)$_2$][Au(GeCl$_3$)$_2$]}$_2$ [59] the

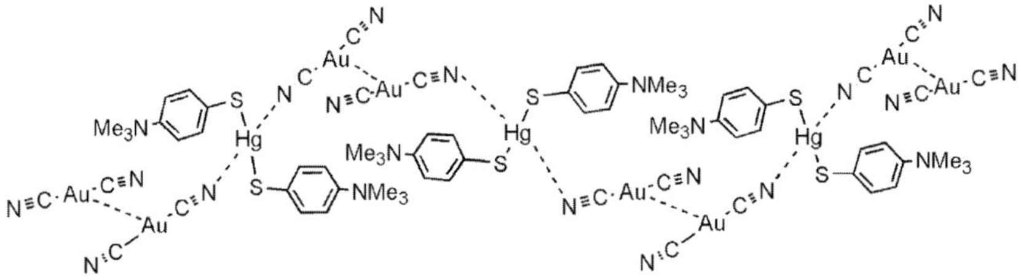

Figure 2.12 Chain structure of [Hg(Tab)$_2$][Au(CN)$_2$]$_2$.

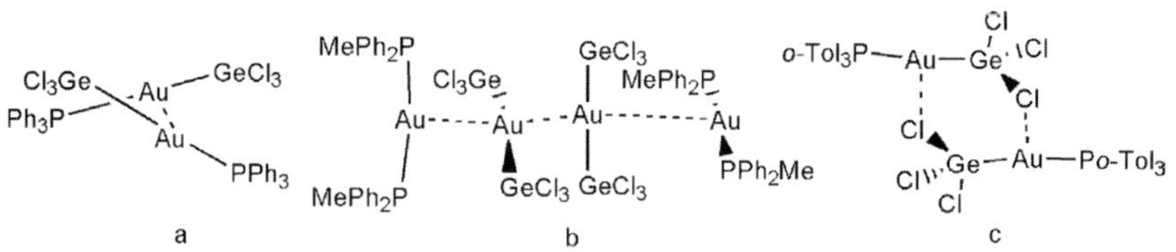

Figure 2.13 Different aggregates of [Au(GeCl$_3$)L].

Figure 2.14 Association of gold monomers in a +/neutral/− pattern.

ions associate in a + − − + pattern with a very short Au–Au distance between the anions of 2.881(1) Å (Figure 2.13b). Dimers of [Au(GeCl$_3$){P(o-Tol)$_3$}] do not exhibit aurophilic interactions and are supported by Au–Cl contacts (Figure 2.13c).

The polymorphism of [AuCl(PPh$_2$Me)], which crystallizes in two different forms, dimers or trimers, as well as its luminescent behavior have been revised as part of a study of polymorphism of gold(I) dicoordinated linear gold(I) complexes [60]. Other example of trinuclear association is obtained by the reaction of [AuCl(SMe$_2$)] with 2-aminopiridine, whose crystal structure consists of a +/neutral/− organization of units (Figure 2.14). A two-dimensional network is constructed by N−H···Cl hydrogen bonds [61].

Reaction of heterocyclic thiones with [AuCl(tht)] (tht = tetrahydrothiophene) affords neutral or cationic species. Among them complex [Au(SNC$_5$H$_5$)$_2$]ClO$_4$ has been studied by X-ray diffraction. Five of the six cations in the cell are linked by short Au–Au contacts and the sixth cation is monomeric [62]. Association of molecules in the previous examples is in most cases linear. An exception to this pattern is the piperidine complex [AuCl(C$_5$NH$_{11}$)]$_4$. The four gold centers connect to form a square instead of a chain [63] (Figure 2.15a). A different motif is found in the crystal structure of [Au(C≡CSiMe$_3$)(CN^{t}Bu)] which displays a tetranuclear structure in which one central gold atom lies on a crystallographic threefold axis and the rest are connected to it through gold–gold interactions (Figure 2.15b, the rectangle represents

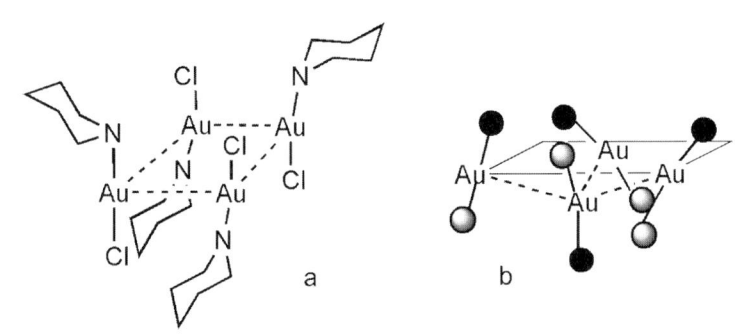

Figure 2.15 Tetrameric association of monomers of [AuCl(C$_5$NH$_{11}$)] (a) and [Au(C≡CSiMe$_3$)(CN^{t}Bu)] (b).

the plane which contains three of the four gold centers). The central gold atom lies 0.61 Å out of the plane formed by the other three gold atoms [64].

2.2.1.3 Formation of Infinite Chains and Supramolecular Arrays

Small changes in substituents may lead to different aurophilic aggregation patterns. Thus, different structures have been described for very similar species and polymorphs of different nuclearities have even been found for one complex. Thus, a classification based only on the nuclearity of the aggregates is not only difficult, but may also hide other interesting comparisons. For these reasons some chain structures have been mentioned above. The following tries to complete those commentaries.

Among the widely studied [AuX(PR$_3$)] (X = halogen, PR$_3$ tertiary phosphine) [AuBrL] (L = 2-methylphenylphosphine) crystallizes as a folded chain [65]. Chains of the tetrahydrothiophene and tetrahydroselenophene complexes [AuX(EC$_4$H$_8$)] (E = S, X = Cl, Br or I; E = Se, X = I) [66, 67] exhibit different patterns. In the crystal structure of the tetrahydrothiophene compounds [AuX(SC$_4$H$_8$)] (X = Cl, Br) neutral units associate, whereas the structure of the iodine complexes consists of [AuI$_2$]$^-$ and [Au(EC$_4$H$_8$)$_2$]$^+$ (E = S, Se) ions connected in the sequence $\cdots + - + - \cdots$. Another example of a polymeric linear chain with a chalcogen donor ligand is that of compound [AuBrL] (L = ω-thiocaprolactam) [68]. Chains of neutral [AuCl(2-picoline)] or ionic [Au(3-picoline)$_2$][AuCl$_2$] units, the latter in a $\cdots + - + - \cdots$ pattern, have been described. Compound [Au(3-picoline)$_2$](SbF$_6$) also crystallizes as chains. Its structure is complicated by the formation of C–H$\cdots$F hydrogen bonds [69]. The two chain polymorphs of [Au(CNC$_6$H$_{11}$)$_2$]PF$_6$, as well as their luminescent behavior have also been analyzed [60]. Upon protonation of [Au$_3$(MeOC = NMe)$_3$] (see Section 2.2.4) different salts of the cation [Au{C(OMe)NMeH}$_2$]$^+$ are produced with different structures. Compound [Au{C(OMe)NMeH}$_2$](C$_7$Cl$_2$NO$_3$)·CHCl$_3$ is a monomer, the unsolvated complex [Au{C(OMe)NMeH}$_2$](C$_7$Cl$_2$NO$_3$) crystallizes as dimers and the luminescent [Au{C(OMe)NMeH}$_2$](CF$_3$SO$_3$) forms infinite chains [70]. These examples evidence that not only the anion, but also the solvent, are important in determining the final structure.

Different structures are found in [AuX(CNR)] (X = halogen) derivatives. [AuCl(CNR)] (R = mesityl, o-xylyl, p-(tosyl)methyl] and [AuI{CN(CH$_2$C(O)OMe}] form dimers. Chain complexes are represented by [AuCl(CNR)] (R = Me, ^{t}Bu, Ph), [AuBr(CNR)] (R = Ph, ^{t}Bu, o-xylyl) and [AuI(CN-o-xylyl)]. Aggregation in sheets is observed for [AuX{CN(CH$_2$C(O)OMe}] (X = Cl, Br). Among [Au(CN)(CNR)] chains are obtained for R = ^{t}Bu, Cy (Cy = cyclohexyl), o-xylyl (gold–gold contacts connect side by side chains for R = ^{n}Bu) and nets for R = ^{i}Pr, Me [71, 72]. Luminescent emissions have been described for many of these derivatives. Comparison of the luminescent properties of [AuX(CNCy)] (X = Cl, Br, I) and [AuCl(CNR)] (R = p-tosylmethyl and ^{t}Bu) could indicate that the emission energies are more sensitive to the association mode than to the ground state Au–Au distance [73]. The nitrate-isocyanide species [Au(ONO$_2$)(CNC$_6$H$_{11}$)] and [Au(ONO$_2$)(CN^{i}Pr)] crystallize as strings. The structure of the former is based on dimers aligned into strings with alternating short and long aurophilic bonding, whereas that of [Au(ONO$_2$)(CN^{i}Pr)] is based on strings of alternating hexa- and octanuclear units which share corners. The edge-sharing Au_3 triangles are connected through short Au–Au contacts [74]. Chain structures are also found in [AuCl(S$_3$)] (S$_3$ = 1,4,7-trithiacyclononane) and the water-soluble compound rac-[Au{C≡CC(Me)(Et)OH}(TPA)] [75, 76]. By treatment of [AuClL] (L=CNB(H)NMe$_3$) with KI compound [AuL$_2$][AuI$_2$] is obtained. Both ions associate through gold–gold contacts to form an infinite linear one-dimensional chain [77]. Neutral units [Au(CN)(PMe$_3$)] aggregate to build chains [78] and the salts of the [Au(CN)$_2$]$^-$ anion display different structural features and luminescent properties, depending on the cation [79–81].

The building block [Au(CN)$_2$]$^-$ has been revealed as a versatile unit in the formation of heteronuclear arrays which, in some cases, display unusual and interesting luminescent properties, such as vapochromic behavior. In many of these complexes the cyanide group uses the carbon and nitrogen ends as Lewis bases able to coordinate two different metal cations (i.e., Au–CN–M) [82–86]. These ideas may be illustrated by the reaction of K[Au(CN)$_2$] with MX$_2$ salts (M = Cu, Ni, Co, Mn; X = ClO$_4$, NO$_3$...) and pyrazine which affords [M(pyrazine)][Au(CN)$_2$]$_2$ compounds. The dimensional patterns or/and properties change upon formation of these heteronuclear derivatives. For example, the structure of the copper derivative consists of a 3D array with chains of Cu(II)-pyrazine units connected by [Au(CN)$_2$]$^-$ bridges; a second identical network connected through aurophilic interactions interpenetrates. No modification is observed in the dimensionality compared with that of the parent gold cyanide complex, but an increment in the thermal stability is described in the final compound. In the structure of [Cu(tmeda)][Au(CN)$_2$]$_2$ (tmeda = N,N,N′,N′-tetramethylenediamine) aurophilic interactions connect 1D chains which contain copper(II)-bridging and pendant [Au(CN)$_2$] units. In this case the aurophilic interactions increase the structural dimensionality from one to three. The compound is synthesized by reaction of Cu(ClO$_4$)$_2$·6H$_2$O, tmeda (tetramethylethylenediamine) and K[Au(CN)$_2$]. The structure of [Au$_2$Cu(CN)$_4$(tacn)]$_n$ (tacn = 1,4,7- triazacyclononane) contains two different sort of chains. One of them contains gold atoms in an ···ABCDABCD··· pattern (Figure 2.16a), and the other is built from two kinds of

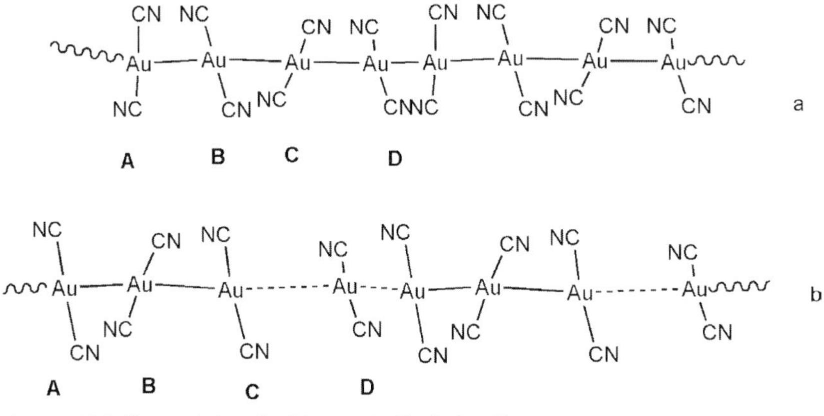

Figure 2.16 Different chains of gold atoms in [Au$_2$Cu(tacn)]$_n$.

centrosymetric trinuclear units and two [Au(CN)$_2$]$^-$ anions (Figure 2.16b). Both chains are bridged by copper atoms to afford a 3D polymer.

The two polymorphs of [Cu(DMSO)$_2$][Au(CN)$_2$]$_2$, one based on 1D chains, the other on 2D corrugated sheets (which form a 3D network through aurophilic interactions), exhibit vapochromic behavior. Both show reversible changes in color when exposed to volatile solvents. The vapochromic behavior of the two polymorphs is essentially identical. Solvent exchange transforms both polymorphs to [Cu(Solvent)$_x$][Au(CN)$_2$]$_2$ in which the solvent molecules coordinate to the copper centers [87]. Polymorphs of [Cu(DMSO)$_2$][Au(CN)$_2$]$_2$ also show similar weak antiferromagnetic coupling, but different thermal stabilities respect the lost of the first DMSO molecules. The dicyanide gold(I) building block has also been useful for the synthesis of hereometalic gold–iron spin crossover complexes. One example is represented by the compound {[Fe(3-CNpy)][Au(CN)$_2$]}·nH$_2$O for which aurophilic interactions are involved in the supramolecular arrangement in two of the three *supramolecular* isomers [88].

2.2.2
Clustering of Gold Atoms at One Center

Aurophilic bonding may induce and support clustering of gold atoms at one center (Y) resulting in unexpected geometries and small Au–Y–Au angles. These compounds represent an interesting subtopic among complexes which contain gold–gold intramolecular interactions. It includes hypervalent species and represents another evidence of the influence of these interactions on gold chemistry. Two facts reveal the importance of the gold–gold interactions in these complexes. One is related to the final structures which in some cases do not follow the VSEPR rules, the other is that some of these gold hypervalent derivatives have not been described for complexes with organic groups instead of AuL$^+$ units coordinated to the central atom.

Homoleptic species have been described in which the central atom is bonded to a different number of "Au(PR$_3$)" (PR$_3$ = monophosphine) or "Au$_2$(P~P)" (P~P = diphosphine) groups. The number of gold atoms which can gather at the central atom in these species can be from two to six. One example of the highest nuclearity obtained is the hypervalent complex [Au$_6$(μ-C)(PPh$_3$)$_6$]$^{2+}$ whose structure has been determined by X-ray crystallography [89]. This compound has been studied by ^{252}Cf plasma desorption mass spectrometry [90]. Analogous hexaaurated complexes have also been synthesized with different diphosphines and monophosphines. Figure 2.17

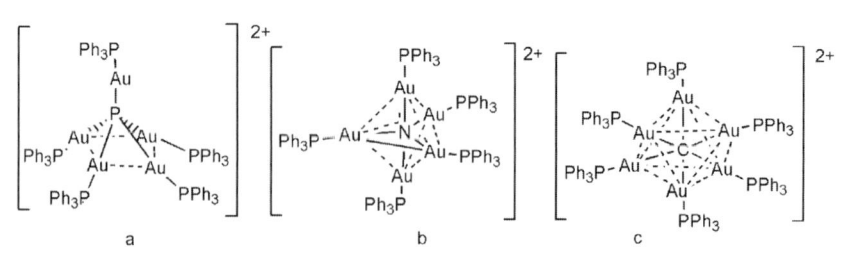

Figure 2.17 Complexes with different centers clustering five or six gold atoms.

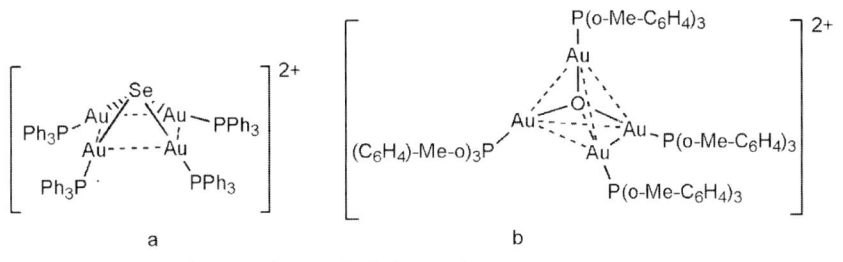

Figure 2.18 Tetranuclear complexes with chalcogenide centers.

shows the highest nuclearity confirmed by X-ray crystallography for a concrete central atom, six for carbon; five for nitrogen or phosphorus. In some cases higher nuclearities than four have been proposed, based on spectroscopic data (e.g., 5 or 6 for sulfur and selenium, and six for phosphorus).

The geometries of these species in general obey the VSEPR rules but interesting exceptions are found, leading in many cases to deficient electronic species stabilized by gold–gold contacts. Some examples are the tetranuclear cations $[Au_4(\mu\text{-}E)(PPh_3)_4]^{n+}$ ($n = 1$, E = As; $n = 2$, E = S, Se) which display square pyramidal geometry with the atom E at the apex of the pyramid [91, 92], whereas nitrogen, phosphorus and oxygen analogous tetranuclear species with different phosphines show tetrahedral geometry towards the central atom (Figure 2.18). Another exception to the VSEPR rules is the square planar pyramid geometry found for $[Au_5(\mu\text{-}P)(PPh_3)_5](BF_4)_2$ (Figure 2.17a).

Although, in general, the highly aurated species commented above may be synthesized by reaction of $[Au(PR_3)]^+$ with species of lower nuclearity, in some cases direct reaction routes are used. Treatment of $C(B(OMe)_2)_4$ with $[AuCl(PPh_3)]$ in the presence of CsF affords $[Au_4(\mu\text{-}C)(PR_3)_4]^+$, but also species of higher nuclearities, depending on the phosphine [93]. A good reaction pathway for the hexaaurated carbon centered species $[Au_6(\mu\text{-}C)(PR_3)_6]^{2+}$, consists in the reaction of TMSCHN$_2$ (TMS = trimethylsilyl) with $[Au_3(\mu\text{-}O)(PR_3)_3]^+$ in the presence of NEt$_3$ [89]. In the crystal structure of $[Au_4(\mu\text{-}HC)(PPh_3)_4]^+$ the cations form dimers (Figure 2.19) through gold–gold contacts [94].

The first members of the series, namely the starting products for the synthesis of complexes with higher nuclearities, are synthesized by different methods, depending

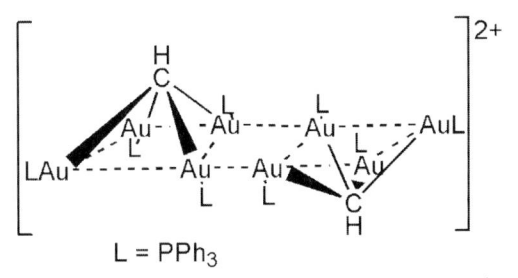

Figure 2.19 Dimeric association of $[Au_4(\mu\text{-}HC)(PPh_3)_4]^+$.

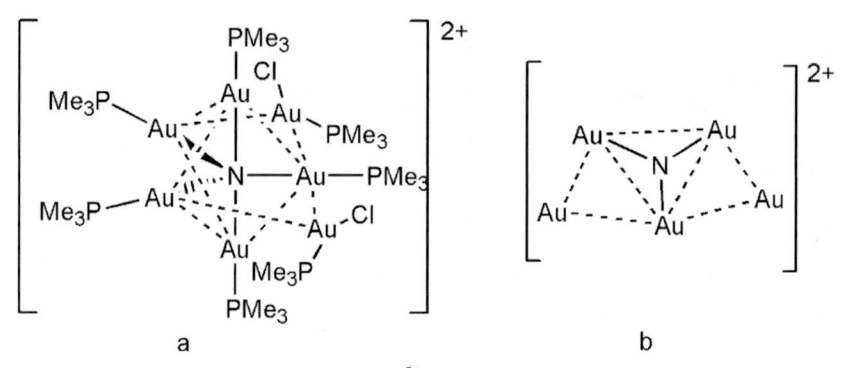

Figure 2.20 $\{[Au_5(\mu\text{-}N)(PPh_3)_5][AuCl(PMe_3)]_2\}^{2+}$ (a). View of the cation from one apical vertex of the pyramid. The phosphine and chloride ligands have been omitted for clarity (b).

on the central atom. The tetraaurated $[Au_4(\mu\text{-}N)(PPh_3)_4]^+$ has been prepared from the reaction of $[Au_3(\mu\text{-}O)(PPh_3)_3]BF_4$ and ammonia at $-60\,°C$, the reaction of $[AuX(PR_3)]$ ($X = BF_4$, PF_6) and Au^- in liquid ammonia, from tris(trimethylsilyl)amine or hexamethyldisilazane with $[AuCl(PR_3)]$ and CsF or by reaction of NH_4ClO_4 with $[Au(acac)(PR_3)]$ (acac=acetylacetonate). The tetrahedral environment at the nitrogen atom is highly distorted [95–99]. The trigonal bipyramidal cation $[Au_5(\mu\text{-}N)(PPh_3)_5]^{2+}$ may be obtained by addition of $[Au(PPh_3)]^+$ to $[Au_4(\mu\text{-}N)(PPh_3)_4]^+$ [100]. Theoretical studies at different levels have been carried out in order to analyze the stability and structures of these clusters [101–103]. Aurophilic interactions connect two $[AuCl(PMe_3)]$ units to the cation $[Au_5(\mu\text{-}N)(PMe_3)_5]^{2+}$ in $[Au_5(\mu\text{-}N)(PPh_3)_5][AuCl(PMe_3)]_2[BF_4]_2$ (Figure 2.20) [104].

Complexes with phosphorus or arsenic as the central atom have been synthesized from silylpricogenides and $[Au_3(\mu\text{-}O)(PR_3)_3]^+$ or $[Au(PR_3)]BF_4$ [95, 105–108]. As commented above, the structure of $[Au_4(\mu\text{-}As)(PPh_3)_4]^+$ does not display the classical tetrahedral geometry towards the As center [95]. The arsenic atom is situated in the apex of a square pyramid, with the lone electron pair allocated to the As apex. This fact would be in agreement with other situations in which the final geometry of the molecules is influenced by the aurophilic attraction, and supports the idea that the stabilization of the final compound is based on the presence of aurophilic interactions which probably are less favorable in a tetrahedral geometry for arsenic, in comparison with nitrogen, due to the different size of these atoms.

A mixture of complexes is obtained by reaction of PH_3 and $[Au_3(\mu\text{-}O)(PPh_3)_3]BF_4$ including $[Au_5(\mu\text{-}P)(PPh_3)_6](BF_4)_2$ in which the phosphorus atom bridges four "Au(PPh_3)" and one "$Au(PPh_3)_2$" fragments (Figure 2.21). In the crystal structure the asymmetric unit contains two crystallographically independent dications with the same trigonal bipyramidal arrangement at the phosphorus center and different orientations of the $[Au(PPh_3)_2]^+$ unit. Other species obtained from the same reaction are $[Au_5(\mu\text{-}P)(PPh_3)_5](BF_4)_2$, which displays a square pyramidal environment at the phosphorus center, and $[Au_9(\mu\text{-}P)_2(PPh_3)_8](BF_4)_3$ [109]. In these examples the maximum number of gold atoms at the phosphorus center is five, nevertheless the

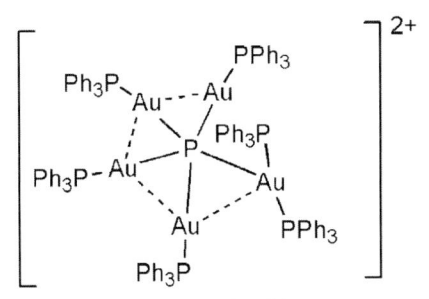

Figure 2.21 $[Au_5(\mu\text{-}P)(PPh_3)_6]^{2+}$.

hexanuclear species $[Au_6(\mu\text{-}P)(PPh_3)_6]^{3+}$ has been proposed, based on ^{31}P NMR studies.

The oxonium cation $[Au_3(\mu\text{-}O)(PPh_3)_3]^+$ can be obtained by reaction of [AuX(PPh_3)] with Ag_2O in the presence of NaBF_4, from $[Au(PPh_3)]^+$ with water in different media, or through hydrolysis of acetylacetonate gold(I) derivatives [Au(acac)(PR_3)] with ammonium salts $(NH_2R'_2)$OTf (OTf = trifluoromethylsulfonate). Lower yields are obtained by the last method. These pathways, or slightly modifications of them, have given a great variety of trinuclear derivatives with different monophosphines as ligands [33, 96, 110–117]. The trinuclear cation $[Au_3(\mu\text{-}O)(PR_3)_3]^+$ displays interesting structural features. Bulky monophosphines [R = o-Tol or p-OMe(C_6H_3)] prevent association in the solid state and monomers are obtained (Figure 2.22a). For less sterically demanding phosphines formation of dimers has been described. The six gold atoms of two trigonal-pyramidal units can adopt two patterns in these dimers. The first consists of a six-membered ring with a chair conformation (Figure 2.22b). This geometry has been found for medium size phosphines, such as PPh_3, PPh_2Me and AsPh_3. The second is found in $[Au_3(\mu\text{-}O)(PMe_3)_3]^+$, which contains a small phosphine. In this pattern both units share edges to give a tetrahedral subunit Au_4 (Figure 2.22c). From theoretical studies it has been proposed that the tetrahedral arrangement (Figure 2.22c) for gold centers in this dimerization process is slightly favorable for the cationic species "OAu_3^+" and the rectangular dimerization is preferred upon addition of phosphine [118]. These trinuclear species are used in the synthesis of highly aurated compounds and in catalysis [13]. Their use as potential gold atoms precursors in organic media, either by thermal or under H_2 conditions, has also been reported. In these processes the phosphine ligands seem to act as reducing agents as well as oxygen traps and gold nanoparticles are produced [119].

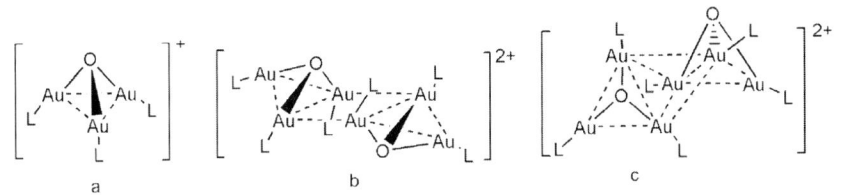

Figure 2.22 Structural patterns of $[Au_3(\mu\text{-}O)L_3]^+$ cations.

Figure 2.23 E = S, Se.

The tetrahedral environment of the oxygen center in the tetranuclear cation [Au$_4$(μ-O){P(o-Tol)$_3$}$_4$]$^{2+}$ has been confirmed by X-ray crystallography [120] (Figure 2.18b).

Dinuclear derivatives [Au$_2$(μ-E)(PR$_3$)$_2$] with E = sulfur or selenium have been obtained by reaction of Na$_2$S or (NH$_2$)$_2$CSe with the corresponding halogold complex [AuX(PR$_3$)]. Their crystal structures and those of analogous species with diphosphines have been reported (Figure 2.23) and display very narrow Au–E–Au angles [121–124]. The dinuclear sulfur compounds can also be synthesized in higher yield starting from Li$_2$S [125].

Coordination of one more "Au(PR$_3$)" fragment to these dinuclear compounds leads to the trinuclear species, that may also be obtained from the reaction of E(SiR$_3$)$_2$ (E = S, Se, Te) and the corresponding gold precursor. The reaction of H$_2$S with Me$_4$N[Au(C$_6$F$_5$)Cl] leads to [Au$_3$(μ-S)(C$_6$F$_5$)$_3$]$^{2-}$, whereas the same reaction with Et$_4$N[Au(C$_6$F$_5$)Cl] gives, depending on the workup, [Au(SH)(C$_6$F$_5$)]$^-$ or [Au$_3$(μ-S)(C$_6$F$_5$)$_3$]$^{2-}$ [126]. The structure of the NEt$_4^+$ salt of the trinuclear derivative has been determined by X-ray diffraction. The structures observed for trinuclear species resemble those of the oxonium derivatives discussed before with the size of the ligand being important in the final association. The cation [Au$_3$(μ-S)(PR$_3$)$_3$]$^+$ (R = methyl 4,6-O-benzilydene-3-deoxy-3-(diphenylphosphino)-α-D-altropyranoside) [127], and the tetrafluoroborate salts of [Au$_3$(μ-S)(P^{i}Pr$_3$)$_3$]$^+$ and [Au$_3$(μ-S)(PPh$_3$)$_3$]$^+$ are monomeric [113, 128]. Nevertheless, the anion also plays a role and the hexafluorophosphate salt of [Au$_3$(μ-S)(PPh$_3$)$_3$]$^+$ is dimeric, as the structure of [Au$_3$(μ-S)(PPh$_2$Me)$_3$]$^+$. In the salt [Au$_3$(μ-S)(PPh$_3$)$_3$]$_2$[AuFe$_3$(μ-S)(CO)$_9$(PPh$_3$)]$_2$ the cations form dimers with very dissimilar gold–gold distances (2.985(2) and 4.194(2) Å) [129]. A different structure is shown for the trinuclear sulfur centered derivative with PMe$_3$ [Au$_3$(μ-S)(PMe$_3$)$_3$]$^+$. Two monomeric units (A and B) are present in the crystal and associate in dimers (A$_2$ and B$_2$). These dimers are aggregated into strings through gold–gold contacts (Figure 2.24) [128].

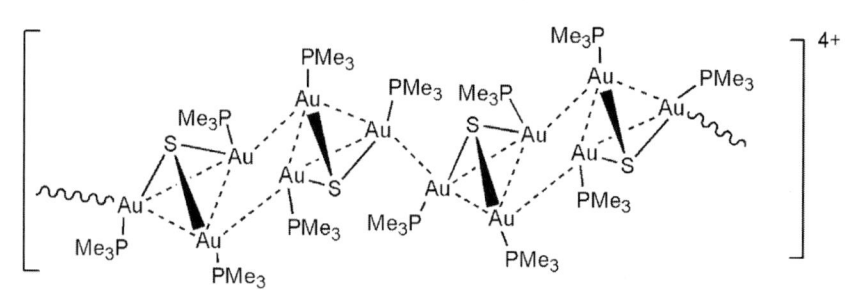

Figure 2.24 Association of pairs of dimers in [Au$_3$(μ-S)(PMe$_3$)$_2$]$^+$.

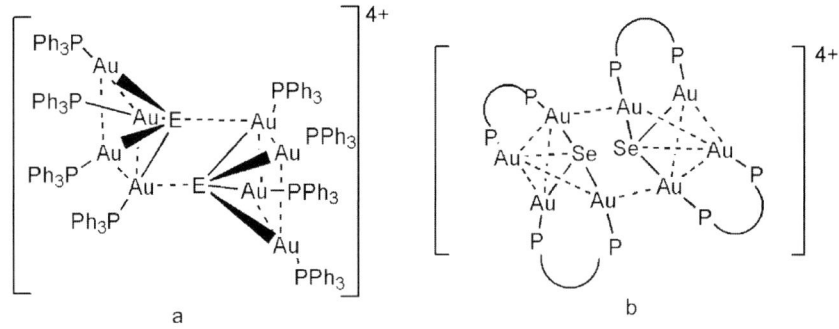

Figure 2.25 [Au$_4$(μ-E)(PPh$_3$)$_4$]$_2^{4+}$ (E = S, Se) (a), [Au$_4$(μ-Se)(dppf)$_2$]$_2^{4+}$ (b).

Crystal structures of the trinuclear [E{Au(PPh$_3$)}$_3$]$^+$ (E = Se, Te) have also been described [130, 131] and consist of the association of monomers (EAu$_3$) through two gold–gold contacts. Theoretical studies for [Au$_n$(μ-Se)(PH$_3$)$_n$]$^{(n+2)+}$ (n = 2–6) models, using quasi-relativistic pseudopotentials at the MP2 level are in good agreement with experimental geometries [124]. Recently, a density functional study of closed-shell attraction has been carried out on [M$_3$(μ-X)L$_3$]$^+$ (X = O, S, Se, M = Au, Ag, Cu) systems [132].

The tetranuclear cations [Au$_4$(μ-E)(PPh$_3$)$_4$]$^{2+}$ (E = S, Se) and other tetranuclear species with diphosphine ligands exhibit square pyramidal geometry with the chalcogenide atom at the apex of the pyramid [91, 92, 123–125]. Dimerization of the cations is also observed for these tetranuclear species. In the isostructural [Au$_4$(μ-E)(PPh$_3$)$_4$](OTf)$_2$ (E = S, Se) the molecules form dimers through Au–Se or Au–S interactions (Figure 2.25a), the shorter intermolecular gold–gold distances are 4.45 and 4.25 Å, respectively. In the crystal structure of compound [Au$_2$(μ-S)(PPh$_2$Me)$_2$(dppf)]$^{2+}$ (dppf 1,1′-bis(diphenylphosphino)ferrocene) two units are connected through intermolecular gold–gold contacts of 2.9520(12) Å, in this case the tetranuclear units could be regarded as a trigonal bipyramid with the apical positions occupied by one gold atom and the lone pair of electrons at the sulfur atom. Intermolecular aurophilic interactions (3.298 Å) are also present in [Au$_4$(μ-Se)(dppf)$_2$]$^{2+}$ (Figure 2.25b).

Dinuclear or trinuclear complexes can be used as building blocks to obtain complexes with higher nuclearities and interesting structures. The crystal structure of compound [Au$_5$(μ-S)$_2$(PPh$_3$)$_4$][SnMe$_3$Cl] has been reported and confirms the triply bridging nature of the sulfur ligand. In the cation one gold atom bridges two "Au$_2$(μ-S)(PPh$_3$)$_2$" units but there is no gold–gold interaction between the bridging gold atom and any other in the molecule (Figure 2.26) [133].

The use of chalcogen atoms and diphosphines as building blocks in the same molecule allows the synthesis of derivatives with higher nuclearities. The crystal structures of the hexanuclear species [Au$_6$(μ-Se)$_2$(dppf)$_3$]$^{2+}$ (Figure 2.27a) and [Au$_6$(μ-S)$_2${Ph$_2$PN(p-CH$_3$C$_6$H$_4$)PPh$_2$}$_3$]$^{2+}$ have been reported. Attempts to crystallize [Au$_5$(μ-Se)$_2$(dppf)$_2$]NO$_3$ afford a decanuclear structure (Figure 2.27b) which consists of two [Au$_5$(μ-Se)$_2$(dppf)$_2$]$^+$ units connected through aurophilic interactions [123, 124, 134].

Figure 2.26 $[Au_5(\mu\text{-}S)_2(PPh_3)_4]^+$.

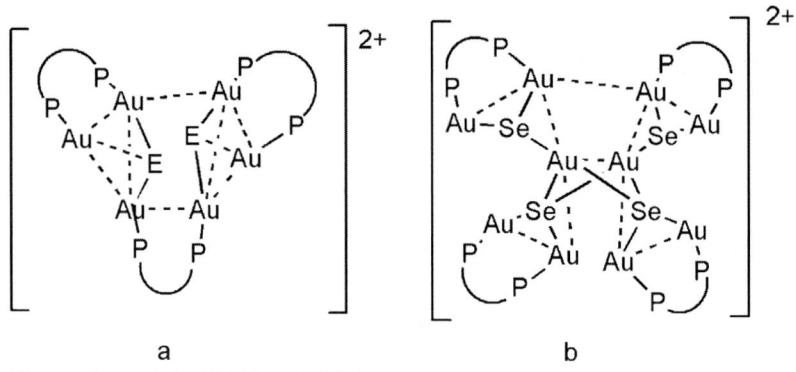

Figure 2.27 E = S, Se (a); P–P = dppf (b).

Several gold–gold contacts are present in the structure of $[Au_{12}(\mu_3\text{-}S)_4(dppm)_6]X_4$ (dppm = bis(diphenylphosphino)methane) (Figure 2.28a), most in the range 3.001(1) to 3.342(1) Å [135]. This complex and other highly aurated derivatives of stoichiometry $[Au_{10}(\mu_3\text{-}S)_4\{(PPh_2)_2N(^nPr)\}_4]X_2$, $[Au_6(\mu\text{-}S)_2\{(Ph_2P)_2N(p\text{-}CH_3C_6H_4)\}_3]^{2+}$ exhibit intense emissions in the green and orange region upon excitation at $\lambda > 350$ nm. Examples of selenide compounds are $[Au_{10}(\mu_3\text{-}Se)_4(dppm)_4]^{2+}$ (Figure 2.28b) and $[Au_{18}(\mu_3\text{-}Se)_8(dppm)_6]^{2+}$ [136].

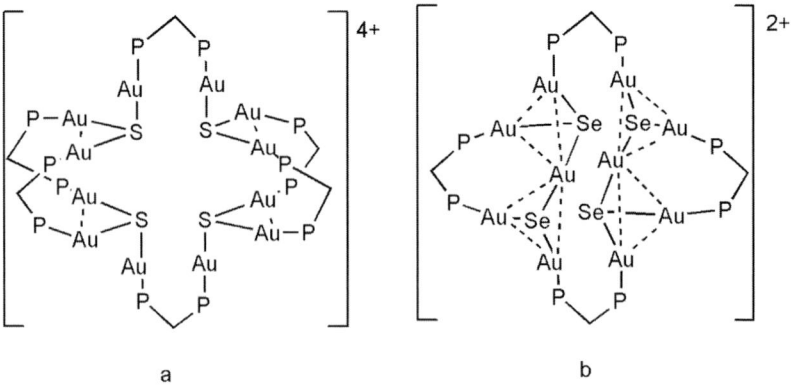

Figure 2.28 $[Au_{12}(\mu_3\text{-}S)_4(dppm)_6]^{4+}$ (a) and $[Au_{10}(\mu_3\text{-}Se)_4(dppm)_4]^{2+}$ (b).

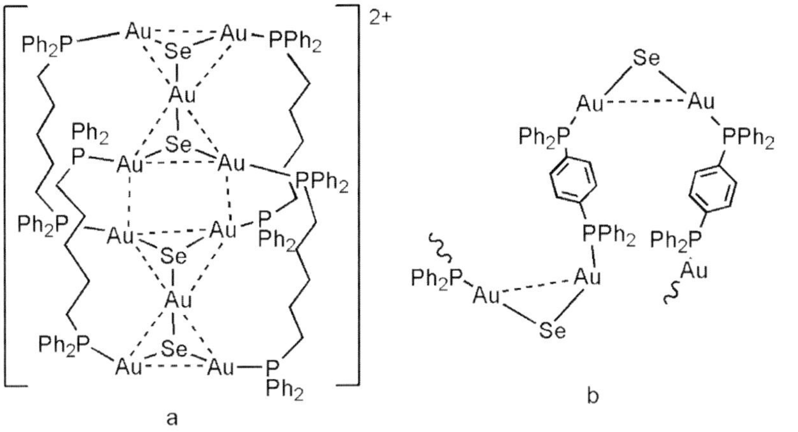

2.2 Gold(I)–Gold(I) Interactions

Figure 2.29 Examples of polynuclear gold complexes with selenide and diphosphine ligands.

Different diphosphines may lead to different structures. Figure 2.29a represents the structure of a decanuclear compound with the ligand 1,5-bis(diphenylphosphino)pentane. A chain structure is obtained by using the diphosphine p-bis(diphenylphosphino)benzene (Figure 2.29b). The complicated crystal structures of $[Au_6(\mu-Se)_2(dppbp)_3]Cl_2$ (dppbp = 4,4'-bis(diphenylphosphino)biphenyl), $[Au_{34}(\mu-Se)_{14}(tpep)_6(tpep^{Se})_2]Cl_6$ (tpep = 1,1,1,-tris(diphenyphosphinoethyl)phosphine; $tpep^{Se}$ = 1,1,-bis(diphenyphosphinoethyl)-1-(diphenylselenophosphinoethyphosphine) have also been reported [137].

Heteropolymetallic compounds in which one central chalcogen atom connects two gold and one heterometal atoms have also been described. Oxygen connecting gold and rhodium [138] (Figure 2.30a) or sulfur bridging gold and platinum are some examples. In complexes $[Au_2Pt(\mu_3-S)_2(\mu-P \sim P)]$ [P $\sim$ P = dppm or dppf] (Figure 2.30b) it is possible to control the presence of aurophilic interactions between the two gold centers by changing the diphosphine; the gold–gold distance changes from 2.916(1) (P$\sim$P = dppm) to 3.759(3) Å (P$\sim$P = dppf) [139].

Compound $[Au_2Pt_4(\mu_3-S)_4(PPh_3)_8]^{2+}$ has been prepared from $[AuCl(SMe_2)]$ and $[Pt_2(\mu-S)_2(PPh_3)_4]$. Its structure shows a short gold–gold interaction (2.837 Å) (Figure 2.31) [140].

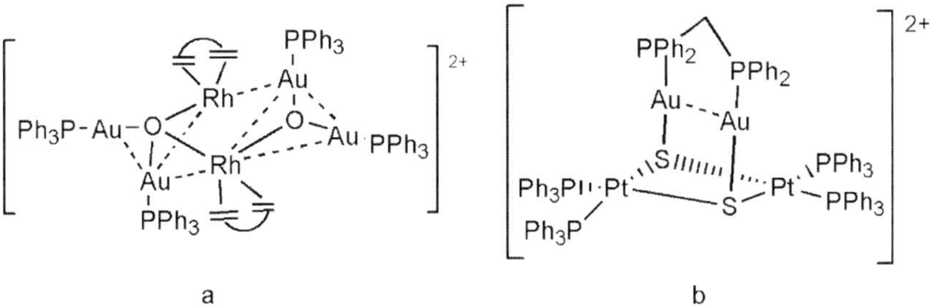

Figure 2.30 Heterometallic complexes with chalcogenide bridging ligands.

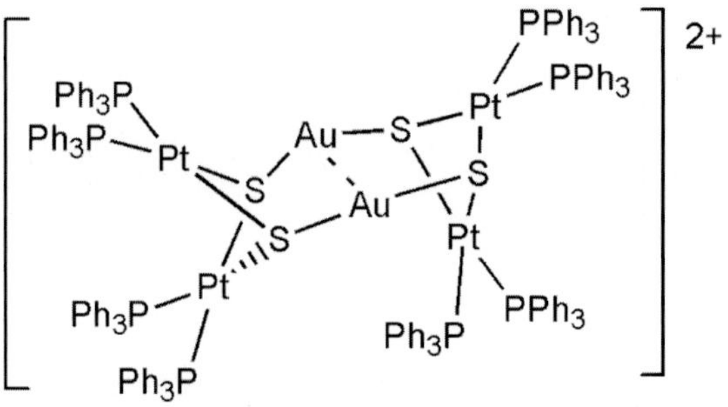

Figure 2.31 $[Au_2Pt_4(\mu_3\text{-}S)_4(PPh_3)_8]^{2+}$.

Reaction of [AuX(PPh$_3$)] (X = halogen) with [Au(PPh$_3$)]A (A = ClO$_4$, BF$_4$), starting from [AuX(PPh$_3$)] and AgBF$_4$ or through reaction of [AuX(PPh$_3$)] and AgSbF$_6$ in 2:1 molar ratio affords complexes in which a halide atom bridges two gold centers [141, 142]. Association of dinuclear units leads to tetranuclear species supported by four gold–gold contacts (Equation 2.2). This association is observed for salts of large anions such as SbF$_6^-$ and not for smaller ions such as ClO$_4^-$ or BF$_4^-$.

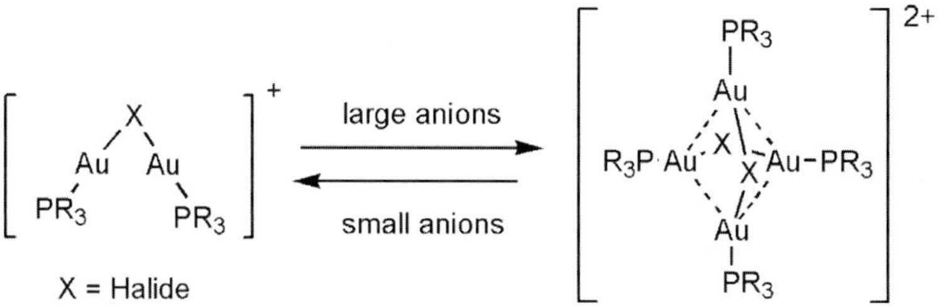

(2.2)

2.2.3
Complexes with Bridging Monodentate Ligands

Monodentate ligands may gather more than one gold atom. In this section some polydentate ligands in which only one of the donor atoms connect various gold centers will also be included, although Sections 2.2.4 and 2.2.5 will be dedicated to complexes containing bi- or polydentate ligands.

Several monodentate building blocks have been used in which the donor atom gathers two or more "Au(PR$_3$)" fragments. Different compounds with nitrogen donor ligands are known. The species [Au$_3$(μ-O)(PR$_3$)$_3$]$^+$ is frequently used as the aurating agent in the synthesis of these polyaurated derivatives. Reaction of 1,2-diphenylhidrazine with [Au$_3$(μ-O)(PPh$_3$)$_3$]$^+$ affords a mixture of complexes [Au$_3$(μ-N-1,4-C$_6$H$_4$-NHPh)(PPh$_3$)$_3$]$^+$ and [Au$_3$(μ-N-1,2-C$_6$H$_4$-NHPh)(PPh$_3$)$_3$]$^+$, the latter in low yield (Equation 2.3). The crystal structure of the former has been determined. Both species can be obtained independently by reaction of the gold oxo complex with the corresponding *ortho* or *para*-semidine [143].

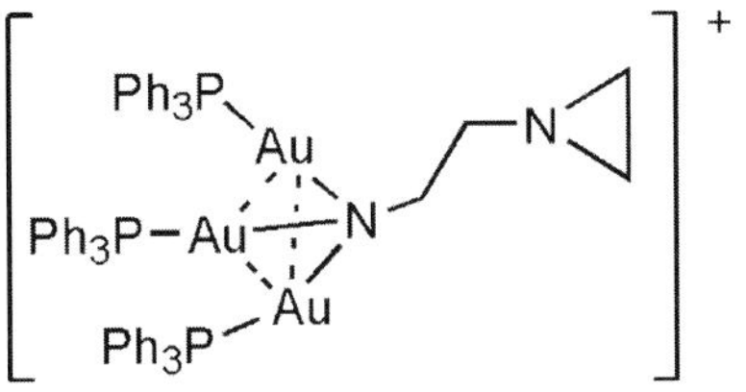

Figure 2.32 2-N-(aziridilydin)ethylimido compound.

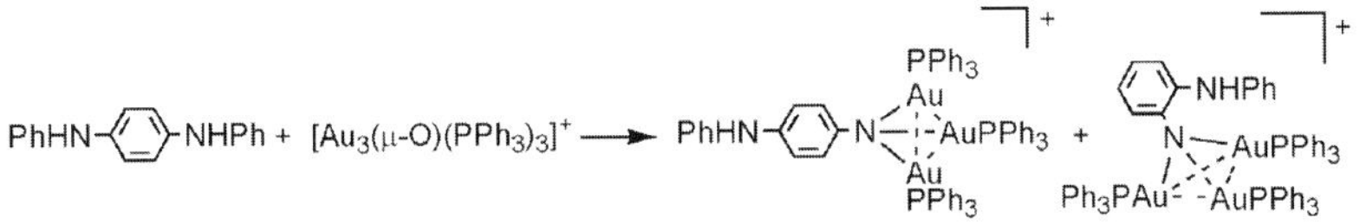

$$PhHN-\langle\rangle-NHPh + [Au_3(\mu-O)(PPh_3)_3]^+ \longrightarrow [PhHN-\langle\rangle-N(AuPPh_3)_3]^+ + [\langle\rangle-NHPh\cdot N(AuPPh_3)_3]^+ \quad (2.3)$$

The crystal structures of the dinuclear compounds $[Au_2(\mu-NNPh_2)(PPh_3)_2]$ and $[Au_2\{\mu-N(SiMe_2)_2\}(PEt_3)_2]$ have been determined. Trinuclear complexes of stoichiometries $[Au_3(\mu-NX)L_3]^+$ (X = Ph, OH, OSiMe$_3$, Cl, SiMe$_3$, NR$_2$ (R = Ph, ^{t}Bu, Me, Bz); L = phosphine) have also been synthesized and many of them structurally characterized. The nitrogen atom is in a distorted tetrahedral environment as shown for the 2-N-(aziridilydin)ethylimido ligand (Figure 2.32) [144–151].

In solution many of the species $[Au_3(\mu-NNR_2)L_3]^+$ are unstable and decompose to the hexanuclear clusters $[Au_6L_6]^{2+}$ (Equation 2.4). Addition of ethyltriflate to $[Au_3(\mu-NMe_2)(PPh_3)_3]^+$ affords $[Au_3(\mu-NNMe_2Et)(PPh_3)_3]^{2+}$ whose crystal structure has been determined.

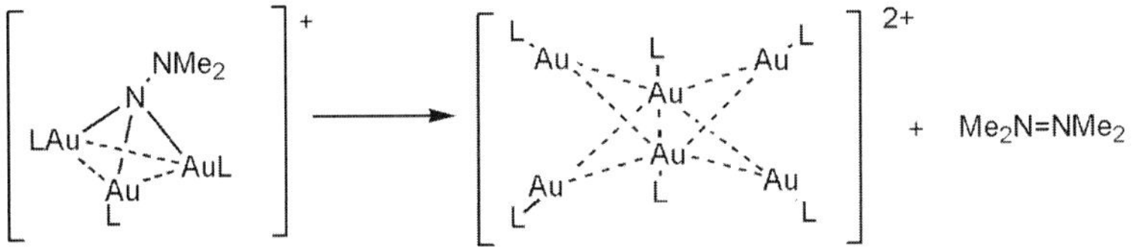

$$2[Au_3(\mu-NMe_2)L_3]^+ \longrightarrow [Au_6L_6]^{2+} + Me_2N=NMe_2 \quad (2.4)$$

Different situations are represented by $[Au_5(\mu-N^tBu)_2(PPh_3)_4]^+$ (Figure 2.33a) [152] in which two trigonal NAu$_3$ pyramids share one vertex or $[Au\{N(SiMe_3)_2\}]_4$, which is obtained by the metathesis reaction of gold(I) chloride with lithium bis(trimethylsylil)amide. In its crystal structure the four gold atoms form a square through short gold–gold distances of 3.0100(3) and 3.0355(3) Å [153] (Figure 2.33b). $[Au_3(\mu-NC_9H_6N)(PPh_3)_3]^+$ coordinates one more "Au(PPh$_3$)" fragment leading to $[Au_4(\mu-NC_9H_6N)(PPh_3)_4]^{2+}$ in which the quinolinaminate ligand acts as bidentate (Figure 2.33c) [154]. A striking example is the dinitrogen compound of Figure 2.33d. The N–N distance is 1.475(14) Å, typical of an N–N single bond. This species and others with different phosphines can be reduced and protonated to ammonia with different yields, depending on the phosphine [155].

Thiolate ligands also afford an interesting variety of species [156–161]. Because or their relation with the antiarthritic drug Auranofin, thiolate-goldphosphine derivatives have received great interest. The simplest structures are those in which the thiolate bridges two gold atoms and thus different compounds of stoichiometry

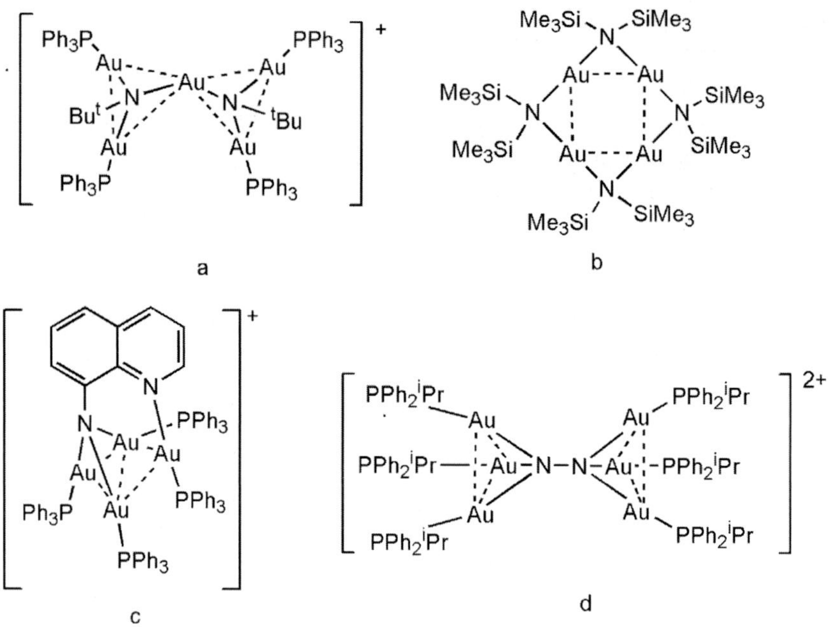

Figure 2.33 $[Au_5(\mu-N^tBu)_2(PPh_3)_4]^+$ (a), $[Au\{N(SiMe_3)_2\}]_4$ (b) $[Au_4(\mu-NC_9H_6N)(PPh_3)_4]^+$ (c) and $[Au_6(\mu-N_2)(PPh_2^iPr)_6]^{2+}$ (d).

$[Au_2(\mu-SR)(PR'_3)_2]^+$ or $[Au_2(\mu-SR)(P\sim P)\}]^+$ ($P\sim P$ = diphosphine) have been synthesized. Frequently, the crystal structures are built by the aggregation of two units through intermolecular aurophilic interactions (Figure 2.34) although, for example, $[Au_2\{\mu-S-(2-COOH-C_6H_4)\}(PPh_3)_2]^+$ cystallizes as a dinuclear monomer with an intramolecular gold–gold contact. Compound $[Au_2(\mu-SMe)(PPh_3)_2]_2(SO_3CF_3)_2$ cocrystallizes with the trinuclear $[Au_3(\mu-SMe)(PPh_3)_3](SO_3CF_3)_2$ [27].

Examples of selenolate and telurolate ligands bridging two or more gold atoms are $[Au_2\{\mu-Se-(p-ClC_6F_4)\}(PPh_3)_2]$ [162] $[Au_2(\mu-SeR)(PPh_3)_2]^+$ (R = benzyl, naphthyl, Ph) [163], $[Au_2\{\mu-Te-2,4,6-Ph_3-C_6H_2\}(PPh_3)_2]^+$ [164] or the tetrameric $[Au\{\mu-Te-C(SiMe_3)_3\}]_4$ [165]. The crystal structure of $[Au_2(\mu-SePh)(dppf)]^+$ contains two molecules in the asymmetric unit which exhibit intramolecular aurophilic interactions.

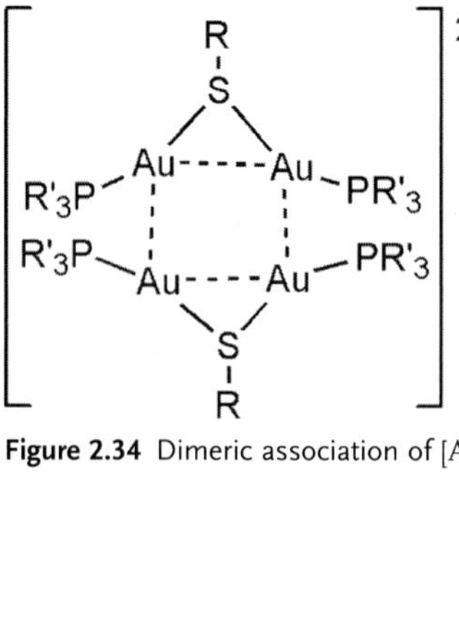

Figure 2.34 Dimeric association of $[Au_2(\mu-SR)(PR'_3)_2]^+$.

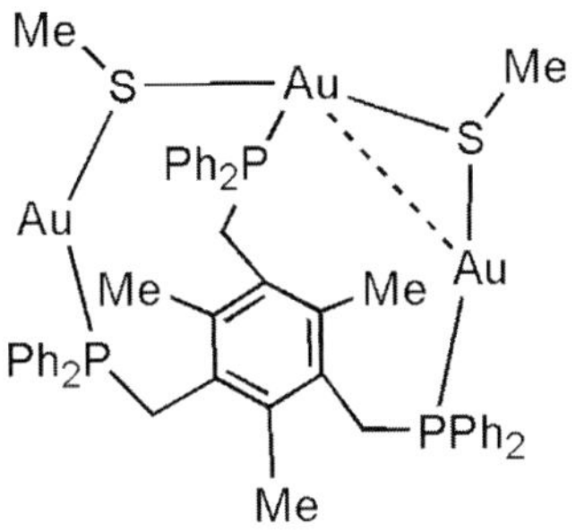

Figure 2.35 Trinuclear gold compound with bridging thiolate and triphosphine ligands.

These molecules are connected through only one intermolecular gold–gold contact [166].

The use of both di- or triphosphines and chalcogenolates as building blocks in the same molecule leads to different species. The trinuclear derivative shown in Figure 2.35 is an example [167]. A more complicated structure is that of [Au$_9$(μ-p-tc)$_6$(μ-dppm)$_4$]$^{3+}$ (p-tc = p-thiocresolate) which exhibits short gold–gold interactions. An emission at 540 nm is observed upon excitation at 450 nm in the solid state at room temperature that increases in intensity at 77 K [168].

Examples of dinuclear complexes with carbon donor ligands bridging the two gold centers connected through aurophilic interactions include cyclopentadienyl [169] dicyanomethane [170], cyano(phenylthiolato)methylidene (Figure 2.36a), nitro (phenylsulfonyl)methylidene [171], 2,3,6-trifluorophenyl [172], ethoxycarbonyl-(triphenylphosphoranylidene)-methyl [173], (2-pyridyl)-tri-p-tolylphosphinemethylene [174], triphenylphosphonio(2-pyridyl)methylene [175], bis(diphenylphoshino)methylidene (Figure 2.36b) [176] or 1-methoxycarbonyl-2,2-diphenyl-3-diphenylphosphino-2-phosphapropane (Figure 2.36c) [177]. Compound [Au$_4$(μ-CS(=O)Me$_2$)(PPh$_3$)$_4$]$^{2+}$ (Figure 2.36d) contains a hypercoordinated ylidic carbon [178]. The ligand 1,3-bis(triphenylphosphine)-2-oxo-1,3-propanediylideneylide coordinates two gold atoms to each ylidic carbon atom in [Au$_4${μ-(CPPh$_3$)$_2$CO}(PPhMe$_2$)$_4$](ClO$_4$)$_2$ (Figure 2.36e) [179]. 1,2-bis(phenylene)ethane coordinates two gold atoms connected through aurophilic interactions to one carbon or each phenyl fragment [180]. In [SPPh$_2$C(AuPPh$_3$)$_2$PPh$_2$CH(AuPPh$_3$)COOMe]ClO$_4$ (Figure 2.36f) one of the carbon atoms is coordinated to two gold centers connected through aurophilic interactions [181]. In [Au$_5$(Mes)$_5$] (Mes = 2,4,6-Me$_3$C$_6$H$_2$) (Figure 2.36g) each mesitylene group coordinates to two gold centers. The gold atoms form a pentanuclear cycle through gold–gold contacts [182]. The mesitylene fragments are nearly perpendicular to the mean plane formed by the gold atoms.

Phosphorus (Figure 2.37a and b), boron (Figure 2.37c) or arsenic donor ligands may gather gold atoms connected through short interactions [183–186]. The hexanuclear [Au$_6$(μ-1,2-P$_2$-C$_6$H$_4$)(C$_6$F$_5$)$_6$(dppe)$_2$] contains four gold(I) atoms. Each dppe bridges two gold(I) centers which are connected through aurophilic interactions. Another gold–gold interaction joins two gold(I) atoms, each of them bonded to a different dppe [187].

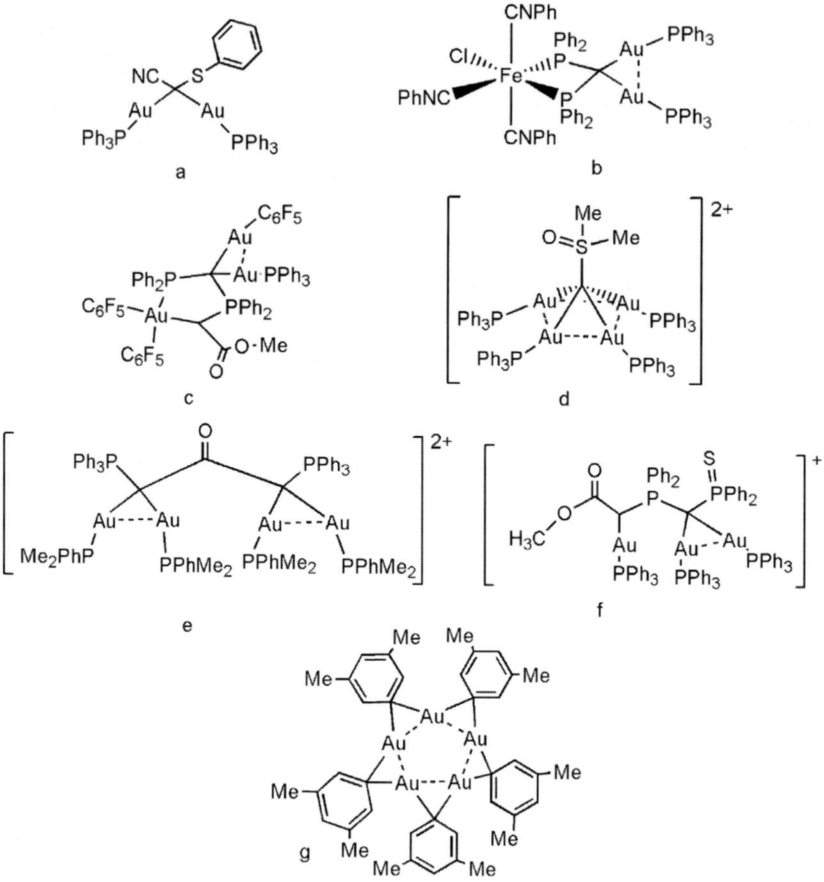

Figure 2.36 Some complexes with carbon donor ligands.

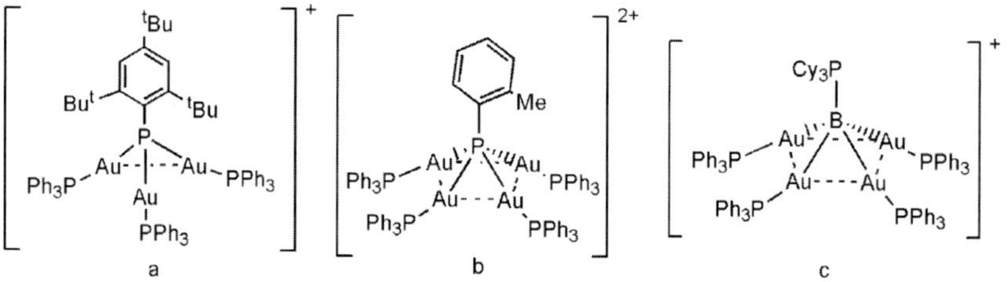

Figure 2.37 Complexes with phosphorus or boron donor ligands bridging three or four gold atoms.

2.2.4
Complexes with Bridging Bidentate Ligands

A wide variety of complexes in which a bidentate ligand bridges two gold centers have been described. The dinuclear units may exhibit intramolecular, intermolecular or

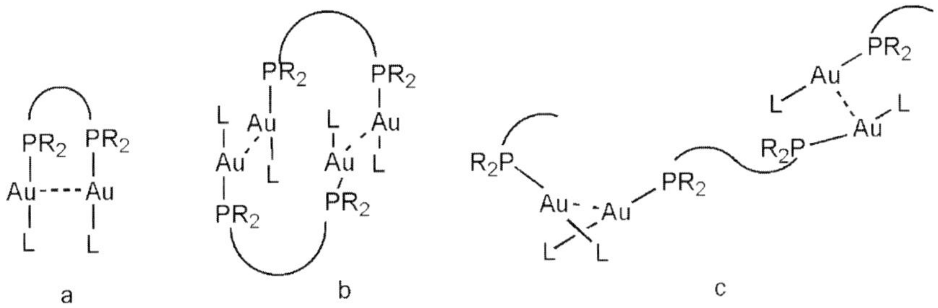

Figure 2.38 Different patterns in gold derivatives with stoichiometries [Au$_2$L$_2$(μ-P~P)].

both sorts of interactions. Many of these species are based on phosphorus donor ligands (P~P), mostly diphosphines, which include a wide range of spacers between the two PR$_2$ units. Different patterns have been described, as shown in Figure 2.38.

Examples of stoichiometry [Au$_2$X$_2$(P~P)] (X = halogen) are well known. Among them complexes with alkylic phosphines PR$_2$(CH$_2$)$_n$PR$_2$ have been widely studied and exhibit different structural patterns. Although it seems that long bridging diphosphine ligands lead to prominence of polymeric chains, this is not the unique factor. Other aspects seem to be relevant in an efficient packing of solvent and gold molecules to afford a favorable conformation [188]. In complexes [Au$_2$Cl$_2${PPh$_2$(CH$_2$)$_n$PPh$_2$}] for $n = 1$ intramolecular aurophilic interactions (Figure 2.38a) [189] are present, whereas for $n = 2$ two polymorphs have been described, in one of them the molecules associate in dimers, the other form is chain polymeric [190, 191]. For $n = 3$ two polymorphs have also been described, in one of them intermolecular interactions afford a chain, the other displays intramolecular interactions [192]. DFT calculations using a large relativistic basis have been carried out in order to analyze the aurophilic interaction with the aim of reproducing the observed molecular structure in the second linkage isomer. In the structure of compound [Au$_2$Cl$_2${PPh$_2$(CH$_2$)$_5$PPh$_2$}] intermolecular interactions afford a chain-like structure and for $n = 4$ neither intra- or intermolecular interactions are present [193]. For $n = 6$ and X = I polymeric chains through gold–gold contacts are obtained [194], these are interwoven to afford discrete layers. Another example of chain structure is obtained when X = Cl and P~P = bis(diisopropyl) ethane [195].

[Au$_2$X$_2$(PTP)] (PTP = 2,5-bis(diphenylphosphino)thiophene; X = Cl, I) crystallize as dimers with intermolecular gold–gold interactions. Variable temperature NMR experiments have been carried out in order to analyze the presence of aurophilic interactions in solution which are consistent with shorter gold–gold contacts for the iodo complex in the solid state. Solution luminescence studies show dual emission for these species that has been interpreted in terms of a monomer–dimer equilibrium. In the solid state dimer emission is dominant [196]. [Au$_2$Cl$_2$(PR$_2$-Z-PR$_2$)] with Z = Me$_3$P=C, bicyclopropyl, CH$_2$=C–C=CH$_2$, cis-HC=CH, NH, CF$_2$, 2,3-dimethyl-1,4-diphosphabuta-1,3-diene, 1,2-diphenyl-3,4-diphosphinidenecyclobutene or azanorbornane (Figure 2.39a) show intramolecular interactions in the dinuclear structures. Among the oligothiophene complexes [Au$_2$Cl$_2$(PR$_2$-Z-PR$_2$)] (PR$_2$-Z-PR$_2$ = 3,3'-bis(diphenylphosphino)-2,2'-bithiophene and 3,3'''-dihexyl-3',3''-

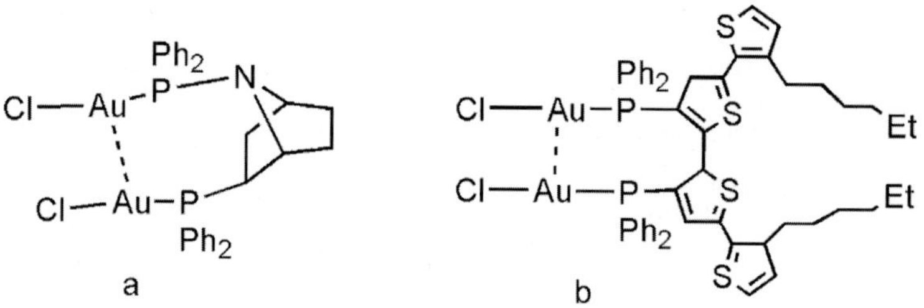

Figure 2.39 Azanorbornane (a) and quaterthiophene (b) compounds.

bis(diphenylphosphino)-2,5′ : 2′,2″ : 5″,2‴-quaterthiophene) the former displays different features depending on the crystallization media. Aurophilic intramolecular interactions are observed in the crystals grown from CH_2Cl_2/Et_2O (Figure 2.39b) whereas from CH_2Cl_2/toluene a toluene adduct crystallizes which does not exhibit gold–gold contacts. Shorter intramolecular aurophilic interactions have been found in the quaterthiophene than in the bithiophene derivative. Chains are obtained for $Z = C \equiv C$ (whose luminescent properties have been studied), trans- CH=CH, and $CH_2C(=CH_2)CH_2$ [197–208].

The luminescent properties of complexes $[Au_2Cl_2(P{\sim}P)]$ $[P{\sim}P = bis(2\text{-diphenyl-}$ phosphino)phenylether (dpephos), 9,9-dimethyl-4,5-bis(diphenylphosphino)xanthene (xantphos), and 4,6-bis(diphenylphosphino)dibenzofuran (dbfphos)] have been studied. Aurophilic intramolecular interactions are present in $[Au_2Cl_2(\text{dephos})]$ and $[Au_2Cl_2(\text{xantphos})]$ which are probably responsible for the presence of a low energy band [209].

Not only halogen atoms but other anionic ligands have also been used in the synthesis of $[Au_2X_2(P{\sim}P)]$ complexes which exhibit the pattern shown in Figure 2.38a. Some of these complexes, such as $[Au(C_6F_5)(PPh_2C_6H_4C_6H_4PPh_2)Au(C_6F_5)]$ are involved in studies which try to better understand the isomerization of gold(II) species (Equation 2.5) [210].

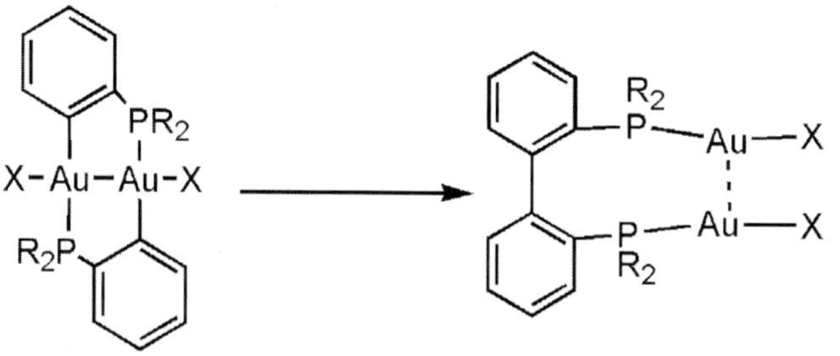

$$(2.5)$$

Other examples of complexes $[Au_2X_2(P{\sim}P)]$ that show similar structures to those in Figure 2.38a include anionic X ligands such as thiolates [211–213] or triphenylsilyl [214], with $P{\sim}P = \text{dppm}$; $X = RC \equiv C$ (R = Ph, 4-OMe-C_6H_4) with $P{\sim}P = N,N$-bis (diphenylphosphino)propylamine [215]; $X = SC(OMe) = NC_6H_4$-4-(NO_2) with $P{\sim}P = (PPh_2)_2(CH_2)_n$ ($n = 1$–4, no aurophilic interactions are found for $n = 4$) or dppf [216]; $X = [S_2P(R)(OR')]$ (R = p-C_6H_4-OCH_3; R′ = c-C_5H_9), with $P{\sim}P = (PPh_2)_2(CH_2)_n$ [$n = 1$–4; only for $n = 1$ intramolecular interactions are present, for the other three

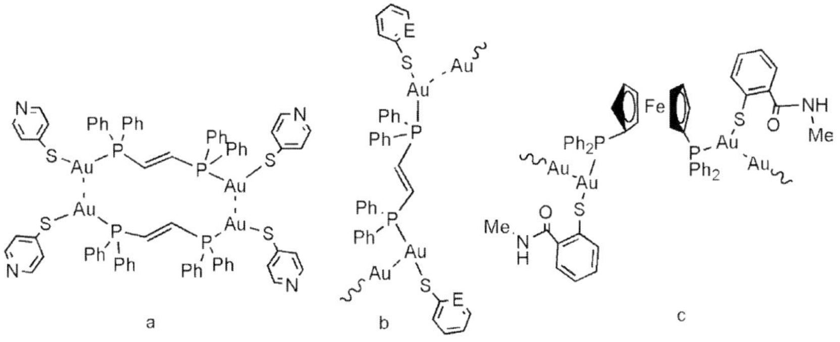

Figure 2.40 Formation of dimers or chains in complexes with diphosphines and thiolates.

diphosphines anti conformations are observed in which the gold atoms show secondary interactions with the second sulfur atom of ∼3.5–3.8 Å [217]. The luminescent properties of many of the above-mentioned derivatives have been analyzed. Complexes [Au$_2$X$_2$(P∼P)] with X = alkynyl, P∼P = N, N-bis(diphenylphosphino)alkyl or aryl amine are luminescent, both in the solid state at room temperature and at 77 K and also in frozen solution with lifetimes in the millisecond range. Upon excitation at λ > 350 nm [Au$_2${SC(OMe)=NC$_6$H$_4$-4-NO$_2$}$_2$(μ-Ph$_2$P-R-PPh$_2$)] (R=CH$_n$, n = 2–4) show emissions in the blue or green regions in solution or in the solid state, respectively.

A systematic structural study of [Au$_2${S-(CH$_2$)$_n$-pyridine}{Ph$_2$P–Z–PPh$_2$}] [Z = CH=CH, (CH$_2$)$_x$ (x = 2–4); n = 0–2] compounds has been carried out with the aim of understanding the influence of the introduction of thiolate instead of halogen ligands, to compare phenylthiolate or pyridilthiolate species and the effect of the presence of hydrogen bonds in the final structures. Independent molecules, formation of dimers or chains through aurophilic interactions are obtained (Figure 2.40a and b). Other derivatives with thiolates as anionic ligands which display chain structures are [Au$_2$(SR)$_2$(P∼P)] (R = Ph, P∼P = 1,2-bis(diphenylphosphino)ethene; R = S-p-4-Me-C$_6$H$_4$, P∼P = PPh$_2$(CH$_2$)$_5$PPh$_2$) and [Au$_2$Cl(SPh)(P∼P)] (P∼P = 1,2-bis(diphenylphosphino)ethene) [26, 218–220]. In complexes [Au$_2${SC$_6$H$_4$-2-(C(=O)Y)}$_2$(μ-dppf)], when Y = NHMe intermolecular interactions lead to the formation of chains (Figure 2.40c). Intramolecular aurophilic interactions are present for Y = NH$_2$ and only hydrogen bonds are present for Y = OH [221]. Nevertheless, all display chain formation in the solid state based on different interactions which depend on Y. Formation of chains is also observed in [Au$_2$(OC(O)CF$_3$)$_2$(dppf)] [222] and [Au$_2$(C≡CPh)$_2${2,6-bis(diphenylphosphino)pyridine}] [223].

Complexes of stoichiometry [Au$_2$L$_2$(E∼E)] (the final charge depending on the ligands) have also been structurally characterized with E a donor atom different from phosphorus. In complexes with dithiolate ligands the sulfur atoms frequently do not coordinate symmetrically to both gold atoms (Figure 2.41a). Thus, one gold is coordinated to a sulfur atom of the dithiolate and also to the other at a longer distance. Further coordination of one more gold fragment in this sort of species leads to trinuclear complexes (Figure 2.41b). In [Au$_2$(PR$_3$)$_2$(MNT)] (MNT = 1,2-dicya-

Figure 2.41 Complexes with dithiolate bridging ligands.

noethene-1-2-dithiolate) each "AuPR$_3$" fragment coordinates to a different sulfur atom. An intramolecular interaction connects both gold atoms. The gold–gold distance follows the pattern: P(OPh)$_3$ < PEt$_3$ < PPh$_3$ < PMe$_3$, ≪ PCy$_3$ and ranges from 2.991(1) to 4.883(3) Å, for PR$_3$ = PCy$_3$ no aurophilic interactions are present [224–226]. A similar structure is obtained for [Au$_2$(PPh$_3$)$_2$(S$_3$PPh)] [227] with an intramolecular gold–gold distance of 3.1793(4) Å, whereas the trithiophosphate analogous derivative [Au$_2$(PPh$_3$)$_2$(S$_3$POMe)] displays intra- and intermolecular interactions of 3.1793(4) and 3.3464(5) Å, respectively, the latter associating the molecules into strings. Further auration affords P–S cleavage. *Ab initio* quantum chemical calculations have been carried out to analyze these results. Emissions which have been assigned to ligand to metal–metal charge transfer (LMMCT) excited states have been observed in [Au$_2${S$_2$C=C(C$_{12}$H$_6$-2,7-R$_2$)}(PR'$_3$)$_2$] (R = H, R' = Ph; R = ^{t}Bu; R' = Me, Ph) (Figure 2.41c) [228].

Formation of dimers has been observed for dinuclear complexes with dichalcogenolate bridging ligands (Figure 2.42a and b) [229, 230]. Twelve-membered cycles formed through aurophilic interactions between two [Au$_2$Cl$_2$L] units (Figure 2.42a) (L = 7-tolenesulfonyl-7-aza-1,4-dithiacyclononane) associate in the crystal through π–π interactions. The luminescence of compound [Au$_3$(μ-i-MNT)$_2$(PEt$_3$)$_2$]$^-$ (i-MNT 1,1-dicyanoethylene-2,2'-thiolate) (Figure 2.42c.) has been studied and the excited state distortions analyzed using Raman data [231].

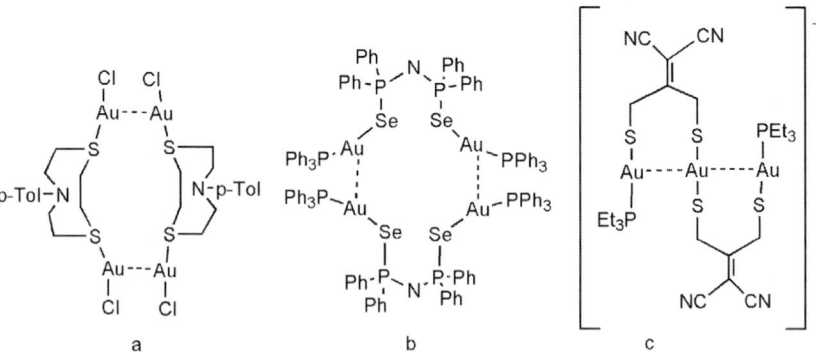

Figure 2.42 Dimer associations of dichalcogenolate complexes (a and b). Trinuclear derivative (c).

Figure 2.43 Chain complexes with dialquinyl ligands.

Dialquinyl groups have been used among bidentate C-donor ligands as interesting building blocks by modulation of the fragment between the two alquinyl fragments. Chain structures have been obtained with different spacers such as substituted benzene (Figure 2.43). Many of these compounds display luminescent properties modulated by the presence of aurophilic interactions. In the crystal structure of [Au$_2$Cl$_2$(1,1'-di-isocyanoferrocene)] the gold centers are connected through intramolecular aurophilic interactions. These molecules form monolayes through intermolecular gold–gold contacts [232–235].

An example of the formation of chains with N-donor bidentate ligands is compound [Au$_2$(diethylenetriamine)$_2$](BF$_4$)$_2$ [236]. Other bidentate ligands afford monomeric structures (Figure 2.44) such as 6-ferrocenylmethyl)amino-2-picoline (Figure 2.44a) [237], 8-aminoquinoline (2.33.c) [154], N-ferrocenecarbaldehide-2-mercaptoethylamino (Figure 2.44b) [238] 3-R-2-sulfanylpropeonate (Figure 2.44c) [239]. The bis-*ortho*-carborane compound shown in Figure 2.44d contains an intramolecular aurophilic interaction (3.119(2) Å). Variable temperature ^{31}P NMR experiments have been carried out in order to estimate the strength of this interaction based on the calculated height of the barrier of interconversion between the *syn* and *anti* forms of the compound [240]. *Nido*-carboranes may act as bridging

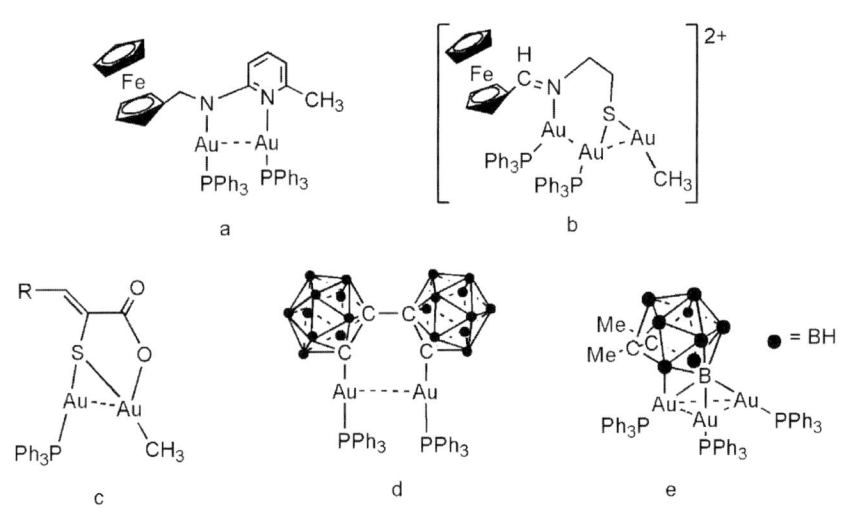

Figure 2.44 Different bidentate ligands connecting two or three gold centers.

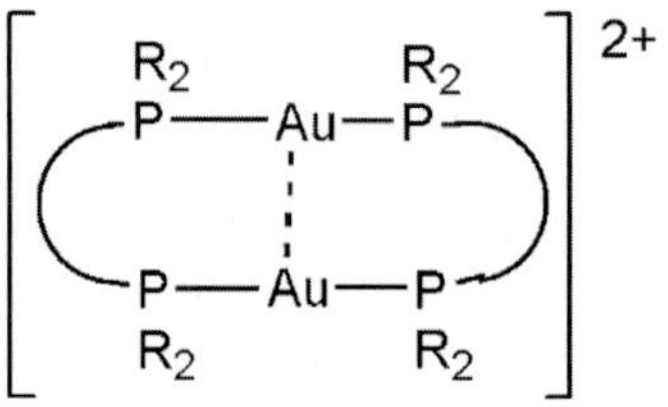

Figure 2.45 $[Au_2(P{\sim}P)_2]^{2+}$.

ligands (Figure 2.44e) [241]. In compound [9-exo-{Au(PPh$_3$)}-9-(μ-H)-10-endo-{Au(PPh$_3$)}-7,8-Me$_2$-nido-7,8-C$_2$B$_9$H$_8$] one of the gold centers is involved in a three-center two-electron B–H–Au bond. Several mixed metal complexes outside the scope of this discussion have also been prepared with these nido-carborane ligands.

Bidentate diphosphines, such as bis(diphenylphosphino)methane (dppm) [242–244], 1,2-bis(diphenylphosphino)ethane (dppe) [245], 1,2-bis(hydroxymethylphosphino)ethane (hmpe) [246], 1,2-bis(dicyclohexyphosphine)ethane (dcype) [247] or bis(diphenylphosphino)isopropane (dppip) [248] build structures of stoichiometry $[Au_2(P{\sim}P)_2]^{2+}$ such as those shown in Figure 2.45. The crystal structure of $[Au_2(dmpe)_2]X_2 \cdot nH_2O$ (dmpe = 1,2-bis(dimethylphosphino)ethane; X = Cl, $n = 2$; Br, $n = 1.5$) consist of polymeric chains anion/water molecule/cation/anion/cation/anion/water molecule [249].

The luminescent properties of some of these species have been studied [250, 251]. Substrate-binding reactions of the 3[dσ∗pσ] excited state of $[Au_2(dcpm)_2]^{2+}$ derivatives (dcpm = bis(dicyclohexylphosphino)methane) have been carried out [252, 253] and the electroluminescence (EL) from the triplet state of $[Au_2(dppm)_2](SO_3CF_3)_2$ has been reported [254]. The synthesis of the water soluble $[Au_2(hmpe)_2]Cl_2$ has been afforded from reaction of [AuCl(PPh$_3$)] and the diphosphine in dichloromethane/water media, whereas the reaction of Na[AuCl$_4$] and hmpe in water affords the homoleptic mononuclear derivative [Au(hmpe)$_2$]Cl. The synthetic results may be compared with those for the diphosphine 1,2-bis(dihydroxymethylphosphino)benzene (hmpb) for which both reactions afford the water soluble mononuclear [Au(hmpb)$_2$]Cl. Complexes $[Au_2(dcype)_2](PF_6)_2$ and $[Au_2(hmpe)_2]Cl_2$ are luminescent, the latter in the solid state and in water solution.

Some structural patterns found in complexes $[Au_2(dppm)_2]^{2+}$ with (pseudo)halide anions are shown in Figure 2.46 [255, 256]. Compounds [Au$_2$X$_2$(dppm)$_2$] (X = Cl, Br) display the structure shown in Figure 2.46a. Two phases (monoclinic and triclinic) are known for X = Cl. This structure is also found in [Au$_2$Cl$_2$(cdpp)] (cdpp = bis(diphenylphosphine)vinylidene) [257]. In the crystal structure of [Au$_2$Cl(P$\sim$P)]Cl (P$\sim$P = bis(diphenylphosphino)cyclopropylidene or bis(diphenylphosphino)-2-etoxyethylidene) one of the chloride atoms is bonded to one gold atom following the pattern shown in Figure 2.46b. [258]. Monodentate coordination of the ligand dithiobenzoate in [Au$_2$(S$_2$CPh)(dppm)$_2$]Cl has also been described [259]. The structure of [Au$_3$Cl$_2$(dppm)$_2$]PF$_6$ is based on a triangular Au$_3$-core (Figure 2.46d) [260]. Reaction of [AuBr(CH$_3$)$_2$(dppm)] with AgNO$_3$ in methanol affords the tetranuclear species shown in Figure 2.46e [261].

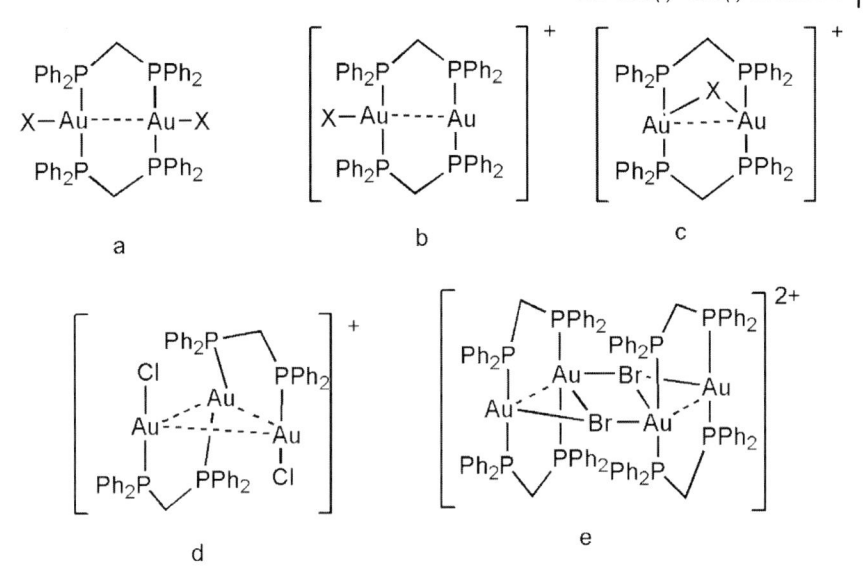

Figure 2.46 Different geometries in gold complexes with bidentate diphosphines and halide ligands.

Longer chains afford different features and thus the crystal structures shown by compounds of empirical formula [Au{PPh$_2$(CH$_2$)$_n$PPh$_2$}]$^+$ consist on rings [Au$_2${PPh$_2$(CH$_2$)$_n$PPh$_2$}$_2$]$^{2+}$ for $n=3$ or 5 but a polymer for $n=4$ [262]. The reactions of these complexes with [Au(CN)$_2$]$^-$ affords a pentanuclear species ($n=3$; Figure 2.47a) or a chain structure ($n=5$; Figure 2.47b). In this case the polymerization does not influence the luminescent emissions which, for all the complexes, show a maximum at about 410 nm.

[Au$_3$(C≡CPh)$_2$(dppm)$_2$][Au(C≡CPh)$_2$] is luminescent, both in the solid state and in solution. Its crystal structure resembles that represented in Figure 2.46d. The three gold centers form a triangle and are connected through aurophilic interactions. One gold center is bonded to both dppm ligands. The other phosphorus atom of each dppm diphosphine coordinates to one gold atom, which is also bonded to one acetylide fragment [263]. The ligand 3,5-di-*tert*-butyl-1,2,4-triphospholyl acts as a mono and bidentate ligand in [Au$_2$(3,5-^{t}Bu–3,5–C$_2$P$_3$)$_2$(PPh$_3$)$_2$] thus, one ligand bridges the two gold centers through one phosphorus atom and the other ligand uses two phosphorus atoms, each of them coordinated to a different gold center [264].

Crystal structures of [Au$_2$(E~E)$_2$]$^{2+}$ (E = chalcogen donor atom) have been described with dithiocarbamate [265–271], xantate [272], MNT [273–275], dithiophosphonate [276], dithiophosphinate [277], dithiophosphate [278, 279] or trithiocarbonate [280] ligands. The gold–gold interaction in complexes shown in Figure 2.48 has been analyzed by *ab initio* Hartree–Fock calculations [281].

In many cases the molecules associate though intermolecular aurophilic interactions leading to the formation of chains, some examples are [Au$_2${S$_2$CNPr$_2$}$_2$],

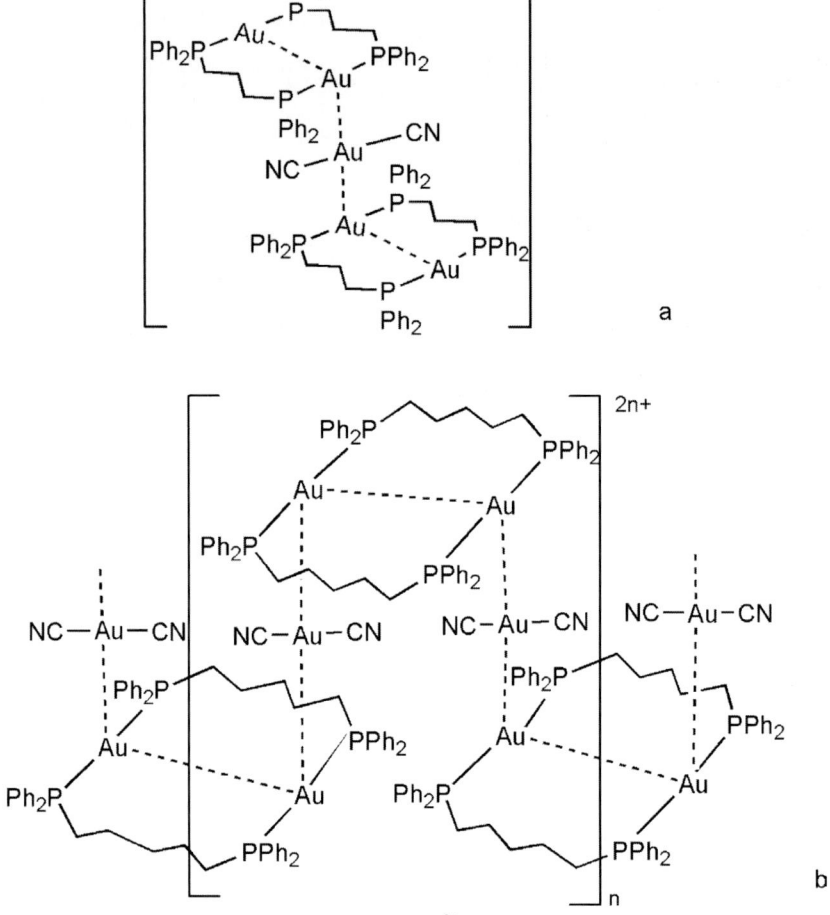

Figure 2.47 Aggregates of [Au$_2$\{PPh$_2$(CH$_2$)$_n$PPh$_2$\}$_2$]$^{2+}$ rings (a) and [Au(CN)$_2$]$^-$ units (b).

[Au$_2$\{S$_2$CN(C$_2$H$_5$)$_2$\}$_2$], [Au$_2$\{S$_2$CN(C$_5$H$_{11}$)$_2$\}$_2$], [Au$_2$\{S$_2$CN(C$_2$H$_4$OMe)$_2$\}$_2$], [Au$_2$(^{n}Bu-xantate)$_2$], [Au$_2$\{S$_2$PR'(OR)\}$_2$] (Figure 2.49) [R = Ph, R' = C$_5$H$_9$; R = Fe(C$_5$H$_4$)(C$_5$H$_5$), R' = (CH$_2$)$_2$O(CH$_2$)$_2$OMe], [Au$_2$\{S$_2$P(OR)$_2$\}$_2$] (R = Me, Et, ^{i}Pr), [Au$_2$\{S$_2$PPh$_2$\}$_2$].

Figure 2.48 [Au$_2$\{SSeC=C(CN)$_2$\}$_2$]$^{2-}$ (a) and [Au$_2$\{Se$_2$C=C(CN)$_2$\}$_2$]$^{2-}$ (b).

Figure 2.49 Chains of [AuS$_2$PR(OR')]$_2$ units.

Luminescent emissions have been reported for many of these species. Interaction of compound [Au$_2$\{S$_2$CN(C$_5$H$_{11}$)$_2$\}$_2$] with different VOC vapors results in an important color change and luminescence. In the absence of VOCs the emission is quenched. Molecular aggregation is required in order to observe luminescence. Emissions at 690 nm and broad profiles tailing to 850 nm are observed for [Au$_2$(^{n}Bu–xantate)$_2$] and [Au$_2$(Et-xantate)$_2$]. Studies of dithiophosphate [Au$_2$\{S$_2$P(OR')$_2$\}$_2$] (R = Me, Et, ^{n}Pr, ^{n}Bu) and dithiophosphonate [Au$_2$\{S$_2$PR(OR')\}$_2$] (R = Ph, R' = C$_5$H$_9$; R = 4-C$_6$H$_4$-OMe, R' = 1S,5R,2S (-) menthyl) complexes reveal an important influence of the aurophilic interactions in the luminescent properties.

Different metallic fragments may associate through gold atoms. One example is compound [Au$_2$\{Co(aet)$_2$(en)\}$_2$]$^{4+}$. In its crystal structure two hexacoordinate cobalt fragments [Co(aet)$_2$(en)]$^+$ (aet = 2-aminoethanethiolate; en = ethylenediamine) bridge two gold centers which exhibit almost linear geometry and intramolecular interactions [282]. Two isomers with the ligands C_2-cis and C_1-cis have been characterized (Figure 2.50). Similar complexes with ruthenium or platinum have been described with different ligands. These complexes also contain "S–Au–S" units connected through aurophilic interactions in which the sulfur atoms are bonded both to gold and to the heterometal [283–285].

Compound [\{Au(PTA)$_3$\}$_2$\{Au$_2$(iso-MNT)\}$_2$] (Figure 2.51a) is an example of unsupported gold–gold interactions. It has been prepared from the water soluble [AuCl(PTA)$_3$] and K$_2$[iso-MNT] [286] and its crystal structure consists on a nonlinear tetranuclear chain. A cyclic disposition of the four gold atoms is found in [Au$_4$(S$_2$CMe)$_4$] (Figure 2.51b) [287].

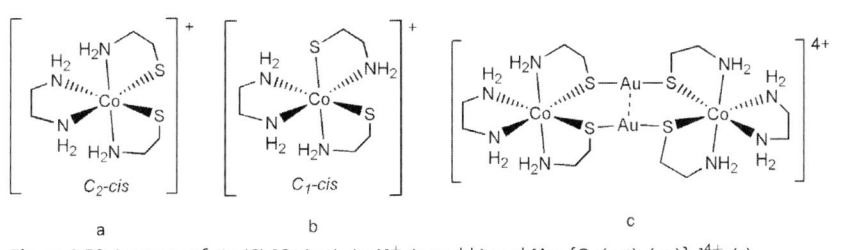

Figure 2.50 Isomers of cis-(S)-[Co(aet)$_2$(en)]$^+$ (a and b) and [Au$_2$\{Co(aet)$_2$(en)\}$_2$]$^{4+}$ (c).

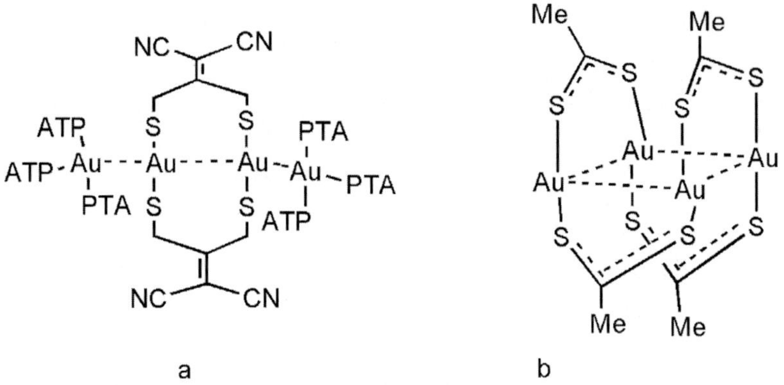

Figure 2.51 Tetranuclear complexes with bridging sulfur donor ligands.

Very short gold–gold distances have been found in [Au$_2$(2,6-Me$_2$Ph-form)$_2$] (Figure 2.52a) (2.711(3) Å). In the bis(amidinate) trinuclear compound [Au$_3$Cl(2,6-Me$_2$-form)$_2$(tht)] (Figure 2.52b) two gold–gold distances (about 3.01 Å) are in the range of aurophilic interactions [288].

Cyclic structures have been described for [Au$_4$(N∼N)$_4$] complexes with N∼N = 1,3-diphenyltriazenide (Figure 2.53a) [289], formiamidinate [290] or 3,5-^{t}Bu$_2$-pz (pz pyrazole) [291] (Figure 2.53b). Emissions have been described for formiamidate complexes [Au$_4$(ArNC(H)NAr)$_4$] (Ar = 4-OMe-C$_6$H$_4$, 3,5-Cl$_2$-C$_6$H$_3$, 4-Me-C$_6$H$_4$) in the solid state, both at room temperature and 77 K, at about 490 cm^{-1} with weaker emissions at about 530 cm^{-1}. Compound [Au$_4$(3,5-^{t}Bu$_2$-pz)$_4$] is luminescent in the solid and solution state at room temperature. The emission maxima appear in the 550–600 nm region for dichloromethane solutions and at 541 or 517 nm for solid or dichloromethane monosolvated samples, respectively.

Reaction of K[4-MePh-form] with [Au$_2$(2,6-Me$_2$Ph-form)$_2$] in 1 : 1 molar ratio in THF leads to the crystallization of [Au$_2$(2,6-Me$_2$Ph-form)$_2$][Au$_4$(4-MePh-form)$_4$]· 2THF, which contains a mixture of dinuclear and tetranuclear derivatives. Compound [Au$_2$(2,6-Me$_2$-form)$_2$]·2[Hg(CN)$_2$]$_2$·2THF (Figure 2.54) is obtained by reaction of [Au$_2$(2,6-Me$_2$-form)$_2$] and [Hg(CN)$_2$]. It crystallizes as a tetrahydrofuran solvate in a

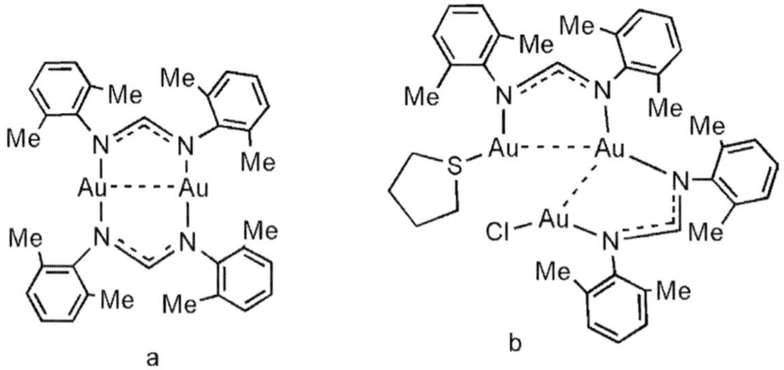

Figure 2.52 Amidinate complexes.

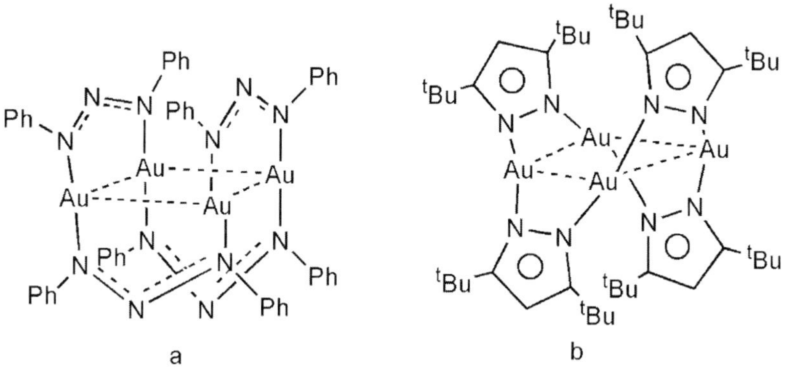

Figure 2.53 [Au$_4$(N$_3$Ph$_2$)$_4$] (a) and [Au$_4$(3,5-^{t}Bu$_2$–pz)$_4$] (b).

2D structure. The gold–gold distance increases from 2.711(3) Å in the starting product, to 2.9047(17) Å in the final chain [292].

[Au(3,5-Ph$_2$pz)(dppp)] [35], [Au$_2$(G)(dmpe)] (G = guaninato dianion; dmpe = 1,2-bis(dimethylphosphino)ethane) [293] or the xantine derivatives [Au$_2$(HX)(dmpe)], [Au$_2$(H$_2$TT)(dmpe)] (H$_3$X = xantine; H$_2$TT = 1,3-dimethyl-8-thioxantine) [294] are examples of dinuclear gold species in which different bidentate ligands bridge both gold atoms in discrete structures with an intranuclear aurophilic interaction. The X-ray structure of the luminescent [Au$_2$(G)(dmpe)] compound (Figure 2.55a) shows an infrequent N3, N9 bridging mode coordination for guanine. The crystal structure also reveals that the AuP$_2$ moieties interact weakly with the Br$^-$ anions (Au–Br distance 3.932(1) Å). In the tetranuclear [Au$_4$(3,5-Ph$_2$pz)$_2$(dppm)$_2$], [Au$_4$(3,5-Ph$_2$pz)$_2$(2,6-Me$_2$Ph-form)$_2$] (Figure 2.55b) and [Au$_4$(3,5-Ph$_2$pz)$_3$(2,6-Me$_2$Ph-form)] species the gold centers connect through gold–gold interactions forming a square [295, 296].

Homoleptic complexes with bidentate C~C donor ylide derivatives are well known. Oxidative addition of some gold(I) [Au$_2${(CH$_2$)$_2$PR$_2$}$_2$] species by addition of halogen, pseudo-halogen, or alkyl halide affords dinuclear Au(II) complexes which

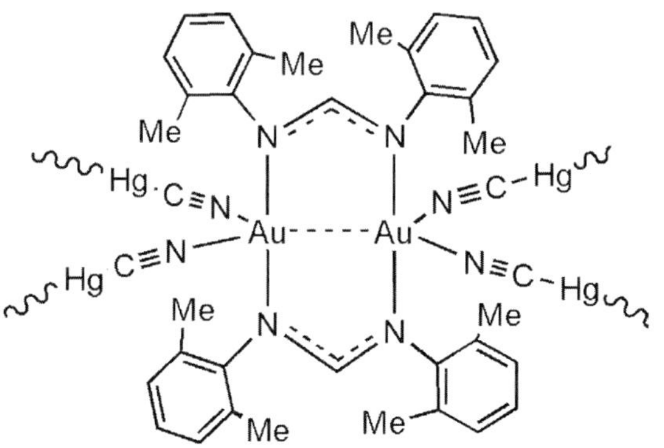

Figure 2.54 [Au$_2$(2,6-Me$_2$Ph-form)$_2$]·2[Hg(CN)$_2$].

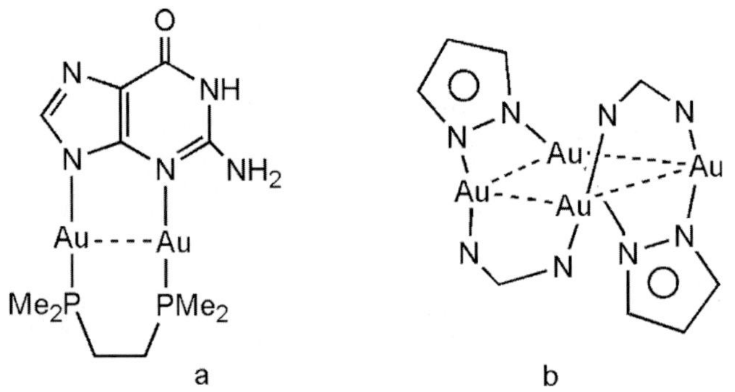

Figure 2.55 [Au$_2$(G)(dmpe)] (a) and [Au$_4$(3,5-Ph$_2$pz)$_2$(2,6-Me$_2$Ph-form)$_2$] (substituents in both nitrogen ligands have been omitted for clarity) (b).

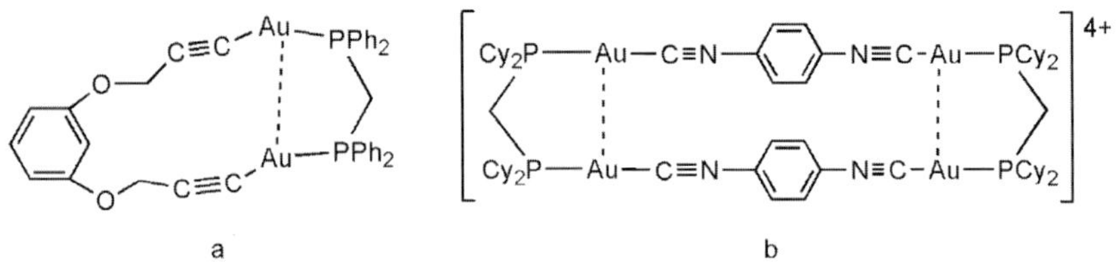

Figure 2.56 Complexes with bis-acethylide (a) and diphosphine ligands (b).

contain a conventional metal–metal bond between the two d^9 gold atoms. Dinuclear species with two different bridging ligands, one an ylide, and the other xantate, dithiocarbamate or pyridine-2-thiolate have been reported. All show intramolecular gold–gold interactions and, in some cases, also intermolecular contacts [297–300].

A large family of analogous compounds to that shown in Figure 2.56a with different numbers of methylene spacers in the diphosphine ligand and different isomers of the benzene ring, have been synthesized and their crystal structures determined. The increment of the methylene groups in the diphosphine chain controls largely the presence and intensity of the gold–gold contacts. The complexes are luminescent, both in solution and in the solid state [301]. It is also possible to design different macrocycles by combining four ligands, frequently two diphosphines and two other ligands (Figure 2.56b) [302].

Complexes of this type have been deeply analyzed. They may be present in solution, an example is shown in Scheme 2.1. In many cases the solid-state structures of the isolated complexes are not representative of the structures in solution. The crystal structures of ring and [2]catenates have been described whereas crystal structures of doubly braided [2]catenates are much less frequent. In these structures gold–gold interactions may combine with hydrogen bonds and π–π interactions [303–306].

Bidentate N∼C donor ligands display head to tail arrangement in homoleptic dinuclear complexes [Au$_2$(N∼C)$_2$] that include [Au$_2${CH$_3$im(CH$_2$py)}$_2$]$^{2+}$ [CH$_3$im

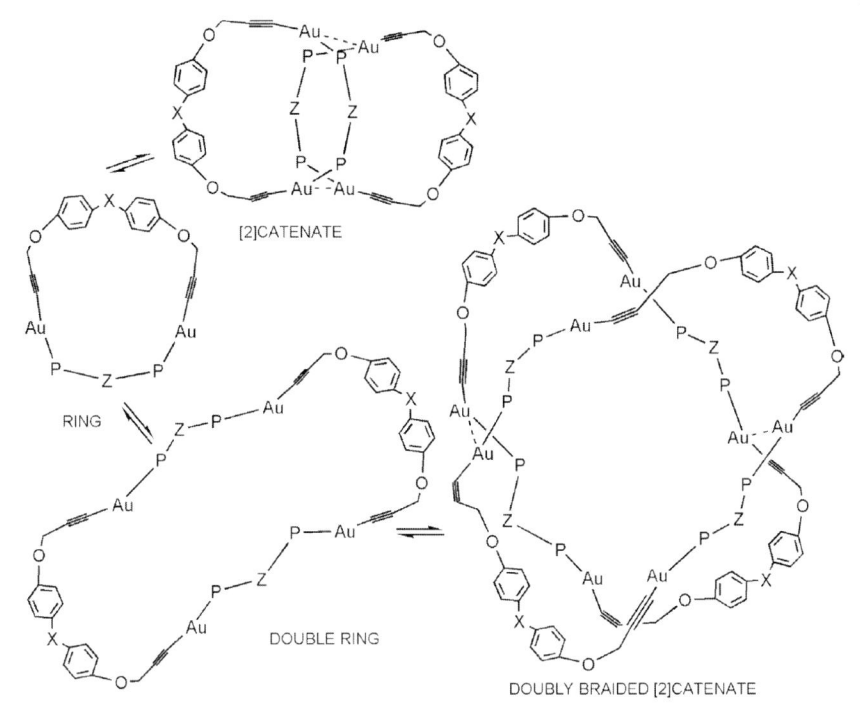

Scheme 2.1 P = PPh$_2$, X = O, S, SO$_2$, CH$_2$, CHR, CR$_2$, cyclohexylidene, Z = (CH$_2$)$_n$ (n = 2–6).

(CH$_2$py) = 1-methyl-3-(2-pyridinylmethyl)-imidazolium] (Figure 2.57a) whose luminescent behavior has been described [307], 2-bis(trimethylsilyl)methylpyridyl [308] or 2-(dimethylaminomethyl)ferrocenyl [309] complexes. In the dinuclear derivative with the 4,5-dihydro-4,4-dimethyl-1,3-oxazol-2-yl)thien-3-yl ligand (Figure 2.57b) the gold–gold distance is 2.8450(6) Å [310].

In the crystal structure of [Au$_2$(4,6-Me$_2$pym-2-S)$_2$] (4,6-Me$_2$pym-2-S = 4,6-dimethylpyrimidinethiolate) (Figure 2.58a) [311] two independent molecules are present. The intermolecular Au–Au distances are 3.544(1) and 3.783(1) Å. The planar molecules interact through π–π interactions which involve the pyrimidine ligands.

Figure 2.57 Complexes with N∼C bridging ligands.

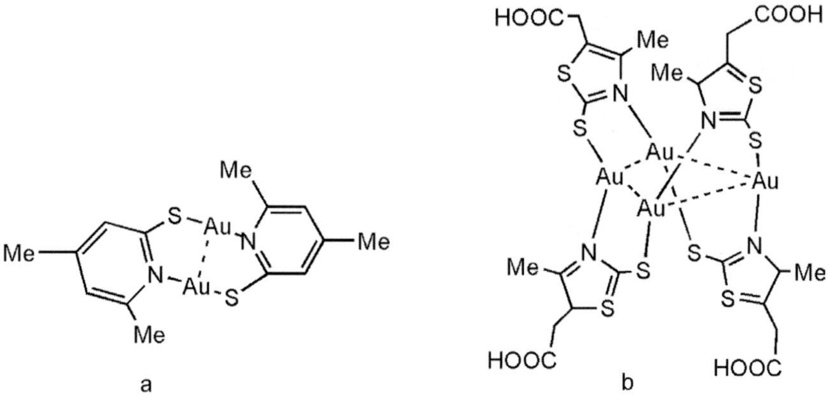

Figure 2.58 Di (a) and tetranuclear (b) complexes with N∼S bridging ligands.

Its luminescent emissions in the solid state and in solution have been analyzed. The spectrum in solution shows one or two bands, depending on the concentration. Reaction of [AuCl(SMe$_2$)] with L = 2-mercapto-4-methyl-5-thiazoleacetic acid in molar ratio 1:1 affords a luminescent tetranuclear [Au$_4$L$_4$] (Figure 2.58b) species [269]. In the emission spectrum two maxima appear at 585 and 720 nm at room temperature but there is only one maximum at 585 nm at 77 K. A similar tetranuclear reorganization is observed in [Au$_4$(etu-H)$_4$] (etu-H = ethylenethiourea deprotonated at the nitrogen atom). In this case the tetranuclear units form chains through intermolecular gold–gold contacts [312].

[Au$_2$(OTf)$_2$(dppm)] (OTf = trifluoromethylsulfonate) reacts with TU = 2-thiouracile or Me-TU = 2-methyl-2-thiouracile leading to [Au$_2$(TU)(dppm)]$^+$ or [Au$_2$(Me-TU)(dppm)]$^+$ in which the gold centers are bridged by both ligands. The molecules display intramolecular gold–gold contacts and also associate through intermolecular aurophilic interactions to build a helix and exhibit luminescence tribochromism. These complexes are non-emissive or weakly luminescent in the solid state and upon crushing they are converted into the bright emissive unprotonated species. The crystal structure of the unprotonated compound with 6-methyl-2-thiouracil consists of the association of two dinuclear units through a unique gold–gold intermolecular interaction [313]. This is a reversible process governed by both acid–base properties and aurophilic interactions. The crystal structures of [M$_2$(PPh$_2$CH$_2$SPh)$_2$]$^{2+}$ (M = Au, Ag) have been determined, and the ligands are in a head to tail arrangement. Ab initio HF/II and MP2/II calculations have been performed in order to analyze the metallophilic attraction in both complexes [314]. A head to tail arrangement is also observed in [Au$_2$(ER$_2$-R'-C$_6$H$_3$)$_2$] (Figure 2.59) (E = P, As, R = Ph, Et, R' = H; E = As, R = Ph; R' = Me) [315–317].

Absorption and emission spectra of [Au$_2$(S$_2$CNEt$_2$)(P∼P)]$^+$ and [Au$_2$\{S$_2$C$_2$(CN)$_2$\}(P∼P)] (P∼P = diphosphine) are dependent on the concentration. These data support aggregation in solution. The crystal structure of compound [Au$_2$\{S$_2$C$_2$(CN)$_2$\}\{PMe$_2$CH$_2$PMe$_2$\}] is helical polymeric, whereas complexes [Au$_2$\{S$_2$C$_2$(CN)$_2$\}\{PPh$_2$(CH$_2$)$_2$PPh$_2$\}], [Au$_2$\{S$_2$C$_2$(CN)$_2$\}\{PPh$_2$(CH=CH)PPh$_2$\}] and [Au$_2$(μ-C$_3$S$_5$)

Figure 2.59 [Au$_2$(ER$_2$-R'-C$_6$H$_3$)$_2$].

(μ-CH$_2$PPh$_2$CH$_2$)] consist of dinuclear units with intramolecular interactions [318–320]. A less common structure is that shown in Figure 2.60a [321]. In each dimercapto ligand one of the sulfur centers is connected to two gold atoms with different distances. Sulfur ylide derivatives of stoichiometry [Au$_2$(dppm){(CH$_2$)$_2$S(O)NMe$_2$}]$^+$ and [Au$_4$(dppm)(PPh$_2$CHPPh$_2$){(CH$_2$)$_2$S(O)NMe$_2$}]$^+$ have been characterized. In the former the dinuclear molecules associate in dimers with shorter intermolecular than intramolecular gold–gold interactions. The compound is luminescent both in solution and in the solid state. In the latter (Figure 2.60b) the tridentate ligand (PPh$_2$CHPPh$_2$)$^-$ bridges three gold centers through the two phosphorus and methylene carbon atoms. Three gold centers are connected in a triangular shape through aurophilic interactions. The fourth gold atom does not exhibit gold–gold contacts [322].

Two isomers of [Au$_2$(dppm)$_2$(dtp)]BF$_4$ (dtp = [S$_2$P(R)(OR')]$^-$; R = p-C$_6$H$_4$-OCH$_3$; R' = c-C$_5$H$_9$) have been isolated and their crystal structures determined (Figure 2.61). The isomer of Figure 2.61a (yellow) changes into that of Figure 2.61b (white) upon heating [217]. The structures [Au$_2$(dppm)$_2$(S$_2$CNEt$_2$)]$^+$ [244], [Au$_2${bis(dimethylphosphino)methane}$_3$]$^{2+}$ [323] and that of the luminescent [Au$_2${(Ph$_2$Sb)$_2$O}$_3$]$^{2+}$ [324] are similar to that in Figure 2.61a.

[Au$_3$(CNC)(dppm)$_2$]$^+$ [CNC = pyridyl-2,6-diphenyl^{2-}] (Figure 2.62a) contains a triangle of gold atoms connected through aurophilic interactions. The cytotoxicity of its chloride salt toward different cancer cells lines has been analysed [325]. 1,1'-bis(diphenylthiophosphoryl)ferrocene (dptpf) and 1,1'-bis(diphenylselenophosphoryl) ferrocene (dpspf) act as trans- ligands in the mononuclear cations [Au(dptpf)]$^+$

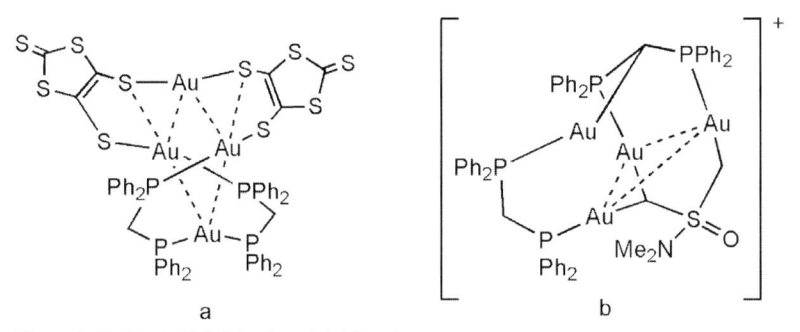

Figure 2.60 [Au$_4$(μ-C$_3$S$_5$)$_2$(μ-dppm)$_2$] (a) and [Au$_4$(PPh$_2$CH$_2$PPh$_2$)(PPh$_2$CHPPh$_2$){(CH$_2$)$_2$S(O)NMe$_2$}]$^+$ (b).

Figure 2.61 Isomers of [Au$_2$(dtp)(dppm)$_2$]$^+$.

or [Au(dpspf)]$^+$. In the crystal structure these cations associate to form dinuclear (1,1′-bis(diphenylthiophosphoryl)ferrocene) or trinuclear (1,1′-bis(diphenylselenophosphoryl)ferrocene) species through gold–gold contacts, the latter is shown in Figure 2.62b [326, 327].

Derivatives of stoichiometry [Au$_2${1,2-S$_2$-1,2-C$_2$B$_{10}$H$_{10}$}(P∼P)] have been structurally characterized for P∼P = dppey (1,2-bis(diphenylphosphino)ethylene) or dppb (o-bis(diphenylphosphino)benzene). In both cases each gold atom is coordinated to one sulfur of the carborane-dithiolate ligand and to a phosphorus of the diphosphine. The gold centers are connected through an intramolecular aurophilic interaction. The bulky nature of the carborane diphosphine 1,2-(PPh$_2$)$_2$-1,2-C$_2$B$_{10}$H$_{10}$ and its tendency to act as chelate directs the formation of a tetranuclear compound (Figure 2.63) [328, 329].

The cycloaurated cation [Au$_5$(C$_6$H$_4$PPh$_2$)$_4$]$^+$ contains a pair of ipso-carbon-digold interactions (Figure 2.64a) [330]. Two [Au$_2$(dppm)$_2$]$^{2+}$ (Figure 2.64b) units are bridged by two dithiolates in [Au^{III}Au$_4^I$(μ–dppm)$_4$(tdb)$_2$] (tdb = toluene-3,4-dithiolate). The sulfur atoms are also bonded to the central gold(III). This compound is luminescent and the emissions appear in the 560 nm region both in the solid state

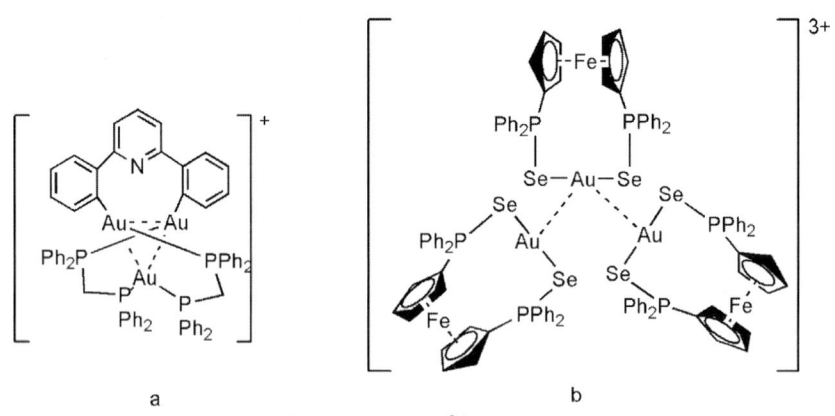

Figure 2.62 [Au$_3$(CNC)(dppm)$_2$]$^+$ and [Au(dpspf)]$_3^{3+}$.

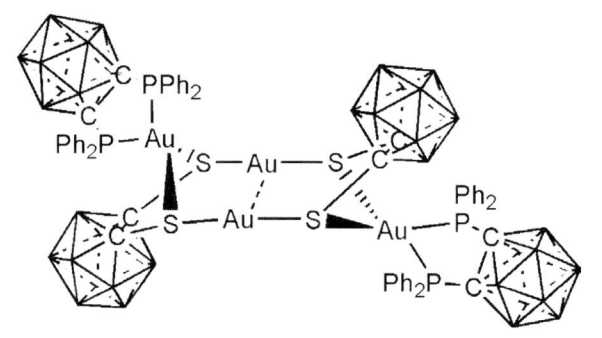

Figure 2.63 [Au$_4${1,2-S$_2$-1,2-C$_2$B$_{10}$H$_{10}$}$_2${1,2-(PPh$_2$)$_2$-1,2-C$_2$B$_{10}$H$_{10}$}$_2$].

and in solution at room temperature [331]. Compound [Au$_5$(dppm)$_2$(2-S-NC$_9$H$_6$)$_3$]$^{2+}$ (Figure 2.64c) exhibits dual emissions at 500 and 605 nm in solution [332].

Examples of higher nuclearities are [Au$_6$(o-CH$_3$C$_6$H$_4$CS$_2$)$_6$] [333] in which the six gold atoms of the central cluster are almost coplanar. In addition to many Au–Au interactions within the hexameric cluster, a close intermolecular contact exists (Au–Au 3.195(1) Å), forming chains of hexanuclear units. The compound [Au$_6$(3,5-diphenylpyrazolate)$_6$] exhibits the same gold:ligand ratio, but it is monomeric. The six gold atoms are in the vertex of two tetrahedra that share one edge [334]. From the non-chiral fragments piperizine-1,4-bis(dithiocarbamate) and bis(diphenylphosphino)methane a luminescent chiral Au$_{16}$ ring complex (Figure 2.65a) has been built. The molecule may be considered as the association of four monomers (Figure 2.65b). From ^{1}H NMR and ESI-MS studies it is proposed that the compound exists in dilute solution as its monomer [335]. Both the tetramer and monomer are luminescent.

Dinuclear units [Au$_2$(CN^{t}Bu)$_2${NC$_3$(4-Cl-C$_6$H$_4$)(O)}$_2$] form infinite chains of parallel orientated molecules through gold–gold contacts (Figure 2.66). Analogous complexes with more demanding groups than CN^{t}Bu, as PEt$_3$ or PPh$_3$, do not exhibit aurophilic interactions [336]. These complexes are interesting because of their properties, the dye ligand diketopyrrolopyrrole is luminescent but the coordination to gold (and also to other metals such as silver, copper, palladium or platinum) leads to

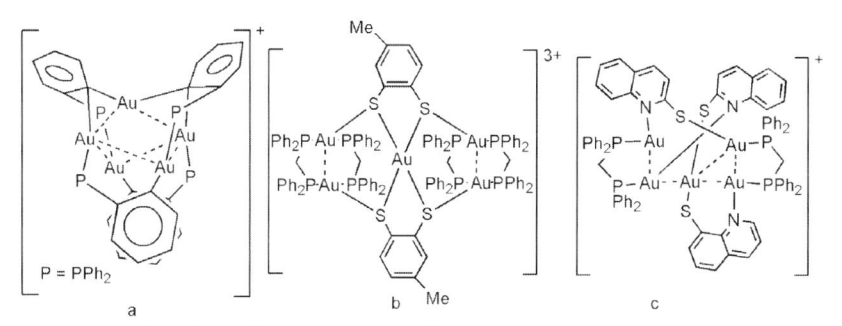

Figure 2.64 Polynuclear gold complexes with different bidentate ligands.

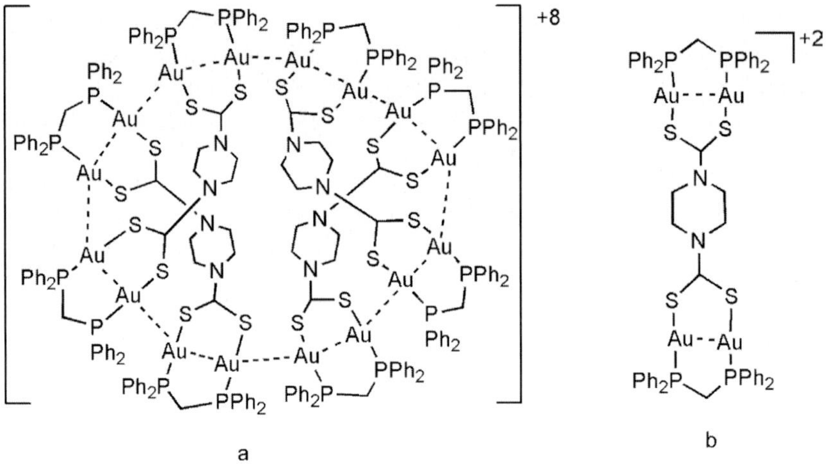

Figure 2.65 Cationic chiral Au$_{16}$ ring (a) and the monomer (b).

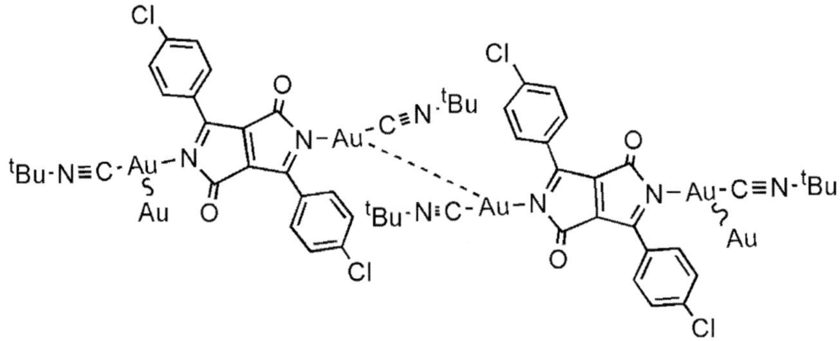

Figure 2.66 Formation of chains in [^{t}BuNCAuN{C$_3$(4-Cl-C$_6$H$_4$)(O)}$_2$NAuCN^{t}Bu].

better solubility, high fluorescence quantum yields and bathochromic absorptions, compared with the free ligand.

Planar nine-membered rings, as shown in Figure 2.67, represent an interesting group of complexes which have been widely studied. Compound [Au$_3$(MeOC=NMe)$_3$]

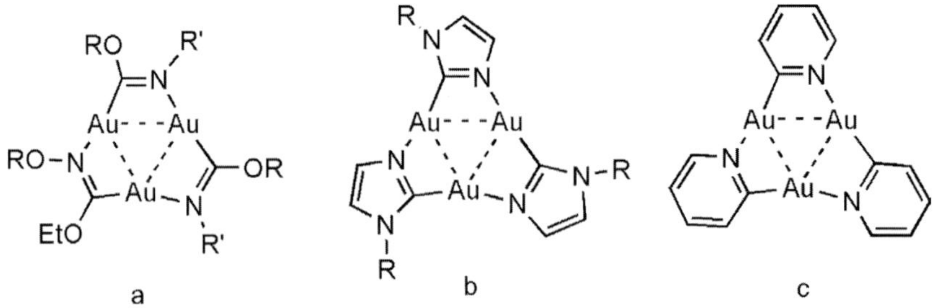

Figure 2.67 Nine-membered triauracycles.

(Figure 2.67a; R = R' = Me) displays solvoluminescence. These compounds also react with acceptor molecules or metal ions M^+ ($M^+ = Ag^+, Tl^+$). Structures of [Au$_3$(MeOC NMe)$_3$] with the acceptors 2,4,7-trinitro-9-fluorenone and 2,3,4,7-tetranitro-9-fluorenone consist of columns of planar molecules showing the pattern ···donor–acceptor–donor–acceptor··· Compound [Au$_3$(EtOC=NpTol)$_3$] interacts with octafluoronaphthalene to form an adduct which crystallizes with the same pattern. The stacking with octafluoronaphthalene completely quenches the photoluminescence. In the structures with 2,7-dinitro-9-fluorenone or 2,4,7-trinitro-9-fluorenone the columns are built from ···donor–donor–acceptor–donor–donor–acceptor··· molecules. Aurophilic interactions produce dimers of nine-membered rings with nearly trigonal prismatic Au$_6$ cores. The same organization ···donor–donor–M–donor–donor–M··· is observed in the adducts formed by [Au$_3$(EtOC = NC$_6$H$_4$-4-Me)$_3$] with M = Tl$^+$ and [Au$_3$(1-benzyl-2-imidazolate)$_3$] (Figure 2.67b; R = CH$_2$C$_6$H$_4$) with M = silver or thallium. The luminescent properties of the silver or thallium species are sensitive to the temperature and the metallic ion M$^+$ [337–339]

In the crystal structure of [Au$_3$(NC$_5$H$_4$)$_3$] the molecules aggregate through gold–gold contacts into two distinct structural motifs. One of the molecules forms an extended chain, the other a dimer through two aurophilic interactions. The disposition of the gold centers in both motifs is shown in Figure 2.68. After months of standing in the atmosphere or immersion in 4 M hydrochloric acid for days hourglass figures form within the crystals [340]. Gold decomposition seems to be the origin of these figures.

Mixed silver–gold trinuclear nine-membered rings which contain different ligands have also been afforded. In the structures of [Au(EtOC=N-pTol)Ag$_2$(μ-3,5-Ph$_2$pz)$_2$]$_2$ and [Au(1-benzyl-2-imidazolate)Ag$_2$(μ-3,5-Ph$_2$pz)$_2$]$_2$ there are no short gold–gold contacts, [Au$_2$(EtOC=N-pTol)$_2$Ag(μ-3,5-Ph$_2$pz)]$_2$ exhibits inter- and intramolecular aurophilic interactions [341] (Figure 2.69).

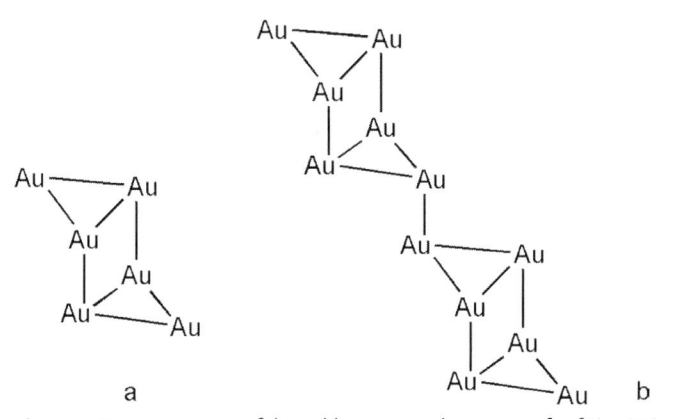

Figure 2.68 Arrangements of the gold centers in the two motifs of [Au$_3$(NC$_5$H$_4$)$_3$].

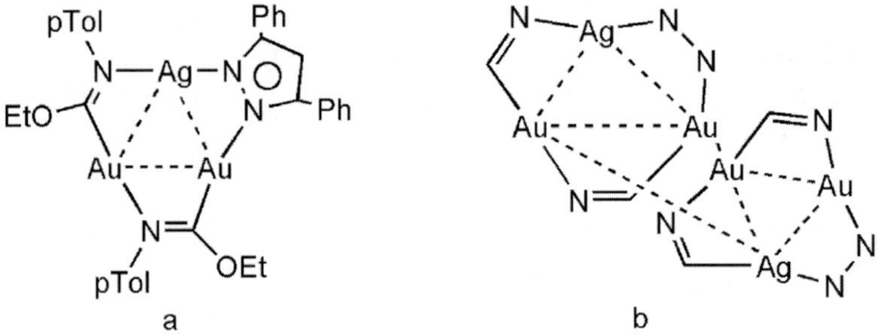

Figure 2.69 Mixed Au–Ag nine-membered ring (a) and association of the molecules (substituents have been omitted for clarity) through gold–gold contacts (b).

2.2.5
Complexes with Polydentate Ligands

Triphosphines can bridge three or four gold atoms. The crystal structures of compounds [Au$_3$Cl$_2$(dpmp)$_2$]Cl (dpmp = bis(diphenylphosphinomethyl)phenylphosphine), [Au$_3$X$_3$(tppm)] (X = Cl, I; tppm = tris(diphenylphosphino)methane), [Au$_3$Cl(tppm)$_2$], [Au$_4$X$_2$(dpmp)$_2$]$^+$, [Au$_4$X$_2$(dpmp)$_2$]$^{2+}$ (X = Cl, SCN with different structures), [Au$_3$(P∼P∼P)$_2$]$^{3+}$ (P∼P∼P = dpmp, bis(2-(diphenylphosphino)ethyl)phenylphosphine), [Au$_3$X(dpmp)$_2$]$^{2+}$ (X = Cl, I), [Au$_4$Cl$_2$(dpma)$_2$]$^{2+}$ (dpma = bis (diphenylphoshinomethyl)phenylarsine) and [Au$_3$Cl$_3$(P∼P∼P)] [P∼P∼P = dpma, dpmp, bis (2-(diphenylphosphino)phenyl)phenylphosphine (TP)] illustrate some of the different patterns obtained (Figure 2.70). Many of these complexes exhibit luminescent properties which have been analyzed. In addition, heterometallic complexes containing gold and iridium or gold and ruthenium with similar patterns, but with two different metal centers, have been described and their luminescent properties also studied. The fluxional behavior in solution of compound [Au$_2$(TP)$_2$]$^{2+}$ has been studied by NMR [245, 342–354].

A layer structure is found for [Au$_3$Cl$_3${MeC(CH$_2$PMe$_2$)$_3$}] (Figure 2.71a), however compound [AuCl$_3${MeC(CH$_2$PPh$_2$)}$_3$] is monomeric [355]. By reaction of BF$_3$OEt$_2$ with [AuCl(Ph$_2$OH)] compound [{FB(OPPh$_2$)$_3$}Au$_3$Cl$_3$Au$_3${(OPPh$_2$)$_3$BF}]$^+$ is obtained (Figure 2.71b) [39]. In the crystal structure each tridentate ligand bridges three gold centers which are connected through gold–gold contacts. The two Au$_3$ units are bridged by three chloride ligands. A more complicated structure is obtained for [Au$_5$(C$_6$F$_5$){C(SPPh$_2$)$_2$}$_2$(PPh$_3$)] (Figure 2.71c) in which the longer gold–gold distances are those which connect the "Au(PPh$_3$)" and "Au(C$_6$F$_5$)" fragments to one of the three gold atoms bridged by two [C(SPPh$_2$)$_2$]$^{2-}$ ligands [181].

The 3,5-bis(N-methylimidazoliumyl)methyl)pyrazole anion [H$_3$L]$^{2+}$ acts as a tetradentate ligand in the intensely luminescent compound [Au$_4$L$_2$]$^{2+}$. Each gold center is coordinated to one nitrogen and one carbon atom. The four gold atoms form a square through gold–gold interactions [356]. Structures of [Au$_4$(μ-tetradentate ligand)L$_4$]$^{n+}$ [tetradentate ligand = 2,3,6,7-tetrakis(diphenylphosphino)tetrathiafulvalene {L = Cl, n = 0 [357]}; 2,2'-dibenzimidazolate {L = PPh$_3$, n = 2 [358]}; or

Figure 2.70 Complexes with triphosphines.

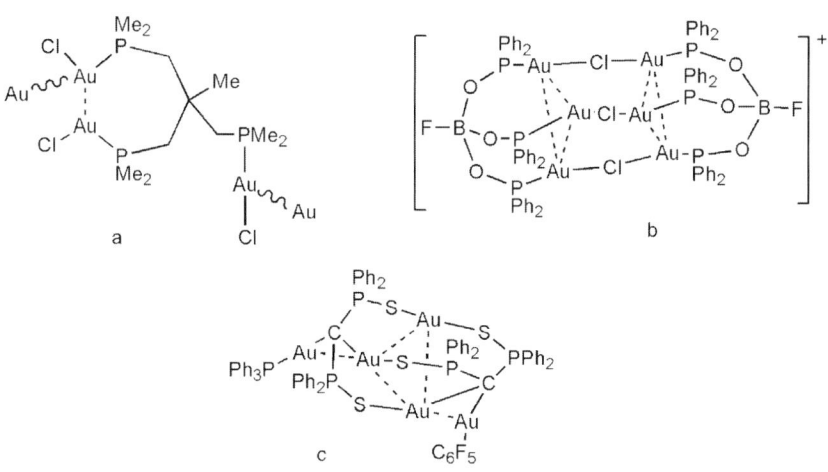

Figure 2.71 [Au$_3$Cl$_3${MeC(CH$_2$PMe$_2$)$_3$}] (a) [{FB(OPPh$_2$)$_3$}Au$_3$Cl$_3$Au$_3${FB(OPPh$_2$)$_3$}] (b) and [Au$_5$(C$_6$F$_5$){C(SPPh$_2$)$_2$}$_2$(PPh$_3$)] (c).

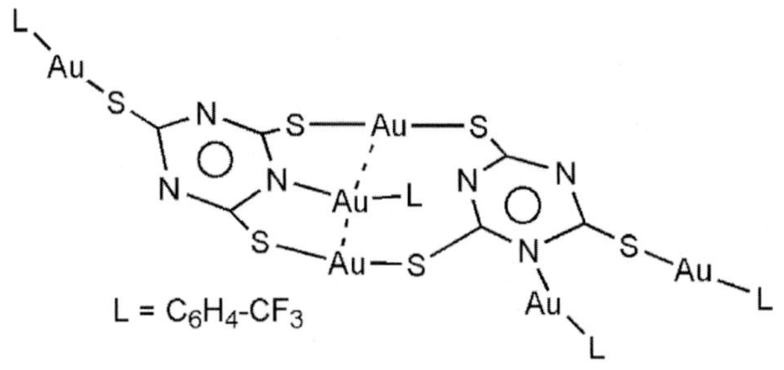

Figure 2.72 [Au$_6${N$_3$C$_3$)S$_3$}$_2${P(3-CF$_3$-C$_6$H$_4$)$_3$}$_4$].

1,4,8,11-(diphenylphosphinomethyl)-1,4,8,11-tetraazacyclotetradecane {L = Cl, n = 0) [359]}] have been determined. Each donor atom of the tetradentate ligand coordinates to one "AuL" unit. The tetrathiafulvalene and benzimidazolate derivatives contain two intramolecular aurophilic contacts which connect different gold centers. The tetraazacyclotetradecane derivative does not exhibit intramolecular gold–gold interactions, instead intermolecular Au–Au contacts build a polymer. The luminescent properties of the tetraazacyclotetradecane and benzimidazolate derivatives have been studied. The luminescent properties of compounds with the ligand (N$_3$C$_3$)S$_3$$^{3-}$ = 1,3,5-triazine-2,4,6-trithiolate have also been described. In [Au$_3${(N$_3$C$_3$)S$_3$}(PR$_3$)$_3$] (R = Ph, py) the three "Au(PR$_3$)" units coordinate to the sulfur atoms and the trinuclear species do not exhibit aurophilic interactions. Attempts to crystallize [Au$_3${(N$_3$C$_3$)S$_3$}L$_3$] (L = PPhMe$_2$, ^{t}BuNC) lead to a reorganization (Equation 2.6) with formation of two linear S–Au–S bonds. The result is a hexanuclear unit in which four gold atoms are connected through aurophilic interactions. Intermolecular aurophilic interactions lead to the formation of two-dimensional stacked layers [360, 361]. For L = 4-CF$_3$-C$_6$H$_4$ the structure is different. Instead of a chain an oligomer built from aurophilic interactions between two [Au$_6${(N$_3$C$_3$)S$_3$}$_2${P(3-CF$_3$-C$_6$H$_4$)$_3$}$_4$] units, such as that shown in Figure 2.72, is formed. The main difference between the hexanuclear units in the oligomer and the chain compounds is that in the oligomer one ligand uses one of the outer nitrogen atoms to coordinate to one gold center and the inner atom is free, whereas in the chain complexes both ligands use inner nitrogen atoms to coordinate to gold [362].

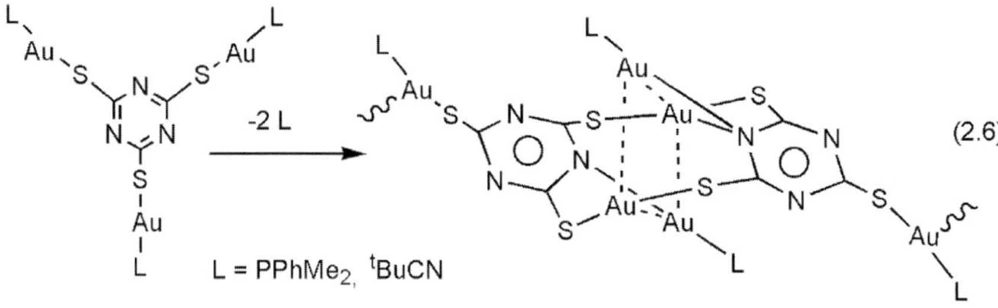

(2.6)

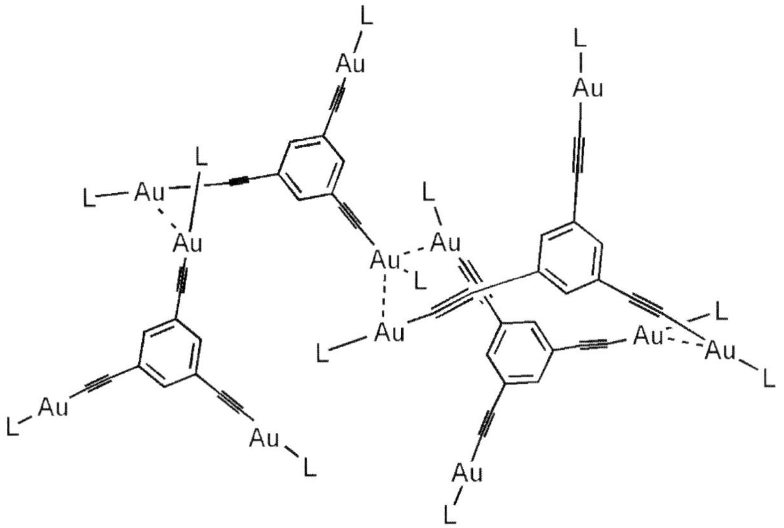

Figure 2.73 Polymeric structure of [{1,3,5(C≡CAuL)}$_3$-C$_6$H$_3$].

Reaction of the triacetylide [{1,3,5(C≡CAu)}$_3$-C$_6$H$_3$]$_n$ and the corresponding monodentate ligand (L) (L = isocyanide, phosphine, phosphite) gives [{1,3,5 (C≡CAuL)}$_3$-C$_6$H$_3$]. In the crystal structures of compounds with L = CN^{t}Bu and P(OMe)$_3$ molecules related by lattice translations are connected through aurophilic interactions to form a polymer (Figure 2.73) [363].

2.3
Gold(I)-Gold(III) Interactions

Aurophilic attraction has been almost synonymous with Au(I)–Au(I) attraction. However some complexes which contain short d^{10}–d^8 Au(I)–Au(III) and d^8–d^8 Au(III)–Au(III) have been reported, suggesting weak contacts. Nevertheless, examples of such interactions are rare. Au(I)–Au(III) short distances have been reported for doubly bridged ylide compexes (in which the ligands support such interactions). On some occasions assignation of oxidation states to the gold atoms is easy and Mössbauer spectroscopy can be used in order to analyze this point. Complexes with chalcogenide, chalcogenolate and polydentate ligands have also been described. Some of these compounds are described here.

[AuCl(terpy)]$_2$[AuCl$_2$]$_3$[AuCl$_4$] is obtained as an insoluble secondary product from the reaction of [HAuCl$_4$] with terpyridine (terpy). Its structure (Figure 2.74) is built from three linear [AuCl$_2$]$^-$ anions which link two [AuCl(terpy)]$^{2+}$ cations through axial gold(I)–gold(III) contacts (3.2997(8) Å). In addition, the [AuCl(terpy)]$^{2+}$ cations are linked through the chloride atoms of square planar [AuCl$_4$]$^-$ anions leading to an infinite chain [364]. The cations of the salt [Au^{III}Cl(BPA-H)]PF$_6$ (BPA = HN(CH$_2$-2-C$_5$H$_4$N)$_2$) form columns in the crystal structure with Au(III)–Au(III) distances of 3.54

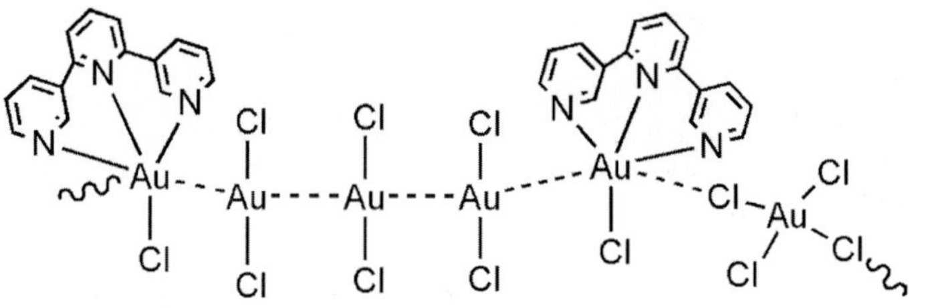

Figure 2.74 [AuCl(terpy)]$_2$[AuCl$_2$]$_3$[AuCl$_4$].

and 3.73 Å. The gold centers are four coordinated to the three nitrogen atoms of the BPA-H and to a chloride ligand. In the salt [Au^{III}Cl(BPA-H)][Au^{I}Cl$_2$] anion and cation alternate building columns with Au(I)–Au(III) distances of 3.34 Å [365].

As commented above, oxidative addition to gold(I) complexes affords gold(II) derivatives in which the gold centers exhibit a *real* bond, nevertheless, in some cases gold(I)–gold(III) species are described which exhibit a longer gold–gold distance. Scheme 2.2 represents an example of the above-mentioned reactivity for methylenethiophosphinate bridging ligands [366]. Compound **B** is the final gold(II)–gold(II) complex, whereas compound **C** corresponds to the final gold(I)–gold(III) derivative. The three compounds have been structurally characterized by X-ray studies. The intramolecular gold–gold distances are 3.040(1) Å (**A**), 2.607(1), 2.611(1) Å (**B**) and 3.050(3) Å (**C**).

Mixed valence oxidation states have also been described for the ylide bidentate ligands shown in Figure 2.75 [367–369]. [Au$_2$Br$_2${(CH$_2$)$_2$PPh$_2$}$_2$] (Figure 2.75a) has been analysed by X-ray photoelectron spectroscopy (XPS) and X-ray diffraction. The

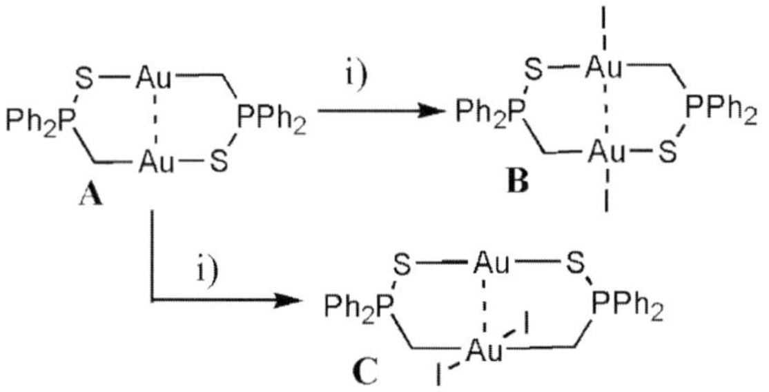

Scheme 2.2 (i) I$_2$.

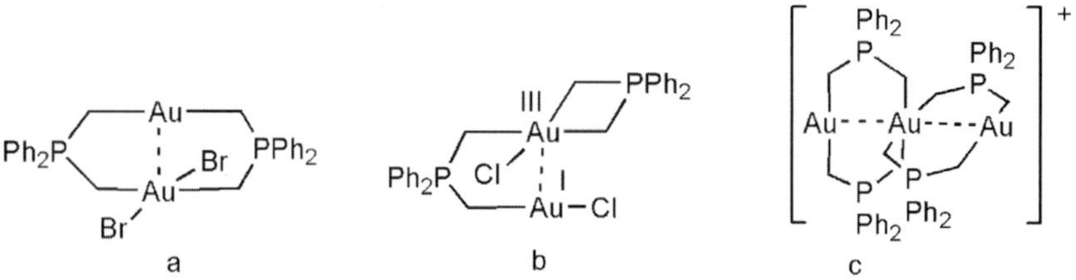

Figure 2.75 Mixed valence complexes with ylide ligands.

2.3 Gold(I)-Gold(III) Interactions

Scheme 2.3 (i) MgMe$_2$, (ii) LiC$_6$F$_5$.

gold(I)–gold(III) distance is 3.061(2) Å. The isomerization of [Au$_2$Cl$_2${(CH$_2$)$_2$PPh$_2$}$_2$] leads to [ClAu^I{μ-(CH$_2$)$_2$PPh$_2$}Au^{III}Cl{(CH$_2$)$_2$PPh$_2$}] (Figure 2.75b) in which the gold (I)–gold(III) distance is 3.184(1) Å. The gold–gold distances in the compound of Figure 2.75c are 3.052(1) and 3.049(1) Å.

The reaction of [Au$_2$(O$_2$CPh){2-(PR$_2$)C$_6$H$_4$}$_2$] (R = Et, Ph) with MgMe$_2$ leads to [Me$_2$Au{μ-(2-(PR$_2$)C$_6$H$_4$}$_2$Au] (Scheme 2.3, **A**). The same reaction with LiC$_6$F$_5$ leads to [(C$_6$F$_5$)Au{μ-(2-(PR$_2$)C$_6$H$_4$}$_2$Au(C$_6$F$_5$)] (Scheme 2.3, **B**) which decomposes upon prolonged heating, leading to the dinuclear gold(I) species [(C$_6$F$_5$)Au{μ-(2-(PR$_2$)C$_6$H$_4$C$_6$H$_4$-2-(PR$_2$)Au(C$_6$F$_5$)] (Scheme 2.3, **C**) and the mixed gold(I)–gold(III) [(C$_6$F$_5$)$_2$Au{μ-2-(PR$_2$)C$_6$H$_4$}$_2$Au] (Scheme 2.3, **D**). The crystal structures of complexes [Me$_2$Au{μ-2-(PPh$_2$)C$_6$H$_4$}$_2$Au] and [(C$_6$F$_5$)$_2$Au{μ-2-(PPh$_2$)C$_6$H$_4$}$_2$Au] have been reported [210] with gold–gold distances of 2.8874(4) and 2.931(1), 2.921(1) Å, respectively.

In [Au$_3$(C$_6$F$_5$)$_2$(PN)(PPh$_3$)$_2$] (PNH$_2$ = 2-(diphenylphosphino)aniline) the nitrogen atoms bridge two gold(I) and one gold(III) centers (Figure 2.76a). The gold(I)–gold (III) distances are 3.3219(3) and 3.4961(3) Å [370]. Different examples with tridentate ligands PhP(C$_6$H$_3$-2-S-3-R)$_2$ (R = SiMe$_3$, H) have been described [371, 372]. The Au(I)–Au(III) distances in [Au$_2${PhP(C$_6$H$_3$-2-S-3-SiMe$_3$)$_2$}$_2$] (Figure 2.76b) and [Au$_2${PhP(C$_6$H$_4$-2-S)$_2$}$_2$] are 2.919(1) and 2.978(2) Å, respectively. Both complexes contain a gold(III) center in a square planar geometry and the gold(I) center exhibits a

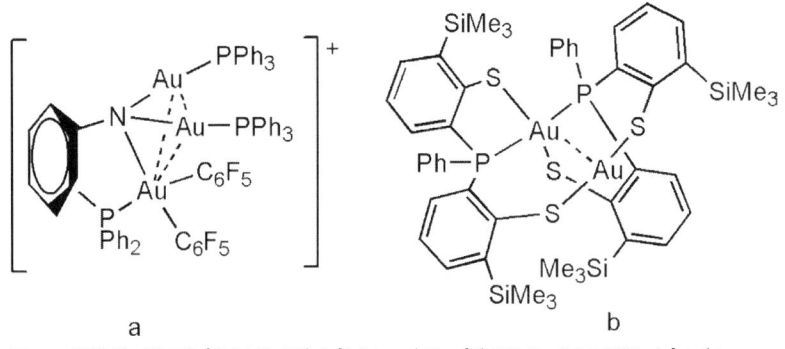

Figure 2.76 [Au(C$_6$F$_5$)$_2${PNH(AuPPh$_3$)$_2$}] (a) and [Au$_2${PhP(C$_6$H$_3$-2-S-3-SiMe$_3$)$_2$}$_2$] (b).

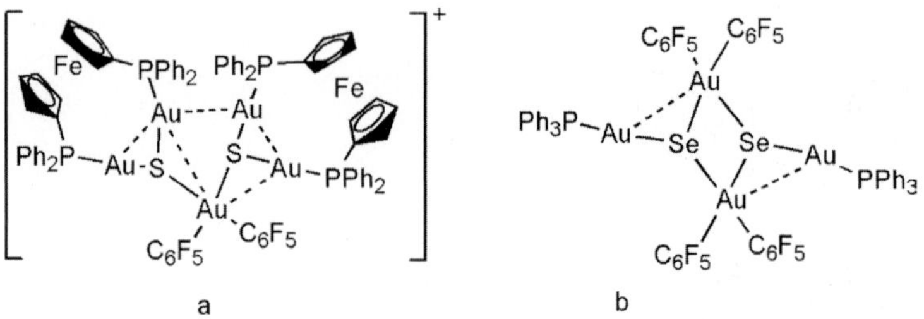

Figure 2.77 [Au$_5$(μ-S)$_2$(C$_6$F$_5$)$_2$(dppf)$_2$] (a) and [Au$_4$(μ-Se)$_2$(C$_6$F$_5$)$_4$(PPh$_3$)$_2$] (b).

linear distorted environment. The former is obtained in very low yield. The crystal structure of [Au$_2${P(C$_6$H$_4$-2-S)$_3$}$_2$] has also been reported. The gold(III)–gold(III) distance is 3.43 Å.

Mixed Au(I)–Au(III) complexes have been described with sulfide and selenido as bridging ligands, some examples are shown in Figure 2.77. Theoretical studies on mixed valence chalcogen species and other compounds support the existence of such gold(I)–gold(III) interactions which are weaker than the gold(I)–gold(I) contacts, involving energies in the range 21–25 kJ mol^{-1} [373–376]. The gold(I)–gold(III) distances in [Au$_5$(C$_6$F$_5$)$_2$(μ-S)$_2$(dppf)$_2$] (Figure 2.77a) are 3.2195(8) and 3.3661 Å. In [E{Au(C$_6$F$_5$)$_3$}{Au$_2$(dppf)}] Au(I)–Au(III) dissimilar distances have been found of 3.5330(8) and 3.8993(8) Å for E = Se and 3.404(1), 3.759(1) Å for E = S. The shorter gold(I)–gold(III) distance is 3.4120(6) Å in [Se$_2${Au(C$_6$F$_5$)$_2$}$_2${Au(PPh$_3$)}$_2$] (Figure 2.77b).

Compound [(C$_6$F$_5$)$_3$Au(SC$_6$H$_4$NH$_2$)(AudppmAu)(SC$_6$H$_4$NH$_2$)Au(C$_6$F$_5$)$_3$] (Equation 2.7, **B**) is obtained by reaction of [Au(C$_6$F$_5$)$_3$(tht)] and [Au$_2$(SC$_6$H$_4$NH$_2$)$_2$(μ-dppm)] (Equation 2.7, **A**). It contains two gold(I) and two gold(III) centers [211]. Coordination of the gold(III) centers results in quenching of the luminescence observed in the precursor **A** which is luminescent in the solid state at 77 K. The gold (I)–gold(I) distance in the precursor **A** is longer than that in the tetranuclear derivative **B**. The asymmetric unit contains two independent molecules, the gold (I)–gold(III) distances are 3.2812(7), 3.3822(7) Å in one and 3.2923(7), 3.4052(7) Å in the other.

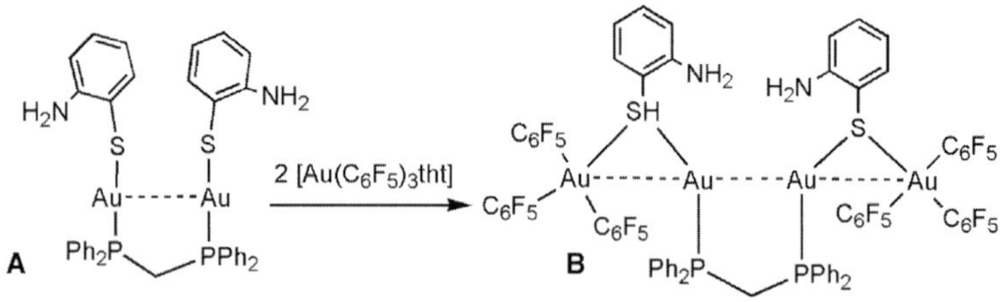

(2.7)

In [Au$_4$(C$_6$F$_5$)$_6$(ER)$_2$(dppf)] (E = S, R = C$_6$F$_5$; E = Se, R = Ph] (Figure 2.78) the gold (I)–gold(III) distances are 3.608(2) Å for the thiolate compound; 3.693 and 3.578 Å in the selenolate species [166,377].

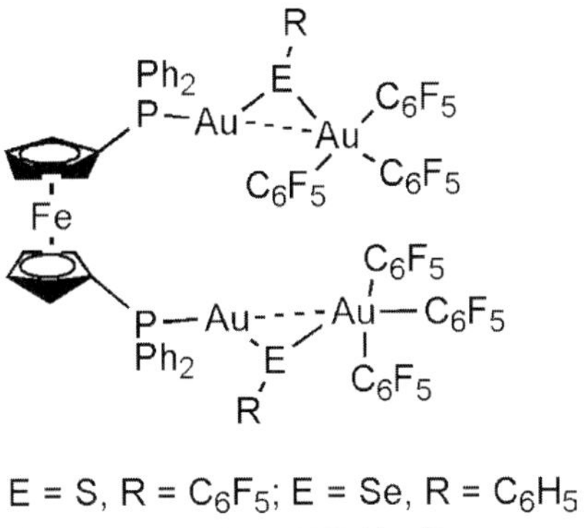

E = S, R = C$_6$F$_5$; E = Se, R = C$_6$H$_5$

Figure 2.78 [Au$_4$(C$_6$F$_5$)$_6$(ER)$_2$(dppf)].

2.4
Concluding Remarks

Eighteen years ago the *aurophilicity* concept was introduced into the chemist's vocabulary. Since then a great number of findings and scientific efforts have revealed its potential. By selecting the ligands, chemists may build a wide variety of structures based on the presence of aurophilic interactions, both at the molecular and supramolecular levels. Some striking findings have been the aggregation of mononuclear dicoordinated gold ions of the same charges or the synthesis of hypervalent species in which one central atom bridges a different number of gold atoms.

In addition to the synthetic advances, the study of the optical properties of these species is a growing field. Many complexes which exhibit these interactions are luminescent. Other interesting points are the induction of rotator phases or the modulation of magnetic properties by designing the appropriate heterometallic complexes. Gold–gold interactions are still a *hot* topic in gold chemistry. This is evidenced in the observation that some complexes which exhibit these interactions, characterized some years ago, are revisited and modified in order to study their properties.

References

1 Schmidbaur, H. (1990) *Gold Bulletin*, **23**, 11.
2 Pyykkö, P. (1997) *Chemical Reviews*, **97**, 597.
3 Schmidbaur, H. (2000) *Gold Bulletin*, **33**, 3.
4 Gimeno, M.C. and Laguna, A. (2003) *Comprehensive Coordination Chemistry II*, 5 (eds S. Patai and Z. Rappoport), Elsevier, New York, p. 911.
5 Schmidbaur, H., Schier, A. (2007) in *Comprehensive Organometallic Chemistry III*, 2 (eds R. Crabtree, M. Mingos and K. Meyer) Elsevier, New York, p. 251.
6 Pyykkö, P. (2004) *Angewandte Chemie (International Edition in English)*, **43**, 4412.
7 Pyykkö, P. *Chemical Society Reviews*, DOI:10.1039/b708613j.

8 Schmidbaur, H., Cronje, S., Djordjevic, B. and Shuster, O. (2005) *Chemical Physics*, **311**, 151.
9 Dorin, D.J. and Toste, F.D. (2007) *Nature*, **446**, 395.
10 Schmidbaur, H. (ed.) (1999) *Gold: Progress in Chemistry, Biochemistry and Technology*, John Wiley & sons.
11 Bachman, R.E., Fioritto, M.S., Fetics, S.K. and Cocker, T.M. (2001) *Journal of the American Chemical Society*, **123**, 5376.
12 Anderson, K.M., Goeta, A.E. and Steed, J.W. (2007) *Inorganic Chemistry*, **46**, 6444.
13 Hashmi, A.S.K. and Hutchings, G.J. (2006) *Angewandte Chemie (International Edition in English)*, **45**, 7896.
14 Papathanasiou, P., Salem, G., Waring, P. and Willis, A.C. (1997) *Journal of The Chemical Society-Dalton Transactions*, 3435.
15 Assefa, Z., McBurnett, B.G., Staples, R.J. and Fackler, J.P., Jr (1995) *Inorganic Chemistry*, **34**, 4965.
16 Stefanescu, D.M., Yuen, H.F., Glueck, D.S., Golen, J.A., Zakharov, L.N., Incarvito, C.D. and Rheingold, A.L. (2003) *Inorganic Chemistry*, **42**, 8891.
17 Attar, S., Bearden, W.H., Alcock, N.W., Alyea, E.C. and Nelson, J.H. (1990) *Inorganic Chemistry*, **29**, 425.
18 Banger, K.K., Banham, R.P., Brisdon, A.K., Cross, W.I., Damant, G., Parsons, S., Prichard, R.G. and Sousa-Pedrares, A. (1999) *Journal of The Chemical Society-Dalton Transactions*, 427.
19 Aucott, S.M., Bhattacharyya, P., Milton, H.L., Slawin, A.M.Z. and Wooliins, J.D. (2003) *New Journal of Chemistry*, **27**, 1466.
20 Monkowius, U., Nogai, S. and Schmidbaur, H. (2003) *Organometallics*, **22**, 145.
21 Nunokawa, K., Onaka, S., Tatematsu, T., Ito, M. and Sakai, J. (2001) *Inorganica Chimica Acta*, **322**, 56.
22 Pyykkö, P., Li, J. and Runeberg, N. (1994) *Chemical Physics Letters*, **218**, 133.
23 Toronto, D.V., Weissbart, B., Tinti, D.S. and Balch, A.L. (1996) *Inorganic Chemistry*, **35**, 2484.

24 Weissbart, B., Toronto, D.V., Balch, A.L. and Tinti, D.S. (1996) *Inorganic Chemistry*, **35**, 2490.
25 Monkowius, U., Nogai, S. and Schmidbaur, H. (2003) *Zeitschrift für Naturforschung. Teil B*, **58**, 751.
26 Onaka, S., Katsukawa, Y., Shiotsuka, M., Kanegawa, O. and Yamashita, M. (2001) *Inorganica Chimica Acta*, **312**, 100.
27 Sladek, A., Angermaier, K. and Schmidbaur, H. (1996) *Chemical Communications*, 1959.
28 Jones, P.G. and Thone, C. (1990) *Chemische Berichte*, **123**, 1975.
29 Canales, S., Crespo, O., Gimeno, M.C., Jones, P.G., Laguna, A. and Romero, P. (2003) *Dalton Transactions*, 4525.
30 Preisenberger, M., Bauer, A., Schier, A. and Schmidbaur, H. (1997) *Journal of The Chemical Society-Dalton Transactions*, 4753.
31 Forward, J.M., Bohmann, D., Fackler, J.P., Jr and Staples, R.J. (1995) *Inorganic Chemistry*, **34**, 6330.
32 Nomiya, K., Noguchi, R., Ohsawa, K. and Tsuda, K. (1998) *Journal of The Chemical Society-Dalton Transactions*, 4101.
33 Kuźmina, L.G., Bagaturyants, A.A., Howard, J.A.K., Grandberg, K.I., Karchava, A.V., Shubina, E.S., Saitkulova, L.N. and Bakhmutova, E.V. (1999) *Journal of Organometallic Chemistry*, **575**, 39.
34 Ovejero, P., Cano, M., Pinilla, E. and Torres, M.R. (2002) *Helvetica Chimica Acta*, **85**, 1686.
35 Mohamed, A.A., Grant, T., Staples, R.J. and Fackler, J.P., Jr (2004) *Inorganica Chimica Acta*, **357**, 1761.
36 Al-Ktaifani, M.M., Hitchcook, P.B. and Nixon, J.F. (2003) *Journal of Organometallic Chemistry*, **665**, 101.
37 Raubenheimer, H.G., Scott, F., Kruger, G.J., Toerien, J.G., Otte, R., van Zyl, W., Taljaard, I., Olivier, P. and Linford, L. (1994) *Journal of The Chemical Society-Dalton Transactions*, 2091.
38 Hollatz, C., Shier, A. and Schmidbaur, H. (1997) *Journal of the American Chemical Society*, **119**, 8115.

39 Hollatz, C., Schier, A., Riede, J. and Schmidbaur, H. (1999) *Journal of The Chemical Society-Dalton Transactions*, 111.

40 Hollatz, C., Schier, A. and Schmidbaur, H. (1999) *Zeitschrift für Naturforschung. Teil B*, **54**, 30.

41 Hao, L., Mansour, M.A., Lachicotte, R.J., Gysling, H.J. and Eisenberg, R. (2000) *Inorganic Chemistry*, **39**, 5520.

42 Römbke, P., Schier, A. and Schmidbaur, H. (2001) *Journal of The Chemical Society-Dalton Transactions*, 2482.

43 Balch, A.L. and Fung, E.Y. (1990) *Inorganic Chemistry*, **29**, 4764.

44 Heathcote, R., Howell, J.A.S., Hennings, N., Cartlidge, D., Cobden, L., Coles, S. and Hursthouse, M. (2007) *Dalton Transactions*, 1309.

45 Wang, H.M.J., Chen, C.Y.L. and Lin, I.J.B. (1999) *Organometallics*, **18**, 1216.

46 Codina, A., Fernández, E.J., Jones, P.G., Laguna, A., López-de-Luzuriaga, J.M., Monge, M., Olmos, M.E., Pérez, J. and Rodriguez, M.A. (2002) *Journal of the American Chemical Society*, **124**, 6781.

47 Hakansson, M., Eriksson, H. and Jagner, S. (2000) *Journal of Organometallic Chemistry*, **602**, 133.

48 Meyer, J., Piana, H., Wagner, H. and Schubert, U. (1990) *Chemische Berichte*, **26**, 791.

49 Stocker, F.B. and Britton, D. (2000) *Acta Crystallographica Section C-Crystal Structure Communications*, **56**, 798.

50 Hussain, M.S., Al-Arfaj, A.R., Akhtar, M.N. and Issab, A.A. (1995) *Polyhedron*, **15**, 2781.

51 Coker, N.L., Bauer, J.A.K. and Elder, R.C. (2004) *Journal of the American Chemical Society*, **126**, 12.

52 Coker, N.L., Bedel, C.E., Krause, J.A. and Elder, R.C. (2006) *Acta Crystallographica Section E-Structure Reports Online*, **62**, m319.

53 Chen, J.-X., Zhang, W.-H., Tang, X.-Y., Ren, Z.-G., Li, H.-X., Zhang, Y. and Lang, J.-P. (2006) *Inorganic Chemistry*, **45**, 7671.

54 Fettouhi, M., Wazeer, M.I.M., Ahmad, S. and Isab, A.A. (2004) *Polyhedron*, **23**, 1.

55 White-Morris, R.L., Olmstead, M.M., Jiang, F. and Balch, A.L. (2002) *Inorganic Chemistry*, **41**, 2313.

56 Conzelmann, W., Hiller, W., Strähle, J. and Sheldrick, G.M. (1984) *Zeitschrift fur Anorganische und Allgemeine Chemie*, **512**, 169.

57 Adams, H.N., Hiller, W. and Strähle, J. (1982) *Zeitschrift fur Anorganische und Allgemeine Chemie*, **485**, 81.

58 Bauer, A., Schier, A. and Schmidbaur, H. (1995) *Journal of The Chemical Society-Dalton Transactions*, 2919.

59 Bauer, A. and Schmidbaur, H. (1996) *Journal of the American Chemical Society*, **118**, 5324.

60 Balch, A.L. (2004) *Gold Bulletin*, **37**, 45.

61 Yip, J.H.K., Feng, R. and Vital, J.J. (1999) *Inorganic Chemistry*, **38**, 3586.

62 Usón, R., Laguna, A., Laguna, M., Jiménez, J., Gómez, M.P., Sainz, A. and Jones, P.G. (1990) *Journal of The Chemical Society-Dalton Transactions*, 3457.

63 Guy, J.J., Jones, P.G., Mays, M.J. and Sheldrick, G.M. (1977) *Journal of The Chemical Society-Dalton Transactions*, 8.

64 Vicente, J., Chicote, M.T., Abrisqueta, M.D., Guerrero, R. and Jones, P.G. (1997) *Angewandte Chemie (International Edition in English)*, **36**, 1203.

65 Schmidbaur, H., Weidenhiller, G., Steigelmann, O. and Muller, G. (1990) *Zeitschrift für Naturforschung. Teil B*, **45**, 747.

66 Ahrland, S., Norén, B. and Oskarson, A. (1985) *Inorganic Chemistry*, **24**, 1330.

67 Ahrland, S., Dreisch, K., Norén, B. and Oskarsson, A. (1993) *Materials Chemistry and Physics*, **35**, 281.

68 Núñez Gaytán, M.-E., Bernès, S., Rodriguez de San Miguel, E. and de Gyves, J. (2004) *Acta Crystallographica Section C-Crystal Structure Communications*, **60**, m414.

69 Jones, P.G. and Ahrens, B. (1998) *Zeitschrift für Naturforschung. Teil B*, 653.

70 Jiang, F., Olmstead, M.M. and Balch, A.L. (2000) *Journal of The Chemical Society-Dalton Transactions*, 4098.

71 White-Morris, R.L., Stender, M., Tinti, D.S., Balch, A.L., Rios, D. and Attar, S. (2003) *Inorganic Chemistry*, **42**, 3237.

72 Ecken, H., Olmstead, M.M., Noll, B.C., Attar, S., Schlyer, B. and Balch, A.L. (1998) *Journal of The Chemical Society-Dalton Transactions*, 3715.

73 Elbjeirami, O., Omary, M.A., Stender, M. and Balch, A.L. (2004) *Dalton Transactions*, 3173.

74 Wilton-Ely, J.D.E.T., Ehlich, H., Schier, A. and Schmidbaur, H. (2001) *Helvetica Chimica Acta*, **84**, 3216.

75 Parker, D., Roy, P.S., Ferguson, G. and Hunt, M.M. (1989) *Inorganica Chimica Acta*, **155**, 227.

76 Mohr, F., Cerrada, E. and Laguna, M. (2006) *Organometallics*, **25**, 644.

77 Kaska, W.C., Mayer, H.A., Elsegood, M.R.J., Horton, P.N., Hursthouse, M.B., Redshaw, C. and Humphrey, S.M. (2004) *Acta Crystallographica Section E-Structure Reports Online*, **60**, m563.

78 Ahrland, S., Aurivillius, B., Dreisch, K., Norén, B. and Oskarson, A. (1992) *Acta Chemica Scandinavica*, **46**, 262.

79 Rawaschdeh-Omary, M.A., Omary, M.A., Patterson, H.H. and Fackler, J.P., Jr (2001) *Journal of the American Chemical Society*, **123**, 11237.

80 Rawaschdeh-Omary, M.A., Omary, M.A. and Patterson, H.H. (2000) *Journal of the American Chemical Society*, **122**, 10371.

81 Stender, M., Olmstead, M.M., Balch, A.L., Rios, D. and Attar, S. (2003) *Journal of The Chemical Society-Dalton Transactions*, 4282.

82 Leznoff, D.B. and Lefebvre, J. (2005) *Gold Bulletin*, **38**, 47.

83 Suarez-Varela, J., Sakiyama, H., Cano, J. and Colacio, E. (2007) *Journal of The Chemical Society-Dalton Transactions*, 249.

84 Zhang, H., Cai, J., Feng, X.-L., Li, T., Li, X.-Y. and Ji, L.-N. (2002) *Inorganic Chemistry Communications*, **5**, 637.

85 Zhou, H.-B., Wang, S.-P., Dong, W., Liu, Z.-Q., Wang, Q.-L., Liao, D.-Z., Jiang, Z.-H., Yan, S.-P. and Cheng, P. (2004) *Inorganic Chemistry*, **43**, 4552.

86 Han, W., Yi, L., Liu, Z.-Q., Gu, W., Yan, S.-P., Cheng, P., Liao, C.-Z. and Jiang, Z.-H. (2004) *European Journal of Inorganic Chemistry*, 2130.

87 Lefebvre, J., Batchelor, R.J. and Leznoff, D.B. (2004) *Journal of the American Chemical Society*, **126**, 16117.

88 Galet, A., Muñoz, M.C., Martínez, V. and Real, J.A. (2004) *Chemical Communications*, 2268.

89 Gabbai, F., Schier, A., Riede, J. and Schmidbaur, H. (1997) *Chemische Berichte*, **130**, 111.

90 McNeal, C.J., Winpenny, R.E.P., Hughes, J.M., Macfarlane, R.D., Pignonelt, L.H., Nelson, L.T.J., Gardner, T.G., Irgens, L.H., Vigh, G. and Fackler, J.P., Jr (1993) *Inorganic Chemistry*, **32**, 5582.

91 Canales, S., Crespo, O., Gimeno, M.C., Jones, P.G. and Laguna, A. (1999) *Chemical Communications*, 679.

92 Canales, F., Gimeno, M.C., Jones, P.G. and Laguna, A. (1994) *Angewandte Chemie (International Edition in English)*, **33**, 769.

93 Schmidbaur, H. and Steigelmann, O. (1992) *Zeitschrift für Naturforschung. Teil B*, **47**, 1721.

94 Schmidbaur, H., Gabbai, F.P., Schier, A. and Riede, J. (1995) *Organometallics*, **14**, 4969.

95 Zeller, E., Beruda, H., Kolb, A., Kissinger, P., Riede, J. and Schmidbaur, H. (1991) *Nature*, **352**, 141.

96 Vicente, J., Chicote, M.T., Guerrero, R., Jones, P.G. and Ramirez de Arellano, M.C. (1997) *Inorganic Chemistry*, **36**, 4438.

97 Sladek, A. and Schmidbaur, H. (1995) *Zeitschrift für Naturforschung. Teil B*, **50**, 859.

98 Slovokhotov, Y.L. and Struchkov, Y.T. (1984) *Journal of Organometallic Chemistry*, **277**, 143.

99 Perevalova, E.G., Smyslova, E.I., Dyadchenko, V.P., Grandberg, K.I. and Nesmeyanov, A.N. (1980) *Akad Nauk SSSR Ser Khim*, 1455.

100 Grohmann, A., Riede, J. and Schmidbaur, H. (1990) *Nature*, **345**, 140.

101 Häberlen, O.D., Schmidbaur, H. and Roesch, N. (1994) *Journal of the American Chemical Society*, **116**, 8241.
102 Pyykkö, P. and Jian, L. (1993) *Inorganic Chemistry*, **32**, 2630.
103 Häberlen, O.D., Chung, S.-C. and Roesch, N. (1994) *International Journal of Quantum Chemistry*, **S28**, 595.
104 Angemaier, K. and Schmidbaur, H. (1995) *Inorganic Chemistry*, **34**, 3120.
105 Zeller, E., Beruda, H. and Schmidbaur, H. (1993) *Chemische Berichte*, **126**, 2033.
106 Schmidbaur, H., Weidenhiller, G. and Steigelmann, O. (1991) *Angewandte Chemie (International Edition in English)*, **30**, 433.
107 Bachman, R.E. and Schmidbaur, H. (1996) *Inorganic Chemistry*, **35**, 1399.
108 Zeller, E. and Schmidbaur, H. (1993) *Journal of The Chemical Society-Dalton Transactions*, 69.
109 Beruda, H., Zeller, E. and Schmidbaur, H. (1993) *Chemische Berichte*, **126**, 2037.
110 Angemaier, K. and Schmidbaur, H. (1994) *Inorganic Chemistry*, **33**, 2069.
111 Nesmeyanov, A.N., Perevalova, E.G., Struchkov, Y.T., Antipin, M.Y., Grandberg, K.I. and Dyadhenko, V.P. (1980) *Journal of Organometallic Chemistry*, **201**, 343.
112 Yang, Yi., Ramamoorthy, V. and Sharp, P.R. (1993) *Inorganic Chemistry*, **32**, 1946.
113 Schmidbaur, H., Kolb, A., Zeller, E., Schier, A. and Beruda, H. (1993) *Zeitschrift fur Anorganische und Allgemeine Chemie*, **619**, 1575.
114 Angemaier, K. and Schmidbaur, H. (1995) *Acta Crystallographica Section C-Crystal Structure Communications*, **51**, 1793.
115 Tripathi, U.M., Schier, A. and Schmidbaur, H. (1998) *Zeitschrift für Naturforschung. Teil B*, **53**, 171.
116 Kuźmina, L.G., Churakov, A.V., Grandberg, K.I. and Howard, J.A.K. (2005) *Journal of Chemical Crystallography*, **35**, 635.
117 Mathieson, T.J., Schier, A. and Schmidbaur, H. (2000) *Zeitschrift für Naturforschung. Teil B*, **55**, 1000.
118 Chung, S.-C., Krüger, S., Schmidbaur, H. and Rösch, N. (1996) *Inorganic Chemistry*, **35**, 5387.
119 Bardají, M., Uznanski, P., Amiens, C., Chaudret, B. and Laguna, A. (2002) *Chemical Communications*, 598.
120 Schmidbaur, H., Hofreiter, S. and Paul, M. (1995) *Nature*, **377**, 503.
121 El-Eltri, M.M. and Scovell, W.M. (1990) *Inorganic Chemistry*, **29**, 480.
122 Jones, P.G. and Thöne, C. (1991) *Chemische Berichte*, **124**, 2725.
123 Canales, F., Gimeno, M.C., Laguna, A. and Jones, P.G. (1996) *Journal of the American Chemical Society*, **118**, 4839.
124 Canales, S., Crespo, O., Gimeno, M.C., Jones, P.G., Laguna, A. and Mendizabal, F. (2000) *Organometallics*, **19**, 4985.
125 Canales, F., Gimeno, M.C., Laguna, A. and Villacampa, M.D. (1996) *Inorganica Chimica Acta*, **244**, 95.
126 Vicente, J., Chicote, M.T., González-Herrero, P., Grünwald, C. and Jones, P.G. (1997) *Organometallics*, **16**, 3381.
127 Shi, J.-C., Kang, B.-S. and Mak, T.C.W. (1997) *Journal of The Chemical Society-Dalton Transactions*, 2171.
128 Angemaier, K. and Schmidbaur, H. (1994) *Chemische Berichte*, **127**, 2387.
129 Albano, V.G., Castellari, C., Femoni, C., Iapalucci, M.C., Longoni, G., Monari, M., Rauccio, M. and Zacchini, S. (1999) *Inorganica Chimica Acta*, **291**, 372.
130 Angemaier, K. and Schmidbaur, H. (1996) *Zeitschrift für Naturforschung. Teil B*, **51**, 879.
131 Lensch, C., Jones, P.G. and Sheldrick, G.M. (1982) *Zeitschrift für Naturforschung. Teil B*, **37**, 944.
132 Fang, H. and Wang, S.-G. (2007) *Journal of Molecular Modeling*, **13**, 255.
133 Jones, P.G., Lensch, C. and Sheldrick, G.M. (1982) *Zeitschrift für Naturforschung. Teil B*, **37**, 141.
134 Yam, V.W.-W., Cheng, E.C.-C. and Zhu, N. (2001) *Angewandte Chemie-International Edition*, **40**, 1763.
135 Yam, V.W.W., Cheng, E.C.C. and Cheung, K.K. (1999) *Angewandte Chemie (International Edition in English)*, **38**, 197.
136 Fenske, D., Langetepe, T., Kappes, M.M., Hampe, O. and Weis, P. (2000)

137 Sevillano, P., Langetepe, T. and Fenske, D. (2003) *Zeitschrift für Anorganische und Allgemeine Chemie*, **629**, 207.
138 Shan, H. and Sharp, P.R. (1996) *Angewandte Chemie (International Edition in English)*, **35**, 635.
139 Li, Z., Loh, Z.-H., Mok, K.F. and Hor, T.S.A. (2000) *Inorganic Chemistry*, **39**, 5299.
140 Li, Z., Mok, K.F. and Hor, T.S.A. (2003) *Journal of Organometallic Chemistry*, **682**, 73.
141 Schmidbaur, H., Hamel, A., Mitzel, N.W., Schier, A. and Nogai, S. (2002) *Proceedings of the National Academy of Sciences of the United States of America*, **99**, 4916.
142 Jones, P.G., Sheldrick, G.M., Usón, R. and Laguna, A. (1980) *Acta Crystallographica Section C-Crystal Structure Communications*, **36**, 1486.
143 Xia, A., James, A.J. and Sharp, P.R. (1999) *Organometallics*, **18**, 451.
144 Ramamoorthy, V. and Sharp, P.R. (1990) *Inorganic Chemistry*, **29**, 3336.
145 Flint, B.W., Yang, Y. and Sharp, P.R. (2000) *Inorganic Chemistry*, **39**, 602.
146 Kolb, A., Bissinger, P. and Schmidbaur, H. (1993) *Zeitschrift für Anorganische und Allgemeine Chemie*, **619**, 1580.
147 Angemaier, K. and Schmidbaur, H. (1995) *Chemische Berichte*, **128**, 817.
148 Lange, P., Beruda, H., Hiller, W. and Schmidbaur, H. (1994) *Zeitschrift für Naturforschung. Teil B*, **49**, 781.
149 Ramamoortly, V., Wu, Z., Yi, Y. and Sharp, P.R. (1992) *Journal of the American Chemical Society*, **114**, 1526.
150 Tripathi, U.M., Schier, A. and Schmidbaur, H. (1998) *Inorganic Chemistry*, **37**, 174.
151 Kuźmina, L.G., Howard, J.A.K., Grandberg, K.I., Alexandrov, G.G. and Kuźmin, V.S. (1999) *Koord Khim (Russ) (Coordination Chemistry)*, **25**, 546.
152 Allan, R.E., Beswick, M.A. and Paver, M.A., Raithby, P.R., Steiner, A. and Wright, D.S. (1996) *Angewandte Chemie (International Edition in English)*, **35**, 208.
153 Bunge, S.D., Just, O. and Rees, W.S., Jr (2000) *Angewandte Chemie (International Edition in English)*, **39**, 3082.
154 Schmidbaur, H., Kolb, A. and Bissinger, P. (1992) *Inorganic Chemistry*, **31**, 4370.
155 Shan, H., Yang, Y., James, A.J. and Sharp, P.R. (1997) *Science* **275**, 1460.
156 Sladek, A., Schneider, W., Angermaier, K., Bauer, A. and Schmidbaur, H. (1996) *Zeitschrift für Naturforschung. Teil B*, **51**, 765.
157 Chen, J., Jiang, T., Wei, G., Mohamed, A.A., Homrighausen, C., Bauer, J.A.K., Bruce, A.E. and Bruce, M.R.M. (1999) *Journal of the American Chemical Society*, **121**, 9225.
158 Mohamed, A.A., Chen, J., Bruce, A.E., Bruce, M.R.M., Bauer, J.A.K. and Hill, D.T. (2003) *Inorganic Chemistry*, **42**, 2203.
159 Wang, S. and Fackler, J.P., Jr (1990) *Inorganic Chemistry*, **29**, 4404.
160 Sladek, A. and Schmidbaur, H. (1995) *Chemische Berichte*, **128**, 907.
161 López-de-Luzuriaga, J.M., Sladek, A., Schneider, W. and Schmidbaur, H. (1997) *Chemische Berichte*, **130**, 641.
162 Jones, P.G. and Thöne, C. (1992) *Zeitschrift für Naturforschung. Teil B*, **47**, 600.
163 Eikens, W., Kienitz, C., Jones, P.G. and Thöne, C. (1994) *Journal of The Chemical Society-Dalton Transactions*, 83.
164 Lang, E.S., Maichle-Mossmer, C. and Strahle, J. (1994) *Zeitschrift für Anorganische und Allgemeine Chemie*, **620**, 1678.
165 Bonasia, D.J., Gindelbert, D.E. and Arnold, J. (1993) *Inorganic Chemistry*, **32**, 5126.
166 Canales, S., Crespo, O., Gimeno, M.C., Jones, P.G. and Laguna, A. (2004) *Inorganic Chemistry*, **43**, 7234.
167 Wilton-Ely, J.D.E.T., Hofreiter, S., Mitzel, N.W. and Schmidbaur, H. (2001) *Zeitschrift für Naturforschung. Teil B*, **56**, 1257.
168 Chen, J., Mohamed, A.A., Abdou, H.B., Bauer, J.A.K., Fackler, J.P., Jr, Bruce, A.E. and Bruce, M.R.M. (2005) *Chemical Communications*, 1575.

169 Perevalova, E.G., Grandberg, K.I., Baukova, T.V., Dyadchenko, V.P., Slowokhotov, Y.L. and Struchkov, Y. (1982) *Koord Khim (Russ)(Coordination Chemistry)*, **8**, 1337.

170 Smyslova, E.I., Perevalova, E.G., Dyadchenko, V.P., Grandberg, K.I., Slovokhotov, Y.L. and Struchkov, Y.T. (1991) *Journal of Organometallic Chemistry*, **215**, 269.

171 Schmidbaur, H., Cronje, S., Djordjevic, B. and Shuster, O. (2005) *Chemical Physics*, **311**, 151.

172 Usón, R., Laguna, A., Fernández, E.J., Mendía, A. and Jones, P.G. (1988) *Journal of Organometallic Chemistry*, **350**, 129.

173 Vicente, J., Chicote, M.T., Cayuelas, J.A., Fernandez-Baeza, J., Jones, P.G., Sheldrick, G.M. and Espinet, P. (1985) *Journal of The Chemical Society-Dalton Transactions*, 1163.

174 Vicente, J., Chicote, M.T., Lagunas, M.C. and Jones, P.G. (1992) *Chemical Communications*, 1730.

175 Vicente, J., Chicote, M.T., Lagunas, M.C., Jones, P.G. and Ahrens, B. (1997) *Inorganic Chemistry*, **36**, 4938.

176 Riera, V., Ruiz, J., Solans, X. and Tauler, E. (1990) *Journal of The Chemical Society-Dalton Transactions*, 1607.

177 Gimeno, M.C., Jones, P.G., Laguna, A., Laguna, M. and Lázaro, I. (1993) *Journal of The Chemical Society-Dalton Transactions*, 2223.

178 Vicente, J., Chicote, M.T., Guerrero, R. and Jones, P.G. (1996) *Journal of the American Chemical Society*, **118**, 699.

179 Vicente, J., Chicote, M.T., Saura-LLamas, I., Jones, P.G., Meyer Base, K. and Erdbrugger, E.F. (1988) *Organometallics*, **7**, 997.

180 Baukova, T.V., Kuźmina, L.G., Oleinikova, N.A., Lemenovskii, D.A. and Blumenfel'd, A.L. (1997) *Journal of Organometallic Chemistry*, **530**, 27.

181 Gimeno, M.C., Laguna, A., Laguna, M. and Sanmartín, F. (1993) *Organometallics*, **12**, 3984.

182 Meyer, E.M., Gambarotta, S., Floriani, C., Chiesi-Villa, A. and Guastini, C. (1989) *Organometallics*, **8**, 1067.

183 Fenske, D. and Simon, F. (1996) *Zeitschrift fur Anorganische und Allgemeine Chemie*, **622**, 45.

184 Zeller, E., Beruda, H., Riede, J. and Schmidbaur, H. (1993) *Inorganic Chemistry*, **32**, 3068.

185 Schmidbaur, H., Zeller, E., Weidenhiller, G., Steigelmann, O. and Beruda, H. (1992) *Inorganic Chemistry*, **31**, 2370.

186 Blumenthal, A., Beruda, H. and Schmidbaur, H. (1993) *Chemical Communications*, 1005.

187 Blanco, M.C., Fernandez, E.J., Olmos, M.E., Pérez, J. and Laguna, A. (2006) *Chemistry – A European Journal*, **12**, 3379.

188 Van Calcar, P.M., Olmstead, M.M. and Balch, A.L. (1997) *Inorganic Chemistry*, **36**, 5231.

189 Schmidbaur, H., Wohlleben, A., Wagner, F., Orama, O. and Huttner, G. (1977) *Chemische Berichte*, **110**, 1748.

190 Bates, P.A. and Waters, J.M. (1985) *Inorganica Chimica Acta*, **98**, 125.

191 Eggelston, D.S., Chodosh, D.F., Girard, G.R. and Hill, D.T. (1985) *Inorganica Chimica Acta*, **108**, 221.

192 Kaim, W., Dogan, A., Klein, A. and Zalis, S. (2005) *Zeitschrift fur Anorganische und Allgemeine Chemie*, **631**, 1355.

193 Schmidbaur, H., Bissinger, P., Lachmann, J. and Steigelmann, O. (1992) *Zeitschrift für Naturforschung. Teil B*, **47**, 1711.

194 Van Calcar, P.M., Olmstead, M.M. and Balch, A.L. (1995) *Journal of The Chemical Society-Dalton Transactions*, 1773.

195 Baker, R.T., Calabrese, J.C. and Wescott, S.A. (1995) *Journal of Organometallic Chemistry*, **498**, 109.

196 Slott, T.L., Wolf, M.O. and Patrick, B.O. (2005) *Inorganic Chemistry*, **44**, 620.

197 Schmidbaur, H., Graf, W. and Müller, G. (1988) *Angewandte Chemie (International Edition in English)*, **27**, 417.

198 Dziwok, K., Lachmann, J., Müller, G., Schmidbaur, H. and Wilkinson, D.L. (1990) *Chemische Berichte*, **123**, 423.

199 Schmidbaur, H., Dziwok, K., Brohmann, A. and Muller, G. (1989) *Chemische Berichte*, **122**, 893.

200 Jones, P.G. (1980) *Acta Crystallographica. Section B-Structural Science*, **36**, 2775.

201 Lange, P., Schier, A. and Schmidbaur, H. (1997) *Zeitschrift für Naturforschung. Teil B*, **52**, 769.

202 Jones, P.G. and Bembenek, E. (1996) *Acta Crystallographica Section C-Crystal Structure Communications*, **52**, 2396.

203 Freytag, M., Ito, S. and Yoshifuji, M. (2006) *Asian Journal of Chemistry*, **1**, 693.

204 Clot, O., Akahori, Y., Moorlag, C., Leznoff, D.B., Wolf, M.O., Batchelor, R.J., Patrick, B.O. and Ishii, M. (2003) *Inorganic Chemistry*, **42**, 2704.

205 Bardaji, M., de la Cruz, M.T., Jones, P.G. and Laguna, A. (2005) *Inorganica Chimica Acta*, **358**, 1365.

206 Eggelston, D.S., McArdke, J.V. and Zuber, G.E. (1987) *Journal of The Chemical Society-Dalton Transactions*, 677.

207 Schmidbaur, H., Paschalidis, C., Steigelmann, O. and Mueller, G. (1989) *Chemische Berichte*, **122**, 1851.

208 Yeo, W.-C., Vittal, J.J., Koh, L.L., Tan, G.-K. and Leung, P.-H. (2004) *Organometallics*, **23**, 3474.

209 Pintado-Alba, A., De la Riva, H., Nieuwhuyzen, M., Bautista, D., Raithby, P.R., Sparkes, H.A., Teat, S.J., López-de-Luzuriaga, J.M. and Lagunas, M.C. (2004) *Dalton Transactions*, 3459.

210 Bennett, M.A., Hockless, D.C.R., Rae, A.D., Welling, L.L. and Willis, A.C. (2001) *Organometallics*, **20**, 79.

211 Bardají, M., Calhorda, M.J., Costa, P.J., Jones, P.G., Laguna, A., Pérez, M.R. and Villacampa, M.D. (2006) *Inorganic Chemistry*, **45**, 1059.

212 Wilton-Ely, J.D.E.T., Schier, A., Mitzel, N.W., Nogal, S. and Schmidbaur, H. (2002) *Journal of Organometallic Chemistry*, **643**, 313.

213 Smyth, D.M. and Tienkink, E.R.T. (2002) *Zeitschrift für Kristallographie*, **217**, 363.

214 Piana, H., Wagner, H. and Schubert, U. (1991) *Chemische Berichte*, **124**, 63.

215 Yip, S.-K., Lam, W.H., Zhu, N. and Yam, V.W.-W. (2006) *Inorganica Chimica Acta*, **359**, 3639.

216 Ho, S.Y., Cheng, E.C.-C., Tiekink, E.R.T. and Yam, V.W.-W. (2006) *Inorganic Chemistry*, **45**, 8165.

217 Maspero, A., Kani, I., Mohamed, A.A., Omary, M.A., Staples, R.J. and Facler, J.P., Jr (2003) *Inorganic Chemistry*, **42**, 5311.

218 Onaka, M.Y.S., Yamauchi, R., Ozeki, T., Ito, M., Sunahara, T., Sugiura, Y., Shiotsuka, M., Nunokawa, K., Horibe, M., Okazaki, J., Lida, A., Chiba, H., Imai, H. and Sako, K. (2005) *Journal of Organometallic Chemistry*, **690**, 57.

219 Nasmayanaswamy, R., Young, M.A., Parkhurst, E., Ouellette, M., Kerr, M.E., Ho, D.M., Elder, R.C., Bruce, A.E. and Bruce, M.R.M. (1993) *Inorganic Chemistry*, **32**, 2506.

220 Onaka, S., Katsukawa, Y. and Yamashita, M. (1998) *Chemistry Letters*, 525.

221 Smyth, D.R., Hester, J., Young, V.G., Jr and Tienkink, E.R.T. (2002) *CrystEngComm*, **4**, 517.

222 Low, P.M.N., Zhang, Z.-Y., Mak, T.C.W. and Hor, T.S.A. (1997) *Journal of Organometallic Chemistry*, **539**, 45.

223 Shieh, S.-J., Hong, X., Peng, S.-M. and Che, C.-M. (1994) *Journal of The Chemical Society-Dalton Transactions*, 3067.

224 Gimeno, M.C., Jones, P.G., Laguna, A., Laguna, M. and Terroba, R. (1994) *Inorganic Chemistry*, **33**, 3932.

225 Nakamoto, M., Kojiman, H., Paul, M., Hiller, W. and Schmidbaur, H. (1988) *Zeitschrift für Anorganische und Allgemeine Chemie*, **619**, 1341.

226 Dávila, R.M., Staples, R.J., Elduque, A., Harlass, M.M., Kyle, L. and Fackler, J.P., Jr (1994) *Inorganic Chemistry*, **33**, 5940.

227 Preisenberg, M., Pyykkö, P., Schier, A. and Schmidbaur, H. (1999) *Inorganic Chemistry*, **38**, 5870.

228 Vicente, J., González-Herrero, P., García-Sanchez, Y., Jones, P.G. and Bardají, M. (2004) *Inorganic Chemistry*, **43**, 7516.

229 Amoroso, A.J., Blake, A.J., Danks, J.P., Fenske, D. and Schroder, M. (1999) *New Journal of Chemistry*, **23**, 345.
230 Canales, S., Crespo, O., Gimeno, M.C., Jones, P.G., Laguna, A., Silvestru, A. and Silvestru, C. (2003) *Inorganica Chimica Acta*, **347**, 16.
231 Hanna, S.D., Khan, S.I. and Zink, J.I. (1996) *Inorganic Chemistry*, **35**, 5813.
232 MacDonald, M.A., Puddephatt, R.J. and Yap, G.P.A. (2000) *Organometallics*, **19**, 2194.
233 Vicente, J., Chicote, M.-T., Alvarez-Falcon, M.M. and Bautista, D. (2004) *Organometallics*, **23**, 5707.
234 Irwin, M.J., Vital, J.J. and Puddephatt, R.J. (1997) *Organometallics*, **16**, 3541.
235 Siemeling, U., Rother, D., Bruhn, C., Fink, H., Weidner, T., Trager, F., Rothenberger, A., Fenske, D., Priebe, A., Maurer, J. and Winter, R. (2005) *Journal of the American Chemical Society*, **127**, 1102.
236 Yau, J., Mingos, D.M.P., Menzer, S. and Williams, D.J. (1995) *Journal of the Chemical Society-Dalton Transactions*, 2575.
237 Barranco, E.M., Crespo, O., Gimeno, M.C., Jones, P.G. and Laguna, A. (2004) *European Journal of Inorganic Chemistry*, 4820.
238 Barranco, E.M., Gimeno, M.C., Jones, P.G., Laguna, A. and Villacampa, M.D. (1999) *Inorganic Chemistry*, **38**, 702.
239 Barreiro, E., Casas, J.S., Couce, M.D., Sanchez, A., Sordo, J., Varela, J.M. and Vazquez-López, E.M. (2003) *Dalton Transactions*, 4754.
240 Harwell, D.E., Mortimer, M.D., Knobler, C.B., Anet, F.A. and Hawthorne, M.F. (1996) *Journal of the American Chemical Society*, **118**, 2679.
241 Feffery, J.C., Jelliss, P.A. and Stone, F.G.A. (1994) *Journal of the Chemical Society-Dalton Transactions*, 25.
242 Cao, Q.-Y., Yin, B. and Liu, J.-H. (2006) *Acta Crystallographica Section E-Structure Reports Online*, **62**, m2730.
243 Wang, J.-C., Khan, M.N.I. and Facker, J.P., Jr (1989) *Acta Crystallographica Section C-Crystal Structure Communications*, **45**, 1482.
244 Khan, M.N.I., King, C., Heinrich, D.D., Fackler, J.P., Jr and Porter, L.C. (1989) *Inorganic Chemistry*, **28**, 2150.
245 Schruh, W., Kopacka, H., Wurst, K. and Peringer, P. (2001) *Chemical Communications*, 2186.
246 Beming, D.E., Katti, K.V., Bames, C.L., Volkert, W.A. and Ketring, A.R. (1997) *Inorganic Chemistry*, **36**, 2765.
247 Shaefer, W.P., Marsh, R.E., McCleskey, T.M. and Gray, H.B. (1991) *Acta Crystallographica Section C-Crystal Structure Communications*, **47**, 2553.
248 Pons, A., Rossell, O., Seco, M., Solans, X. and Font-Bardia, M. (1996) *Journal of Organometallic Chemistry*, **514**, 177.
249 Jaw, H.-R.C., Savas, M.M., Rogers, R.D. and Mason, W.R. (1989) *Inorganic Chemistry*, **28**, 1028.
250 Che, C.-M., Kwong, H.-L., Poon, C.-K. and Yam, V.W.-W. (1990) *Journal of the Chemical Society-Dalton Transactions*, 3215.
251 Leung, K.H., Phillips, D.L. Mao, Z., Che, C.-M., Miskowski, V.M. and Chan, C.-K. (2002) *Inorganic Chemistry*, **41**, 2002.
252 Fu, W.-F., Chan, K.-C., Miskowski, V.M. and Che, C.-M. (1999) *Angewandte Chemie (International Edition in English)*, **38**, 2783.
253 Fu, W.-F., Chan, K.-C., Cheung, K.-K. and Che, C.-M. (2001) *Chemistry – A European Journal*, **7**, 4656.
254 Ma, Y., Che, C.-M., Chao, H.-Y., Zhan, X., Han, W. and Shen, J. (1999) *Advanced Materials*, **11**, 852.
255 Shain, J. and Fackler, J.P., Jr (1987) *Inorganica Chimica Acta*, **131**, 157.
256 Liou, L.-S., Liu, C.-P. and Wang, J.-C. (1994) *Acta Crystallographica Section C-Crystal Structure Communications*, **50**, 538.
257 Schmidbaur, H., Herr, R., Muller, G. and Riede, J. (1985) *Organometallics*, **4**, 1208.
258 Schmidbaur, H., Pollok, T., Herr, R., Wagner, F.E., Bau, R., Riede, J. and Muller, G. (1986) *Organometallics*, **5**, 566.
259 Wang, J.-C. and Liu, L.-K. (1994) *Acta Crystallographica Section C-Crystal Structure Communications*, **50**, 704.

260 Constable, E.C., Housecroft, C.E., Neuburger, M., Schaffner, S. and Shardlow, E.J. (2007) *Acta Crystallographica Section E-Structure Reports Online*, **63**, m1698.
261 Paul, M. and Schmidbaur, H. (1996) *Chemische Berichte*, **129**, 77.
262 Brandys, M.C. and Puddephatt, R.J. (2001) *Chemical Communications*, 1280.
263 Che, C.-M., Yip, H.-K., Lo, W.-C. and Peng, S.-M. (1994) *Polyhedron*, **13**, 887.
264 Hofmann, M., Heinemann, F.W. and Zenneck, U. (2002) *Journal of Organometallic Chemistry*, **643**, 357.
265 Ho, S.Y. and Tiekink, E.R.T. (2002) *Zeitschrift für Kristallographie*, **217**, 589.
266 Hesse, R. and Jennische, P. (1972) *Acta Chemica Scandinavica* **26**, 3855.
267 Heinrich, D.D., Wang, J.-C. and Fackler, J.P., Jr (1990) *Acta Crystallographica Section C-Crystal Structure Communications*, **46**, 1444.
268 Mansour, M.A., Connick, W.B., Lachicotte, R.J., Gysling, H.J. and Eisenberg, R. (1998) *Journal of the American Chemical Society*, **120**, 1329.
269 Tzeng, B.-C., Liu, W.-H., Liao, J.-H., Lee, G.-H. and Peng, S.-M. (2004) *Crystal Growth and Design*, **4**, 573.
270 Bishop, P., Marsh, P., Brisdon, B.J. and Mahon, M.F. (1998) *Journal of The Chemical Society-Dalton Transactions*, 675.
271 Hong, M.-C., Lei, X.-J., Huang, Z.-Y., Kang, B.-S., Jiang, F.-L. and Liu, H.-Q. (1993) *Chinese Science Bulletin*, **38**, 912.
272 Mohamed, A.A., Kani, I., Ramirez, A.O. and Fackler, J.P., Jr (2004) *Inorganic Chemistry*, **43**, 3833.
273 Ribas, X., Mas-Torrent, M., Perez-Benitez, A., Dias, J.C., Alves, H., Lopes, E.B., Henriques, R.T., Molins, E., Santos, I.C., Wurst, K., Foury-Leylekian, P., Almeida, M., Veciana, J. and Rovira, C. (2005) *Advanced Functional Materials*, **15**, 1023.
274 Khan, M.N.I., Wang, S. and Fackler, J.P., Jr (1989) *Inorganic Chemistry*, **28**, 3579.
275 Staples, R.J., Fackler, J.P., Jr and Assefa, Z. (1995) *Zeitschrift für Kristallographie*, **210**, 379.
276 van Zyl, W.E., López-de-Luzuriaga, J.M., Mohamed, A.A., Staples, R.J. and Fackler, J.P., Jr (2002) *Inorganic Chemistry*, **41**, 4579.
277 van Zyl, W.E., López-de-Luzuriaga, J.M., Fackler, J.P., Jr and Staples, R.J. (2001) *Canadian Journal of Chemistry*, **79**, 896.
278 Lee, Y.-A., McGarrah, J.E., Lachicotte, R.J. and Eisenberg, R. (2002) *Journal of the American Chemical Society*, **124**, 10662.
279 Lawton, S.L., Rohrbaugh, W.J. and KoKotailo, G.T. (1972) *Inorganic Chemistry*, **11**, 2227.
280 Vicente, J., Chicote, M.T., Gonzalez-Herrero, P. and Jones, P.G. (1995) *Chemical Communications*, 745.
281 Dietzsch, W., Franke, A., Hoyer, E., Gruss, D., Hummel, H.-U. and Otto, P. (1992) *Zeitschrift für Anorganische und Allgemeine Chemie*, **611**, 81.
282 Konno, T., Hattori, M., Yoshimura, T. and Hirotsu, M. (2002) *Chemistry Letters*, 230.
283 Shin, R.Y., Tan, G.K., Koh, L.L., Vital, J.J., Goh, L.Y. and Webster, R.D. (2005) *Organometallics*, **24**, 539.
284 Vicente, J., Chicote, M.T., Huertas, S., Jones, P.G. and Fischer, A.K. (2001) *Inorganic Chemistry*, **40**, 6193.
285 Robertson, S.D., Slawin, A.M.Z. and Woollins, J.D. (2007) *European Journal of Inorganic Chemistry*, 247.
286 Fackler, J.P., Jr Staples, R.J. and Asefa, Z. (1994) *Chemical Communications*, 431.
287 Piovesana, O. and Zanazzi, P.F. (1980) *Angewandte Chemie (International Edition in English)*, **19**, 561.
288 Abdou, H.E., Mohamed, A.A. and Fackler, J.P., Jr (2005) *Inorganic Chemistry*, **44**, 166.
289 Beck, J. and Strahle, J. (1986) *Angewandte Chemie (International Edition in English)*, **25**, 95.
290 Mohamed, A.A., Abdou, H.E., Irwin, M.D., López-de-Luzuriaga, J.M. and Fackler, J.P., Jr (2003) *Journal of Cluster Science*, **14**, 253.
291 Yang, G. and Raptis, R.G. (2003) *Inorganica Chimica Acta*, **352**, 98.
292 Mohamed, A.A., Abdou, H.E. and Fackler, J.P., Jr (2006) *Inorganic Chemistry*, **45**, 11.

293 Colacio, E., Crespo, O., Cuesta, R., Kivekas, R. and Laguna, A. (2004) *Journal of Inorganic Biochemistry*, **98**, 595.

294 Colacio, E., Cuesta, R., Gutierrez-Zorrilla, J.M., Luque, A., Román, P., Giraldi, T. and Taylor, M.R. (1996) *Inorganic Chemistry*, **35**, 4232.

295 Abdou, H.E., Mohamed, A.A. and Fackler, J.P., Jr (2007) *Inorganic Chemistry*, **46**, 141.

296 Mohamed, A.A., López-de-Luzuriaga, J.M. and Fackler, J.P., Jr (2003) *Journal of Cluster Science*, **14**, 61.

297 Schmidbaur, H., Mandl, J.E., Richter, W., Bejenke, V., Frank, A. and Huttner, G. (1977) *Chemische Berichte*, **110**, 2236.

298 Heinrich, D.D., Staples, R.J. and Fackler, J.P., Jr (1995) *Inorganica Chimica Acta*, **229**, 61.

299 Bardají, M., Jones, P.G., Laguna, A. and Laguna, M. (1995) *Organometallics*, **14**, 1310.

300 Bardají, M., Connelly, N.G., Gimeno, M.C., Jones, P.G., Laguna, A. and Laguna, M. (1995) *Journal of the Chemical Society-Dalton Transactions*, 2245.

301 Hunks, W.J., MacDonald, M.-A., Jennings, M.C. and Puddepahtt, R.J. (2000) *Organometallics*, **19**, 5063.

302 Irwin, M.J., Rendina, L.M., Vital, J.J. and Puddephatt, R.J. (1996) *Chemical Communications*, 1281.

303 McArdle, C.P., Vittal, J.J. and Puddephatt, R.J. (2000) *Angewandte Chemie-International Edition*, **39**, 3819.

304 McArdle, C.P., Jennings, M.C., Vittal, J.J. and Puddephatt, R.J. (2001) *Chemistry – A European Journal*, **7**, 3572.

305 Habemehl, N.C., Jennings, M.C., McArdle, C.P., Mohr, F. and Puddephatt, R.J. (2005) *Organometallics*, **24**, 5004.

306 Habemehl, N.C., Eisler, D.J., Kirby, C.W., Yue, N.L.-S. and Puddephatt, R.J. (2006) *Organometallics*, **25**, 2921.

307 Catalano, V.J. and Moore, A.L. (2005) *Inorganic Chemistry*, **44**, 6558.

308 Papasergio, R.I., Raston, C.L. and White, A.H. (1987) *Journal of The Chemical Society-Dalton Transactions*, 3085.

309 Jacob, K., Voigt, F., Merzweller, K. and Pietzsch, C. (1997) *Journal of Organometallic Chemistry*, **545**, 421.

310 Desmet, M., Raubenheimer, H.G. and Kruger, G.J. (1997) *Organometallics*, **16**, 3324.

311 Hao, L., Lachicotte, R.J., Gysling, H.J. and Eisenberg, R. (1999) *Inorganic Chemistry*, **38**, 4616.

312 Jones, P.G. and Friendrichs, S. (1999) *Chemical Communications*, 1365.

313 Lee, Y. and Eisenberg, R. (2003) *Journal of the American Chemical Society*, **125**, 7778.

314 Fernández, E.J., López-de-Luzuriaga, J.M., Monge, M., Rodriguez, M.A., Crespo, O., Gimeno, M.C., Laguna, A. and Jones, P.G. (1998) *Inorganic Chemistry*, **37**, 6002.

315 Bennett, M.A., Bhargava, S.K., Griffiths, K.D., Robertson, G.B., Wickramasinghe, W.A. and Willis, A.C. (1987) *Angewandte Chemie (International Edition in English)*, **26**, 258.

316 Kitadai, K., Takahashi, M., Takeda, M., Bhargava, S.L., Privér, S.H. and Bennet, M.A. (2006) *Dalton Transactions*, 2560.

317 Bhargava, S.K., Mohr, F., Bennett, M.A., Welling, L.L. and Willis, A.C. (2000) *Organometallics*, **19**, 5628.

318 Tang, S.-S., Cheng, C.-P., Lin, I.J.B., Liou, L.-S. and Wang, J.-C. (1997) *Inorganic Chemistry*, **36**, 2294.

319 Davila, R.M., Elduque, A., Staples, R.J., Harlass, M. and Fackler, J.P., Jr (1994) *Inorganica Chimica Acta*, **217**, 45.

320 Cerrada, E., Jones, P.G., Laguna, A. and Laguna, M. (1996) *Inorganica Chimica Acta*, **249**, 163.

321 Cerrada, E., Laguna, A., Laguna, M. and Jones, P.G. (1994) *Journal of The Chemical Society-Dalton Transactions*, 1325.

322 Lin, I.J.B., Liu, C.W., Liu, L.-K. and Wen, Y.-S. (1992) *Organometallics*, **11**, 1447.

323 Bensh, W., Prelati, M. and Ludwig, W. (1986) *Chemical Communications*, 1762.

324 Bojan, V.R., Fernández, E.J., Laguna, A., López-de-Luzuriaga, J.M., Monge, M., Olmos, M.E. and Silvestru, C. (2005)

Journal of the American Chemical Society, **127**, 11564.

325 Kui, S.C.F., Huang, J.-S., Sun, R.W.-Y., Zhu, N. and Che, C.-M. (2006) *Angewandte Chemie (International Edition in English)*, **45**, 4663.

326 Canales, S., Crespo, O., Gimeno, M.C., Jones, P.G. and Laguna, A. (2000) *Journal of Organometallic Chemistry*, **613**, 50.

327 Canales, S., Crespo, O., Gimeno, M.C., Jones, P.G. and Laguna, A. (1995) *Journal of the Chemical Society-Dalton Transactions*, 3563.

328 Crespo, O., Gimeno, M.C., Jones, P.G. and Laguna, A. (1993) *Chemical Communications*, 1696.

329 Crespo, O., Gimeno, M.C., Jones, P.G., Ahrens, B. and Laguna, A. (1997) *Inorganic Chemistry*, **36**, 495.

330 Bennett, M.A., Welling, L.L. and Willis, A.C. (1997) *Inorganic Chemistry*, **36**, 5670.

331 Chen, Y.-D., Zhang, L.-Y., Qin, Y.-H. and Chen, Z.-N. (2005) *Inorganic Chemistry*, **44**, 6456.

332 Tzeng, B.-C., Che, C.-M. and Peng, S.-M. (1996) *Journal of the Chemical Society-Dalton Transactions*, 1769.

333 Schuerman, J.A., Froncek, F.R. and Selbin, J. (1986) *Journal of the American Chemical Society*, **108**, 336.

334 Raptis, R.G., Murray, H.H., III and Fackler, J.P., Jr (1987) *Chemical Communications*, 737.

335 Yu, S.Y., Zhang, Z.-X., Cheng, E.C.-C., Li, Y.Z., Yam, V.W.-W., Huang, H.-P. and Zhang, R. (2005) *Journal of the American Chemical Society*, **127**, 17994.

336 Lorenz, I.-P., Limmert, M., Mayer, P., Piotrowski, H., Langhals, H., Poppe, M. and Polborm, K. (2002) *Chemistry – A European Journal*, **8**, 4047.

337 Olmstead, M.M., Jiang, F., Attar, S. and Balch, A.L. (2001) *Journal of the American Chemical Society*, **123**, 3260.

338 Burini, A., Bravi, R., Facker, J.P., Jr, Galassi, R., Grant, T.A., Omary, M.A., Rawashdeh-Omary, M., Peitroni, B.R. and Staples, R.J. (2000) *Inorganic Chemistry*, **39**, 3158.

339 Mohamed, A.A., Rawashdeh-Omary, M.A., Omary, M.A. and Fackler, J.P., Jr (2005) *Dalton Transactions*, 2597.

340 Hayashi, A., Olmstead, M.M., Attar, S. and Balch, A.L. (2002) *Journal of the American Chemical Society*, **124**, 5791.

341 Ahmed, A.A., Galassi, R., Papa, F., Burini, A. and Fackler, J.P., Jr (2006) *Inorganic Chemistry*, **45**, 7770.

342 Balch, A.L., Catalano, V.J., Noll, B.C. and Olmstead, M.M. (1990) *Journal of the American Chemical Society*, **112**, 7558.

343 Balch, A.L., Fung, E.Y. and Olmstead, M.M. (1990) *Journal of the American Chemical Society*, **112**, 5181.

344 Yam, V.W.-W., Lai, T.-F. and Che, C.-M. (1990) *Journal of the Chemical Society-Dalton Transactions*, 3747.

345 Kosaka, Y., Shinozaki, Y., Tsutsumi, Y., Kaburagi, Y., Yamamoto, Y., Sunada, Y. and Tatsumi, K.T. (2003) *Journal of Organometallic Chemistry*, **671**, 8.

346 Bardají, M., Laguna, A., Orera, V.M. and Villacampa, M.D. (1998) *Inorganic Chemistry*, **37**, 5125.

347 Xiao, H., Weng, Y.-X., Wong, W.-T., Mak, T.C.W. and Che, C.-M. (1997) *Journal of the Chemical Society-Dalton Transactions*, 221.

348 Li, D., Che, C.-M., Peng, S.-M., Liu, S.-T., Zhou, Z.-Y. and Mak, T.C.W. (1993) *Journal of the Chemical Society-Dalton Transactions*, 189.

349 Tanase, T., Masuda, K., Matsuo, J., Hamaguchi, M., Begum, R.A. and Yano, S. (2000) *Inorganica Chimica Acta*, **299**, 91.

350 Bardají, M., Laguna, A., Jones, P.G. and Fischer, A.K. (2000) *Inorganic Chemistry*, **39**, 3560.

351 Stutzer, A., Bissinger, P. and Schmidbaur, H. (1992) *Chemische Berichte*, **125**, 367.

352 Zank, J., Shrier, A. and Schmidbaur, H. (1998) *Journal of the Chemical Society-Dalton Transactions*, 323.

353 Zank, J., Shrier, A. and Schmidbaur, H. (1999) *Journal of the Chemical Society-Dalton Transactions*, 415.

354 Che, C.-M., Yip, H.-K., Yam, V.W.-W., Cheung, P.-P., Lai, T.-F., Shieh, S.-J. and

Peng, S.-M. (1992) *Journal of The Chemical Society-Dalton Transactions*, 427.

355 Sutzer, A., Bissinger, P. and Schmidbaur, H. (1992) *Zeitschrift für Naturforschung. Teil B*, **47**, 1261.

356 Zhou, Y. and Chen, W. (2007) *Organometallics*, **26**, 2742.

357 Cerrada, E., Díaz, C., Díaz, M.C., Hurtsthouse, M.B., Laguna, M. and Light, M.E. (2002) *Journal of the Chemical Society-Dalton Transactions*, 1104.

358 Tzeng, B.-C., Li, D., Peng, S.-M. and Che, C.-M. (1993) *Journal of the Chemical Society-Dalton Transactions*, 2365.

359 Tzeng, B.-C., Cheung, K.-K. and Che, C.-M. (1996) *Chemical Communications*, 1681.

360 Hunks, W.J., Jennings, M.C. and Puddephatt, R.J. (1999) *Inorganic Chemistry*, **38**, 5930.

361 Tzeng, B.-C., Che, C.-M. and Peng, S.-M. (1997) *Chemical Communications*, 1771.

362 Nunokawa, T.S.K. and Onaka, S. (2004) *Chemistry Letters*, **33**, 1300.

363 Irwin, M.J., Manojlovíc-Muir, L., Muir, K.W., Puddephatt, R.J. and Yufit, D.S. (1997) *Chemical Communications*, 219.

364 Hollins, L.S. and Lippard, S.J. (1983) *Journal of the American Chemical Society*, **105**, 4293.

365 Cao, L., Jennings, M.C. and Puddephatt, R.J. (2007) *Inorganic Chemistry*, **46**, 1361.

366 Manzany, A.M. and Fackler, J.P., Jr (1984) *Journal of the American Chemical Society*, **106**, 801.

367 Raptis, R.G., Porter, L.C., Emrich, R.J., Murray, H.H. and Fackler, J.P., Jr (1990) *Inorganic Chemistry*, **29**, 4408.

368 Fackler, J.P., Jr and Trzcinska-Bancroft, B. (1985) *Organometallics*, **4**, 1891.

369 Schmidbaur, C.H.H., Reber, G. and Müller, G. (1987) *Angewandte Chemie (International Edition in English)*, **26**, 1146.

370 Fernández, E.J., Gil, M., Olmos, M.E., Crespo, O., Laguna, A. and Jones, P.G. (2001) *Inorganic Chemistry*, **40**, 3018.

371 Ortner, K., Hilditch, L., Dilworth, J.R. and Abram, U. (1998) *Inorganic Chemistry Communications*, **1**, 469.

372 Ortner, K., Hilditch, L., Zheng, Y., Dilworth, J.R. and Abram, U. (2000) *Inorganic Chemistry*, **39**, 2801.

373 Calhorda, M.J., Canales, F., Gimeno, M.C., Jiménez, J., Jones, P.G., Laguna, A. and Veiros, L.F. (1997) *Organometallics*, **16**, 3837.

374 Canales, S., Crespo, O., Gimeno, M.C., Jones, P.G., Laguna, A. and Mendizabal, F. (2001) *Organometallics*, **20**, 4812.

375 Mendizabal, F. and Pyykkö, P. (2004) *Physical Chemistry Chemical Physics*, **6**, 900.

376 Mendizabal, F., Zapata-Torres, G. and Olea-Azar, C. (2003) *Chemical Physics Letters*, **382**, 92.

377 Crespo, O., Canales, F., Gimeno, M.C., Jones, P.G. and Laguna, A. (1999) *Organometallics*, **18**, 3142.

3
Gold Nanomaterials

Eduardo J. Fernández and Miguel Monge

3.1
Introduction

The strong development of activities related to the nanoscience and nanotechnology of gold has led to a great number of publications in this research field. In this chapter, we will focus on the synthesis of cluster aggregates displaying gold–gold bonds and the formation of gold nanoparticles. Special consideration will be given to the chemical bottom-up approach for the synthesis of gold nanomaterials using different methods.

Section 3.2 describes the synthesis and structure of small gold clusters taking into account the number of gold atoms, the reducing agents employed and the stabilizing ligands used, such as phosphines, arsines, thiols, and so on. Section 3.3 discusses the synthesis and properties of large size gold clusters such as the Au_{55} cluster series. Finally, in Section 3.4, we describe the synthesis of gold nanoparticles and the development of new nanomaterials based on gold nanoparticles using bottom-up chemical methods and give a short summary of their properties and applications.

3.2
Molecular Gold Clusters

Herein we describe the synthesis and characterization of small-size homonuclear gold cluster compounds with a nuclearity ranging from 3 to 39 gold atoms. These types of clusters are usually characterized through definitive structural techniques such as single-crystal X-ray diffraction, enabling in-depth study of the structural arrangements found in the solid state.

This section is divided into two main subsections. The first describes the synthesis and characterization of phosphine-stabilized gold clusters. The second focuses on the synthesis and characterization of thiol-stabilized gold clusters and the study of other stabilizers such as arsines or boranes are described.

Modern Supramolecular Gold Chemistry: Gold-Metal Interactions and Applications.
Edited by Antonio Laguna
Copyright © 2008 WILEY-VCH Verlag GmbH & Co. KGaA, Weinheim
ISBN: 978-3-527-32029-5

3.2.1
Synthesis and Structural Characterization of Phosphine-Stabilized Gold Clusters

The first tetranuclear gold clusters were reported in the 1980s, namely $[Au_4(PPh_3)_4(\mu\text{-}I)_2]$ (Au–Au distances from 2.649(1) to 2.828(1) Å) [1] and $[Au_4(dppm)_3(\mu_3\text{-}I)I]$ (Au-Au distances from 2.724(1) to 2.947(1) Å) [2]. The first cluster was prepared through a degradation reaction of the nonanuclear cluster $[Au_9(PPh_3)_8]^{3+}$, while the second was obtained by reaction of the already mentioned $[Au_4(PPh_3)_4(\mu\text{-}I)_2]$ cluster and one equivalent of dppm ligand. Both structures display distorted tetrahedral cores of gold atoms with the iodine atoms bridging two different gold centers in the first case and three gold centers in the second (Scheme 3.1).

In the 1990s, tetranuclear gold clusters $[Au_4(PR_3)_4]BF_4$ (R = ^{t}Bu [3] and ($C_6H_3Me\text{-}1,3,5$) [4]) were reported. Both clusters are synthesized from the corresponding oxonium salts $[O\{Au(PR_3)\}_3]BF_4$ and $[HO\{Au(PR_3)\}_2]BF_4$ (R = ^{t}Bu and ($C_6H_3Me\text{-}1,3,5$)). In both clusters, X-ray diffraction analysis showed that gold atoms form a regular tetrahedron and each phosphine ligand is coordinated to one gold center (Scheme 3.2).

The latest examples of tetranuclear gold clusters are based on phosphinocarborane ligands, namely $[Au_4\{(PPh_2)_2C_2B_9H_{10}\}_2\{P(4\text{-}OMeC_6H_4)_3\}_2]$ [5] and, more recently, $[Au_4\{(PPh_2)_2C_2B_9H_{10}\}_2\{PPh_3\}_2]$ [6]. The first cluster is prepared by treating the tetranuclear cluster $[Au_4\{(PPh_2)_2C_2B_9H_{10}\}_2(AsPh_3)_2]$ with the tertiary phosphine P(4-OMeC$_6$H$_4$)$_3$. The second is prepared by reaction of $[Au(NO_3)(PPh_3)]$ with the closo phosphinocarborane ligand $[1,2\text{-}(PPh_2)\text{-}1,2\text{-}C_2B_{10}H_{10}]$ in a degradation reaction of the carborane cage. In both structures, the gold atoms define a tetrahedral core in which two of the gold atoms are chelated by the anionic diphosphinocarborane, and each of the other two atoms is bonded to one arsine or phosphine ligand. Interestingly, all these clusters display intense luminescence, probably arising from gold-centered spin-forbidden transitions (Scheme 3.3).

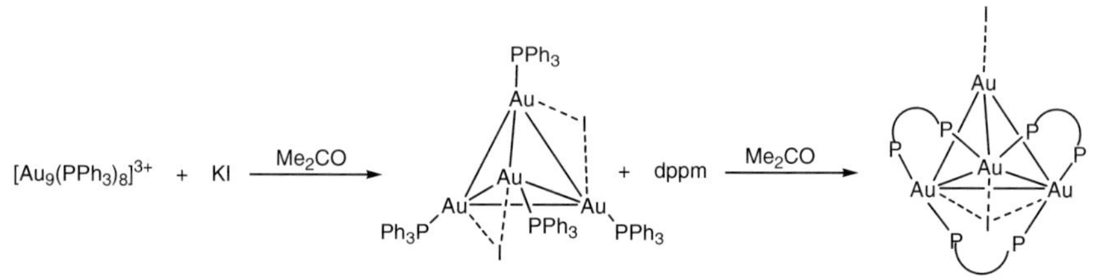

Scheme 3.1 Synthesis and structure of the clusters $[Au_4(PPh_3)_4(\mu\text{-}I)_2]$ and $[Au_4(dppm)_3(\mu_3\text{-}I)I]$.

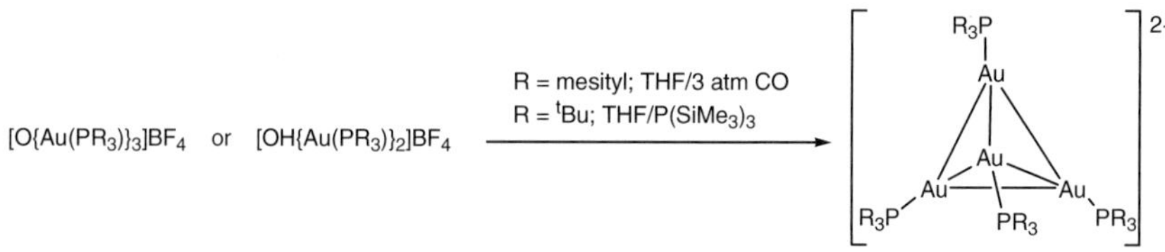

Scheme 3.2 Structure of the clusters $[Au_4(PR_3)_4]BF_4$ (R = ^{t}Bu and ($C_6H_3Me\text{-}1,3,5$)).

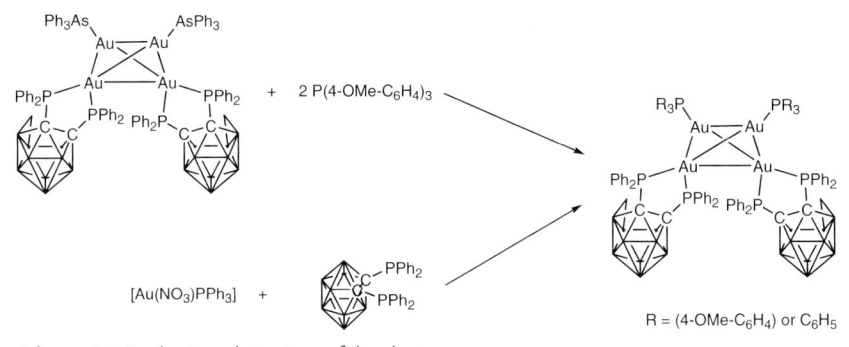

Scheme 3.3 Synthesis and structure of the clusters [Au$_4${(PPh$_2$)$_2$C$_2$B$_9$H$_{10}$}$_2${P(4-OMeC$_6$H$_4$)$_3$}$_2$] and [Au$_4${(PPh$_2$)$_2$C$_2$B$_9$H$_{10}$}$_2${PPh$_3$}$_2$].

To date, there is only one case of a pentanuclear gold cluster [Au$_5$(dppm)$_3$(dppm-H)][NO$_3$]$_2$ [7] in which the metal atoms display a spiked-tetrahedral arrangement. Four of the gold centers form a tetrahedral cluster while the fifth is bonded to one of the former metal centers. Each dppm and (dppm-H)$^-$ ligand acts as a bridging ligand between two gold atoms. This cluster was prepared using a metal evaporation technique in solution and in the presence of (dppm-H)$^-$ ligand.

Among phosphine-stabilized hexanuclear gold clusters, two interesting structures have been analyzed. The first is the cationic [Au$_6$(PPh$_3$)$_6$]$^{2+}$ cluster, prepared through decomposition of the hydrazido complex [R$_2$NN(AuPPh$_3$)$_3$]BF$_4$ [8] or through degradation of the cluster [Au$_8$(PPh$_3$)$_8$]$^{2+}$ in the presence of [Ag(CN)$_2$]$^-$ [9]. The cluster structure displays an edge-sharing bitetrahedral arrangement of the gold centers with a phosphine ligand bonded to each gold atom. In this case, the shortest Au–Au distance corresponds to the gold atoms sharing the tetrahedral units. The second example is the cluster [Au$_6$(dppp)$_4$]$^{2+}$ (dppp = diphenylphosphinopropane), synthesized through degradation of cluster [Au$_9$(PPh$_3$)$_8$]$^{2+}$ with an excess of dppp ligand, in which four gold atoms form a tetrahedron while the two remaining gold centers are placed in a trans-edge bridge arrangement surrounded by four diphosphine ligands [10] (Scheme 3.4).

The structure of the heptanuclear cluster [Au$_7$(PPh$_3$)$_7$]OH has been characterized through X-ray diffraction studies. The gold centers display a pentagonal bipyramid geometry with a phosphine ligand bonded to each of them. This cluster was prepared through the reaction of gold vapor with PPh$_3$ in toluene as solvent [11, 12].

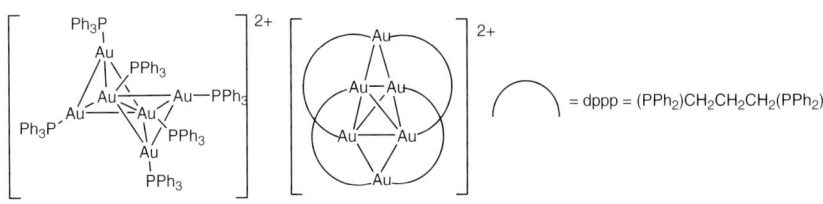

Scheme 3.4 Structure of the clusters [Au$_6$(PPh$_3$)$_6$]$^{2+}$ and [Au$_6$(dppp)$_4$]$^{2+}$.

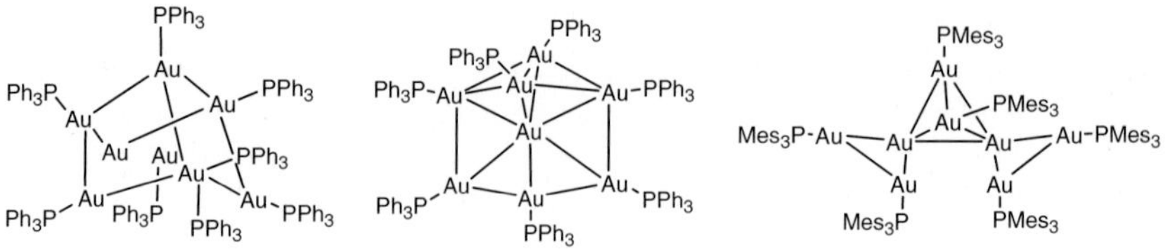

Figure 3.1 Structure of dicationic clusters $[Au_8(PPh_3)_8]^{2+}$, $[Au_8(PPh_3)_7]^{2+}$, $[Au_8\{P(C_6H_2Me_3\text{-}1,3,5)_3\}_6]^{2+}$.

Three octanuclear gold cluster structures must be mentioned, namely $[Au_8(PPh_3)_8]^{2+}$, $[Au_8(PPh_3)_7]^{2+}$ and $[Au_8\{P(C_6H_2Me_3\text{-}1,3,5)_3\}_6]^{2+}$ showing eight, seven or six arylphosphine ligands, respectively [4, 13–15] (Figure 3.1).

Cluster $[Au_8(PPh_3)_8]^{2+}$ can be prepared by reaction of the nonanuclear cluster $[Au_9(PPh_3)_8]^{3+}$ with two equivalents or excess of triphenylphosphine [13], by photolysis of the azide complex $[AuN_3(PPh_3)]$ in THF [16] or by addition of triphenylphosphine to the $[Au_8(PPh_3)_6]^{2+}$ cluster [4]. Two crystal structures with different anions (PF_6^- and alazarinsulfonate) [13, 15] have been determined and both display a core of gold atoms in a centered chair arrangement with an additional gold atom bonded to three of the metal atoms of the chair and a central gold atom. The inner Au–Au distances lie between 2.587(5) and 2.771(5) Å for the alazarinsulfonate salt and between 2.64(1) and 2.72(1) Å for the PF_6^- one. Each gold atom is bonded to a triphenylphosphine ligand.

Cluster $[Au_8(PPh_3)_7]^{2+}$ can be synthesized through two unusual pathways. The first pathway consists of the reaction of cluster $[Au_9(PPh_3)_8]^{2+}$ with $Na_2S_2C_2(CN)_2$ salt [17] and the second consists of the reaction of cluster $[Au_8(PPh_3)_8]^{2+}$ with the rhodium compound $[RhCl(C_8H_{14})_2]_2$ through elimination of one PPh_3 ligand [14]. The crystal structure of this octanuclear cluster displays a butterfly shape with four gold centers and another three metal atoms bridging the opposite atoms of the butterfly part. An eighth gold atom lies in the center of the cluster and seven triphenylphosphine ligands are bonded to the peripheral metal atoms. The central-to-peripheral Au–Au distances are shorter (2.63(1)–2.72(1) Å) than the peripheral ones (2.80(1)–2.94(1) Å).

Octanuclear cluster $[Au_8\{P(C_6H_2Me_3\text{-}1,3,5)_3\}_6]^{2+}$ is prepared by reduction of the oxonium salt $[O\{Au\{P(C_6H_2Me_3\text{-}1,3,5)_3\}_3\}_3]$ in the presence of 3 atm of CO [4]. The structure shows a cluster core composed of a distorted Au_4 tetrahedron in which two of the gold vertexes are also bonded to two other gold atoms. Six phosphine ligands are linked to the gold atoms of the unshared vertexes.

The synthesis of nonanuclear gold clusters is carried out mainly through the reduction of a mononuclear gold(I) complex. Using $NaBH_4$ as reducing agent against $[Au(SCN)(PCy_3)]$ or $[Au(NO_3)(PPh_3)]$ precursor complexes the gold clusters $[Au_9(SCN)_3(PCy_3)_5]$ [18] and $[Au_9(PPh_3)_8][NO_3]_3$ [19] are obtained, respectively. Also, when the molecular gold precursor is complex $[AuCl(PPh_3)]$ and the reducing agent is the titanium compound $[Ti(\eta\text{-}C_5H_5Me)_2]$ the cluster $[Au_9(PPh_3)_8]^{3+}$ can be synthesized [20]. Finally, synthesis of cluster $[Au_9(PPh_3)_8]^{3+}$ can also be afforded by an aggregation reaction of the octanuclear cluster $[Au_8(PPh_3)_8]^{2+}$ with complex

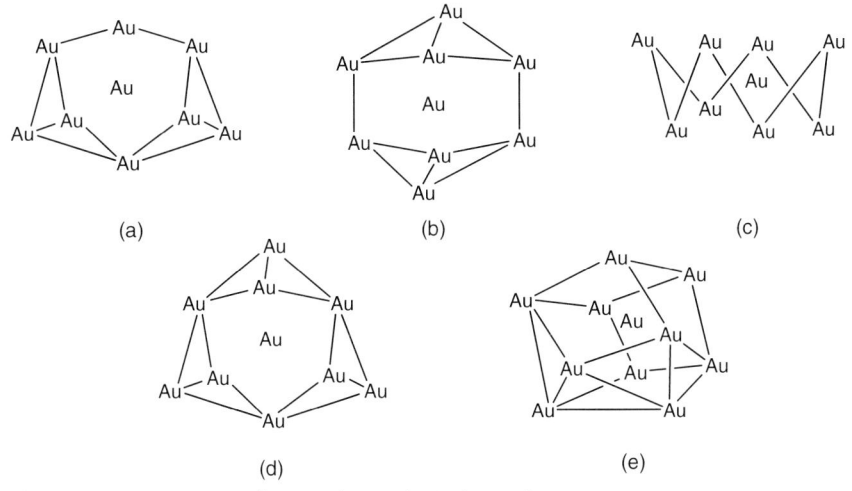

Figure 3.2 Core structures of nonanuclear (a, b, c), decanuclear (d) and undecanuclear (e) gold clusters.

[Au(NO$_3$)(PPh$_3$)] [21]. Similar clusters with different triarylphosphines, such as P(p-C$_6$H$_4$Me)$_3$ or P(p-C$_6$H$_4$OMe)$_3$, have also been described [22, 23].

The structure of the heteroleptic cluster [Au$_9$(SCN)$_3$(PCy$_3$)$_5$] [18] displays a six-membered ring of metal atoms with one gold atom in the center of the chair and two other peripheral gold atoms each bonded to three metals (Figure 3.2a). Interestingly, cluster [Au$_9${P(p-C$_6$H$_4$OMe)$_3$}$_8$]$^{3+}$ displays two different structural arrangements depending on the counteranion. When the anion is PF$_6^-$ and the phosphine is P(p-C$_6$H$_4$Me)$_3$ [22] or the anion is NO$_3^-$ and the phosphine is P(p-C$_6$H$_4$OMe)$_3$ [24], the cluster structure can be defined as a centered chair of gold atoms with two additional edge bridging gold atoms with phosphine ligands linked to all gold centers with the exception of the central one. (Figure 3.2b) If the anion is BF$_4^-$ [23] or NO$_3^-$ [24] and the phosphine is P(p-C$_6$H$_4$OMe)$_3$, the cluster structure adopts a centered crown arrangement of gold atoms. (Figure 3.2c)

The decanuclear cluster [Au$_{10}$(PCy$_2$Ph)$_6$Cl$_3$]$^+$ can be synthesized by reducing the mononuclear precursor [AuCl(PCy$_2$Ph)] with NaBH$_4$ [25]. Other decanuclear gold clusters such as [Au$_{10}$(PPh$_3$)$_3$(C$_6$F$_5$)$_4$] [26] and [Au$_{10}$(PPh$_3$)$_7${S$_2$C$_2$(CN)$_2$}$_2$] [27] can be obtained through aggregation reactions to the nonanuclear cluster [Au$_9$(PPh$_3$)$_8$]$^{3+}$ and three equivalents of [Au(C$_6$F$_5$)$_2$]$^-$ or one equivalent of [S$_2$C$_2$(CN)$_2$]$^-$, respectively. The gold cluster core is similar in the three decanuclear clusters and consists of three trigonal bipyramidal Au$_5$ clusters that share a gold central atom and an edge (Figure 3.2d).

There are several synthetic pathways that permit the synthesis of undecanuclear clusters. The reduction of the mononuclear precursor [AuI(PPh$_3$)] or the dinuclear precursor [Au$_2$X$_2$(BINAP)] with NaBH$_4$ leads to clusters [Au$_{11}$(PPh$_3$)$_7$I$_3$] [28] and [Au$_{11}$(BINAP)$_4$X$_2$] [29] (X = Cl or Br; BINAP = 2,2′-bis(diphenylphosphino)-1,1′-binaphthyl), respectively; reaction of gold vapor with complex [Au(SCN)(PPh$_3$)] in ethanol leads to cluster [Au$_{11}$(PPh$_3$)$_7$(SCN)$_3$]$^{3+}$ [30]; the photolytic reduction of

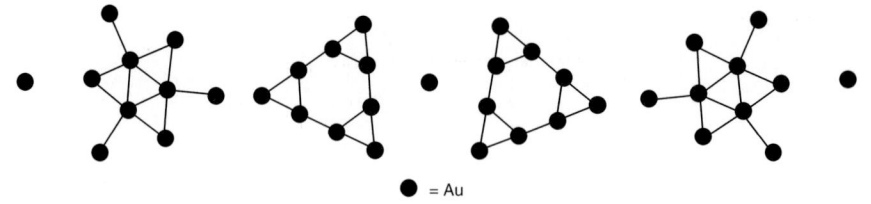

Figure 3.3 Molecular structure of the cluster $[Au_{39}(PPh_3)_{14}Cl_6]Cl_2$: individual 1:9:9:1:9:9:1 gold layers.

complex $[AuN_3(PPh_3)]$ in CH_2Cl_2 gives rise to cluster $[Au_{11}(PPh_3)Cl_2]^+$ [16]; cluster $[Au_{11}(PMe_2Ph)_{10}]^{3+}$ is obtained from the degradation reaction of cluster $[Au_{13}(PMe_2Ph)_{10}Cl_2]^{3+}$ in the presence of four equivalents of free PMe_2Ph; [31] aggregation of Cl^- or SCN^- to cluster $[Au_9(PPh_3)_8]^{3+}$ gives rise to clusters $[Au_{11}(PPh_3)_8X_2]^+$ (X = Cl or SCN) [21, 32] and, finally, ligand substitution reactions yield clusters $[Au_{11}(dppp)_5][SCN]_3$ [33] and $[Au_{11}(PPh_3)_7(SCN)_3]$ [34].

As regards their solid state structures, most undecanuclear gold clusters display a C_{3v} metal core arrangement deriving from a centered icosahedron with a trinuclear face substituted by a single metal atom (Figure 3.2e). In the case of $[Au_{11}(BINAP)_4X_2]^+$ clusters, the origin of their chiroptical activity, observed through their circular dichroism spectra, could be related to the deformation of the Au_{11}^{3+} core, since the starting materials did not display this activity.

The tredecanuclear cluster $[Au_{13}(PMe_2Ph)_{10}Cl_2]^+$ [31] can be obtained through reduction of the gold(I) precursor $[AuCl(PMe_2Ph)]$ using the titanium complex $[Ti(\eta\text{-}C_6H_5Me)_2]$ as reducing agent, or through photolysis of the complex $[AuN_3(PPh_3)]$ in the presence of PMe_2Ph ligand [16]. The core structure of this cluster consists of a centered icosahedron of gold atoms with chlorine ligands bonded to the gold atoms in apical positions and phosphine ligands bonded to the remaining metal centers.

The largest gold cluster characterized by X-ray diffraction studies is the triphenylphosphine-based cluster $[Au_{39}(PPh_3)_{14}Cl_6]Cl_2$ [35]. This derivative is obtained by reduction of $HAuCl_4$ with $NaBH_4$ in the presence of PPh_3. The crystal structure determined by X-ray diffraction studies reveals a 1:9:9:1:9:9:1 layered hcp/hcp' structure. The encapsulated gold atom displays an Au—Au mean distance with the other 12 gold atoms of 3.040 Å. Of the 39 gold atoms, only the encapsulated atom can be considered as a "bulk" atom. There are three types of 38 peripheral gold atoms: 14 of them are coordinated to PPh_3; 6 to chloride ligands; and 14 are not coordinated (Figure 3.3).

3.2.2
Gold Clusters with Other Ligand Stabilizers

Although most of the structurally characterized gold clusters are phosphine-based systems, in recent years some studies have focused on the synthesis of gold clusters with other stabilizers coexisting with phosphines such as thiols, arsines and boranes. Certain heteroleptic gold clusters stabilized with these ligands and arylphosphines in the same molecule have been structurally characterized.

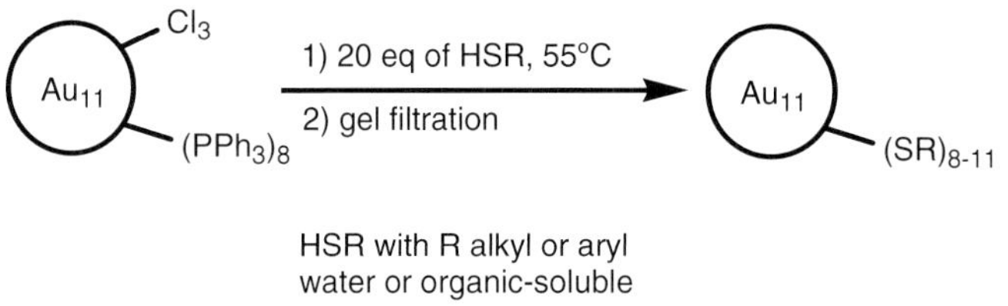

Scheme 3.5 Ligand exchange reaction of the cluster [Au$_{11}$(PPh$_3$)$_8$Cl$_3$] with ω-functionalized alkanethiols.

Thiol-stabilized gold clusters have been prepared by reaction of the previously commented undecagold cluster [Au$_{11}$(PPh$_3$)$_8$Cl$_3$] with ω-functionalized alkanethiols [36, 37] (Scheme 3.5). In these studies the XPS, TGA and elemental analysis confirm that the undecagold core is preserved during the exchange reaction between triphenylphosphine and the thiol ligands. The TEM analysis also suggests that the Au$_{11}$ core is present in all clusters since the size analysis of the TEM micrographs reveals average core sizes for all clusters, including the [Au$_{11}$(PPh$_3$)$_8$Cl$_3$] precursor, of 0.8 ± 0.2 nm. In contrast to larger size gold nanoparticles, in this case the NMR mechanistic studies provide strong evidence that the Au$_{11}$ core remains intact during the ligand exchange reaction.

The thiol-modified gold-phosphine cluster [Au$_{11}$(S-4-NC$_5$H$_4$)$_3$(PPh$_3$)$_7$] [38] has been synthesized both by reduction of the mononuclear precursor [Au(S-4-NC$_5$H$_4$)(PPh$_3$)] with NaBH$_4$ and by reacting the phosphine-stabilized cluster [Au$_9$(PPh$_3$)$_8$][NO$_3$]$_3$ with HS-4-NC$_5$H$_4$. The molecular structure of this cluster has been determined via a synchrotron radiation source and it displays an incomplete icosahedral skeleton with an almost C_{3v} symmetry axis. The gold distances are in the 2.6295–3.2693 Å range, depending on the position of the gold centers (center or peripheral) in the structure. The TEM analysis shows particles with diameter of approximately 0.6 nm that could reflect the existence of the undecanuclear core. Finally, the XPS spectrum of this cluster exhibits three peaks corresponding to three peripheral gold atoms each bonded to a thiol ligand, seven peripheral gold atoms each bonded to PPh$_3$ ligands and one central gold atom (Figure 3.4).

Tredecanuclear clusters [Au$_{13}$(PPh$_3$)$_4${S(CH$_2$)$_{11}$CH$_3$}$_2$Cl$_2$] and [Au$_{13}$(PPh$_3$)$_4${S(CH$_2$)$_{11}$CH$_3$}$_4$] have been prepared via exchange of dodecanethiol onto phosphine-halide gold clusters. These clusters were characterized using quantitative high-angle annular dark-field scanning transmission electron microscopy (HAADF-STEM) to determine the atomic composition of the cluster cores and confirm their monodispersity. The core atom counts and size for each gold cluster are 13.6 ± 3.4 atoms and 0.8 ± 0.1 nm for cluster [Au$_{13}$(PPh$_3$)$_4${S(CH$_2$)$_{11}$CH$_3$}$_2$Cl$_2$] and 13.4 ± 3.8 atoms and 0.8 ± 0.1 nm for cluster [Au$_{13}$(PPh$_3$)$_4${S(CH$_2$)$_{11}$CH$_3$}$_4$] [39].

Similar reactions between the undecagold cluster [Au$_{11}$(PPh$_3$)$_8$Cl$_3$] and n-alkanethiols of different chain-lengths surprisingly lead to biicosahedral gold clusters [Au$_{25}$(PPh$_3$)$_{10}$(SC$_n$H$_{2n+1}$)$_5$Cl$_2$]$^{2+}$ ($n = 2, 8, 10, 12, 14, 16$ and 18). In all cases, the chemical composition of the clusters was determined through ESI mass spectrometry and the crystal structure of the cluster [Au$_{25}$(PPh$_3$)$_{10}$(SC$_2$H$_5$)$_5$Cl$_2$]$^{2+}$ was characterized by X-ray diffraction studies. The Au$_{25}$ cluster core can be defined as a dimer of

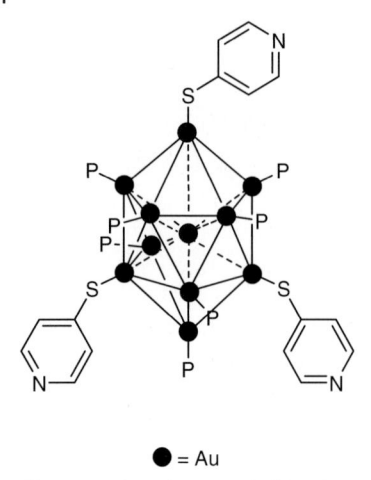

Figure 3.4 Crystal structure of the cluster $[Au_{11}(S\text{-}4\text{-}NC_5H_4)_3(PPh_3)_7]$.

Au_{13} icosahedrons sharing a vertex gold atom with Au–Au bond lengths that are in the range 2.70–3.00 Å. Five thiolate ligands bridge two Au_5 rings of each Au_{13} cluster subunit; two chloride ligands are bonded to the Au centers in the apical position of the biicosahedron cluster and, finally, the rest of the peripheral Au atoms are each bonded to a PPh_3 ligand [40] (Figure 3.5).

Higher nuclearity homoleptic thiolate-protected gold clusters do exist but they have not been characterized by X-ray diffraction studies. This type of cluster will be discussed below in Section 3.4.1.1.

Although most of the molecular gold clusters were synthesized using triarylphosphines as stabilizing ligands, some examples using triphenylarsine $AsPh_3$ have been reported. The first example of an arsine stabilized gold cluster is the tetranuclear cluster $[Au_4\{(PPh_2)_2C_2B_9H_{10}\}_2(AsPh_3)_2]$ [41], prepared by reaction between the diphosphinocarborane ligand $[1,2\text{-}(PPh_2)\text{-}1,2\text{-}C_2B_{10}H_{10}]$ and two equivalents of the gold precursor $[AuCl(AsPh_3)]$. In this reaction, degradation in the carborane cage of the

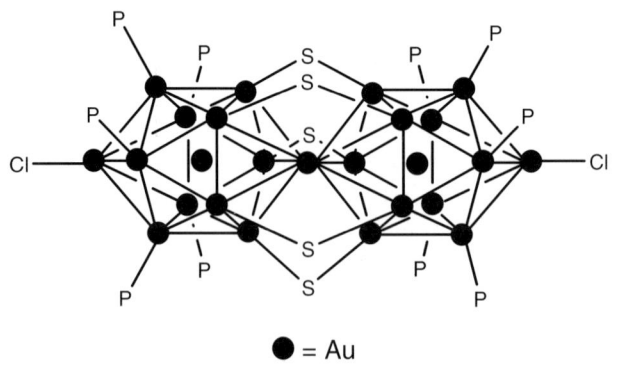

Figure 3.5 Crystal structure of the cluster $[Au_{25}(PPh_3)_{10}(SC_2H_5)_5Cl_2]^{2+}$.

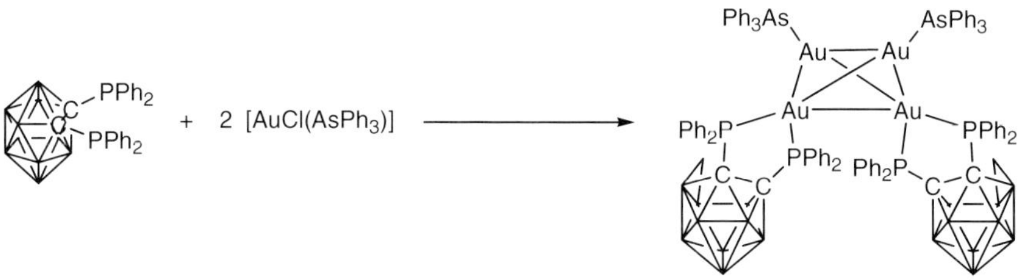

Scheme 3.6 Synthesis of the cluster [Au$_4${(PPh$_2$)$_2$C$_2$B$_9$H$_{10}$}$_2$(AsPh$_3$)$_2$].

diphosphine ligand occurred to afford the anionic nido-ligand. The crystal structure displayed an Au$_4$ tetrahedral core in which two of the gold centers were each chelated by a phosphinocarborane ligand while the two others were bonded to one arsine ligand. The Au–Au bond lengths were classified into three groups: 2.9148(8) Å (long); 2.8371(8) and 2.8062(7) Å (medium) and 2.6036(7)–2.6751(8) Å (short) (Scheme 3.6).

The hexadecanuclear homoleptic gold cluster [Au$_{16}$(AsPh$_3$)$_8$Cl$_{16}$] [42] has been synthesized by reduction of [AuCl(AsPh$_3$)] with NaBH$_4$. The crystal structure of this cluster consists of a centered Au$_{13}$ icosahedron in which one of the gold vertexes was bonded to three metal centers forming a pendent tetrahedron.

There are few trinuclear gold compounds in which the gold–gold distances and the oxidation state of the metal atoms justify their proposal as homonuclear clusters. An interesting example is the trinuclear cluster [Au$_2$(PPh$_3$)$_2$Au(PPh$_3$)(C$_2$B$_9$H$_9$Me$_2$)] reported by Stone et al. [43]. This cluster is synthesized by deprotonation with NaH and addition of [AuCl(PPh$_3$)] to the dinuclear complex [Au(PPh$_3$)(μ-H)Au(PPh$_3$)(C$_2$B$_9$H$_9$Me$_2$)]. Its structure consists of a triangle, defined by the gold centers with Au–Au bond lengths between 2.691(1) and 3.010(1) Å, in which each PPh$_3$ ligand binds a gold center and the carborane cage is bonded to the three gold centers through a boron atom.

3.3
Large-Size Gold Clusters. The Au$_{55}$ Case

The fuzzy frontier between the molecular and the nanometric level can be elucidated from an electronic point of view. Molecules and small clusters can be described as systems in which the metal atoms form well-defined bonding and antibonding orbitals. Large clusters or small nanoparticles (quantum dots) with dimensions of a few nanometers are intermediate between the size of molecules and bulk material, presenting discrete energy levels with a small band gap owing to quantum-mechanical rules. Finally, larger particles tend to lose this trend and display a typical band structure similar to that of the bulk material.

Large gold clusters belong to a class of materials that display special features due to their intermediate size; interesting studies in this respect include the work by Schmid and coworkers on phosphine-stabilized gold clusters. These studies mainly focused on cluster [Au$_{55}$(PPh$_3$)$_{12}$Cl$_6$] based on its synthesis and characterization in the 1980s and 1990s to study their properties and potential applications in recent years.

Synthesis of $[Au_{55}(PPh_3)_{12}Cl_6]$ is carried out by reduction of $[AuCl(PPh_3)]$ with diborane [44] and since this synthesis was described for the first time, several studies have focused on its structural characterization. The calculated stoichiometry is in good agreement with the proposed Au_{55} cluster composition. The EXAFS measurements display the coordination environment of gold atoms and this is in accordance with a mean coordination of 7 by other gold atoms that is consistent with a cuboctahedral structure for the Au_{55} cluster. The Mössbauer spectrum also agrees with the expected structural model since it displays the existence of four classes of gold atoms in the cluster structure, consistent with a two layer 55-atom cuboctahedron in which 13 inner atoms are surrounded by 24 naked gold atoms, 12 phosphine bonded gold atoms and 6 chlorine bonded gold centers [45]. Nevertheless, a further analysis of the structure of this Au_{55} cluster through cryoHRTEM, X-ray scattering, analytical ultracentrifugation and thermoanalytical measurements contradicts the assumption of monodispersity for this system since the observed particles are heterogeneous in size and structure. The size distribution displays a maximum, in agreement with the diameter of a hypothetical Au_{55} cluster (1.4 ± 0.4 nm) [46]. Schmid and coworkers subsequently succeeded in obtaining microcrystals of cluster $[Au_{55}(PPh_3)_{12}Cl_6]$ in preparative amounts to overcome the above-mentioned problems of decomposition or lack of monodispersity. These microcrystals are prepared by the fast evaporation of concentrated CH_2Cl_2 solutions of the Au_{55} cluster, leading to microcrystals of about 1 μm in size. The HRTEM image of a selected Au_{55} microcrystal displays a monocrystalline particle containing about 10^8 clusters [47] (Figure 3.6).

Since the late 1990s, most research on Au_{55} clusters has focused on two main areas: first, the reactivity of Au_{55} clusters with other ligands and the study of this effect on the structure and properties of the resulting clusters; and second, taking into account the attractive properties of these small size clusters due to their quantum size behavior, part of the research has focused on studying highly ordered nanostructures

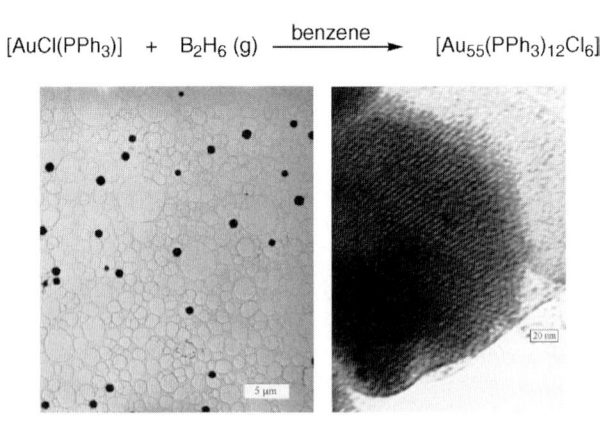

$$[AuCl(PPh_3)] + B_2H_6 (g) \xrightarrow{benzene} [Au_{55}(PPh_3)_{12}Cl_6]$$

Figure 3.6 Synthesis of $[Au_{55}(PPh_3)_{12}Cl_6]$ and TEM and HRTEM micrographs of microcrystals of the Au_{55} cluster obtained from concentrated CH_2Cl_2 solutions. Reproduced from reference [47] by permission of The Royal Society of Chemistry.

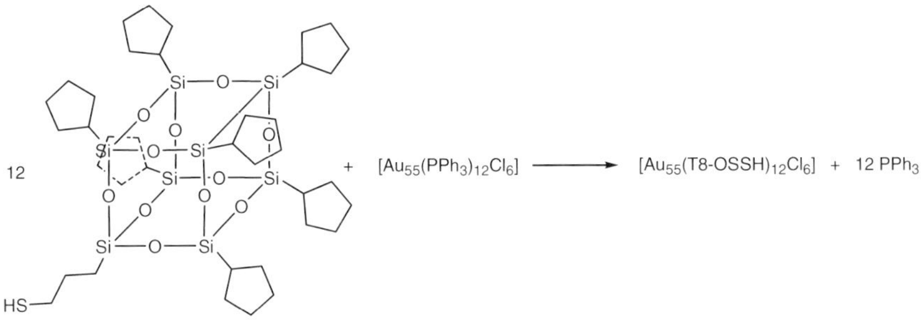

Scheme 3.7 Synthesis of the [Au$_{55}$(T8-OSS-SH)$_{12}$Cl$_6$] cluster.

or new nanomaterials based on the Au$_{55}$ cluster. These new nanomaterials are very interesting from an applied point of view in the nanoelectronics field.

A Au$_{55}$ cluster has been synthesized by the exchange of PPh$_3$ ligands in [Au$_{55}$(PPh$_3$)$_{12}$Cl$_6$] by the thiol functionalized silsesquioxane derivative (cyclopentyl)$_7$Si$_8$O$_{12}$(CH$_2$)$_3$SH. This ligand exchange leads to an increase in total cluster diameter from 2.1 to 4.4 nm and confers great stability to the Au$_{55}$ cluster core. Furthermore, TEM images of a [Au$_{55}$(T8-OSS-SH)$_{12}$Cl$_6$] sample display monodisperse particles, whose size (1.4 ± 0.1 nm) is in agreement with the Au$_{55}$ core size, leading to a ligand shell of 3.0 nm. This ligand shell may be responsible for the particular properties of this cluster, such as its solubility in non-polar solvents. This thick ligand shell may also prevent cluster agglomeration and good affinity of thiol groups and gold atoms, which leads to a highly stable Au$_{55}$ cluster [48] (Scheme 3.7).

The exchange of PPh$_3$ in cluster [Au$_{55}$(PPh$_3$)$_{12}$Cl$_6$] by Na$_2$[B$_{12}$H$_{11}$SH] allows the formation of [Au$_{55}${(B$_{12}$H$_{11}$SH)Na$_2$}$_{12}$Cl$_6$] after six weeks of reaction. This cluster has been characterized spectroscopically by means of ^{1}H and ^{11}B NMR spectroscopy and IR spectroscopy. The monodispersity of the cluster has been confirmed through TEM analysis. Further reaction of cluster [Au$_{55}${(B$_{12}$H$_{11}$SH)Na$_2$}$_{12}$Cl$_6$] with (octyl)$_4$NBr yields cluster [Au$_{55}${(B$_{12}$H$_{11}$SH){(N(octyl)$_4$)}$_2$}$_{12}$Cl$_6$]. An interesting property of these clusters is that they display electromigration in an electric field due to the very high concentration of ionic charges [49].

The reaction of [Au$_{55}$(PPh$_3$)$_{12}$Cl$_6$] with thiols R–SH (R = C$_6$H$_{13}$, C$_3$H$_7$ and PhC$_7$H$_4$) gives rise to Au$_{75}$ monolayer thiol protected clusters. A combined study using mass spectrometry, HPLC analysis, UV–Vis spectroscopy and thermogravimetric results produced a preliminary formulation [Au$_{75}$(SC$_6$H$_{13}$)$_{40}$]. In this case, the TEM images were not consistent since size estimates varied significantly [50].

The nanomaterials based on the Au$_{55}$ cluster can be described according to their dimensionality. Starting from zero-dimensional nanomaterials a novel class of a nanocontainer has been recently reported. It consists of an inorganic [Au$_{55}${SCH$_2$(4-Py)}$_{16}$(PPh$_3$)$_3$] cluster confined within an organic cage composed of zinc metalloporphyrins. The external cage is built up using six zinc-tetraaryl porphyrin molecules with four terminal olefin groups, each assembled around the Au$_{55}$ cluster via intermolecular olefin-methatesis. Ligand exchange inside the cage is possible since the porphyrin nanocage has holes for the irreversible interpenetration

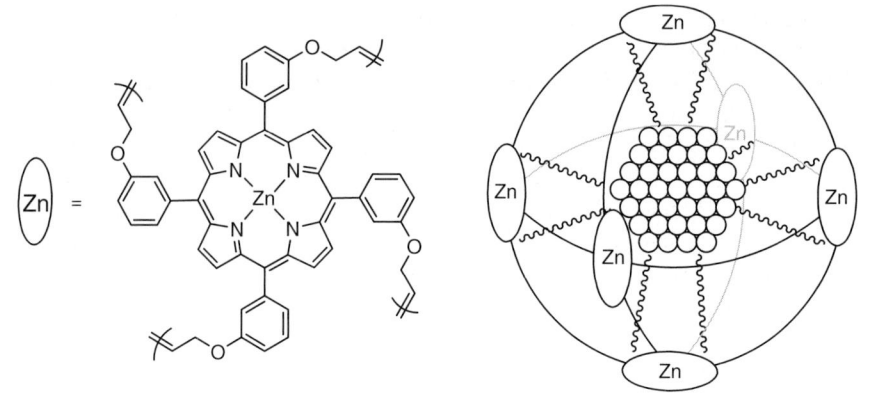

Figure 3.7 [Au$_{55}${SCH$_2$(4-Py)}$_{16}$(PPh$_3$)$_3$] cluster confined within a six-zinc porphyrin cage.

of small substrates. This class of materials may have potential for application in different fields such as molecular recognition [51] (Figure 3.7).

It is also possible to build up one-dimensional short chains of [Au$_{55}$(PPh$_3$)$_{12}$Cl$_6$] clusters between tungsten tips on a SiO$_2$ surface. For this, a CH$_2$Cl$_2$ solution of the cluster is deposited by applying a voltage at the tips, leading to quantum dot wires that can act as single transistors at room temperature. Intensity–voltage curves display pronounced coulomb blockades between −0.5 and 0.5 V [52].

Two-dimensional arrays of [Au$_{55}$(Ph$_2$PC$_6$H$_4$SO$_3$H)$_{12}$Cl$_6$] clusters can also be achieved. For example, both 2D hexagonal and 2D cubic lattices of Au$_{55}$ clusters on poly(ethyleneimine) (PEI) films through classic acid–base reactions between the –NH groups of the PEI film and the –SO$_3$H functions of the cluster ligand shell. The choice for a hexagonal or a cubic 2D arrangement could be attributed to the influence of the polymer film [53] (Figure 3.8).

Multilayer systems of [Au$_{55}$(PPh$_3$)$_{12}$Cl$_6$] double layers and SiO$_2$ films can also be synthesized and characterized. For this purpose, a combination of techniques is

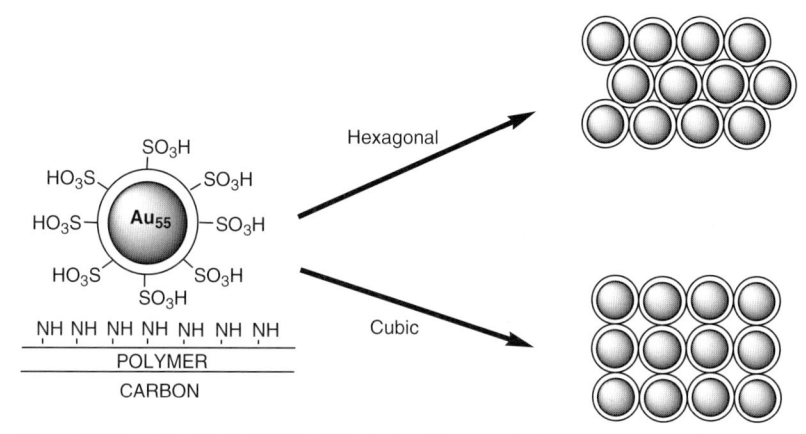

Figure 3.8 [Au$_{55}$(Ph$_2$PC$_6$H$_4$SO$_3$H)$_{12}$Cl$_6$] cluster, 2D superstructures.

needed, namely, spin-coating for the preparation of the double cluster layer and plasma-assisted physical vapor deposition (PAPVD) for the generation of SiO_2 films. These multilayer systems have been characterized through AFM and SEM microscopy. A similar approach has been used for the preparation of $[Au_{55}(PPh_3)_{12}Cl_6]$ monolayers embedded between SiO_2 films. The cluster monolayers have also been obtained by spin coating while SiO_2 films have been prepared by a special anodic plasma arc technique. This system has been characterized by AFM and TEM microscopy. Both the multilayer and monolayer systems present interesting electrical properties [54, 55].

A further step has been the synthesis and characterization of Au_{55} clusters within mesoporous silica. The cluster $[Au_{55}(PPh_3)_{12}Cl_6]$ can be prepared via reduction of $[AuCl(PPh_3)]$ with diborane and then incorporated in SBA-15 silica or directly prepared inside the mesoporous silica. TEM micrographs reveal that Au_{55} clusters are located inside the pores rather than in the outer surface, and the individual clusters are arranged within the mesopores [56].

The preparation of polymer 2D networks of Au_{55} clusters can be achieved by two different procedures. First, $[Au_{55}(PPh_3)_{12}Cl_6]$ cluster monolayers are prepared using the Langmuir–Blodgett (LB) technique and, in a subsequent step, a toluene solution of a dithiol linker is sprayed. The second option is to perform a light-induced polymerization of vinyl-terminated fragments that belong to the cluster ligand p-styryl-diphenylphosphine [57].

Finally, naked Au_{55} clusters can coalesce into well-crystallized $(Au_{55})_\infty$ superstructures by adding a thiol-terminated fourth-generation dendrimer (96 SH groups) to cluster $[Au_{55}(PPh_3)_{12}Cl_6]$. In this reaction, the dendrimer is able to display a multiple role since it is responsible for removing PPh_3 and Cl ligands from the Au_{55} cluster and it also acts as a matrix for superstructure crystallization [58] (Scheme 3.8).

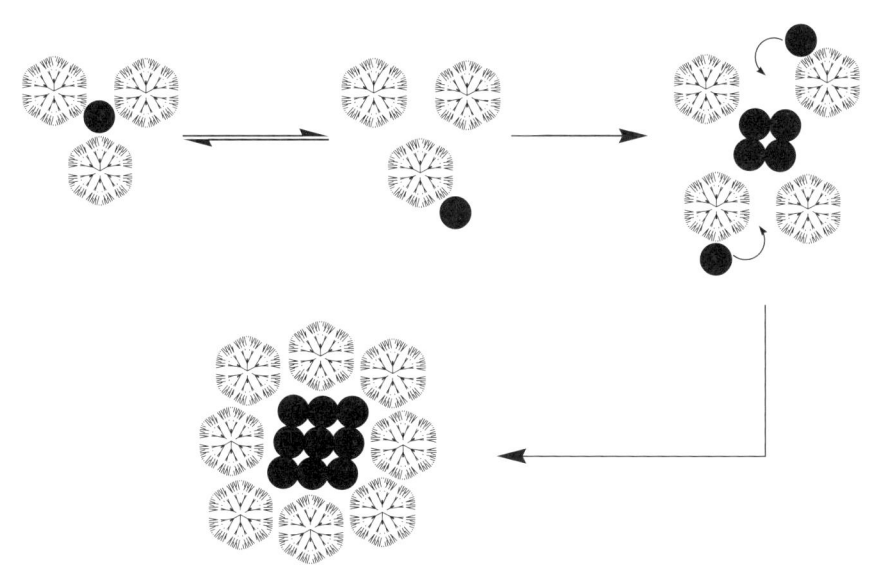

Scheme 3.8 Proposed $(Au_{55})_\infty$ superstructure formation in the presence of dendrimer excess.

3.4
Gold Nanoparticles

Although the term "small nanoparticle" was used in the previous section to describe large gold clusters, in this section we will focus on the synthesis and characterization of gold nanoparticles (Au NPs) from the sub-nanometer scale up to some hundreds of nm. The Au_{55} species can be classified as stoichiometric clusters due to the high degree of monodispersity observed, while small gold nanoparticles, also termed gold colloids, are characterized by some dispersity in size.

In this section, we will emphasize how the different chemical approaches allow control over the size, shape and assembly of Au NPs. The first subsection describes the most common synthetic approaches for the synthesis of gold nanoparticles, based on historically known methods from Faraday and Turkevitch. The surface chemistry of Au NPs, Au nano-bio hybrids synthesis and certain properties and applications of gold nanoparticles will be briefly summarized.

3.4.1
Synthesis of Gold Nanoparticles

The first preparation of gold colloids was described by Faraday in 1857 and it consisted of a two-phase reduction of an aqueous solution of $Na[AuCl_4]$ with a solution of carbon disulfide in which the reducing agent phosphorus was dissolved [59]. HRTEM characterization of these particles later showed that Faraday's method led to a wide distribution of particle sizes from the smallest of 3 nm to the largest of around 30 nm in diameter. Almost one century later, in 1951, Turkevitch reported one of the most popular synthetic approaches for the synthesis of gold colloids, that is the citrate method [60]. This method consists of reducing $HAuCl_4$ using trisodium citrate both as reducing agent and stabilizing ligand and leads to gold nanoparticles of about 20 nm. Further studies have enabled particle size to be controlled by varying the gold:citrate ratio [61].

3.4.1.1 Monolayer Protected Clusters
A very important method is that reported by Brust and Schiffrin in 1994 [62], relating to the synthesis of thiol-derivatized gold nanoparticles in a two-phase liquid–liquid system. Briefly, $AuCl_4^-$ is transferred from an aqueous solution to the organic toluene phase using tetraoctylammonium bromide. On addition of dodecanethiol (DDT) and aqueous $NaBH_4$, the organic phase changes color to deep brown, forming very small DDT-protected gold nanoparticles with diameters between 1 and 3 nm (Figure 3.9). This starting point has given rise to an entire family of thiolate-protected gold nanoparticles, also called monolayer protected clusters (MPCs) due to their small size.

Following this report, several modifications of the method with regard to the stabilizing ligands or reaction conditions have been described. Some interesting examples are described in the following paragraphs.

The first attempted modifications focused on the use of highly polar thiols to obtain very small water soluble MPCs. Thus, for example, the synthesis of gluthatione

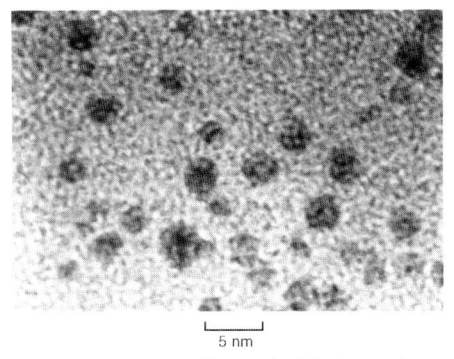

Figure 3.9 Brust–Schiffrin method for the synthesis of monolayer protected gold clusters. Reproduced from reference [62] by permission of The Royal Society of Chemistry.

protected gold clusters [63] of about 0.9 nm and a $Au_{28}(SG)_{19}$ composition has been proposed in view of the MALDI-MS, ESI-MS and XRD spectra. This type of glutathione-protected gold cluster has also been studied through ESI mass spectrometry of previously electrophoretically fractioned Au : SG mixtures of clusters, leading to the isolation of magic-numbered Au_n clusters ($n = 18, 21, 25, 28, 32, 39$) [64].

Tiopronin water soluble Au clusters have been described and their characterization by HRTEM and TGA has shown clusters of about 1.8 nm and a composition of $Au_{201}(Tiopronin)_{85}$ [65]. Polymer protected water soluble gold MPCs have also been synthesized using thiolated polyethyleneglycol (SH-PEG). HRTEM analysis displays a polydisperse population of clusters of 2.8 ± 1 nm size and TGA results yields an average $Au_{807}(SH\text{-}PEG)_{98}$ composition [66].

The Brust–Schiffrin method has been modified using arenethiol ligands. The gold MPCs core sizes vary, depending on the ligand employed, and they are larger and less stable than the alkanethiol protected clusters. For example, when 4-phenylbutane-1-thiol (PhC_4S) is used as ligand, an $Au_{314}(PhC_4S)_{143}$ composition, based on the TGA results, is proposed [67]. Interesting results have also been achieved with benzenethiolate (S-C_6H_5). The most abundant MPCs display compositions $[Au_n(SPh)_m]^{z-}$ like $[140, 78]^{3-}$, $[110, 62]^{6-}$ and $[44, 28]^{2-}$ estimated by mass spectrometry correlated with ^{1}H NMR and elemental analysis [68].

Other modifications to the reaction conditions of the Brust–Schiffrin method, such as a reduction temperature of $-78\,°C$ and the use of a hyperexcess of hexanethiol, results in an $Au_{38}(thiolate)_{24}$, based on TEM observations, LDI-TOF mass spectrometry, TGA analysis and elemental analysis [69]. The influence of preparation temperature on the size and monodispersity of dodecylthiol monolayer protected gold clusters has also been reported. Both TEM and SAXS measurements show that higher temperatures increase polydispersity. This modification of polydispersity may be related to the existence of a dynamic exchange of thiols at the particle surface with thiols in the solvent [70].

Another Au_{38} cluster core has been reported using phenylethanethiolate as protecting ligand, giving rise to 1.1 nm core diameter nanoparticles, as revealed by

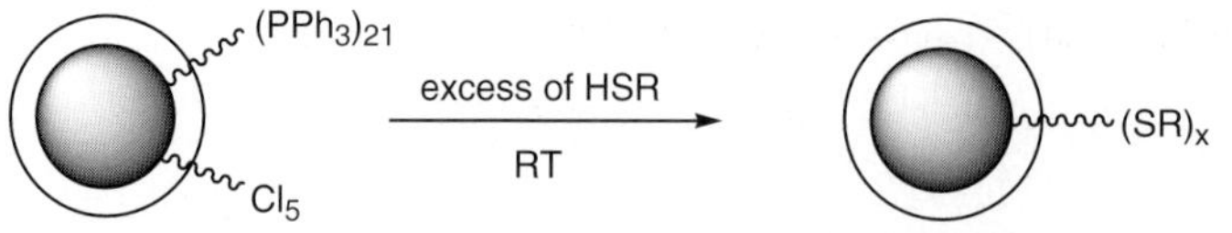

Scheme 3.9 Thiol ligands used in the ligand exchange reaction with 1.5 nm PPh$_3$-stabilized Au NPs.

$$R = -(CH_2)_nCH_3 \ (n = 2,5,7,11,15,17)$$
$$-p\text{-}Ph\text{-}OH$$
$$-p\text{-}Ph\text{-}CH_3$$
$$-p\text{-}Ph\text{-}Ph$$
$$-(CH_2)_3Si(OCH_3)_3$$
$$-(CH_2)_nCOOH \ (n = 1,2,6,11)$$
$$-(CH_2CH_2O)_nCH_2CH_2OH \ (n = 1,2)$$
$$-(CH_2)_2PO(OH)_2$$
$$-(CH_2)_2NHMe_2^+Cl^-$$
$$-(CH_2)_2NMe_3^+Cl^-$$
$$-(CH_2CH_2O)_nCH_2CH_2NMe_3^+Cl^- \ (n = 1,2)$$
$$-(CH_2)_2SO_3^-Na^+$$

TEM and ATG measurements. These Au$_{38}$(PhCH$_2$CH$_2$S)$_{24}$ MPCs undergo place exchange reactions with other thiolate ligands or even PPh$_3$ ligand [71, 72]. It is possible to synthesize thiol-functionalized gold nanoparticles through ligand exchange reactions. Thus, the use of ω-functionalized thiols with a phosphine stabilized [Au$_{11}$(PPh$_3$)$_8$Cl$_3$] cluster or 1.5 nm triphenylphosphine-stabilized gold nanoparticles is a versatile approach since a wide variety of technologically important functional groups can be employed [73, 74] (Scheme 3.9). Recently, the undecagold cluster [A$_{11}$(PPh$_3$)$_8$Cl$_3$] has been used as a precursor of the Au$_{25}$(SG)$_{18}$ cluster, providing a large-scale synthesis of this thiolate-protected gold cluster [75].

A simple method for preparing gold MPCs using dimercaptosuccinic acid (DMSA) as reducing and stabilizing agent has been reported. The core sizes are in the range 10 to 13 atoms (about 0.8 nm), as observed in mass spectroscopic analysis [76].

3.4.1.2 Other Ligands

Although thiol ligands are the most widely used as stabilizers in the synthesis of Au NPs, other organic ligands bearing amine, phosphine, phosphine oxide, carboxylate or other functional groups have been employed alone or in mixtures.

Amines have been used scarcely compared to thiols but they can be employed to achieve good dispersity control. Gold nanoparticles can be synthesized from self-assembled gold(I) amine precursors like [AuCl(NH$_2$R)] (R = C$_8$H$_{17}$, C$_{12}$H$_{25}$ and C$_{16}$H$_{33}$). These Au(I) complexes tend to self-organize in the solid state into a supramolecular organization as a fibrous material. Upon decomposition through air exposure of the samples or in THF under 3 bar H$_2$ at RT, Au NPs of 23 ± 6 nm and 10 ± 3 nm are obtained, respectively. This method enables good control of the dispersity and a self-organization of the particles [77] (Scheme 3.10).

The modification to the Brust–Schiffrin method through the addition of laurylamine (LAM) or octadecylamine (ODA) instead of thiols to colloidal gold particles

[AuCl(tht)] + NH$_2$(CH$_2$)$_n$-CH$_3$ ⟶ [AuCl(NH$_2$-(CH$_2$)$_n$-CH$_3$)]

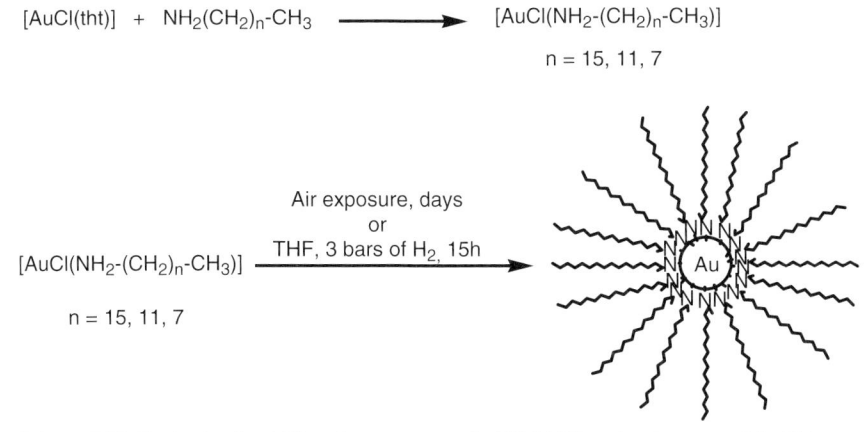

n = 15, 11, 7

[AuCl(NH$_2$-(CH$_2$)$_n$-CH$_3$)] $\xrightarrow{\text{Air exposure, days or THF, 3 bars of H}_2\text{, 15h}}$

n = 15, 11, 7

Scheme 3.10 Synthesis of gold(I) amine precursors [AuCl(NH$_2$R)] and amine-capped Au NPs.

leads to very monodispersed particles of 5.3 ± 0.8 nm in size that assemble into close-packed structures upon solvent evaporation. The study also shows that the surface-bound alkylamine monolayer may be exchanged with other amines [78].

Using amine chemistry for reduction and for surface stabilization at the same time, gold nanoparticles can be prepared in water directly by addition of oleyl amine (OLA) to a solution of AuCl$_4^-$. The XRD measurements show the peaks that confirm the face centered cubic (fcc) lattice of gold. The TEM analysis of the obtained nanoparticles displays narrow size distributions and, for example, when high concentrations of oleyl amine are used an average core size of 10 ± 0.6 nm is achieved [79].

Triphosphinogold oxonium salt [O(AuPPh$_3$)$_3$]BF$_4$ in the presence of amine and dioxygen is also a versatile gold atom source in mesitylene. The amine participates in the reduction of gold(I) atom precursor. An excess of amine enables the synthesis of ligand-protected nanoparticles, which also display a great tendency to self-organize [80].

A one-pot method for the synthesis of amine-stabilized gold nanoparticles using 3-(trimethoxysilylpropyl)diethylenetriamine has been reported. In this case, the amine acts as a reducing agent and capping ligand in the reaction with HAuCl$_4$ and highly stable Au NPs with sizes between 8 and 20 nm can be obtained by changing the ligand/Au mole ratio. The incorporated siloxy functionality is further used to form silica shells around the particles [81].

Amine ligands have also been used as stabilizers in the presence of other ligands such as, for example, phosphine oxides. The use of tri-n-octylphosphine oxide (TOPO) ligand as both a reaction medium and passivating ligand in the presence of octadecylamine in a 1:0.57 molar ratio at 190 °C, enables the growth of Au NPs (8.59 ± 1.09 nm) to be controlled in the reduction of HAuCl$_4$ with sodium borohydride. The good control on the dispersity enables the formation of ordered hexagonal close-packed 2D lattices [82].

A combination of octylamine (OA) and hexadecylamine (HDA) has been used for the synthesis of gold nanoparticles using the organometallic complex

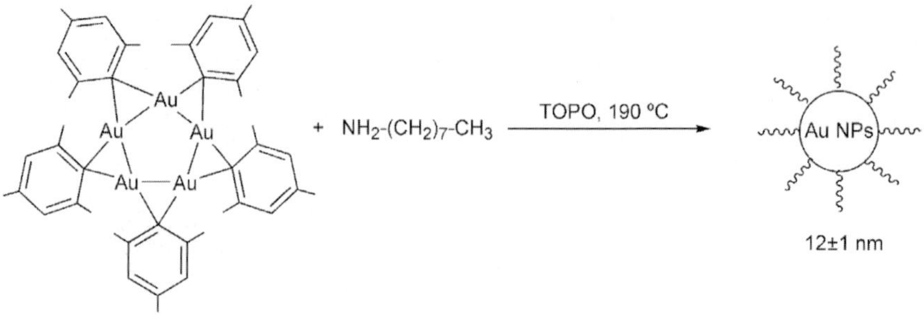

Scheme 3.11 Synthesis of Au NPs from the organometallic precursor [Au(μ-mesityl)]$_5$.

[Au(μ-mesityl)]$_5$ (mesityl = 2,4,6-Me$_3$C$_6$H$_2$) as precursor. This Au(I) complex is dissolved first in OA and injected into a heated (300 °C) solution of HDA leading to nanoparticles ranging in size from 10 to 80 nm. Nevertheless, when a more diluted solution of complex [Au(μ-mesityl)]$_5$ in OA is injected into a TOPO solution at 190 °C, highly monodisperse gold nanoparticles of 12 ± 1 nm are obtained [83] (Scheme 3.11).

In a more recent report, the single step reduction of complex [Au(acac)(PPh$_3$)] (acac = acetylacetonate) at high temperatures in solutions of varied TOPO and HDA concentrations leads to 4–15 nm diameter Au NPs with low polydispersity. A 50:50 mixture of TOPO and HDA leads to Au NPs with a particle size of 10.2 ± 1.4 nm [84].

Phosphines are less used as capping ligands for Au NPs except those already commented in the Au$_{55}$ cluster types. A modification of the Brust–Schiffrin method enables the synthesis of phosphine-stabilized gold nanoparticles of small size (1.5 nm). This biphasic synthesis involves the use of a phase transfer reagent tetraoctylammonium bromide (TOAB) to facilitate the transfer of chloroaurate ions from an aqueous solution to an organic phase (toluene) containing PPh$_3$. The reduction is carried out using NaBH$_4$ instead of diborane gas used by Schmid and coworkers for the synthesis of Au$_{55}$ clusters. The average empirical formula was determined using the complementary techniques XPS and TGA and was estimated to be Au$_{101}$(PPh$_3$)$_{21}$Cl$_5$ [85].

Another noteworthy result is the synthesis of phosphinine stabilized gold nanoparticles in which the reduction of the Au(I) precursor [AuCl(SMe$_2$)] is carried out using sodium naphthalenide in the presence of two different phoshinine ligands. Depending on the ligand used and its amount, different nanoparticle sizes and degrees of polydispersity are achieved with nanoparticles ranging from 8.3 ± 2.0 nm to 20.1 ± 7.5 nm in size. These nanoparticles have been included in silica and titania matrixes [86] (Scheme 3.12).

There is an interesting study in which the long-chain carboxylate ligand myristate was used for the synthesis of a gold(I) precursor complex [Au(C$_{13}$H$_{27}$COO)(PPh$_3$)] which, upon thermolysis at 180 °C under N$_2$, led to monodisperse 12 nm Au NPs stabilized by myristate and PPh$_3$ ligands. An increase in reaction time or thermolysis temperature resulted in larger size nanoparticles [87].

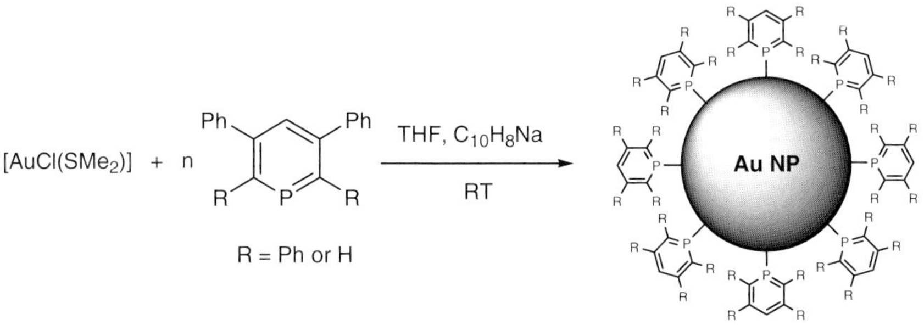

Scheme 3.12 Synthesis of Au NPs using phosphinine as capping ligand.

3.4.1.3 Polymers

Polymer-stabilized Au NPs have been the object of intense research in recent years because, apart from their intrinsic synthetic interest, these hybrid materials have interesting properties and potential applications. There are also several advantages in the use of polymers as stabilizers: enhancement of long-term stability; adjustment of solubility or amphiphilicity of nanoparticles, tuning of the properties of Au NPs, and the promotion of compatibility and processability.

The large number of synthetic methods can be classified into four types: the first technique that groups several methods is the covalent "grafting from", where polymer chains grow from the small initiators that have been pre-anchored to the Au NPs; the second method is the covalent "grafting to" technique that allows one-pot synthesis of Au NPs by using polymers that bear functional groups at the end or in the middle of the polymer chain; the third group of synthetic methods includes the protection of Au NPs with polymers that do not contain specific atoms or groups, a method known as physisorption, and which includes the use of water-soluble polymers, block copolymer micelles (nanoreactors) or star block copolymers; and finally, a fourth method is the "post-modification of pre-formed Au NPs" in which common methods such as the citrate reduction or the Brust–Schiffrin method are used to obtain very monodisperse Au NPs in the first stage, followed by a second step comprising facile exchange of weakly-bound citrate ions with polymers or modification of end-functionalized thiols with polymers. Recent examples of each synthetic method are described in the following paragraphs.

A dense polymer brush is obtained using the "grafting from" techniques. Surface-initiated polymerization in conjunction with a living polymerization technique is one of the most useful synthetic routes for the precise design and functionalization of the surfaces of various solid materials with well-defined polymers and copolymers. Above all, surface-initiated living radical polymerization (LRP) is particularly promising due to its simplicity and versatility and it has been applied for the synthesis of Au NPs.

For example, Mandal *et al.* have reported the synthesis of Au core–shell NPs containing a gold core and poly(methyl methacrylate) (PMMA) shells by surface-confined living radical polymerization on gold nanoparticles. The synthesis of Au NPs has been carried out in the presence of 11-mercaptoundecanol (MUD) and subsequent esterification with 2-bromoisobutyryl bromide (BIB). Atom transfer living radical polymerization has been conducted on the Br-terminated Au NPs using copper(I) bromide/1,4,8,11-tetramethyl-1,4,8,11-tetraazacyclotetradecane (Me$_4$Cyclam) as a catalyst. The TEM analysis displays Au NPs of 50–70 nm size and this particle size is in good agreement with the size calculated using the Mie theory from the plasmon band energy observed in the UV–Vis measurements [88].

Au NPs coated with an initiator group consisting of an alkanethiol chain bearing a 2-bromoisobutyryl group suitable for living radical polymerization have been prepared by the simple one-pot reduction of tetrachloroaurate with NaBH$_4$, in the presence of an initiator group containing disulfide. In this case, a surface-initiated living radical polymerization of methyl methacrylate (MMA) mediated by a copper complex was performed. TEM observations of the AuNPs coated with PMMA (PMMA-AuNPs) revealed that the particles were well dispersed in the polymer matrix without forming aggregates [89] (Scheme 3.13).

Spherical gold nanoparticles coated with poly(N-isopropylacrylamide) (PNIPAM) grafts have been synthesized by controlled radical polymerization. The polymerization of N-isopropylacrylamide was initiated from the surface of the nanoparticles modified with 4-cyanopentanoic acid dithiobenzoate for reversible addition–fragmentation chain-transfer polymerization. The mean diameter of the Au core was 3.2 nm, as observed by means of high-resolution transmission electron microscopy [90].

Kim *et al.* used the exchange reaction to synthesize cross-linked AuNP–PNIPAM core–shell hybrid structures, as well as a brush-type AuNP/PNIPAM hybrid through surface-initiated ATRP in an aqueous medium. The disulfide initiators, [BrC(CH$_3$)$_2$COO(CH$_2$)$_{11}$S]$_2$, were bound to AuNPs synthesized by citrate reduction. They have studied the effect of cross-linking on the thermo-responsiveness of the AuNP/PNIPAM hybrids for potential use as a stimuli responsive optical device, such as surface plasmon resonance-based sensing materials [91].

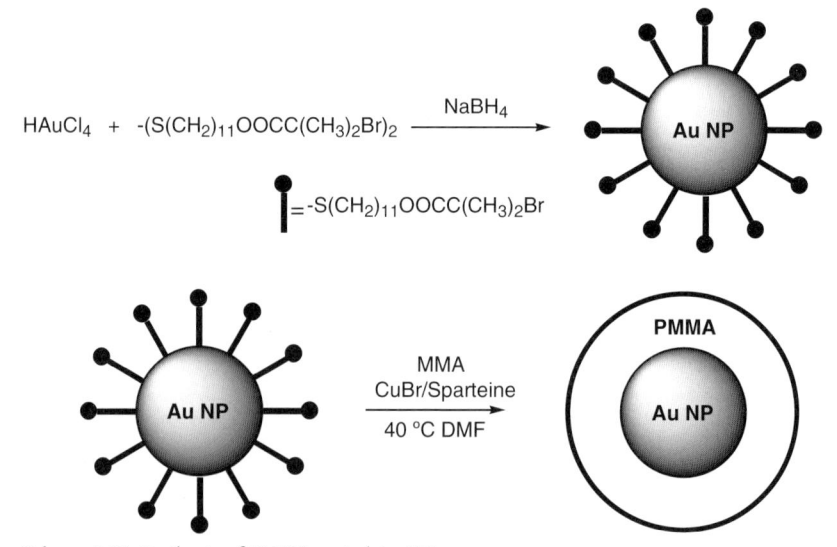

Scheme 3.13 Synthesis of PMMA-coated Au NPs.

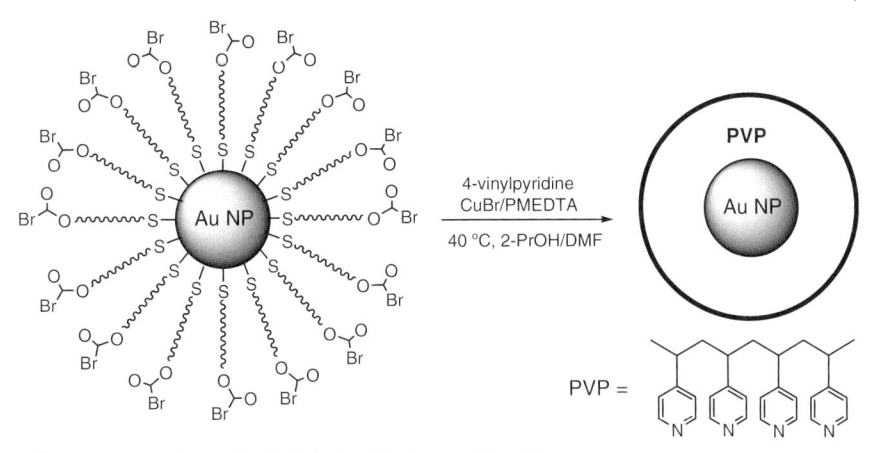

Scheme 3.14 Synthesis of poly(4-vinylpyridine)-coated Au NPs.

In a recent report, new nanocomposites of Au NPs and poly(4-vinylpyridine) were obtained through surface-initiated atom-transfer radical polymerization (SI-ATRP). The citrate-stabilized gold nanoparticles were first modified by the disulfide initiator [BrC(CH$_3$)$_2$COO(CH$_2$)$_{11}$S]$_2$ for ATRP initiation, and the subsequent polymerization of 4-vinylpyridine occurred on the surface of the gold particles. The assembled Au@PVP nanocomposites are pH-responsive because of the pyridyl groups, which are facially protonated and positively charged. The TEM micrographs show Au NPs of around 15 nm size [92] (Scheme 3.14).

As mentioned above, the "grafting to" technique enables in a one-pot reaction the synthesis of Au NPs stabilized by sulfur-containing polymers, which bear functional groups such as dithioester, trithioester, thiol, thioether and disulfide at the end of a polymer chain or in the middle. This method leads to nanoparticles similar to those obtained by the Brust–Schiffrin method in which alkanethiol-protected Au NPs of small size are obtained. This "grafting to" technique leads to very stable nanomaterials that also present a high surface graft density of polymer brush on the Au NP surface.

For example, Wuelfing et al. reported on the synthesis of Au NPs using the thiolated polymer, α-methoxy-ω-mercapto-poly(ethylene glycol) (PEG-SH), as stabilizer in a modification of the Brust–Schiffrin method using a 1/12 polymer thiol/AuCl$_4^-$ ratio. Transmission electron microscopy showed that the product had modestly polydisperse Au cores of average diameter 2.8 ± 1 nm. This nanomaterial led to characteristics uniquely different from alkanethiolate MPCs, notably aqueous solubility, thermal and chemical stability, ligand footprint size, and ionic conductivity [66].

Using the "grafting to" technique, thiol-terminated polystyrene (PS-SH) and poly(ethylene oxide) (PEO-SH) polymers were found to form dense brushes on the faceted gold nanoparticle surfaces. Depending on the polymer, the ligand grafting densities on the gold nanoparticles are 1.2- to 23.5-fold greater than those available via self-assembled monolayer formation of the corresponding two-dimensional gold surface. In terms of particle size, when PS-SH polymers were used, the nanoparticle diameters were between 4.4 and 7.7 nm; when PEO-SH was used, the size was between 3.6 and 9.2 nm [93].

Au NPs protected with a thermo-responsive polymer such as PNIPAM by the covalent "grafting to" technique with different end-functional PNIPAMs and various ratios between PNIPAM and HAuCl$_4$ has been studied. PNIPAM samples were synthesized through either conventional radical polymerization or living/controlled radical polymerization. With this approach, very small and quite monodisperse Au NPs are obtained with diameters ranging from 1.5 to 2.3 nm [94].

Gold nanoparticles coated by short thiol end functional polystyrene homopolymers (PS-SH) can be incorporated into a poly(styrene-b-2-vinylpyridine) diblock copolymer template (PS-b-P2VP). It has been found that the surface density of PS chains on the gold particles is critical in controlling their location in block copolymer templates [95].

Template core–shell particles with cores comprised mainly of poly(glycidyl methacrylate) (GMA) and shells consisting mainly of PNIPAM and amino or thiol-functionalized have been used for the synthesis of Au NPs. The obtained hybrid particles exhibited a reversible color change from red to purple, which originated from the surface plasmon resonance of gold nanoparticles and was temperature-dependent in the range 25–40 °C [96] (Scheme 3.15).

Near monodisperse Au NPs in the size range of 1–4 nm can be obtained using dodecylthioether end-functionalized PMMA as stabilizer. Particle size is controlled by varying the concentration of the stabilizing polymer, which can be readily displaced by thiol ligands to yield MPCs of the usual type [97].

Methylstyrenic polymer containing thioether side groups, PMS-(CH$_2$SCH$_3$)$_n$, has also been used to stabilize Au NPs using the "grafting to" technique, leading to small size Au NPs of approximately 3.5 nm diameter [98].

Gold nanoparticles can also be stabilized using polymers that do not have specific functional groups through physisorption. Among the possible stabilizers, the polymers used most often to stabilize Au NPs are the water soluble polymers poly(N-vinylpyrrolidone) (PVP), poly(ethylene glycol) (PEG), poly(vinyl pyridine), poly(vinyl alcohol) (PVA), poly(vinyl methyl ether) (PVME), and polyelectrolytes such as PAA, chitosan, polyethyleneimine (PEI) or poly(diallyl dimethylammonium) chloride (PDDA) [99].

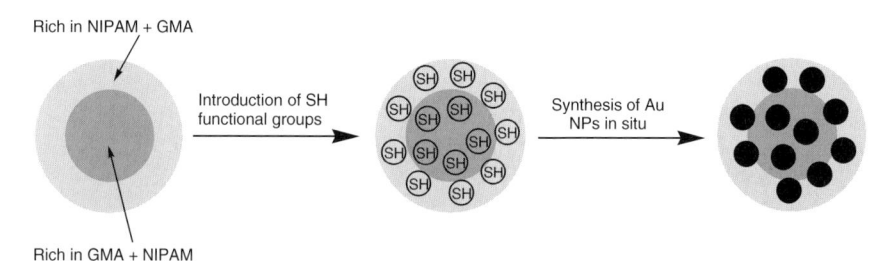

Scheme 3.15 Au NPs synthesized on template core–shell particles based on GMA and PNIPAM polymers and functionalized with thiol groups.

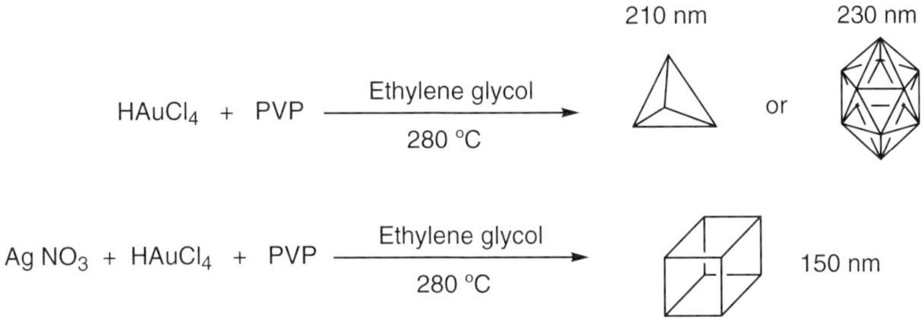

3.4 Gold Nanoparticles

$$HAuCl_4 + PVP \xrightarrow[280\,°C]{Ethylene\ glycol} \text{210 nm (tetrahedron)} \text{ or } \text{230 nm (icosahedron)}$$

$$AgNO_3 + HAuCl_4 + PVP \xrightarrow[280\,°C]{Ethylene\ glycol} \text{150 nm (cube)}$$

Scheme 3.16 Platonic gold nanocrystals of tetrahedral, cubic and icosahedral shape.

Among these polymers, the one used most often in the synthesis of Au NPs is PVP. Its use has increased most in recent years, with the focus on the less effective protection of this polymer by physisorption that permits the synthesis of particles of large size and broad size distribution but displaying various shapes such as spherical, polyhedral, rod-like, triangular and hexagonal plate-like, and so on.

For example, Kim et al. have described the synthesis of platonic gold nanocrystals by injecting PVP and $HAuCl_4$ into boiling ethylene glycol. Ethylene glycol acts as a solvent and reducing agent, while PVP acts as a particle stabilizer and controls the shape of the particles, leading to cubic (150 ± 14 nm size), tetrahedral (210 ± 20 nm size) and icosahedral nanocrystals (230 ± 20 nm size) [100] (Scheme 3.16).

In a similar reaction but changing the conditions, especially with very short reaction times (37–100 s), small Au NPs can be obtained with sizes ranging from 3.6 to 11.8 nm [101].

When the reactants are the same ($HAuCl_4$ and PVP) but the solvent is changed to water and a reduced amount of sodium citrate as reducing agent is added, gold nanoplates of triangular or truncated triangular shape are formed. These nanoplates are single crystals with a planar width of 80–500 nm and thickness 10–40 nm [102].

When changing the solvent and the reductant to N,N-dimethylformamide (DMF), the reduction of $HAuCl_4$ in the presence of PVP leads to polyhedral Au NPs of different shapes. This shape has been tuned among penta-twinned decahedrons, truncated tetrahedrons, cubes, octahedrons and hexagonal thin plates by introducing a small amount of salt (NaOH or NaCl) into a DMF solution containing PVP, and changing the temperature or the concentration of the gold precursor [103].

Large-scale icosahedral gold nanocrystals with good uniformity can be obtained in a one-step thermal process using PVP and $HAuCl_4$ as reactants in a reflux bath at 100 °C. Observation by TEM shows that most of the nanoparticles have a projected quasi-spheroid shape of 220 ± 40 nm in size [104] (Figure 3.10).

Stable gold hydrosols can also be obtained using a one-step method based on the reduction at low temperature of an aqueous solution of $HAuCl_4$ using PVP as capping and reducing agent. The PVP/Au ratio has an important effect on both the size and shape of the obtained nanoparticles. High PVP/Au ratios favor the formation of spherical nanoparticles with narrow size distributions [105].

When glycerol is used as solvent and reducing agent, the formation of Au NPs from $HAuCl_4$ and PVP can be tuned using different reaction conditions. Thus, if the

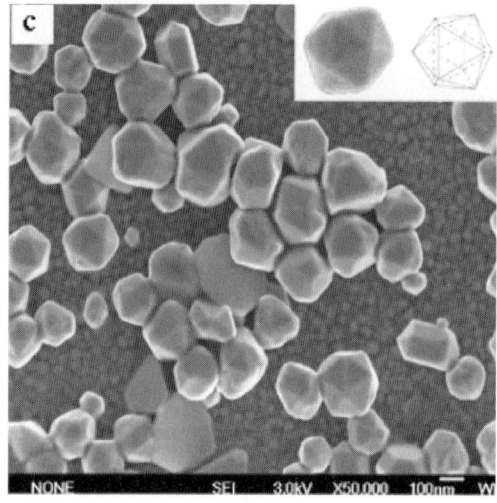

Figure 3.10 Icosahedral gold nanocrystals obtained from $HAuCl_4$ and PVP. Reprinted with permission from reference [104]. Copyright 2006 American Chemical Society.

reaction is carried out under reflux, spherical Au NPs of 7.9 ± 0.2 nm are obtained. Under microwave heating conditions, in addition to spherical NPs of 6.9 ± 0.1 nm, trigonal nanoprisms of 7.1 ± 0.2 nm are obtained [106].

Gold nanoparticles and nanoplates have also been synthesized using a continuous method under UV irradiation in the presence of citric acid and PVP and using $HAuCl_4$ as precursor. Spherical gold nanoparticles were synthesized at a higher solution flow rate (6 mL min^{-1}). Particle sizes decreased from more than 10 nm to 2 nm, with increasing concentrations of PVP in solution [107] (Figure 3.11).

Polyethylene glycol has been used as reducing and stabilizing agent for Au NPs. The stability of the resulting Au colloids and the reaction rates are dependent on polymer molar mass. The Au NPs are characterized using UV–Vis, analyzing the plasmon bands [108].

Among the water soluble polymers capable of stabilizing Au NPs through physisorption, chitosan is a widely used capping–reducing agent. This polymer is a naturally occurring polysaccharide that displays very good biodegradable and biocompatible characteristics, which prevent any environmental toxicity or biological

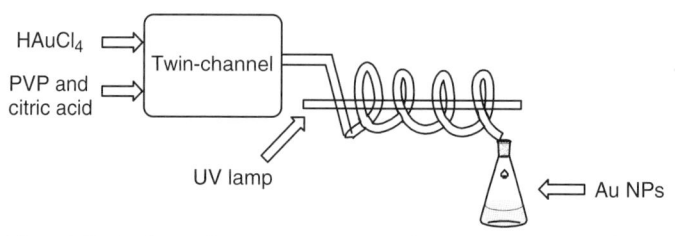

Figure 3.11 Synthesis of Au NPs using UV irradiation in a continuous flow.

hazards in the synthesis of Au NPs. When Au NPs are synthesized using $HAuCl_4$ and chitosan in the absence of tripolyphosphate (TPP), the particles obtained are spherical in shape, and particle size and size distributions vary slightly with the concentration and molecular weight of chitosan. TPP can ionically cross-link with chitosan leading to a gel-like suspension in which Au NPs can form. In this case, the TEM images show bimodal size distribution with spherical and polygonal shapes [109].

Chitosan-stabilized Au NPs can be selectively synthesized on surfaces like poly (dimethylsiloxane) (PDMS) films using $HAuCl_4$ as precursor. The computation of surface plasmon bands (SPBs) based on Mie theory and experimental results indicates that the particles are partially coated by chitosan. The proposed mechanism implies that chitosan acts as a reducing/stabilizing agent. Furthermore, PDMS films patterned with chitosan could induce localized synthesis of gold nanoparticles in regions capped with chitosan only [110].

Another class of polymers capable of stabilizing Au NPs through physisorption is amphiphilic block copolymers. Initial reports describe the formation of Au NPs in the presence of different amounts of diblock copolymers like PS-P2VP (polystyrene-block-poly-2-vinylpyridine) [111] or PS-PEO (polystyrene-block-polyethyleneoxide) [112].

Au NPs have been synthesized in polymeric micelles composed of amphiphilic block copolymers. Poly(styrene)-block-poly(2-vinylpyridine) in toluene has been used as nanocompartments loaded with a defined amount of $HAuCl_4$ and reduced with anhydrous hydrazine. The metal ions can be reduced in such a way that exactly one Au NP is formed in each micelle, where each particle is of equal size between 1 and 15 nm [113]. In another example, the addition of $HAuCl_4$ to the triblock copolymer (PS-b-P2VP-b-PEO) (polystyrene-block-poly-2-vinyl pyridine-block-polyethylene oxide) permits the synthesis of Au NPs using two different routes, such as the reduction of $AuCl_4^-$ by electron irradiation during TEM observation or by addition of an excess of aqueous $NaBH_4$ solution [114].

The use of a star block copolymer such as poly(styrene)-block-poly(2-vinylpyridine) (PS-block-P2VP) prepared from the coupling of the diblock copolymer PS-block-P2VP and ethylene glycol dimethacrylate(EGDMA) yields small Au NPs. This star-block copolymer enables the conversion of $HAuCl_4$ into single gold nanoparticles of size 4.1 ± 0.06, 5.6 ± 0.1, and 6.8 ± 0.25 nm in which the P2VP part remains at the core of the star-block copolymer [115] (Scheme 3.17).

The micellization of polystyrene-block-poly(4-vinylpyridine) (PS-block-P4VP) in chloroform is carried out through the interaction between P4VP blocks and $HAuCl_4$, leading to micelles with PS as the shell and the $P4VP/HAuCl_4$ complex as the core. The subsequent reduction of $HAuCl_4$ using hydrazine gives rise to Au NPs of about 15 nm in the core. Addition of methanol to the chloroform solution gives rise to a core–shell reversion forming vesicle-like reversed hybrid polymeric micelles (RHPMs), where PS is the wall and protonated P4VP/Au NPs is the shell [116].

Using PS-block-P2VP block copolymer in toluene, monolayers containing $HAuCl_4$ can be obtained through spin-coating. Highly-ordered monodisperse Au NPs over a large surface area can be obtained directly after oxygen plasma treatment, used to

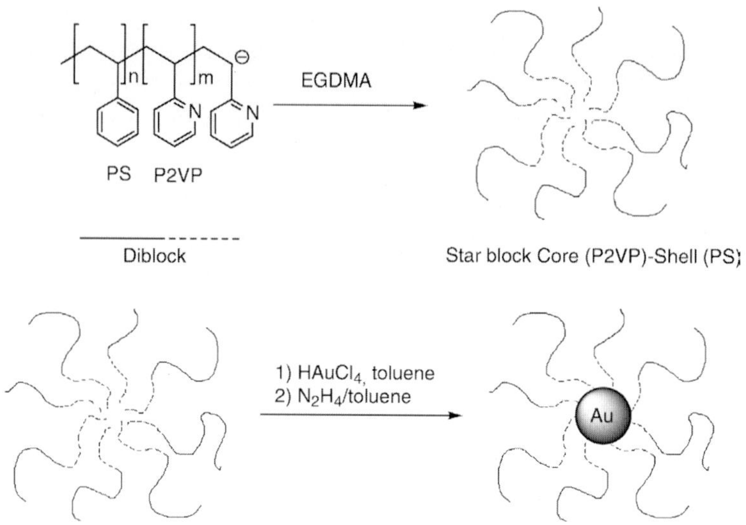

Scheme 3.17 Synthesis of Au NPs using the star-block copolymer PS-b-P2VP.

remove the block copolymer template. These Au NPs prepared from the PS-b-P2VP block copolymer template are excellent catalysts for growing silicon nanowires [117].

Poly(ethylene oxide)-poly(propylene oxide)-poly(ethylene oxide) (PEO-PPO-PEO) (Pluronic) block copolymer is a very efficient reducing agent and nanoparticle stabilizer. Au NPs of about 10 nm can be stabilized with PEO-PPO-PEO block copolymer solutions in water and at room temperature and using $HAuCl_4$ as precursor. The formation of gold nanoparticles is controlled by the overall molecular weight and relative block length of the block copolymer [118].

Au NPs have been obtained using a thermoresponsive and pH-responsive triblock copolymer such as poly(ethylene glycol)-b-poly(4-vinylpyridine)-b-poly(N-isopropylacrylamide) PEG-P4VP-PNIPAM with $HAuCl_4$ as the source of gold atoms and $NaBH_4$ the reducing agent. This approach yields discrete 2 nm size gold nanoparticles, 10 nm size gold@polymer core-shell nanoparticles, and 7 nm gold nanoparticle clusters by simply changing the pH or temperature of the aqueous solution of the PEG-P4VP-PNIPAM triblock copolymer [119].

The fourth method for the preparation of polymer stabilized Au NPs is the "postmodification of pre-formed Au NPs". This method is used to avoid broad distribution of sizes of Au NPs stabilized with polymers through any of the methods described previously. As we have mentioned before, in a first step very monodisperse Au NPs are obtained by common methods, such as the citrate reduction or the Brust–Schiffrin method. In a second step, the exchange of weakly bound citrate ions with polymer or modification of end-functionalized thiols with polymers is performed.

The first report on the use of pre-made Au NPs was the blending of these particles within a polystyrene matrix [120]. The study focused on the synthesis of novel gold nanoparticles (PS-Au) decorated with covalently bound thiol-capped polystyrene macromolecules (PS-SH), and their successful dispersion in a PS matrix. The Au

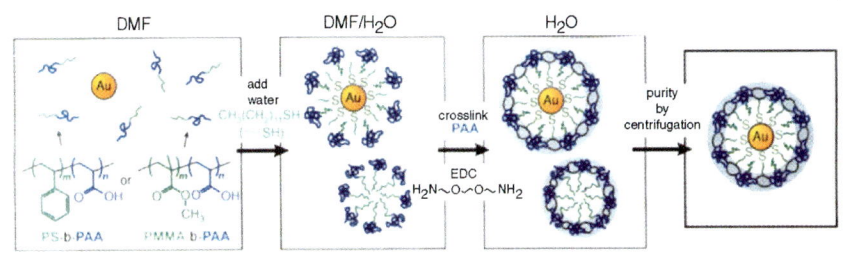

Scheme 3.18 Dodecanethiol-capped Au NPs stabilized with diblock copolymers PS-b-PAA and PMMA-b-PAA. Copyright 2005 Wiley-VCH Verlag GmbH & Co. KGaA. Reproduced with permission from reference [121].

NP diameters observed through TEM are 6.2 ± 1.7 nm showing some degree of polydispersity.

In another recent example, citrate-capped Au NPs are modified with 1-dodecanethiol in a first step. These premade nanoparticles were encapsulated with block copolymers such as poly(styrene-block-acrylic acid) (PS-b-PAA) and poly(methylmethacrylate-block-acrylic acid) (PMMA-b-PAA) leading to core–shell hybrid materials. The Au NP diameters are 12 and 31 nm with average shell thickness of about 15 nm [121] (Scheme 3.18).

Citrate-capped Au NPs have been coated with a layer composed of the double hydrophilic block copolymer poly(ethylene oxide)-block-poly(2-(dimethylamino)ethyl methacrylate)-SH (PEO-b-PDMA-SH) leading to core–shell, almost spherical, Au NPs of about 18 nm. The shell cross-linking of these hybrid Au NPs gives rise to high colloidal stability [122].

3.4.1.4 Dendrimers

Dendrimers are a special class of arborescent monodisperse nanometer sized molecules that have been used in the synthesis of Au NPs as surface stabilizers or nanoreactor/templates for nanoparticle growth. Moreover, these hybrid nanomaterials have great potential for application in different fields such as sensors, imaging in cells, electrooptical devices, catalysis, drug delivery agents, and so on.

Dendrimers can stabilize Au NPs in several ways, namely, dendrimer stabilized nanoparticles (DSNs), in which the dendrimers act as ligands using their peripheral functional groups as surface-nanoparticle stabilizers; dendrimer encapsulated nanoparticles (DENs), in which the dendrimer sequesters gold ions in a first step and a chemical reduction is carried out in a second step, leading to very small Au NPs inside the dendrimer skeleton; the third class are nanoparticle cored dendrimers (NCDs) that are built up using dendrons as ligand stabilizers for small Au NPs; finally, it is also possible to synthesize dendrimer-nanoparticle composites (DNCs) in which several Au NPs are stabilized inside the dendrimer (Figure 3.12).

In an early report by Esumi *et al.*, Au NPs were prepared by reducing $HAuCl_4$ using UV irradiation in the presence of poly(amidoamine) (PAMAM) dendrimers of generations 0 to 5. The average particle size decreased with increase in the

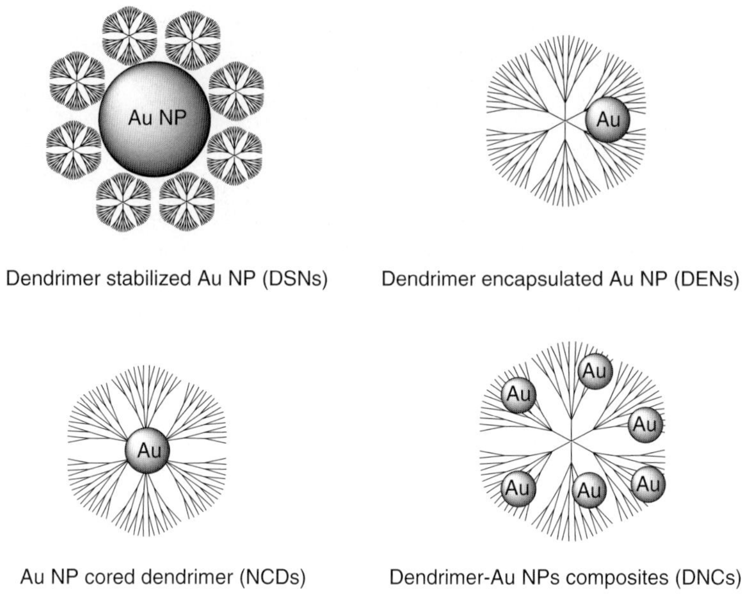

Figure 3.12 Stabilization of Au NPs using dendrimers.

concentration of surface amino groups of the dendrimers. In particular, Au NPs with a diameter less than 1 nm were obtained in the presence of the later generation dendrimers (G3-5). In this study, the authors suggest that when the generation of the dendrimers increases, particle growth is prevented by the three-dimensional structure of the dendrimer, resulting in a smaller particle size that may be in agreement with DENs [123].

Thiol-functionalized PAMAM dendrimers have also been used as surface stabilizers to obtain dendrimer stabilized Au NPs (DSNs) of very small size (1.5–2.1 nm). The study shows that dendrimer molecules are highly flexible and can undergo a conformational change to accumulate thiol terminal groups on one side of the molecule interacting with the nanoparticle surface [124] (Scheme 3.19).

Another report by Garcia *et al.* describes the synthesis of amine-functionalized PAMAM dendrimers–gold colloid nanocomposites. Synthesis was carried out using HAuCl$_4$ as metal precursor and NaBH$_4$ as reducing agent in the presence of PAMAM dendrimer. The obtained nanoparticles were very small (2–3 nm) and the authors suggest that the factor driving the interaction of the colloids with the dendrimers is an association of Au with the primary amine terminal groups and, perhaps, the interior

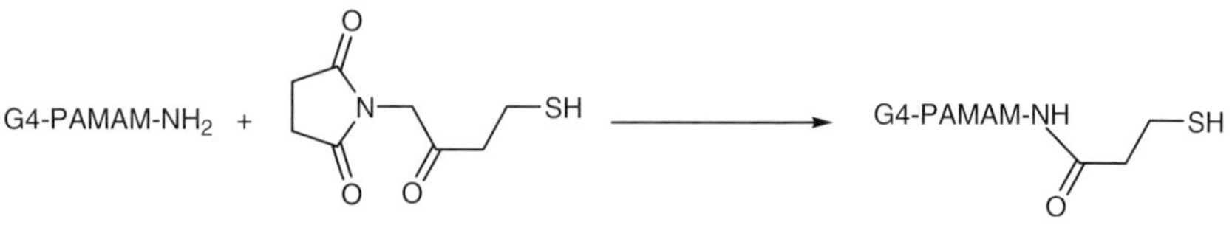

Scheme 3.19 Synthesis of a thiol-terminated PAMAM dendrimer.

secondary and tertiary amines of the dendrimer, suggesting the encapsulation of the Au NPs within the dendrimers [125].

Monodisperse, dendrimer-stabilized Au NPs (DSNs) have been obtained by reducing $HAuCl_4$ in the presence of amine-functionalized PAMAM dendrimers (G2–G6), using hydrazine as reducing agent and with the same molar ratios of dendrimer terminal nitrogen ligands/gold atoms. The sizes of the synthesized Au DSNs observed through TEM analysis decreased with the increase in the number of dendrimer generations (15.4 ± 5.8 nm for PAMAM-G2, 12.0 ± 2.8 nm for PAMAM-G3, 9.1 ± 3.2 nm for PAMAM-G4, 8.6 ± 2.8 nm for PAMAM-G5, and 7.1 ± 1.9 nm for PAMAM-G6). The authors suggest that the slow reduction process with hydrazine favors aggregation and Ostwald ripening of NPs. As a result, the formed Au DSNs were large and covered with a monolayer of dendrimer molecules. In contrast, dendrimer-encapsulated Au NPs with diameters less than 5 nm can be prepared by means of a fast reduction process using $NaBH_4$ as described below [126].

Positively charged PAMAM dendrimers are obtained in aqueous solution upon addition of $HAuCl_4$. Subsequent addition of $NaBH_4$ enables the synthesis of Au NPs. In this respect, the influence of reaction conditions and dendrimer generation on the resulting nanostructures has been studied. The characterization techniques used were TEM, small-angle neutron scattering (SANS), and small-angle X-ray scattering (SAXS), showing that lower generation dendrimers (2–4) aggregate when stabilizing the metal particles giving rise to DSNs and generation 6–9 dendrimers can template one gold colloid per dendrimer molecule leading to DENs [127].

Highly monodisperse, 1–2 nm diameter Au DENs have been synthesized using PAMAM dendrimers, having both quaternary ammonium groups and primary amines on their periphery, as templates in a workup of less than 30 min and using $NaBH_4$ as reducing agent and $HAuCl_4$ as the Au NPs source. The high monodispersity of these Au DENs is due to three reasons: a template-based approach; the high permanent charge on the surface of the dendrimers; and the use of "magic number" ratios of Au precursor:dendrimers, since Au NPs with 55 or 140 metal atoms form energetically favorable structures [128].

The third class of AuNPs-dendrimer nanomaterials is Nanoparticle Cored Dendrimers (NCDs). The first report on this topic describes a new strategy in which Fréchet-type dendrons with a single thiol group at the focal point were used as a surface stabilizer of Au NPs. These dendronized Au NPs were synthesized using a modification of the two-phase Brust–Schiffrin method, giving rise to highly stable and very monodisperse Au NPs of small size (about 2.4–3.1 nm) [129] (Figure 3.13).

This type of Au NCDs can also be obtained using similar Fréchet-type polyaryl ether dendritic disulfide wedges from generations 1 to 5 by the reduction of $HAuCl_4$ phase-transferred into toluene in the presence of the dendrons. TEM analysis shows that the NCDs are small (2.01–3.93 nm) and not very monodisperse [130].

By using a similar synthetic approach, G1-G3 4-pyridone-based dendrons have been used to stabilize monodisperse Au NPs of 2.0 ± 1.0, 3.3 ± 1.1, and 5.1 ± 1.7 nm, respectively [131].

Daniel *et al.* have also reported on the use of thiol-functionalized dendrons as stabilizers for the synthesis of NCDs of about 2.1 nm. The synthetic procedure

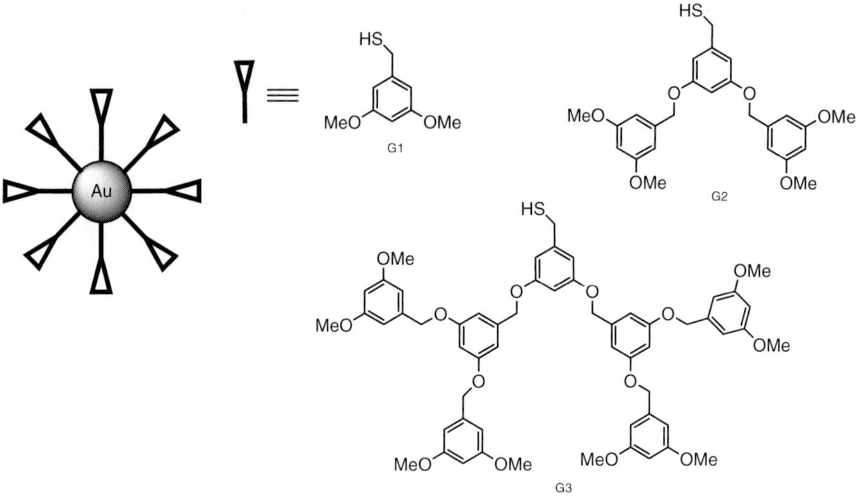

Figure 3.13 Synthesis of gold nanoparticle-cored dendrimers using Fréchet-type dendrons.

consists of a ligand exchange reaction between dodecanethiol–stabilized Au NPs prepared using the Brust–Schiffrin method and the thiol-functionalized dendron, keeping the size of the Au NPs constant during the reaction. An interesting feature of these NCDs is that the dendrons bear peripheral sylylferrocene groups that enable $H_2PO_4^-$, HSO_4^-, and adenosine-5'-triphosphate (ATP^2) to be identified by means of cyclic voltammetry [132, 133].

Dendrimer-Au NPs composites can be obtained using Au NPs with a single carboxyl group on the surface coupled to a generation 5 PAMAM dendrimer through covalent bonding. This approach can be used to obtain Au NPs with an average core diameter of 2.8 nm. Thus, in the presence of an activation agent such as di-iso-propylcarbodiimide (DIPCDI), the monocarboxylic-Au NPs are expected to form amide bonds with the amino groups on the dendrimer surface [134] (Scheme 3.20).

Bielinska et al. define DNCs as hybrid nanoparticles formed by the dispersion and immobilization of guest atoms or small clusters in dendritic polymer matrices. These authors describe the synthesis of 5–25 nm DNCs of Au NPs and G5-PAMAM dendrimers through UV decomposition of the PAMAM-HAuCl$_4$ precursor. These nanoparticles have been imaged by TEM in both *in vitro* and *in vivo* conditions [135].

3.4.1.5 Seeded Growth

The use of monodisperse colloid seed particles for the synthesis of larger core diameter nanoparticles has been reported for a long time. Nanoparticles of more than 5 nm size can be grown from seed nanoparticles by the epitaxial addition of metal salts. These metal salts can be reduced in a surface-catalyzed reaction with a mild reducing agent, in the presence or absence of surfactants, leading to larger Au NPs. The use of well-defined seeds is very important to obtain nanoobjects of narrow size dispersity or defined shape such as large spherical Au NPs or nanorods. This

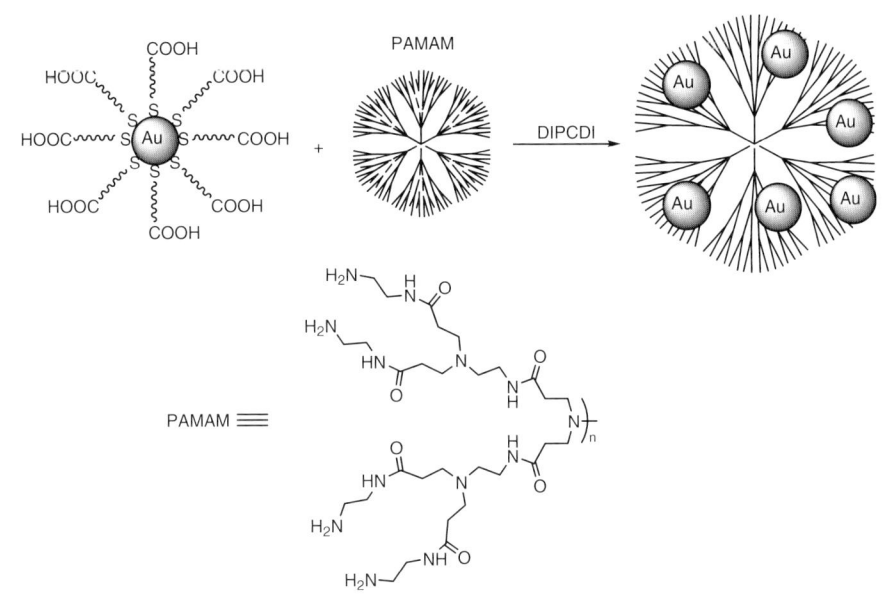

Scheme 3.20 Dendrimer-gold nanoparticle composite prepared from G5-PAMAM dendrimer and Au NPs bearing −COOH groups.

subsection describes the seeded growth of gold nanoparticles in terms of the shape of the particles, spherical or anisotropic rods.

The synthesis of spherical nanoparticles using the mediated seeded-growth method has been carried out using different mild reducing agents such as citrate, organic acids or hydroxylamine and well-defined monodisperse seed particles.

When hydroxylamine is used as reducing agent, it is possible to achieve a surface-catalyzed reduction of Au^{3+} ions on small seed Au NPs prepared using the citrate method, leading to spherical nanoparticles ranging from 30 to 100 nm in size [136]. It was noted that the reduction of Au^{3+} ions was greatly catalyzed on any surface and, therefore, stronger reducing agents could be used. Growth conditions using boiling citrate led to the growth of seed nanoparticles at the expense of nucleating new particles [137].

Citrate-stabilized Au NPs of 12 nm size have been used as seed particles in the growth of larger Au NPs using a wide range of reducing agents and conditions. The results show that a step-by-step particle enlargement procedure allows a large seed particle to Au^{3+} ratio to be maintained throughout successive growth steps [138]. Taking all these aspects into account, fairly uniform spherical Au NPs of 5–40 nm in size can be prepared using 3.5 nm size Au seeds. Primary nuclei have been prepared using $NaBH_4$ as reducing agent and citrate as stabilizing ligand. During seeding growth, ascorbic acid is used as a mild reducing agent and cetyltrimethylammonium bromide (CTAB) as the surfactant in the growth medium [139].

Au nanorods with tunable aspect ratio can also be synthesized in aqueous solution using surfactants and controlling the growth conditions of seed nanoparticles. High

Step 1 ○ ⇨ ● ⇨ ⬠

Nucleation — Growth — Faceted Nanocrystals

Step 2

Preferential binding of CTAB to {100} face

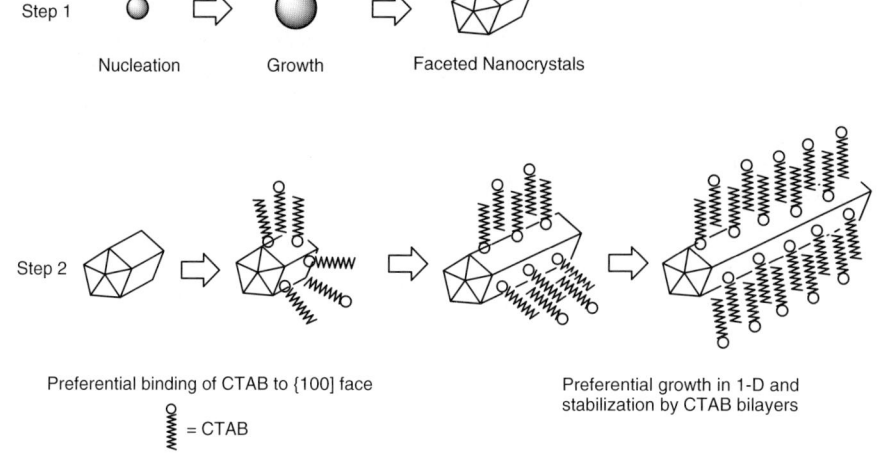

⸲ = CTAB

Preferential growth in 1-D and stabilization by CTAB bilayers

Scheme 3.21 Surfactant-directed growth of gold nanorods.

aspect ratio cylindrical Au nanorods can be obtained from 3.5 nm Au seed NPs prepared by reduction with NaBH$_4$, using citrate as capping ligand, in a two- or three-step seeding process and with careful control of the concentration of hexadecyltrimethylammonium bromide C$_{16}$TAB and ascorbic acid [140]. By changing the pH of this process, the reaction yield was improved from 4% to 90% and gold nanorods with aspect ratios of 18.8 ± 1.3 and 20.2 ± 1.2 were obtained. It seems that the growth mechanism of gold nanorods is governed by the adsorption of the surfactant (C$_{16}$TAB) to different crystal faces rather than to the tips of the rods. It is also possible that the surfactants are adsorbed onto the nanorod crystal faces in a bilayer manner, promoting a "zipping" mechanism in which longer nanorods can be obtained [141]. More recently, Murphy and coworkers have proposed a general mechanism, called surfactant-directed growth of gold nanorods, only relevant to fcc metals. It has been highlighted that the sterics of the ammonium head group of the surfactant CTAB are most compatible with the lattice arrangement observed along the crystal faces that promotes nanorod growth [142] (Scheme 3.21).

In a recent report, Pérez-Juste et al. have reported short aspect ratio (1 to 6) Au nanorods with a yield of about 50%. The control of the aspect ratio and monodispersity of the nanorods is achieved by reducing the temperature and concentration of surfactant C$_{16}$TAB. The authors also observed that C$_{16}$TAB-stabilized seed Au NPs improved nanorod yield with respect to that obtained from citrate-stabilized seed particles. An electric-field-directed growth of Au nanorods mechanism has been proposed. Thus, ascorbic acid can reduce Au^{3+} ions to Au$^+$ in the presence of C$_{16}$TAB leading to AuCl$_2^-$ C$_{16}$TAB micelles. These gold ions are preferentially delivered to the ends of a growing nanorod since this part of the rod has a higher electric field gradient than the rest of the structure and this may lead to anisotropic growth [143] (Scheme 3.22).

It is also known that the presence of AgNO$_3$ enables better control of the shape and yield of Au nanorods. The proposed mechanism is related to an adsorption of Ag$^+$

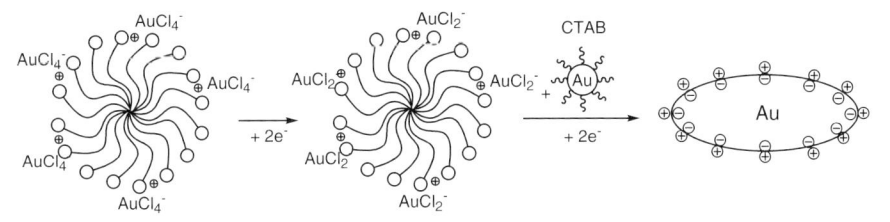

Scheme 3.22 Electric-field-directed growth of Au nanorods mechanism.

ions in the form of AgBr (Br$^-$ originating from C$_{16}$TAB) preventing the growth of passivated Au crystal facets [144]. If the seed Au particles are stabilized with C$_{16}$TAB instead of citrate and a co-surfactant such as benzyldimethylammonium bromide (BDAB) is used, the addition of Ag$^+$ ions prompts the formation of 99% Au nanorods with aspect ratios ranging from 1.5 to 5 [145].

3.4.1.6 Nanoparticle–Biomolecule Hybrids

Biomaterials such as proteins/enzymes or DNA display highly selective catalytic and recognition properties. Au nanoparticles or nanorods show electronic, photonic and catalytic properties. The convergence of both types of materials gives rise to Au NP–biomolecule hybrids that represent a very active research area. The combination of properties leads to the appearance of biosensors due to the optical or electrical transduction of biological phenomena. Moreover, multifunctional Au NP–peptide hybrids can be used for targeting nuclear cells where genetic information is stored and could be useful for biomedical applications [146].

Au NPs functionalized with biomolecules can be synthesized using different methods, depending on factors inducing the interactions promoted between the nanoparticle and the biomolecule. These interactions can be classified as electrostatic adsorption, chemisorption and covalent binding and, finally, specific affinity interactions. Some examples are given in the following paragraphs (Scheme 3.23).

Electrostatic interactions play a significant role when Au NPs are stabilized with anionic ligands, like citrate, and adsorb positively charged proteins like immunoglobulin G (IgG) at a pH above the isoelectric point of the citrate ligand. Electrostatic binding is produced between the negatively charged citrate groups at the NP surface and the amino acid side chains of the protein bearing the positive charges [147].

The photoreduction of AuCl$_4$$^-$ in the presence of dimyristoyl-L-alpha-phosphatidyl-DL-glycerol, which is a negatively charged phospholipid, gives rise to Au NPs coated through electrostatic adsorption of a biomolecule, leading to a nanosized model of a biomembrane [148].

The second group of interactions between Au NPs and biomolecules includes both the chemisorption of thiolated biomolecules such as L-cysteine to Au NPs and the covalent binding between Au NPs and biomolecules using bifunctional linkers. Naka *et al.* describe an example of the use of L-cysteine molecule-bearing thiol groups, which are used for the functionalization of Au NPs. The oligomerization of the cysteine units in solution leads to oligopeptide capping-agents for the Au NPs [149].

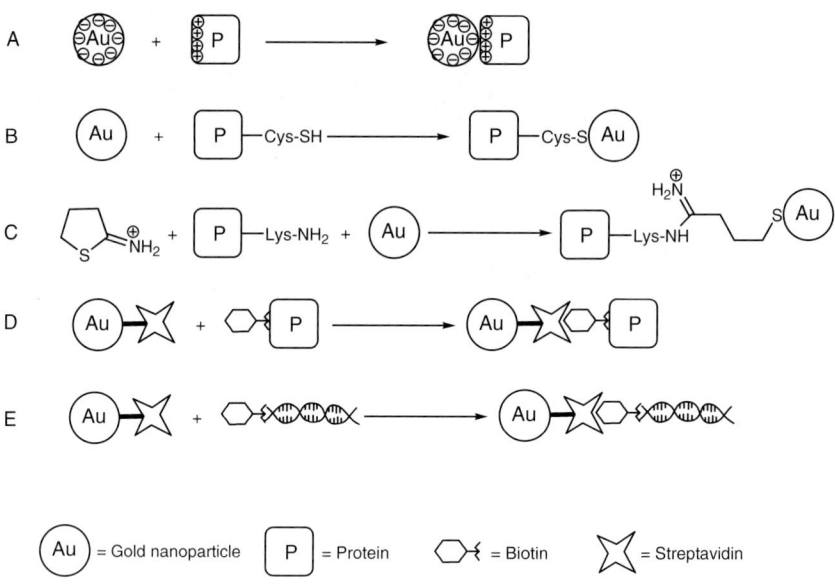

Scheme 3.23 Different types of Au NP–biomolecule hybrids depending on biomolecule–nanoparticle interaction.

Proteins like serum albumin have cysteine residues that can also stabilize Au NPs through chemisorption of the thiol groups [150].

In a recent report, 19- or 20-residue thiolated oligonucleotides were used in an exchange reaction with glutathione monolayer-protected gold clusters. After thiol substitution, the resulting DNA–nanoparticle conjugates can be separated by gel electrophoresis on the basis of the number of bound oligonucleotides and assembled with one another by DNA–DNA hybridization [151].

Bifunctional linkers bearing functional groups like thiols, disulfides or phosphines are widely used for covalent binding between Au NPs and biomolecules. For example, Au NPs stabilized with a functionalized polysiloxane shell have been used for covalent binding to biomolecules [152].

Au NP-peptide chain hybrids can be obtained by covalent binding of L-lysine residues to the surface of very small (2 nm) Au NPs prepared through solid-phase reactions [153].

DNA-functionalized Au NPs have been prepared using nucleic acids bearing n-alkylthiol groups [154]. Another study showed that thiolated single-stranded nucleic acids of different lengths (8 to 135 bases) can be attached to 10 nm size Au NPs at different Au:nucleic acid ratios [155].

The third group of Au NP–biomolecule interactions are referred to as specific affinity interactions and include nanoparticles functionalized with groups that provide affinity sites for the binding of biomolecules such as proteins and oligonucleotides.

Biotinylated proteins such as immunoglobulins and serum albumins or biotinylated oligonucleotides can bind through affinity interactions streptavidin-functionalized Au NPs [156, 157].

Recently, biotin–streptavidin-mediated aggregation of long gold nanorods has been reported. In an initial study, Caswell *et al.* built end-to-end linkages of gold nanorods driven by biotin–streptavidin interaction. These linkages are due to the preferred binding of thiol molecules to the Au{111} surface as opposed to the gold nanorod side faces [158]. In a more recent report, the same authors describe a simple procedure to biotinylate the entire gold nanorod surface and subsequently form a 3D assembly by addition of streptavidin [159].

The use of carbohydrate-stabilized Au NPs has produced a number of examples of affinity-binding interactions with biomolecules. For example, Au NPs stabilized with mannose have been used to visualize FimH proteins on type 1 Pili of *E. coli* through TEM [160]. Au NPs modified with galactosyl and glycosyl head-groups recognize the HIV-associated glycoprotein gp120 through multivalent interactions [161].

Au NPs coated with a monolayer of the synthetic drug vancomycin have shown multivalent affinity to membranes of Gram-negative bacteria leading to antimicrobial activity [162].

3.4.1.7 Nanoparticle Assembly

One of the most important objectives of nanoscience and nanotechnology is to take advantage of the unique properties of nanomaterials. Nanomaterials can be synthesized from the assembly of individual Au NPs to give more or less ordered nanostructures possessing one-, two- or three-dimensional architectures. As described in the final section of this chapter, these Au nanomaterials can be applied in different fields such as optics, electronics, sensing, catalysis, biology-related applications, and so on.

Gold nanoparticle assembly can be induced in several ways such as binding of nanoparticles to other nanoparticles or nanomaterials, deposition of nanoparticles on functionalized surfaces, or the use of templates. This subsection presents some examples of the different types of assemblies of Au NPs.

Controlled assemblies consisting of the binding of Au NPs to other Au NPs or nanomaterials are usually induced in the solution phase and can be driven through covalent bond formation, electrostatic interactions, crystallization, hydrophobic or van der Waals interactions, and so on.

Multifunctional ligands are able to link nanoparticles, for example in small assemblies like extended architectures. In early reports, Osifchin *et al.* and Brust *et al.* reported the self-assembly of alkyl-thiol capped gold nanoparticles using dithiol ligands [163, 164].

Extended networks of nanoparticles can be created using electrostatic interactions. In a recent study, fullerenes bearing piperazynyl groups were used to assemble Au NPs capped with citrate groups [165] (Figure 3.14).

It is also possible to induce self-assembly of Au NPs by changing solution conditions to induce precipitation. The addition of pentadecylamine to triphenylphosphine-stabilized gold nanoparticles in a ligand exchange reaction gives rise to two types of 2D and 3D nanocrystal superlattices (NCSs) of Au NPs: a first NCS occurring early in the ligand exchange giving Au NPs of 1.8 nm size and a second NCS at the end of the ligand exchange reaction giving Au NPs of 8 nm size [166].

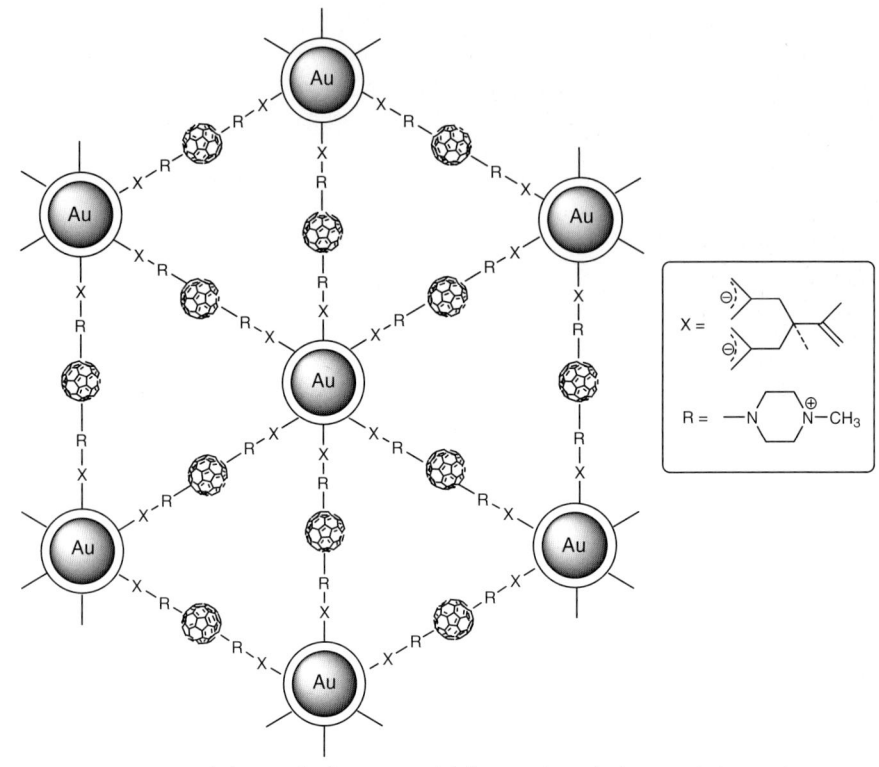

Figure 3.14 Extended network of Au NPs and fullerenes through electrostatic interactions.

Alkanethiol-capped Au NPs can spontaneously self-assemble into 2D arrays. It is worth mentioning that a bimodal population of Au NPs of well-defined sizes self-aggregate in a hexagonal-closed-packed arrangement [167].

Another interesting study describes the synthesis of large hexagonal close-packed Au NCSs from water-soluble mercaptosuccinic acid (MSA)-coated Au nanoparticles used as the building units. Au NPs that form the superlattices can be interconnected by interparticle chemical bonding thanks to the mercaptosuccinic acid molecules over the Au nanoparticle surface [168] (Figure 3.15).

Self-assemblies of gold nanorods can be generated with an aspect ratio of about 4.6 (12 nm in diameter and 50–60 nm in length) in one, two and three dimensions, leading to the formation of superlattices upon concentration from aqueous solutions [169]. Moreover, Au nanorods of high aspect ratio (13–18; mean length ∼200–290 nm), prepared using the seeding growth method, spontaneously self-assemble in concentrated solutions to produce liquid crystals based on ordered arrays of metallic nanoparticles [170].

A different approach can be used to induce nanoparticle self-assembly on surfaces or matrixes both by electrostatic interactions and chemical bonding between a functionalized nanoparticle and a surface. This is a vast area of research in which many types of substrates like Si, SiO_2, TiO_2, Al_2O_3, MgO, carbon nanotubes, and so

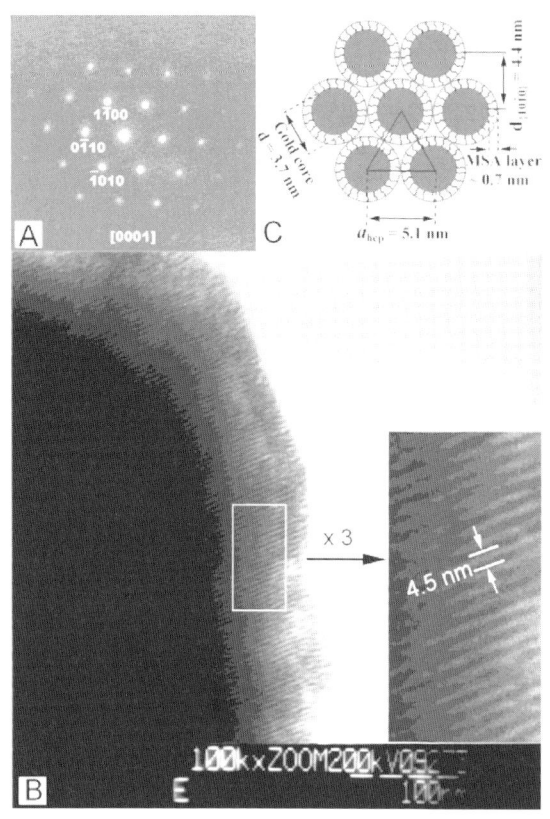

Figure 3.15 (A) Small-angle electron diffraction pattern recorded from an individual Au NCS; (B) HRTEM micrograph showing the supperlattice fringes in the hcp system. (C) Stacking model of Au NPs in the hcp system. Reprinted with permission from reference [168]. Copyright 2003 American Chemical Society.

on, have been used. A number of examples are presented below to give a representative idea of this field of research.

Self-organized 2D arrays of Au NPs on a silicon substrate modified with an aminopropyltriethoxylsilane (APS) monolayer have been reported. The Au NPs in the arrays have been capped with alkanethiols [171] (Figure 3.16).

Ultrathin TiO_2 films and modified gold nanoparticles have been used to prepare 2D Au NP monolayers. The Au NP alkylthiol capping ligands bear surface hydroxy groups that enable the attachment of the NPs to the TiO_2 surface through a sol–gel process. The average diameter of the Au NPs is 4.7 ± 0.64 nm [172] (Figure 3.16).

Recently, the chemical synthesis of micrometer-scale mirrors based on Au nanoparticles and $Ca_2Nb_3O_{10}$ nanoplates has been reported. Thus, the reaction of $Ca_2Nb_3O_{10}$ nanoplates with 3-aminopropyltrimethoxysilane produces 3-aminopropylsilyl-perovskite plates. The nanoparticle mirror is produced upon addition of a

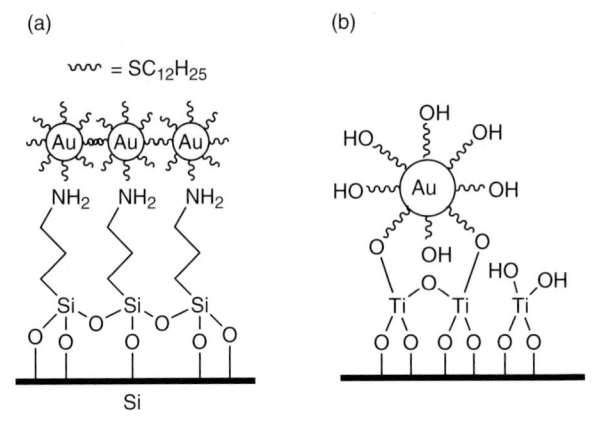

Figure 3.16 2D arrangement of Au NPs on a silicon substrate modified with an aminopropyltriethoxylsilane monolayer (a). 2D monolayer of Au NPs on ultrathin TiO_2 (b).

suspension of the perovskite colloid to an aqueous dispersion of citrate-capped Au NPs of 12.3 nm [173].

SiO_2 is a very important substrate for the preparation of gold-based nanomaterials. Depending on the choice of chemical method, different types of nanomaterials can be designed. One option is to cover Au NPs with a SiO_2 shell (Au@SiO_2 NPs) using sol–gel methods that can produce, in some cases, well ordered superstructures; Au NPs can be introduced into mesoporous SiO_2 materials bearing channels in which Au NPs can be inserted; sol–gel methods can be used to create silicate matrixes around Au NPs and it is also possible to decorate SiO_2 large nanoparticles with small Au NPs at their surface.

In an early work, Liz-Marzán et al. reported the coating of Au NPs with silica using the silane coupling agent (3-aminopropyl)-trimethoxysilane (APS) that binds the Au surface through its amine functional groups and bears the trimethoxysilane group ready for condensation with SiO_3^{2-}. The Au@SiO_2 nanoparticles can be transferred into ethanol for further growth, enabling precise control of the thickness of the silica layer from a few to several hundred nanometers. As shown in Figure 3.17, some of the Au@SiO_2 nanoparticles tend to self-assemble [174]. In a subsequent work, 3D-ordered artificial opals were prepared by sedimentation of Au@SiO_2 core–shell colloid spheres [175].

Synthetic encapsulation of 2 or 5 nm Au NPs into the pores of mesoporous silica materials MCM-41 and MCM-48 has been performed by growing the porous structure in the presence of the metal particles in an aqueous solution [176].

A two-step procedure based on the attachment of an organogold precursor within the channels of thiol-functionalized ordered mesoporous silica HMS-C16 and SBA-15 has been carried out. In a second step, the chemical reduction yields Au NPs exclusively located within the pore channels [177].

$AuCl_4^-$ ions have been stabilized by organically modified sol–gel monomers of the type $(MeO)_3Si-X-NH_2$ (X = alkyl spacer) containing amine functional groups. The

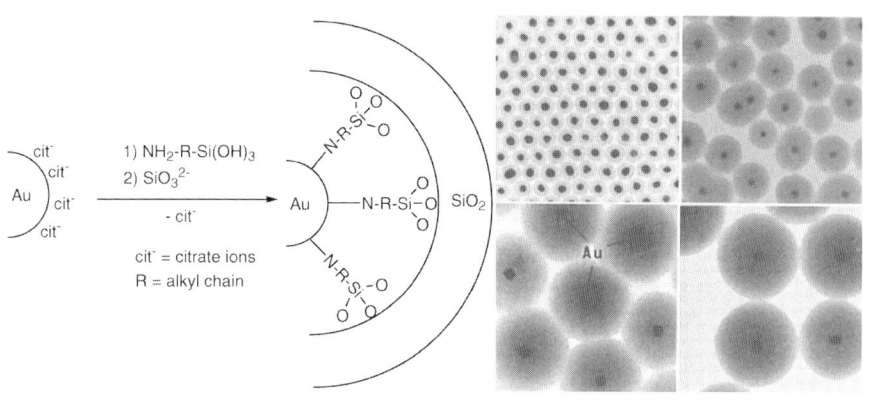

Figure 3.17 Formation of a silica shell on the surface of citrate-stabilized Au NPs (left). TEM micrographs of silica-coated Au NPs (right). Reprinted with permission from reference [174]. Copyright 1996 American Chemical Society.

hydrolysis of the material in a polycondensation step and the subsequent reduction of the Au ions with $NaBH_4$ led to aproximately 5 nm Au NPs surrounded by an organically modified silica-sol [178].

More recently, similar materials have been synthesized taking advantage of a similar sol–gel approach but using thiol functionalized sol–gel monomers such as $(MeO)_3Si–(CH_2)_3–SH$ and preformed thiol-capped Au NPs. After calcination of the materials TEM analysis showed that most of the Au NPs in the silica matrix display sizes between 1.5 and 5 nm. These nanomaterials have been employed as catalysts in CO oxidation due to their high-surface-area [179].

A new liquid-phase approach for the synthesis of small gold particles by the adsorption of complex $[Au(en)_2]^{3+}$ on a SiO_2 surface allows small Au NPs (<5 nm) to be introduced onto SiO_2. Working at high pH, SiO_2 bears a high number of negatively charged species favoring interaction with the positively-charged gold precursor. Further thermal treatment in the presence of H_2 or air gives rise to the formation of small Au NPs with sizes between 3 and 5 nm [180].

Decorated Au NPs-SiO_2 nanospheres can be prepared by reaction of 3-aminopropylsilane-modified SiO_2 nanospheres (470 nm) with citrate-coated gold nanoparticles (9.7 nm) in water. Reaction of the Au-SiO_2 nanomaterial with different chain-length alkanethiols causes the Au NPs to associate with string-like structures for short-chain thiols and hexagonally packed arrays of Au NPs for thiols with long-chain thiols on the silica surfaces [181].

3.5
Properties and Applications of Gold Nanoparticles

Although the main purpose of this chapter is to describe the most important chemical methods for the synthesis of gold nanoparticles and nanomaterials, it seems essential to at least mention the most important properties and applications of gold nanoparticles. Initially, some of the most important properties of Au NPs can be classified into two main areas, namely photophysical and catalytic properties. Since some examples of catalysis with gold are treated in another chapter of this book, we will describe here the most important photophysical properties of Au NPs. We will also emphasize the associated applications of these properties, especially in biology.

Among the photophysical properties, Au NPs have been the object of intense research due to their surface plasmon resonance (SPR). Gold nanospheres display a characteristic red color that is due to the collective oscillation of electrons in the conduction band, known as surface plasmon oscillation. The oscillation frequency usually appears in the visible for gold, leading to a characteristic SPR absorption. The SPR is absent for Au NPs with core sizes less than 2 nm and for bulk gold and it is size- and shape-dependent. For example, while 9 nm size Au NPs display an SPR absorption at 520 nm, 48 and 99 nm size Au NPs display the SPR bands at 533 and 575 nm, respectively. Gold nanorods have also attracted much attention since it is fairly easy to control their size, shape and dispersity, factors which are responsible for the changes in the SPR bands. These nanorods display two plasmon resonances, a transverse SPR that is similar to that of the nanospheres and a longitudinal SPR band that increases with larger aspect ratios [182].

Theoretically, SPR absorption can be estimated by solving Maxwell's equations. Gustav Mie rationalized this for spherical particles in 1908. Nowadays these equations can be solved to predict the corresponding SPR bands for spheres, concentric spherical shells, spheroids and infinite cylinders, and an approximation is required for other geometries. The routine measurement of the SPR absorption of most reported processes of synthesis of Au NPs is, indeed, one of the key points for the characterization of new nanomaterials [183].

This strong plasmon absorption and its sensitivity to the local environment have made Au NPs and nanomaterials attractive candidates as colorimetric sensors. Colorimetric response can be due to the metal particle aggregation, which affects the SPR band of the isolated particles due to plasmon coupling and induced dipoles.

For example, Au NPs functionalized with 15-crown-5-alkanethiol chemically recognize K^+ from an aqueous matrix containing physiologically important cations, such as Li^+, Cs^+, NH_4^+, Mg^{2+}, Ca^{2+}, and an excess amount of Na^+. The colloidal solution changes from red to blue upon addition of K^+ and this is due to the selective formation of Au NP aggregates [184] (Figure 3.18).

The SPR of Au NPs has also been applied to nanobiotechnology and Au NPs probes can be used for the colorimetric detection of oligonucleotides [185], polynucleotides with single base imperfections [186], antigens [187] or adenosine [188].

Another interesting photophysical property of Au NPs and nanorods is surface-enhanced Raman scattering (SERS), which is a powerful tool for relaying information on molecules placed on metallic substrates in the 10–200 nm size scale. Raman vibrations of isolated molecules are very weak but it is possible to take advantage of nanosized metals since the molecular Raman vibrations excited by visible light are enhanced by several orders of magnitude.

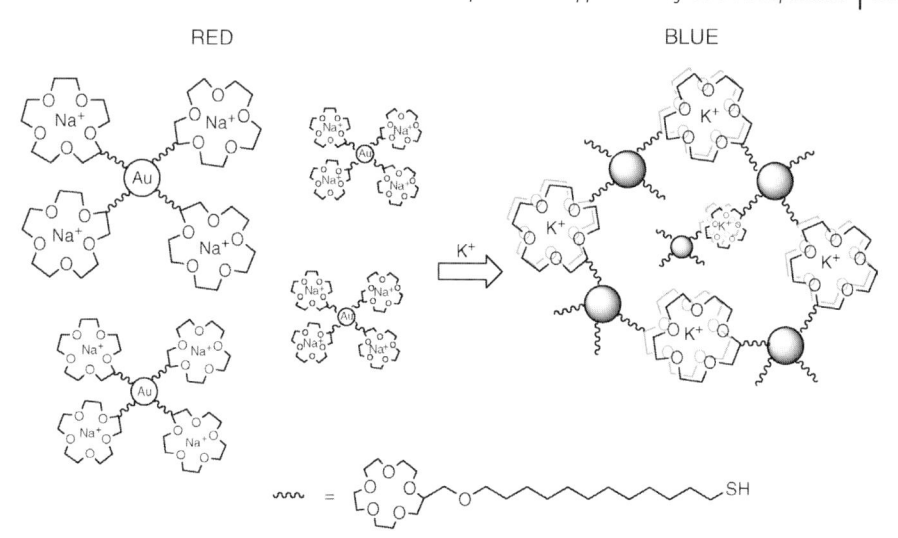

Figure 3.18 Colorimetric sensor based on Au NPs for the detection of K^+ ions.

Small Au NPs [189] and nanorods [190] display enhanced fluorescent emissions over the bulk metal that can be tuned as the size or the aspect ratio increases.

Enhanced Rayleigh (Mie) scattering can be used for the imaging of biological systems. Au NP surface plasmon resonance scattering increases as nanoparticle size is enhanced. By using this property, conjugated anti-EGFR antibody Au NP can distinguish between cancer and non-cancer cells based on the strong scattering images that the nanoparticle conjugate produces when it binds to cancer cells [191].

A very interesting and promising physical property of Au NPs is the photothermal effect. Au nanocages can be bioconjugated with antibodies for selective targeting of cancer cells. Light exposure can convert the absorbed photons into a temperature increase of the nanocage lattice. Heat dissipation may cause selective damage to cancer cells [192].

Au NPs functionalized with biomolecules present changes in the properties of the isolated nanoparticles and in their interaction with the environment. For example, Au NPs stabilized with long-chain alkanethiols are only soluble in low-polarity organic solvents, while tiopronin or coenzyme A stabilized Au NPs are completely soluble in water [193]. As is known, the use of electrical or optical signals can change the chemical properties of biomolecules. When these biomolecules are linked to Au NPs, the use of the electrical or optical signals can be employed to control the interaction of the Au NP–biomolecule hybrids with the environment. For example, it is possible to study the reaction between flavin and a diaminopyridine derivative at the surface of an Au NP through the electrochemical control of the reduced form of flavin that favors hydrogen bonding with the pyridine [194].

The chemical reactivity of biomolecules can be controlled through the study of the interactions of Au NPs with the biomaterials. Moreover, this study may report on the state of a reaction through transduction of the biomolecular transformations into an output signal, taking advantage of the unique photophysical and electronic properties of Au NPs. For example, DNA reactivity, such as the enzymatic extension of the DNA chain, can be controlled when DNA is bound to Au NPs through an alkylthiol linker [195].

The DNA hybridization of an Au NP–DNA hybrid can be controlled through inductive coupling of a radiofrequency (1 GHz) electromagnetic field to the Au NP (1.4 nm), which acts as an antenna, increasing the local temperature of the bound DNA and inducing denaturation. The switch on–off of the electromagnetic field induces a reversible denaturation–rehybridization of the DNA double-helix [196].

Au NPs can be used as reporter probes of DNA reactivity to develop nanobiosensors that are able to detect and recognize specific DNA sequences. Thus, DNA functionalized with an organic dye at the opposite end of the nucleic acid chain can bind the Au NP in a flexible arrangement leading to a short distance between the organic dye and the Au NP, thus promoting the quenching of dye fluorescence. Hybridization between this DNA and a complementary DNA leads to a rigidified spacer between the dye and the Au NP, thus inhibiting quenching [197].

As mentioned before, the changes in the properties of Au NPs upon their aggregation provides a way to identify the association process and design optical nanobiosensors through the detection of SPR absorption changes [198]. For example, Au NPs have been used as colorimetric sensors that show color changes with different aggregation states of DNA [199, 200].

The controlled aggregation of Au NPs stabilized with nucleic acids has been used as an optical probe for the detection of Pb^{2+} ions or adenosine by DNAzymes [201].

When the Au NPs–biomolecule hybrids assemble on different surfaces, a new field of properties and applications is opened. If the surfaces are transducers such as electrodes, piezoelectric crystals or field-effect transistors, biosensors and bioelectronic devices can be designed [202]. Biocatalytic electrodes for applications such as biosensors have been designed by the co-deposition of redox enzymes/proteins and Au NPs on electrode supports. For example, a single efficient electrical contact between Au NP and the redox enzyme glucose oxidase (GOx). This bioconjugate can be attached to a thiolated monolayer associated with an electrode. This display can be used to generate conductive domains and surfaces, and the conductivity properties of the system transduce a biosensing process such as the transformation of glucose into gluconic acid and the release of one electron that can be detected [203].

3.6
Concluding Remarks

From this survey it can be seen that the chemical synthesis of gold clusters, nanoparticles and nanomaterials has been applied with great success. These advances in chemical synthesis have generated a whole range of gold-based nanomaterials presenting different sizes (number of gold atoms) and shapes by using different types of stabilizing agents as ligands, polymers, dendrimers, and so on. The synthesis of Au_{55} clusters and thiolate-protected gold clusters shows the possibility of accessing even sub-nm sized gold particles with special properties related to their small size. Larger, almost monodispersed, Au NPs can be obtained using organic ligands such as amines, carboxylic acids or phosphorus-donor ligands as capping agents. The synthesis of polymer stabilized Au NPs has given rise to new core–shell structures, presenting a variety of sizes and shapes, that combine the properties of the metal nanoparticles and organic macromolecules. The design of anisotropic nanomaterials through seeded-growth, self-assembled gold nanomaterials and nanoparticle–biomolecule hybrids has also given rise to specific materials with remarkable properties and applications. In this chapter we have cited some of the most interesting properties and applications of nanoparticles, especially those related to the surface plasmon resonance photophysical properties and the chemical and biochemical sensing applications related to them.

References

1 Demartin, F., Manassero, M., Naldini, L., Ruggeri, R. and Sansoni, M. (1981) *Journal of the Chemical Society. Chemical Communications*, 222.

2 van der Velden, J.W.A., Bour, J.J., Pet, R., Bosman, W.P. and Noordik, J.H. (1983) *Inorganic Chemistry*, **22**, 3112.

3 Zeller, E., Beruda, H. and Schmidbaur, H. (1993) *Inorganic Chemistry*, **32**, 3203.

4 Yang, Y. and Sharp, P.R. (1994) *Journal of the American Chemical Society*, **116**, 6983.

5 Calhorda, M.J., Crespo, O., Gimeno, M.C., Jones, P.G., Laguna, A., Lopez-de-Luzuriaga, J.M., Perez, J.L., Ramón, M.A. and Veiros, L.F. (2000) *Inorganic Chemistry*, **39**, 4280.

6 Zhang, D., Dou, J., Li, D. and Wang, D. (2007) *Journal of Coordination Chemistry*, **60**, 825.

7 van der Velden, J.W.A., Bour, J.J., Vollenbroek, F.A., Beurskens, P.T. and Smits, J.M.M. (1979) *Journal of the Chemical Society. Chemical Communications*, 1162.

8 Ramamoorthy, V., Wu, Z., Yi, Y. and Sharp, P.R. (1992) *Journal of the American Chemical Society*, **114**, 1526.

9 Briant, C.E., Hall, K.P., Mingos, D.M.P. and Wheeler, A.C. (1986) *Journal of the Chemical Society Dalton Transactions*, 687.

10 van der Velden, J.W.A., Bour, J.J., Steggerda, J.J., Beurskens, P.T., Roseboom, M. and Noordik, J.H. (1982) *Inorganic Chemistry*, **21**, 4321.

11 van der Velden, J.W.A., Beurskens, P.T., Bour, J.J., Bosman, W.P., Noordik, J.H., Kolenbranden, M. and Buskens, J.A.K.M. (1984) *Inorganic Chemistry*, **23**, 146.

12 Marsh, R.E. (1984) *Inorganic Chemistry*, **23**, 3682.

13 Vollenbroek, F.A., Bosman, W.P., Bour, J.J., Noordik, J.H. and Beurskens, P.T. (1979) *Journal of the Chemical Society. Chemical Communications*, 387.

14 van der Velden, J.W.A., Bour, J.J., Bosman, W.P. and Noordik, J.H. (1983) *Inorganic Chemistry*, **22**, 1913.

15 Manassero, M., Naldini, L. and Sansoni, M. (1979) *Journal of the Chemical Society. Chemical Communications*, 385.

16 Strähle, J. (1995) *Journal of Organometallic Chemistry*, **488**, 15.

17 Cheetham, G.M.T., Harding, M.M., Haggitt, J.L., Mingos, D.M.P. and Powell, H.R. (1993) *Journal of the Chemical Society. Chemical Communications*, 1000.

18 Cooper, M.K., Dennis, G.R., Hendrick, K. and McPartlin, M. (1980) *Inorganica Chimica Acta*, **45**, L151.

19 Cariati, F. and Naldini, L. (1972) *Journal of the Chemical Society Dalton Transactions*, 2286.

20 Dyson, P.J. and Mingos, D.M.P. (1999) in *Homonuclear Clusters and Colloids of Gold:*

Synthesis, Reactivity, Structural and Theoretical Considerations in Gold: Progress in Chemistry, Biochemistry and Technology, (ed. H. Schmidbaur), Wiley, Chichester.

21 Hall, K.P. (1981) Part II Thesis, University of Oxford.

22 Bellon, P.L., Cariati, F., Manassero, M. Naldini, L. and Sansoni, M. (1971) *Journal of the Chemical Society. Chemical Communications*, 1423.

23 Hall, K.P., Theobald, B.R.C., Gilmour, D.I., Mingos, D.M.P. and Welch, A.J. (1982) *Journal of the Chemical Society. Chemical Communications*, 529.

24 Briant, C.E., Hall, K.P. and Mingos, D.M.P. (1984) *Journal of the Chemical Society. Chemical Communications*, 291.

25 Briant, C.E., Hall, K.P., Wheeler, A.C. and Mingos, D.M.P. (1984) *Journal of the Chemical Society. Chemical Communications*, 248.

26 Laguna, A., Laguna, M. and Gimeno, M.C. (1992) *Organometallics*, **11**, 2759.

27 Cheetham, G.M.T., Harding, M.M., Haggitt, J.L., Mingos, D.M.P. and Powell, H.R. (1993) *Journal of the Chemical Society. Chemical Communications*, 1000.

28 McCleverty, J.A. and da Motta, M.M.M. (1973) *Journal of the Chemical Society Dalton Transactions*, 2571.

29 Yanagimoto, Y., Negishi, Y., Fujihara, H. and Tsukuda, T. (2006) *Journal of Chemical Physics B*, **110**, 11611.

30 Vollenbroek, F.A., Bouten, D.C.P., Trooster, J.M., van der Berg, J.P. and Bour, J.J. (1978) *Inorganic Chemistry*, **17**, 1345.

31 Briant, C.E., Theobald, B.R.C., White, J.W., Bell, L.K. and Mingos, D.M.P. (1981) *Journal of the Chemical Society. Chemical Communications*, 201.

32 Bos, W., Kanters, R.P.F., van Halen, C.J., Bosman, W.P., Behm, H., Smits, J.M.M., Beurskens, P.T., Bour, J.J. and Pignolet, L.H. (1986) *Journal of Organometallic Chemistry*, **307**, 385.

33 Smits, J.M.M., Bour, J.J., Vollenbroek, F.A. and Beurskens, P.T. (1983) *Journal of Crystallographic and Spectroscopic Research*, **13**, 355.

34 Vollenbroek, F.A., van der Velden, J.W.A., Bour, J.J. and Trooster, J.M. (1981) *Recueill des Travaux Chimique des Pays-Bas*, **100**, 375.

35 Teo, B.K., Shi, X. and Zhang, H. (1992) *Journal of the American Chemical Society*, **114**, 2743.

36 Woehrle, G.H., Warner, M.G. and Hutchison, J. (2002) *The Journal of Physical Chemistry. B*, **106**, 9979.

37 Warner, M.G. and Hutchison, J. (2005) *Inorganic Chemistry*, **44**, 6149.

38 Nunokawa, K., Onaka, S., Ito, M., Horibe, M., Yonezawa, T., Nishihara, H., Ozeki, T., Chiba, H., Watase, S. and Nakamoto, M. (2006) *Journal of Organometallic Chemistry*, **691**, 638.

39 Menard, L.D., Gao, S.-P., Xu, H., Twesten, R.D., Harper, A.S., Song, Y., Wang, G., Douglas, A.D., Yang, J.C., Frenkel, A.I., Nuzzo, R.G. and Murray, R.W. (2006) *The Journal of Physical Chemistry. B*, **110**, 12874.

40 Shichibu, Y., Negishi, Y., Watanabe, T., Chaki, N.K., Kawaguchi, H. and Tsukuda, T. (2007) *The Journal of Physical Chemistry. C*, **111**, 7845.

41 Crespo, O., Gimeno, M.C., Jones, P.G., Laguna, A. and Villacampa, M.D. (1997) *Angewandte Chemie (International Edition in English)*, **36**, 993.

42 Richter, M. and Strähle, J. (2001) *Zeitschrift für Anorganische und Allgemeine Chemie*, **627**, 918.

43 Jeffery, J.C., Jelliss, P.A. and Stone, F.G.A. (1994) *Journal of the Chemical Society Dalton Transactions*, 25.

44 Schmid, G., Boese, R., Pfeil, R., Bandermann, F., Meyer, S., Calis, G.H.M. and van der Velden, J.W.A. (1981) *Chemische Berichte*, **114**, 3634.

45 Thiel, R.C., Benfield, R.E., Zanoni, R., Smit, H.H.A. and Dirken, M.W. (1993) *Structure and Bonding*, **81**, 1, and references therein.

46 Rapoport, D.H., Vogel, W., Cölfen, H. and Schlögl, R. (1997) *The Journal of Physical Chemistry. B*, **101**, 4175.

47 Schmid, G., Pugin, R., Sawitowski, T., Simon, U. and Marler, B. (1999) *Chemical Communications*, 1303.
48 Schmid, G., Pugin, R., Malm, J.-O. and Bovin, J.-O. (1998) *European Journal of Inorganic Chemistry*, 813.
49 Schmid, G., Pugin, R., Meyer-Zaika, W. and Simon, U. (1999) *European Journal of Inorganic Chemistry*, 2051.
50 Balasubramanian, R., Guo, R., Mills, A.J. and Murray, R.W. (2005) *Journal of the American Chemical Society*, **127**, 8126.
51 Inomata, T. and Konishi, K. (2003) *Chemical Communications*, 1282.
52 Schmid, G., Liu, Y.-P.-., Schumann, M., Raschke, T. and Radehaus, C. (2001) *Nano Letters*, **1**, 405.
53 Schmid, G., Bäumle, M. and Beyer, N. (2000) *Angewandte Chemie-International Edition*, **39**, 181.
54 Reuter, T., Neumeier, S., Schmid, G., Koplin, E. and Simon, U. (2005) *European Journal of Inorganic Chemistry*, 3670.
55 Neumeier, S., Reuter, T. and Schmid, G. (2005) *European Journal of Inorganic Chemistry*, 3679.
56 Lotz, A.R. and Fröba, M. (2005) *Zeitschrift fur Anorganische und Allgemeine Chemie*, **631**, 2800.
57 Schmid, G., Vidoni, O., Torma, V., Pollmeier, K., Rehage, H. and Vassiliev, A. (2005) *Zeitschrift fur Anorganische und Allgemeine Chemie*, **631**, 2792.
58 Schmid, G., Meyer-Zaika, W., Pugin, R., Sawitowski, T., Majoral, J.-P., Caminade, A.-M. and Turrin, C.-O. (2000) *Chemistry – A European Journal*, **6**, 1693.
59 Faraday, M. (1857) *Philosophical Transactions of the Royal Society of London*, **147**, 145.
60 Turkevich, J., Stevenson, P.C. and Hillier, J. (1951) *Discussions of the Faraday Society*, **11**, 55.
61 Frens, G. (1973) *Nature: Physical Science*, **241**, 20.
62 Brust, M., Walker, M., Bethell, D., Schiffrin, D.J. and Whyman, R. (1994) *Journal of the Chemical Society. Chemical Communications*, 801.
63 Schaaff, T.G., Knight, G., Shafigullin, M.N., Borkman, R.F. and Whetten, R.L. (1999) *The Journal of Physical Chemistry. B*, **102**, 10643.
64 Negishi, Y., Takasugi, Y., Sato, S., Yao, H., Kimura, K. and Tsukuda, T. (2004) *Journal of the American Chemical Society*, **126**, 6518.
65 Templeton, A.C., Chen, S., Gross, S.M. and Murray, R.W. (1999) *Langmuir*, **15**, 66.
66 Wuelfing, W.P., Gross, S.M., Miles, D.T. and Murray, R.W. (1998) *Journal of the American Chemical Society*, **120**, 12696.
67 Chen, S. and Murray, R.W. (1999) *Langmuir*, **15**, 682.
68 Price, R.C. and Whetten, R.L. (2005) *Journal of the American Chemical Society*, **127**, 13750.
69 Jimenez, V.L., Georganopoulou, D.G., White, R.J., Harper, A.S., Mills, A.J., Lee, D. and Murray, R.W. (2004) *Langmuir*, **20**, 6864.
70 Jørgensen, J.M., Erlacher, K., Pedersen, J.S. and Gothelf, K.V. (2005) *Langmuir*, **21**, 10320.
71 Donkers, R.L., Lee, D. and Murray, R.W. (2004) *Langmuir*, **20**, 1945.
72 Wang, W. and Murray, R.W. (2005) *Langmuir*, **21**, 7015.
73 Woehrle, G.H., Warner, M.G. and Hutchison, J.E. (2002) *The Journal of Physical Chemistry. B*, **106**, 9979.
74 Woehrle, G.H., Brown, L.O. and Hutchison, J.E. (2005) *Journal of the American Chemical Society*, **127**, 2172.
75 Shichibu, Y., Negishi, Y., Tsukuda, T. and Teranishi, T. (2005) *Journal of the American Chemical Society*, **127**, 13464.
76 Negishi, Y. and Tsukuda, T. (2003) *Journal of the American Chemical Society*, **125**, 4046.
77 Gomez, S., Philippot, K., Collière, V., Chaudret, B., Senocq, F. and Lecante, P. (2000) *Chemical Communications*, 1945.
78 Kumar, A., Mandal, S., Selvakannan, P.R., Pasricha, R., Mandale, A.B. and Sastry, M. (2003) *Langmuir*, **19**, 6277.
79 Aslam, M., Fu, L., Su, M., Vijayamohanan, K. and Dravid, V.P.

(2004) *Journal of Materials Chemistry*, **14**, 1795.
80 Uznanski, P., Amiens, C., Chaudret, B. and Bryszewska, E. (2006) *Polish Journal of Chemistry*, **80**, 1845.
81 Zhu, H., Pan, Z., Hagaman, E.W., Liang, C., Overbury, S.H. and Dai, S. (2005) *Journal of Colloid and Interface Science*, **287**, 360.
82 Green, M. and O'Brien, P. (2000) *Chemical Communications*, 183.
83 Bunge, S.D., Boyle, T.J. and Headley, T.J. (2003) *Nano Letters*, **3**, 901.
84 Fleming, D.A. and Williams, M.E. (2004) *Langmuir*, **20**, 3021.
85 Weare, W.W., Reed, S.M., Warner, M.G. and Hutchison, J.E. (2000) *Journal of the American Chemical Society*, **122**, 12890.
86 Moores, A., Goettmann, F., Sanchez, C. and Le Floch, P. (2004) *Chemical Communications*, 2842.
87 Yamamoto, M. and Nakamoto, M. (2003) *Chemistry Letters*, **32**, 452.
88 Mandal, T.K., Fleming, M.S. and Walt, D.R. (2002) *Nano Letters*, **2**, 3.
89 Ohno, K., Koh, K., Tsujii, Y. and Fukuda, T. (2002) *Macromolecules*, **35**, 8989.
90 Raula, J., Shan, J., Nuopponen, M., Niskanen, A., Jiang, H., Kauppinen, E.I. and Tenhu, H. (2003) *Langmuir*, **19**, 3499.
91 Kim, D.J., Kang, S.M., Kong, B., Kim, W.-J., Paik, H., Choi, H. and Choi, I.S., (2005) *Macromolecular Chemistry and Physics*, **206**, 1941.
92 Li, D., He, Q., Cui, Y. and Li, J. (2007) *Chemistry of Materials*, **19**, 412.
93 Corbierre, M.K., Cameron, N.S. and Lennox, R.B. (2004) *Langmuir*, **20**, 2867.
94 Shan, J., Nuopponen, M., Jiang, H., Kauppinen, E. and Tenhu, H. (2003) *Macromolecules*, **36**, 4526.
95 Kim, B.J., Bang, J., Hawker, C.J. and Kramer, E.J. (2006) *Macromolecules*, **39**, 4108.
96 Suzuki, D. and Kawaguchi, H. (2005) *Langmuir*, **21**, 8175.
97 Hussain, I., Graham, S., Wang, Z., Tan, B., Sherrington, D.C., Rannard, S.P., Cooper, A.I. and Brust, M. (2005) *Journal of the American Chemical Society*, **127**, 16398.
98 Huang, H.-M., Chang, C.-Y., Liu, I.-C., Tsai, H.-C., Lai, M.-K. and Tsiang, R.C.-C. (2005) *Journal of Polymer Science Part A: Polymer Chemistry*, **43**, 4710.
99 Shan, J. and Tenhu, H. (2007) *Chemical Communications*, 4580, and references therein.
100 Kim, F., Connor, S., Song, H., Kuykendall, T. and Yang, P. (2004) *Angewandte Chemie-International Edition*, **43**, 3673.
101 Salvati, R., Longo, A., Carotenuto, G., De Nicola, S., Pepe, G.P., Nicolais, L. and Barone, A. (2005) *Applied Surface Science*, **248**, 28.
102 Ah, C.S., Yun, Y.J., Park, H.J., Kim, W.-J., Ha, D.H. and Yun, W.S. (2005) *Chemistry of Materials*, **17**, 5558.
103 Chen, Y., Gu, X., Nie, C.-G., Jiang, Z.-Y., Xie, Z.-X. and Lin, C.-J. (2005) *Chemical Communications*, 4181.
104 Zhou, M., Chen, S. and Zhao, S. (2006) *The Journal of Physical Chemistry. B*, **110**, 4510.
105 Hoppe, C.E., Lazzari, M., Pardiñas-Blanco, I. and López-Quintela, M.A. (2006) *Langmuir*, **22**, 7027.
106 Grace, A.N. and Pandian, K. (2006) *Colloid Surface A*, **290**, 138.
107 Yang, S., Zhang, T., Zhang, L., Wang, S., Yang, Z. and Ding, B. (2007) *Colloid Surface A*, **296**, 37.
108 Longenberger, L. and Mills, G. (1995) *The Journal of Physical Chemistry*, **99**, 475.
109 Huang, H. and Yang, X. (2004) *Biomacromolecules*, **5**, 2340.
110 Wang, B., Chen, K., Jiang, S., Reincke, F., Tong, W., Wang, D. and Gao, C. (2006) *Biomacromolecules*, **7**, 1203.
111 Spatz, J.P., Mössmer, S. and Möller, M. (1996) *Angewandte Chemie (International Edition in English)*, **35**, 1510.
112 Möller, M., Spatz, J.P. and Roescher, A. (1996) *Advanced Materials*, **8**, 337.
113 Spatz, J.P., Mössmer, S., Hartmann, C., Möller, M., Herzog, T., Krieger, M., Boyen, H.-G., Ziemann, P. and Kabius, B. (2000) *Langmuir*, **16**, 407.

114 Gohy, J.-F., Willet, N., Varshney, S., Zhang, J.-X. and Jérôme, R. (2001) *Angewandte Chemie-International Edition*, **40**, 3214.
115 Youk, J.H., Park, M.-K., Locklin, J., Advincula, R., Yang, J. and Mays, J. (2002) *Langmuir*, **18**, 2455.
116 Hou, G., Zhu, L., Chen, D. and Jiang, M. (2007) *Macromolecules*, **40**, 2134.
117 Lu, J.Q. and Yi, S.S. (2006) *Langmuir*, **22**, 3951.
118 Sakai, T. and Alexandridis, P. (2004) *Langmuir*, **20**, 8426.
119 Zheng, P., Jiang, X., Zhang, X., Zhang, W. and Shi, L. (2006) *Langmuir*, **22**, 9393.
120 Corbierre, M.K., Cameron, N.S., Sutton, M., Mochrie, S.G.J., Lurio, L.B., Rühm, A. and Bruce Lennox, R. (2001) *Journal of the American Chemical Society*, **123**, 10411.
121 Kang, Y. and Taton, T.A. (2005) *Angewandte Chemie-International Edition*, **44**, 409.
122 Luo, S., Xu, J., Zhang, Y., Liu, S. and Wu, C. (2005) *The Journal of Physical Chemistry. B*, **103**, 22159.
123 Esumi, K., Suzuki, A., Aihara, N., Usui, K. and Torigoe, K. (1998) *Langmuir*, **14**, 3157.
124 Chechik, V. and Crooks, R.M. (1999) *Langmuir*, **15**, 6364.
125 Garcia, M.E., Baker, L.A. and Crooks, R.M. (1999) *Analytical Chemistry*, **71**, 256.
126 Shi, X., Ganser, T.R., Sun, K., Balogh, L.P. and Baker, J.R., Jr (2006) *Nanotechnology*, **17**, 1072.
127 Gröhn, F., Bauer, B.J., Akpalu, Y.A., Jackson, C.L. and Amis, E.J. (2000) *Macromolecules*, **33**, 6042.
128 Kim, Y.-G., Oh, S.-K. and Crooks, R.M. (2004) *Chemistry of Materials*, **16**, 167.
129 Kim, M.-K., Jeon, Y.-M., Jeon, W.S., Kim, H.-J., Hong, S.G., Park, C.G. and Kim, K. (2001) *Chemical Communications*, 667.
130 Gopidas, K.R., Whitesell, J.K. and Fox, M.A. (2003) *Journal of the American Chemical Society*, **125**, 6491.
131 Wang, R., Yang, J., Zheng, Z., Carducci, M.D., Jiao, J. and Seraphin, S. (2001) *Angewandte Chemie-International Edition*, **40**, 549.
132 Daniel, M.-C., Ruiz, J., Nlate, S., Palumbo, J., Blais, J.-C. and Astruc, D. (2001) *Chemical Communications*, 2000.
133 Daniel, M.-C., Ruiz, J., Nlate, S., Blais, J.-C. and Astruc, D. (2003) *Journal of the American Chemical Society*, **125**, 2167.
134 Worden, J.G., Dai, Q. and Huo, Q. (2006) *Chemical Communications*, 1536.
135 Bielinska, A., Eichman, J.D., Lee, I., Baker, J.R., Jr and Balogh, L. (2002) *Journal of Nanoparticle Research*, **4**, 395.
136 Brown, K.R. and Natan, M.J. (1998) *Langmuir*, **14**, 726.
137 Brown, K.R., Walter, D.G. and Natan, M.J. (2000) *Chemistry of Materials*, **12**, 306.
138 Jana, N.R., Gearheart, L. and Murphy, C.J. (2001) *Chemistry of Materials*, **13**, 2313.
139 Jana, N.R., Gearheart, L. and Murphy, C.J. (2001) *Langmuir*, **17**, 6782.
140 Jana, N.R., Gearheart, L. and Murphy, C.J. (2001) *The Journal of Physical Chemistry. B*, **105**, 4065.
141 Gao, J., Bender, C.M. and Murphy, C.J. (2003) *Langmuir*, **19**, 9065.
142 Murphy, C.J., Sau, T.K., Gole, A.M., Orendorff, C.J., Gao, J., Gou, L., Hunyadi, S.E. and Li, T. (2005) *The Journal of Physical Chemistry. B*, **109**, 13857.
143 Pérez-Juste, J., Liz-Marzán, L.M., Carnie, S., Chan, D.Y.C. and Mulvaney, P. (2004) *Advanced Functional Materials*, **14**, 571.
144 Jana, N.R., Gearheart, L. and Murphy, C.J. (2001) *Advanced Materials*, **13**, 1389.
145 Nikoobakht, B. and El-Sayed, M.A. (2003) *Chemistry of Materials*, **15**, 1957.
146 Tkachenko, A.G., Xie, H., Coleman, D., Glomm, W., Ryan, J., Anderson, M.F., Franzen, S. and Feldheim, D.L. (2003) *Journal of the American Chemical Society*, **125**, 4700.
147 Shenton, W., Davis, S.A. and Mann, S. (1999) *Advanced Materials*, **11**, 449.
148 Ibano, D., Yolota, Y. and Tominaga, T. (2003) *Chemistry Letters*, **32**, 574.
149 Naka, K., Itoh, H., Tampo, Y. and Chujo, Y. (2003) *Langmuir*, **19**, 5546.
150 Hayat, M.A. (1989) *Colloidal Gold: Principles, Methods and Applications*, Academic Press, New York.

151 Ackerson, C.J., Sykes, M.T. and Kornberg, R.D. (2005) *Proceedings of the National Academy of Sciences of the United States of America*, **102**, 13383.

152 Schroedter, A. and Weller, H. (2002) *Angewandte Chemie-International Edition*, **41**, 3218.

153 Sung, K.-M., Mosley, D.W., Peelle, B.R., Zhang, S. and Jacobson, J.M. (2004) *Journal of the American Chemical Society*, **126**, 5064.

154 Mirkin, C.A., Letsinger, R.L., Mucic, R.C. and Storhoff, J.J. (1996) *Nature*, **382**, 607.

155 Parak, W.J., Pellegrino, T., Micheel, C.M., Gerion, D., Williams, S.C. and Alivisatos, A.P. (2003) *Nano Letters*, **3**, 33.

156 Niemeyer, C.M. (2001) *Angewandte Chemie-International Edition*, **40**, 4128.

157 Gestwicki, J.E., Strong, L.E. and Kiessling, L.L. (2000) *Angewandte Chemie-International Edition*, **39**, 4567.

158 Caswell, K.K., Wilson, J.N., Bunz, U.H.F. and Murphy, C.J. (2003) *Journal of the American Chemical Society*, **125**, 13914.

159 Gole, A. and Murphy, C.J. (2005) *Langmuir*, **21**, 10756.

160 Lin, C.-C., Yeh, Y.-C., Yang, C.-Y., Chen, C.-L., Chen, G.-F., Chen, C.-C. and Wu, Y.-C. (2002) *Journal of the American Chemical Society*, **124**, 3508.

161 Nolting, B., Yu, J.-J., Liu, G., Cho, S.-J., Kauzlarich, S. and Gervay-Hague, J. (2003) *Langmuir*, **19**, 6465.

162 Gu, H., Ho, P.L., Tong, E., Wang, L. and Xu, B. (2003) *Nano Letters*, **3**, 1261.

163 Osifchin, R.G., Andres, R.P., Henderson, J.I., Kubiak, C.P. and Domine, R.N. (1996) *Nanotechnology*, **7**, 412.

164 Brust, M., Fink, J., Bethell, D., Schiffrin, D.J. and Kiely, C. (1995) *Journal of the Chemical Society, Chemical Communications*, 1655.

165 Lim, I.-I.S., Ouyang, J., Luo, J., Wang, L., Zhou, S. and Zhong, C.-J. (2005) *Chemistry of Materials*, **17**, 6528.

166 Brown, L.O. and Hutchison, J.E. (2001) *The Journal of Physical Chemistry. B*, **105**, 8911.

167 Kiely, C.J., Fink, J., Brust, M., Bethell, D. and Schiffrin, D.J. (1998) *Nature*, **396**, 444.

168 Wang, S., Sato, S. and Kimura, K. (2003) *Chemistry of Materials*, **15**, 2445.

169 Nikoobakht, B., Wang, Z.L. and El-Sayed, M.A. (2000) *The Journal of Physical Chemistry. B*, **104**, 8635.

170 Jana, N.R., Gearheart, L.A., Obare, S.O., Johnson, C.J., Edler, K.J., Mann, St. and Murphy, C.J. (2002) *Journal of Materials Chemistry*, **12**, 2909.

171 Liu, S., Zhu, T., Hu, R. and Liu, Z. (2002) *Physical Chemistry Chemical Physics*, **4**, 6059.

172 Yonezawa, T., Matsune, H. and Kunitake, T. (1999) *Chemistry of Materials*, **11**, 33.

173 Kim, J.Y. and Osterloh, F.E. (2006) *Journal of the American Chemical Society*, **128**, 3868.

174 Liz-Marzán, L.M., Giersig, M. and Mulvaney, P. (1996) *Langmuir*, **12**, 4329.

175 García-Santamaría, F., Salgueiriño-Maceira, V., López, C. and Liz-Marzán, L.M. (2002) *Langmuir*, **18**, 4519.

176 Kónya, Z., Puntes, V.F., Kiricsi, I., Zhu, J., Ager, J.W., III, Ko, M.K., Frei, H., Alivisatos, P. and Somorjai, G.A. (2003) *Chemistry of Materials*, **15**, 1242.

177 Guari, Y., Thieuleux, C., Mehdi, A., Reyé, C., Corriu, R.J.P., Gomez-Gallardo, S., Philippot, K. and Chaudret, B. (2003) *Chemistry of Materials*, **15**, 2017.

178 Bharathi, S. and Lev, O. (1997) *Chemical Communications*, 2303.

179 Budroni, G. and Corma, A. (2006) *Angewandte Chemie-International Edition*, **45**, 3328.

180 Zanella, R., Sandoval, A., Santiago, P., Basiuk, V.A. and Saniger, J.M. (2006) *The Journal of Physical Chemistry. B*, **110**, 8559.

181 Osterloh, F., Hiramatsu, H., Porter, R. and Guo, T. (2004) *Langmuir*, **20**, 5553.

182 Hu, M., Chen, J., Li, Z.-Y., Au, L., Hartland, G.V., Li, X., Marquez, M. and Xia, Y. (2006) *Chemical Society Reviews*, **35**, 1084.

183 Mie, G. (1908) *Annals of Physics*, **25**, 377.

184 Lin, S.-Y., Liu, S.-W., Lin, C.-M. and Chen, C. (2002) *Analytical Chemistry*, **74**, 330.

185 Reynolds, R.A., III, Mirkin, C.A. and Letsinger, R.L. (2000) *Journal of the American Chemical Society*, **122**, 3795.

186 Storhoff, J.J., Elghanian, R., Mucic, R.C., Mirkin, C.A. and Letsinger, R.L. (1998) *Journal of the American Chemical Society*, **120**, 1959.

187 Thanh, N.T.K. and Rosenzweig, Z. (2002) *Analytical Chemistry*, **74**, 1624.

188 Liu, J. and Lu, Y. (2004) *Analytical Chemistry*, **76**, 1627.

189 Nie, S. and Emory, S.R. (1997) *Science*, **275**, 1102.

190 Eustis, S. and El-Sayed, M. (2005) *The Journal of Physical Chemistry. B*, **109**, 16350.

191 El-Sayed, I.H., Huang, X. and El-Sayed, M.A. (2005) *Nano Letters*, **5**, 829.

192 Chen, J., Wiley, B., Li, Z.Y., Campbell, D., Saeki, F., Cang, H., Au, L., Lee, J., Li, X. and Xia, Y. (2005) *Advanced Materials*, **17**, 2255.

193 Templeton, A.C., Chen, S., Gross, S.M. and Murray, R.W. (1999) *Langmuir*, **15**, 66.

194 Boal, A.K. and Rotello, V.M. (1999) *Journal of the American Chemical Society*, **121**, 4914.

195 Peña, S.R.N., Raina, S., Goodrich, G.P., Fedoroff, N.V. and Keating, C.D. (2002) *Journal of the American Chemical Society*, **124**, 7314.

196 Hamad-Schifferli, K., Schwartz, J.J., Santos, A.T., Zhang, S. and Jacobson, J.M. (2002) *Nature*, **415**, 152.

197 Maxwell, D.J., Taylor, J.R. and Nie, S. (2002) *Journal of the American Chemical Society*, **124**, 9606.

198 Elghanian, R., Storhoff, J.J., Mucic, R.C., Letsinger, R.L. and Mirkin, C.A. (1997) *Science*, **277**, 1078.

199 Reynolds, R.A., III, Mirkin, C.A. and Letsinger, R.L. (2000) *Journal of the American Chemical Society*, **122**, 3795.

200 Souza, G.R. and Miller, J.H. (2001) *Journal of the American Chemical Society*, **123**, 6734.

201 Liu, J. and Lu, Y. (2004) *Analytical Chemistry*, **76**, 1627.

202 Katz, E. and Willner, I. (2004) *Angewandte Chemie-International Edition*, **43**, 6042.

203 Xiao, Y., Patolsky, F., Katz, E., Hainfeld, J.F. and Willner, I. (2003) *Science*, **299**, 1877.

4
Gold–Heterometal Interactions and Bonds
Cristian Silvestru

4.1
Introduction

The chemistry of compounds containing heterometallic gold–metal bonds or interactions has raised considerable interest in the last decades due to their potential application in solid-state chemistry, catalysis, the electronics industry and biochemistry besides the traditional uses of this metal [1]. The development of this topic in gold chemistry was largely supported by the progress achieved in the single-crystal X-ray technique which provided an extremely powerful tool for the structural characterization of a continuously increasing number of gold compounds.

This chapter is dedicated to the structural chemistry of molecular compounds containing gold–heterometal bonds and/or interactions, supermolecules (*"well-defined discrete oligomolecular species that result from the intermolecular association of a few components"* [2–4]) and supramolecular assemblies achieved in the solid state. The intermolecular bond types for supramolecular assemblies which are considered in this chapter will be only gold–heterometal dative bonding (electron-pair donor–acceptor bonding or Lewis acid–base interactions), secondary bonds or ionic interactions [5]. The supramolecular nature of gold complexes based on *aurophilic* [that is the gold (I)–gold(I) bonding interaction] (see also Chapter 2) or on gold–nonmetal (see also Chapter 5) interactions will be considered only for compounds that also contain gold–heterometal bonds.

A "secondary bond", as defined by Alcock [6–8], is an interaction between two atoms characterized by a distance longer than the sum of the covalent radii but shorter than the sum of the van der Waals radii of the corresponding atoms. Such secondary interactions are weaker than normal covalent or dative bonds, but strong enough to connect individual molecules and to modify the coordination geometry of the atoms involved. They are often present in a crystal, thus resulting in self-assembled supermolecules or supramolecular architectures. For gold complexes,

Table 4.1 Atomic radii for gold and other metals and the corresponding gold–metal distances (Å) [9a].

Atom X	r_{cov}	Σr_{cov} (Au,M)	r_{vdW}	Σr_{vdW} (Au,M)	Atom X	r_{cov}	Σr_{cov} (Au,M)	r_{vdW}	Σr_{vdW} (Au,M)
Au	1.34	2.68	1.70[a]	3.40	Ti	1.32	2.66	2.00[b]	3.70
					V	1.32	2.66	2.00[b]	3.70
Li	1.23	2.57	1.80[a]	3.50	Nb	1.34	2.68	2.00[b]	3.70
Na	1.54[a]	2.88	2.31	4.01	Cr			2.00[b]	3.70
K	2.03	3.37	2.31	4.01	Mo	1.29	2.63	2.00[b]	3.70
Ga	1.25	2.59	1.90[a]	3.60	W	1.30	2.64	2.00[b]	3.70
In	1.50	2.84	1.90[a]	3.60	Mn	1.17	2.51	2.00[b]	3.70
Tl	1.55	2.89	2.00[a]	3.70	Tc			2.00[b]	3.70
Si	1.17	2.51	2.10[a]	3.80	Re	1.28	2.62	2.00[b]	3.70
Ge	1.22	2.56	2.00[b]	3.70	Fe	1.16	2.50	2.00[b]	3.70
Sn	1.40	2.74	2.20[a]	3.90	Ru	1.24	2.58	2.00[b]	3.70
Pb	1.54	2.88	2.00[a]	3.70	Os	1.26	2.60	2.00[b]	3.70
Sb	1,41	2.75	2.20	3.86	Co	1.16	2.50	2.00[b]	3.70
Bi	1.52	2.86	2.40	4.06	Rh	1.25	2.59	2.00[b]	3.70
					Ir	1.26	2.60	2.00[b]	3.70
Hg	1.44	2.78	1.50[a]	3.20	Ni	1.15	2.49	1.60[a]	3.30
					Pd	1.28	2.62	1.60[a]	3.30
					Pt	1.29	2.63	1.70[a]	3.40
					Cu	1.17	2.51	1.40[a]	3.10
					Ag	1.34	2.68	1.70[a]	3.40

[a] Ref. [9b].
[b] Ref. [9c].

the interatomic distances for these additional interactions cover quite a large range. Table 4.1 shows the covalent and van der Waals radii for Au and metals involved in bonds to the gold center [9], as well as the gold–element distances estimated from these values, which were used for comparison with the interatomic distances actually found in the discussed compounds.

The organization of the chapter will follow the groups containing metals in the Periodic Table. A search based on the Cambridge Structure Database as well as the original literature up to 2007 was used to collect all the available structure determinations by X-ray diffractometry. A table format has been chosen in some cases to summarize the compounds belonging to a particular class, also including some common structural features (e.g., monomeric, dimeric, or supramolecular associations) as well as bond distances and interatomic angles. The structural diagrams were redrawn, using the program DIAMOND [10], on the basis of reported atomic coordinates, and the original atom numbering scheme was usually preserved. When the molecular parameters for the structure discussed were not available in the original articles or the deposited CIF files, they were also calculated using the program DIAMOND. In some cases, for clarity and a better view of the coordination environment around gold or heterometal atoms, only parts of the organic groups are represented.

4.2
Main Group Metal–Gold Compounds

In contrast with other topics in coordination gold chemistry which grew rapidly in the last decades, for example, gold complexes with nitrogen, phosphorus or chalcogen donor ligands [11–17], organogold compounds [14, 18–22] or homo- and heteronuclear gold clusters [23–27], the chemistry of main group metal–gold compounds is very limited and is still waiting to be developed in the future. This section will review the literature on the structure of stoichiometrically defined molecular compounds containing Au–main group metal and Au–Hg bonds or interactions. Alkali metal aurides or solid-state materials with interactions between gold and alkali-earth metals are not included since they do not contain discrete molecular clusters. To the best of our knowledge there are no reports so far on species with Au–alkali earth, Au–Cd, Au–Zn or Au–Al bonds.

4.2.1
Gold–Group 1 Metal Compounds

The structures of only four complexes containing Au–Li, Au–Na or Au–K interactions have been described. Reactions of Li[C_6H_4(CH_2NMe_2)-2] with Au(PPh$_3$)Br or Au$_2$[C_6H_4(CH_2NMe_2)-2]$_2$ (2 : 1 molar ratio) gave the tetranuclear species Au$_2$Li$_2$[C_6H_4-(CH_2NMe_2)-2]$_4$ **1** whose dimeric nature is preserved in benzene solution. On the basis of NMR and ^{197}Au Mössbauer data it was suggested that the aryl ligands bridge gold and lithium atoms [28]. The solid-state structure was established by single-crystal X-ray diffraction [29]. The crystal contains discrete molecular units built from two almost linear [Ar$_2$Au]$^-$ anions [C(1)–Au(1)–C(19) 167(1)°] which coordinate two Li$^+$ cations through the nitrogens of the pendant arms. The result is a planar *trans*-Au$_2$Li$_2$ core with Au–Li interactions in the range 2.848(2)–2.903(2) Å, Li–Au–Li and Au–Li–Au angles close to 90° and distorted tetrahedral coordination of the alkali metal (Figure 4.1). No further inter-dimer metal–metal interactions are present in the crystal.

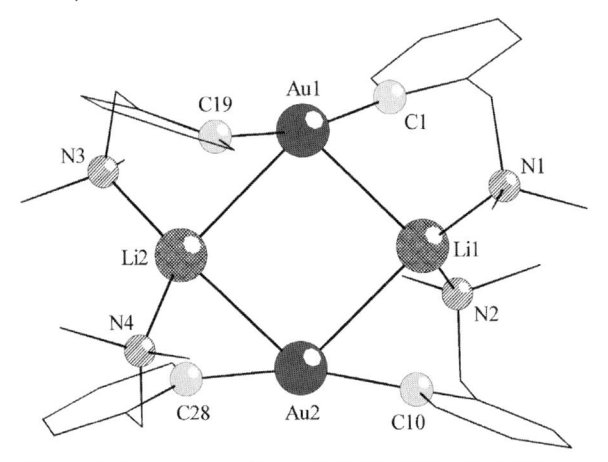

Figure 4.1 The structure of Au$_2$Li$_2$[C_6H_4(CH_2NMe_2)-2]$_4$ **1** [29].

$$[Au_2Na(Ph_2phen)_3](PF_6)_3 \rightleftharpoons [Au_2(Ph_2phen)_3](PF_6)_2 + NaPF_6$$

$$\updownarrow$$

$$[Au_2(Ph_2phen)_2](PF_6)_2 + P_2phen$$

Scheme 4.1

A gold-based metallocryptate, $[Au_2Na(P_2phen)_3](PF_6)_3$ **2**, was prepared from P_2phen, $NaPF_6$ and $Au(tht)Cl$ (3:2:3 molar ratio) as a yellow solid. Its ^{31}P NMR spectrum in $CDCl_3/CH_2Cl_2$ solution is consistent with a fast loss of Na^+ ion resulting in an equilibrium mixture according to Scheme 4.1. Addition of $TlNO_3$ to a solution of **2** resulted in exchange of Na by Tl and formation of a new metallocryptate, $[Au_2Tl(P_2phen)_3](PF_6)_3$ [30].

The solid state structure of **2** revealed a Na^+ ion loosely held inside a cage capped by two trigonal AuP_3 fragments (Figure 4.2a). The position of the alkali atom is disordered over three partially occupied sites, with an average Au—Na—Au angle of 163.4°. The average Au—Na distance (2.845 Å) is much smaller than the sum of the van der Waals radii of the corresponding atoms [$\Sigma r_{vdW}(Au,Na)$ 4.0 Å] [9]. Taking into account the expected repulsive interaction between the Na^+ ion and the Au(I) centers,

(a)

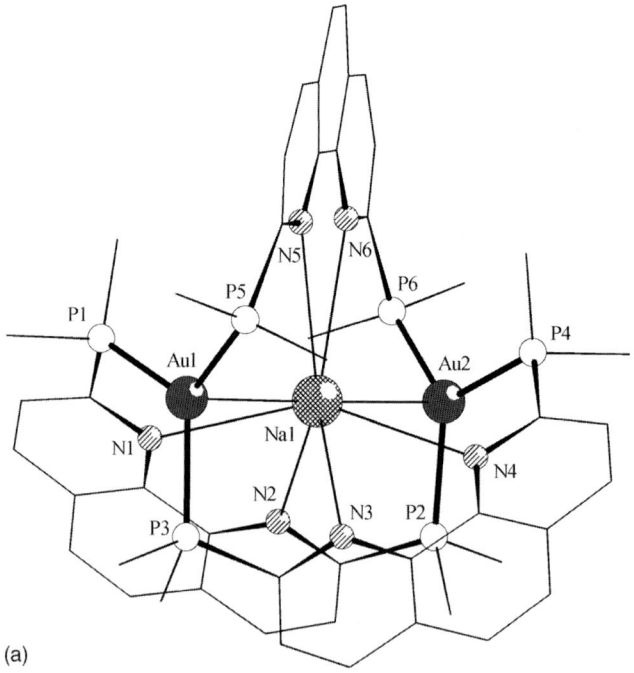

Figure 4.2 The structure of the cation of (a) $[Au_2Na(P_2phen)_3](PF_6)_3$ **2** [only Na(1) atom is shown] [30], and (b) $[Au_2K(P_2napy)_3](ClO_4)_3$ **3** [32] (the phenyl rings, less *ipso* carbons, on phosphorus have been omitted for clarity).

(b)

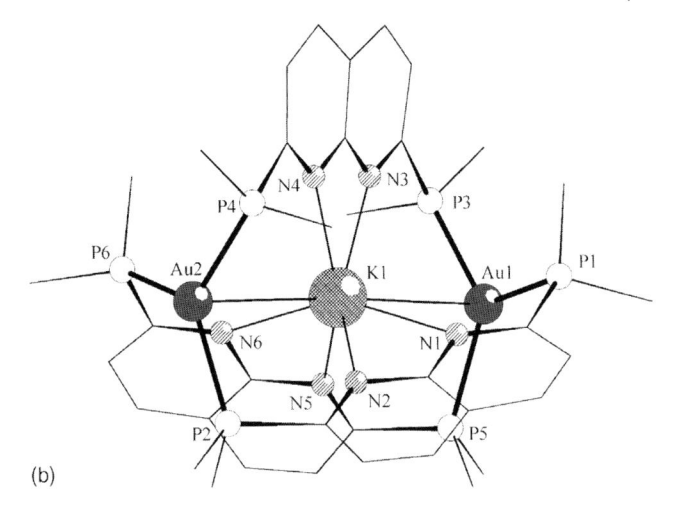

Figure 4.2 (Continued).

it is likely that the alkali metal is kept inside the cavity based on the Na−N (average 3.21 Å) interactions which are, however, significantly weaker than found in the related phenantroline-based cryptate [Na{N(CH$_2$-phen-CH$_2$)$_3$N}]Br (2.70 Å) [31].

A related yellow potassium metallocryptate, [Au$_2$K(P$_2$napy)$_3$](ClO$_4$)$_3$ **3**, was isolated after LiClO$_4$ treatment of the reaction mixture obtained from P$_2$napy and an Au(I) complex prepared *in situ* by reduction of K[AuCl$_4$] with 2,2-thioethanol, in methanol [32]. The structure of the cation in **3** (Figure 4.2b) is similar to that of the Na complex **2**. The Au−K distances are slightly different [Au(1)−K(1) 3.615(5), Au(1)−K(2) 3.575(5) Å] and the Au(1)−K(1)−Au(2) angle is close to linearity [177.1(1)°]. The gold atoms of the trigonal AuP$_3$ fragments in **3** are deviated from the P$_3$ plane outside the cage [Au(1) 0.178, Au(2) 0.201 Å] resulting in an Au−Au separation of 7.19 Å. This contrasts with the behavior found for **2**, that is, gold atoms displaced from the P$_3$ planes towards the central Na atom [Au(1) 0.230, Au(2) 0.237 Å] and Au−Au separation of 5.61 Å, the effect being at least in part a contribution from the larger size of the K$^+$ cation.

The structure of the first coordinating polymer of tetrachloroaurate, [K(4,4'-bipy)(AuCl$_4$)]$_n$ **4**, was reported recently [33]. The compound was obtained from K[AuCl$_4$] and 4,4'-bipyridine (2 : 1 molar ratio) in ethanol/water, in open atmosphere. The crystal contains linear chains of K atoms bridged by the nitrogen atoms of the aromatic ligand [N(1)−K(1)−N(1ab) 180°], which extend along the *a* axis. These chains are further bridged through planar [AuCl$_4$]$^−$ anions. This results in linear −Au−K−Au−K− chains along the *c* axis [K(1)−Au(1)−K(1") 180°], with K atoms in the axial positions of the tetragonally elongated K$_2$AuCl$_4$ octahedron. The Au^{III}−K distances [Au(1)−K(1) 3.680(1) Å] in **4** are close to the average value (3.595 Å) of the Au^I−K distances in **3**. The overall arrangement is of a layer with alternating metal atoms in the corners of rectangular cavities (Figure 4.3). In the crystal the layers stack parallel to each other, with an interlayer separation of 3.916(2) Å.

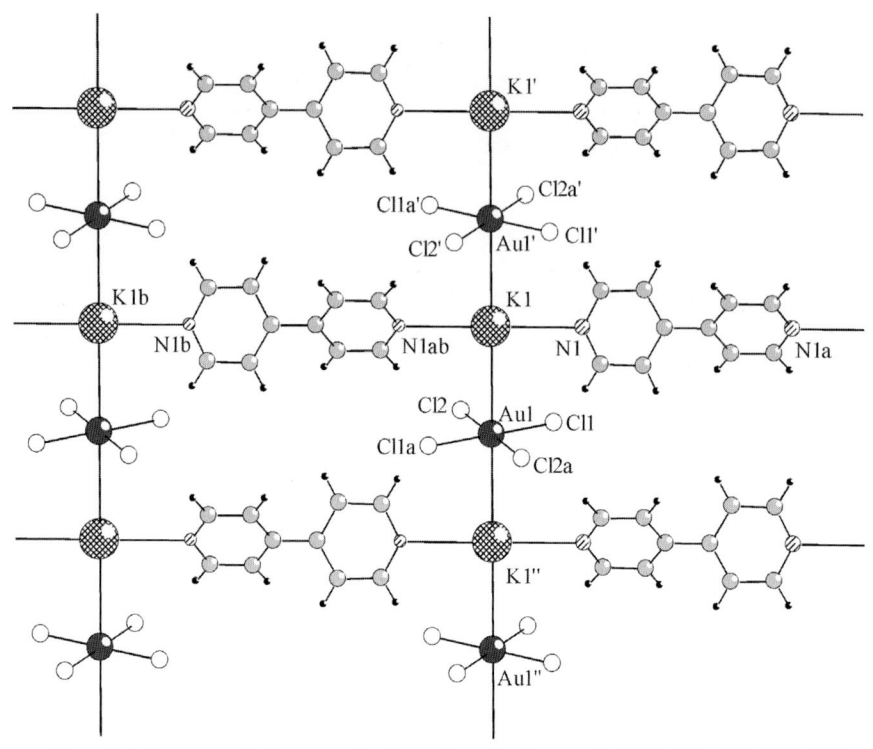

Figure 4.3 Fragment of a layer coordination polymer in the crystal of [K(4,4′-bipy)(AuCl$_4$)]$_n$ **4** [33].

4.2.2
Gold–Mercury Compounds

In view of the filled d shells, the elements belonging to Group 12, as expected, exhibit few of the characteristic properties of transition metals. For this reason the compounds containing Au—Hg bonds or interactions will be treated together with those dedicated to main group metals. No molecular compounds containing bonds or interactions between gold and zinc or cadmium atoms were found in a search on the Cambridge Structure Database.

Most compounds with Au—Hg bonds whose structure was established by X-ray diffraction fit into two main groups: (i) ylide complexes, and (ii) M-centered, Au-rich clusters.

The chemistry of metal–gold-containing ylide bimetallic complexes based on methylenethiophosphinate ligand was developed by Fackler and coworkers [34–39]. The chemical similarities of Au(I) and Hg(II) ions were used to synthesize a series of complexes, as shown in Scheme 4.2.

Compounds **5** and **7** contain the same [AuHg{CH$_2$P(S)Ph$_2$}$_2$]$^+$ cation (Figure 4.4a) in which two carbon atoms coordinate linearly to the Hg(II) center and the Au(I) center is bonded linearly to the sulfur atoms, thus resulting in an eight-membered

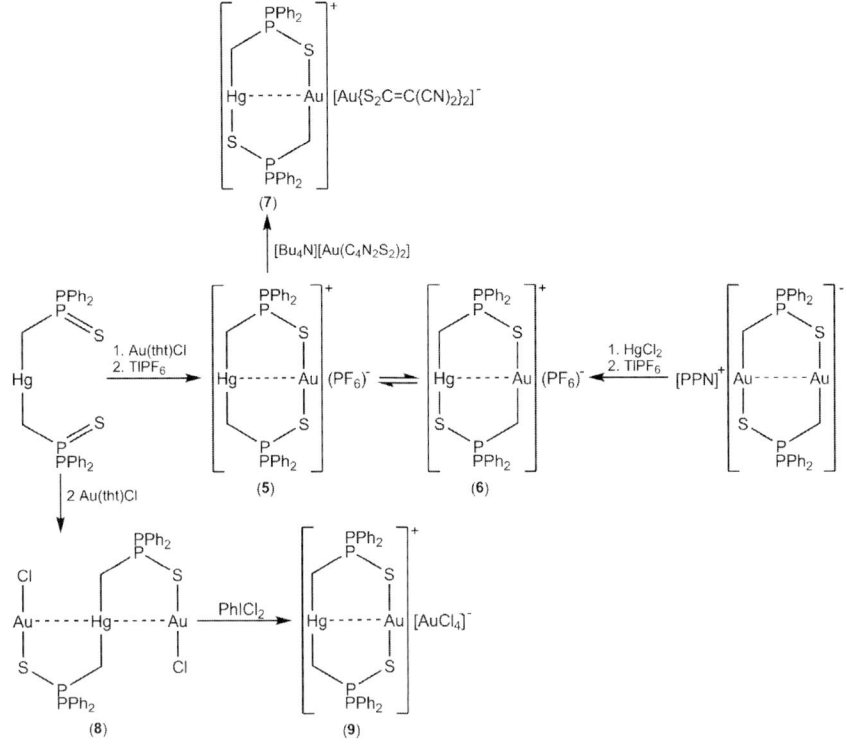

Scheme 4.2

ring. A weak *transannular* gold–mercury interaction is established in each compound, the interatomic Au–Hg distance being shorter than the sum of the van der Waals radii of the corresponding atoms [Σr_{vdW}(Au,Hg) 3.2 Å] [9] (Table 4.2). No further intermolecular metal–metal interactions are present in the crystals of 5 and 7.

The reaction of [PPN][Au$_2${CH$_2$P(S)Ph$_2$}$_2$] with HgCl$_2$, followed by TlPF$_6$ treatment, resulted in a rearrangement of the Au–C bonds and compound **6** was isolated. It is an isomer of compound **5** in the cation of which both gold and mercury are linearly coordinated by a carbon and a sulfur atom. The isomers **5** and **6** do not interconvert under reflux in THF [36]. The *transannular* gold–mercury distance in the cation of **6** [2.989(1) Å] is about 0.1 Å shortest than in the cations containing the S–Au–S/C–Hg–C structural units. In the crystal the cations of **6** dimerize and form a tetranuclear species (Figure 4.4b). The Hg and Au atoms are experimentally crystallographically indistinguishable and the intermolecular metal–metal distance [3.150(2) Å] is similar to the intermolecular Au–Au separations in the related Au$_2$[CH$_2$P(S)Ph$_2$]$_2$ (3.04 Å) [40]. Based on the tendency of Au(I) complexes to associate through short intermolecular gold–gold contacts, this intermolecular metal–metal distance was assigned to an Au(I)–Au(I) interaction.

When the reaction of Hg[CH$_2$P(S)Ph$_2$]$_2$ with Au(tht)Cl was performed in a 1:2 molar ratio the trinuclear species **8** was quantitatively isolated [38]. The molecule

188 | *4 Gold–Heterometal Interactions and Bonds*

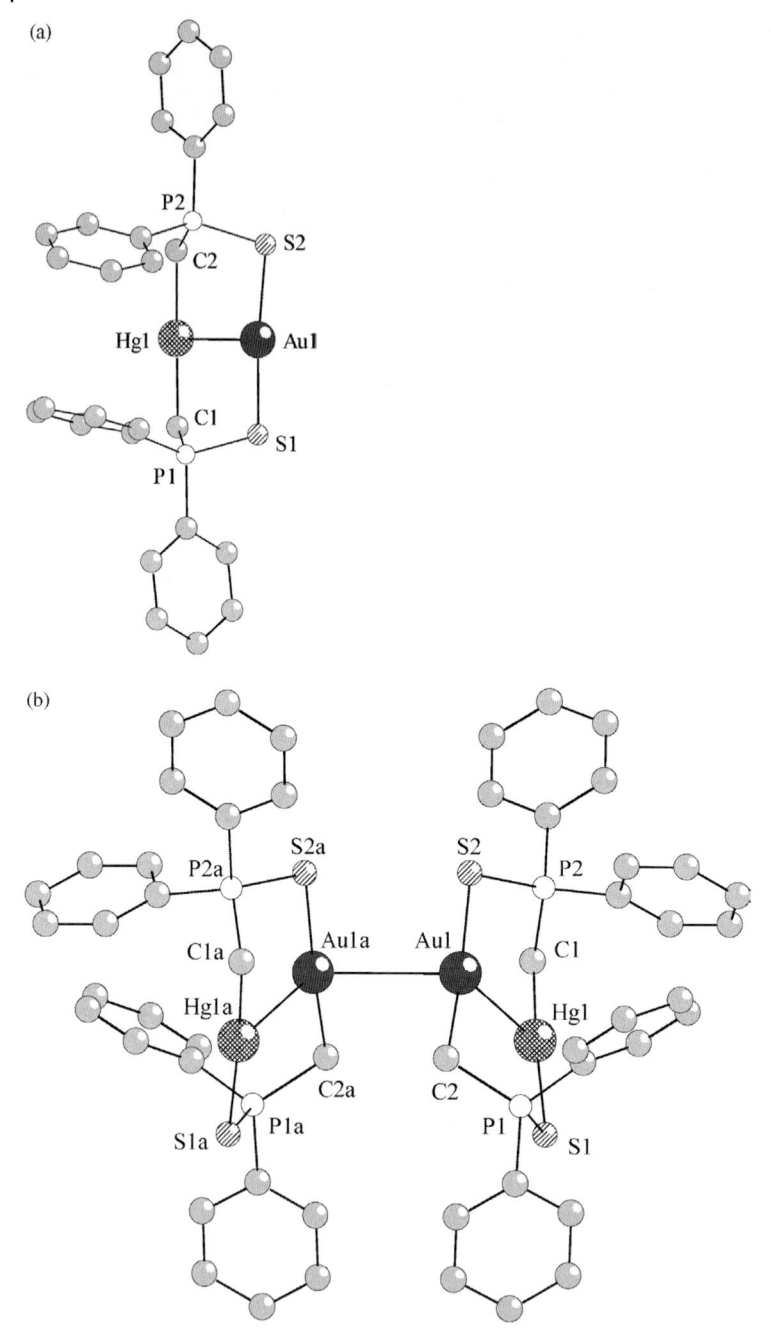

Figure 4.4 The structure of the cation of (a) [AuHg{CH$_2$P(S)Ph$_2$}$_2$](PF$_6$) **5** [36], and (b) [Au$_2$Hg$_2${CH$_2$P(S)Ph$_2$}$_4$](PF$_6$)$_2$ **6** [36].

Table 4.2 Selected molecular parameters for gold–mercury ylide complexes.

Compound	Au–Hg (Å)	Au–S (Å)	X–Au–Y (°)	X–Hg–Y (°)	Ref.
5	3.088(1)	2.299(6)/2.313(6)	175.0(2)^a	175.0(7)^b	[36]
6	2.989(1)	2.337(5)	175.2(7)^c	174.8(7)^c	[36]
7	3.085(1)	2.297(3)/2.288(4)	173.7(2)^a	178.1(5)^b	[39]
8	3.310(1)/3.361(1)	2.270(6)/2.260(5)	174.6(2)/174.1(2)	178.1(8)^b	[38]
9	3.079(2)	2.308(8)/2.313(8)	176.1(3)^d	179.1(9)^b	[38]

^{a}X = Y = S.
^{b}X = Y = C.
^{c}X = S, Y = C.
^{d}X = S, Y = Cl.

contains a central mercury atom doubly bridged to two gold atoms. The latter are linearly coordinated to one chlorine atom and one sulfur atom (Table 4.2). The three metal atoms in the molecular unit are not colinear (Au–Hg–Au 151.8°) and in this case the intramolecular Au–Hg distances [3.310(1)/3.361(1) Å] are significantly long to be considered as intermetallic interactions [cf. Σr_{vdW}(Au,Hg) 3.2 Å] [9]. In the crystal the molecules are packed in a way featuring a zigzag one-dimensional chain of metal atoms, but the intermolecular Au(I)–Au(I) distance of 4.21 Å excludes a supramolecular association through gold–gold interactions.

The oxidation of **8** resulted in the mixed-valence compound **9** [38] which contain the same cation as the complexes **5** and **7** described above and an [Au^{III}Cl$_4$]$^-$ anion. The main interesting feature is the one-dimensional chain arrangement in the crystal, with alternating cations and anion separated by intermetallic distances of 3.404(2) Å (Au^I–Au^{III}) and 4.658 Å (Hg^{II}–Au^{III}) (Figure 4.5).

The development of the second class of compounds with gold–mercury bonds, that is, Hg-containing, Au-rich clusters, was mainly the contribution of Pignolet and coworkers [41–43]. Such compounds were generally obtained by addition of mercury

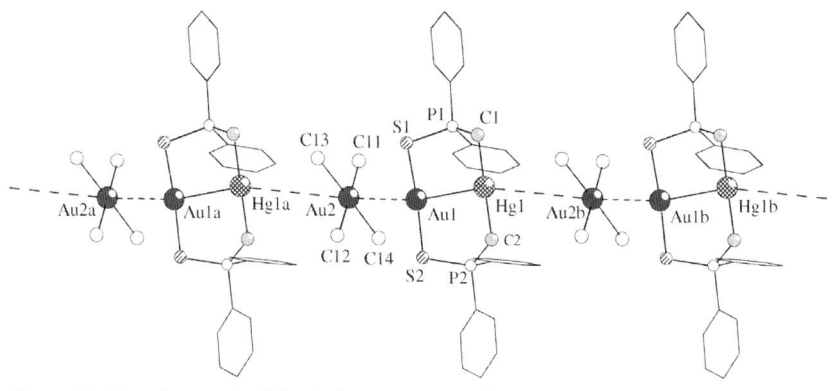

Figure 4.5 View along c axis of the chain arrangement of alternating cations and anion in the crystal of [AuHg{CH$_2$P(S)Ph$_2$}$_2$][AuCl$_4$] **9** [38].

to preformed Pt/Au or Pd/Au clusters. Thus, the treatment of the 16-electron cluster [{(Ph$_3$P)Au}$_6$Pt(PPh$_3$)](NO$_3$)$_2$ with Hg$_2$(NO$_3$)$_2$ resulted in isolation of the 18-electron cluster [{(Ph$_3$P)Au}$_5$Pt(PPh$_3$)(HgNO$_3$)$_2$](NO$_3$) **10** after an oxidative addition with elimination of an Au(PPh$_3$) fragment. The cluster core of the cation of **10** is best described as a fragment of a Pt-centered icosahedron. It contains two Hg atoms each μ$_4$-connected to folded PtAu$_3$ systems which have a common PtAu$_2$ fragment [Au–Hg 2.775(4)–3.072(3) Å] (Figure 4.6a) [41]. When the same Pt/Au cluster was reacted with metallic mercury simple addition occurred with isolation of [{(Ph$_3$P)Au}$_6$Pt(PPh$_3$)(HgNO$_3$)](NO$_3$) **11** [42]. In this case the cation contains only one Hg atom μ$_3$-connected to platinum and two gold atoms of a folded PtAu$_3$ system [Au–Hg 2.945(3)/

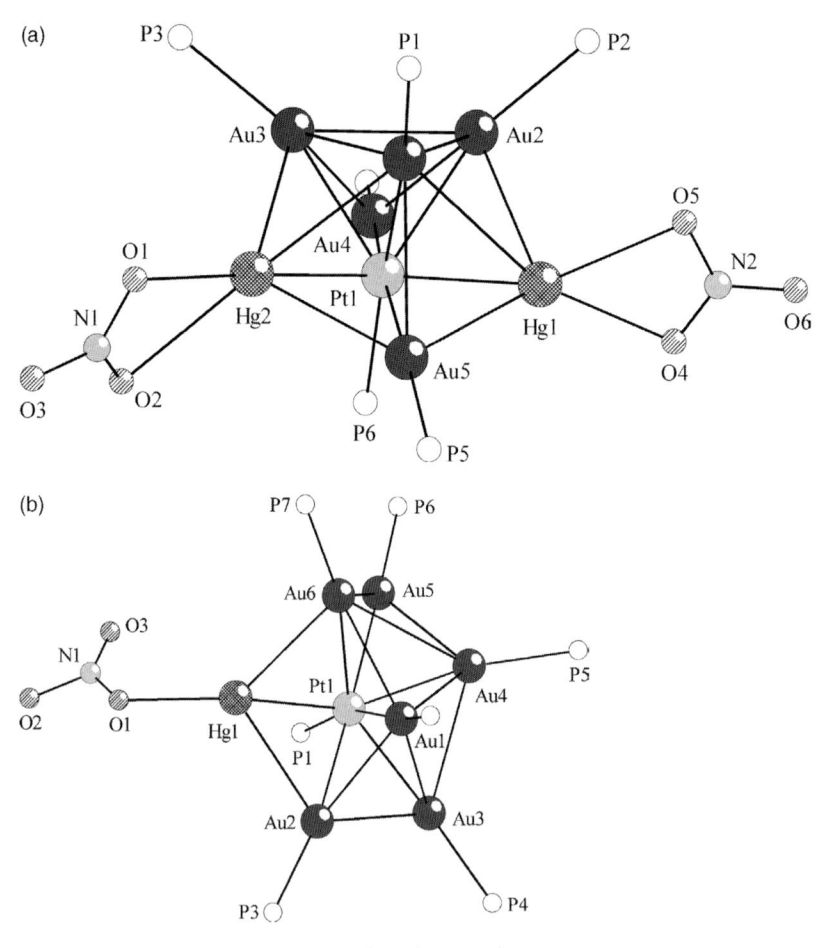

Figure 4.6 The structure of the cations of (a) [{(Ph$_3$P)Au}$_5$Pt(PPh$_3$)(HgNO$_3$)$_2$](NO$_3$) **10** [41]; (b) [{(Ph$_3$P)Au}$_6$Pt(PPh$_3$)(HgNO$_3$)](NO$_3$) **11** [42], (c) [{(Ph$_3$P)Au}$_6$Pd(Hg)$_2$](NO$_3$)$_2$ **12** [43], and (d) [{(Ph$_3$P)Au}$_8$Pt(Hg)](NO$_3$)$_2$ **14** [43] (for clarity, phenyl groups are not shown).

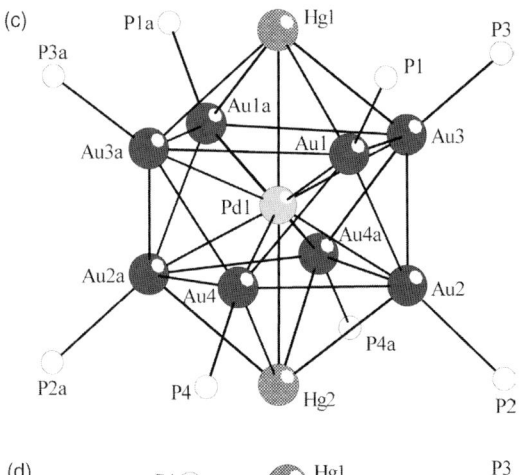

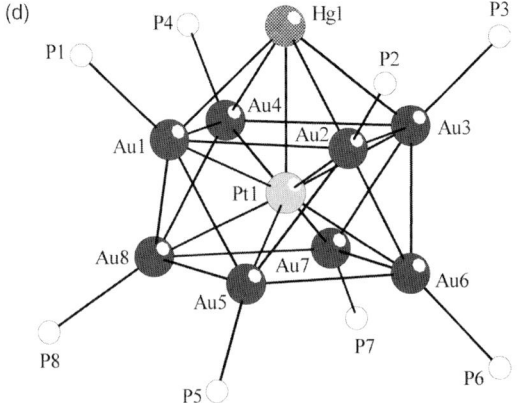

Figure 4.6 (Continued).

2.988(2) Å] (Figure 4.6b). In both compounds **10** and **11** the nitrate ligands are bound to mercury atoms in a bidentate or monodentate fashion, respectively.

The reaction of the preformed 16-electron clusters [{(Ph$_3$P)Au}$_8$M](NO$_3$)$_2$ (M = Pt, Pd) with metallic Hg resulted in isolation of the isomorphous 20-electron clusters [{(Ph$_3$P)Au}$_8$M(Hg)$_2$](NO$_3$)$_2$ in which two mercury atoms cap in a μ_4-fashion the gold atoms of two opposite square faces in an M-centered square-antiprism [Au–Hg 2.915(7)–3.056(7) Å for M = Pd **12**] (Figure 4.6c) [43]. The same structure of the cation [Au–Hg 3.0011(5)/3.0070(5) Å] was established for the 18-electron cluster [{(Ph$_3$P)Au}$_8$Pt(Hg)$_2$](NO$_3$)$_4$ **13** [44]. The compound was isolated as red cubes by slow diffusion of diethyl ether into a methanol/CH$_2$Cl$_2$ solution of [{(Ph$_3$P)Au}$_8$Pt(HgNO$_3$)$_2$](NO$_3$)$_2$, a cluster obtained by reacting [{(Ph$_3$P)Au}$_8$Pt](NO$_3$)$_2$ with Hg$_2$(NO$_3$)$_2$ [43, 44].

The isomorphous, 18-electron, monomercury clusters [{(Ph$_3$P)Au}$_8$M(Hg)](NO$_3$)$_2$ (M = Pt, Pd) were isolated either as intermediates in the synthesis of [{(Ph$_3$P)Au}$_8$M(Hg)$_2$](NO$_3$)$_2$ or by reaction of the dimercury clusters with one equivalent of [{(Ph$_3$P)

Au}$_8$M](NO$_3$)$_2$ [43]. The structure of [{(Ph$_3$P)Au}$_8$Pt(Hg)](NO$_3$)$_2$ **14** was determined and the cation was found to be similar [Au–Hg 2.979(5)–3.079(5) Å] to that observed in cluster **12**, one of the capping Hg atoms being removed (Figure 4.6d). In compounds **12–14** the mercury atoms also interact with the central metal atom (Pd or Pt) while, in contrast to compounds **10** and **11**, the capping Hg atoms have no close contacts with NO$_3^-$ anions.

4.2.3
Gold–Group 13 Metal Compounds

To the best of our knowledge no species with Au–Al bonds have been reported yet. By contrast, a few compounds with Au–Ga and Au–In have been described, while the chemistry of compounds with Au–Tl bonds was quite largely developed by the contributions of Laguna and coworkers.

4.2.3.1 Gold–Gallium Compounds

The gold cluster complex, Au$_3$(GaI$_2$)$_3$(Cp*Ga)$_5$ **15**, in which the gold atoms are bonded only to an electropositive main-group metal, was recently reported to be obtained as an air-sensitive orange–yellow solid by adding Au(PPh$_3$)X (X = Cl, I) to a solution of "[Cp*Ga/GaI]" in CH$_2$Cl$_2$, along with [Au(PPh$_3$)$_n$]$^+$ (n = 2, 3) species (as suggested by spectroscopic data). It is worthwhile to note that the reverse addition of the gallium reagent to the solution of the gold starting material resulted in isolation of the cluster [Au$_6$(PPh$_3$)$_6$]$^{2+}$, which does not contain Ga fragments coordinated to gold atoms. The cluster **15** was also prepared from Au(PCy$_3$)I and "[Cp*Ga/GaI]" (along with [Au(PCy$_3$)$_2$]$^+$[Cp*GaI$_3$]$^-$) or Au(PPh$_3$)I and a mixture of Cp*Ga and GaI$_3$. In the latter case the [Au(PPh$_3$)$_n$]$^+$ species was not formed [45]. The X-ray crystal structural analysis of **15** revealed an almost planar Au$_3$Ga$_3$ system with a trigonal Au$_3$ fragment [Au(1)–Au(2) 2.726(1), Au(1)–Au(1a) 2.804(1) Å] μ$_2$-bridged by three GaI$_2$ units [Au–Ga 2.528(1)–2.532(1) Å]. Two symmetry-related Au atoms are each coordinated by two Cp*Ga units [Au(1)–Ga(1) 2.384(1), Au(1)–Ga(3) 2.620(1) Å], whereas the third unique gold atom is coordinated by only one Cp*Ga unit [Au(2)–Ga(2) 2.377(2) Å] (Figure 4.7). The DFT calculations on the Cp analog of **1** supported the presence of an Au$_3$ cluster with Au–Au bonds between near-charge-neutral gold atoms and strongly polarized CpGa–Au bonds.

The gallium(I) derivative Ga(DDP) was found to stabilize Au(I) centers. Indeed, treatment of Au(PPh$_3$)Cl with a stoichiometric amount of Ga(DDP) resulted in its insertion into the Au–Cl bond and isolation of (Ph$_3$P)Au[Ga(DDP)Cl] **16**, while the use of an excess of Ga(DDP) resulted in further substitution of the PPh$_3$ and Au[Ga(DDP)][Ga(DDP)Cl] **17** was produced [46]. The chlorine atom in **17** can be substituted by a methyl following reaction with MeLi or abstracted with Na[B{C$_6$H$_3$(CF$_3$)$_2$-3,5}$_4$] in THF, to produce Au[Ga(DDP)][Ga(DDP)Me] **18** or the ionic species [Au{Ga(DDP)(THF)}$_2$][B{C$_6$H$_3$(CF$_3$)$_2$-3,5}$_4$] **19** (Scheme 4.3) [46, 47]. It should also be noted that the THF-free complex [Au{Ga(DDP)}$_2$][B{C$_6$H$_3$(CF$_3$)$_2$-3,5}$_4$] **20** was also isolated from **17** and Na[B{C$_6$H$_3$(CF$_3$)$_2$-3,5}$_4$] in C$_6$H$_5$F as a noncoordinating solvent [47].

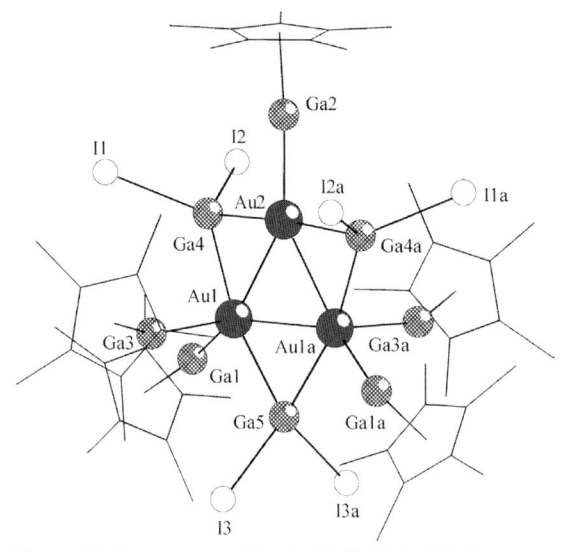

Figure 4.7 The structure of Au$_3$(GaI$_2$)$_3$(Cp*Ga)$_5$ **15** [45].

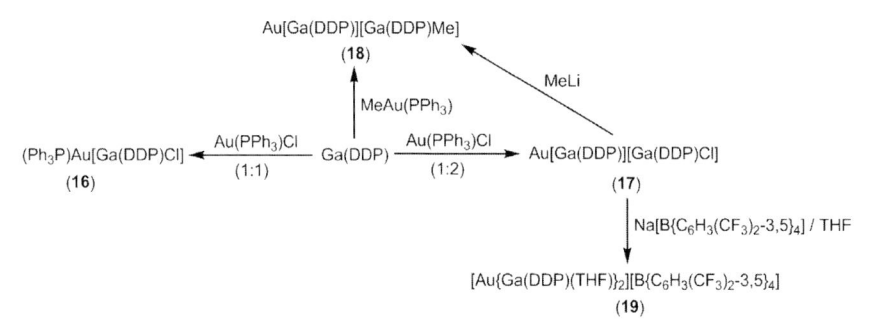

Scheme 4.3

The molecular structure of compounds **16** (Figure 4.8a), **17** (Figure 4.8b) and **19** (Figure 4.8c) consists of a central gold atom almost linearly coordinated [P(1)–Au(1)–Ga(1) 174.64(5)° in **16**; Ga(1)–Au(1)–Ga(2) 177.88(2)° and 170.93° in **17** and **19**, respectively). The Au–Ga bonds in these three compounds [Au(1)–Ga(1) 2.411(1) Å in **16**; Au(1)–Ga(1) 2.4117(7) and Au(1)–Ga(2) 2.4125(7) Å in **17**; Au(1)–Ga(1) 2.393 and Au(1)–Ga(2) 2.391 Å in **19**, respectively] are slightly longer than those observed for the equatorial GaCp* units in the cluster complex Au$_3$(GaI$_2$)$_3$(Cp*Ga)$_5$ **15**. The coordination of the Ga(DDP) unit to gold resulted in an increased electrophilicity of the gallium atom, a behavior supported by the weak coordination of THF to each of the gallium centers in the cation of **19** [47]. No further intermolecular metal–metal interactions are present in the crystals of these compounds.

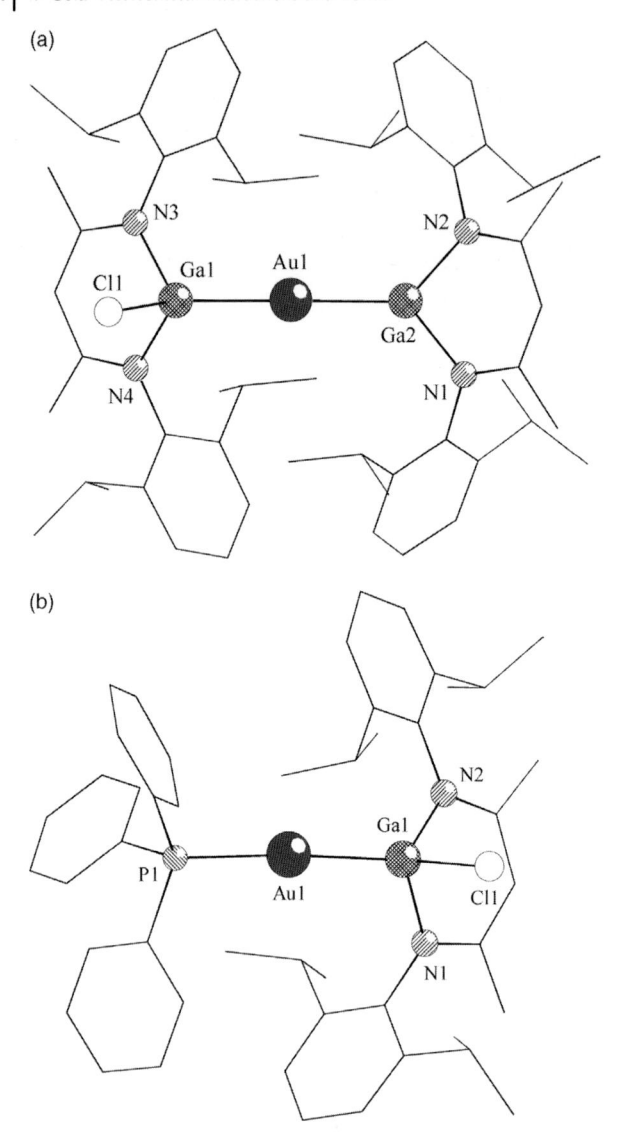

Figure 4.8 The structure of (a) (Ph$_3$P)Au[Ga(DDP)Cl] **16**; (b) Au[Ga(DDP)][Ga(DDP)Cl] **17** [46], and (c) of the cation in [Au{Ga(DDP)(THF)}$_2$][B{C$_6$H$_3$(CF$_3$)$_2$-3,5}$_4$] **19** (the methyls of the isopropyl groups are not shown for clarity) [47].

4.2.3.2 Gold–Indium Compounds

Only two compounds containing Au–In bonds have been structurally characterized to date. They were prepared following reactions which include ligand redistribution and redox processes. Thus, the treatment of Au(PPh$_3$)Cl with InCl (molar ratio 1:1) in THF leads only to isolation of (THF)$_2$Cl$_2$In-InCl$_2$(THF)$_2$ and no gold species were

(c)

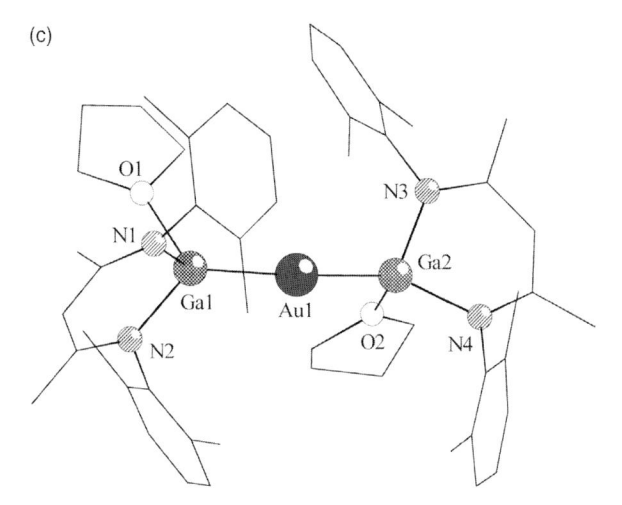

Figure 4.8 (Continued).

identified. However, when the reaction was carried out in the presence of dppe the lemon-yellow cluster (dppe)$_2$Au$_3$In$_3$Cl$_6$(THF)$_3$ **21** (Figure 4.9a) was obtained [48]. A similar reaction between Au(PPh$_3$)Br with InBr in the presence of dppe produced an ionic orange species, [(dppe)$_2$Au][(dppe)$_2$Au$_3$In$_3$Br$_7$(THF)] **22** (Figure 4.9b) [49]. Both **21** and the anion of **22** contain a basically similar Au$_3$In$_3$ cluster core, that is, a Au$_3$In$_2$ trigonal bipyramid with the In atoms in apical positions [Au–In 2.829(1)–3.004(1) Å in **21**; 2.828(2)–2.973(1) Å in **22**] and a third In atom bridging [Au–In 2.778(1)/2.839 (1) Å in **21**; 2.761(2)/2.826(2) Å in **22**] the shortest edge of the gold triangle [Au(1)–Au (3) 2.562(1) Å in **21**; 2.575(1) Å in **22**]. The two long Au–Au edges [Au–Au 2.939(1)/

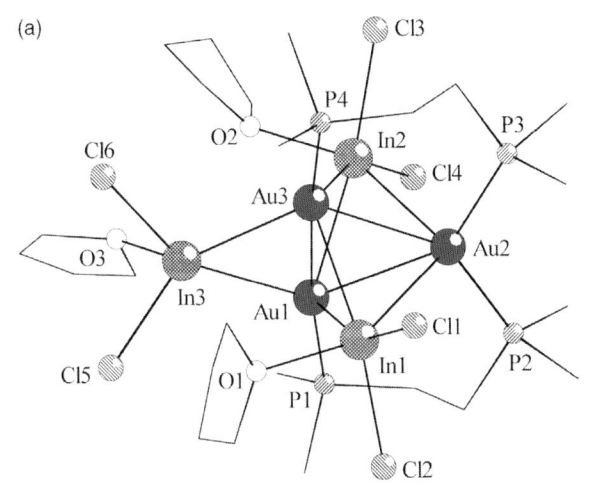

Figure 4.9 The structure of (a) (dppe)$_2$Au$_3$In$_3$Cl$_6$(THF)$_3$ **21** [48], and (b) of the anion in [(dppe)$_2$Au][(dppe)$_2$Au$_3$In$_3$Br$_7$(THF)] **22** (the phenyl rings, less *ipso* carbons, on phosphorus have been omitted for clarity) [49].

(b)

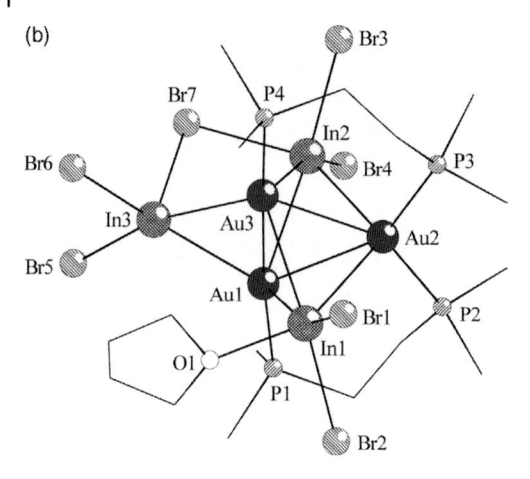

Figure 4.9 (Continued).

2.931(1) Å in **21**; 2.860(1)/2.858(1) Å in **22**] are spanned by the dppe ligands. There is no straightforward way to assign an oxidation state to the metal atoms in compounds **21** and **22** using standard qualitative rules of cluster chemistry. Based on the linearity of the P(1)–Au(1)–Au(3)–P(4) system, the short Au(1)–Au(3) distance and the DFT calculations an oxidation state Au^0 was assigned for these gold atoms. The third gold atom, Au(2), was then assigned oxidation state Au^{+1}. For the indium atoms an oxidation state of In^{+2} for two atoms and In^{+1} for the third one was assumed [49].

4.2.3.3 Gold–Thallium Compounds

The last years have revealed a continuous interest in the synthesis and investigation of compounds containing gold–thallium interactions, mainly due to their luminescent properties which suggest potential applications as LEDs or selective VOCs sensors [22, 50–52]. Three basic strategies were developed to prepare such compounds: (i) the use of bridging ligands containing different donor centers with affinity for these metal atoms and placed in the ligand skeleton at distances which bring the two metal atoms close enough to each other to result in supported intramolecular gold–thallium interactions; (ii) the use of gold-based metallocryptands which produce unsupported gold–thallium interactions by encapsulation of a thallium atom; (iii) the use of basic gold(I) complexes which can react with thallium(I) salts leading, via acid–base stacking, to supramolecular networks which contain unsupported gold–thallium interactions.

The number of compounds with Au–Tl interactions for which the structure has been investigated by single-crystal X-ray diffraction has increased considerably in the last years and some important molecular dimensions are collected in Table 4.3.

The first Au–Tl compound, $[AuTl\{CH_2P(S)Ph_2\}_2]_n$ **23**, was obtained by Fackler and coworkers by reacting $[PPN][Au\{CH_2P(S)Ph_2\}_2]$ with Tl_2SO_4 [35, 37]. In the molecular unit the Au(I) atom is linearly coordinated by two carbon atoms, while the sulfur atoms are coordinated to the Tl(I) center. In the solid state an extended one-dimensional chain polymer is formed along the crystallographic *b* axis of the

Table 4.3 Selected molecular parameters for gold–thallium compounds.

Compound	Au–Tl (Å)	C–Au–E (°)	Tl–Au–Tl (°)	M–Tl–Au (°)	Association degree	Ref.
[AuTl{CH$_2$P(S)Ph$_2$}$_2$]$_n$ **23**	2.959(2)/3.003(2)	173.4(10)[a]	162.9(1)	162.7(1)	1D chain polymer	[35, 37]
[Au$_2$Tl(P$_2$phen)$_3$](ClO$_4$)$_3$ **24**	2.917(1)/2.911(1)			174.47(2)[c]	discrete cation	[30]
[AuPdTl(P$_2$phen)$_3$](PF$_6$)$_2$ **25**	2.811(9)			177.0(4)[d]	discrete cation	[53]
[AuPtTl(P$_2$phen)$_3$](PF$_6$)$_2$ **28**	2.900(1)			172.30(2)[e]	discrete cation	[53]
[Tl{Au$_3$(bzim)$_3$}$_2$]$_n$(PF$_6$)$_n$ **29**	2.971(1)–3.045(1)	171.5(5)–176.4(4)[b]			columnar polymer of cations	[54]
[Tl{Au$_3$(carb)$_3$}$_2$]$_n$(PF$_6$)$_n$ **30**	3.067(1)–3.107(1)	174.5(2)–176.9(3)[b]			columnar polymer of cations	[54]
[Tl[Au(C$_6$Cl$_5$)$_2$]]$_n$ **31**	2.9726(5)/3.0044(5)	177.9(3)[a]			1D chain polymer	[56]
Tl(η^6-toluene)[Au(C$_6$Cl$_5$)$_2$] **34**	2.9115(2)	176.9(2)[a]	180.0	180.0[c]	discrete dinuclear unit	[60]
[Tl(η^6-toluene){Au(C$_6$Cl$_5$)$_2$}]$_2$ (dioxane) **35**	2.8935(3)	177.6(2)[a]			discrete tetranuclear unit	[61]
Tl$_2$Au$_2$(C$_6$Cl$_5$)$_4$(acetone) **36**	3.0331(6)–3.1887(6)	176.9(4)/176.5(3)[a]	70.72(1)/71.19(1)	103.57(2)/100.99(2)[c]	discrete tetranuclear unit	[62, 63]
Tl$_2$Au$_2$(C$_6$Cl$_5$)$_4$(PhMeCO) **37**	3.0167(4)–3.2313(4)	178.0(2)/173.7(2)[a]	72.78(1)/73.29(1)	100.07(1)/104.92(1)[c]	discrete tetranuclear unit	[63]
Tl$_2$Au$_2$(C$_6$Cl$_5$)$_4$(Hacac) **38**	3.0852(4)–3.1326(4)	178.0(2)/174.6(2)[a]	73.38(1)/73.31(1)	101.26(1)/102.90(1)[c]	discrete tetranuclear unit	[63]
Tl$_2$Au$_2$(C$_6$Cl$_5$)$_4$(Hacac)(4,4'-bipy) **39**	3.0314(3)/3.2414(3)	174.5(2)/177.7(2)[a]	68.36(1)/70.63(1)	99.60(1)/104.73(1)[c]	tetranuclear unit	[63]
[Tl(THF)$_{0.5}${Au(C$_6$Cl$_5$)$_2$}]$_n$ **40**	2.9078(3)–3.0918(3)	176.0(2)/179.0(2)[a]	130.14(1)/175.57(1)	138.71(1)/157.06(1)[c]	1D chain polymer	[59]
[Tl(THF)$_2${Au(C$_6$Cl$_5$)$_2$}]$_n$ **41**	3.0764(4)/3.1981(4)	179.7(2)[a]	156.51(1)	164.08(1)[c]	1D chain polymer	[59]
[Tl(Hacac)$_2${Au(C$_6$Cl$_5$)$_2$}]$_n$ **47**	2.9684(5)/2.9894(5)	178.4(3)[a]	170.02(2)	155.10(2)[c]	1D chain polymer	[59]
[Tl(OPPh$_3$)$_2${Au(C$_6$F$_5$)$_2$}]$_n$ **48**	3.0358(8)/3.0862(8)	180[a]	180	163.16(1)[c]	1D chain polymer	[64]
[{Tl(OPPh$_3$)}{Tl(OPPh$_3$)(THF)} Au(C$_6$Cl$_5$)$_2$]$_2$]$_n$ **49**	3.0529(3)–3.3205(3)	178.1(2)/176.7(2)[a]	142.61(1)/ 168.53(1)	156.56(1)/131.60(1)[c]	1D chain polymer	[55]
[{Tl(OPPh$_3$)}{Tl(OPPh$_3$)(acetone)} Au(C$_6$Cl$_5$)$_2$]$_2$]$_n$ **50**	3.0937(3)–3.2705(4)	178.2(2)/178.6(2)[a]	143.88(1)/ 166.74(1)	158.31(1)/135.72(1)[c]	1D chain polymer	[55]

(Continued)

Table 4.3 (Continued)

Compound	Au–Tl (Å)	C–Au–E (°)	Tl–Au–Tl (°)	M–Tl–Au (°)	Association degree	Ref.
[Tl(2,2′-bipy){Au(C$_6$F$_5$)$_2$}]$_n$ 52	3.0120(6)/3.4899(6)	173.2(4)^a	120.19(2)	120.19(2)^c	1D chain polymer	[65]
[Tl(1,10-phen){Au(C$_6$F$_5$)$_2$}]$_n$ 53	3.0825(4)/3.1397(4)	171.9(1)^a	140.52(1)	176.08(1)^c	1D chain polymer	[65]
[Tl(H$_2$NCH$_2$CH$_2$N=CMe$_2$){Au(C$_6$Cl$_5$)$_2$}]$_n$ 56	2.9571(11)–3.0365(11)	180.0/178.9(2)^a	180.0/160.15(2)	144.26(1)/139.43(4)^c	1D chain polymer	[66]
[Tl(Me$_2$C=NCH$_2$CH$_2$N=CMe$_2$){Au(C$_6$Cl$_5$)$_2$}]$_n$ 58	3.0519(3)/3.0877(3)	180.0^a	180.0	150.59(1)^c	1D chain polymer	[66]
[Tl(Me$_2$C=NCH$_2$CH$_2$CH$_2$N=CMe$_2$){Au(C$_6$F$_5$)$_2$}]$_n$ 59	3.0479(2)/3.1427(2)	180.0^a	180.0	139.20(1)^c	1D chain polymer	[66]
[Tl$_2$(μ-DMSO)$_2${Au(C$_6$Cl$_5$)$_2$}$_2$]$_n$ 60	3.1220(5)/3.1794(4) 3.1337(6)/3.2839(6)	180.0^a 174.3(3)^a	180.0 71.60(1)	126.23(2)/153.78(2)^c	1D chain polymer	[67]
[Tl$_2$(μ-DMSO)$_3${Au(C$_6$F$_5$)$_2$}$_2$]$_n$ 61	3.2225(6)/3.2465(6) 3.5182(8)	179.1(4)^a 177.9(4)^a	178.82(2)	67.09(2)^c	1D chain polymer	[67]
[NBu$_4$][Tl$_2${Au(C$_6$Cl$_5$)$_2$}{μ-Au(C$_6$Cl$_5$)$_2$}$_2$]$_n$ 62	3.0940(3)/3.1001(3)	166.7(2)/ 168.6(2)^a	82.66(1)/ 84.88(1)	91.66(1)/92.31(1)^c	1D chain polymer	[58]
[NBu$_4$][Tl{Au(C$_6$Cl$_5$)$_2$}{Au(3,5-C$_6$Cl$_2$F$_3$)$_2$}]$_n$ 63	3.0559(4)–3.1678(4) 3.0108(2)/3.0914(2)	180^a 180.0^a	180 180.0	108.83(1)–158.20(1)^c 164.44(1)^c	1D chain polymer	[58]
[NBu$_4$][Tl{Au(3,5-C$_6$Cl$_2$F$_3$)$_2$}$_2$]$_n$ 64	3.0317(3) 3.0329(5)/3.1196(5) 2.9704(7)	170.1(2)^a 180.0^a 168.9(4)^a	180.0	75.87(1)/119.33(1)^c 168.83(2)^c 71.78(1)/117.89(2)^c	1D chain polymer	[58]
[Tl$_3$(acac)$_2${Au(C$_6$F$_5$)$_2$}$_2$]$_n$ 65	3.0653(4)	177.3(3)^a			1D double-chain polymer	[57]
[Tl$_2$(acac){Au(C$_6$Cl$_5$)$_2$}$_2$]$_n$ 66	3.0963(7)/3.2468(7)	177.0(4)^a	126.60(2)	131.34(2)^c	2D network	[57]
[Tl(4,4′-bipy)$_{0.5}${Au(C$_6$Cl$_5$)$_2$}]$_n$ 67	3.0561(3)/3.1743(3)	180.0^a	180.0	147.38(1)^c	2D network	[68]
[{Tl(4,4′-bipy)}$_2${Au(C$_6$F$_5$)$_2$}$_2$]$_n$ 68	3.0161(2)	177.6(1)^a			2D network	[69]
[{Tl(4,4′-bipy)(THF)}{Au(C$_6$F$_5$)$_2$}]$_n$ 69	3.2155(3)/3.4800(3)	177.2(2)^a	163.24(1)	163.24(1)^c	2D network	[65]

[Tl(4,4′-bipy){Au(C$_6$Cl$_5$)$_2$}]$_n$ **70**	2.9647(2)/3.0052(2)	180.0[a]		163.40(1)[c]	[68]
[Tl(4,4′-bipy){Au(C$_6$Cl$_5$)$_2$}· 0.5toluene]$_n$ **71**	3.0797(2)/3.1110(2)	180.0[a]		149.02(1)[c]	[68]
[{Tl(4,4′-bipy)}{Tl(4,4′-bipy)$_{0.5}$ (THF)}{Au(C$_6$Cl$_5$)$_2$}$_2$]$_n$ **72**	3.0323(4)–3.0540(4)	173.9(3)/176.1(3)[a]	133.77(1)/ 160.99(1)	163.16(1)/ 129.97(1)[c]	[69]
[{Tl(4,4′-bipy)}{Tl(4,4′-bipy)$_{0.5}$(THF)} {Au(C$_6$Cl$_5$)$_2$}$_2$·THF]$_n$ **73**	2.9874(6)–3.1221(6)	176.5(3)/177.4(3)[a]	160.60(2)/ 178.89(2)	154.08(2)/ 164.52(2)[c]	[68]

[a] E = C.
[b] E = N.
[c] M = Au.
[d] M = Pd.
[e] M = Pt.

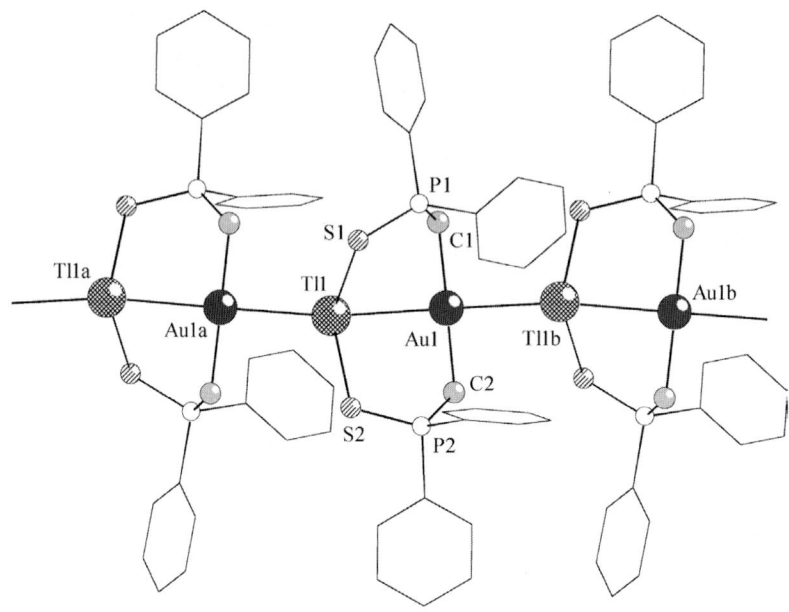

Figure 4.10 View of the one-dimensional chain polymer in the crystal of [AuTl{CH$_2$P(S)Ph$_2$}$_2$]$_n$ **23** [37].

lattice (Figure 4.10), with both supported intramolecular [Au(1)–Tl(1) 2.959(2) Å] and unsupported intermolecular [Au(1a)–Tl(1) 3.003(2) Å] gold–thallium interactions and angular Tl(1)–Au(1)–Tl(1b) [162.9(1)°] and Au(1a)–Tl(1)–Au(1) [162.7(1)°] systems. The overall geometry around the gold and thallium atoms is distorted square planar (AuC$_2$Tl$_2$ core) and distorted trigonal bipyramidal with a vacant equatorial coordination site (TlS$_2$Au$_2$ core), respectively. Although no formal metal–metal bond is present in this heterometallic compound, molecular orbital calculations suggest that the Au–Tl interactions are associated with relativistic effects on electrons in these heavy-metal atoms. The strong luminescence observed in the solid state was attributed to these Au–Tl interactions.

Gold-based metallocryptates of the type [Au$_2$Tl(P$_2$phen)$_3$](ClO$_4$)$_3$ **24**, [AuPdTl(P$_2$phen)$_3$]X$_2$ [X = PF$_6^-$ **25**, BF$_4^-$ **26**, Cl$^-$ **27**] or [AuPtTl(P$_2$phen)$_3$](PF$_6$)$_2$ **28** were reported by Catalano and coworkers [30, 53]. Compounds **25** and **28** were isolated as deep-red, air-stable solids by reacting P$_2$phen, Au(tht)Cl and Pd$_2$(dba)$_3$ or Pt(dba)$_2$ (3:1:1 molar ratio) with excess thallium(I) acetate, followed by metathesis with excess NH$_4$PF$_6$. Exchange of the anion was achieved by treatment with the corresponding sodium salts [50]. All compounds feature the same structure of the cation as observed for the related Na derivative, that is, a Tl$^+$ ion held inside a cage capped by two trigonal AuP$_3$ fragments in **24** or one AuP$_3$ and one MP$_3$ fragment in **25–27** (M = Pd) or **28** (M = Pt), mainly due to Au–Tl interactions. The shortened intermetallic Au–Tl separations observed for mixed-metal species **25–28** compared to the symmetrical **24** derivative seem to be the result of stronger closed-shell, metal–metal interaction due to a dipole moment introduced by the dissimilar capping metals.

The electron-rich trinuclear Au$_3$(bzim)$_3$ and Au$_3$(carb)$_3$ react with TlPF$_6$ to yield the "sandwich" clusters [Tl{Au$_3$(bzim)$_3$}$_2$](PF$_6$) **29** and [Tl{Au$_3$(carb)$_3$}$_2$](PF$_6$) **30** [54]. The cation of both compounds contains a Tl atom which interacts in a distorted trigonal prismatic coordination with six Au(I) atoms from two cyclic Au$_3$C$_3$N$_3$ moieties at Au–Tl distances ranging from 2.971(1) to 3.107(1) Å, indicative of appreciable metal–metal interaction. Two Au(I) atoms on each planar trinuclear unit are involved in intermolecular aurophilic bonding interactions [Au–Au 3.109(1)/3.066(1) Å in **29**; 3.059(1)/3.052(1) Å in **30**], thus resulting in an infinite columnar chain with a ···BBABBA··· pattern (Figure 4.11).

Based on the ability of relativistic heavy atoms to exhibit a marked tendency to form aggregates with metal–metal distances shorter than the sum of their van der Waals radii an exciting chemistry of compounds containing interactions between gold(I) and thallium(I) atoms, that is, centers with d^{10} and s^2 configurations, was developed by Laguna and coworkers [22, 50–52]. They employed a synthetic strategy based on reactions between basic pentahalophenyl gold(I) complexes of the type [AuR$_2$]$^-$ (R = C$_6$Cl$_5$, C$_6$F$_5$, C$_6$Cl$_2$F$_3$-3,5) and Tl(I) salts as a Lewis acid, the resulting products being highly dependent on reaction conditions such as solvent, presence of ligands, and so on. Many of these complexes were investigated by single-crystal X-ray diffraction and were found to involve a large variety of assemblies in the crystals, such as discrete molecular units, 1D linear or zigzag polymeric chains, or 2D and 3D networks (for selected molecular parameters see Table 4.3). In most cases they are built by means of what are generally considered to be "weak" unsupported metal–metal interactions. Long X–Tl (X = halogen) contacts between the thallium center and halogen atoms of the almost linear [AuR$_2$]$^-$ unit seem also to contribute to the stability of many of these systems. It is worth noting that theoretical studies revealed for the Au–Tl interaction in these systems a surprising calculated strength of about 276 kJ mol^{-1}, from which 80% is due to an ionic contribution and 20% to dispersion (van der Waals) [55].

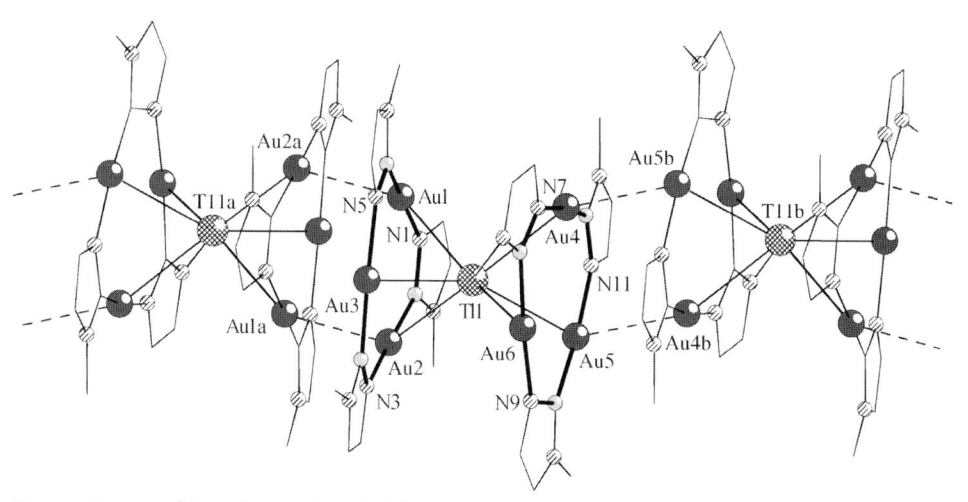

Figure 4.11 View of the columnar chain built by cations in the crystal of [Tl{Au(bzim)$_3$}$_2$](PF$_6$) **29** [54].

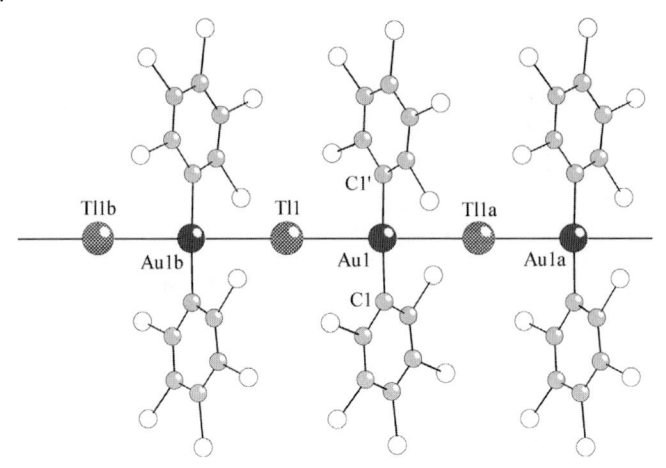

Figure 4.12 View of the chain polymer in the crystal of [Tl{Au(C$_6$Cl$_5$)$_2$}]$_n$ **31**.

The reaction of equimolecular amounts of [NBu$_4$][AuR$_2$] and TlPF$_6$ in THF afforded the isolation of the precursor polymeric complexes [Tl(AuR$_2$)]$_n$ [R = C$_6$Cl$_5$ (**31**), C$_6$F$_5$ **32**, C$_6$Cl$_2$F$_3$-3,5 **33**] [56–58]. The structure of **31** consists of 1D linear polymer chains with unsupported Au–Tl interactions between alternating [Au(C$_6$Cl$_5$)$_2$]$^-$ anions and Tl$^+$ cations [Au(1)–Tl(1) 2.9726(5), Au(1)–Tl(1a) 3.0044(5) Å; Tl(1)–Au(1)–Tl(1a) and Au(1)–Tl(1)–Au(1b) 180°] (Figure 4.12) [56]. Each Au(I) atom is almost linearly coordinated by two pentachlorophenyl groups, a pattern common for all the other compounds listed in Table 4.3. The Tl center is not further coordinated, which allows adjacent chains to get close to each other and to establish Tl–Cl interactions. This results in empty channels that run parallel to the crystallographic z axis, with a hole diameter as large as 10.471 Å (the shortest nonbonding interchain Au–Au distance). These channels are large enough to accommodate different volatile organic compounds (VOCs) which can enter the lattice. Indeed, exposure of **31** in the solid state to VOCs (e.g., tetrahydrofuran, acetone, tetrahydrothiophene, 2-fluoropyridine, acetonitrile, acetylacetone, pyridine) resulted in a perceptible change in the color of the solid, which reverts to that of the starting material upon heating the solid adducts [56, 59]. In all complexes the process is completely reversible with no detectable degradation, even after 10 exposure/heating cycles [59].

A particular feature of the polymeric complexes [Tl(AuR$_2$)]$_n$ **31–33** is that the Tl(I) centers still keep their acid properties, despite having a lone pair. They react easily with neutral donor ligands to give compounds with a variety of structures in the solid state. A summary of the reactions providing complexes for which the structure was established in the solid state is given in Schemes 4.4–4.6. In most cases the polymeric structure in the solid state is preserved, with the donor ligands coordinated to the thallium centers. In other cases the polymeric structure is destroyed and discrete dinuclear or tetranuclear units are present in the crystal. Generally, secondary interactions between metal atoms and the halogen atoms of perhalophenyl groups are present, which may contribute to the stability of the compounds.

4.2 Main Group Metal–Gold Compounds

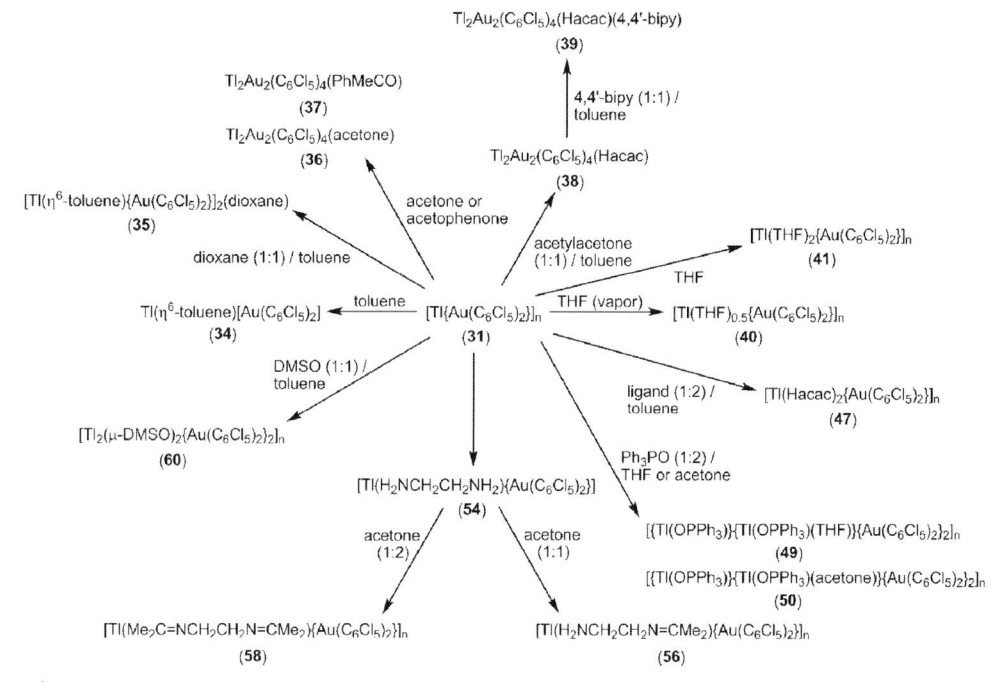

Scheme 4.4

Treatment with toluene of the yellow solid obtained from the reaction of a dichloromethane suspension of TlPF$_6$ with [NBu$_4$][Au(C$_6$Cl$_5$)$_2$] afforded the isolation of an ocher-colored solid. The single-crystal X-ray diffraction studies revealed that the polymeric structure found for **31** was destroyed and the crystal contains discrete molecules of Tl(η^6-toluene)[Au(C$_6$Cl$_5$)$_2$] **34** (Figure 4.13a) [intermolecular metallophilic contacts are ruled out, the shortest Au–Au distance being 3.4625(2) Å] [60]. The

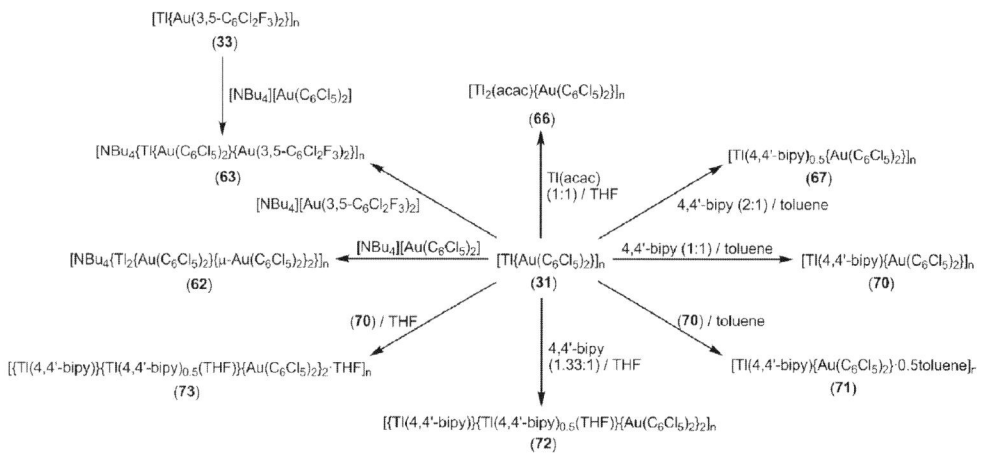

Scheme 4.5

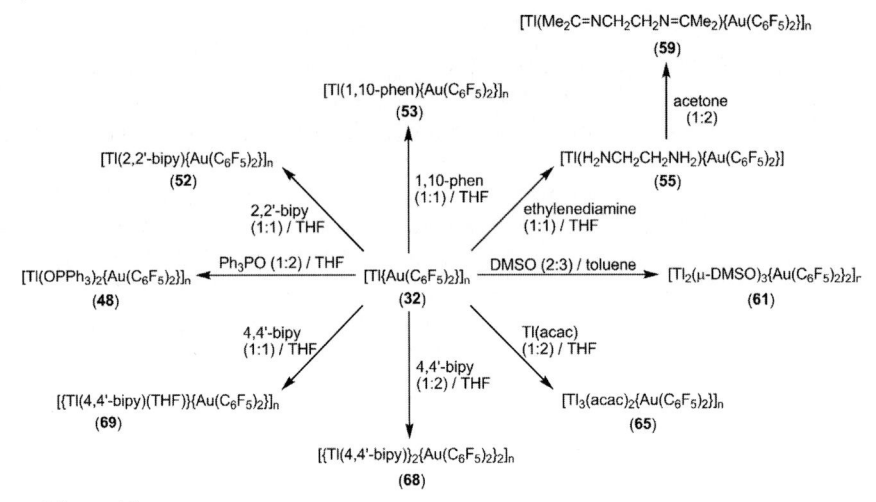

Scheme 4.6

molecule exhibits one of the shortest gold–thallium unsupported interaction distances [Au(1)–Tl(1) 2.9115(2) Å] and the coordination hemisphere of thallium(I) is completed by an unusual η^6-toluene contact, with Tl–C distances in the range 3.281(5)–3.458(5) Å. When a suspension of **31** in toluene is reacted with 1,4-dioxane the coordination number of thallium(I) is further increased and the white tetranuclear species [Tl(η^6-toluene){Au(C$_6$Cl$_5$)$_2$}]$_2$(dioxane) **35** was formed. It contains a molecule

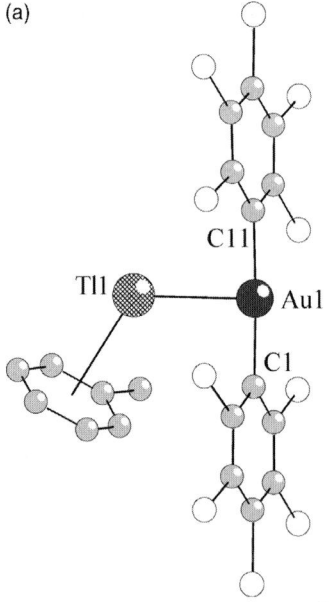

Figure 4.13 The structure of (a) Tl(η^6-toluene)[Au(C$_6$Cl$_5$)$_2$] **34** [60], and (b) [Tl(η^6-toluene){Au(C$_6$Cl$_5$)$_2$}]$_2$(dioxane) **35** [61].

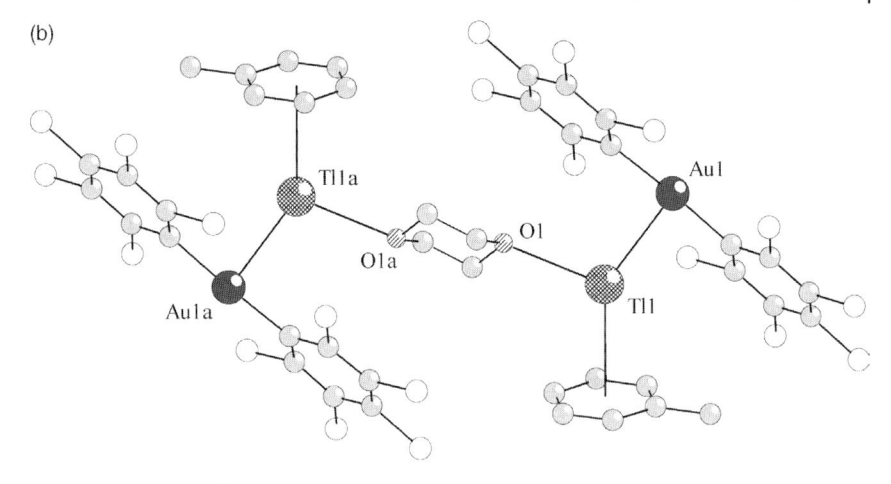

Figure 4.13 (Continued).

of 1,4-dioxane weakly bridging the thallium(I) centers [Tl(1)–O(1) 2.827(4) Å] of two Tl(η^6-toluene)[Au(C$_6$Cl$_5$)$_2$] units (Figure 4.13b). The main striking feature of **35** is the shortest Au–Tl distance [2.8935(3) Å] found to date in gold–thallium complexes [61].

When the reaction between [NBu$_4$][AuR$_2$] and TlPF$_6$ was carried out in acetone the unexpected, loosely bound butterfly cluster Tl$_2$Au$_2$(C$_6$Cl$_5$)$_4$(acetone) **36** was formed [62]. The same complex was also obtained from a solution of the polymeric [Tl{Au(C$_6$Cl$_5$)$_2$}]$_n$ **31** in acetone [63]. In the tetranuclear species the metal atoms are held together by four unsupported Au(I)–Tl(I) interactions [3.0331(6)–3.1887(6) Å] which result in a four-membered Au$_2$Tl$_2$ ring folded along an additional transannular Tl–Tl interaction [3.6027(6) Å]. The acetone molecule weakly bridges, through its oxygen, the thallium atoms [Tl–O 2.968(9)/2.903(9) Å] (Figure 4.14a) [62]. Similar central Tl$_2$Au$_2$(C$_6$Cl$_5$)$_4$ cores were found for the related pale yellow Tl$_2$Au$_2$(C$_6$Cl$_5$)$_4$(PhMeCO) **37** and Tl$_2$Au$_2$(C$_6$Cl$_5$)$_4$(Hacac) **38** [63]. Compared to the molecular structure of **36** one main difference is a longer transannular Tl–Tl interaction [3.7110(4) Å in **37**; 3.7152(4) Å in **38**], which might be due to the more sterically demanding acetophenone and acetylacetone ligands. On the other hand, the coordination of the organic molecules is different: in **37** the oxygen of the ketone bridges asymmetrically the two Tl atoms [2.713(5)/3.086(6) Å], while in **38** the acetylacetone molecule is coordinated through its oxygen in the keto form to only one of the Tl atoms [2.826(5) Å]. The reaction of **38** with 4,4′-bipyridine (1:1 molar ratio) in toluene afforded the isolation of the salmon pink complex Tl$_2$Au$_2$(C$_6$Cl$_5$)$_4$(-Hacac)(4,4′-bipy) **39** in which the Tl$_2$Au$_2$(C$_6$Cl$_5$)$_4$ core is preserved, but the diketone exhibits a bimetallic triconnective coordination pattern [Tl–O 2.707(4)/2.904(4) and 2.959(4) Å] and the aromatic amine is coordinated to a Tl atom [Tl(1)–N(1) 2.874(5) Å]. Moreover, the other nitrogen atom of 4,4′-bipyridine is involved in an intermolecular N···H–O interaction with the enol form of the acetylacetone ligand, leading to a dimer association (Figure 4.14b) [63].

Exposure of the solid [Tl{Au(C$_6$Cl$_5$)$_2$}]$_n$ **31** to vapors of THF provided crystals of [Tl(THF)$_{0.5}${Au(C$_6$Cl$_5$)$_2$}]$_n$ **40** for which an X-ray diffraction study revealed that part of the

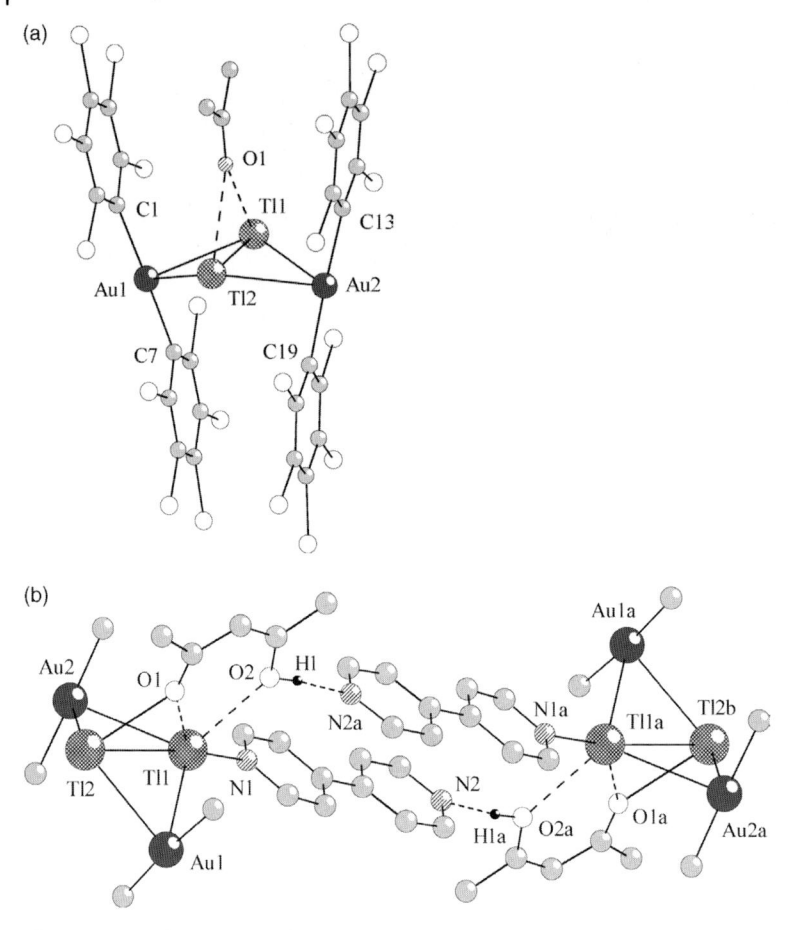

Figure 4.14 (a) The structure of Tl$_2$Au$_2$(C$_6$Cl$_5$)$_4$(acetone) **36** [62], and (b) the dimer association through hydrogen bonding in the crystal of Tl$_2$Au$_2$(C$_6$Cl$_5$)$_4$(Hacac)(4,4'-bipy) **39** [63].

empty channels available in the crystal of the parent compound **31** are occupied by solvent molecules [59]. The polymeric structure is preserved but the alternating Au and Tl atoms form a zigzag chain. In the asymmetric unit there are two thallium atoms with a different environment (Figure 4.15a). Thus, while Tl(1) remained uncoordinated to THF with only intermetallic interactions and an angular Au–Tl–Au fragment [138.71(1)°], the Tl(2) binds to the oxygen atom of the THF molecule [Tl(2)–O(1) 2.697(6) Å] in a nearly planar trigonal geometry [Au(1)–Tl(2)–Au(2a) 157.06(1)°]. In addition to the Au–Cl and Tl–Cl contacts within a chain polymer, some Tl–Cl contacts between adjacent chains are also preserved. In **40** there are still vacant coordination sites at thallium and the crystal contains empty channels that might allow the entrance of additional THF molecules. Indeed, **31** reacts with THF and the green complex [Tl(THF)$_2${Au(C$_6$Cl$_5$)$_2$}]$_n$ **41** could be isolated. The crystal of **41** consists of similar zigzag chains of alternating Au and Tl atoms [Au(1)–Tl(1) 3.1981(4), Au(1)–Tl(1a) 3.0764

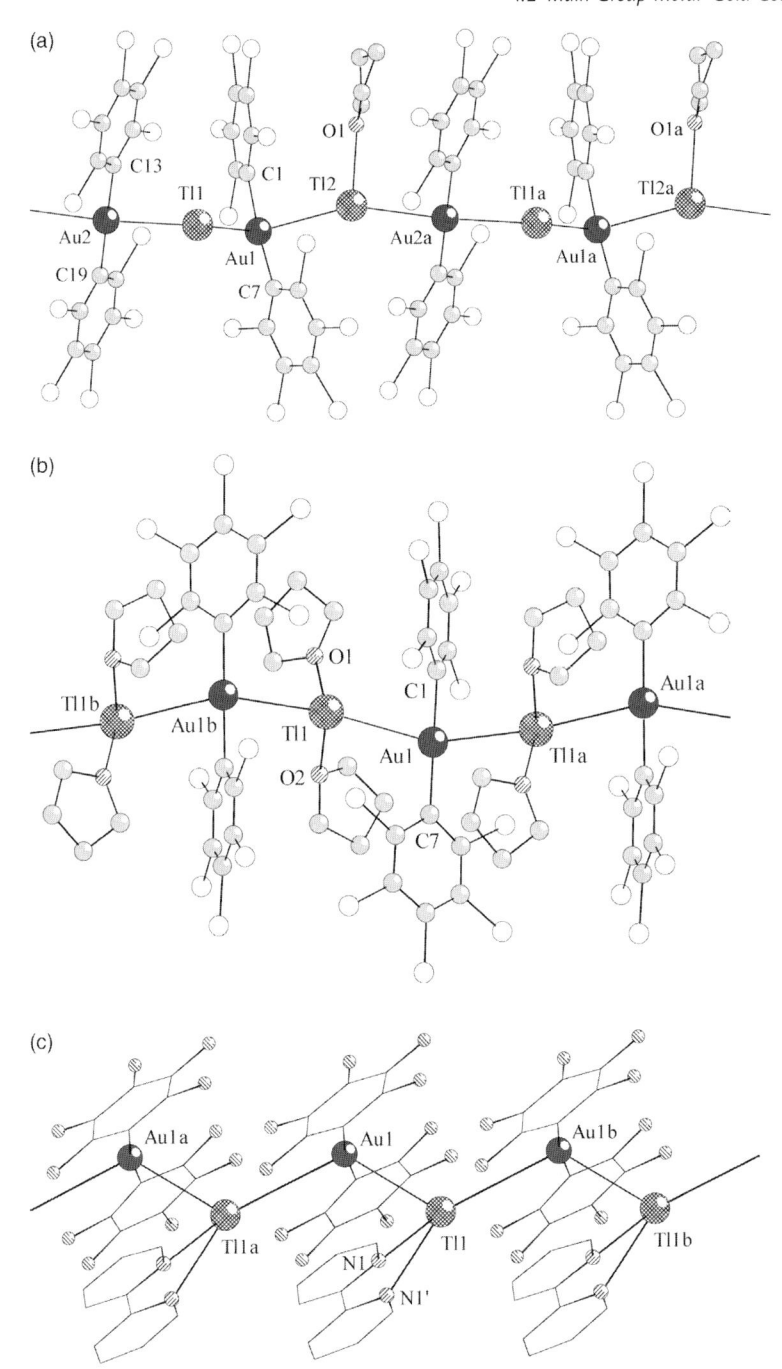

Figure 4.15 View of the chain polymer in the crystal of (a) [Tl(THF)$_{0.5}$\{Au(C$_6$Cl$_5$)$_2$\}]$_n$ **40**, (b) [Tl(THF)$_2$\{Au(C$_6$Cl$_5$)$_2$\}]$_n$ **41** [59], (c) [Tl(2,2′-bipy)\{Au(C$_6$F$_5$)$_2$\}]$_n$ **52** and (d) [Tl(1,10-phen)\{Au(C$_6$F$_5$)$_2$\}]$_n$ **53** [65].

(d)

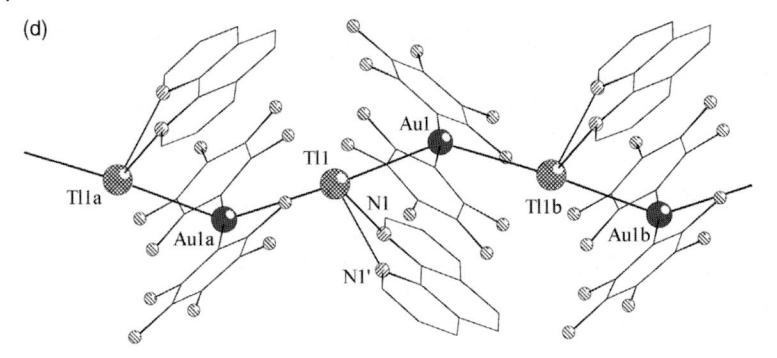

Figure 4.15 (Continued).

(4) Å], but in this case each thallium atom is coordinated by two THF molecules [Tl–O 2.653(7)/2.648(6) Å] in a distorted *pseudo*-trigonal bipyramidal geometry (Figure 4.15b) [59]. The coordination sphere of each Tl(I) being complete, the crystal of **41** no longer presents interchain Tl–Cl contacts and free channels [59]. The yellow complex [Tl(MeCN)$_2$$\{$Au(C$_6Cl_5$)$_2$$\}$]$_n$ **42** was prepared similarly [59]. Compounds of the same stoichiometry, that is [Tl(L)$_2$$\{$Au(C$_6Cl_5$)$_2$$\}$]$_n$, but different colors, were also obtained when **31** was reacted in a 1:2 molar ratio with other pure organic ligands as solvents [59]: yellow – THT **43**, 2-fluoropyridine **44**, NEt$_3$ **45**; red – pyridine **46** [59]. The reaction of **31** with acetylacetone as solvent resulted in the red complex [Tl(Hacac)$_2$$\{$Au(C$_6Cl_5$)$_2$$\}$]$_n$ **47** [59], behavior which contrasts with the cluster compound **38** obtained when the same reaction is performed in toluene [63]. The structure of **47** revealed a polymeric chain similar to that found in **41**, each acetylacetone molecule being coordinated in its enol form, stabilized by intramolecular O–H···O hydrogen bonding which gives rise to a six-membered ring [59].

A yellow–green complex of the same stoichiometry as found for **41**, that is [Tl(OPPh$_3$)$_2$$\{$Au(C$_6F_5$)$_2$$\}$]$_n$ **48**, was obtained by reacting Li[Au(C$_6$F$_5$)$_2$] with TlNO$_3$ and triphenylphosphine oxide in Et$_2$O/MeOH [64] or [NBu$_4$][Au(C$_6$F$_5$)$_2$], TlPF$_6$ and Ph$_3$PO in THF [55]. Surprisingly, treatment of the gold(III) complex [Au(C$_6$F$_5$)$_2$Cl(PPh$_3$)] with equimolecular amounts of Tl(acac) gave the same complex **48**. Its structure is similar to that of **41**, but the Tl–Au–Tl angle in the polymeric chain is 180°. In contrast, the reaction of TlPF$_6$ with [NBu$_4$][Au(C$_6$Cl$_5$)$_2$] in THF or acetone, in the presence of two equivalents of Ph$_3$PO, resulted in isolation of a new type of green complexes [$\{$Tl(OPPh$_3$)$\}$$\{$Tl(OPPh$_3$)(THF)$\}$$\{$Au(C$_6Cl_5$)$_2$$\}$$_2$]$_n$ **49** and [$\{$Tl(OPPh$_3$)$\}$$\{$Tl(OPPh$_3$)(acetone)$\}$$\{$Au(C$_6Cl_5$)$_2$$\}$$_2$]$_n$ **50** [55]. In the solid state, as in the case of **40**, in both compounds **49** and **50** the asymmetric unit contains four metal atoms with a Tl–Au–Tl–Au arrangement. Each thallium atom of the asymmetric unit has a different environment, that is, one Tl is coordinated by the oxygen atoms of one OPPh$_3$ ligand and one solvent molecule in a *pseudo*-trigonal-bipyramidal geometry, while the second Tl atom only bears an OPPh$_3$ ligand and has a strongly distorted tetrahedral geometry with a vacant coordination site. This asymmetric unit is repeating in the one-dimensional, zigzag polymers.

4.2 Main Group Metal–Gold Compounds

The precursor polymeric complex [Tl{Au(C$_6$F$_5$)$_2$}]$_n$ **32** exhibits a similar reactivity to that of the related compound **31**. Thus, **32** reacts with two equivalents of pyridine, in THF, to give a yellow solid formulated as a 1 : 1 electrolyte in solution, [Tl(Py)$_2$][Au(C$_6$F$_5$)$_2$] **51**, on the basis of conductivity measurements [65]. The reactions of **32** with one equivalent of the chelating ligands 2,2′-bipyridine or 1,10-phenanthrolin in THF, gave yellow solids which also behave as 1 : 1 electrolytes in acetone. In the solid state, both compounds were found to exhibit chain polymeric structures, [Tl(2,2′-bipy){Au(C$_6$F$_5$)$_2$}]$_n$ **52** (Figure 4.15c) and [Tl(1,10-phen){Au(C$_6$F$_5$)$_2$}]$_n$ **53** (Figure 4.15d), with the Tl centers coordinated by the nitrogen atoms of a chelating aromatic amine. The major differences between the two polymeric chains built via unsupported Au–Tl interactions are (i) the asymmetry in the Au–Tl distances, that is, 3.0825(4)/3.1397(4) Å in **52** versus 3.0120(6)/3.4899(6) Å in **53**, and (ii) the acute angular Au–Tl–Au angle [120.19(2)°] in **52** versus the almost linear Au–Tl–Au angle [176.08(1)°] in **53**.

An interesting behavior was reported for the [Tl(H$_2$NCH$_2$CH$_2$NH$_2$)(AuR$_2$)] complexes obtained as white [R = C$_6$Cl$_5$ **54**] or green [R = C$_6$F$_5$ **55**] solids [66]. These complexes react at room temperature with ketones in THF to give new amine-imine or diimine complexes, depending on the stoichiometry. They also react in the solid state with ketone vapors at room temperature in a few seconds. Thus, with equimolecular amounts of acetone the orange compounds [Tl(H$_2$NCH$_2$CH$_2$N=CMe$_2$)(AuR$_2$)]$_n$ [R = C$_6$Cl$_5$ **56**; C$_6$F$_5$ **57**] were obtained, while reactions with acetone in 1 : 2 molar ratio, in THF, gave the complexes [Tl(Me$_2$C=NCH$_2$CH$_2$N=CMe$_2$)(AuR$_2$)]$_n$ [R = C$_6$Cl$_5$ **58** – red; C$_6$F$_5$ **59** – yellow]. It is worthwhile to mention that the use of the stronger acceptor acetophenone in the reaction with the precursors **54** and **55** resulted in conversion of both amine groups into imine groups, regardless of the molar ratio employed. The 1D polymeric nature of compounds **56**, **58** and **59** was proved by single-crystal X-ray diffraction. The chains contain alternating gold(I) and thallium(I) centers similar to other related polynuclear Au/Tl systems with unsupported metal–metal interactions. The Au(I) atoms display square-planar coordination, while the Tl(I) atoms are chelated by two nitrogen atoms of a chelating amine-imine in **56** or diimine ligand in **58** and **59**, respectively, in a distorted *pseudo*-trigonal bipyramidal environment [66]. It should be noted here that, even with a degree of disorder, the structure of **54** was also determined and the crystal contains only discrete dinuclear molecules, with an intramolecular Au–Tl distance of about 3.1 Å and an intermolecular Au–Tl distance of about 4.04 Å between neighboring molecules [66].

The reaction of the polymeric complexes [Tl(AuR$_2$)]$_n$ [R = C$_6$Cl$_5$ **31**, C$_6$F$_5$ **32**] with DMSO resulted in compounds with different stoichiometry, that is, the yellow [Tl$_2$(μ-DMSO)$_2${Au(C$_6$Cl$_5$)$_2$}$_2$]$_n$ **60** and the white [Tl$_2${μ-DMSO}$_3${Au(C$_6$F$_5$)$_2$}$_2$]$_n$ **61**, regardless of the molar ratio of the reagents [67]. The X-ray diffraction studies revealed, in both cases, the formation of infinite chains formed via both unsupported Au–Tl interactions and bridging DMSO molecules between the Tl centers. For compound **60** the monodimensional polymer is formed from repeating Au–Tl(μ-DMSO)$_2$Tl units with symmetrically bridged solvent molecules. An additional Au(C$_6$Cl$_5$)$_2$ also bridges the Tl atoms of the repeating unit [Au(1)–Tl(1) 3.2839(6), Au(1)–Tl(2) 3.1337(6) Å] and the Tl–Tl distance of 3.7562(6) Å suggests a metal–metal interaction (Figure 4.16a). In contrast, the repeating unit in **61** is Au–Tl(μ-DMSO)$_3$Tl in which

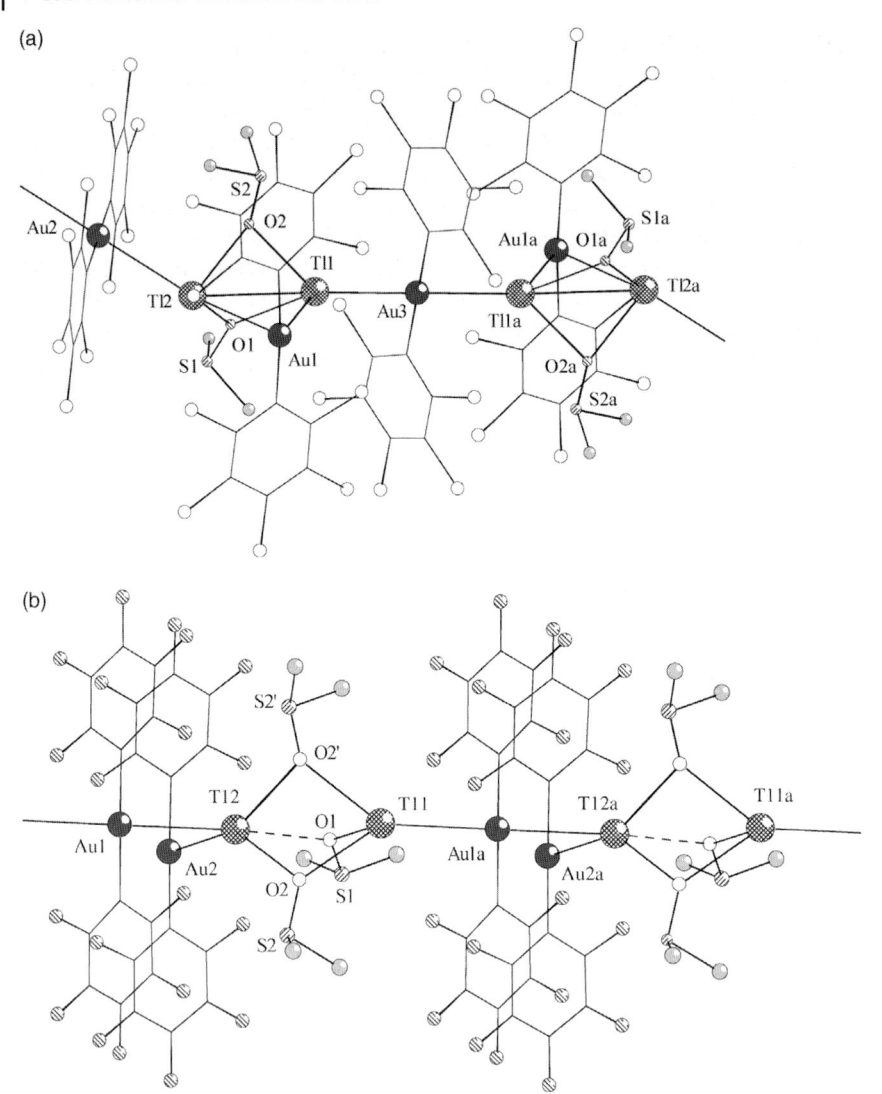

Figure 4.16 View of the chain polymer in the crystal of
(a) [Tl$_2$(μ-DMSO)$_2${Au(C$_6$Cl$_5$)$_2$}$_2$]$_n$ **60**, and (b) [Tl$_2${μ-DMSO}$_3${Au
(C$_6$F$_5$)$_2$}$_2$]$_n$ **61** [67].

two of the DMSO molecules are symmetrically bridged, while the third one acts as an asymmetrical bridge. An additional terminal Au(C$_6$F$_5$)$_2$ is connected to a thallium center [Au(2)–Tl(2) 3.5182(8) Å] (Figure 4.16b). For **61** the Tl–Tl distance of 4.10 Å is too long to consider any bonding between the thallium centers.

Taking advantage of the acid properties of the Tl(I) centers in the polymeric [Tl(AuR$_2$)]$_n$ derivatives, the neutral derivatives **31** and **33** were reacted with [NBu$_4$][AuR$_2$] (R = C$_6$Cl$_5$, C$_6$Cl$_2$F$_3$-3,5) in different molar ratios, in toluene as noncoordinating solvent, to yield the ionic compounds [NBu$_4${Tl$_2${Au(C$_6$Cl$_5$)$_2$}{μ-Au(C$_6$Cl$_5$)$_2$}$_2$}]$_n$ **62** (yellow), [NBu$_4${Tl{Au(C$_6$Cl$_5$)$_2$}{Au(3,5-C$_6$Cl$_2$F$_3$)$_2$}}]$_n$ **63** (yellow), and [NBu$_4${Tl{Au(3,5-C$_6$Cl$_2$F$_3$)$_2$}$_2$}]$_n$ **64** (green) [58]. The resulting anionic heterometallic chains are different. Thus, only half [Au(C$_6$Cl$_5$)$_2$]$^-$ per thallium(I) atom was incorporated in **62** with respect to the starting precursor **31**, resulting in a chain which contains folded tetranuclear Au$_2$Tl$_2$ cores connected by AuR$_2$ units (Figure 4.17a). In contrast to the butterfly clusters **36–39** [62, 63] no transannular Tl–Tl interactions are present in **62** and the Tl–Au–Tl fragment between two Au$_2$Tl$_2$ cores is linear. For the complexes **63** (Figure 4.17b) and **64** each thallium(I) center of the polymetallic Au/Tl chain binds a terminal [Au(C$_6$Cl$_2$F$_3$-3,5)$_2$]$^-$ fragment. In all three complexes the coordination around the Tl atoms is almost planar and, therefore, the inert pair is apparently stereochemically inactive.

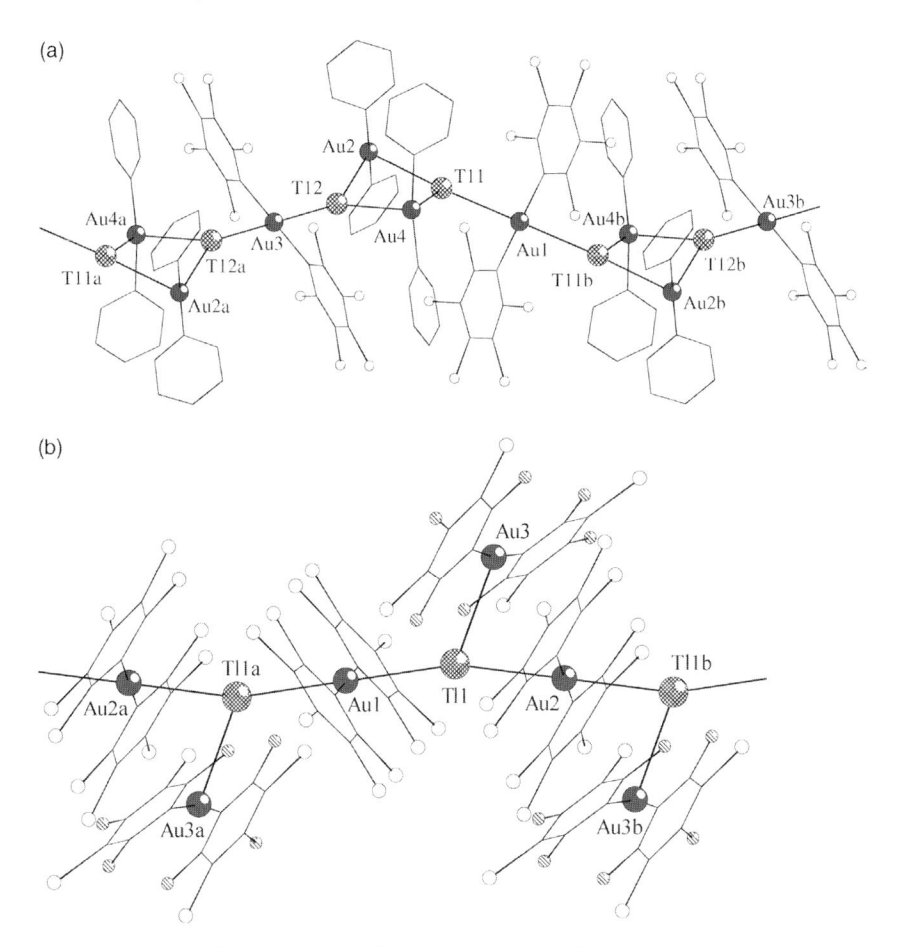

Figure 4.17 View of the chain polymer of anions in the crystal of (a) [NBu$_4${Tl$_2${Au(C$_6$Cl$_5$)$_2$}{μ-Au(C$_6$Cl$_5$)$_2$}$_2$}]$_n$ **62** (for clarity, the Cl atoms on the phenyl groups of the Tl$_2$Au$_2$ core are omitted), and (b) [NBu$_4${Tl{Au(C$_6$Cl$_5$)$_2$}{Au(3,5-C$_6$Cl$_2$F$_3$)$_2$}}]$_n$ **63** [58].

The heteropolynuclear complexes [Tl(AuR$_2$)]$_n$ [R = C$_6$Cl$_5$ **31**, C$_6$F$_5$ **32**] were also found to react with Tl(acac) in 1 : 1 or 1 : 2 molar ratio, in toluene, to afford isolation of yellow [Tl$_2$(acac){Au(C$_6$Cl$_5$)$_2$}]$_n$ **66** and white [Tl$_3$(acac)$_2${Au(C$_6$F$_5$)$_2$}]$_n$ **65**, regardless of the molar ratio used and recovering the excess of starting materials when the molar ratio was not adequate [57]. In both cases dimeric Tl$_2$(acac)$_2$ act as bridging units between Tl[Au(C$_6$X$_5$)$_2$] fragments, but in **66** Au–Tl contacts prevail, while in **65** the most important interactions are the Tl–Tl contacts. Thus, the crystal of **65** contains a double-chain unidimensional polymer built from isolated dinuclear Tl[Au(C$_6$F$_5$)$_2$] and Tl$_2$(acac)$_2$ units via two Tl–O bonds [2.577(3) Å], with additional unsupported Tl–Tl contacts [Tl(3)–Tl(2a) 3.7200(4), Tl(3)–Tl(2′) 3.7607(4) Å] (Figure 4.18a). In contrast, in **66** a two-dimensional supramolecular architecture is achieved; zigzag infinite chains formed via unsupported Au–Tl interactions are joined by Tl$_2$(acac)$_2$

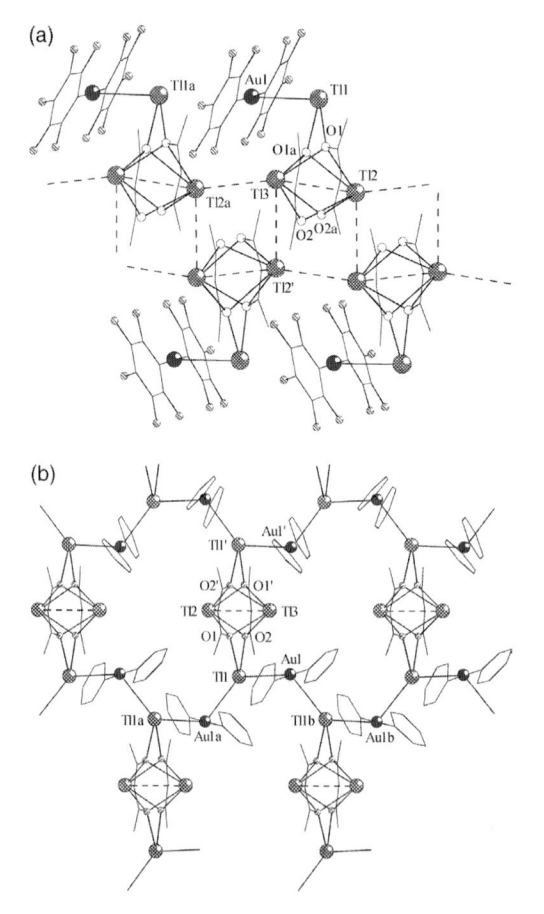

Figure 4.18 (a) Double-chain 1D polymer in the crystal of [Tl$_3$(acac)$_2${Au(C$_6$F$_5$)$_2$}]$_n$ **65**, and (b) 2D network in the crystal of [Tl$_2$(acac){Au(C$_6$Cl$_5$)$_2$}]$_n$ **66** (for clarity, the Cl atoms on the phenyl groups are omitted) [57].

units bridging thallium centers of different chains through Tl–O bonds [Tl(1)–O(1) 2.685(8), Tl(1)–O(2) 2.676(9) Å] (Figure 4.18b).

The use of 4,4′-bipyridine as spacer between the thallium centers of the polymeric precursors [Tl(AuR$_2$)]$_n$ [R = C$_6$Cl$_5$ **31**, C$_6$F$_5$ **32**] gives rise to a variety of materials which contain different amounts of the aromatic amine and display two- or three-dimensional networks in the solid state [65, 68, 69]. Thus, the reaction of **31** with 4,4′-bipyridine (2 : 1 molar ratio) in toluene led to the orange complex [Tl(4,4′-bipy)$_{0.5}${Au(C$_6$Cl$_5$)$_2$}]$_n$ **67**. Its crystal contains zigzag infinite chains formed via unsupported Au–Tl interactions [Tl–Au–Tl 180°, Au–Tl–Au 147.38(1)°]. The thallium atoms from adjacent polymetallic chains are connected through bridging aromatic amine ligands, resulting in a two-dimensional polymer (Figure 4.19) [68], with trigonal planar coordination environment of the thallium centers.

The reaction of TlPF$_6$ with [NBu$_4$][Au(C$_6$F$_5$)$_2$] and 4,4′-bipyridine, in THF, gave the pale yellow complex [{Tl(4,4′-bipy)}$_2${Au(C$_6$F$_5$)$_2$}$_2$]$_n$ **68**, which also exhibits a 2D network (Figure 4.20a) [69]. In this case almost linear tetranuclear units, [Tl$_2${Au

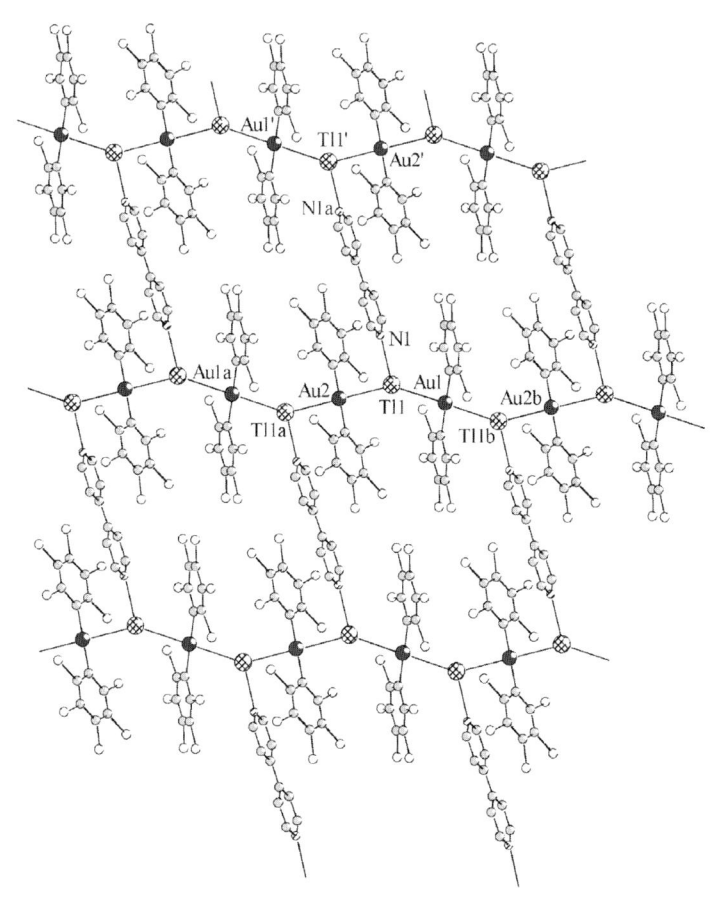

Figure 4.19 View of the 2D network in the crystal of [Tl(4,4′-bipy)$_{0.5}${Au(C$_6$Cl$_5$)$_2$}]$_n$ **67** [68].

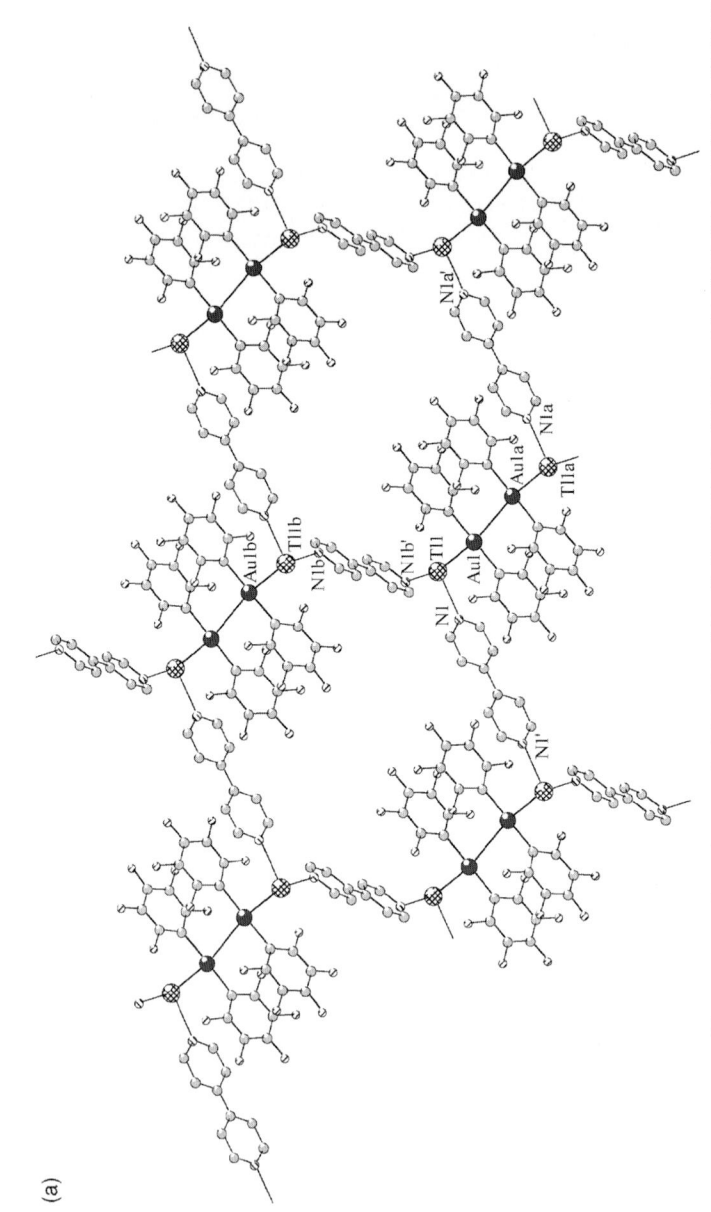

Figure 4.20 View of the 2D networks in the crystal of (a) [{Tl(4,4'-bipy)}₂{Au(C₆F₅)₂}₂]ₙ **68** [69], and (b) [{Tl(4,4'-bipy)(THF)}{Au(C₆F₅)₂}]ₙ **69** (for clarity, the F atoms on the phenyl groups are omitted) [65].

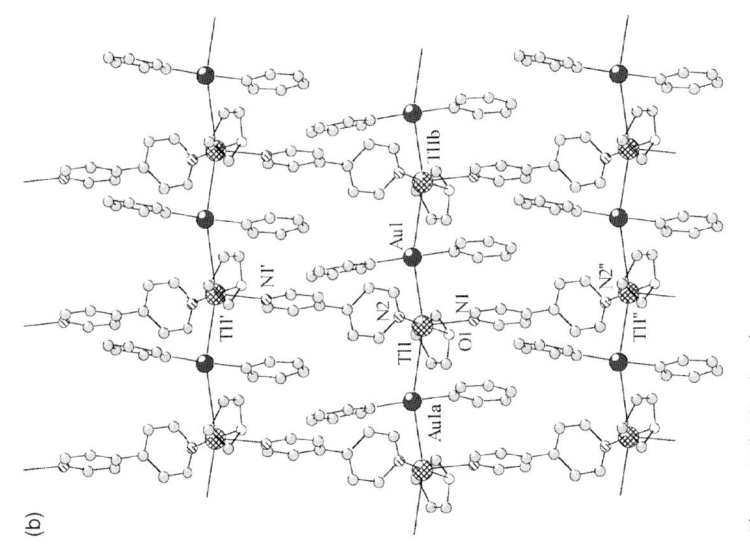

Figure 4.20 (*Continued*).

$(C_6F_5)_2\}_2]$ [Tl(1)–Au(1)–Au(1a) 164.92(1)°], are formed through a gold–gold contact [Au(1)–Au(1a) 3.4092(3) Å]. The terminal thallium centers of a tetranuclear unit are linked to four other different tetranuclear moieties through bridging 4,4'-bipyridine ligands. When the reaction was performed between preformed $[Tl\{Au(C_6F_5)_2\}]_n$ **32** and 4,4'-bipyridine, in THF, the complex $[\{Tl(4,4'\text{-bipy})(THF)\}\{Au(C_6F_5)_2\}]_n$ **69** was isolated [65]. Its crystal contains chains of alternating Au(I) and Tl(I) centers linked via unsupported Au–Tl interactions. Each Tl atom is coordinated by the oxygen of a THF molecule and by two nitrogen atoms from two different 4,4'-bipyridine molecules which act as bridging ligands between adjacent chains, leading to a 2D network (Figure 4.20b). The environment of a thallium center, considering both the Tl–N/Tl–O bonds and the intermetallic contacts, is trigonal planar in **68** and *pseudo*-octahedral in **69**. In both compounds adjacent layers are connected by additional Tl–F interactions [3.170(2) and 3.174(2) Å in **68**; 3.289(4) Å in **69**] and including these contacts the structures can be described as three-dimensional supramolecular associations.

The related pentachlorophenyl derivative, $[Tl(4,4'\text{-bipy})\{Au(C_6Cl_5)_2\}]_n$ **70**, was obtained as an orange solid, either by reacting the polymeric precursor **31** with an equimolecular amount of 4,4'-bipyridine or from complex **67** and half an equivalent of the aromatic amine [68]. Its solid-state structure is completely different from that of **68**. It features polymeric chains of alternating Au(I) and Tl(I) centers linked via unsupported Au–Tl interactions, which grow along the *a* axis. Molecules of 4,4'-bipyridine bridge the thallium atoms (*pseudo*-trigonal bipyramidal TlN_2Au_2 core, with gold atoms in apical positions) from parallel chains resulting in a 3D supramolecular architecture with channels which might be filled by solvent molecules (Figure 4.21) [68]. Indeed, this is the case for the deep brown crystals of $[Tl(4,4'\text{-bipy})\{Au(C_6Cl_5)_2\}\cdot 0.5\text{toluene}]_n$ **71**, obtained by recrystallization of **70** from toluene, where the channels are occupied by toluene molecules. The main difference between these two structures resides in the orientation of the aromatic rings of the bridging ligands that is rotated to form an angle of 36.2(3)° in **70** and coplanar in **71**, a behavior which results in larger holes in **71** (minimum distance 3.550 Å) than in **70** (minimum distance 2.992 Å) [68].

Using a different molar ratio (2 : 1.5) between **31** and 4,4'-bipyridine, in THF, the red complex $[\{Tl(4,4'\text{-bipy})\}\{Tl(4,4'\text{-bipy})_{0.5}(THF)\}\{Au(C_6Cl_5)_2\}_2]_n$ **72** was isolated [68, 69], while attempts to grow single crystals of **70** from THF resulted in the related red complex $[\{Tl(4,4'\text{-bipy})\}\{Tl(4,4'\text{-bipy})_{0.5}(THF)\}\{Au(C_6Cl_5)_2\}_2\cdot THF]_n$ **73** [68]. The molecular structure of both compounds was established and was found to exhibit some common features. For both compounds the asymmetric unit contains four metal centers with a Tl–Au–Tl–Au arrangement, with one thallium coordinated by nitrogens from two bipyridine ligands and the other coordinated by the nitrogen atom from the third bipyridine ligand and the oxygen atom from a THF molecule. The Au_2Tl_2 units are associated into chains which grow along the *a* axis. For **72** neighboring chains are connected into an infinite two-dimensional structure through bipyridine molecules which bridge Tl atoms of different types, while the remaining bipyridine moieties bridge Tl centers of the same type from parallel layers, thus resulting in a 3D supramolecular association (Figure 4.22a) [69]. In

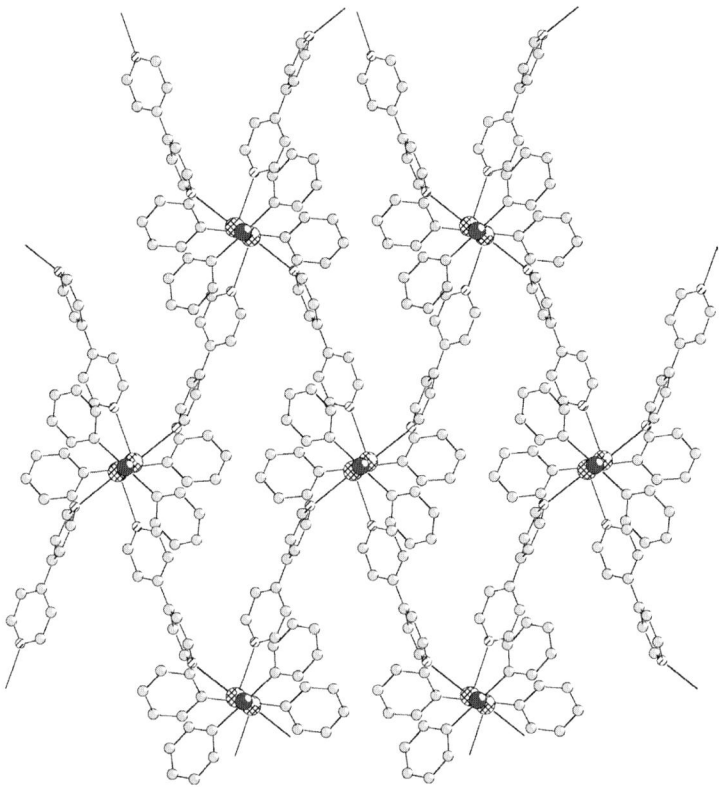

Figure 4.21 View of the 3D network in the crystal of [Tl(4,4′-bipy){Au(C$_6$Cl$_5$)$_2$}]$_n$ **70** (for clarity, the Cl atoms on the phenyl groups are omitted) [68].

contrast, the bridging pattern of the 4,4′-bipyridine molecules in **73** results in a different 3D architecture with channels occupied by additional uncoordinated THF molecules (Figure 4.22b) [68]. This results in different coordination cores for the thallium centers: while in **72** distorted tetrahedral TlONAu$_2$ and *pseudo*-trigonal bipyramidal TlN$_2$Au$_2$ cores are present, in **73** all thallium atoms exhibit distorted *pseudo*-trigonal bipyramidal TlN$_2$Au$_2$ cores.

4.2.4
Gold–Group 14 Metal Compounds

A few compounds containing gold–silicon, gold–germanium and gold–tin bonds and one compound with a gold–lead interaction have so far been structurally characterized by single-crystal X-ray diffraction and they are listed in Table 4.4. Two strategies were generally used for the preparation of the majority of the compounds: (i) reaction of gold(I) halide complexes with alkali metal salts of [MR$_3$]$^-$ anions, or

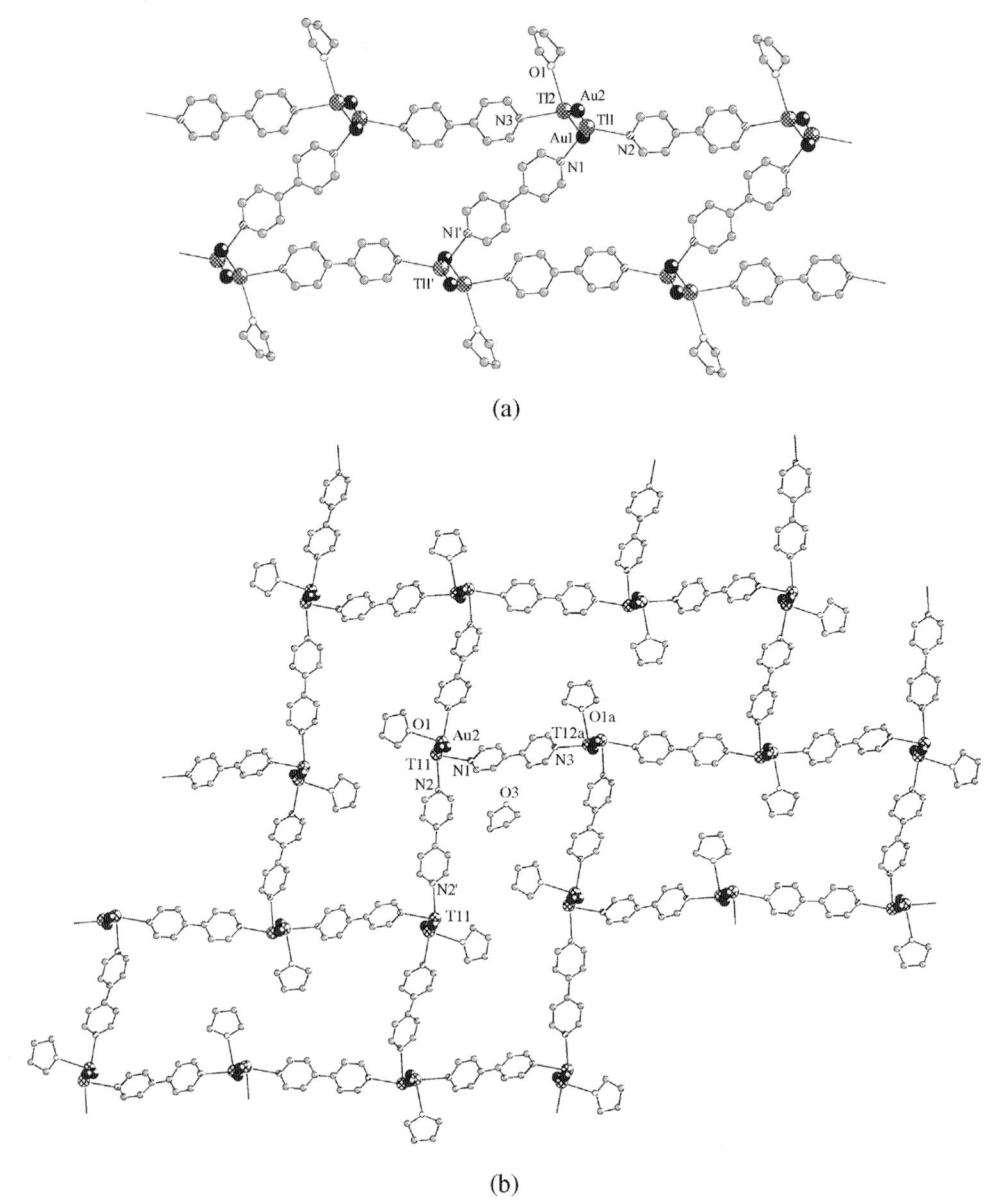

Figure 4.22 View of the 3D networks in the crystal of (a) [{Tl(4,4′-bipy)}{Tl(4,4′-bipy)$_{0.5}$(THF)}{Au(C$_6$Cl$_5$)$_2$}$_2$]$_n$ **72** [69], and (b) [{Tl(4,4′-bipy)}{Tl(4,4′-bipy)$_{0.5}$(THF)}{Au(C$_6$Cl$_5$)$_2$}$_2$·THF]$_n$ **73** (for clarity, the perchlorophenyl groups are omitted) [68].

Table 4.4 Selected molecular parameters for gold–Group 14 metal compounds.

Compound	Au–M (Å)	P–Au–M (°)	Ref.
Compounds with Au–Si bonds/interactions			
(MePh$_2$P)AuSiPh$_3$ 74	2.354(4)	173.6(2)	[70]
(MePh$_2$P)AuSi(SiMe$_3$)$_3$ 75	2.356(2)	170.15(9)	[70]
(Me$_3$P)AuSiPh$_3$ 76	2.362(1)	178.68(5)	[71]
(Ph$_3$P)AuSiCl(Cp*)$_2$ 77	2.363(2)	171.31(6)	[72]
(dppm)(AuSiPh$_3$)$_2$ 78	2.344(1)/2.358(1)	161.28(5)/172.74(5)	[73]
(dppe)(AuSiPh$_3$)$_2$ 79	2.352(2)	173.12(9)	[73]
[(Ph$_3$P)$_2$Au][Ph$_3$SiAuCl] 80	2.291(2)	177.96(8)^a	[74]
Compounds with Au–Ge bonds/interactions			
(Ph$_3$P)AuGeCl$_3$ 81	2.406(1)	168.3(1)	[75]
(Ph$_3$As)AuGeCl$_3$ 82	2.389(1)	169.35(1)^b	[76]
[(2-MeC$_6$H$_4$)$_3$P]AuGeCl$_3$ 83	2.376(1)	167.9(1)	[77]
(Ph$_3$P)AuGeCl[N(SiMe$_3$)$_2$] 84	2.4136(11)	169.48(3)	[78]
(Cy$_3$P)AuGeCl[N(SiMe$_3$)$_2$] 85	2.4106(5)	176.66(2)	[78]
(Et$_3$P)AuGeCl[N(SiMe$_3$)$_2$] 86	2.4169(5)	175.22(3)	[78]
(Ph$_3$P)AuGeF[N(SiMe$_3$)$_2$] 87	2.4085(3)	177.447(18)	[78]
(Ph$_3$P)$_3$AuGeCl$_3$ 88	2.563(1)	101.7(1)–105.9(1)	[75]
[(Me$_3$SiN=PPh$_2$)$_2$C=Ge(I)Au]$_2$ 89	2.330(1)/2.341(1)	175.7(3)/173.7(3)^c	[79]
IAuGe[N(SiMe$_3$)C(Ph)C(SiMe$_3$)(C$_5$H$_4$N-2)]Cl 90	2.346(2)	172.8(6)^d	[80]
CH$_2$(AsPh$_2$)$_2$Au]$_2$(Cl$_3$GeAuCl)$_2$ 91	2.323(1)	173.86(6)^e	[76]
[PPh$_4$][Au{Ge(NSiMe$_3$CH$_2$)$_3$CMe}$_2$] 92	2.423(2)	180.0^f	[81]
[(Me$_2$PhP)$_2$Au][Au(GeCl$_3$)$_2$] 93	2.409(1)–2.417(1)	169.98(4)/169.20(4)^f	[77]
[(MesNC)$_2$Au][Au(GeCl$_3$)$_2$] 94	2.3866(7)	180	[82]

(Continued)

Table 4.4 (Continued)

Compound	Au–M (Å)	P–Au–M (°)	Ref.
[(dppm)Au]₂[Au(GeCl₃)₃] 96	2.4150(6)–2.5351(7)	101.72(2)/104.97(2)/153.12(2)[f]	[83]
[K{[2.2.2]crypt}]₅[Au₃Ge₁₈] 97	2.437(2) – 2.469(2)		[84]
[K{[2.2.2]crypt}]₈K[Au₃Ge₄₅] 98	2.479–2.565		[84, 85]
[Li(THF)₆][AuGe₁₈{Si(SiMe₃)₃}₆] 99	2.6876(6)		[86]
Compounds with Au–Sn bonds/interactions			
[(Me₂PhP)₂AuSnCl₃ 100	2.881(1)	102.2(1)/104.0(1)	[87]
(Ph₃P)₂AuSn[N(tolyl-p)SiMe₂]₃SiMe 101	2.5651(13)	179.91(9)	[88]
(dppb)Au₂[Sn{N(tolyl-p)SiMe₂}₃SiMe]₂ 102	2.6804(13)	92.0(4)/85.2(4)[g]	[88]
[(Ph₃P)₇Au₈(SnCl₃)]₂[SnCl₆] 103	2.625(3)		[89]
(Ph₃P)₄Au₄(μ-SnCl₃)₂ 104	2.972(1)/2.819(1)		[90]
[p-tolyl₃P)Au]₂Mn₂(SnCl₂)(CO)₆(dppm) 105	2.813(4)/2.815(4)		[91]
[NBu₃H]₅[Au(SnB₁₁H₁₁)₄] 106	2.601(1)/2.589(1)		[93]
[NBu₃H]₂[{(Ph₃P)Au}₂(SnB₁₁H₁₁)₂] 107	2.737(1)/2.761(1)	116.86(3)/120.17(3)	[94]
[NBu₃Me]₄[{(Ph₃P)Au}₂(SnB₁₁H₁₁)₃] 108	2.711(1) – 3.032(1)	108.45(6)–122.72(6)	[94]
[NBu₃H]₃[{(Et₃P)Au}₃(SnB₁₁H₁₁)₃] 109	2.7329(8)–3.1204(9)		[95]
[NBu₃Me]₄[{(dppm)Au₂}₂(SnB₁₁H₁₁)₄] 110	2.7105(7)–3.3078(9)		[95]
Compounds with Au–Pb bonds/interactions			
Pb[Au{CH₂P(S)Ph₂}₂]₂ 111	2.896(1)/2.963(2)	180.0[h]	[37]

[a] Cl–Au–Si angle.
[b] As–Au–Ge angle.
[c] Ge–Au–N angle.
[d] Ge–Au–I angle.
[e] Ge–Au–Cl angle.
[f] Ge–Au–Ge angle.
[g] Sn–Au–C angle.
[h] Pb–Au–Au angle.

(ii) insertion of MX$_2$ (X=halogen) or MR$_2$ into the Au–halide bond of tertiary phosphine complexes.

4.2.4.1 Gold–Silicon Compounds

Most of the gold-silicon compounds whose molecular structures have been reported so far were obtained by reacting phosphine-gold(I) chlorides with triorganosilyl lithium salts [70, 71, 73]. Typical examples are (MePh$_2$P)AuSiPh$_3$ **74**, (MePh$_2$P)AuSi(SiMe$_3$)$_3$ **75** and (Me$_3$P)AuSiPh$_3$ **76** (Figure 4.23a), whose crystals contain discrete molecules with an Au–Si bond of about 2.36 Å and almost linear P–Au–Si fragments (Table 4.4).

The reaction of decamethylsilicocene, Si(Cp*)$_2$, with Au(PPh$_3$)Cl, in toluene, resulted in insertion of the silylene into the Au–Cl bond and isolation of (Ph$_3$P)AuSiCl(Cp*)$_2$ **77**. The compound is molecular, with a π-interaction between a silicon-bound Cp* substituent and the central gold atom [Au(1)-Cp*$_{centroid}$ 3.317 Å] (Figure 4.23b) which increases the distortion of the P–Au–Si from linearity [P(1)–Au(1)–Si(1) 171.31(6)°].

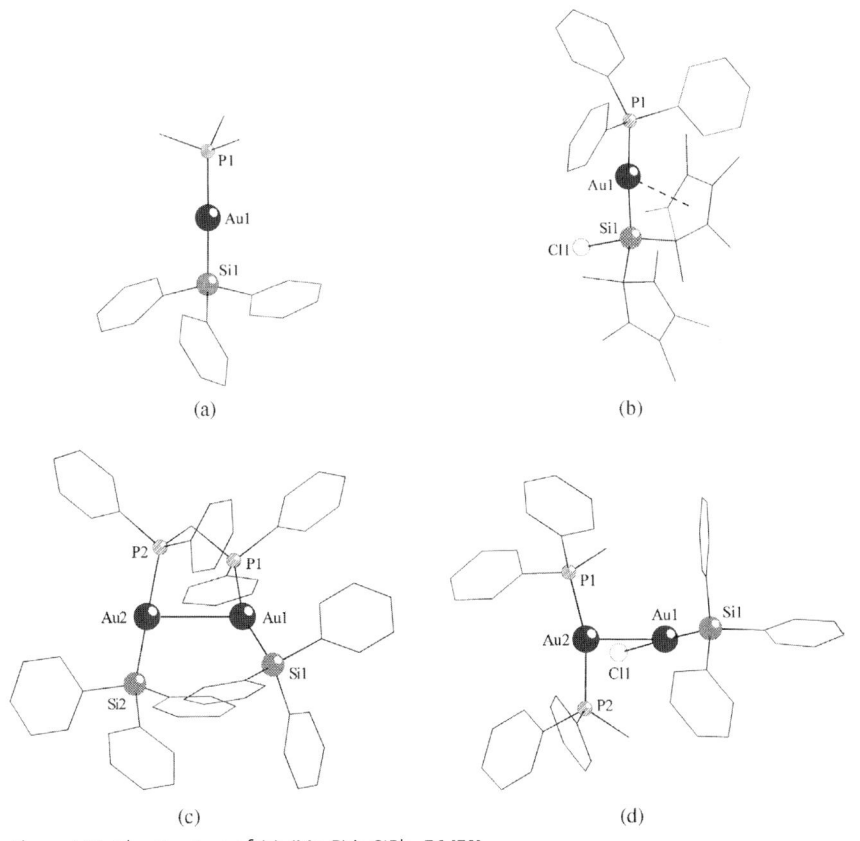

Figure 4.23 The structure of (a) (Me$_3$P)AuSiPh$_3$ **76** [71], (b) (Ph$_3$P)AuSiCl(Cp*)$_2$ **77** [72], (c) (dppm)(AuSiPh$_3$)$_2$ **78** [73], and (d) [(MePh$_2$P)$_2$Au](Ph$_3$SiAuCl) **80** [74].

The dinuclear complexes (dppm)(AuSiPh$_3$)$_2$ **78** and (dppe)(AuSiPh$_3$)$_2$ **79** were also found to be molecular [73], but their structures are different. While in **78** an intramolecular gold–gold interaction is established [Au(1)–Au(2) 3.1680(3) Å] (Figure 4.23c), in the related **79** the backbone of the bridging ligand is twisted to bring the gold centers far from each other [73] and steric congestion rules out any intermolecular gold–gold interaction.

Compound (Ph$_2$MeP)AuSiPh$_3$ **74** was found to react with Au(PPh$_2$Me)Cl to undergo ligand redistribution reaction which resulted in the ionic complex [(Ph$_3$P)$_2$Au](Ph$_3$SiAuCl) **80** [74]. Its crystal contains dinuclear units in which the linear-distorted cation [P(1)–Au(2)–P(2) 164.90(6)°] and anion [Cl(1)–Au(1)–Si(1) 177.96(8)°] are orthogonal to each other and connected through a gold–gold interaction [Au(1)–Au(2) 2.9807(4) Å] (Figure 4.23d).

No compounds with more than one silicon atom bound to gold have yet been structurally characterized.

4.2.4.2 Gold–Germanium Compounds

The chemistry of compounds with gold–germanium bonds has developed considerably since Schmidbaur and coworkers reported the first derivatives obtained by the insertion reaction of GeCl$_2$ (as its dioxane complex) into a gold–halogen bond [75]. Both neutral and ionic compounds have been reported and a larger structural diversity in comparison with the silicon analogs has been established (Table 4.4).

Most compounds were obtained by the insertion of germylene derivatives into a gold-halogen bond. Thus, reaction of GeCl$_2$(1,4-dioxane) with Au(L)Cl [L = Ph$_3$P, Ph$_3$As, (2-MeC$_6$H$_4$)$_3$P] gave the the compounds (Ph$_3$P)AuGeCl$_3$ **81** [75], (Ph$_3$As)AuGeCl$_3$ **82** [76] and [(2-MeC$_6$H$_4$)$_3$P]AuGeCl$_3$ **83** [77]. For **81** and **82** (Figure 4.24a) the crystals contain dimer associations via gold–gold contacts [Au(1)–Au(1a) 2.960(1) and 2.941(1) Å, respectively], with orthogonal E–Au–Ge fragments (E = P, As). In contrast, the bulkier *ortho*-tolyl groups on phosphorus in the related **83** prevent the aggregation via gold–gold interaction, but a dimer in the solid state is achieved through weak intermolecular gold-chlorine interactions [Au(1)–Cl(2a) 3.299(3) Å]

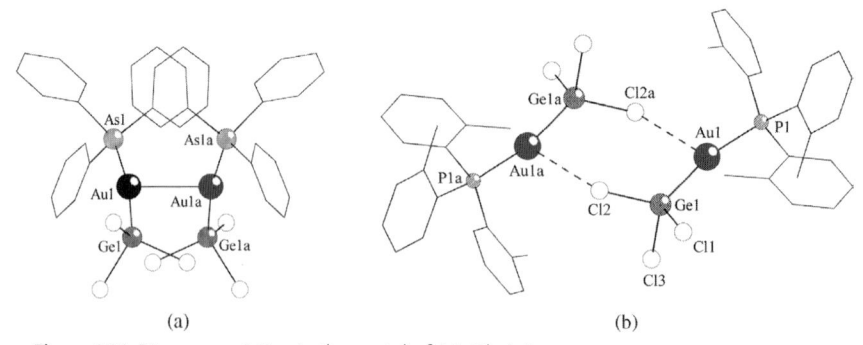

Figure 4.24 Dimer association in the crystal of (a) (Ph$_3$As)AuGeCl$_3$ **82** [76], and (b) [(2-MeC$_6$H$_4$)$_3$P]AuGeCl$_3$ **83** [77].

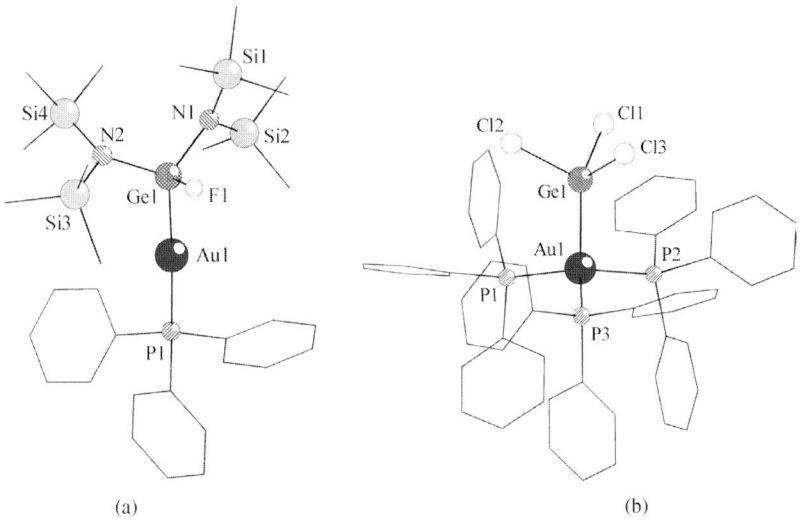

Figure 4.25 Molecular structure of (a) (Ph$_3$P)AuGe(F)[N(SiMe$_3$)$_2$] **87** [78], and (b) (Ph$_3$P)$_3$AuGeCl$_3$ **88** [75].

(Figure 4.24b). These intermolecular interactions may be the reason why the molecular E–Au–Ge fragments deviate significantly from linearity (Table 4.4).

The presence of bulkier groups on the germanium atom can result in molecular compounds. This is the case with compounds of the type (R$_3$P)AuGeCl[N(SiMe$_3$)$_2$] [R = Ph **84**, Cy **85**, Et **86**] which were obtained by insertion of the germylene Ge[N(SiMe$_3$)$_2$]$_2$ into the Au–Cl bond of Au(L)Cl [78]. Treatment of [{(Ph$_3$P)Au}$_3$(μ-O)][BF$_4$] with Ge[N(SiMe$_3$)$_2$]$_2$ results in fluoride abstraction from the [BF$_4$]$^-$ anion to yield the unexpected (Ph$_3$P)AuGeF[N(SiMe$_3$)$_2$] **87** (Figure 4.25a) [78]. Although gold atom is not involved in additional intermolecular interaction, for compound **84** the P–Au–Ge fragment is considerably bent [169.48(3)°], probably due to packing forces.

The reaction of Au(PPh$_3$)Cl with GeCl$_2$(1,4-dioxane) in the presence of two equivalents of triphenylphosphine leads to (Ph$_3$P)$_3$AuGeCl$_3$ **88** [75]. The compound crystallizes in two modifications which differ only in the formula units in the unit cell. The compound is molecular (Figure 4.25b), with a *quasi*-tetrahedrally co-ordinated gold atom and a considerable longer Au–Ge bond [Au(1)–Ge(1) 2.563(1) Å] compared to the related two-coordinated gold(I) derivatives. This suggests that the compound **88** might be the result of a weakly co-ordinating [GeCl$_3$]$^-$ anion and a [Au(PPh$_3$)$_3$]$^+$ cation.

The bisgermavinylidene [(Me$_3$SiN=PPh$_2$)$_2$C=Ge → Ge=C(PPh$_2$=NSiMe$_3$)$_2$] was used as a source of unstable germavinylidene and its reaction with two equivalents of AuI gave, following an insertion reaction, the complex **89** (Equation 4.1) [79]. A dimer association is achieved in this case through bridging halogermenes, showing that the germavinylidene underwent insertion into the Au–I bond instead of forming a donor–acceptor interaction. The Au–Ge bond distances [2.329(1) and 2.341(1) Å] are shorter than in (R$_3$P)AuGeCl$_3$ and a transannular Au–Au interaction of 3.092(7) Å is established.

$$\text{(4.1)}$$

Recently, the first Ge^{II}–Au^{I} adduct was reported, IAuGe[N(SiMe$_3$)C(Ph)C(SiMe$_3$)(C$_5$H$_4$N-2)]Cl **90**, obtained by reacting Ge[N(SiMe$_3$)C(Ph)C(SiMe$_3$)(C$_5$H$_4$N-2)]Cl with gold(I) iodide, in THF [80]. This result demonstrated the Lewis base behavior of the germanium(II) starting material. The compound is monomeric (Figure 4.26), with one of the shortest gold–germanium bonds [Au(1)–Ge(1) 2.346(2) Å] found in neutral compounds.

Ionic compounds with anions containing one, two- or three Au–Ge bonds have also been reported. They exhibit a surprising structural diversity (Figure 4.27). Thus, treatment of CH$_2$(AsPh$_2$AuCl)$_2$ with GeCl$_2$(1,4-dioxane) resulted only in the mono-insertion product [CH$_2$(AsPh$_2$)$_2$Au]$_2$(Cl$_3$GeAuCl)$_2$ **91**, which contains linear anions with only one Au–Ge bond, the shortest found so far [Au(2)–Ge(1) 2.323(1) Å] [76]. In its lattice a tetranuclear unit with a sequence (−)(++)(−) is formed; each gold atom of a central dication [transanular Au(1)–Au(1a) 3.1613(8) Å] is connected to an orthogonally oriented anion through Au–Au contacts [Au(1)–Au(2) 3.0499(6) Å] (Figure 4.27a).

Discrete anions with two Au–Ge bonds were found in the crystal of [PPh$_4$][Au{Ge(NSiMe$_3$CH$_2$)$_3$CMe}$_2$] **92**, obtained by LiBr elimination between the tris(amido)germanate [MeC(CH$_2$NSiMe$_3$)$_3$Ge]Li(THF)$_2$ and [PPh$_4$][AuBr$_2$], in THF [81]. The Ge–Au–Ge fragment is linear (180.0°) (the gold atom resides upon a crystallographic center of symmetry) and is effectively shielded by the peripheral SiMe$_3$ groups of the two triamidogermanate units (Figure 4.27b).

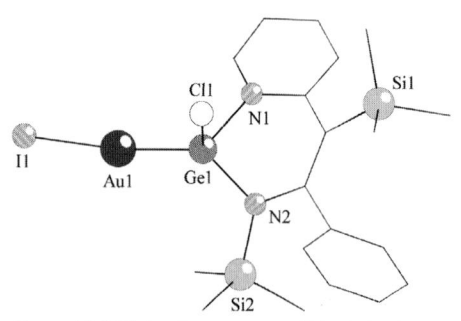

Figure 4.26 Molecular structure of IAuGe[N(SiMe$_3$)C(Ph)C(SiMe$_3$)(C$_5$H$_4$N-2)]Cl **90** [80].

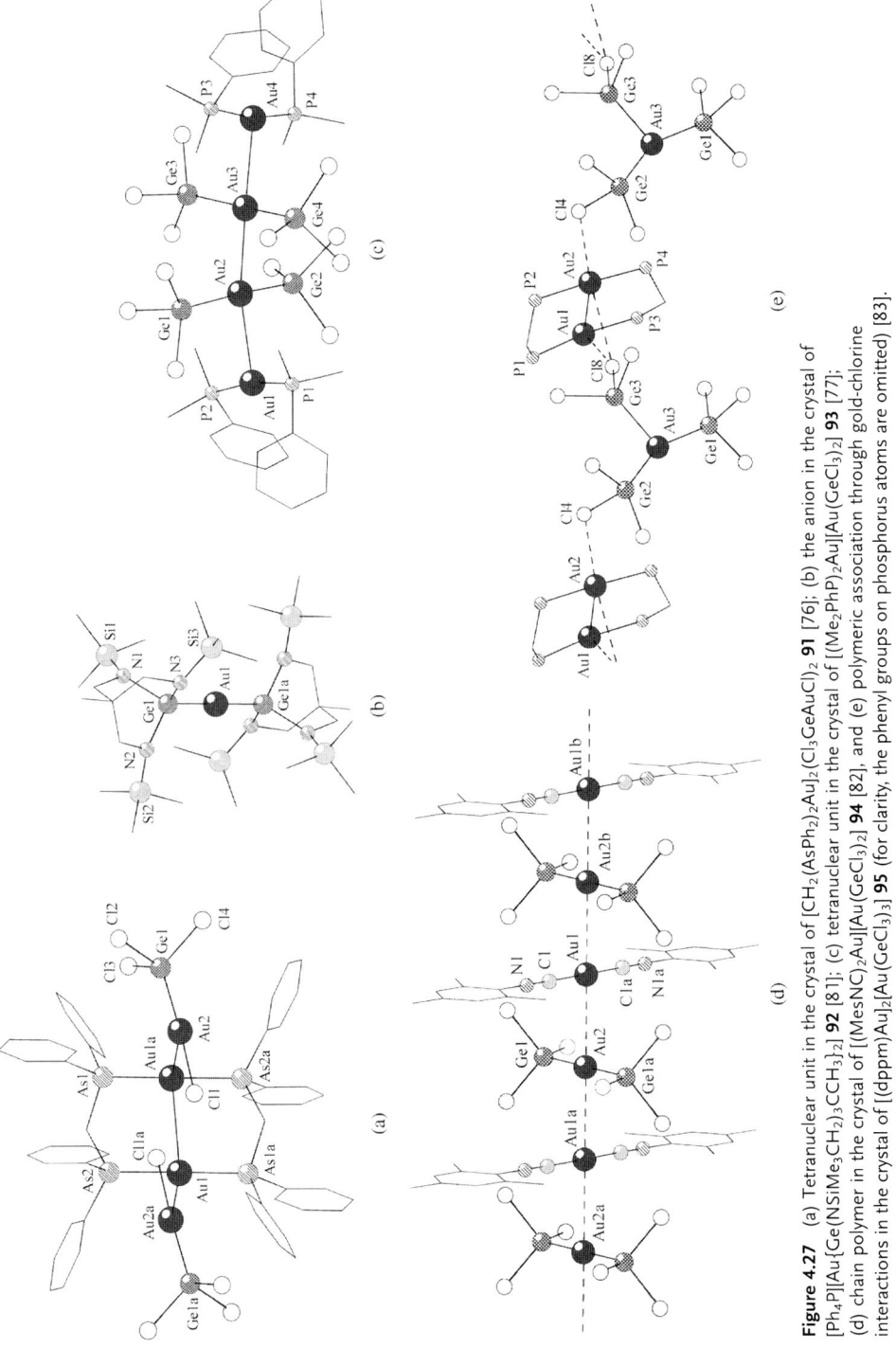

Figure 4.27 (a) Tetranuclear unit in the crystal of [CH$_2$(AsPh$_2$)$_2$Au]$_2$(Cl$_3$GeAuCl)$_2$ **91** [76]; (b) the anion in the crystal of [Ph$_4$P][Au{Ge(NSiMe$_3$CH$_2$)$_3$CCH$_3$}$_2$] **92** [81]; (c) tetranuclear unit in the crystal of [(Me$_2$PhP)$_2$Au][Au(GeCl$_3$)$_2$] **93** [77]; (d) chain polymer in the crystal of [(MesNC)$_2$Au][Au(GeCl$_3$)$_2$] **94** [82], and (e) polymeric association through gold-chlorine interactions in the crystal of [(dppm)Au]$_2$[Au(GeCl$_3$)$_3$] **95** (for clarity, the phenyl groups on phosphorus atoms are omitted) [83].

The reaction of Au(PPhMe$_2$)Cl with GeCl$_2$(1,4-dioxane) (1:1 molar ratio) in THF surprisingly resulted in the ionic derivative [(Me$_2$PhP)$_2$Au][Au(GeCl$_3$)$_2$] **93** [77]. Distortion from linearity was observed for both the [(Me$_2$PhP)$_2$Au]$^+$ cations [P–Au–P 170.06(9)/168.38(8)°] and the [Au(GeCl$_3$)$_2$]$^-$ anions [Ge–Au–Ge 169.98(4)/169.20(4)°]. The ionic components are associated in *quasi*-linear tetranuclear units [Au(1)–Au(2)–Au(3) 166.92(2)°, Au(2)–Au(3)–Au(4) 172.98(2)°] with an all-staggered conformation in the unexpected sequence (+)(−)(−)(+) (Figure 4.27c). The gold–gold distance between the central anions [Au(2)–Au(3) 2.881(1) Å] is even shorter than those established with the metal atoms of the terminal cations [Au(1)–Au(2) 2.981(1), Au(3)–Au(4) 2.976(1) Å]. This suggests that auriophilicity-based bonding is strong enough to overrule coulomb repulsion between two anions [77].

The complex [(MesNC)$_2$Au][Au(GeCl$_3$)$_2$] **94**, obtained by insertion of GeCl$_2$ into the Au–Cl bond of (mesitylisonitrile)gold(I) chloride, is also ionic [82]. However, a different solid-state structure was established, that is, in the crystal linear strings [Au–Au–Au 180°] parallel to the crystallographic *b* axis are formed from alternating, perpendicular, [(MesNC)$_2$Au]$^+$ cations and [Au(GeCl$_3$)$_2$]$^-$ anions, with quite large separation between the metal centers [Au(1)–Au(2) 3.402(1) Å] (Figure 4.27d). A temperature dependence of the Au–Au distance (21 °C, 3.435 Å; −74 °C, 3.402 Å; −110 °C, 3.390 Å) was noted.

Only one compound containing an anion with three gold–germanium bonds, [(dppm)Au]$_2$[Au(GeCl$_3$)$_3$] **96**, has been structurally characterized so far [83]. Crystals of **96** were isolated from a CD$_3$CN solution of [(dppm)Au]$_2$[Au(GeCl$_3$)$_2$]$_2$ **95**, following a decomposition process. The homoleptic dianion [Au(GeCl$_3$)$_3$]$^{2-}$ has a three-coordinate gold atom in a planar geometry between a T- and a Y-shaped pattern [Ge(1)–Au(3)–Ge(2) 153.12(2), Ge(1)–Au(3)-Ge(3) 104.97(2), Ge(2)–Au(3)–Ge(3) 101.72(2)°] and the three Au–Ge distances fall into two categories, that is, two short bonds [Au(3)–Ge(1) 2.4150(6), Au(3)–Ge(2) 2.4284(6) Å] and a longer one [Au(3)–Ge(3) 2.5351(7) Å]. This distortion might suggest that this dianion is the result of an interaction between a [GeCl$_3$]$^-$ group and the gold atom of a linear [Au(GeCl$_3$)$_2$]$^-$ anion [83]. Weak interionic contacts between gold atoms of the cation and chlorine atoms of the counter ion [Au(2)–Cl(4) 3.349, Au(1)–Cl(8) 3.468, Au(2)–Cl(8) 3.408 Å; cf. Σr_{vdW}(Au,Cl) 3.5 Å] result in a chain polymeric association (Figure 4.27e).

Recently gold–germanium clusters were reported to be obtained by reacting Zintl ions of the type [Ge$_9$]$^{4-}$ with gold(I) compounds. Thus, treatment of K$_4$Ge$_9$ with Au(PPh$_3$)Cl in ethylenediamine in the presence of [2.2.2]crypt produced plate-shaped crystals of [K([2.2.2]crypt)]$_5$[Au$_3$Ge$_{18}$] **97** as well as diamond-shaped crystals of [K([2.2.2]crypt)]$_8$K[Au$_3$Ge$_{45}$] **98** [84, 85]. In **97** the anion contains two Ge$_9$ cluster units linked through the germanium atoms of a trigonal face to a trigonal Au$_3$ fragment (Figure 4.28) [84]. The anion of **98** contains four [Ge$_9$] cluster units covalently linked by another nine Ge atoms. The individual clusters are connected through Ge-atom bridges and are further coordinated to three gold atoms. Each gold atom is coordinated by four Ge atoms in a trapezoid fashion [85]. Another gold–germanium cluster, the red derivative [Li(THF)$_6$][AuGe$_{18}${Si(SiMe$_3$)$_3$}$_6$] **99**, was obtained by reacting [Li(THF)$_4$][Ge$_9${Si(SiMe$_3$)$_3$}$_3$] with Au(PPh$_3$)Cl, in THF [86]. In

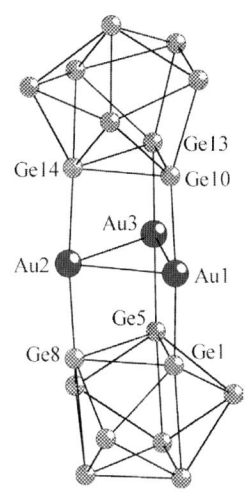

Figure 4.28 Structure of the anion in the crystal of [K([2.2.2]crypt)]$_5$[Au$_3$Ge$_{18}$] **97** [84].

this case the cluster anion consists of a central gold atom connected in a trigonal-antiprismatic fashion to three germanium atoms from each of the two [Ge$_9${Si(SiMe$_3$)$_3$}$_3$] units. Owing to the high coordination number, the Au–Ge bond in **99** [2.6876(6) Å] is the longest observed in a molecular gold–germanium compound so far. The steric protection offered by the bulky Si(SiMe$_3$)$_3$ substituents on germanium atoms prevents any other interaction with the gold atom.

4.2.4.3 Gold–Tin Compounds

Quite a few gold–tin compounds have been structurally characterized and the interatomic Au–Sn distances in these compounds cover a large range from 2.56 to 3.31 Å.

Gold–tin homologs of the gold–silicon or gold–germanium compounds are (Me$_2$Ph-P)$_2$AuSnCl$_3$ **100**, (Ph$_3$P)AuSn[N(tolyl-*p*)SiMe$_2$]$_3$SiMe **101** and (dppb)Au$_2$[Sn{N(tolyl-*p*)SiMe$_2$}$_3$SiMe]$_2$ **102**. The complex **100** was prepared by insertion of SnCl$_2$ into the gold–chlorine bond of Au(PPhMe$_2$)Cl, in the presence of Me$_2$PhP, in acetone [87]. The gold atom has a distorted trigonal planar coordination (Figure 4.29a), reflected in the wide opened P(1)–Au(1)–P(2) angle [153.8(2)°]. The Au–Sn distance of 2.881(1) Å is considerably longer than the sum of the covalent radii for the corresponding atoms, Σr_{cov}(Au,Sn) 2.74 Å [9]. Much stronger gold–tin bonds were found for compounds **101** [2.5651(13) Å] and **102** [2.6804(13) Å], which were obtained from tris(amido)stannate [MeSi{Me$_2$Si(tolyl-*p*)N}$_3$Sn]Li(Et$_2$O) and Au(PPh$_3$)Cl or the gold(II) bis(ylide) complex Au$_2$(CH$_2$PPh$_2$CH$_2$)$_2$Cl$_2$ [88]. The crystal of **101** contains discrete molecules with the expected linear geometry for the gold(I) atom (Figure 4.29b), the "lamp shade" arrangement of the peripheral tolyl groups preventing aurophilic interactions. For **102** each metal atom of the central gold(II) eight-membered diaurocycle is bound to a tripodal tris(amido)tin fragment, resulting in a unique, almost linear, Sn–Au–Au–Sn system [Sn–Au–Au' 174.62(3)°] [88]. The

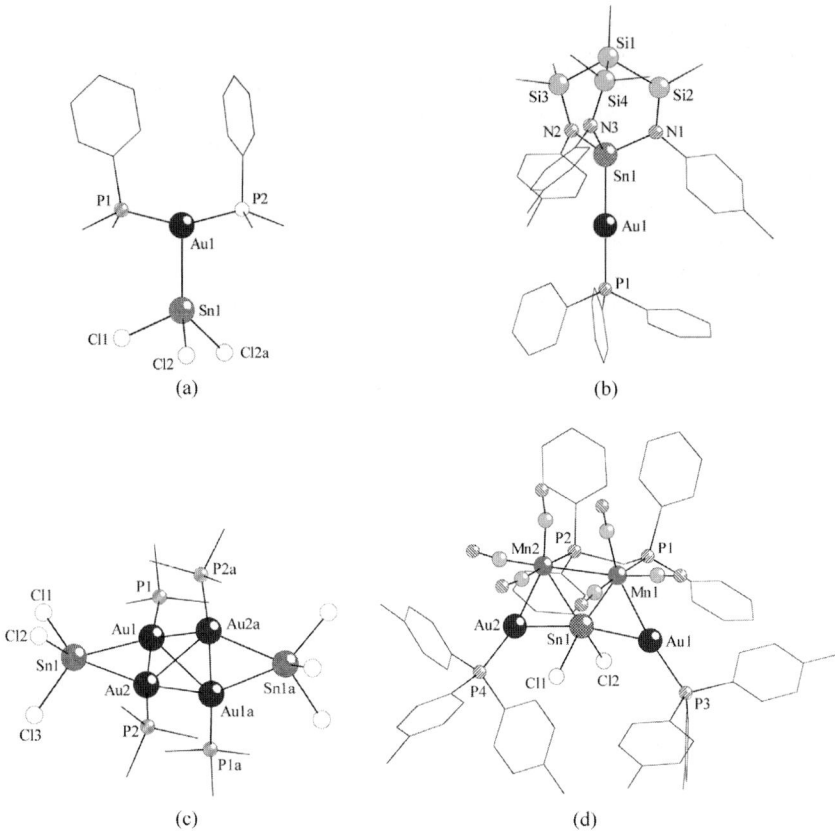

Figure 4.29 The structures of (a) (Me$_2$PhP)$_2$AuSnCl$_3$ **100** [87], (b) (Ph$_3$P)AuSn[N(tolyl-*p*)SiMe$_2$]$_3$SiMe **101** [88], (c) Au$_4$(PPh$_3$)$_4$(μ-SnCl$_3$)$_2$ **104** [90], and (d) [(*p*-tolyl$_3$P)Au]$_2$Mn$_2$(SnCl$_2$)(CO)$_6$(μ-dppm) **105** [91].

coordination geometry of the gold(II) centers is square planar as usually exhibited in gold(II) derivatives with gold–gold bonds [Au^{II}-Au^{II} 2.7492(13) Å].

The structures of a few gold–tin clusters have also been determined. The compound [(Ph$_3$P)$_7$Au$_8$(SnCl$_3$)]$_2$[SnCl$_6$] **103** was obtained from [Au$_8$(PPh$_3$)$_7$][NO$_3$]$_2$ or [Au$_9$(PPh$_3$)$_8$][NO$_3$]$_3$ and excess of SnCl$_2$·2H$_2$O or from [Au$_8$(PPh$_3$)$_8$][NO$_3$]$_2$ and [NEt$_4$][SnCl$_3$], in acetone. The gold core of the cation can be described as a fragment of an Au-centered icosahedron, with each of the peripheral gold atoms coordinated by a phosphine ligand. The tin atom of the trichlorostannyl unit is strongly bound to the central gold atom of the Au$_8$ core [Au–Sn 2.625(3) Å] [89]. In compound (Ph$_3$P)$_4$Au$_4$(μ-SnCl$_3$)$_2$ **104** a tetrahedral Au$_4$ cluster has two opposite edges asymmetrically bridged by a trichlorostannyl group [Au(1)–Sn(1) 2.972(1), Au(2)–Sn(1) 2.819(1) Å] (Figure 4.29c). Each gold atom is also coordinated by a terminal PPh$_3$ ligand [90].

A pentanuclear manganese–gold–tin cluster, [(p-tolyl$_3$P)Au]$_2$Mn$_2$(SnCl$_2$)(CO)$_6$(μ-dppm) **105**, was prepared by reacting the tetranuclear digold cluster [(p-tolyl$_3$P)Au]$_2$Mn$_2$(CO)$_6$(μ-dppm) with SnCl$_2$ in THF [91]. The compound contains a planar Au$_2$Mn$_2$Sn skeleton, with the gold atoms bridging the Mn–Sn edges of the central Mn$_2$Sn core (Figure 4.29d). The magnitude of the Sn–Au distances [2.813(4)/2.815(4) Å], similar to that observed for (PhMe$_2$P)$_2$AuSnCl$_3$ **100**, is indicative of a rather weak Sn–Au bonding interaction.

Investigations on gold chemistry of the stannaborate ligand, [SnB$_{11}$H$_{11}$]$^{2-}$, were recently reported and revealed a remarkable gold-stannaborate clusters diversity [92–95]. Thus, red-orange crystals of [NBu$_3$H]$_5$[Au(SnB$_{11}$H$_{11}$)$_4$] **106** were obtained from the reaction of AuCl$_3$ with [Bu$_3$NH]$_2$[SnB$_{11}$H$_{11}$] in methylene chloride [93]. It features a homoleptic square-planar Au(III) atom, with linear Sn–Au–Sn units and strong Au–Sn bonds [2.601(1)/2.589(1) Å] (Figure 4.30a). Treatment of Au(PPh$_3$)Cl with one equivalent of [SnB$_{11}$H$_{11}$]$^{2-}$ resulted in a mixture of two cluster species whose crystals could be obtain by selective crystallization, [NBu$_3$H]$_2$[{(Ph$_3$P)Au}$_2$(SnB$_{11}$H$_{11}$)$_2$] **107** and [NBu$_3$Me]$_4$[{(Ph$_3$P)Au}$_2$(SnB$_{11}$H$_{11}$)$_3$] **108** [94]. The anions of both compounds contain a dinuclear fragment with very short Au–Au interatomic distance [2.625(1) and 2.590(1) Å in **107** and **108**, respectively] which are bridged by two (Figure 4.30b) or three (Figure 4.30c) stannaborate anions. While the Au–Sn–Au bridges in **107** are symmetrical [Au–Sn 2.737(1)/2.761(1) Å], those in **108** are unsymmetrical, with short Au–Sn bonds [2.711(1)–2.727(1) Å] of the same magnitude as in **107** and long Au–Sn distances in the range 2.947(1)–3.032(1) Å. The bonding situation was studied in detail by theoretical calculations and it was concluded that the stability of the dimeric

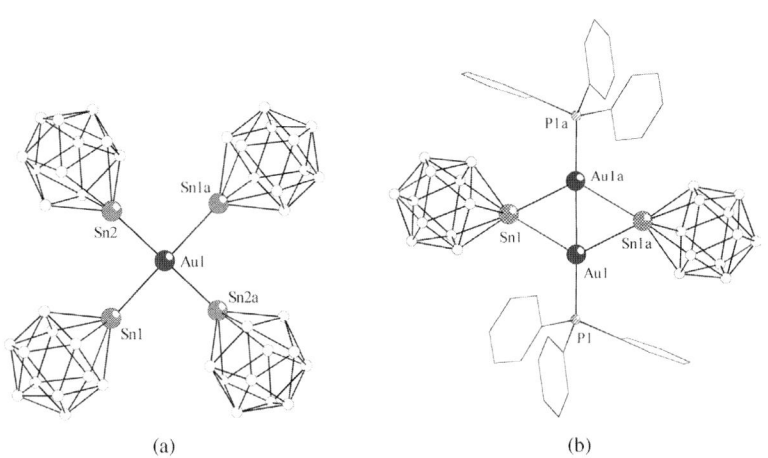

Figure 4.30 The structure of the anion in the crystal of (a) [NBu$_3$H]$_5$[Au(SnB$_{11}$H$_{11}$)$_4$] **106** [93], (b) [NBu$_3$H]$_2$[{(Ph$_3$P)Au}$_2$(SnB$_{11}$H$_{11}$)$_2$] **107** [94], (c) [NBu$_3$Me]$_4$[{(Ph$_3$P)Au}$_2$(SnB$_{11}$H$_{11}$)$_3$] **108** [94], (d) [NBu$_3$H]$_3$[{(Et$_3$P)Au}$_3$(SnB$_{11}$H$_{11}$)$_3$] **109** (ethyl groups on phosphorus are not shown) [95], and (e) [NBu$_3$Me]$_4$[{(dppm)Au$_2$}$_2$(SnB$_{11}$H$_{11}$)$_4$] (**110**) (phenyl groups on phosphorus are not shown) [95].

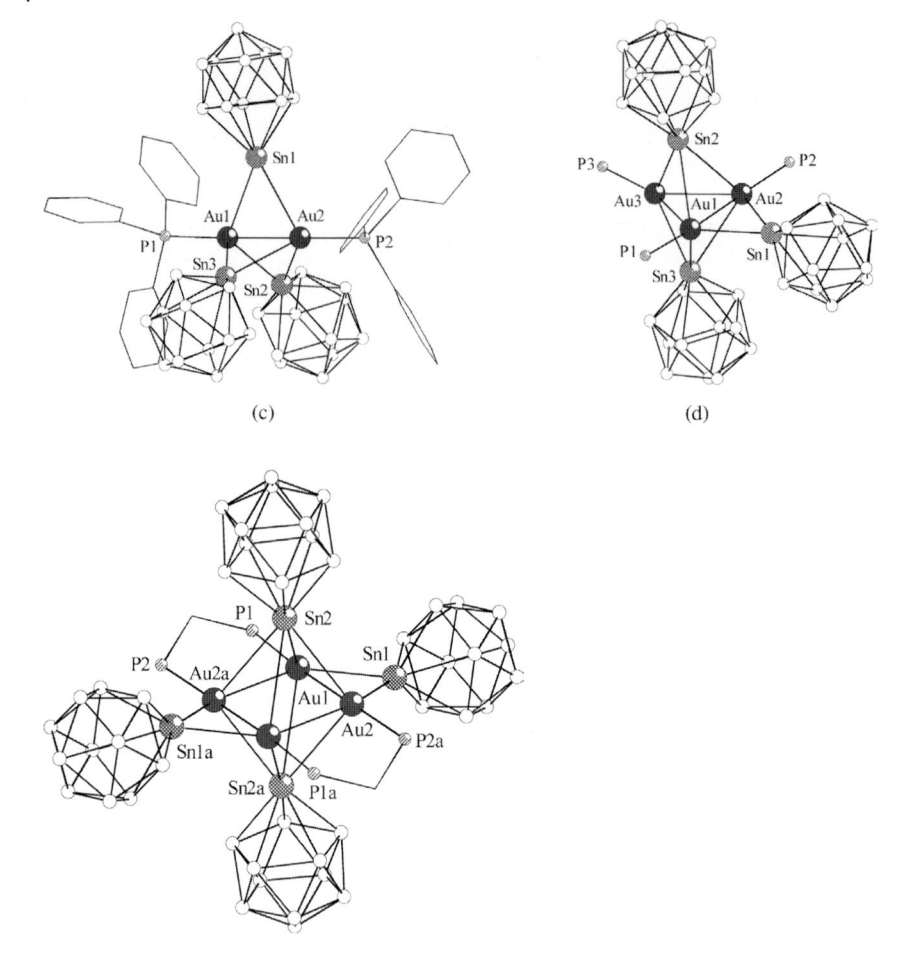

Figure 4.30 (Continued).

species over the respective monomers is mainly due to dispersive interactions between the valence electrons of tin and gold. Using Au(PEt$_3$)Cl as the gold starting material a trinuclear gold cluster, [NBu$_3$H]$_3$[{(Et$_3$P)Au}$_3$(SnB$_{11}$H$_{11}$)$_3$] **109**, was obtained whose anion can be considered as derived from a dinuclear anion as found in **107** by coordinating a (R$_3$P)Au$^+$ unit to the Au–Au bond [95]. In this anion two of the stannaborate moieties act as μ$_3$-ligands on both sides of the triangular gold core [Au–Sn range 2.7329(8)–3.1204(9) Å], while the third heteroborate unit bridges [Au(1)–Sn(1) 2.7353(8), Au(2)–Sn(1) 2.8113(9) Å] the shortest Au(1)–Au(2) edge [2.606(1) Å] (Figure 4.30d). An even larger tetranuclear gold cluster, [NBu$_3$Me]$_4$[{(dppm)Au$_2$}$_2$(SnB$_{11}$H$_{11}$)$_4$] **110**, could be obtained by using an appropriate gold precursor, that is, Au$_2$Cl$_2$(dppm) [95]. The central Au$_4$ core is a rectangle with short linear P–Au–Au–P edges [Au(1)–Au(2a) 2.622(1) Å] bridged symmetrically by stannaborate ligands [Au(1)–Sn(1) 2.7289(7), Au(2)–Sn(1) 2.7105(7) Å], and longer Au–Au edges

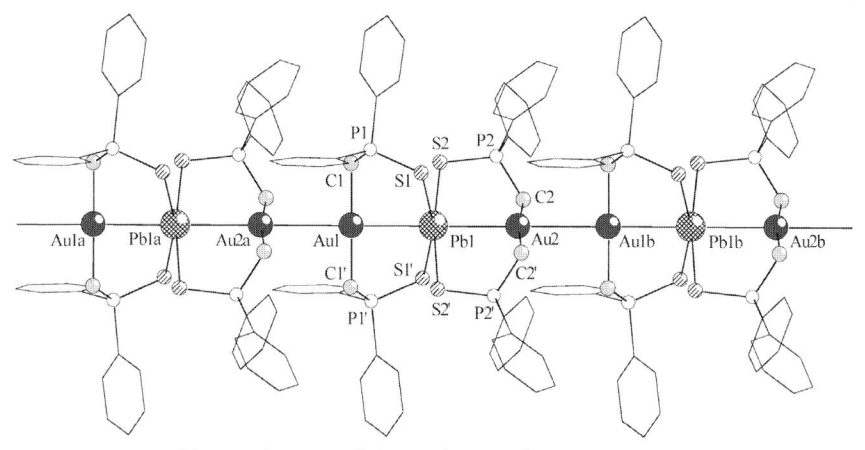

Figure 4.31 View of the one-dimensional chain polymer in the crystal of Pb[Au{CH$_2$P(S)Ph$_2$}$_2$]$_2$ **111** [37].

[Au(1)–Au(2) 2.848(1) Å] bridged by dppm ligands. The resulting six-membered Au$_4$Sn$_2$ ring has a chair conformation, with tin atoms in the apices. The planar Au$_4$ rectangle is μ$_4$-capped by two additional stannaborate moieties [Au–Sn range 2.8396(8)–3.3078(9) Å] (Figure 4.30e).

4.2.4.4 Gold–Lead Compounds

No lead homologs of the above compounds of lighter Group 14 elements have been reported so far. Only the structure of the gold–lead complex based on methylenethiophosphinate ligand, Pb[Au{CH$_2$P(S)Ph$_2$}$_2$]$_2$ **111** has been reported [37]. The molecular unit contains two gold(I) atoms linearly coordinated by carbon atoms, while the four sulfur atoms of the ylide moieties are coordinated to the central Pb(II) atom. This results in a *spiro*-bicyclic system with the lead atom as the *spiro* atom and an orthogonal arrangement of the two eight-membered rings. Weak *transannular* gold–lead interactions are established [Au(1)–Pb(1) 2.896(1), Au(2)–Pb(2) 2.963(2) Å], the interatomic Au–Pb distances being shorter than the sum of the van der Waals radii of the corresponding atoms [Σr_{vdW}(Au,Pb) 3.7 Å] [9]. In the crystal an extended one-dimensional chain polymer is formed with intermolecular Au–Au interactions [Au(1)–Au(2a) 3.149(2) Å] and linear Au–Pb–Au and Pb–Au–Au fragments (Figure 4.31).

4.2.5
Gold–Group 15 Metal Compounds

In contrast to the extensive use of phosphorus-based ligands in the coordination chemistry of gold, there are very few examples of gold complexes containing heavier Group 15 metal compounds as ligands. The complexes containing gold–antimony or gold–bismuth bonds or interactions which have been characterized by single-crystal X-ray diffraction are listed in Table 4.5.

Table 4.5 Selected molecular parameters for gold–Group 15 metal compounds.

Compound	Au–M (Å)	Sb–Au–Sb (°)	Ref.
Compounds with Au–Sb bonds/interactions			
[Au(SbPh$_3$)$_4$][Au(C$_6$F$_5$)$_2$] **112**	2.585(1)–2.669(1)	108.47(1)–110.48(1)	[96]
[Au(SbPh$_3$)$_4$][Au{C$_6$H$_2$(NO$_2$)$_3$-2,4,6}$_2$] **113**	2.647(1)–2.655(1)	107.8(1)–111.0(1)	[97]
[Au(SbPh$_3$)$_4$](ClO$_4$) **114**	2.656(2)–2.658(2)	108.8(1)–110.1(1)	[98]
[Au$_2${(Ph$_2$Sb)$_2$O}$_3$](ClO$_4$)$_2$ **115**	2.6048(5)–2.6173(5)	118.04(2)–121.29(1)	[99]
[Au(SbMes$_3$)$_2$](ClO$_4$) **116**	2.5853(9)/2.5856(9)	173.08(3)	[100]
[Au(SbMes$_3$)$_2$](O$_3$SCF$_3$) **117**	2.5617(8)/2.5590(8)	174.74(3)	[100]
Au(SbMes$_3$)Cl **118**	2.5100(2)	177.39(3)^a	[100]
(Et$_3$P)$_6$Au$_8$(SbPh)$_2$(SbPh$_2$)$_4$ **119**	2.577(1)–2.735(1)		[101]
Compounds with Au–Bi bonds/interactions			
[Bi{C$_6$H$_4$(CH$_2$NMe$_2$)-2}$_2$][Au(C$_6$F$_5$)$_2$] **120**	3.7284(5)		[103]

^{a} Sb–Au–Cl angle.

4.2.5.1 Gold–Antimony Compounds

Until a few years ago this class of compounds included only three ionic compounds containing the same complex cation, that is [Au(SbPh$_3$)$_4$]X (X = [Au(C$_6$F$_5$)$_2$]$^-$ **112** [96], [Au{C$_6$H$_2$(NO$_2$)$_3$-2,4,6}$_2$]$^-$ **113** [97] and (ClO$_4$)$^-$ **114** [98]). Such complexes were obtained by addition of SbPh$_3$ to solutions of AuR(tht) and an equilibrium between the neutral three-coordinate AuR(SbPh$_3$)$_2$ and the ionic complex [Au(SbPh$_3$)$_4$][AuR$_2$] species in solution has been proposed [97]. In the solid state, regardless of the nature of the counter anion, the [Au(SbPh$_3$)$_4$]$^+$ cations show an almost tetrahedral coordination at the gold center (cf. Figure 4.32a for [Au(SbPh$_3$)$_4$][Au{C$_6$H$_2$(NO$_2$)$_3$-2,4,6}$_2$] **113** [97]), with Sb–Au–Sb bond angles in the range 107.8 to 111.0° and Au–Sb bonds between 2.585(1) and 2.669(1) Å.

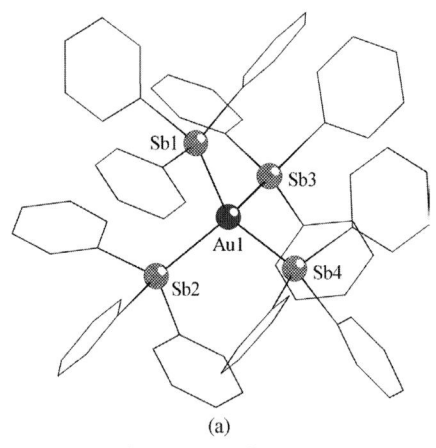

Figure 4.32 The structure of the cation in the crystal of (a) [Au(SbPh$_3$)$_4$][Au{C$_6$H$_2$(NO$_2$)$_3$-2,4,6}$_2$] **113** [97], (b) [Au$_2${(Ph$_2$Sb)$_2$O}$_3$](ClO$_4$)$_2$ **115** [99], and (c) [Au(SbMes$_3$)$_2$](ClO$_4$) **116** [100]; (d) the structure of Au(SbMes$_3$)Cl **118** [100].

(b)

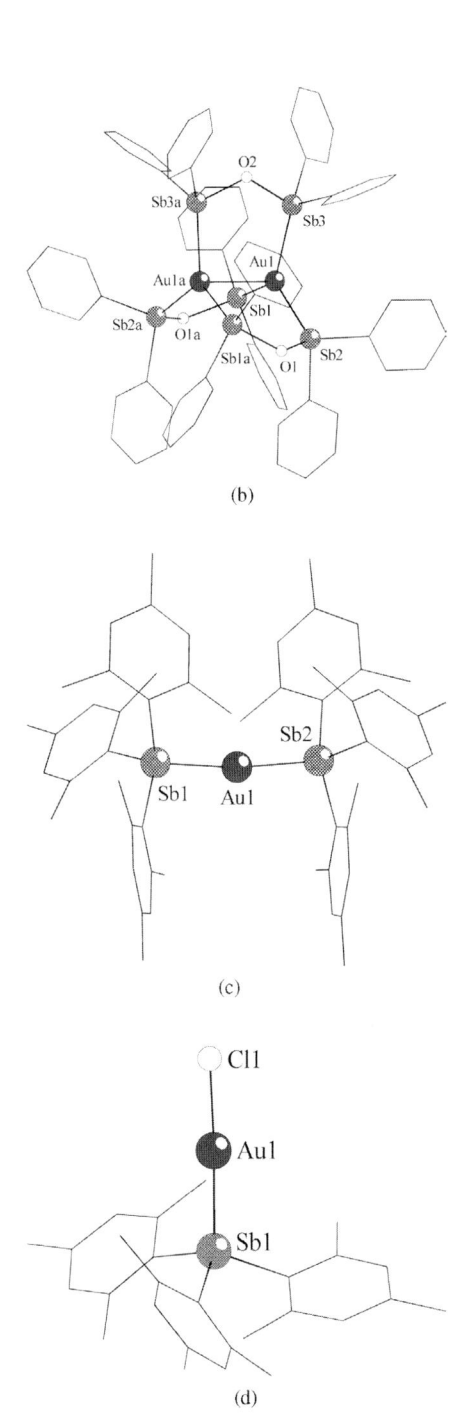

(c)

(d)

Figure 4.32 (*Continued*).

Recently, the ionic complex [Au$_2$\{(Ph$_2$Sb)$_2$O\}$_3$](ClO$_4$)$_2$ **115** was prepared by reaction between the Au(I) precursor [Au(tht)$_2$](ClO$_4$) and bis(diphenylstibine)oxide in a 2 : 3 molar ratio [99]. Its cation (Figure 4.32b) shows two gold centers in a trigonal-planar environment as a result of the bridging nature of the Sb donor ligands. The Au–Sb bond distances, which range from 2.6048(5) to 2.6173(5) Å, are slightly shorter than those found in the tetracoordinate [Au(SbPh$_3$)$_4$]$^+$ cations, as expected for a lower coordination number. As in other dinuclear gold(I) compounds, an intramolecular Au–Au contact [3.0320(4) Å] is established between the two Au(I) centers of the complex cation.

Changing the phenyl group on antimony for a bulkier, more electron-withdrawing, mesityl group allowed the isolation of the gold–stibine complexes containing two-coordinate gold atoms. Thus, treatment of [Au(tht)$_2$]X with SbMes$_3$ produced crystals of the ionic complexes [Au(SbMes$_3$)$_2$](ClO$_4$) **116** and [Au(SbMes$_3$)$_2$](O$_3$SCF$_3$) **117**. The cation contains a gold atom linearly coordinated by antimony atoms (Figure 4.32c), with Au–Sb bonds shorter than in **115**.

Reaction of Au(tht)Cl with SbMes$_3$ resulted, regardless of the molar ratio used, in the isolation of the neutral complex Au(SbMes$_3$)Cl **118**, which also contains a linear Sb–Au–Cl system (Figure 4.32d), with the shortest Au–Sb bond [2.5100(2) Å] found so far in gold–stibine complexes. For none of these complexes was further association of the molecules observed in the solid state.

The [SbPh]$^{2-}$ and [SbPh$_2$]$^-$ anions were also found to behave as ligands for gold. The polynuclear complex (Et$_3$P)$_6$Au$_8$(SbPh)$_2$(SbPh$_2$)$_4$ **119** was obtained by the reaction of Au(PEt$_3$)Cl with a mixture of PhSb(SiMe$_3$)$_2$ and Ph$_2$SbSiMe$_3$ in the presence of bis(diphenylphosphanyl)methane [101]. It consists of a distorted heterocubic central [Au$_6$Sb$_2$] unit connected on two opposite edges by two Sb–Au–Sb fragments (Figure 4.33). The SbPh units act as μ$_3$-ligands [2.5883(15) Å] and are placed in opposite corners of the central cube. The SbPh$_2$ units act as asymmetric bridges [2.577(1)/2.735(1) and 2.580(1)/2.730(2) Å] between a gold atom from the central cube and the peripheral rings. All gold atoms exhibit tetrahedral coordination, but as part of different cores, that is, AuP$_2$Sb$_2$, AuPSb(Au)$_2$ and AuSb$_2$(Au)$_2$.

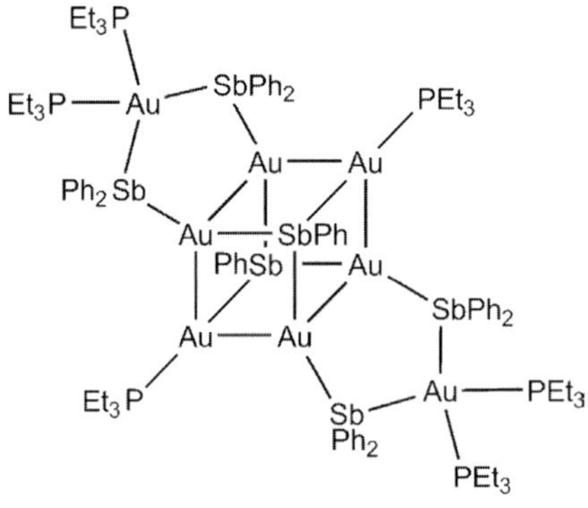

Figure 4.33 Schematic structure of compound (Et$_3$P)$_6$Au$_8$(SbPh)$_2$(SbPh$_2$)$_4$ **119** [101].

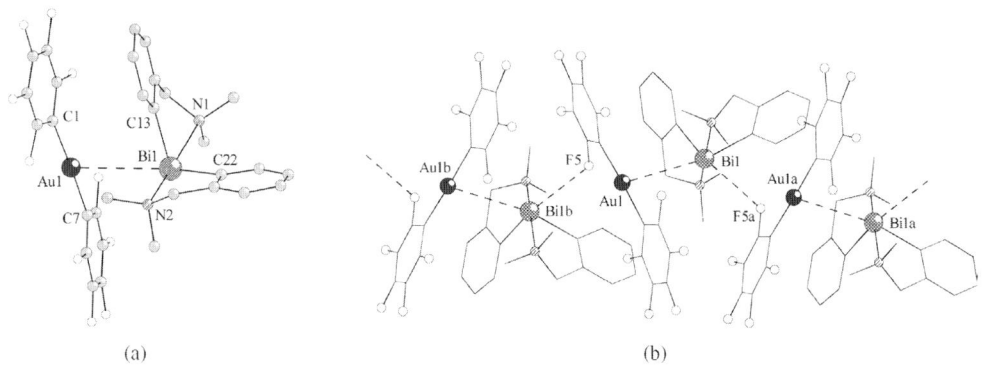

Figure 4.34 (a) The structure of [Bi{$C_6H_4(CH_2NMe_2)$-2}$_2$][Au$(C_6F_5)_2$] **120** [101], and (b) view of the chain polymer through Bi–F contacts in the crystal [103].

4.2.5.2 Gold–Bismuth Compounds

The bismuth ligands have received much less attention in the coordination chemistry of gold. Schmidbaur *et al.* reported that tertiary bismuthine (BiR$_3$) ligands cannot be employed as donor ligands for Au(I) complexes due to rapid transorganylation processes that give rise to organogold compounds [102].

Taking advantage of the Lewis acidic site at the bismuth atom in the hypervalent diorganobismuth compound [2-(Me$_2$NCH$_2$)C$_6$H$_4$]$_2$BiCl, a transmetallation reaction with the gold precursor [AuAg(C$_6$F$_5$)$_2$]$_n$ was performed and the complex [Bi{C$_6$H$_4$(CH$_2$NMe$_2$)-2}$_2$][Au(C$_6$F$_5$)$_2$] **120** was isolated [103]. The structure determination revealed a bismuth center (C,N)-chelated by the pendant arm ligands and a linear diorganogold anion. The most interesting feature of the structure of **120** is the presence of the first Au–Bi interaction described to date, with an Au(1)–Bi(1) distance of 3.7284(5) Å (Figure 4.34a). The nature of this interaction was shown to be consistent with the presence of a high ionic contribution (79%) and a dispersion type (van der Waals) interaction (21%). Additional Bi–F contacts [3.404(5) Å] between adjacent molecules results in a monodimensional polymer (Figure 4.34b).

4.3
Transition Metal–Gold Compounds

The topic of molecular transition metal–gold bond-containing compounds, especially of cluster compounds, has developed considerably in recent years (Table 4.6).

Several reviews of the chemistry of heteronuclear gold clusters are available and the reader is referred to these works for earlier studies in the field [23–27]. Since the number of well structurally characterized compounds is quite large, this section will be focused only on the structures of basic compounds as well as the most recent results on molecular compounds containing Au–transition metal bonds or interactions. In many cases analogous compounds are available for the metals belonging to the same group and therefore, when appropriate, the common structural patterns will be presented for

Table 4.6 Literature references for structurally characterized compounds with Au–transition metal bond/interactions.

Metal	Au–M range (Å)	Ref.	Metal	Au–M range (Å)	Ref.
Ti	2.719–3.007	[104–106]	Ru	2.633–3.169	[229–232, 245–247, 256, 261–278, 291, 295, 298, 304, 310–325, 328, 344–364, 369, 374–390, 395–399, 403, 404]
V	2.690–2.836	[107–109]	Os	2.646–3.173	[230, 233, 248, 276, 279–290, 326–335, 365–369, 391–393, 399, 400, 405–414, 420–422]
Nb	2.910–3.033	[110–112]	Co	2.450–2.873	[126, 146, 161, 166, 169, 172, 173, 178, 208, 257, 298, 307, 315, 389, 423–438]
Cr	2.629–2.770	[113–119]	Rh	2.531–3.147	[136, 216, 297, 310, 311, 344, 352, 376–379, 404, 432, 439–460, 488]
Mo	2.691–3.242	[113, 120–128, 147–152, 158, 160, 161, 162, 164, 166, 168–170]	Ir	2.593–3.132	[209, 221, 247, 296, 376, 446, 461–487]
W	2.697–3.156	[113, 126, 127, 129–146, 150–157, 159, 163, 164, 167]	Ni	2.632–3.015	[489–494]
Mn	2.566–3.084	[91, 127, 171–174, 178–186, 196–198, 210, 213, 214, 218, 222]	Pd	2.611–3.267	[43, 495–510]
Tc	2.589	[223]	Pt	2.575–3.349	[34, 41–44, 131, 144, 145, 159, 217, 258, 396, 403, 490, 491, 494, 498, 511–572]
Re	2.566–3.059	[175–178, 187–195, 199–209, 211, 212, 215–217, 219–221]	Cu	2.526–3.215	[201, 202, 345, 346, 402, 556, 564, 565, 571, 577–587]
Fe	2.509–3.108	[160, 225–228, 234–244, 249–260, 292–294, 296–309, 336–343, 370–373, 394, 401, 402, 415–419]	Ag	2.702–3.311	[54, 194, 201, 202, 401, 494, 556, 562, 563, 566, 568–570, 576, 578, 582, 584, 587–628]

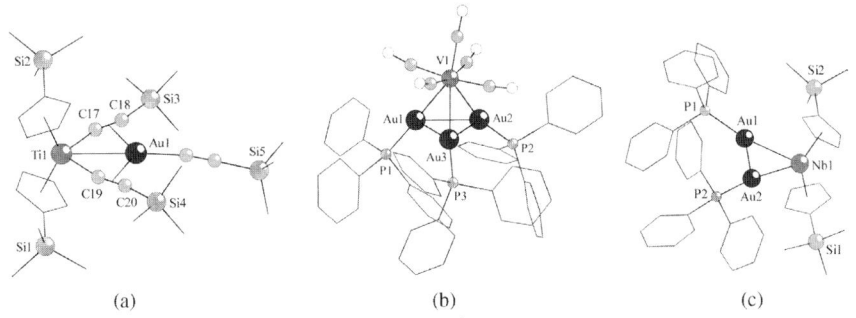

Figure 4.35 The structure of (a) (Me$_3$SiC≡C)AuTi(η^5-C$_5$H$_4$SiMe$_3$)$_2$(C≡CSiMe$_3$)$_2$ **121** [104], (b) [(Ph$_3$P)Au]$_3$V(CO)$_5$ **125** [108], and (c) the cation in [{(Ph$_3$P)Au}$_2$Nb(η^5-C$_5$H$_4$SiMe$_3$)$_2$](PF$_6$) **127** [110].

such series of derivatives. Also, in some cluster compounds substitutional disorder was reported for heavy metal atoms and in these cases the Au–metal interatomic distances were not considered for the data presented in Table 4.6.

To the best of our knowledge there are no structure reports on species with Au–Group 3 metals (Sc, Y, La, Ac) as well as Au–lanthanoids and Au–actinoids bonds, or compounds with Au–Hf, Au–Zr or Au–Ta bonds or interactions.

4.3.1
Gold–Titanium Compounds

Few compounds with Au–Ti bonds have been structurally characterized and all of them are molecular compounds. The neutral complexes (RC≡C)AuTi(η^5-C$_5$H$_4$SiMe$_3$)$_2$(C≡CR)$_2$ [R = ^{t}Bu **120**, SiMe$_3$ **121** (Figure 4.35a)] and [(η^5-C$_5$H$_4$SiMe$_3$)$_2$Ti(C≡CSiMe$_3$)$_2$]Au[C$_6$H$_2$(CF$_3$)$_3$-2,4,6] **122** are monomeric, with the organogold(I) moiety in a trigonal-planar environment, stabilized by the chelating effect of the organometallic π-tweezer bis(alkynyl) titanocene [104]. The strong η^2-coordination of both alkynyl ligands brings the Au(I) atom in close proximity to the formally Ti(IV) center, at distances [2.975(1), 3.007(2) and 2.995(1) Å in **120**, **121** and **122**, respectively] which suggests a direct Au–Ti interaction. Reaction of Cp*$_2$Ti=C=CH$_2$ with Au(PPh$_3$)Cl afforded the complex Cp*$_2$Ti(Cl)(μ-C=CH$_2$)Au(PPh$_3$), which contains a titanium–gold bond [2.9547(8) Å] asymmetrically bridged by the vinylidene ligand [105].

The shortest, unsupported Au–Ti bond [2.719(1) Å] was found in the linear anion of [K(15-crown-5)$_2$][(Et$_3$P)AuTi(CO)$_6$] **123**, obtained from Au(PEt$_3$)Cl and [K(15-crown-5)$_2$]$_2$[Ti(CO)$_6$] [106].

4.3.2
Gold–Group 5 Metal Compounds

Few gold compounds contain vanadium-carbonyl fragments or dicyclopentadienyl-niobium fragments, respectively, and no additional interactions between discrete heterometalic cores are present in the crystal of any of them.

4.3.2.1 Gold–Vanadium Compounds

The neutral, dinuclear (Ph$_3$P)AuV(CO)$_6$ **124** contains the shortest, unsupported Au–V bond [2.690(3) Å] established by a gold atom placed in the capping position with respect to an octahedral V(CO)$_6$ core [107].

Two cluster compounds have also been described. The neutral [(Ph$_3$P)Au]$_3$V(CO)$_5$ **125** was prepared by treatment of alkali metal salts of the [V(CO)$_5$]$^{3-}$ with Au(PPh$_3$)Cl [108]. It contains a V(CO)$_5$ group μ$_3$-bridged to a trigold unit [Au–V 2.709(1)–2.756(1) Å] (Figure 4.35b) [108]. In the heptanuclear cluster cation of [{(Ph$_3$P)Au}$_6$V(CO)$_4$](PF$_6$)(OH) **126** the vanadium atom of the V(CO)$_4$ group is connected to all six gold atoms [Au–V 2.725–2.836 Å] [109].

4.3.2.2 Gold–Niobium Compounds

The molecular structure has been established for only two cluster compounds with Au–Nb bonds, [{(Ph$_3$P)Au}$_2$Nb(η^5-C$_5$H$_4$SiMe$_3$)$_2$](PF$_6$) **127** and Au$_3$(μ-H)$_6$[Nb(η^5-C$_5$H$_4$SiMe$_3$)$_2$]$_3$ **128**. In the cation of **127** a Nb(η^5-C$_5$H$_4$SiMe$_3$)$_2$ bridges symmetrically [Nb(1)–Au(1) 2.9139(8) Å, Nb(1)–Au(2) 2.9098(8) Å] the gold atoms of a P–Au–Au–P fragment, *trans* to the phosphorus atoms (Figure 4.35c) [110]. The structure of the hexanuclear raft cluster **128** consists of a central, almost equilateral, gold triangle with each Au–Au bond bridged by a diorganoniobium fragment. In this case the Au–Nb bonds [range 2.967(2)–3.033(3) Å] are considerably longer than in **127**. Additionally, each Au–Nb unit is bridged by one hydride ligand. The most interesting feature of **127** is the planarity of the Au$_3$Nb$_3$ core [111, 112].

4.3.3
Gold–Group 6 Metal Compounds

While only a few compounds with Au–Cr bonds have been structurally characterized, for the heavier elements of Group 6 there are many more complexes for which the molecular structure have been established by X-ray diffraction and the diversity of the compounds is also greater.

4.3.3.1 Gold–Chromium Compounds

All characterized compounds contain either Cr(CO)$_n$ or Cr(L)(CO)$_n$ (L = cyclopentadienyl derivative, PPh$_3$) fragments bound to gold. Unsupported Au–Cr bonds are found in the neutral (Ph$_3$P)AuCr(CO)$_3$(η^5-C$_5$H$_4$CH$_2$CH$_2$NMe$_2$) **129** [2.6291(11) Å] [113] or the bent anion of [NBu$_4$][Au{Cr(η^5-Cp)(CO)$_3$}$_2$] **130** [Au–Cr 2.641(9)/2.635(8) Å; Cr–Au–Cr 162.2(3)°] [114]. Additional hydride or carbonyl bridges between the metal centers are present in complexes (Ph$_3$P)AuCr(μ-H)(CO)$_5$ **131** [Au–Cr 2.770(2) Å] [115, 116] and (Ph$_3$P)AuCr(μ-CO)$_2$(CO)(η^5-C$_5$H$_4$CHO) **132** [Au–Cr 2.632(2) Å] [117].

The trinuclear cluster [(Ph$_3$P)Au]$_2$Cr(CO)$_5$(PPh$_3$) **133** contains an organochromium group bridging the gold atoms of the P–Au–Au–P fragment [Au–Cr 2.6932(6)/2.7038(7) Å] [118] in a similar fashion as in the niobium complex **127**. Recently, CrCp(CO)$_2$ groups were also reported to bridge each Au–P edge of the planar Au$_3$P$_3$ six-membered ring in the cluster Au$_3$P$_3$[Cr(η^5-Cp)(CO)$_2$]$_6$ **134** [Au–Cr range 2.707(2)–2.748(3) Å] [119].

4.3.3.2 Gold–Molybdenum and Gold–Tungsten Compounds

There are both differences and similarities in the structural chemistry of compounds with Au–Mo and Au–W bonds. Heterometallic compounds with an unsupported Au–metal bond are common for molybdenum derivatives [120–124] and recent examples are complexes related to the chromium derivative **129**, that is (Ph$_3$P)AuMo(CO)$_3$(η^5-C$_5$H$_4$CH$_2$CH$_2$NMe$_2$) **135** [Au–Mo 2.7208(6) Å] and the protonated species [(Ph$_3$P)AuMo(CO)$_3$(η^5-C$_5$H$_4$CH$_2$CH$_2$NHMe$_2$)]Cl **136** [Au–Mo 2.7063(5) Å] [113], or the almost linear anion of [PPN][2,2'-μ-Au-{1,2-μ-NH^{t}Bu-2,2,2-(CO)$_3$-*closo*-2,1-MoCB$_{10}$H$_{10}$}$_2$] **137** [Au-Mo 2.7377(9)/2.7422(8) Å; Mo–Au–Mo 170.80(2)°] [125]. Dinuclear compounds with an Au–Mo bond and additional ligands bridging the metal centers are not known. However, such systems are present in some clusters containing μ-*p*-tolylmethylidyne or carbonyl groups [126, 127] or μ$_3$-hydride bridging the AuMo$_2$ core in (Ph$_3$P)Au(μ$_3$-H)[Mo(CO)$_4$]$_2$(μ-dppm) **138** [Au–Mo 2.900(3)/2.914(2) Å] [128]. In contrast, dinuclear complexes with an unsupported Au–W bond are much rarer [113, 129], a recent example being the tungsten analog **139** of compound **136** [Au–W 2.7121(5)/2.7060(5) Å] [113]. Much more common are dinuclear and polynuclear compounds in which the metal centers of an Au–W bond are bridged by μ-arylmethylene [130, 131], μ-arylmethylidyne [132–140], μ-C≡C^{t}Bu [141], μ-CN(Et)Me [142], μ-CO or μ-CS ligands [127, 143–145]. A recent example, (Ph$_3$P)AuW(μ-CO)(μ-C≡C^{t}Bu)(η^5-Cp)(NO) **140**, features both alkynyl and carbonyl groups supporting the Au–W bond [2.803(3) Å] (Figure 4.36a) [141]. Both neutral and ionic cluster compounds containing μ$_3$-*p*-tolylmethylidyne or μ$_3$-ethylidyne groups bridging a trimetallic core with at least one gold atom are well represented in the chemistry of Au–W compounds [126, 131, 135, 136, 145, 146].

The affinity of gold for sulfur ligands has prompted the development of the chemistry of gold complexes based on [ME$_4$]$^{2-}$ ligands (M = Mo, W; E = S, Se) and related ligands [147–157]. Many compounds of the type (L)AuS$_2$MS$_2$Au(L) (L = PR$_3$, AsPh$_3$) have been reported for both molybdenum [147–152] and tungsten [150–153]. These trinuclear heterobimetallic species contain an almost linear L–Au–M–Au–L

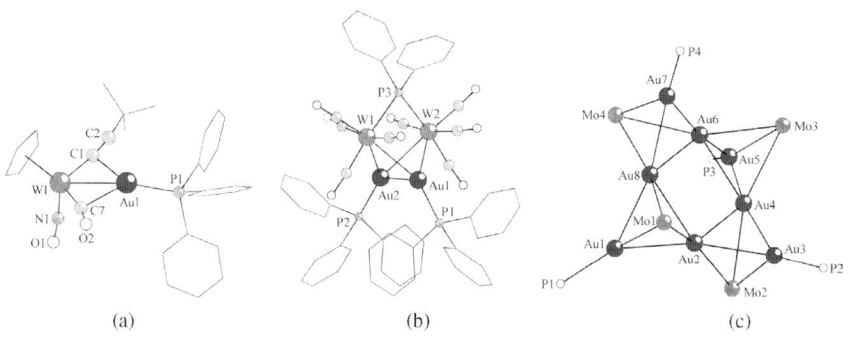

Figure 4.36 The structure of (a) (Ph$_3$P)AuW(μ-CO)(μ-C≡C^{t}Bu)(η^5-Cp)(NO) **140** [141], (b) the anion in [NEt$_4$][{(Ph$_3$P)Au}$_2$W$_2$(CO)$_8$(μ-PPh$_2$)] **150** [163], and (c) Au$_4$[(Ph$_3$P)Au]$_4$Mo$_4$(CO)$_{20}$ **152** (for clarity, the phenyl groups on phosphorus and the CO ligands are not shown) [170].

fragment, with tricoordinate gold atoms. The formation of the *spiro*-bicyclic system with orthogonal four-membered AuS$_2$M rings brings the metal centers in close proximity, for example, Au–Mo and Au–W distances of 2.804(1) and 2.822 (8) Å in MS$_4$[Au(PPh$_2$Py)]$_2$ [M = Mo **141**, W **142**] [152]. Treatment of [NH$_4$]$_2$[WS$_4$] with the gold(I)-ylide complex AuCl(CH$_2$PPh$_3$) resulted in WS$_4$[Au(CH$_2$PPh$_3$)]$_2$ **143** [149], which contains one three- and one two-co-ordinate gold center (CAuS$_2$ and CAuS cores, if the Au–W contacts are not taken into account). The selenium derivative WSe$_4$[Au(PPh$_2$Me)]$_2$ **144** [154] exhibits the similar common structure as the thio analogs. The related WSe$_4$[Au(PPh$_3$)][Au(PPh$_3$)$_2$] **145** has a longer transannular Au–W distance [3.138(5) Å] for the tetrahedral coordinated gold atom than for the gold atom in the trigonal planar environment [2.854(5) Å] [155]. A similar pattern, that is, an Mo–Au distance significantly greater for the four-coordinate gold atom [3.133(1) Å] than for the three-coordinate gold atom [2.838 (1) Å], is observed in the MoOS$_3$[Au(PPh$_3$)][Au(PPh$_3$)$_2$] **146** derivative. In this case the oxotrithiomolybdate has one triply bridging and two doubly bridging sulfur atoms, resulting in an angular Au–Mo–Au fragment [94.5(1)°] [148]. The use of diphosphine ligands as spacers afforded the isolation of ionic compounds of the type [Et$_4$N]$_2$[S$_2$WS$_2$Au(μ-L)AuS$_2$WS$_2$] (L = dppe, dppf), which contain terminal tetrathiotungstate groups [156].

A number of neutral or ionic clusters of different nuclearity have also been structurally characterized. A (Ph$_3$P)Au unit can bridge two metal centers of which at least one is a Mo or W atom in trinuclear or tetranuclear species with AuMo$_2$ [158] and AuMoMn [127] or (AuPtW)W [159] cores, respectively. Recently, the mixed-metal clusters [NEt$_4$][{(Me$_3$P)Au}MoFe$_5$(μ$_6$-C)(CO)$_{17}$] **147** [160] and [PPh$_4$][{(Ph$_3$P)Au}MoCo$_5$(μ$_6$-N)(CO)$_{14}$] **148** [161] were reported. Both anions contain a μ$_3$-(R$_3$P)Au unit capping a trigonal MoFe$_2$ [Au–Mo 2.963(3) Å] or MoCo$_2$ [Au–Mo 3.159(1) Å] face of the octahedral C- or N-centered cluster cores.

Fragments containing a P–Au–Au–P skeleton can be connected to a metal center as in [{(Ph$_3$P)Au}$_2$Mo(η^5-Cp)(CO)$_2$(PMe$_3$)](BF$_4$) **149** [Au-Mo 2.7859(8)/2.7791(8) Å] [162] or can bridge two or three metal centers as in [NEt$_4$][{(Ph$_3$P)Au}$_2$W$_2$(CO)$_8$(μ-PPh$_2$)] **150** [Au-W 2.872(1)–2.952(1) Å] (Figure 4.36b) [163], [(Ph$_3$P)Au]$_2$M$_3$(CO)$_9$[Au(PPh$_3$)] (μ-OEt)(μ$_3$-OEt)$_2$ (M = Mo, W) [164] and [NEt$_4$][{(dppm)Au$_2$}MoFe$_5$(μ$_5$-C)(CO)$_{17}$] **151** [160].

Finally, polynuclear species which can be described as [(Ph$_3$P)Au]$_n$ clusters with gold atoms bridged by an Mo or W-containing fragment are also known. Examples are [(Ph$_3$P)Au]$_3$WH$_4$(PMe$_3$)$_3$ [165], [{(Ph$_3$P)Au}$_4$Mo(η^5-Cp)(CO)$_2$](PF$_6$) [122], [{(Ph$_3$P)Au}$_5$M(CO)$_4$](PF$_6$) (M = Mo [166], W [167]) and [{(Ph$_3$P)Au}$_7$M(CO)$_3$]X (M = Mo, X = OH [168, 169]; M = W, X = PF$_6$ [167]). A recent report describes the structure of the new cluster type Au$_4$[(Ph$_3$P)Au]$_4$Mo$_4$(CO)$_{20}$ **152** whose structure can be described as a square gold cluster, whose edges are all bridged by Mo(CO)$_5$ and Au(PPh$_3$) units, with alternate orientation [170]. The four Au atoms that define the square are unsupported by ligands and are connected to six metal atoms in a distorted trigonal prismatic geometry. The unshared four Au atoms are three-connected to metal atoms and carry a PPh$_3$ ligand, thus exhibiting a tetrahedral environment (Figure 4.36c).

4.3.4
Gold–Group 7 Metal Compounds

The structures of several compounds with Au–Mn and Au–Re bonds have been reported and all are neutral or ionic species without further association through metal–metal interaction resulting in supramolecular architectures.

4.3.4.1 Gold–Manganese and Gold–Rhenium Compounds

Few dinuclear compounds with Au–Mn [171–174] and Au–Re [175–177] bonds are known. Typical examples for unsupported Au–metal bonds are (μ-dppf)[AuMn(CO)$_5$]$_2$ **153** [Au–Mn 2.56(3)/2.58(3) Å] [172] and (Ph$_3$P)AuRe(CO)(η^5-Cp)(N=NC$_6$H$_4$OMe-4) **154** [Au–Re 2.615(1) Å] [175], while for supported Au–metal bonds the compounds 3,8-[(Ph$_3$P)Au]-8-μ-H-3,3,3-(CO)$_3$-*closo*-3,1,2-MnC$_2$B$_9$H$_{10}$ **155** [Au–Mn 2.9284(4) Å] [174] and (Ph$_3$P)AuRe(μ-H)(CO)(PPhMe$_2$)$_3$ **156** [Au–Re 2.779(1) Å] [176], with either an additional Au–B interaction or a bridging hydrido ligand, can be considered. The cluster [(Ph$_3$P)Au]$_4$[AuMn(CO)$_5$]$_2$ **157** also contains one of the shortest unsupported Au–Mn bonds [2.566(3)/2.569(2) Å] [178].

Clusters containing (Ph$_3$P)Au units bridging symmetrically or asymmetrically a single bond M–M fragment (M = Mn [179–186], Re [187–195]) are numerous. Typical compounds are (Ph$_3$P)AuM$_2$(μ-PPh$_2$)(CO)$_8$ (Au–Mn 2.678(3)/2.710(3) Å in **158** [180]; Au–Re 2.810(2)/2.764(2) Å in **159** [190]), (μ-dppe)AuMn$_2$(μ-PPh$_2$)(CO)$_7$ **160** [Au–Mn 2.601(2)/2.756(2) Å] [181], or (Ph$_3$P)AuRe$_2$(μ-S-naphthyl-2)(CO)$_8$ **161** [Au–Re 2.7853(8)/2.8110(7) Å] [193]. Related compounds in which an Au(PPh$_3$) unit bridges a heterometallic Mn–M (M = Sn (**105**) [91], Mo [127]) bond are also known.

Neutral or ionic clusters with one or two μ-(R$_3$P)Au units connecting an unsaturated Mn=Mn system are known [91, 196, 197]. The tetranuclear species [(*p*-tolyl$_3$P)Au]$_2$Mn$_2$(μ-dppm)(CO)$_6$ **162** (Figure 4.37a) [91] and [{(Ph$_3$P)Au}$_2$Mn$_2$(μ$_3$-H)(μ-dppm)(CO)$_6$](PF$_6$) **163** [196] contain an Mn$_2$Au$_2$ core in a flattened-butterfly arrangement with diagonal geometry for the Au atoms. In **163** a hydride ligand

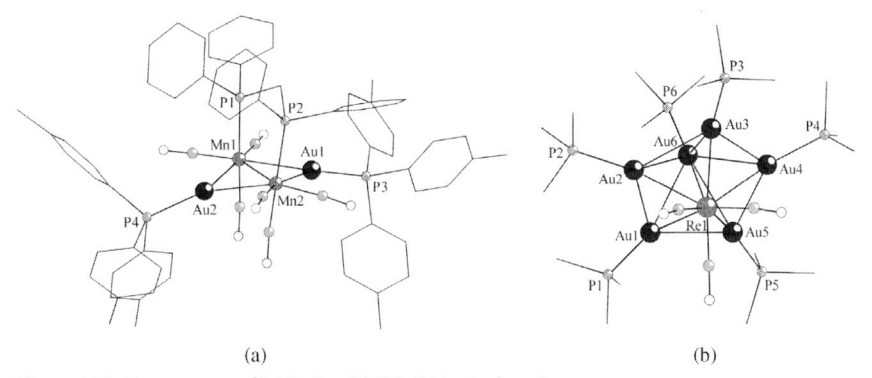

Figure 4.37 The structure of (a) [μ-(*p*-tolyl$_3$P)Au]$_2$Mn$_2$(μ-dppm)(CO)$_6$ **162** [91], (b) the cation in [{(Ph$_3$P)Au}$_6$Re(CO)$_3$](PF$_6$) **182** (for clarity, only *ipso* carbon of the phenyl groups on phosphorus are shown) [211], (c) the anion of [PPN][Au{Mn$_2$(CO)$_8$(μ-PPh$_2$)}$_2$] **187** [182], and (d) (Ph$_3$P)AuTc(=NMes)$_3$ **190** [223].

(c)   (d)

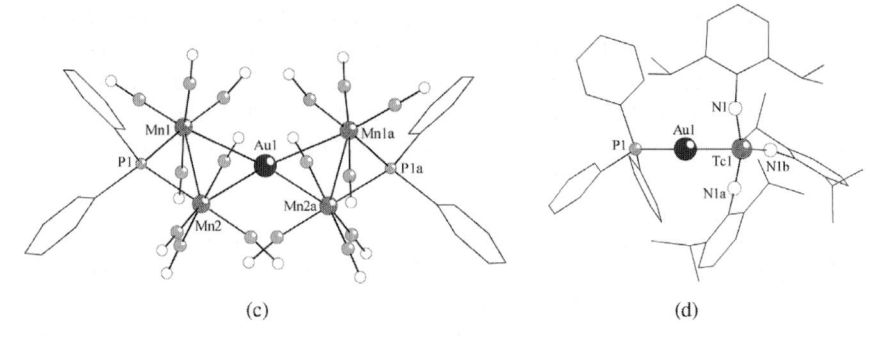

Figure 4.37 (Continued).

bridges a Mn_2Au core and the corresponding Au–Mn bonds [2.768(4)/2.845(3) Å] are significantly longer than those exhibited by the second gold atom [Au–Mn 2.665(4)/2.703(3) Å] or those found in the neutral cluster **162** [Au–Mn 2.669(3)–2.696(3) Å]. The structure of the unsaturated pentanuclear cluster $[(Ph_3P)Au]_3Mn_2[\mu\text{-}P(OEt)_2](CO)_6$ **164** was also reported and it features a slightly distorted trigonal bipyramidal Au_3Mn_2 core, with the manganese atoms at equatorial sites [198]. A similar structure with an M=M double bond was established for the related $[(Ph_3P)Au]_3M_2(\mu\text{-}PPh_2)(CO)_6$ (M = Mn **165** [180]; Re **166** [190]).

The tetranuclear clusters exhibits Au_nM_{4-n} ($n = 1$–3) cores of different geometries. Most of them contain tetrahedral $AuRe_3$ [199, 200] or $AuRe_2M$ (M = Cu, Ag) [194, 201, 202], Au_2Mn_2 [180], Au_2Re_2 [190, 194, 202–205] and Au_3Re [188] skeletons. Recent examples are $[(Ph_3P)Au][(Ph_3P)M]Re_2(\mu\text{-}PCy_2)(CO)_7(C\equiv CPh)$ [M Cu **167**, Ag **168**] [201], $[(Ph_3P)Au]_2Mn_2(\mu\text{-}PPh_2)[\mu\text{-}C(Ph)O](CO)_6$ **169** [Au–Mn 2.702(2)–3.084(2) Å] [180] or $[(Ph_3P)Au]_2Re_2(\mu\text{-}PPh_2)_2(CO)_6$ **170** [Au–Re 2.862(1)–3.059(1) Å] [180]. Tetranuclear clusters with a more or less flattened butterfly $AuRe_3$ [206, 207], Au_2Re_2 [208], Au_2ReIr [209] and AuM_3 (M = Mn [210], Re [178, 211]) cores, folded about a Re–Re or Au–M (M = Re, Ir) bond, are also known. Examples of such compounds are $(Ph_3P)AuRe_3(\mu_3\text{-}H)(\mu\text{-}H)_3(CO)_9(PPh_3)$ **171** [Au–Re 2.926(1)/2.967(1) Å] [207], $(\mu\text{-}dppf)Au_2Re_2(CO)_9$ **172** [Au–Re 2.730(2)/2.757(2) Å] [208], $[(Ph_3P)Au]_3Mn(CO)_4$ **173** [Au–Mn 2.584(4)–2.620(4) Å] [210] and $[(Ph_3P)Au]_3Re(CO)_4$ **174** [Au–Re 2.710(1)–2.784(1) Å] [211].

Tetranuclear clusters with planar, opened Au–Au–M–M (M = Mn, Re) or Au–Au–Re–Mn cores and a μ_4-PCy ligand have also been structurally characterized, for example $[(Ph_3P)Au]_2Mn_2(\mu_4\text{-}PCy)(CO)_8$ **175** [Au–Mn 2.762(2) Å] [183] and $[(Et_3P)Au]_2Re_2(\mu\text{-}H)(\mu\text{-}PCy_2)(\mu_4\text{-}PCy)(CO)_6$ **176** [Au–Re 2.919(1) Å] [212].

Different clusters with higher nuclearity which contain a Mn or Re atom connected to aggregates of four or more $Au(PR_3)$ units are known. These include cations with a $[(Ph_3P)Au]_4M$ core containing two $(R_3P)Au$ units μ_3-bridged on both sides of a trigonal Au_2M fragment [211, 213, 214], for example $[\{(Ph_3P)Au\}_4Re(CO)_4](PF_6)$ **177** [Au–Re 2.743(1)–2.805(1) Å] [211]. A different pentanuclear core was found in $[\{(Ph_3P)Au\}_4(H)_4Re\{P(tolyl\text{-}p)_3\}_2](BPh_4)$ **178**, which consists of an Au_3Re tetrahedron with an Au–Re edge bridged by a gold atom [215].

The heterometallic core in the hexanuclear clusters of type [{(Ph₃P)Au}₅(H)₄Re(PR₃)₂](PF₆) (R = Ph **179** [216], *p*-tolyl **180** [217]) features two Au₃Re tetrahedrons with a common Au–Re edge, while the inner Au₆M core in the heptanuclear clusters of type [{(Ph₃P)Au}₆M(CO)₃](PF₆) (M = Mn **181** [218], Re **182** (Figure 4.37b) [211]) is a pentagonal bipyramid with the M atom in an axial position. The shortest distance being between the heterometal atom to the axial Au atom [Au–Mn 2.618(1) Å in **181**; Au–Re 2.722(1) Å in **182**].

A class of ionic cluster compounds without analogy in the manganese cluster chemistry is based on octahedral C-centered metal cores, with one to three faces μ₃-bridged by Au(PPh₃) or Re(CO)₃ groups [219–221]. Examples are [NEt₄]₂[{(Ph₃P)Au}HRe₆(μ₆-C)(CO)₁₈] **183** [219] with one trigonal face capped by gold, [NEt₄]₂[{(Ph₃P)Au}₂Re₆(μ₆-C)(CO)₁₈] **184** [219] and [AsPh₄]₂[{(Ph₃P)Au}Re₇(μ₆-C)(CO)₂₁] **185** [220], with two opposite faces capped by two gold atoms or one gold and one rhenium atom, respectively. An IrRe₆ octahedron tricapped by two Au and one Re atom was found in the anion of [PPh₄][{(Ph₃P)Au}₂IrRe₆(μ₆-C)(CO)₂₀] **186** [221].

Crystallographically characterized clusters in which a four-coordinated gold(I) center connects two polynuclear fragments are rare and few examples are known for compounds with Au–Mn or Au–Re bonds. In the anion of [PPN][Au{Mn₂(CO)₈(μ-PPh₂)}₂] **187** a gold atom acts as a *spiro* atom between two planar AuMn₂P metallacycles [Au–Mn 2.800(1)/2.812(1) Å] (Figure 4.37c) [182], while in the neutral complex AuMn₄(μ-H)₄(μ₃-H)(CO)₁₂[μ-(EtO)₂POP(OEt)₂]₂ **188** two AuMn₂ triangles are fused at gold [Au–Mn 2.771(2)–2.841(2) Å] and hydride bridges are present between metal centers [222]. In [{(Ph₃P)Au}₆AuRe₂(CO)₈](PF₆) **189** two Au₄Re trigonal bipyramids are connected by a common axial Au atom [Au_spiro–Re 2.831(1) Å] [211].

4.3.4.2 Gold–Technetium Compounds

The unique Au–Tc compound, (Ph₃P)AuTc(=NMes)₃ **190**, was obtained by reacting the *in situ* prepared Na[Tc(=NMes)₃] with Au(PPh₃)Cl, in THF. The green crystals contain discrete molecules with a linear P–Au–Tc fragment and an Au–Tc bond of 2.589(1) Å (Figure 4.37d) [223].

4.3.5
Gold–Group 8 Metal Compounds

The structural chemistry of compounds containing Au–Group 8 metal bonds, and especially that of clusters, is by far the best represented in the literature in comparison with those of the other transition metals [224]. There are many similarities between the heterometallic cluster cores in compounds with Au–Fe, Au–Ru and Au–Os bonds, but also some differences which will be emphasized in the subsequent discussion.

There are relatively few dinuclear compounds with Au–Fe [225–228] and Au–Ru [229–232] bonds and most of them contain Au–C_carbonyl interactions or a hydride bridging the two metal atoms. Examples are [(*p*-tolyl)Ph₂P]AuFe(SiMePh₂)(PMe₃)(CO)₃ **191** [Au–Fe 2.527(3) Å] [228] or [(Ph₃P)AuRu(μ-H)₂(CO)(PPh₃)₃](PF₆) **192** [Au–Ru 2.788(1) Å] [230]. The compound IAuOs(μ-CH₂)Cl(NO)(PPh₃)₂ **193** is the unique dinuclear gold–osmium species characterized by X-ray diffraction and contains metal centers connected at an Au–Os distance of 2.788(1) Å [233]. The trimetallic complex [NEt$_4$][(Ph$_3$P)AuFe(CO)$_4$W(CO)$_5$] **194** contains an unsupported Au–Fe bond [2.520(3) Å] [234], while in the cation of [{(Ph$_3$P)Au}$_2$Os(μ-H)$_3$(PPh$_3$)$_3$] (PF$_6$) **195** [Au–Os 2.696(1)/2.709(1) Å] the absence of an Au–Au bond was suggested to be due to the presence of hydride ligand bridging gold and osmium atoms [230].

Trinuclear clusters are much more common for iron and both AuFe$_2$ [235–239] and Au$_2$Fe [240–244] cores have been described. They include neutral compounds such as (Ph$_3$P)AuFe$_2$(μ-S^{i}Pr)(μ-CO)(CO)$_6$ **196** [Au–Fe 2.644(1)/2.695(1) Å] [238] or [(Ph$_3$P)Au]$_2$Fe(CO)$_3$[P(OEt)$_3$] **197** [Au–Fe 2.509(3)/2.561(3) Å] [242], the latter containing the shortest Au–Fe bond, or ionic species as [NEt$_4$][(Ph$_3$P)AuFe$_2$(μ-CO)$_2$(CO)$_6$] **198** [Au–Fe 2.622(1)/2.698(1) Å] [236]. Complexes with gold atoms from two Au$_2$Fe or three AuFe$_2$ units bridged by bidentate or tridentate phosphine ligands, that is (μ-dppe)$_2$[Au$_2$Fe(CO)$_4$] **199** [241] and [μ$_3$-MeC(CH$_2$PPh$_2$)$_3$][AuFe$_2$(μ-HC=CHPh)(μ-CO)(CO)$_6$]$_3$ **200** [239], have also been described. The cation of the unusual, unique compound [{(Ph$_3$P)AuAu(PPh$_3$)}(η^5-C$_5$H$_4$)Fe(η^5-Cp)](BF$_4$) **201** contains a chain Au–Au–Fe [Au–Fe 2.818(9) Å], while one of the cyclopentadienyl ligands bridges the two Au atoms through one of its carbon atoms [240]. By contrast, so far the structures of only two trinuclear clusters with Au–Ru bonds, [(Ph$_3$P)AuRu$_2${μ-EtN[P(OMe)$_2$]$_2$}$_2$(μ-CO)(CO)$_4$](SbF$_6$) **202** [Au–Ru 2.741(1)/2.749(1) Å] [245, 246] and [{(Ph$_3$P)Au}$_2$Ru(μ-H)$_2$(dppm)$_2$](NO$_3$)$_2$ **203** [Au–Ru 2.776(0)/2.786(0) Å] [247], and the neutral species [(Ph$_3$P)Au]$_2$Os(CO)$_4$ **204** [Au–Os 2.667(1)/2.646(1) Å] [248], containing the shortest Au–Os bonds, have been reported.

By far, most of the tetranuclear species exhibit a butterfly structure folded over a transannular metal–metal bond. For Au/Fe clusters this includes one (R$_3$P)Au unit bridging the metal atoms of a Fe–Fe, Fe–Ru or Fe–Co bond in compounds with (AuFe$_2$)Fe [249–254], (AuFe$_2$)Co [255], (AuFeRu)Co [256] and (AuFeCo)Co [257] cores, or two (R$_3$P)Au units bridging the Fe–Pt bond in an (AuFePt)Au core [258]. A similar butterfly structure was observed in [(μ-dppm)$_2$Au$_3$Fe(CO)$_4$]Cl **205**, which contains a Fe(CO)$_4$ fragment bridging two Au atoms [Au–Fe 2.518(2)/2.538(1) Å], with the other two edges of the triangular Au$_3$ system being each spanned by the dppm ligands [259]. A particular nonplanar Au$_2$Fe$_2$ ring formed from a (μ-dppm)$_2$Au$_2$ unit connected to a Fe$_2$(CO)$_8$ fragment [Au–Fe 2.527(2)/2.534(2) Å] was also described. No transannular metal–metal bond is present in (μ-dppm)$_2$Au$_2$Fe$_2$(CO)$_8$ **206** [260].

Similar (AuM$_2$)M cores were found for the Au–Ru [261–278] and Au–Os [276, 279–290] tetranuclear species and in most cases they exhibited a folded butterfly structure. Recent examples are (Ph$_3$P)AuRu$_3$[μ$_3$,η^2-C$_2$C≡CAu(PPh$_3$)](CO)$_9$ **207** [Au–Ru 2.7693(4)/2.7524(3) Å] [278] and (Ph$_3$P)AuOs$_3$(μ-SC$_5$H$_4$N-4)(CO)$_{10}$ **208** [Au–Os 2.7755(8)/2.7744(7) Å] [290]. Bis-cluster species in which two (AuRu$_2$)Ru cores are connected through buta-1,3-diyne derivatives as bridging ligands have been described [290]. In two cases, [(Ph$_3$P)AuRu$_3${μ$_3$-HC(PMe$_2$)$_3$}(CO)$_9$](O$_3$SCF$_3$) **209** [Au–Ru 2.7608(6) Å] [271] and (Et$_3$P)AuOs$_3$(μ$_2$,η^3-C$_3$H$_5$)(CO)$_{10}$ **210** [Au–Os 2.746(1)/2.816(1) Å] [284], the (AuM$_2$)M core was found to be planar.

Tetranuclear clusters with a μ$_3$-Au(PPh$_3$) unit leading to tetrahedral AuM$_3$ (M = Fe, Ru) cores are rare, for example [NEt$_4$][(Ph$_3$P)AuFe$_3$(μ$_3$-O)(μ-CO)$_3$(CO)$_6$] **211** [Au–Fe 2.708(2)–2.736(2) Å] [253] and (Ph$_3$P)AuRu$_3$(μ-H)(μ$_3$-HC$_2$H)(CO)$_9$] **212** [Au–Ru

2.742(2)–3.114(2) Å [272]. For **212** the isomer with butterfly (AuRu$_2$)Ru core was also reported [272]. The only tetrahedral cluster with an Au$_3$Ru core whose structure has been reported so far is [{(Ph$_3$P)Au}$_3$Ru{μ$_3$-MeC(CH$_2$PPh$_2$)$_3$}(μ-H)$_3$](PF$_6$)$_2$ **213** [Au–Ru 2.679(2) Å] [291].

An unusual open-chain Fe–Au–Au–Fe with alternating metal atoms bridged by dppm ligands was found in the compound Au$_2$Fe$_2$(μ-dppm)$_2$[Si(OMe)$_3$]$_2$(CO)$_6$ **214** [Au–Fe 2.535(3)/2.562(3) Å] [292]. Recently, examples of tetranuclear species have been reported in which a (Ph$_3$P)Au unit bridges a metal atom of a triangular M$_3$ fragment and a carbon [293, 294] or boron atom [295] of a ligand attached to it.

A structural diversity of heterodimetallic and -trimetallic cores was observed for the pentanuclear species containing Au–Fe [296–309], Au–Ru [267, 272, 310–325] and Au–Os [326–335]. A (Ph$_3$P)Au unit can bridge in a μ$_3$-fashion a face of a tetrahedral heterodimetallic (Fe$_2$Ir$_2$, Fe$_2$Rh$_2$, Ru$_2$Rh$_2$, Ru$_3$Rh, Ru$_3$Co) core [296, 297, 310, 311, 315] to result in a trigonal bipyramidal geometry, for example, in the ionic [PPh$_4$][(Ph$_3$P)AuFe$_2$Ir$_2$(CO)$_{12}$] **215** [Au–Fe 2.806(1) Å] [296] or the neutral (Ph$_3$P)AuRu$_3$Rh(μ$_3$-H)(μ$_3$-COCH$_3$)(μ-CO)(CO)$_9$(PPh$_3$) **216** [Au–Ru 2.813(1)/2.735(1) Å] [311]. For Ru and Os clusters a common pattern is the edge-bridging of a (R$_3$P)Au fragment to a homonuclear M$_4$ (M = Ru [312], Os [326–330]) or heteronuclear M$_3$W (M = Ru [313], Os [331]), Ru$_3$Co [315] and Os$_3$Ni [332], for example, in (Ph$_3$P)AuRu$_3$W(η^5-Cp)(μ$_3$-H)(μ-H)(CO)$_{11}$ **217** [Au–Ru 2.759(1)/2.722(1) Å] [313] and (Ph$_3$P)AuOs$_4$(μ-H)$_3$(CO)$_{11}$(NMe$_3$) **218** [Au–Os 2.768(2)/2.753(2) Å] [327]. For clusters with an Au$_2$M$_3$ core (M = Fe [298–300], Ru [267, 314, 316–318]) the distorted trigonal bipyramidal geometry can be considered as arising from a butterfly AuM$_3$ fragment μ$_4$-bridged by a second Au atom, for example [(Ph$_3$P)Au]$_2$Fe$_3$(μ$_3$-O)(CO)$_9$ **219** [Au–Fe 2.669(2)–2.782(2) Å] [300] and [(Ph$_3$P)Au]$_2$Ru$_3$(μ$_3$-S)(CO)$_9$ **220** [Au–Ru 2.783(2)–2.867(2) Å] [267]. When the second Au atom acts as a μ$_3$-bridge for the butterfly AuRu$_3$ fragment [272, 314, 319, 320] the resulting pentanuclear core adopts a distorted square-based pyramidal structure, with the basal plane defined by the two Au atoms and two of the Ru atoms and the third Ru atom forming the apex of the pyramid, for example [(Ph$_3$P)Au]$_2$Ru$_3$(μ-H)(μ$_3$-CMe)(CO)$_9$ **221** [Au–Ru 2.7174(4)–2.7737(5) Å] [319].

An edge-bridged (R$_3$P)Au unit was also found in pentanuclear clusters in which all four metal atoms of the butterfly tetrametallic core are coordinated by a carbon [301, 302] or a nitrogen [303, 321] atom. The gold atom bridges the hinge Fe–Fe or Ru–Ru bond, for example, in neutral (Et$_3$P)AuFe$_4$(μ$_4$-COMe)(CO)$_{12}$ **222** [Au–Fe 2.666(2)/2.675(3) Å] [301] or (Ph$_3$P)AuFeRu$_3$(μ$_4$-N)(CO)$_{12}$ **223** [Au–Ru 2.766(1)/2.755(1) Å] [321]. In the crystal of [K(18-crown-6)][(Et$_3$P)AuFe$_4$(μ$_4$-CO)(CO)$_{12}$] **224** a polymeric chain is formed with alternating anions and cations bridged by K···OC(Fe) interactions [302]. The gold atom of an (Et$_3$P)Au fragment was also found to bridge symmetrically the hinge Os–Os bond in [μ-(Et$_3$P)Au]$_2$Os$_3$(CO)$_{10}$ **225** [Au–Os 2.760(1)/2.762(1) Å] [333]. The neutral cluster (Ph$_3$P)AuOs$_4$(μ-H)[μ$_3$-NC(O)Me](CO)$_{12}$ **226** has a butterfly Os$_4$ framework μ$_3$-capped by the Au(PPh$_3$) fragment [Au–Os 2.762(1)–2.940(1) Å] and μ$_3$-bridged by the four-electron-donor amido group [334, 335].

Clusters containing a butterfly tetrametallic core μ$_4$-coordinated by a boron atom (BFe$_4$ [304], BRu$_3$W [322], or BRu$_3$Rh [323]) and an Au(PR$_3$) fragment as a bridge between the boron and a Fe/Ru atom have been reported, for example, (Ph$_3$P)AuRu$_3$W

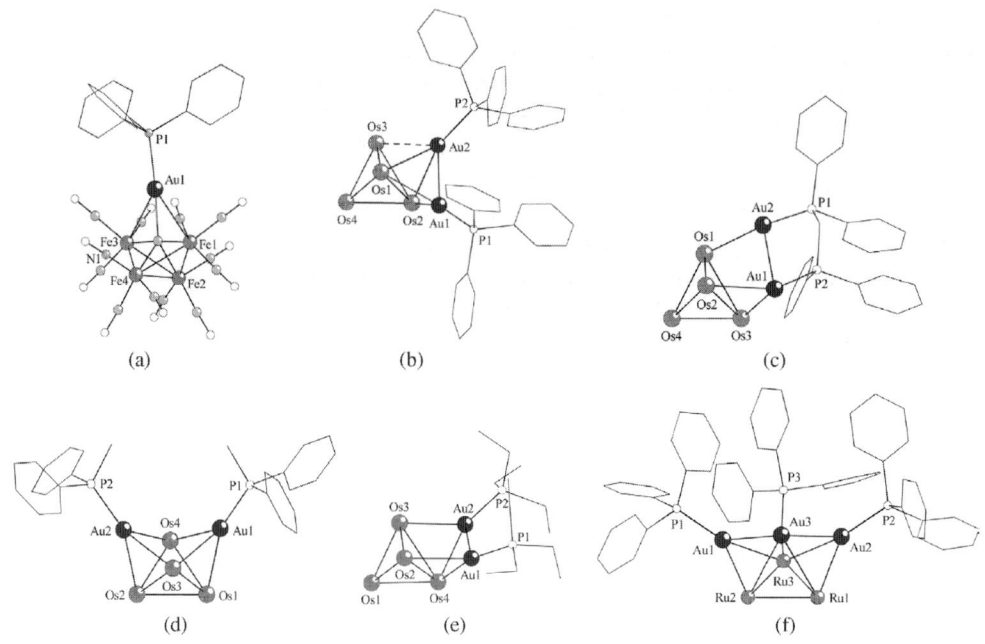

Figure 4.38 (a) The structure of $(Ph_3P)AuFe_4(\mu_4\text{-}C)(CO)_{12}(NO)$ **228** [306], and metal frameworks in (b) $[(Ph_3P)Au]_2Os_2(\mu\text{-}H)_2(CO)_{12}$ **232** [367], (c) $[(dppe)Au_2]Os_4(\mu\text{-}H)_4(CO)_{11}$ **233** [369], (d) $[(Ph_2MeP)Au]_2Os_4(CO)_{12}$ **234** [368], (e) $[(Et_3P)Au]_2Os_4(\mu\text{-}CO)(CO)_{12}$ **235** [368], and (f) $[(Ph_3P)Au]_3Ru_3[\mu_3\text{-}\eta^1\text{:}\eta^1\text{:}\eta^1, \eta^2\text{-}CCHCPh_2OC(O)](CO)_8$ **236** [362] (for clarity, ligands on Os and Ru are not shown).

$(\mu_5\text{-}B)(H)(\eta^5\text{-}Cp)(CO)_{11}$ **227** [Au–Ru 2.692(2) Å] [322]. Related clusters in which the Au (PR_3) unit bridges in a μ_3-M,C,M' fashion a butterfly CM_4 (M = Fe [305, 306], Ru [324]) or CFe_3Co (M′ = Co [307], Fe [308]) core are also known, for example $(Ph_3P)AuFe_4(\mu_5\text{-}C)(CO)_{12}(NO)$ **228** [Au–Fe 2.820(2)/2.866(2) Å] (Figure 4.38a) [306].

Unusual heterometallic cores were found in $[(Ph_3P)Au]_2Fe_3(\mu_5\text{-}P)(\mu\text{-}H)(CO)_9$ **229**, which contains a PFe_3 tetrahedron connected in a Fe,P,P-pattern by a $(Ph_3P)Au$-Au (PPh_3) fragment [309], or in $[(Ph_3P)Au]_2Ru_3(\mu_3\text{-}CEtCMeCPhCPh)(CO)_7$ **230**, whose $Au_2Ru_3C_4$ cluster consists of the $(Ph_3P)Au$–$Au(PPh_3)$ unit interacting as a μ_3-bridge with two metal atoms of the C_4Ru_3 pentagonal bipyramid [325].

Different geometries of the heterometallic core were reported in hexanuclear clusters with Au–Fe [241, 305, 309, 336–343], Au–Ru [262, 264, 298, 315, 324, 344–364] and Au–Os bonds [327, 365–369]. Common structures feature a trigonal bipyramidal fragment capped by a μ_3-$Au(PR_3)$, for example, in $(Ph_3P)AuOs_5(\mu\text{-}H)(\eta^6\text{-}C_6H_6)(CO)_{12}$ **230** [Au–Os 2.819(2)–3.020(2) Å] [365] or $(\mu\text{-}dppe)AuRu_4Cu(\mu_3\text{-}H)_2(CO)_{12}$ **231** [Au–Ru 2.823(2) Å] [345]. Similar metal frameworks are also observed for some hexanuclear systems containing two $Au(PR_3)$ fragments which might be considered to arise from a M_4 tetrahedron with one face capped by one Au atom and one M_2Au face thus formed further capped by the second Au atom. Examples are known both for bimetallic Au_2Ru_4 [354, 355] or trimetallic $Au_2Ru_2Co_2$ [298] and Au_2Ru_3Co [315] species.

Depending on the saturated or the unsaturated nature of the cluster and the phosphine ligands used, different geometries of the hexanuclear Au_2Os_4 metal frameworks were observed: (i) a $(Ph_3P)Au–Au(PPh_3)$ unit asymmetrically bridging one edge of an Os_4 tetrahedron in $[(Ph_3P)Au]_2Os_4(\mu\text{-}H)_2(CO)_{12}$ **232** [Au(1,2)–Os(1,2) 2.839(4)–3.004(5) Å, Au(2)–Os(3) 3.159 Å] (Figure 4.38b) [367]; (ii) one of the Au atoms of the $(dppe)Au_2$ unit bridges an edge of the Os_4 tetrahedron and the second Au atom coordinated to a third Os atom in $[(dppe)Au_2]Os_4(\mu\text{-}H)_4(CO)_{11}$ **233** [Au–Os 2.699(3)–2.825(3) Å] (Figure 4.38c) [369]; (iii) "open" butterfly Os_4 core bicapped by two isolated μ_3-$Au(PR_3)$ units in $[(Ph_2MeP)Au]_2Os_4(CO)_{12}$ **234** [Au–Os 2.784(2)–2.831(2) Å] (Figure 4.38d) [368], or (iv) diedge-bridged tetrahedral Os_4 core in $[(Et_3P)Au]_2Os_4(\mu\text{-}CO)(CO)_{12}$ **235** [Au–Os 2.739(1)–2.863(1) Å] (Figure 4.38e) [368]. Metal frameworks similar to that observed in cluster **235** were found in several other Au_2Ru_4 [346, 356–358] or Au_2Os_4 [327] derivatives.

Most Au_3M_3 species (M = Fe [342], Ru = [262, 264, 362–364]) exhibit a capped-trigonal bipyramidal metal framework, for example, in $[(Ph_3P)Au]_3Ru_3[\mu_3\text{-}\eta^1{:}\eta^1{:}\eta^1, \eta^2\text{-}CCHCPh_2OC(O)](CO)_8$ **236** [Au–Ru 2.7906(9)–2.9226(8) Å] (Figure 4.38f) [362], while for $[(Ph_3P)Au]_3Ru_3(\mu_3\text{-}C_2Ph)(CO)_8$ **237** a distortion towards a capped square pyramid was observed [364].

The cluster $(Ph_3P)AuRu_5(\mu_5\text{-}C_2PPh_2)(\mu\text{-}C_2Ph)(\mu\text{-}PPh_2)(CO)_{13}$ **238** has a scorpion geometry with a Ru_3–Ru–Ru skeleton, the Au atom bridging an edge of a Ru_3 triangle [Au–Ru 2.736(1)/2.767(1) Å] [353]. Some unusual metal frameworks were also observed for the clusters $[(Me_3P)Au]_2Ru_4(\mu_4\text{-}PCF_3)(CO)_{12}$ **239** and $[(Ph_3P)Au]_2Ru_4(\mu_3\text{-}PCF_3)_2(CO)_{12}$ **240** [359]. The former contains an octahedral $AuRu_4P$ core, with apical P and Ru atoms and $Au(PMe_3)$ unit capping an $AuRu_2$ face, while the latter has a Ru_4 metal skeleton with open chain geometry, the long terminal Ru–Ru bonds being each bridged by a $Au(PPh_3)$ moiety.

Hexanuclear clusters with more than three gold atoms have been structurally characterized for iron. Compound $[(dppm)Au_2]_2Fe_2(CO)_8$ **241** [359] contains an almost planar rhomboid of gold atoms. Opposite edges are bridged by dppm and $Fe(CO)_4$ units, with *cis–trans* orientation with respect to the Au_4 system, thus resulting in well separated Au_2Fe triangles. The cation of the cluster $[\{(Ph_3P)Au\}_5\{\mu_5\text{-}Fe(CO)_3\}](PF_6)$ (**242**) has an $FeAu_3$ tetrahedron with two Au_2Fe faces capped by $Au(PPh_3)$ units [343].

A class of hexanuclear clusters is based on a butterfly framework of iron or ruthenium atoms μ_4-connected by a carbon [305, 324, 336, 347–350] or boron [337–341, 351, 352] atom. In some $AuRu_5(\mu_5\text{-}C)$ compounds a Ru-containing moiety (*Ru, C,Ru*)-connected to the wingtip metal atoms, while the $Au(PPh_3)$ bridges the hinge Ru atoms [347–349], as in $(Ph_3P)AuRu_5(\mu_5\text{-}C)(\eta^5\text{-}Cp)(CO)_{13}$ **243** [Au–Ru 2.780(1)/2.750(1) Å] [353]. Two isomers are present in the crystal of $(Et_3P)AuRu_5(\mu_5\text{-}C)(NO)(CO)_{13}$ **244**; both contain a square pyramidal Ru_5 arrangement and differ in the bonding mode of the $Au(PEt_3)$ group, that is, μ_3-bonded to a Ru_3 face [Au–Ru 2.783(2)–3.033(2) Å] in **244a** and spanning a Ru–Ru bond between apical and basal atoms [Au–Ru 2.748(2)/2.792(2) Å] in **244b** [350]. A similar square pyramidal Ru_5 or Ru_4Rh core was also found in related μ_5-B derivatives [351, 352], but in these cases the Au(PPh_3) moiety is μ_3-bridged to the boron and the metal atoms of a Ru–Ru or Ru–Rh bond of the basal plane.

Unique structures were found in the clusters [(PhMe$_2$P)Au]$_2$Ru$_4$(μ_5-C)(CO)$_{12}$ **245**, in which one gold atom acts as a (*Ru,C,Ru*)-bridge and the other gold atom bridges the hinge metal atoms [324], in [(Et$_3$P)Au]$_2$Fe$_4$(μ_6-C)(CO)$_{12}$ **246**, in which a P–Au–Au–P fragment is connected to the butterfly Fe$_4$C core to result in a C-centered octahedral species [305], or in [(dppm)Au$_2$]Fe$_4$(μ_5-C)(CO)$_{12}$ **247**, where one metal atom of the P–Au–Au–P fragment bridges the wingtip atoms in a (*Fe,C,Fe*) fashion, while the second gold is connected to only one hinge iron atom. Clusters with the Au$_2$M$_4$B framework fit into two structure classes: (i) an Au–Au unit bridges an Fe$_{wing}$–B edge through one gold atom and an Fe$_{hinge}$–B edge through the other one, the latter being also involved in an additional weaker interaction with an Fe$_{wing}$ atom [337–340], as in [(Ph$_3$As)Au]$_2$Fe$_4$H(μ_6-B)(CO)$_{12}$ **248** [Au–Fe 2.590(1)/2.613(1); 2.860(2) Å] [340], and (ii) each gold atom of the Au–Au unit bridges one M$_{wing}$–B edge [339, 341, 360, 361], as in [(dppf)Au]$_2$Ru$_4$H(μ_6-B)(CO)$_{12}$ **249** [Au–Ru 2.695(2) Å] [360]. The related [(Ph$_3$P)Au]$_3$Fe$_3$H(μ_6-P)(CO)$_9$ **250** exhibits a similar Au$_3$Fe$_3$P as in **248**, one Fe$_{wing}$ atom being replaced by a gold atom [Au–Fe 2.665(5)–2.772(5) Å], thus resulting in an Au$_3$ triangle [309].

Heptanuclear clusters with Au–Fe [160, 306, 370–373, 394], Au–Ru [264, 319, 328, 374–390] and Au–Os bonds [327, 391–393] have also been reported. Most common structures are based on C-centered [160, 306, 374, 375] or B-centered [376–379] octahedral frameworks with a trigonal metal face capped by a gold atom, for example, in the anion of [NEt$_4$][{(Me$_3$P)Au}MoFe$_5$(μ_6-C)(CO)$_{17}$] **147** (Figure 4.39a) [160]

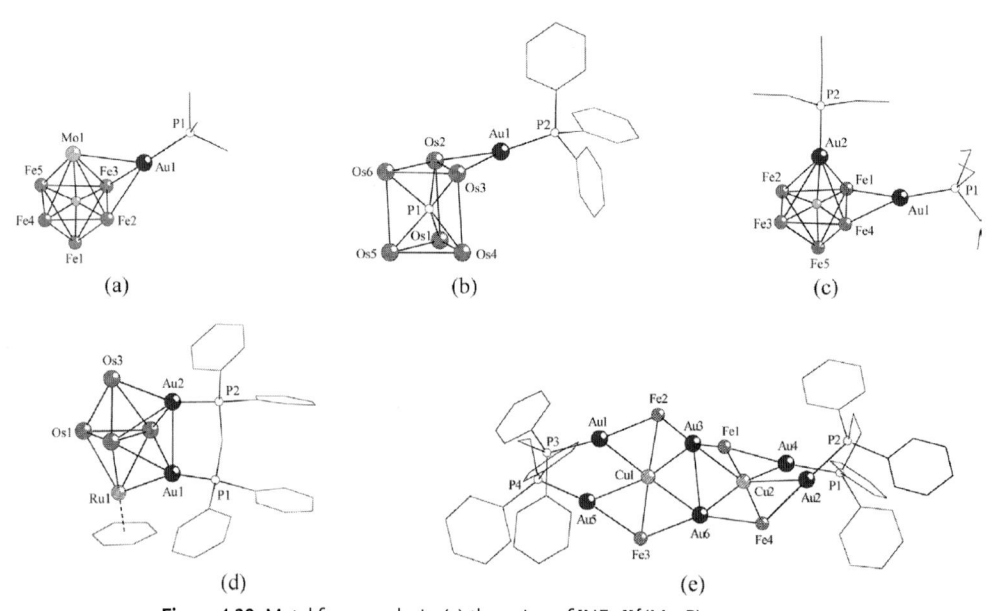

Figure 4.39 Metal frameworks in (a) the anion of [NEt$_4$][{(Me$_3$P)Au}MoFe$_5$(μ_6-C)(CO)$_{17}$] **147** [160], (b) (Ph$_3$P)AuOs$_6$(μ_6-P)(CO)$_{18}$ **253** [391], (c) [(Et$_3$P)Au]$_2$Fe$_5$(μ_6-C)(CO)$_{14}$ **254** [370], (d) (dppm)Au$_2$Os$_4$Ru(η^6-C$_6$H$_6$)(CO)$_{12}$ **265** [392], and (e) [(dppe)Au$_2$]$_2$Au$_2$Cu$_2$[Fe(CO)$_4$]$_4$ **282** [402] (for clarity, ligands on Fe, Ru and Os are not shown).

or neutral [μ-Au(dppb)Au][*trans*-Rh$_2$Ru$_4$(μ$_6$-B)(CO)$_{16}$]$_2$ **251** [379]. A gold atom edge-bridged to an Ru–Ru bond in B-centered BRu$_6$ and *cis*-BRh$_2$Ru$_4$ cores were found in some compounds [378, 380, 381] or in the cluster (Ph$_3$P)AuRu$_6$(μ$_6$-C)(μ-SePh)(CO)$_{15}$ **252** [Au–Ru 2.7253(6)/2.8515(6) Å] [382]. In contrast, compound (Ph$_3$P)AuOs$_6$(μ$_6$-P)(CO)$_{18}$ **254** contains a P-centered trigonal prismatic metallic core with one basal Os–Os edge bridged symmetrically by an Au(PPh$_3$) group [Au–Os 2.793(1)/2.775(1) Å] (Figure 4.39b) [391]. A related digold cluster, [(Et$_3$P)Au]$_2$Fe$_5$(μ$_6$-C)(CO)$_{14}$ **253**, contains a μ$_4$-Au(PEt$_3$) moiety as part of the C-centered metal framework [Au–Fe 2.828(3)–3.036(3) Å] and an additional Fe–Fe edge-bridged Au(PEt$_3$) group [Au–Fe 2.696(2)/2.701(3) Å] (Figure 4.39c) [370]. Other heptanuclear clusters based on a square pyramidal EM$_5$ core include systems in which (i) two isolated Au(PPh$_3$) groups are edge-bridged to opposite Os$_{apical}$–Os$_{basal}$ bonds as in [(Ph$_3$P)Au]$_2$Os$_5$(μ$_5$-C)(CO)$_{14}$ **255** [393]; (ii) gold atoms of an Au–Au fragment bridge the boron atom and opposite Ru–Ru and Ru–Rh bonds, respectively, of the basal plane as in (dppm)Au$_2$RhRu$_4$(μ$_7$-B)(CO)$_{14}$ **256** [379], or (iii) one gold atom of an Au–Au fragment is capping, while the other spans the basal M–M bond of a trigonal face as in (dppm)Au$_2$Fe$_5$(μ$_5$-C)(CO)$_{14}$ **257** [371] or (dppe)Au$_2$Ru$_5$(μ$_5$-C)(CO)$_{14}$ **258** [328]. The structures of clusters containing an angular Au–Au–Au fragment connected through all gold atoms to a butterfly BM$_4$ unit, for example, in [(Ph$_3$P)Au]$_3$Fe$_4$(μ$_7$-B)(CO)$_{12}$ **259** [372] and (ClAu)[(Ph$_3$P)Au]$_2$RhRu$_3$(μ$_6$-B)(μ-H)(η^5-Cp)(CO)$_9$ **260** [386], as well as an ionic cluster with a square planar Au$_4$(PPh$_3$)$_4$ fragment connected through three gold atoms to a face of a PFe$_3$ tetrahedron, that is [{(Ph$_3$P)Au}$_4$Fe$_3$(μ$_7$-P)(CO)$_9$][B{C$_6$H$_3$(CF$_3$)$_2$-3,5}$_4$] **261**, have also been described [373].

The cluster (Ph$_3$P)AuOs$_5$Ru(η^5-Cp)(CO)$_{15}$ **262** has a metal core resulting from a trigonal bipyramidal Os$_5$ with two faces including the same apical metal atom capped by a Au(PPh$_3$) unit and an organoruthenium moiety [392]. Similar overall metal frameworks are found in related heptanuclear clusters containing three gold atoms, that is, two Au(PPh$_3$) units capping AuRu$_2$ or AuRuCo faces of trigonal bipyramidal AuRu$_4$ [264, 387, 388] or AuCoRu$_3$ [389] cores with axial gold atom. By contrast, the gold atoms of an Au$_2$(dppm) ligand are μ$_3$-capping adjacent faces which include both axial atoms of the trigonal bipyramidal M$_5$ or Os$_4$Ru skeleton in (dppm)Au$_2$M$_5$(CO)$_{15}$ (M = Ru **263** [384], Os **264** [392]) or (dppm)Au$_2$Os$_4$Ru(η^6-C$_6$H$_6$)(CO)$_{12}$ **265** (Figure 4.39d) [392] to give a distorted, axial compressed pentagonal bipyramid with an Au–Au bond in the equatorial plane.

Uncommon patterns in which Au$_3$ fragments are connected to an M$_4$ core have also been described. The metal skeleton of (dppm)(Ph$_3$P)Au$_3$Ru$_4$(μ-H)(CO)$_{12}$ **266** may be described as a distorted Au$_2$Ru$_3$ square pyramid with two adjacent Ru$_3$ and AuRu$_2$ triangular faces capped by the metal atoms of an Ru(CO)$_3$ and an Au(PPh$_3$) unit [390], while in [(Ph$_3$P)Au]$_3$Os$_4$(μ-H)$_3$(CO)$_{11}$ **267** [327] two Au$_3$Os and Os$_4$ tetrahedrons are joined through a common osmium atom. An AuRu$_3$ tetrahedron with all three AuRu$_2$ faces capped by two Au(PPh$_3$) groups and a further AuBr unit were found in the cluster (BrAu)[(Ph$_3$P)Au]$_3$Ru$_3$(μ$_3$-CMe)(CO)$_9$ **268** [319].

Most octanuclear clusters are based on C-centered octahedral frameworks, contain two gold atoms and exhibit a variety of overall geometries [160, 318, 384, 395–397]. Two gold-phosphine units can bridge opposite Ru–Ru edges of the equatorial plane [318, 395], as in [(Ph$_3$P)Au]$_2$Ru$_6$(μ$_6$-C)(CO)$_{16}$ **269** [318], or non-adjcent Ru–Pt bonds in [(Ph$_3$P)Au]$_2$PtRu$_5$(μ$_6$-C)(CO)$_{15}$ **270** [396]. The gold atoms of a diphosphine-supported Au–Au bond can span adjacent Ru–Ru bonds [384, 396, 399], as in [PhCH$_2$N(CH$_2$PPh$_2$)$_2$]Au$_2$Ru$_6$(μ$_6$-C)(CO)$_{16}$ **271** [397], while in the anion of [NEt$_4$][{(dppm)Au$_2$}MoFe$_5$(μ$_6$-C)(CO)$_{17}$] **151** [160] an Fe–Fe edge is bridged by one gold atom, whereas the other Au is directly attached to the molybdenum center. Finally, one gold atom of a (R$_3$P)Au–Au(PR$_3$) system is μ$_3$-capping, while the other one bridges an Ru–Ru bond of the same triangular Ru$_3$ face in [(Et$_3$P)Au]$_2$Ru$_5$W(μ$_6$-C)(CO)$_{17}$ **272** [395], or both are μ$_3$-capping adjacent Ru$_3$ faces in the related [(Ph$_3$P)Au]$_2$Ru$_6$(μ$_6$-B)(μ-H)(CO)$_{16}$ **273** [380].

The cluster [(Ph$_3$P)Au]$_3$Ru$_5$(μ$_7$-B)(CO)$_{14}$ **274** contains a square-based pyramidal Ru$_5$ core with a triangular face capped by an Au(PPh$_3$) unit, while a (R$_3$P)Au–Au(PR$_3$) unit is positioned under the square face, with each of the gold atoms connected to boron and spanning opposite Ru–Ru bonds [398].

Octanuclear clusters with a higher number of gold atoms are [(Ph$_3$P)Au]$_4$Os$_4$(μ-H)$_2$(CO)$_{11}$ **275** [327] and [(dppm)Au$_2$]$_2$Os$_4$(μ-H)$_2$(CO)$_{11}$ **276** [400], which exhibit different core geometries. In the former compound the metal framework may be viewed as an Os$_4$ tetrahedron fused with an OsAu$_4$ trigonal bipyramid by sharing the osmium atom and further connected by an additional Au–Os bond. The metal core of **276** also consists of an Os$_4$ tetrahedron, one face of which is capped by three Au atoms while the fourth Au atom caps a resulting OsAu$_2$ face.

Clusters with nuclearity higher than eight are more common for ruthenium [369, 380, 399, 403, 404] and osmium [399, 405–414]. They include species based on a M$_{10}$ tetracapped C-centered octahedron. One Os–Os bond of the capping tetrahedra is bridged by a gold atom, as in [PPh$_3$Me][(Ph$_3$P)AuOs$_{10}$(μ$_6$-C)(CO)$_{24}$] **277** [405, 406], or a via a vertex Au$_4$ tetrahedron, as in [(Cy$_3$P)Au]$_3$AuOs$_{10}$(μ$_6$-C)(CO)$_{24}$ **278** [413]. In (dppm)Au$_2$Ru$_{10}$(μ$_6$-C)(CO)$_{24}$ **279** the Au atoms cap two adjacent, vertex sharing, triangular Ru$_3$ faces [369]. Ruthenium clusters based on B-centered octahedral frameworks with two adjacent trigonal faces capped by a gold atom and an additional Ru–Ru bond bridged by a third gold atom are known [380, 404], for example, [(Ph$_3$P)Au]$_3$Ru$_6$(μ$_6$-B)(CO)$_{16}$ **280** [380]. The structures of some other clusters containing Au$_2$Os$_7$ [407, 408], Au$_2$Os$_8$ [409, 410], Au$_2$Os$_9$ [411], Au$_4$Os$_6$ [412] and Au$_4$Os$_{10}$ [414] cores, as well as of some trimetallic species with Au$_2$Os$_6$Ru$_2$ [399] and AuRu$_3$Pt$_3$Ru$_3$ [403] frameworks, have also been reported.

Unusual clusters with Au–Fe bonds are [(dppe)Au$_2$]$_2$Ag$_4$[Fe(CO)$_4$]$_4$ **281** [401], [(dppe)Au$_2$]$_2$Au$_2$Cu$_2$[Fe(CO)$_4$]$_4$ **282** (Figure 4.39e) and [(dppe)Au$_2$]$_2$Au$_4$[Fe(CO)$_4$]$_4$ **283** [402]. They contain (dppe)Au$_2$ units with gold atoms bridging the bonds of Fe–M–Fe (M = Ag, Cu, Au) systems as well as naked gold atoms bridging Fe–Cu or Fe–Au bonds. The resulting metal atom framework can be described as a two-dimensional triangulated ribbon, twisted around the elongation direction.

Clusters which contain nacked gold atoms attached to a Group 8 metal atom are not very common, but a few structurally characterized compounds are available [304, 415–422]. One gold atom acts as a trimetallic triconnective ligand in [(Ph$_3$P)(OC)$_4$Os]AuOs$_3$(μ-Cl)(CO)$_{10}$ **284**, bridging one metal–metal bond of the Os$_3$ triangle [Au–Os 2.706(2)–2.746(2) Å] [420]. Planar AuM$_4$ fragments with a tetracoordinated gold atom spanned between two M–M bonds are found in the anions of [NBzMe$_3$]$_2$[Au

{Fe$_2$(CO)$_8$}$_2$]Cl **285** [Au–Fe 2.583(1)/2.607(1) Å] [415, 416] and [PPN][Au{Os$_3$(μ-H)(CO)$_{10}$}$_2$] **286** [Au–Os 2.806(1)/2.802(1) Å] [421]. In the neutral cluster AuOs$_2$(CO)$_8$[Os$_4$(μ$_5$-C)(μ-OMe)(CO)$_{12}$] **287** [422] the gold atom links an Os$_2$ fragment [Au–Os 2.664(3)/2.669(3) Å] with the wingtip metal atoms of a COs$_4$ butterfly fragment [Au–Os 2.839(3)/2.822(3) Å]. The gold atom bridges the M$_{wing}$–B edge from two BM$_4$ butterfly clusters to result in a *cis*- or *trans*-planar AuM$_2$B$_2$ core in [Au(PPh$_2$Me)$_2$][Au{Fe$_4$(μ$_5$-BH)(μ-H)(CO)$_{12}$}$_2$] **288** [304, 417] and [PPN][Au{Ru$_4$(μ$_5$-BH)(μ-H)(CO)$_{12}$}$_2$] **289** [304], respectively. An isosceles Au$_3$ triangle has two shorter edges spanned by Fe(CO)$_4$ groups in [NEt$_4$][{(dppm)Au$_2$}Au{Fe(CO)$_4$}$_2$] **290** [418], thus resulting in a naked gold atom in a trapezoidal planar arrangement [Fe–Au–Fe 172.7°]. A similar coordination geometry was found for each gold atom of the planar Au$_4$Fe$_4$ core of the anion in [NEt$_4$]$_4$[Au$_4${Fe(CO)$_4$}$_4$] **291** [419]. The cation of [{(dppm)Au$_2$}$_2$Au{Fe(CO)$_4$}$_2$](BF$_4$) **292** [418] has a metal framework consisting of a "bow tie" of gold atoms whose tips are bridged by dppm ligands, whereas the Fe(CO)$_4$ units triply bridge the central and two apical gold atoms, thus resulting in a hexacoordinated, naked central gold atom.

4.3.6
Gold–Group 9 Metal Compounds

Compounds with Au–metal bonds or interactions are known for all three metals of Group 9. Most dinuclear Au/Co species whose structure has been reported so far [172, 173, 423, 424] contain an unsupported Au–Co bond, for example (Ph$_3$P)AuCo(CO)$_3$(PPh$_3$) **293** with the shortest Au–Co bond of 2.450(1) Å [424]. Unsupported Au–Co bonds are also present in the linear anion of [PPN][Au{Co(CO)$_4$}$_2$] (**294**) [2.509(2) Å] [438], as well as in neutral clusters, for example, the tetranuclear (μ-cis-dpen)[AuCo(CO)$_4$]$_2$ **295** [2.482(7)/2.511(5) Å] [208], the hexanuclear [(OC)$_4$CoAu][(Ph$_3$P)Au]$_2$Co(CO)$_3$ **296**, with a Co(CO)$_4$ unit connected to an axial Au atom [2.536(2) Å] of the trigonal bipyramidal Au$_4$Co core [208], and the octanuclear [(Ph$_3$P)Au]$_4$[AuCo(CO)$_4$]$_2$ **297**, with Co(CO)$_4$ units connected to gold atoms [2.46(2) Å] of an edge-fused di-tetrahedron Au$_6$ core [436, 437]. Supported Au–Co bonds might be considered for the tetranuclear species (μ-dppm)$_2$Au$_2$Co$_2$(CO)$_6$ **298** [Au–Co 2.55(2)/2.45(2) Å], which contains a Co–Au···Au–Co skeleton as part of a dicyclic system with a common Au–Au bond (Figure 4.40a) [425, 426].

Dinuclear Au–Rh [439–441] and Au–Ir [247, 446, 461–469] compounds with both supported and unsupported Au–metal bonds have also been reported. An unsupported Au–Ir bond [2.607(2) Å] can also be considered for the linear, trinuclear complex [(Ph$_3$P)AuIr$_2$(dimen)$_4$(PPh$_3$)](PF$_6$)$_3$, in which an Au(PPh$_3$) unit occupies one of the axial sites of the Ir$_2$(dimen)$_4$ core, while a PPh$_3$ ligand is placed in the second axial position [470]. It was suggested that mono-hydrogen-bridged Au(μ-H)M bonds are longer than unbridged Au–M bonds [247, 463]. Indeed the Au–M bond is longer in [(Ph$_3$P)AuRh(μ-H){(2-Ph$_2$PCH$_2$CH$_2$)$_3$P}](BPh$_4$) **299** [Au–Rh 2.690(1) Å] than in [(Ph$_3$P)AuRh(H){(2-Ph$_2$PCH$_2$CH$_2$)$_3$P}](PF$_6$) **300** [Au–Rh 2.531(1) Å] [439], or in [(Ph$_3$P)AuIr(μ-H)(H)$_2$(PPh$_3$)$_3$](BF$_4$) **301** [Au–Ir 2.751(1) Å] [461, 462] and [(Ph$_3$P)AuIr(μ-H)$_2$(H){(2-Ph$_2$PCH$_2$)$_3$CMe}](PF$_6$) **302** [Au–Ir 2.6845(5) Å] [469] than in [(Ph$_3$P)AuIr(H)(CO)(PPh$_3$)$_3$](PF$_6$) **303** [Au–Ir 2.662(1) Å] [465] and [(Ph$_3$P)AuIr

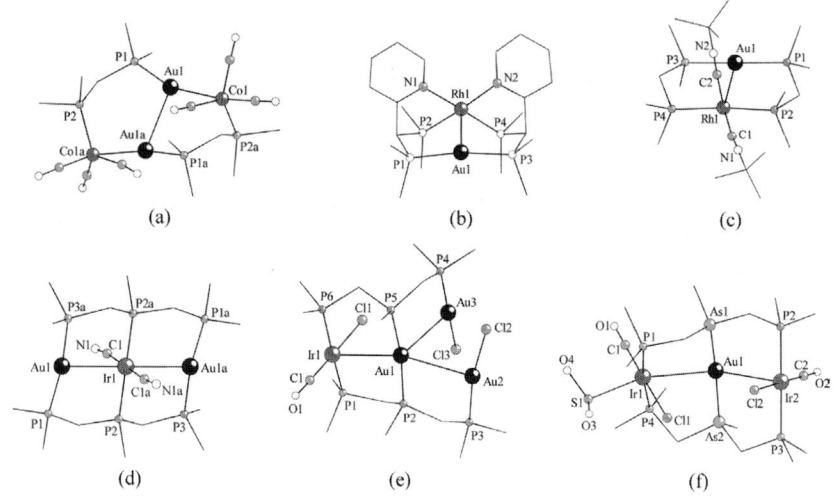

Figure 4.40 Metal frameworks in (a) (μ-dppm)$_2$Au$_2$Co$_2$(CO)$_6$ **298** [425], and the cation of (b) [AuRh(PNP)$_2$](BF$_4$)(NO$_3$) **305** [440], (c) the cation of [AuRh(dppm)$_2$(CN^{t}Bu)$_2$]Cl$_2$ **306** [488], (d) [Au$_2$Ir(μ$_3$-dpmp)$_2$(CN)$_2$](PF$_6$) **313** [476], (e) [(ClAu)$_2$AuIr(μ$_3$-dpmp)$_2$(CO)Cl](PF$_6$) **314** [476], and (f) [AuIr$_2$(μ$_3$-dpma)$_2$(CO)$_2$(SO$_2$)Cl$_2$](HSO$_4$) **316** [475] (for clarity, only *ipso* carbon of phenyl groups are shown for phosphine ligands).

(dppe)$_2$](BPh$_4$)$_2$ **304** [Au–Ir 2.625(1) Å] [464]. Examples of compounds with supported Au–M (M = Rh [440, 441, 488], Ir [466, 467]) are also known. The compound [AuRh(PNP)$_2$](BF$_4$)(NO$_3$) **305** was the first example of a supported, transannular Au–Rh interaction [2.850(2) Å] between a T-shaped gold atom (AuRhP$_2$ core, with *trans* phosphorus atoms) and a square pyramidal rhodium atom (RhAuP$_2$N$_2$ core, with *cis* phosphorus atoms) (Figure 4.40b) [440]. Other d^8–d^{10} compounds with supported Au–M interactions have also been reported; they are based on bridging diphosphine ligands, [AuRh(dppm)$_2$(CN^{t}Bu)$_2$]Cl$_2$ **306** [Au–Rh 2.9214(9) Å] (Figure 4.40c) [488] and [AuRh(tfepma)$_2$(CN^{t}Bu)$_2$]Cl$_2$ **307** [Au–Rh 2.8181(3) Å] [441] or [AuIr(dppm)$_2$(CNMe)$_n$](PF$_6$)$_2$ [Au–Ir 2.944(1) and 2.817(1) Å for n = 2 **308** and 3 **309**, respectively] [467]. Recently, oxidation of **307** afforded the first structurally characterized d^9–d^7 Au^{II}–Rh^{II} singly bonded metal complex [AuRh(tfepma)$_2$(CN^{t}Bu)$_2$Cl$_2$][AuCl$_2$] **310** [Au–Rh 2.6549(4) Å] [441]. Related systems are the trinuclear Au$_2$M species based on the tridentate dpma and dpmp ligands: [Au$_2$M(μ$_3$-dpma)$_2$(CO)Cl](PF$_6$)$_2$ (Au–Rh 3.006(2)–3.074(2) Å **311** [445]; Au–Ir 2.985(2)–3.025(2) Å **312** [472]), with bent Au-M-Au framework, and [Au$_2$Ir(μ$_3$-dpmp)$_2$(CN)$_2$](PF$_6$) **313** [Au–Ir 2.835 Å] (Figure 4.40d) [476], with a linear Au–Ir–Au skeleton. The latter was obtained from the tetranuclear species [(ClAu)$_2$AuIr(μ$_3$-dpmp)$_2$(CO)Cl](PF$_6$) **314**, the cation of which contains a roughly T-shaped, almost planar Au$_3$Ir core [Au–Ir 3.115(2) Å] with the metal atoms bridged by two dpmp, as shown in Figure 4.40e [476]. Bent Ir–Au–Ir frameworks were also found in [AuIr$_2$(μ$_3$-dpma)$_2$(CO)$_2$Cl$_2$](BPh$_4$) **315** [Au–Ir 3.012(1)/3.059(1) Å; Ir–Au–Ir 149.0(1)°] [473, 474] or the sulfur dioxide adduct [AuIr$_2$(μ$_3$-

dpma)$_2$(CO)$_2$(SO$_2$)Cl$_2$](HSO$_4$) **316** [Au–Ir 2.953(1)/3.132(1) Å; Ir–Au–Ir 153.2(1)°] (Figure 4.40f) [475]. Oxidation of the iridium centers resulted in shortening of the Au–Ir bonds [2.812(2)/2.806(2) Å] and an almost liniar Ir–Au–Ir [173.1(1)°] fragment in [AuIr$_2$(μ$_3$-dpma)$_2$(CO)$_2$Cl$_4$]Cl **317** [474].

Trinuclear species also include compounds with triangular metal cores, AuRh$_2$ [442, 443] and Au$_2$M (M = Rh [444], Ir [463, 471]), or open metallic skeletons, Rh–Au–Rh [444], Rh–Au–W [136] and Au–Ir–Au [469]. In the case of [(Ph$_3$P)AuRh$_2$(μ-Cl)$_2$(dmpi)$_4$](PF$_6$) **318** two isomers were obtained, which both contain dimer association of cluster cations through a weak Rh–Rh interaction, that is, 3.008(4) Å in the green modification **318a** and 3.262(1) Å in the red modification **318b**. The two cluster units adopt a *gauche*, staggered conformation in **318a** and a *trans*, eclipsed conformation, in **318b**, with respect to the vector of the Rh–Rh interaction [444].

Most tetranuclear clusters exhibit butterfly metallic cores, for example, Au(CoFe)Co [257], Au(Ir$_2$)Ir [478] and Au(AuIr)Re [209], or tetrahedral metallic cores, for example, AuIr$_3$ [477] and Au$_3$M (M = Co [166], Rh [216, 448], Ir [469]). An uncommon, planar Au$_3$Ir framework was found in the cation of [(Ph$_3$P)Au]$_3$Ir(PPh$_3$)$_2$(NO$_3$)](PF$_6$) **319**, with a transannular Au–Ir bond of 2.675 Å [477].

The common heterometallic framework in gold-containing pentanuclear clusters of Group 9 metals is the trigonal bipyramid, with a phosphine-gold unit in the axial position: AuRh$_4$ [449], Au(Co$_3$)M (M = Fe [427, 428], Ru [429–431]), Au(Rh$_3$)Ru [450], Au(Co$_2$Rh)Ru [432], Au(Rh$_2$M)M (M = Fe [297], Ru [310]), Au(Ir$_2$Fe)Fe [296], Au(MRu$_2$)Ru (M = Co [315], Rh [311]) and Au(Au$_2$M)Au (M = Co [433], Rh [451], Ir [481]). Some particular metal cores were also observed, that is (Ph$_3$P)AuFe$_3$Co(μ$_5$-C)(CO)$_{12}$ **320**, with a gold atom as a μ$_3$-Fe,C,Co bridge for the butterfly CFe$_3$Co fragment [Au–Co 2.873(1) Å] [307], or (Ph$_3$P)AuIr$_4$(μ-PPh$_2$)(CO)$_{10}$ **321**, with a gold atom spanning an Ir–Ir bond of the Ir$_4$ tetrahedron [Au–Ir 2.788(2)/2.731(2) Å] [479, 480].

The structures of few hexanuclear clusters have been reported. Compounds with one or two μ$_3$-Au(PR$_3$) units resulting in capped trigonal bipyramidal cores are common for all three metals, that is, Au$_2$Co$_2$Ru$_2$ [298], Au$_2$CoRu$_3$ [315], AuRh$_2$Ru$_3$ [344] or Au$_2$Ir$_4$ [484]. The metallic framework of the high-nuclearity gold cluster, [{(Ph$_3$P)Au}$_5$Rh(dmpi)$_3$](ClO$_4$)$_2$ **322**, was described as bicapped tetrahedral with two Au(PPh$_3$) groups capping two faces of an Au$_3$Rh tetrahedron, or, alternatively, as derived from a rhodium-centered Au$_{12}$ icosahedron [Au–Rh 2.669(4)–2.731(4) Å] by removal of a hemispherical bowl of seven Au atoms [454]. Unusual cores were found in some Rh clusters [352, 452, 453]. Thus the dication of [{(Ph$_3$P)Au}$_4$Rh$_2$(μ$_4$-O)$_2$(nbd)$_2$](BF$_4$)$_2$ **323** represents a rare example of a double oxygen centered cluster. It consists of a planar Rh(μ-O)$_2$Rh framework with two Au(PPh$_3$) coordinated to each oxygen atom, thus resulting in a trigonal pyramidal coordination geometry for the oxygen atoms. One gold atom from each P–Au–Au–P fragment also bridges the Rh–Rh segment (Au–Rh 3.020/2.984 Å) (Figure 4.41a) [453]. In the hexanuclear cation of [{(Ph$_3$P)Au}$_2$Rh$_4$(μ$_4$-PyS$_2$-2,6)$_2$(tfbb)$_2$](ClO$_4$)$_2$ **324**, the shortest Au–Rh separation of 3.063(1) Å was suggested to be indicative of an intermetallic interaction as a consequence of the metallophilic attraction between both closed-shell d^8–d^{10} metals [452].

Clusters with more than six metal atoms, with few exceptions, fit into some common structural patterns. Metallic frameworks based on a trigonal bipyramidal

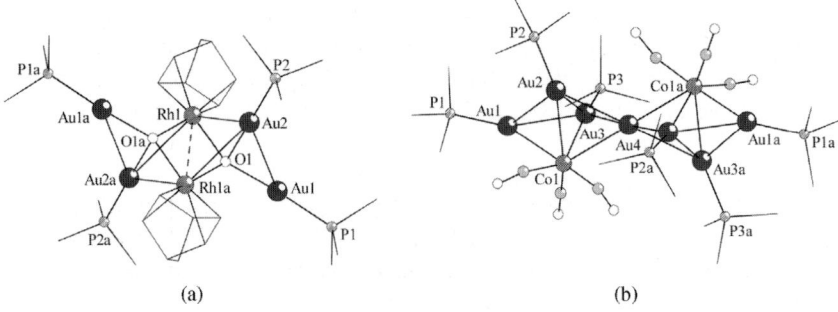

Figure 4.41 Metal frameworks in the cation of (a) [{(Ph$_3$P)Au}$_4$Rh$_2$(μ_4-O)$_2$(nbd)$_2$](BF$_4$)$_2$ **323** [453], and (b) [{(Ph$_3$P)Au}$_6$AuCo(CO)$_6$](NO$_3$) **327** [169] (for clarity, only *ipso* carbon of phenyl groups are shown for phosphine ligands).

core were found for some Co-containing compounds. In all cases reported the cobalt atom is in the equatorial plane. Two additional μ_3-Au(PPh$_3$) units are present in [(Ph$_3$P)Au]$_3$CoRu$_3$(CO)$_{12}$ **325** [389] and [{(Ph$_3$P)Au}$_6$Co(CO)$_2$](PF$_6$) **326** [435], while in [{(Ph$_3$P)Au}$_6$AuCo(CO)$_6$](NO$_3$) **327** two Au$_4$Co trigonal bipyramids share one axial Au atom [Au–Co 2.576(1)–2.656(1) Å] (Figure 4.41b) [169].

A common feature for some Au–Ir clusters is the presence of an octahedral Ir$_6$ core [485–487], with trigonal faces capped by metal atoms. Examples are [NBzMe$_3$][(Ph$_3$P)AuIr$_6$(μ-CO)$_3$(CO)$_{12}$] **328** [485] or [PPh$_4$][(Ph$_3$P)AuIr$_6$Ru$_3$(CO)$_{21}$] **329** [487], in the latter anion three additional faces being capped by Ru(CO)$_3$ groups. The related neutral cluster (Ph$_3$P)AuIr$_7$Ru$_3$(CO)$_{23}$ **330** also has an octahedral Ir$_6$ framework tetrahedrally capped by three Ru and an Ir atom, while an Au(PPh$_3$) group coordinates to the apical iridium atom [487].

A considerable number of clusters based on C-centered [161, 221, 434, 456, 457] or B-centered [376–379, 404] octahedral homo- and heterometal frameworks have been structurally characterized for all three Group 9 metals. Heptanuclear clusters include species with a μ_3-Au(PPh$_3$) group, for example [NEt$_4$][(Ph$_3$P)AuCo$_6$(μ_6-C)(μ-CO)$_5$(CO)$_8$] **331** [434], (Ph$_3$P)Au(*cis*-Rh$_2$)Ru$_4$(μ_6-B)(μ-CO)$_3$(CO)$_{11}$(μ_4-nbd) **332** [377] or (Ph$_3$P)Au(*trans*-Rh$_2$)Ru$_4$(μ_6-B)(μ-CO)$_4$(CO)$_{12}$ **333** [378], or a phosphine-gold unit spanning the Ir–Ir bond in (Cy$_3$P)Au(*cis*-Ir$_2$)Ru$_4$(μ_6-B)(μ-CO)(CO)$_{15}$ **334** [376]. Examples of octanuclear species with an Au–Au fragment connected to the octahedral core are [(Ph$_3$P)Au]$_2$Co$_6$(μ_6-C)(μ-CO)$_4$(CO)$_9$ **335** [434] or [(Ph$_3$P)Au]$_2$Rh$_6$(μ_6-C)(μ-CO)$_7$(CO)$_6$ **336** [456], while as nonanuclear clusters with two or three isolated phosphine-gold units the clusters [PPh$_4$][{(Ph$_3$P)Au}$_2$IrRe$_6$(μ_6-C)(CO)$_{20}$] **186** [221] and [(Cy$_3$P)Au]$_3$(*trans*-Rh$_2$)Ru$_4$(μ_6-B)(μ-CO)$_3$(CO)$_{12}$ **337** [404] can be considered.

For Au–Rh clusters some unique metallic frameworks have also been reported. They include (dppm)Au$_2$RhRu$_4$(μ_7-B)(CO)$_{14}$ **256**, with gold atoms of an Au–Au fragment bridging the boron atom and opposite Ru–Ru and Ru–Rh bonds of the basal plane [379], or [PPh$_4$][(Ph$_3$P)AuRh$_6$(μ_6-C)(μ-CO)$_9$(CO)$_6$] **338** and [(Ph$_3$P)Au]$_2$Rh$_6$(μ_6-C)(μ-CO)$_9$(CO)$_6$ **339**, which contain a C-centered trigonal prismatic metallic core with one or two basal Rh$_3$ faces capped by Au(PPh$_3$) groups [455]. Species with fused CRh$_6$

cores are [(Ph$_3$P)Au]$_4$Rh$_{10}$(μ_6-C)$_2$(μ-CO)$_8$(CO)$_{10}$ **340** and [(Ph$_3$P)Au]$_4$Rh$_{10}$(μ_6-C)$_2$(μ-CO)$_{12}$(CO)$_8$ **341**, which consist of two octahedrons sharing a Rh–Rh edge and four μ_3-Au(PPh$_3$) groups [459], or [NEt$_4$][(Ph$_3$P)AuRh$_{12}$(μ_6-C)$_2$(μ-CO)$_{11}$(CO)$_{12}$] **342**, with a layered aggregation of three trigonal prisms sharing rectangular faces, two additional Rh atoms capping contiguous rectangular faces and a μ_4-Au(PPh$_3$) fragment bridging a Rh$_4$ butterfly-shaped face of the Rh-atom skeleton [460].

The rich-gold clusters [{(Ph$_3$P)Au}$_7$Co(CO)$_2$](PF$_6$)$_2$ **343** [435], [{(Ph$_3$P)Au}$_7$Rh(CO)$_2$](NO$_3$)$_2$ **344** [454] and [{(Ph$_3$P)Au}$_6$(ClAu)$_2$Rh(dmpi)$_2$](PF$_6$) **345** [454, 458] may be described as derivatives of a metal-centered Au$_{12}$ icosahedron formed by removing five and seven vertices from the icosahedral cage.

4.3.7
Gold–Group 10 Metal Compounds

While the structures of only a few compounds containing Au–Ni bonds or interactions have been reported, for Au/Pd compounds and, especially, Au/Pt compounds the structural chemistry is quite rich and both neutral or ionic species have been characterized. In some few Au/Pt compounds metal–metal interactions resulted in supramolecular 1D polymers.

4.3.7.1 Gold–Nickel Compounds
In the dinuclear d^{10}–d^{10} complex [AuNi(dppm)$_2$(CNMe)$_2$]Cl **346** a separation of 3.015 Å was found between the metal centers bridged by dppm ligands, thus suggesting only week intramolecular Au–Ni interaction [489]. The supported Au–Ni interaction [2.8614(8) Å] was considerably strengthened in the d^8–d^{10} derivative [AuNi(dcpm)$_2$(CN)$_2$](ClO$_4$) **347** [490].

A few clusters with Au–Ni bonds have also been described. The trimetallic species, [(Ph$_3$P)Au]$_2$Pt$_3$(Pt$_{1-x}$Ni$_x$)(μ-CO)$_4$(CO)(PPh$_3$)$_3$ and [(Ph$_3$P)Au]$_2$Pt$_2$(Pt$_{2-y}$Ni$_y$)(μ-CO)$_4$(CO)$_2$(PPh$_3$)$_2$ (with octahedral metallic cores) [491] and [{(Ph$_3$P)Au}$_{10}$Au$_2$NiAg$_{12}$(μ-Cl)$_5$Cl$_2$](SbF$_6$) **349** (two Ni- and Au-centered Au$_6$Ag$_6$ icosahedra, sharing a common Au atom) [494], possess substitutional Ni/Pt and Ni/Au disorder, respectively. The dimetallic cluster [PPh$_3$Me]$_2$[Au$_6${Ni$_3$(μ-CO)$_3$(CO)$_3$}$_4$] **350** [Au–Ni 2.632(4)–2.757(4) Å] exhibits a central Au$_6$ octahedron with alternate Au$_3$ triangular faces capped by the four Ni$_3$ triangles [492, 493].

4.3.7.2 Gold–Palladium Compounds
Supported intramolecular d^8–d^{10} interactions are present in doubly bridged [AuPd(dcpm)$_2$(CN)$_2$]Cl **351** [Au–Pd 2.954(1) Å] [490] and monobridged [(Ph$_2$PCH$_2$SPh)AuCl]$_2$PdCl$_2$ **352** [Au–Pd 3.1418(8) Å] (Figure 4.42a) [495], the latter containing both T-shaped and linearly coordinated gold atoms. No further intermolecular metal–metal interactions are present in these compounds.

A few Au–Pd clusters with less than six metal atoms have been structurally characterized. They include the gold-capped palladium *triangulo*-cluster (Ph$_3$P)AuPd$_3$(μ-SO$_2$)(μ-N$_3$)(PPh$_3$)$_3$ **353** [Au–Pd 2.725(1)–2.757(1) Å] [496] and [(Ph$_3$P)Au]$_2$Pd$_2$(μ_3-Se)$_2$(SeH)$_2$(PPh$_3$)$_2$ **354** [497], in which Au atoms are bonded to the planar

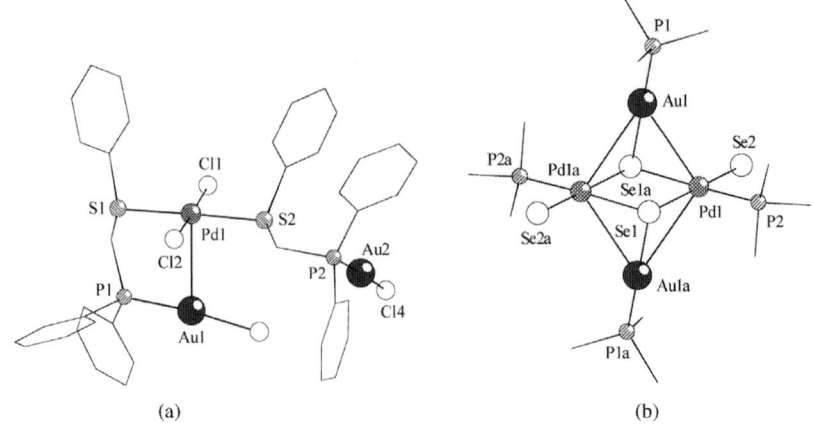

Figure 4.42 Metal frameworks in (a) [(Ph$_2$PCH$_2$SPh)AuCl]$_2$PdCl$_2$ **352** [495], and (b) [(Ph$_3$P)Au]$_2$Pd$_2$ (μ_3-Se)$_2$(SeH)$_2$(PPh$_3$)$_2$ **354** [497] (for clarity, only *ipso* carbon of phenyl groups are shown for phosphine ligands).

Pd$_2$Se$_2$ unit via a formal Se–Au single bond and establish weak d^8–d^{10} interactions [Au–Pd 3.067(2)/3.300(2) Å] (Figure 4.42b). Similar weak Au–Pd metallophilic contacts were recently observed in complexes [(Cy$_3$P)Au]$_2$Pd[S$_2$C=(t-Bu-fy)]$_2$ [3.0479 (1) Å] **355** and [{(Cy$_3$P)Au}$_2$Pd{S$_2$C=(t-Bu-fy)}(dbbpy)](ClO$_4$)$_2$ **356** [3.0658(3)/3.2670 (3) Å] [498].

The high nuclearity, mixed-metal clusters are quite well represented and most compounds fit into two main classes: (i) Pd-rich clusters derived from Pd- or Au-centered palladium polyhedra [499–503], and (ii) Au-rich clusters derived from Pd-centered Au$_{12}$ icosahedron [504–509]. Recent examples for the first type of clusters are the [(R$_3$P)Au]AuPd$_{21}$(μ_3-CO)$_4$(μ-CO)$_{16}$(PR$_3$)$_9$ [R = Me **357**, Et **358**], [PPh$_4$] [AuPd$_{22}$(μ_3-CO)$_4$(μ-CO)$_{16}$(PMe$_3$)$_6$(PPh$_3$)$_4$] **359** [500], Au$_2$Pd$_{41}$(μ_3-CO)$_{15}$(μ-CO)$_{12}$ (PEt$_3$)$_{15}$ **360** [501], Au$_4$Pd$_{28}$(μ_3-CO)$_9$(μ-CO)$_{13}$(PMe)$_{16}$ **361** [502] and Au$_4$Pd$_{32}$(μ_3-CO)$_8$(μ-CO)$_{20}$(PMe)$_{14}$ **362** [503]. Basic representatives for the second type of clusters are [(Ph$_3$P)Au]$_6$[(dppe)Au$_2$](AuCl)$_4$Pd **363** [504] and [(Ph$_3$P)Au]$_8$(AuCl)$_4$Pd **364** [505], which contains a complete (μ_{12}-Pd)Au$_{12}$ core. Several species whose bimetallic core can be described as fragments of the Pd-centered Au$_{12}$ icosahedron resulting from removing six, four or three vertices, are also reported, for example [{(Ph$_3$P)Au}$_6$Pd (PPh$_3$)](PF$_6$)$_2$ **365** [507], [{(Ph$_3$P)Au}$_8$Pd(CO)](NO$_3$)$_2$ **366** [508], or [(Ph$_3$P)Au]$_7$(AuI)$_2$ Pd(I) **367** [509].

Other geometries for Pd-centered Au$_8$ frameworks were found in [{(Ph$_3$P)Au}$_8$Pd] (NO$_3$)$_2$ **368** [510] or in its reaction product with metallic mercury, [{(Ph$_3$P)Au}$_8$Pd (Hg)$_2$](NO$_3$)$_2$ **12** (see Section 2.2, Figure 4.6c) [43], which both exhibit a Pd-centered square antiprism arrangement of the gold atoms.

4.3.7.3 Gold–Platinum Compounds

The structural chemistry of Au–Pt bond or interaction-containing compounds is by far better exemplified in the literature. Dinuclear compounds with the Au(μ-H)Pt

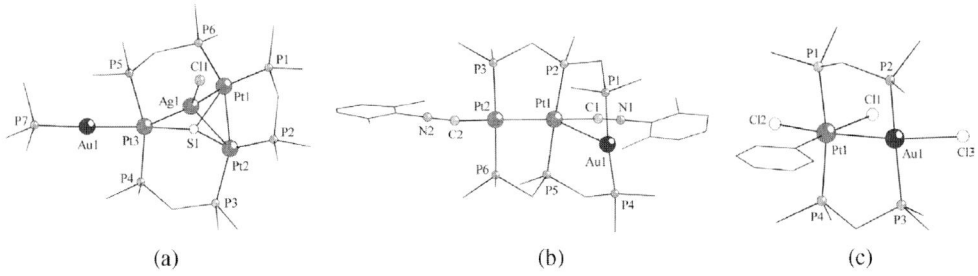

Figure 4.43 Metal frameworks in the cation of (a) [(Ph$_3$P)AuPt$_3$(μ_3-S)(μ_3-AgCl)(dppm)$_3$](PF$_6$) **371** [516], (b) [AuPt$_2$(dpmp)$_2$(dmpi)$_2$](PF$_6$)$_3$ **374** [525], and (c) [AuPt(dppm)$_2$(Ph)(Cl)$_3$](PF$_6$) **379** [522] (for clarity, only *ipso* carbon of phenyl groups are shown for phosphine ligands).

fragment [511, 512] exhibit a longer Au–Pt bond than in the absence of bridging hydride in compounds with Au(PPh$_3$) σ-bonded to Pt, for example [(Ph$_3$P)AuPt(μ-H)(C$_6$Cl$_5$)(PPh$_3$)$_2$](ClO$_4$) **369** [2.792(1) Å] [512] and [(Ph$_3$P)AuPt(PPh$_3$)$_3$](BF$_4$) **370** [2.6158(7) Å] [513]. The shortest, unsupported, Au–Pt bond distance was found in the cluster cation of [(Ph$_3$P)AuPt$_3$(μ_3-S)(μ_3-AgCl)(dppm)$_3$](PF$_6$) **371** [2.575(3) Å] (Figure 4.43a) [516]. No intermolecular metal–metal interactions are present in these compounds.

Several dinuclear compounds with supported intramolecular d^8–d^{10} interactions have been reported [490, 517–522] and most of them are ionic compounds, [AuPt(μ-L)$_2$R$_2$]X, with discrete cations containing Au and Pt atoms doubly bridged by dppm or dcpm ligands, liniar P–Au–P and square planar PtP$_2$R$_2$ fragments. The transannular Au–Pt distances range from 2.910(1) Å in [AuPt(dppm)$_2$(C≡CPh)$_2$](PF$_6$) **372** [519] to 3.082(1) Å in [AuPt(dppm)$_2$(C≡C^{t}Bu)(Cl)]Cl **373** [517]. An Au(I) ion is trapped by two uncoordinated phosphine units in *syn*-[Pt$_2$(dpmp)$_2$(dmpi)$_2$](PF$_6$)$_2$ to give the trinuclear [AuPt$_2$(dpmp)$_2$(dmpi)$_2$](PF$_6$)$_3$ **374**, with an angular Pt–Pt–Au skeleton [110.68(5)°] and a d^8–d^{10} interaction of 3.045(2) Å (Figure 4.43b) [525]. Similar weak Au–Pt metallophilic contacts were recently observed in complexes [(Cy$_3$P)Au]$_2$Pt[S$_2$C=(t-Bu-fy)]$_2$ **375** [3.0529(2) Å] and [{(Cy$_3$P)Au}Pt{S$_2$C=(t-Bu-fy)}(dbbpy)](ClO$_4$) **376** [3.3108(2) Å] [498]. A considerably longer Au–Pt interatomic distance of 3.446(1) Å was found in the neutral, trinuclear derivative (ClAu)$_2$Pt(dppm)$_2$(C≡C^{t}Bu)$_2$ **377**, which contains a *trans*-Pt(C≡C^{t}Bu)$_2$ and two AuCl fragments bridged by two dppm ligands [517]. By contrast, oxidation of the [AuPt(dppm)$_2$(Ph)(Cl)](PF$_6$) **378** [Au–Pt 2.9646(3) Å] resulted in the isolation of the first Au^{II}/Pt^{III} derivative [AuPt(dppm)$_2$(Ph)(Cl)$_3$](PF$_6$) **379** (Figure 4.43c), with a corresponding shortening of the intermetallic distance in the d^7–d^9 complex to 2.6457(3) Å [522].

The ylide complex Pt[Au{CH$_2$P(S)Ph$_2$}$_2$]$_2$ **380** exhibits a structure similar to that of the analogous Pb derivative (see Section 2.4.4), with a linear Au–Pt–Au skeleton, transannular d^8–d^{10} Au–Pt interaction of 3.034(1) Å, and extended one-dimensional chain polymer formation through weak intermolecular d^{10}–d^{10} interactions [Au–Au 3.246 Å] [34]. Oxidation of **380** or adduct formation resulted in discrete molecular

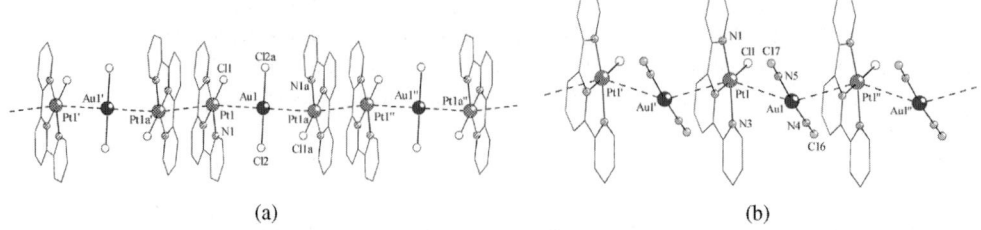

Figure 4.44 View of the 1D chain polymer in the crystal of (a) [{Pt(terpy)Cl}$_2${AuCl$_2$}][AuCl$_4$] **383**, and (b) [Pt(terpy)Cl][Au(CN)$_2$] **386** [523].

compounds with a linear Au–Pt–Au skeleton and stronger Au–Pt bonds, Pt[Au(Cl) {CH$_2$P(S)Ph$_2$}$_2$]$_2$ **381** [2.668(1)/2.662 (1) Å] [34], or Au–Pt interactions, Pt[Au(SO$_2$) {CH$_2$P(S)Ph$_2$}$_2$]$_2$ **382** [2.868(l) Å] [524], as a consequence of removal of electron density from the axial orbitals of these d^{10}–d^8–d^{10} compounds.

Recently, a strategy to obtain extended linear chains of unsupported metallophilic interactions was developed [523]. It is based on the use of polypyridyl-containing square-planar Pt cations, for example [(terpy)PtX]$^+$, and linear [AuX$_2$]$^-$ or planar [AuX$_4$]$^-$ anions. The crystal of [{Pt(terpy)Cl}$_2${AuCl$_2$}][AuCl$_4$] **383** contains a 1D polymeric chain of cations (Figure 4.44a), with an Au–Pt interaction of 3.284(1) Å in the cation unit and an inter-cation Pt–Pt distance of 3.447(1) Å. The Pt(1)–Au(1)–Pt (1a) angle is linear, while the Au(1)–Pt(1)–Pt(1a′) is close to linearity [170.58(3)°]. Similar chains of cations were found in the crystals of related [{Pt(terpy)Cl}$_2${AuBr$_2$}] [AuBr$_4$] **384** and [{Pt(terpy)Cl}$_2${AuCl$_2$}][AuCl$_2$] **385**. The [AuX$_4$]$^-$ anions in **383** and **384**, and half of the [AuCl$_2$]$^-$ anions in **385** maintain the charge balance but do not exhibit any metallophilic interactions. A different type of zigzag 1D polymeric chain was observed in the crystal of [Pt(terpy)Cl][Au(CN)$_2$] **386**. It consists of alternating cations and anions (Figure 4.44b), with an Au–Pt distance of 3.349(1) Å and a nonlinear Pt(1)–Au(1)–Pt(1a″) fragment [147.83(3)°]. A similar 1D polymeric chain was found in the crystal of the isomeric compound [Pt(terpy)CN][AuCl(CN)] **387**.

Both bimetallic and trimetallic clusters with trigonal AuPt$_2$ [526–533], Au$_2$Pt [217, 534] and AuPtW cores [131, 134] have been reported. Examples include ionic or neutral compounds, for example [AuPt$_2$(μ$_3$-S)(dppm)$_3$]Cl **388** [Au–Pt 3.259 (2)/3.113(2) Å] [531], (μ-XAu)Pt$_2$(dppm)$_2$(C≡CFc)$_2$ [X=Cl **389**: Au–Pt 2.6395(3)/ 2.6712(3) Å; X=Br **390**: Au–Pt 2.6786(4)/2.6413(4) Å], obtained by addition of d^{10} AuX to the Pt^I–Pt^I σ-bond in Pt$_2$(dppm)$_2$(C≡CFc)$_2$ [533], [{(Ph$_3$P)Au}$_2$Pt(Cl)(PEt$_3$)$_2$] (O$_3$SCF$_3$) **391** [Au–Pt 2.600(3)/2.601(4) Å] [534], or [(Me$_3$P)AuPtW(μ$_3$-Ctolyl-p) (CO)$_2$(η^5-Cp)(PMe$_3$)$_2$](PF$_6$) **392** [Au–Pt 2.956(2) Å] [131].

Ionic tetranuclear clusters with tetrahedral AuPt$_3$ framework are quite common [535–540], recent examples being [(Ph$_3$P)AuPt$_3$(dppm)$_3$](PF$_6$)$_{0.75}$(BF$_4$)$_{0.25}$ **393** [539] or [1,4-C$_6$H$_4$(CH$_2$PPh$_2$)$_2$Au$_2${Pt$_3$(μ-CO)$_3$(PCy$_3$)$_3$}$_2$](PF$_6$)$_2$ **394** [Au–Pt 2.728 (1)–2.767(1) Å] [540]. An example of a cluster with an Au$_3$Pt core is [{(Ph$_3$P)Au}$_3$Pt (triphos)]Cl **395** [Au–Pt 2.630(1) Å] [541]. A few compounds with butterfly Au$_2$Pt$_2$, folded about the Au–Au bond, [{(Ph$_3$P)Au}$_2$Pt$_2$(dmpi)$_4$(PPh$_3$)$_4$](PF$_6$)$_2$ **396** [Au–Pt 2.718 (2)–3.028(2) Å] [542, 543], Au$_2$PtFe, folded about the Pt–Fe bond [258], or even planar

Au$_2$Pt$_2$ core, with a transannular Au–Au bond, [{(Ph$_3$P)Au}$_2$Pt$_2$(dimen)$_2$(PPh$_3$)$_2$](PF$_6$)$_2$ **397** [Au–Pt 2.8082(5)/2.84229(5) Å] [544], are also known. Unusual tetrametallic cores were reported for some compounds, for example [{(Ph$_3$P)Au}$_2$Pt$_2$(μ-PPh$_2$)$_2$(PPh$_3$)$_2$](PF$_6$)$_2$ **398**, with a Au$_2$Pt$_2$P$_6$ "hammock" skeleton resulting from the "external" addition of Au(PPh$_3$) units to opposite Pt–P bonds of a planar Pt$_2$(μ-P)$_2$ core [545], or [{(Ph$_3$P)Au}Pt{W(η^5-Cp)(CO)$_2$}$_2$(μ-PPh$_2$)$_2$(μ-CO)](PF$_6$) **399**, with an Au(PPh$_3$) unit spanning one Pt–W bond of a linear W–Pt–W system [159].

All pentanuclear Au–Pt clusters whose structure has been reported so far exhibit a trigonal bipyramidal skeleton with Au–phosphine units μ$_3$-bridging a Pt$_3$ triangle [546, 547], as in [{(Me$_3$P)Au}$_2$Pt$_3$(μ-dppm)$_3$](PF$_6$)$_2$ **400** [547], or an Au$_4$Pt core with an equatorial Pt atom, as in [{(Ph$_3$P)Au}$_4$Pt(dppe)](PF$_6$)$_2$ **401** [548]. Octahedral Au$_2$Pt$_3$(Pt$_{1-x}$Ni$_x$) and Au$_2$Pt$_2$(Pt$_{2-y}$Ni$_y$) cores, with substitutional Ni/Pt disorder (see Section 3.7.1) have also been described [491]. The AuPt$_6$ core consisting of a μ$_6$-Au atom sandwiched between two *triangulo* Pt$_3$ rings was found in [Au{Pt$_3$(μ-CO)$_3$(PPh$_3$)$_3$}$_2$](PF$_6$) **402** [549] and [Au{Pt$_3$(μ-CO)$_3$(PMe$_3$)$_4$}$_2$]Cl **403** [550]. Some Au/Pt clusters with unique metallic frameworks have also been reported. They include [(Ph$_3$P)Au]$_2$PtRu$_5$(μ$_6$-C)(CO)$_{15}$ **270** (gold–phosphine units bridging non-adjacent Ru–Pt bonds of a C-centered CPtRu$_5$ octahedron) [396], [{(^{i}PrP)Au}$_2$Pt$_6$(μ-CO)$_6$(dppm)$_3$](Cl)(PF$_6$) **404** (a distorted trigonal prismatic Pt$_6$ core with each triangular face capped by a μ$_3$-Au-phosphine group), [μ$_4$-(Ph$_3$P)Au]$_2$Pt$_7$(μ-CO)$_8$(PPh$_3$)$_4$ **405** [552], and [(Et$_3$P)Au]$_2$Pt$_3$Ru$_6$(μ$_3$-H)$_2$ (CO)$_{21}$ **406** (two fused Pt$_3$Ru$_3$ octahedra with a common Pt$_3$ face and two μ$_3$-Au-phosphine units bridging PtRu$_2$ faces on opposite sides of the Pt$_3$ triangle) [403].

The cation of [{(Ph$_3$P)Au}$_6$Pt(μ-C≡C^{t}Bu)(PPh$_3$)][Au(C≡C^{t}Bu)$_2$] **407** exhibits a central core resulting from two PtAu$_4$ square pyramids fused about a common PtAu$_2$ triangular face [553]. Alternatively, it might be described as derived from a distorted Pt-centered Au$_8$ cube with two missing vertices. A similar metallic framework, that is, Pt-centered Au$_8$ cube with one missing vertex, was found in the related [{(Ph$_3$P)Au}$_7$Pt(H)(PPh$_3$)](PF$_6$)$_2$ **408** [554]. Trimetallic Hg–Au–Pt clusters based on a Pt-centered Au$_8$ square antiprism were also reported (see Section 2.2) [41–44].

As for palladium, the structures of several Au-rich clusters have been reported. For most of them the metallic frameworks can be derived from a Pt-centered Au$_{12}$ icosahedron by removing seven [41], six [555, 556], four [556, 558], three [548, 559–561], or two [556] vertices. Clusters with similar cores in which up to three gold atoms are replaced by atoms of another metal have also been reported, for example, Pt–Au$_5$Pt$_2$ [550], Pt–Au$_6$Hg [42], Pt–Au$_5$Hg$_2$ [41], Pt–Au$_8$Ag [556, 562, 563], Pt–Au$_8$Ag$_2$ [563], Pt–Au$_8$Cu [556, 564] and Pt–Au$_8$Cu$_2$ [564]. In another few compounds additional metal atoms are connected to an M$_3$ face of trimetallic frameworks derived from Pt-centered icosahedra, for example, Cu(Pt-Au$_6$Cu$_3$) [565] and Ag$_2$(Pt-Au$_6$Ag$_6$) [566]. Clusters with a toroidal geometry of the metal core are also known, for example [{(Ph$_3$P)Au}$_8$Pt(CuCl)](NO$_3$)$_2$ **409** [571], [{(Et$_3$P)Au}$_9$Pt](PF$_6$)$_3$ **410** [572] or [{(Ph$_3$P)Au}$_8$Pt(AgNO$_3$)](NO$_3$)$_2$ **410a** [562].

Recently, the first nanosized Au/Pt cluster, [(Ph$_3$P)$_2$Au$_2$]$_2$Pt$_{13}$(CO)$_{10}$(PPh$_3$)$_4$ **411**, was reported; it contains a Pt-centered Pt$_{12}$ icosahedral cage capped by two (Ph$_3$P)Au-Au(PPh$_3$) units [567]. Finally, high nuclearity Au–Pt–Ag clusters, with two metal-centered icosahedra sharing a vertex, have also been described [494, 568–570].

4.3.8
Gold–Group 11 Metal Compounds

Heterometallic Au–Cu and Au–Ag compounds include cluster species but also compounds based on heterometallophilic Au–Cu and Au–Ag interactions. Both theoretical studies and experimental data suggested that metallophilic M(I)–M(I) attraction is present in homometallic dimers of all three coinage metals as a correlation effect and such an interaction is strengthened by the relativistic effects for gold [24, 573–575]. Recently, besides heterometallic Au–Cu and Au–Ag cluster species, compounds with heterometallophilic Au–Cu and Au–Ag interactions have attracted great attention due to their theoretical interest and potential applications. Theoretical calculations were consistent with the fact that the presence of only one gold atom is enough to induce metallophilic attractions in the group congeners and the ligand on gold atom can be used to modulate this effect [576]. A considerable similarity is revealed in the structural chemistry of compounds with Au–Cu and Au–Ag bonds and interactions. However there are many Au–Ag species without analogy with the Au–Cu compounds characterized so far.

4.3.8.1 Gold–Copper Compounds

Compounds with unsupported Au–Cu interaction are rare. The direct reaction between basic $[AuR_2]^-$ (R = perhalophenyl) complexes and M(I) inorganic salts, successfully used to prepare gold(I) compounds with Ag(I) or Tl(I), failed to produce Au(I)–Cu(I) complexes [577]. The unique organometallic chain polymer $[Cu(NCMe)\{Au(C_6F_5)_2\}(\mu_2\text{-}C_4H_4N_2)]_n$ **412** was prepared through a transmetallation reaction between $[Au_2Ag_2(C_6F_5)_4(NCMe)_2]_n$ and CuCl, in the presence of the pyrimidine ligand. It displays isolated dinuclear units with an unsupported, short Au(I)–Cu(I) interaction [2.8216(6) Å], joined together by bridging pyrimidine ligands that bind two Cu(I) ions (Figure 4.45a) [577]. The same transmetallation reaction, in the absence of the pyrimidine, afforded the isolation of $[Cu_2(NCMe)_2\{Au(C_6F_5)_2\}_2]_n$ **413** which consists of planar, tetranuclear Au_2Cu_2 units associated into a

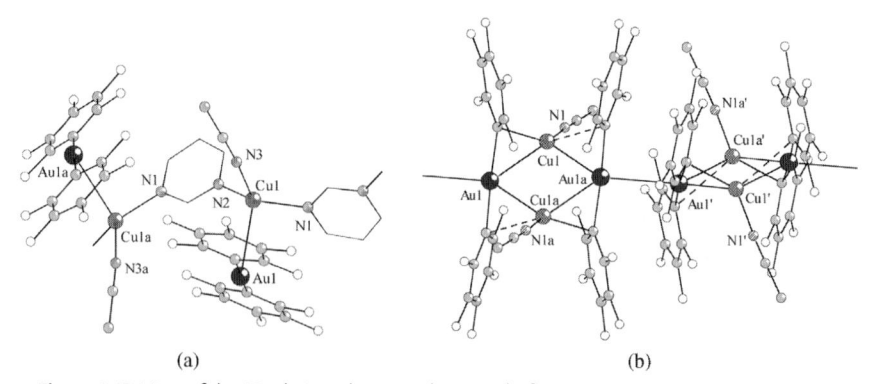

Figure 4.45 View of the 1D chain polymer in the crystal of (a) $[Cu(NCMe)\{Au(C_6F_5)_2\}(\mu_2\text{-}C_4H_4N_2)]_n$ **412** [577], and (b) $[Cu_2(NCMe)_2\{Au(C_6F_5)_2\}_2]_n$ **413** [578].

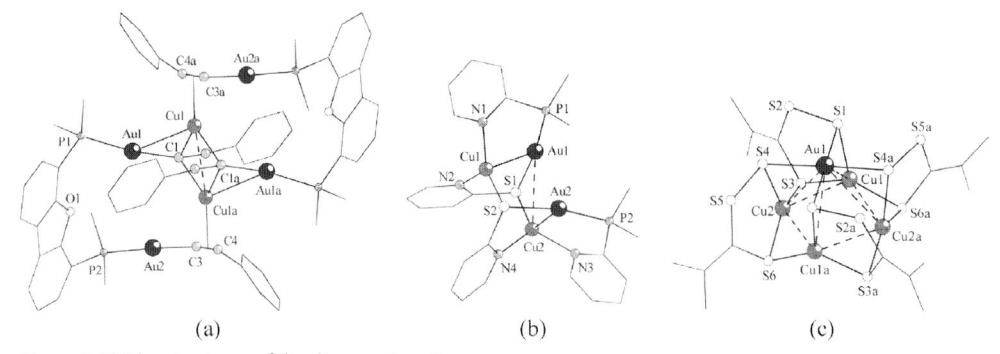

Figure 4.46 The structures of the dimer cations in
(a) [(μ-dbfphos){Au(η^2-C≡CPh)}$_2$Cu]$_2$(PF$_6$)$_2$ **414** [579],
(b) [AuCu(μ-Spy)(μ-PPh$_2$py)]$_2$(PF$_6$)$_2$ **417** [583] (for clarity, only *ipso* carbon of phenyl groups are shown for phosphine ligands), and
(c) the structure of the anion in [NBu$_4$]$_2$[AuCu$_4$(S^{t}BuDED)$_4$] **418** [585] (oxygen atoms and the *tert*-Bu groups of the thio ligands are not shown).

monodimensional polymer through short aurophilic contacts [Au(1a)–Au(1′) 2.9129 (3) Å] (Figure 4.45b) [578]. The Au–Cu distances within a tetranuclear unit [Au–Cu 2.5741(6)/2.5876(5) Å] are considerably shorter than those observed in **412** and even shorter than found in some cluster species, for example [(Ph$_3$P)Au][(Ph$_3$P)Cu]Re$_2$(μ-PCy$_2$)(CO)$_7$(C≡CPh) **167** [Au–Cu 2.644(2) Å] [201].

Another strategy to obtain Au(I)–Cu(I) interactions is the use of organometallic gold (I)–alkynyl complexes able to trap an additional metal ion through π-alkynyl coordination. Thus, treatment of the dinuclear (μ-dbfphos)[Au(C≡CPh)]$_2$ with [Cu(NCMe)$_4$](PF$_6$) resulted in the dimeric species [(μ-dbfphos){Au(η^2-C≡CPh)}$_2$Cu]$_2$(PF$_6$)$_2$ **414** (Figure 4.46a), which displays an unexpected hexanuclear Au$_4$Cu$_2$ core held through Au(1)–Cu(1) [2.852(2) Å] and Cu(1)–Cu(1a) [2.898(3) Å] contacts. In addition, the alkynyl groups contribute to the stabilization of the whole structure, one bridging the Cu(I) atoms of the dimer, the other coordinating only to one copper center [579]. Ionic complexes containing pentanuclear [{Au(η^2-C≡CR)$_2$}$_3$Cu$_2$]$^-$ (R = Ph [580, 582], *p*-tolyl [581]) were also reported, for example [NBu$_4$][{Au(η^2-C≡CPh)$_2$}$_3$Cu$_2$]. For example, reaction of [NBu$_4$][Au(C≡CPh)$_2$] with [Au(C≡CPh)]$_n$ and [Cu(C≡CPh)]$_n$ affords isolation of [NBu$_4$][{Au(η^2-C≡CPh)$_2$}$_3$Cu$_2$] **415**. The structure of the anion revealed a trigonal bipyramidal arrangement of metal atoms. The gold atoms of three linear [Au(C≡CR)$_2$]$^-$ anions described the triangular base, with no Au–Au contacts (3.44–3.53 Å), while each Cu(I) atom from the apical positions is π-coordinated by three alkynyl ligands and establishes contacts with the gold atoms of the basal plane [Au–Cu 2.783–3.016(3) Å] [580]. A particular case is compound [PPN]$_2$[{Au(η^2-C≡CPh)$_2$}$_3$Cu$_2$] [{Au(η^2-C≡CPh)$_2$}$_3$Ag$_2$] **416**, which contains two similar Cu- and Ag-containing anions [582].

Treatment of the mononuclear complex Au(Spy)(PPh$_2$py) with [Cu(NCMe)$_4$](PF$_6$) resulted in the isolation of the heterobimetallic system [AuCu(μ-Spy)(μ-PPh$_2$py)]$_2$(PF$_6$)$_2$ **417**. Its tetranuclear cation can be considered as a dimer resulting

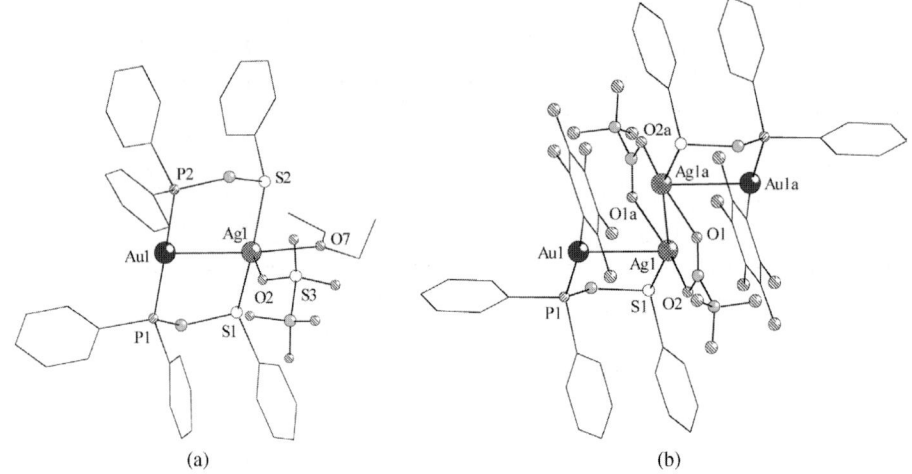

Figure 4.47 The structures of (a) the cation in [AuAg(Ph$_2$PCH$_2$SPh)$_2$(OEt$_2$)](O$_3$SCF$_3$)$_2$ **425**, and (b) [(C$_6$F$_5$)Au]$_2$Ag$_2$(μ-Ph$_2$PCH$_2$SPh)$_2$(μ-O$_2$CCF$_3$)$_2$ **426a** [576].

from the bridging nature of the sulfur atom from the 2-pyridinethiolato ligand. In addition to strong supported d^{10}–d^{10} interactions [Au(1)–Cu(1) 2.634(1), Au(2)–Cu(2) 2.646(1) Å], a weaker Au(1)–Cu(2) interaction [3.108(1) Å] is also present (Figure 4.46b) [583].

The affinity of gold and copper for sulfur ligands and the ability of sulfur to act as a bridging atom between metallic centers was used to prepare heterometallic thiolates in which Au–Cu interactions are present. Thus, treatment of the homonuclear Na$_2$[Cu$_4$(SCH$_2$CH$_2$S)$_3$] with [NEt$_4$][AuBr$_2$] and [PPh$_4$]Br resulted in the isolation of the ionic complex [PPh$_4$]$_6$[Au$_2$Cu$_4$(SCH$_2$CH$_2$S)$_4$]$_2$[Au$_3$Cu$_3$(SCH$_2$CH$_2$S)$_4$]. The disordered heterometallic frameworks in the anions are based on linear S–Au–S and trigonal CuS$_3$ units and metal–metal interactions are present [584]. The structures of two mixed-valence, mixed-coinage metal clusters based on dithiolato ligands have also been described. The compound [NBu$_4$]$_2$[AuCu$_4$(S^{t}BuDED)$_4$](CuCl$_2$) **418** was obtained when the homonuclear [NBu$_4$]$_4$[Cu$_8$(S^{t}BuDED)$_6$] was treated with HAuCl$_4$. The heterometallic anion contains a rectangular pyramidal AuCu$_4$ core stabilized by trimetallic tetraconnective thio ligands. The apical position is occupied by a gold(III) atom four coordinate in a planar arrangement by sulfur atoms from different ligands, and exhibiting weak Au^{III}–Cu^I interactions [Au(1)–Cu(1) 3.051(3), Au(1)–Cu(2) 3.087(3) Å] (Figure 4.46c) [585]. Recently, the structure of the Cu-rich cluster, [AuCu$_8$(μ-dppm)$_3$(tdt)$_5$](ClO$_4$)$_{0.5}$(SbF$_6$)$_{0.5}$ **419**, isolated from a reaction mixture obtained from [NBu$_4$][Au(tdt)$_2$], [Cu$_2$(dppm)$_2$(MeCN)$_2$](ClO$_4$)$_2$ and Na(SbF$_6$), was described [586]. The complex cation in **419** may be regarded as resulting from the attachment of an [Au^{III}(tdt)$_2$]$^-$ anion to the octanuclear [Cu$^I_8$(μ-dppm)$_3$(tdt)$_3$]$^{2+}$ cation through four Au–S–Cu bridges in *syn* conformation. The shortest Au–Cu distances within the cluster cation are 3.216(4) and 3.376(4) Å, which indicates weak Au^{III}–Cu^I interactions.

Trimetallic clusters with tetrahedral AuRe$_2$Cu [201, 202] or capped-trigonal bipyramidal AuRu$_4$Cu [345, 346] skeletons have also been reported. Examples are [(Ph$_3$P)Au][(Ph$_3$P)Cu]Re$_2$(μ-PCy$_2$)(CO)$_7$(C≡CPh) **167** [Au–Cu 2.644(2) Å] [201] or (μ-dppe)AuRu$_4$Cu(μ$_3$-H)$_2$(CO)$_{12}$ **231** [Au–Cu 2.614(2) Å] [345].

Gold-rich clusters include the unique Au-centered icosahedral species [{(Ph$_2$MeP)Au}$_8$Au(CuCl)$_4$](C$_2$B$_9$H$_{12}$) **420** [587]. In the cation of **420** a difference was observed in the lengths of the heterometallic bonds, the radial Au–Cu bonds [2.564(2) – 2.605(2) Å] being significantly shorter than the tangential ones [2.706(2)–2.906(2) Å] [587]. Several other clusters with lower nuclearity display metallic frameworks which can be derived from a Pt-centered icosahedron by removing three or two vertices [556, 564, 565]. They include species with Pt–Au$_8$Cu [556, 564] and Pt–Au$_8$Cu$_2$ cores [564], for example, [(dppp)$_4$Au$_8$Pt(PPh$_3$)(CuCl)](NO$_3$)$_2$ **421** [Au–Cu 2.690(2)–2.830(2) Å] [556] or [{(Ph$_3$P)Au}$_8$Pt(μ$_3$-H)(CuCl)$_2$](NO$_3$) **422** [Au–Cu 2.636(5)–2.887(5) Å] [564]. In another two compounds an additional Cu atom is μ$_3$-connected to a Cu$_3$ face of the trimetallic framework derived from the Pt-centered icosahedron, [{(Ph$_3$P)Au}$_6$Pt(PPh$_3$)(CuCl)$_3$\{μ$_3$-Cu(PPh$_3$)\}](NO$_3$) **423** [Au–Cu 2.632(3)–2.815(3) Å] or [{(Ph$_3$P)Au}$_6$Pt(PPh$_3$)(CuI)$_3$(μ$_3$-Cu)](NO$_3$) **424** [Au–Cu 2.695(10)–2.854(9) Å] [565]. The compound [{(Ph$_3$P)Au}$_8$Pt(CuCl)](NO$_3$)$_2$ **409** [Au–Cu 2.629(4)–2.682(4) Å] [571] exhibits a toroidal geometry of the metal core.

An unusual trimetallic cluster is [(dppe)Au$_2$]$_2$Au$_2$Cu$_2$[Fe(CO)$_4$]$_4$ **282** (see also Section 4.3.5) [402]. In the two-dimensional, triangulated ribbon-like metal framework the Au$_{naked}$–Cu bonds [2.629(4)–2.682(4) Å] are shorter than the Au$_{phosphine}$–Cu bonds [2.717(3)–2.742(2) Å].

4.3.8.2 Gold–Silver Compounds

Although there is an analogy in many aspects between the structural chemistry of heterometallic Au–Ag and Au–Cu compounds, the interest shown in recent years has resulted in a considerable increase in the structural diversity for the gold–silver species resulting from the different strategies used to obtain compounds with a pre-established design, but also due to many unexpected compounds.

Dinuclear derivatives with gold and silver atoms are not common compounds due to their tendency to associate, through metallophilic d^{10}–d^{10} interactions, into polymers. However, in a few cases, the use of appropriate synthetic strategies and bridging ligands resulted in monomeric species which contain supported Au–Ag interactions. The first reported example was [AuAg(PPh$_2$py)$_2$](ClO$_4$)$_2$ **424** in which the gold atom is bonded to both phosphorus atoms, whereas the silver center is bonded to the two sulfur atoms of the bidentate ligands [588]. The two perchlorate anions are each coordinated to silver through one of the oxygen atoms and the mixed-metal dication displays an Au–Ag distance of 2.820(1) Å, corresponding to a d^{10}–d^{10} bonding interaction. A weaker Au–Ag interaction was found in the dication of the related [AuAg(Ph$_2$PCH$_2$SPh)$_2$(OEt$_2$)](O$_3$SCF$_3$)$_2$ **425** [2.9314(5) Å] (Figure 4.47a), obtained from [Au(Ph$_2$PCH$_2$SPh)$_2$](O$_3$SCF$_3$) and AgO$_3$SCF$_3$ [576], or in the neutral AuAg[CH$_2$P(S)Ph$_2$]$_2$ **426** [2.912(1) Å], resulting after treatment of [PPN][Au\{CH$_2$P(S)Ph$_2$\}$_2$] with AgNO$_3$ [589]. In both cases a silver atom is trapped by the pendant sulfur atoms of the ligands attached to gold. The mixed-metal units in **426** are arranged in the crystal to form an infinite linear chain with alternating gold and silver atoms at intermolecular distances of 3.635 Å, which, although long, were assigned to weak metal–metal interactions [589].

When the starting gold complex contains only one Ph_2PCH_2SPh ligand coordinated to the metal through its phosphorus atom, that is $(C_6F_5)Au(Ph_2PCH_2SPh)$, the resulting heterometallic species depends on the nature of the counterion of the silver salt and the molar ratio used. Thus, reaction with AgO_2CCF_3, in a 1:1 or 2:1 molar ratio, produces a tetranuclear species, $[(C_6F_5)Au]_2Ag_2(\mu\text{-}Ph_2PCH_2SPh)_2(\mu\text{-}O_2CCF_3)_2$ **426a**, with two gold and two silver atoms forming an Au–Ag–Ag–Au zigzag chain. The gold and silver atoms are bridged by the phosphorus ligand to afford an Au–Ag interaction of 3.0335(8) Å, while the two central silver atoms are bridged by the trifluoroacetate ligands [Ag–Ag 2.8155(9) Å] (Figure 4.47b) [576]. The complex $[\{(4\text{-}Me_2NC_6H_4)Ph_2P\}Au]_2Ag_2(\mu\text{-}O_2CC_2F_5)_4$ **427**, obtained from $Au[PPh_2(C_6H_4NMe_2\text{-}4)]Cl$ and $Ag(O_2CC_2F_5)$, exhibits a similar structure [Au–Ag 3.0253(3), Ag–Ag 2.9439(6) Å] [590].

Recently, Catalano and coworkers reported a useful strategy to prepare compounds with supported Au–Ag interactions, that is, Au(I)–Ag(I) N-heterocyclic carbene coordination species based on N-substituted imidazole backbones (Figure 4.48) [591–594]. Generally, starting materials of the type $[Au(L)_2](BF_4)$, in which the gold(I) center is bound to the carbene portion of the ligand, were reacted with $AgBF_4$, resulting in species with Ag(I) ions coordinated to the nitrogen atoms of the pyridyl groups. The nuclearity of the isolated heterometallic compounds, as well as the degree of association, was dependent on the nature of the carbene ligand.

Treatment of $[Au\{CH_3im(CH_2py)\}_2](BF_4)$ with one equivalent of $AgBF_4$ produced the dinuclear species $[AuAg\{CH_3im(CH_2py)\}_2](BF_4)_2$ **428**. Both dangling pyridyl groups coordinate the same Ag(I) center and are slightly hinged at the methylene linkages to allow a weak transannular Au(I)–Ag(I) interaction [3.0318(5) Å] [592]. A trimetallic species, $[AuAg_2\{(pyCH_2)_2im\}_2(NCMe)_2](BF_4)_3$ **429**, was isolated when $[Au\{(pyCH_2)_2im\}_2](BF_4)$ was treated with two equivalents of $AgBF_4$. The cation contains two silver ions, each coordinated to the pyridine moieties on the same carbene ligand in a *trans* fashion and to an acetonitrile molecule (T-shaped AgN_3

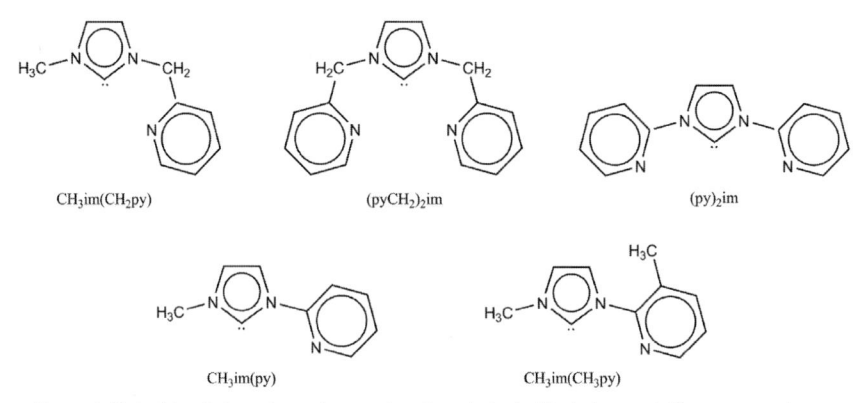

Figure 4.48 Imidazole-based species used as ligands in Au/Ag heterometallic compounds.

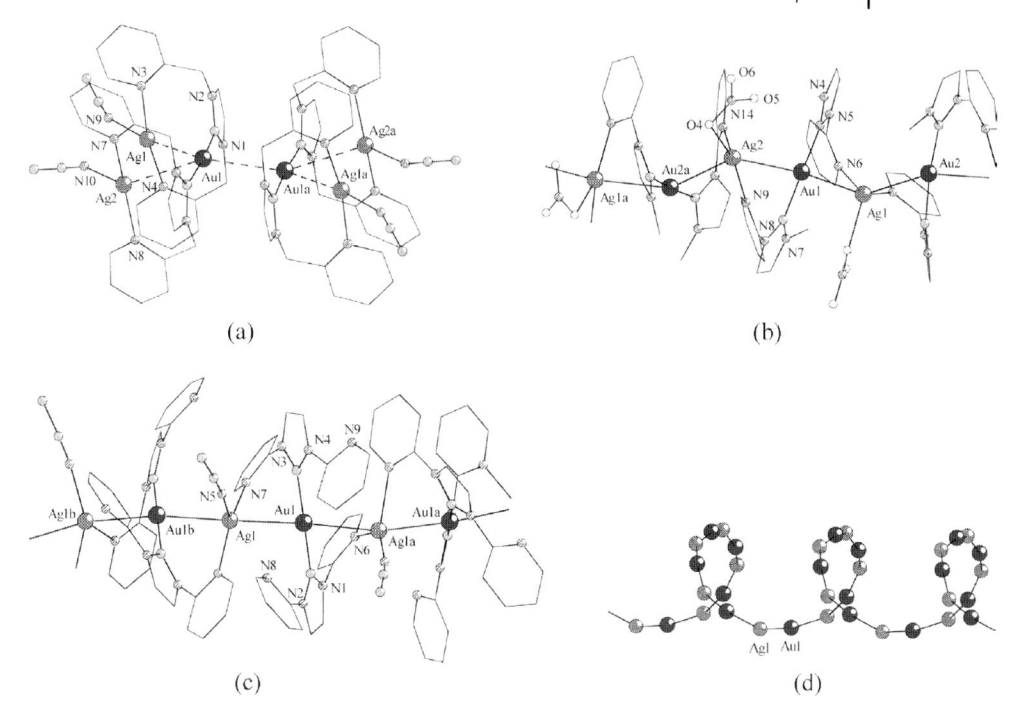

Figure 4.49 (a) Dimer association of the cations in the crystal of [AuAg$_2${(pyCH$_2$)$_2$im}$_2$(NCMe)$_2$](BF$_4$)$_3$ **428** [591]; (b) polymeric chain in the crystal of [{AuAg(CH$_3$impy)$_2$(NO$_3$)}NO$_3$]$_n$ **435** [593]; (c) polymeric chain in the crystal of [{AuAg{(py)$_2$im}$_2$(NCMe)}(BF$_4$)$_2$]$_n$ **430**, and (d) helical metal core in polymer **430** [591].

core). The silver centers also interact weakly with the Au(I) center [Au(1)–Ag(1) 3.2197(17), Au(1)–Ag(2) 3.2819(17) Å]. Two such cations form a dimer association through a weak aurophilic interaction [Au(1)–Au(1a) 3.326(1) Å] (Figure 4.49a) [591].

In contrast, reaction of [Au{(py)$_2$im}$_2$](BF$_4$) with AgBF$_4$ resulted in the heterometallic coordination polymer, [{AuAg{(py)$_2$im}$_2$(NCMe)}(BF$_4$)$_2$]$_n$ **430**, regardless of the reaction conditions and stoichiometry used [591]. The chain polymer consists of Au(I) centers linearly coordinated to two carbene ligands alternating with Ag(I) centers bound to two of the pyridine rings from alternating Au(I) centers and to an acetonitrile molecule (distorted trigonal AgN$_3$ core). One pyridyl group on each py$_2$im ligand remains uncoordinated in the polymeric structure (Figure 4.49c). The metal–metal separations are short with Au(I)–Ag(I) linkages of 2.8359(4) and 2.9042(4) Å. A unique feature of this mixed-metal polymer is that it crystallizes in a helical fashion, which was suggested to be induced by the formation of the six-membered AuAgC$_2$N$_2$ metallacycles (Figure 4.49d). Replacement of the dangling, uncoordinated pyridyl ring by a methyl group (CH$_3$impy ligand) as well as the presence of a methyl substituent on the remaining one [CH$_3$im(CH$_3$py) ligand] produced the related polymers [{AuAg(CH$_3$impy)$_2$(L)}(BF$_4$)$_2$]$_n$ [L = NCMe **431**, NCPh **432**, NCCH$_2$Ph **433**] [Au–Ag 2.833(1)–2.924(1) Å] [593] and [{AuAg{CH$_3$im(CH$_3$py)}$_2$(NCEt)}(BF$_4$)$_2$]$_n$ **434** [Au–Ag

2.9845(5)/2.9641(5) Å] [594]. They are built in a similar way as observed for compound **430**, with Au(I)–Ag(I) interactions in the range 2.833(1)–2.985(1) Å, but with a distorted zigzag arrangement of the metal atoms. Replacement of the nitrile molecule by a coordinated nitrate anion in [{AuAg(CH$_3$impy)$_2$(NO$_3$)}(NO$_3$)]$_n$ **435** (Figure 4.49b) also results in a zigzag polymer with alternating short [2.8125(2) Å] and long [2.9482 (2) Å] Au(I)–Ag(I) interactions [593].

Supported Au–Ag interactions were also found in several trinuclear species with Au$_2$Ag [595–598] or AuAg$_2$ [597, 598] cores. They include the "snake"-type complex [(MesAu)$_2$Ag(μ-dppm)$_2$](ClO$_4$) **436**, with equivalent Au–Ag distances [2.944(2)/2.946 (2) Å] and dppm ligands bridging the central silver and two gold atoms [596], or the triangular species Au(carb)Ag$_2$(μ-3,5-Ph$_2$pz)$_2$ **437**, Au$_2$(carb)$_2$Ag(μ-3,5-Ph$_2$pz) **438** and Au(bzim)Ag$_2$(μ-3,5-Ph$_2$pz)$_2$ **439**, which show intramolecular Au–Ag contacts (3.21–3.28 Å), corresponding to very weak metal–metal interactions, and are further associated into dimers through intermetallic contacts [Au–Ag 3.083(2)/3.310(2) Å for **437**; Au–Au 3.335(1), Au–Ag 3.42 Å for **438**; Au–Ag 3.38/3.53 Å for **439**] [597, 598]. Two related, electron-rich, trinuclear Au$_3$(bzim)$_3$ units can trap a silver cation in a "sandwich" cluster [Ag{Au$_3$(bzim)$_3$}$_2$](BF$_4$) **440** [54, 599], similar to the thallium analog **29**. Within the heterometallic cation the Ag atom interacts strongly with six Au (I) atoms from two cyclic Au$_3$C$_3$N$_3$ moieties [Au–Ag 2.731(2)–2.922(2) Å] and an infinite columnar polymer of the type ···BBABBA··· is formed through weak intercation aurophilic interactions [Au–Au 3.268(1) Å]. The reaction of (μ-PAnP) (AuX)$_2$ with AgSbF$_6$, instead of precipitating AgX, afforded the isolation of isostructural complexes [{(μ-PAnP)(AuX)$_2$}$_2$Ag(SbF$_6$)]$_n$ [X = Cl **441**, Br **442**], which display unprecedented Au(I)–X–Ag(I) halonium cations [600]. The compounds also consist of helical coordination polymers, with silver(I) ions encapsulated at the center of a distorted Au$_4$X$_4$ dodecahedron described by the Au and X atoms of four neighboring (μ-PAnP)(AuX)$_2$ units. In addition to the weakly coordinated halide atom, the Au–Ag separations [3.2180(4) Å in **441**, and 3.2701(5) Å in **442**] suggest weak d^{10}–d^{10} metallophilic interactions, which contribute together with Ag–halide and π–π interactions to the stabilization of the halonium cations [600].

Compounds in which the silver center is coordinated solely by gold atoms, thus exhibiting unsupported Au–Ag bonds, are rare. In addition to compound **440**, only two other complexes have been reported. Reaction of (μ-dppm)(AuCH$_2$SiMe$_3$)$_2$ with AgO$_3$SCF$_3$ (2:1 molar ratio) resulted in the complex [{(μ-dppm)(AuCH$_2$SiMe$_3$)$_2$}$_2$Ag](O$_3$SCF$_3$) **443**, in which two units of the gold(I) starting material directly bond to a silver center through four gold–silver bonds [2.718(1)/2.782(1) Å] in a distorted tetrahedral Au$_4$Ag core [601]. Treatment of [PPN][Au(trip)Cl] with AgO$_3$SCF$_3$ afforded the isolation of a unique green complex, [{(μ-trip)Au}$_6$Ag](O$_3$SCF$_3$) **444**, which exhibits an almost planar, "wheel"-like Au$_6$Ag core, with silver(I) atom in the center of a regular hexagon of gold atoms and *ipso* carbon of the aromatic groups bridging between two consecutive gold atoms. The gold-silver distances [2.797(1), 2.802(1) and 2.809(1) Å] are consistent with formal heterometallic bonds [602].

The phosphorus ylide complex [(Ph$_3$PCH$_2$)$_2$Au]$_2$Ag$_2$(ClO$_4$)$_4$ **445** also contains Au–Ag bonds [(2.783(2)/2.760(2) Å] unsupported by any covalent bridge within a Au$_2$Ag$_2$ ring, but each silver atom is further bonded to two oxygen atoms from two

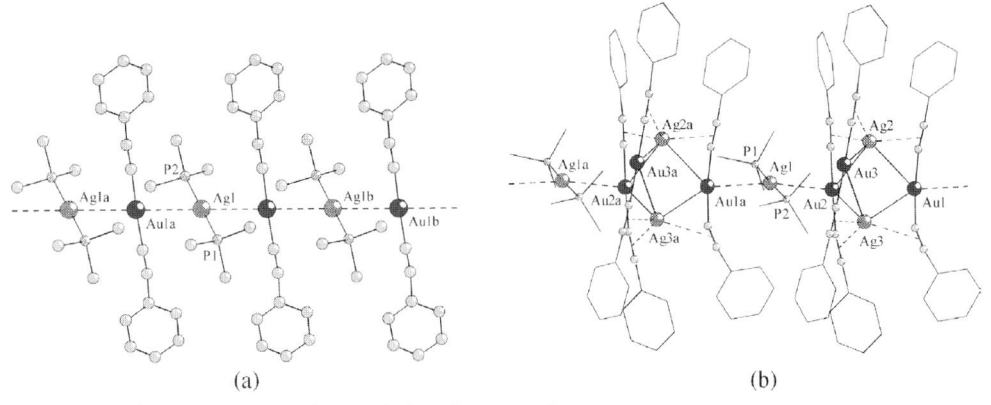

Figure 4.50 Polymer association in the crystal of (a) [{(Me$_3$P)$_2$Ag}{Au(C≡CPh)$_2$}]$_n$ **447**, and (b) [{(Me$_3$P)$_2$Ag}{Au(η^2-C≡CPh)$_2$}$_3$Ag$_2$}]$_n$ **448** [606].

perchlorato ligands which results in a distorted tetrahedral environment for silver [603].

A further few ionic compounds based on unsupported heterometallic interactions between metal centers from the anion and the cation were reported only recently [604–606]. Thus, treatment of [Ag(Tab)$_2$](PF$_6$) with K[Au(CN)$_2$] generated the complex [{(Tab)$_2$Ag}{Au(CN)$_2$}]$_2$ **446** [605]. In the solid state [Ag(Tab)$_2$]$^+$ cations and [Au(CN)$_2$]$^-$ anions are held together via ionic interactions [Au–Ag 2.9598(7)/2.9185(7) Å] to form [(Tab)$_2$M][Au(CN)$_2$] units. Two such fragments are further connected by one gold–gold aurophilic bonding interaction [Au–Au 3.0140(5) Å] to form an uncommon Ag–Au–Au–Ag linear string structure with three ligand–unsupported metal–metal bonds [605].

A linear –Ag–Au–Ag–Au– chain built from alternating cations and anions via weak heterometallophilic bonding [Au(1)–Ag(1) 3.206(2), Au(1)–Ag(1b) 3.224(2) Å] (Figure 4.50a) was found in the crystal of the ionic mixed-metal compound [{(Me$_3$P)$_2$Ag}{Au(C≡CPh)$_2$}]$_n$ **447**, isolated from a mixture of the molecular (Me$_3$P)AuC≡CPh and ionic [(Me$_3$P)$_2$Ag]$^+$[Ag(C≡CPh)$_2$]$^-$ (Au : Ag 1 : 1) [606]. Partial loss of tertiary phosphines from **447** leads to a product which contains [(Me$_3$P)$_2$Ag]$^+$ cations and [{Au(η^2-C≡CPh)$_2$}$_3$Ag$_2$]$^-$ anions. The cluster anion consists of three [PhC≡CAuC≡CPh]$^-$ anions associated via two Ag$^+$ cations [Au–Ag 2.854(2)–3.039(1) Å]. Each Ag(I) atom from the apical positions is π-coordinated by three alkynyl ligands. Density functional calculations proved the trigonal bipyramidal Ag$_2$Au$_3$ core to be largely ionic in nature. In the crystal, the cluster anions are associated with the [(Me$_3$P)$_2$Ag]$^+$ cations via heterometallophilic contacts in which two of the three gold atoms are involved [Au(1a)–Ag(1) 2.9945(8), Au(2)–Ag(1) 2.9996(8) Å] to give the 1D polymeric [{(Me$_3$P)$_2$Ag}{Au(η^2-C≡CPh)$_2$}$_3$Ag$_2$}]$_n$ (**448**) (Figure 4.50b) [606]. Similar cluster anions were found in other ionic species [582, 607], for example, [NBu$_4$][{Au(η^2-C≡CPh)$_2$}$_3$Ag$_2$] [607]. Related cluster compounds are [Au(η^2-C≡CPh)$_2$]$_2$Ag$_2$(PPh$_3$)$_2$ **449**, which contains a planar Au$_2$Ag$_2$ [Au–Ag 2.894(1)–3.028(1) Å] and each silver atom asymmetrically π- bonded to two triple bonds and one phosphine

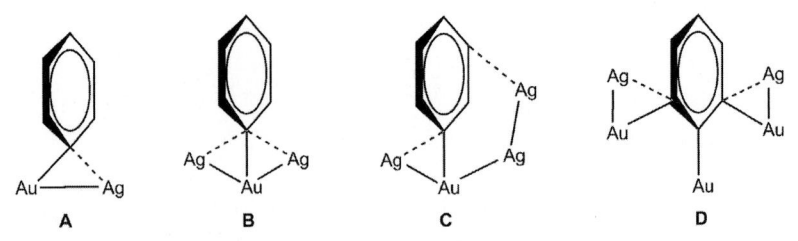

Figure 4.51 Heterometallic bridging pattern of an aryl group in Au–Ag species.

ligand [608], and the pentanuclear [{{(p-tolyl$_3$P)Au}$_2$ {μ-(η^2-C≡C)$_2$Ar}}$_2$Ag](ClO$_4$) **450** (Ar = C$_6$Me$_4$–3,4,5,6) which cation contains a silver ion encased by four AuC≡C fragments and exhibiting weak contacts with the gold atoms [Au–Ag separations in the range 3.166(1)–3.333(1) Å] [609].

In most cases, when arylgold(I) starting materials were used to obtain compounds with gold–silver bonds or interactions a common pattern in the resulting complexes is the bridging nature of the aryl group between Au and Ag atoms (Figure 4.51), which seems to provide further stabilization of the complexes [578, 602, 610–616]. A unique bridging patern of type **D** was also recently observed in some Au–Ag cluster compounds [617–619].

The treatment of Au(mes)(PPh$_3$) with Ag(tht)(O$_3$SCF$_3$) (1 : 1 molar ratio) resulted in the tetranuclear dimer [{(Ph$_3$P)Au(μ-mes)Ag(μ-tht)}$_2$](O$_3$SCF$_3$)$_2$ **451**, the tht ligands acting as bridges between the silver atoms of two isolated Au(μ-mes)Ag fragments [Au–Ag 2.8245(6) Å] [610]. The use of a 1 : 2 molar ratio between Au(mes)(AsPh$_3$) and AgClO$_4$ afforded the isolation of a trinuclear species, [{(Ph$_3$As)Au(μ-mes)}$_2$Ag](ClO$_4$) **452**. Its cation contains a linear Au–Ag–Au system [Au–Ag 2.7758(8) Å], with each of the heterometallic bonds supported by a bridging mesityl group of type **A** (Figure 4.51). In contrast with **451**, which contains discrete tetranuclear cations, the trinuclear cations of **452** are further associated through Au–Au contacts of 3.132(2) Å to form a unidimensional chain polymer (Figure 4.52a) [611].

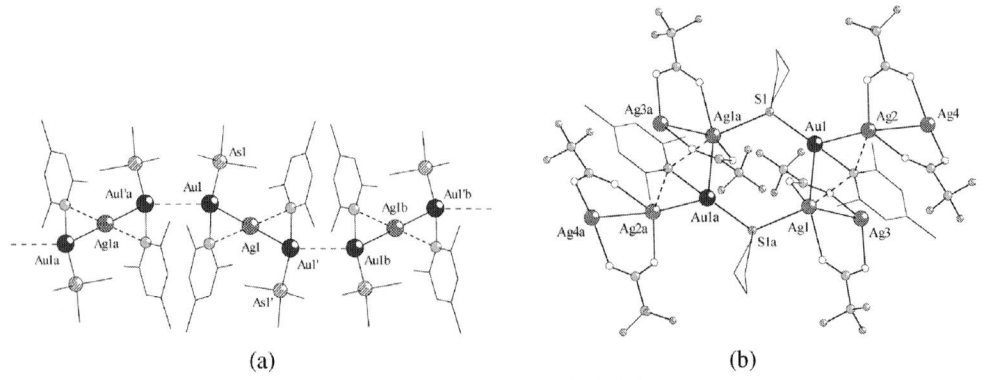

Figure 4.52 (a) Polymer association in the crystal of [{(Ph$_3$As)Au(μ-mes)}$_2$Ag](ClO$_4$) **452** [611] (for clarity, only *ipso* carbon of phenyl groups are shown for arsine ligands), and (b) dimer association in the crystal of [Au(μ$_3$-mes){Ag$_2$(μ-O$_2$CCF$_3$)$_2$}$_2$(tht)]$_n$ **453** [612].

The polymeric compounds [Au(μ$_3$-mes){Ag$_2$(μ-O$_2$CR)$_2$}$_2$(μ-tht)]$_n$ [R = CF$_3$ **453**, CF$_2$CF$_3$ **454**] display a monomeric unit in which the Au(mes) moiety bridges silver atoms from two different Ag$_2$(μ-O$_2$CR)$_2$ units [Au–Ag 2.8226(4)/2.8993(4) Å in **453**; 2.8140(8)/2.8166(8) Å in **454**]. The main difference resides in the bridging pattern of the mesityl group which is of type **B** and **C** (Figure 4.51) in **453** and **454**, respectively. In both complexes two pentanuclear units build a dimer through the tht ligands which bridge the gold atoms (see Figure 4.52b) and further Ag–O–Ag linkages result in two-dimensional polymers [612]. In the related complex [(tht)Au(μ$_3$-mes){Ag$_2$(μ-O$_2$CCF$_2$CF$_3$)$_2$}$_2$(μ-tht)$_2$]$_n$ **455** the mesityl group also exhibits a type **B** bridging pattern, but the gold atom in the monomer has contacts with three silver atoms [Au–Ag 2.8540(6)/2.8845(6)/3.0782(6) Å]. One tht ligand coordinates the gold atom in a terminal fashion, while the other two are involved in bridges between silver atoms of different monomers to build a polymeric layer structure [612]. The reaction of **453** with water (1:2 molar ratio) resulted in the isolation of a partial hydrolysis product of composition [AuAg$_4$(mes)(O$_2$CCF$_3$)$_4$(tht)(H$_2$O)]$_n$ **456**, with three Au–Ag interactions per asymmetric unit [2.8503(6)/2.8708(6)/3.1347(7) Å] and a monodimensional polymeric structure based on Ag–O contacts [612].

The synthetic strategy based on the direct reaction of basic [AuR$_2$]$^-$ (R = perhalophenyl) salts with silver(I) salts was successfully used to prepare complexes with unsupported Au–Ag interactions or with such interactions supported only by aryl bridging ligands. Thus, treatment of [NBu$_4$][Au(C$_6$F$_5$)$_2$] with AgClO$_4$ leads to the complex [Au$_2$Ag$_2$(C$_6$F$_5$)$_4$]$_n$ which can further react with neutral ligands to give neutral complexes of the type [Au$_2$Ag$_2$(C$_6$F$_5$)$_4$(L)$_2$]$_n$ [578, 613–616]. The first compound of this class which was characterized by single-crystal X-ray diffraction was [Au$_2$Ag$_2$(C$_6$F$_5$)$_4$(tht)$_2$]$_n$ **457** which proved to exhibit a linear chain structure built from planar Au$_2$Ag$_2$ rings with unsupported Au–Ag bonds [2.718(2)/2.726(2) Å] connected through strong aurophilic interactions [Au(1)–Au(2a) 2.889(2) Å] (Figure 4.53a) [613, 614]. Similar polymeric chains were found for other related [Au$_2$Ag$_2$(C$_6$F$_5$)$_4$(L)$_2$]$_n$ in which the tht ligands on silver are replaced by L = benzene (coordinated by one edge) **458** [Au–Ag 2.702(2)/2.792(2), Au–Au 3.013(2) Å] [614], acetone **459** [Au–Ag 2.7829(9)/2.7903(9), Au–Au 3.167(1) Å] [615] or acetonitrile **460** [Au–Ag 2.7577(5)/2.7267(5), Au–Au 2.8807(4) Å] (Figure 4.53b) [578]. There

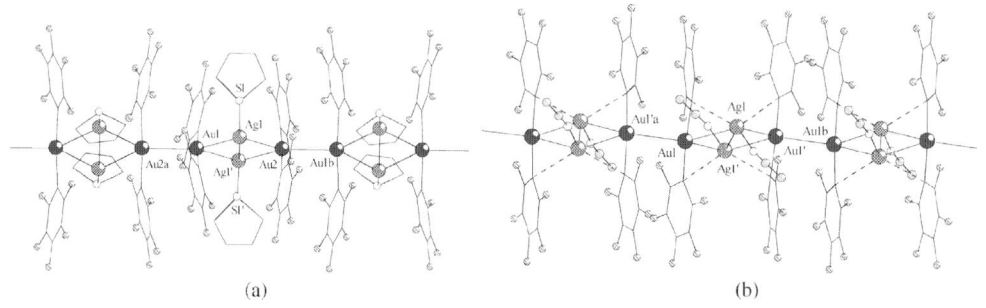

Figure 4.53 Polymer associations in the crystal of
(a) [Au$_2$Ag$_2$(C$_6$F$_5$)$_4$(tht)$_2$]$_n$ **457** [614], and
(b) [Au$_2$Ag$_2$(C$_6$F$_5$)$_4$(NCMe)$_2$]$_n$ **460** [578].

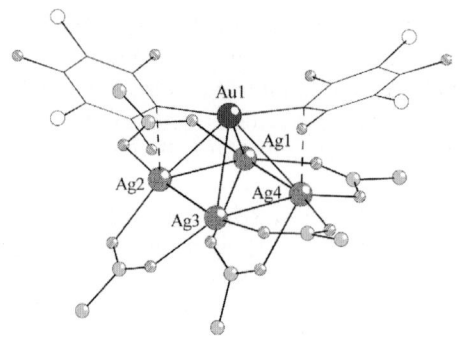

Figure 4.54 Structure of [NBu$_4$]$_2$[Au(μ-3,5-C$_6$Cl$_2$F$_3$)$_2$Ag$_4$ (μ-O$_2$CCF$_3$)$_5$] **461** [616] (for clarity, fluorine atoms of the CF$_3$ groups are not shown).

are also some particular differences: (i) the inter-ring Au–Au distances, consistent with much stronger interactions in **457** and **460** than in **458** and **459**; (ii) the coplanarity of the Au$_2$Ag$_2$ rings in **458**, while they are twisted along the polymeric chain in the other compounds; (iii) stronger transannular Ag–Ag interactions in **457** (3.046 Å) and **458** (3.070 Å) than in **459** [3.181(1) Å] and **460** [3.108(1) Å], and (iv) absence of the Ag–C$_{ipso}$ interaction in **457** and **458**, while in the other two compounds bridging aryl groups support the Au–Ag bonds of the Au$_2$Ag$_2$ ring.

The related precursor [NBu$_4$][Au(C$_6$F$_5$)$_2$] was used to prepare the unique pentanuclear, ionic species [NBu$_4$]$_2$[Au(μ-3,5-C$_6$Cl$_2$F$_3$)$_2$Ag$_4$(μ-O$_2$CCF$_3$)$_5$] **461**, with a square pyramidal AuAg$_4$ core [616]. In this case two of the Au–Ag interactions are unsupported [Au(1)–Ag(1) 3.0134(6), Au(1)–Ag(3) 2.9364(6) Å], while the other two are supported by μ-aryl groups [Au(1)–Ag(2) 2.9019(6), Au(1)–Ag(4) 2.9421(6) Å] (Figure 4.54).

Using reactions between the polymeric silver arylacetylide (AgC≡CC$_6$H$_4$R-4)$_n$ and the dinuclear gold(I) precursor [Au$_2$(dppm)$_2$](SbF$_6$)$_2$, with exclusion of light, a new class of metastable green clusters of the type [Au$_6$Ag$_{13}$(μ-dppm)$_3$(μ$_3$-η^1-C≡CC$_6$H$_4$R-4)$_{14}$](SbF$_6$)$_5$ [R = H **462**, Me **463**] was prepared and the X-ray structure analysis revealed a partial exchange of ligands: the gold atoms are coordinated by two acetylide C donors in quasi-linear arrangements, while the dppm ligands bridge silver atoms. The cations contain an Au$_6$Ag$_{13}$ core (Au–Ag contacts in the range 2.94–3.22 Å) in which two arylacetylide units cap three silver(I) centers, whereas the others are bound to one Au(I) and two Ag(I) centers in an asymmetric mode, respectively [617, 618]. Exposure of **462** to light resulted in a stable red cluster isolated as [Au$_5$Ag$_8$(μ-dppm)$_4$(μ-C≡CC$_6$H$_5$)$_4$(μ$_3$-C≡CC$_6$H$_5$)$_3${μ$_5$-1,2,3-C$_6$(C$_6$H$_5$)$_3$}](PF$_6$)$_3$ **464**. In this case the Au$_5$Ag$_8$ core [Au–Ag contacts in the range 2.743(1)–3.149(1) Å] is supported by phenylacetylide ligands σ-bound to gold and π-coordinated to silver atoms and by a [1,2,3-C$_6$(C$_6$H$_5$)$_3$]$^{3-}$ anion, derived from cyclotrimerization of phenylacetylide, in an unusual μ$_5$-bridging pattern of type **D** (Figure 4.51) [617]. The cluster [Au$_5$Ag$_8$(μ-dppm)$_4$(μ-C≡CC$_6$H$_4$Me-4)$_4$(μ$_3$-C≡CC$_6$H$_4$Me-4)$_3${μ$_5$-1,2,3-C$_6$(C$_6$H$_4$Me-4)$_3$}](PF$_6$)$_3$ **465** has a similar structure [618]. The reaction of the polymeric (AgC≡CFc)$_n$ with [Au$_2$(Ph$_2$PNHPPh$_2$)$_2$(MeCN)$_2$]

(SbF$_6$)$_2$ affords the isolation of another type of high-nuclearity cluster, [Au$_6$Ag$_8$(μ-C≡CFc)$_{12}$Cl](OH) **466**, in which again the acetylide ligands were transferred to gold atoms [619]. The Au$_6$Ag$_8$ core, which traps a chlorine atom, consists of a cubic arrangement of the Ag(I) atoms and the Au(I) atoms capping each Ag$_4$ square plane exhibit Au–Ag contacts in the range 2.92–3.31 Å, supported by π-coordinated ferrocenylacetilide ligands.

Trimetallic clusters with a tetrahedral AuRe$_2$Ag skeleton [194, 201, 202] were also reported, for example [(Ph$_3$P)Au][(Ph$_3$P)Ag]Re$_2$(μ-PCy$_2$)(CO)$_7$(C≡CPh) **168** [Au–Ag 2.756(1) Å] [201].

The unusual trimetallic derivative, [(dppe)Au$_2$]$_2$Ag$_4$[Fe(CO)$_4$]$_4$ **281** (see also Section 4.3.5), exhibits a metal atom framework described as a two-dimensional triangulated twisted ribbon, with gold atoms of (dppe)Au$_2$ units connected to the same silver atom [Au–Ag 2.767(2)–2.824(3) Å] [401].

Gold-rich clusters include the unique Au-centered icosahedral species [{(Ph$_2$MeP)Au}$_8$Au(AgCl)$_4$](C$_2$B$_9$H$_{12}$) **467** [587, 620]. As in the case of the analogous Au/Cu cluster **420** (see Section 3.8.1) for the cation of **467** a difference was noted in the lengths of the radial Au–Ag bonds [2.814(2)–2.853(2) Å] and the tangential ones [2.814(2)–3.078(2) Å]. The Pd-centered icosahedral cluster, [(Ph$_3$P)Au]$_6$PtAg$_6$(AgI$_3$)$_2$ **468**, contains AgI$_3$ units capping two opposite triangular Ag$_3$ faces [566]. Several other trimetallic clusters of lower nuclearity with metallic frameworks derived from a Pt-centered icosahedron by removing three or two vertices are also known. They include species with Pt–Au$_8$Ag [556, 562, 563] and Pt–Au$_8$Ag$_2$ cores [563], for example [(dppp)$_4$Au$_8$Pt(PPh$_3$)(AgCl)](NO$_3$)$_2$ **469** [556] or [{(Ph$_3$P)Au}$_8$Pt(μ$_3$-H)(AgNO$_3$)$_2$](NO$_3$) **470** [563].

A series of high nuclearity Au–Ag clusters, referred to as "clusters of clusters", have also been reported. Most of them display a metal framework based on Au-centered vertex-sharing bi-icosahedra [621–625]. They exhibit different types of arrangements of the four adjacent metal pentagons, that is, either staggered–eclipsed–staggered (*ses*) or staggered–staggered–staggered (*sss*) as well as some intermediate metal configurations. The two middle silver pentagons are bridged by a number of halide ligands, which are referred to as 5-, 6- or 7-ring. Examples are [{(*p*-tolyl$_3$P)Au}$_{10}$Au$_3$Ag$_{10}$(AgCl)$_2$Cl$_5$](SbF$_6$)$_2$ (*ses* – 5-ring) **471** [622], [{(Ph$_3$P)Au}$_{10}$Au$_3$Ag$_{10}$(AgBr)$_2$Br$_6$](SbF$_6$) (*ses* – 6-ring) **472** [621] or [{(MePh$_2$P)Au}$_{10}$Au$_2$Ag$_{11}$(AgBr)$_2$Br$_7$ (*ses* – 7-ring) **473** [625], the latter exhibiting a silver atom joining the icosahedral cores. Similar metal skeletons were found in the trimetallic clusters [{(Ph$_3$P)Au}$_{10}$Au$_2$Ag$_{10}$(AgCl)$_2$NiCl$_5$](SbF$_6$) **474** [494] and [{(Ph$_3$P)Au}$_{10}$Au$_2$Ag$_{10}$(AgCl)$_2$PtCl$_5$]Cl **475** [494, 570], which display two Au$_6$Ag$_6$ icosahedra, one M-centered (M = Ni, Pt) and one Au-centered, sharing a common Au atom. Related trimetallic species are [(Ph$_3$P)Au]$_{10}$Ag$_{11}$(AgCl)$_2$Pt$_2$Cl$_5$ **476** [568] and [(Ph$_3$P)Au]$_{10}$AuAg$_{10}$(AgCl)$_2$Pt$_2$Cl$_5$ **477** [569], in which Pt-centered icosahedra are joind by a silver or gold atom, respectively. A larger Au–Ag cluster is the species [(*p*-tolyl$_3$P)Au]$_{12}$Au$_6$Ag$_{11}$(AgCl)$_2$Pt$_2$Cl$_5$ **478** [626], based on three Au-centered icosahedra sharing three vertices in a triangular array and two additional, capping Ag atoms on the *pseudo*-threefold axis [626]. Recently, two new Au–Ag clusters, that is [(MePh$_2$P)Au]$_{10}$Au$_2$Ag$_5$[Ag(NO$_3$)$_2$]$_2$(NO$_3$)$_7$ **478** and [(MePh$_2$P)Au]$_{10}$Au$_7$[Ag(NO$_3$)$_2$]$_2$(NO$_3$)$_7$ **479**, were reported [627]. Their metal cores can be best described as two Au-centered pentagonal antiprisms sharing an Ag_5 or Au_5 ring and bicapped by silver atoms.

As in the case of copper, the reaction of the homonuclear $Na_3[Ag_9(SCH_2CH_2S)_6]$ with $[NEt_4][AuBr_2]$ and $[PPh_4]Br$ resulted in the isolation of the ionic complex $[PPh_4]_2[Au_2Ag_4(SCH_2CH_2S)_4]$. The $Au_2Ag_4S_8$ framework of the anion consists of linear AuS_2 and trigonal planar AgS_3, with sulfur atoms bridging silver atoms [Ag–Ag 3.170 Å] or gold–silver fragments [Au–Ag 2.996 Å] [584].

Finally, the affinity of coinage metals for sulfur ligands was recently used to create a unique metallosupramolecular architecture containing Au(I), Ag(I) and Cu(II) in the same metal frameworks [628]. Thus, the reaction of $(NH_4)[Au(D-Hpen)_2]$ with $AgNO_3$ (1:1 molar ratio), followed by treatment with excess of $CuCl_2$, afforded the isolation of blue crystals of a heterometallic aggregate with a rock-salt-like lattice structure. It consists of two types of supramolecular ionic cages, $[Au_6Ag_8Cu_6Cl(H_2O)_5(D\text{-pen})_{12}]^+$ **480a** and $[Au_6Ag_9Cu_6Cl_4(H_2O)_6(D\text{-pen})_{12}]^-$ **480b**, each of which accommodates a Cl^- ion. The metal ions are connected by D-penicillaminate ligands through amino, carboxylate and thiolate groups and 18 metallophilic Au–Ag interactions with distances of 2.880(2)–3.249(2) Å are present in the 21-nuclear monoanionic cage cluster.

4.4
Concluding Remarks

Mainly in recent years the use of appropriate synthesis strategies has resulted in a considerable development of the chemistry of gold–heterometal compounds and particularly of molecular, oligomeric and supramolecular species in which the Au(I)–metal interactions are strengthened by the relativistic effects for gold, as is best revealed by the results reported on Au–Tl and Au–Ag compounds. This review of the published data revealed that, besides Au-containing cluster compounds, for most metals there are examples of species with gold–heterometal bonds and/or interactions characterized by single-crystal X-ray diffraction. There also some gaps which are still waiting to be filled, that is, species with Au–metal (alkali earth, Cd, Zn, Al, Sc, Y, Ln, An, Hf, Zr or Ta) bonds or interactions. Much more effort should be concentrated on the chemistry of gold–main group metal compounds which is still poorly developed, although some important achievements have recently been reported.

References

1 Schmidbaur, H. (ed.) (1999) *Gold – Progress in Chemistry, Biochemistry and Technology*, John Wiley & Sons, New York.

2 Lehn, J.M. (1988) *Angewandte Chemie (International Edition in English)*, 27, 89.

3 Lehn, J.M. (1995) *Supramolecular Chemistry Concepts and Perspectives*, VCH, Weinheim.

4 Lehn, J.M., Atwood, J.L., Davies, J.E.D., MacNicol, D.D. and Vögtle, F. (eds) (1996)

Comprehensive Supramolecular Chemistry, Vol. 1–11, Pergamon Press, Oxford.
5   Haiduc, I. and Edelmann, F.T. (1999) *Supramolecular Organometallic Chemistry*, Wiley-VCH, Weinheim.
6   Alcock, N.W. (1972) *Advances in Inorganic Chemistry and Radiochemistry*, **15**, 1.
7   Alcock, N.W. and Countryman, R.M. (1977) *Journal of the Chemical Society-Dalton Transactions*, 217.
8   Alcock, N.W. (1990) *Bonding and Structure: Structural Principles in Inorganic and Organic Chemistry*, Ellis Horwood, Chichester.
9   (a) Emsley, J. (1994) *Die Elemente*, Walter de Gruyter, Berlin. (b) Huheey, J., Keiter, E. and Keiter, R. (1995) *Anorganische Chemie: Prinzipien von Struktur und Reaktivität*, Walter de Gruyter, Berlin. (c) http://www.americanelements.de
10 DIAMOND – Visual Crystal Structure Information System (2001) CRYSTAL IMPACT, Postfach 1251, D-53002 Bonn, Germany.
11 Puddephatt, R.J. (1987) in *Comprehensive Coordination Chemistry* (eds G. Wilkinson, R.D. Gillard and J.A. McCleverty), **Vol. 5**, Pergamon Press, Oxford, pp. 861–923.
12 Akrivos, D., Katsikis, H.J. and Koumoutsi, A. (1997) *Coordination Chemistry Reviews*, **167**, 95.
13 Gimeno, M.C. and Laguna, A. (1997) *Chemical Reviews*, **97**, 511.
14 Gimeno, M.C. and Laguna, A. (2003) in *Comprehensive Coordination Chemistry II* (eds J.A. McCleverty and T.J. Meyer), **Vol. 6**, Elsevier, New York, pp. 911–1145.
15 Strähle, J. (1999) in *Gold – Progress in Chemistry, Biochemistry and Technology* (ed. H. Schmidbaur), John Wiley & Sons, New York, pp. 311–347.
16 Laguna, A. (1999) in *Gold – Progress in Chemistry, Biochemistry and Technology* (ed H. Schmidbaur), John Wiley & Sons, New York, pp. 349–427.
17 Fackler, J.P., Jr, van Zyl, W.E. and Prihoda, B.A. (1999) in *Gold – Progress in Chemistry, Biochemistry and Technology* (ed. H. Schmidbaur), John Wiley & Sons, New York, pp. 795–839.
18 Grohmann, A. and Schmidbaur, H. (1995) in *Comprehensive Organometallic Chemistry II* (eds E.W. Abel, F.G.A. Stone and G. Wilkinson), **Vol. 3**, Pergamon, Oxford, pp. 1–56.
19 Patai, S. and Rappoport, Z. (eds) (1999) *The Chemistry of Organic Derivatives of Gold and Silver*, John Wiley & Sons, Chichester.
20 Schmidbaur, H., Grohmann, A., Olmos, M.E. and Schier, A. (1999) in *The Chemistry of Organic Derivatives of Gold and Silver* (eds S. Patai and Z. Rappoport), John Wiley & Sons, Chichester, pp. 227–311.
21 Schmidbaur, H., Grohmann, A. and Olmos, M.E. (1999) in *Gold – Progress in Chemistry, Biochemistry and Technology* (ed. H. Schmidbaur), John Wiley & Sons, New York, pp. 648–746.
22 Fernández, E.J., Laguna, A. and Olmos, M.E. (2005) *Advances in Organometallic Chemistry*, **52**, 77.
23 Mingos, M.P. and Watson, M.J. (1992) *Advances in Inorganic Chemistry*, **39**, 327.
24 Pyykkö, P. (1997) *Chemical Reviews*, **97**, 597.
25 Pignolet, L.H. and Krogstad, D.A. (1999) in *Gold – Progress in Chemistry, Biochemistry and Technology* (ed. H. Schmidbaur), John Wiley & Sons, New York, pp. 429–493.
26 Dyson, D.J. and Mingos, M.P. (1999) in *Gold – Progress in Chemistry, Biochemistry and Technology* (ed. H. Schmidbaur), John Wiley & Sons, New York, pp. 511–555.
27 Pyykkö, P. (2004) *Angewandte Chemie (International Edition in English)*, **43**, 4412.
28 van Koten, G. and Noltes, J.G. (1979) *Journal of Organometallic Chemistry*, **174**, 367.
29 van Koten, G., Jastrzebski, J.T.B.H., Stam, C.H. and Niemann, N.C. (1984) *Journal of the American Chemical Society*, **106**, 1880.
30 Catalano, V.J., Bennett, B.L., Kar, H.M. and Noll, B.C. (1999) *Journal of the American Chemical Society*, **121**, 10235.

31 Caron, A., Guilhelm, J., Riche, C., Pascard, C., Alpha, B., Lehn, J.-M. and Rodriguez-Ubis, J.-C. (1985) *Helvetica Chimica Acta*, **68**, 1577.

32 Uang, R.-H., Chan, C.-K., Peng, S.-M. and Che, C.-M. (1994) *Chemical Communications*, 2561.

33 Song, C., Li, Y.-Z., Chen, Y.-Q., Xiao, S.-J. and You, X.-Z. (2004) *Acta Crystallographica E*, **60**, m1741.

34 Murray, H.H., Briggs, D.A., Garzon, G., Raptis, R.G., Porter, L.C. and Fackler, J.P. Jr. (1987) *Organometallics*, **6**, 1992.

35 Wang, S., Fackler, J.P., Jr. King, C. and Wang, J.C. (1988) *Journal of the American Chemical Society*, **110**, 3308.

36 Wang, S. and Fackler, J.P. Jr. (1988) *Organometallics*, **7**, 2415.

37 Wang, S., Garzon, G., King, C., Wang, J.-C. and Fackler, J.P. Jr. (1989) *Inorganic Chemistry*, **28**, 4623.

38 Wang, S. and Fackler, J.P. Jr. (1990) *Organometallics*, **9**, 111.

39 Wang, S. and Fackler, J.P. Jr. (1990) *Acta Crystallographica C*, **46**, 2253.

40 Mazany, A.M. and Fackler, J.P. Jr. (1984) *Journal of the American Chemical Society*, **106**, 801.

41 Ito, L.N., Felicissimo, A.M.P. and Pignolet, L.H. (1991) *Inorganic Chemistry*, **30**, 387.

42 Gould, R.A.T. and Pignolet, L.H. (1994) *Inorganic Chemistry*, **33**, 40.

43 Gould, R.A.T., Craighead, K.L., Wiley, J.S. and Pignolet, L.H. (1995) *Inorganic Chemistry*, **34**, 2902.

44 Bour, J.J., Berg, W.v.d., Schlebos, P.P.J., Kanters, R.P.F., Schoondergang, M.F., Bosman, W.P., Smits, J.M.M., Beurskens, P.T., Steggerda, J.J. and van der Sluis, P. (1990) *Inorganic Chemistry*, **29**, 2971.

45 Anandhi, U. and Sharp, P.R. (2004) *Angewandte Chemie (International Edition in English)*, **43**, 128.

46 Kempter, A., Gemel, C. and Fischer, R.A. (2005) *Inorganic Chemistry*, **44**, 163.

47 Kempter, A., Gemel, C., Hardman, N.J. and Fischer, R.A. (2006) *Inorganic Chemistry*, **45**, 3133.

48 Gabbai, F.P., Schier, A., Riede, J. and Schmidbaur, H. (1995) *Inorganic Chemistry*, **34**, 3855.

49 Gabbai, F.P., Chung, S.-C., Schier, A., Kruger, S., Rosch, N. and Schmidbaur, H. (1997) *Inorganic Chemistry*, **36**, 5699.

50 Fernández, E.J., Laguna, A. and López-de-Luzuriaga, J.M. (2005) *Coordination Chemistry Reviews*, **249**, 1423.

51 Fernández, E.J., Laguna, A. and López-de-Luzuriaga, J.M. (2007) *Dalton Transactions*, 1969.

52 Fernández, E.J., Laguna, A. and Olmos, M.E. (2007) *Coordination Chemistry Reviews*, **252**, 1630.

53 Catalano, V.J. and Malwitz, M.A. (2004) *Journal of the American Chemical Society*, **126**, 6560.

54 Burini, A., Bravi, R., Fackler, J.P., Jr., Galassi, R., Grant, T.A., Omary, M.A., Pietroni, B.R. and Staples, R.J. (2000) *Inorganic Chemistry*, **39**, 3158.

55 Fernández, E.J., Laguna, A., López-de-Luzuriaga, J.M., Mendizabal, F., Monge, M., Olmos, M.E. and Perez, J. (2003) *Chemistry – A European Journal*, **9**, 456.

56 Fernández, E.J., López-de-Luzuriaga, J.M., Monge, M., Olmos, M.E., Perez, J., Laguna, A., Mohamed, A.A. and Fackler, J.P. Jr. (2003) *Journal of the American Chemical Society*, **125**, 2022.

57 Fernández, E.J., Laguna, A., López-de-Luzuriaga, J.M., Monge, M., Montiel, M., Olmos, M.E. and Perez, J. (2004) *Organometallics*, **23**, 774.

58 Fernández, E.J., Laguna, A., López-de-Luzuriaga, J.M., Montiel, M., Olmos, M.E. and Perez, J. (2005) *Organometallics*, **24**, 1631.

59 Fernández, E.J., López-de-Luzuriaga, J.M., Monge, M., Montiel, M., Olmos, M.E., Perez, J., Laguna, A., Mendizabal, F., Mohamed, A.A. and Fackler, J.P. Jr. (2004) *Inorganic Chemistry*, **43**, 3573.

60 Fernández, E.J., Laguna, A., López-de-Luzuriaga, J.M., Monge, M., Montiel, M. and Olmos, M.E. (2007) *Inorganic Chemistry*, **47**, 2953.

61 Fernández, E.J., Laguna, A., López-de-Luzuriaga, J.M., Olmos, M.E. and Perez, J. (2003) *Chemical Communications*, 1760.

62 Fernández, E.J., López-de-Luzuriaga, J.M., Monge, M., Olmos, M.E., Perez, J. and Laguna, A. (2002) *Journal of the American Chemical Society*, **124**, 5942.

63 Fernández, E.J., López-de-Luzuriaga, J.M., Olmos, M.E., Perez, J., Laguna, A. and Lagunas, M.C. (2005) *Inorganic Chemistry*, **44**, 6012.

64 Crespo, O., Fernández, E.J., Jones, P.G., Laguna, A., López-de-Luzuriaga, J.M., Mendia, A., Monge, M. and Olmos, M.E. (1998) *Chemical Communications*, 2233.

65 Fernández, E.J., Jones, P.G., Laguna, A., López-de-Luzuriaga, J.M., Monge, M., Montiel, M., Olmos, M.E. and Perez, J. (2004) *Zeitschrift für Naturforschung*, **59b**, 1379.

66 Fernández, E.J., Laguna, A., López-de-Luzuriaga, J.M., Montiel, M., Olmos, M.E. and Perez, J. (2006) *Organometallics*, **25**, 1689.

67 Fernández, E.J., Laguna, A., López-de-Luzuriaga, J.M., Montiel, M., Olmos, M.E. and Perez, J. (2005) *Inorganica Chimica Acta*, **358**, 4293.

68 Fernández, E.J., Laguna, A., López-de-Luzuriaga, J.M., Olmos, M.E. and Perez, J. (2004) *Dalton Transactions*, 1801.

69 Fernández, E.J., Jones, P.G., Laguna, A., López-de-Luzuriaga, J.M., Monge, M., Perez, J. and Olmos, M.E. (2002) *Inorganic Chemistry*, **41**, 1056.

70 Meyer, J., Willnecker, J. and Schubert, U. (1989) *Chemische Berichte*, **122**, 223.

71 Oroz, M.M., Schier, A. and Schmidbaur, H. (1999) *Zeitschrift für Naturforschung*, **54b**, 26.

72 Theil, M., Jutzi, P., Neumann, B., Stammler, A. and Stammler, H.-G. (2002) *Journal of Organometallic Chemistry*, **662**, 34.

73 Piana, H., Wagner, H. and Schubert, U. (1991) *Chemische Berichte*, **124**, 63.

74 Meyer, J., Piana, H., Wagner, H. and Schubert, U. (1990) *Chemische Berichte*, **123**, 791.

75 Bauer, A., Schier, A. and Schmidbaur, H. (1995) *Journal of the Chemical Society-Dalton Transactions*, 2919.

76 Tripathi, U.M., Wegner, G.L., Schier, A., Jockisch, A. and Schmidbaur, H. (1998) *Zeitschrift für Naturforschung*, **53b**, 939.

77 Bauer, A. and Schmidbaur, H. (1996) *Journal of the American Chemical Society*, **118**, 5324.

78 Anandhi, U. and Sharp, P.R. (2006) *Inorganica Chimica Acta*, **359**, 3521.

79 Leung, W.-P., So, C.-W., Kan, K.-W., Chan, H.-S. and Mak, T.C.W. (2005) *Organometallics*, **24**, 5033.

80 Leung, W.-P., So, C.-W., Chong, K.-H., Kan, K.-W., Chan, H.-S. and Mak, T.C.W. (2006) *Organometallics*, **25**, 2851.

81 Contel, M., Hellmann, K.W., Gade, L.H., Scowen, I.J., McPartlin, M. and Laguna, M. (1996) *Inorganic Chemistry*, **35**, 3713.

82 Bauer, A., Schneider, W. and Schmidbaur, H. (1997) *Inorganic Chemistry*, **36**, 2225.

83 Bauer, A. and Schmidbaur, H. (1997) *Journal of the Chemical Society-Dalton Transactions*, 1115.

84 Spiekermann, A., Hoffmann, S.D., Kraus, F. and Fassler, T.F. (2007) *Angewandte Chemie (International Edition in English)*, **46**, 1638.

85 Spiekermann, A., Hoffmann, S.D., Fassler, T.F., Krossing, I. and Preiss, U. (2007) *Angewandte Chemie (International Edition in English)*, **46**, 5310.

86 Schenk, C. and Schnepf, A. (2007) *Angewandte Chemie (International Edition in English)*, **46**, 5314.

87 Clegg, W. (1978) *Acta Crystallographica B*, **34**, 278.

88 Findeis, B., Contel, M., Gade, L.H., Laguna, M., Gimeno, M.C., Scowen, I.J. and McPartlin, M. (1997) *Inorganic Chemistry*, **36**, 2386.

89 Demidowicz, Z., Johnston, R.L., Machell, J.C., Mingos, D.M.P. and Williams, I.D. (1988) *Journal of the Chemical Society-Dalton Transactions*, 1751.

90 Mingos, D.M.P., Powell, H.R. and Stolbert, T.L. (1992) *Transition Metal Chemistry*, **17**, 334.

91 Liu, X.-Y., Riera, V., Ruiz, M.A., Lanfranchi, M. and Tiripicchio, A. (2003) *Organometallics*, **22**, 4500.
92 Gädt, T. and Wesemann, L. (2007) *Organometallics*, **26**, 2474.
93 Marx, T., Mosel, B., Pantenburg, I., Hagen, S., Schulze, H. and Wesemann, L. (2003) *Chemistry – A European Journal*, **9**, 4472.
94 Hagen, S., Pantenburg, I., Weigend, F., Wickleder, C. and Wesemann, L. (2003) *Angewandte Chemie (International Edition in English)*, **42**, 1501.
95 Hagen, S., Wesemann, L. and Pantenburg, I. (2005) *Chemical Communications*, 1013.
96 Jones, P.G. (1982) *Zeitschrift für Naturforschung*, **37**, 937.
97 Vicente, J., Arcas, A., Jones, P.G. and Lautner, J. (1990) *Journal of the Chemical Society-Dalton Transactions*, 451.
98 Jones, P.G. (1992) *Acta Crystallographica C*, **48**, 1487.
99 Bojan, V.R., Fernández, E.J., Laguna, A., López-de-Luzuriaga, J.M., Monge, M., Olmos, M.E. and Silvestru, C. (2005) *Journal of the American Chemical Society*, **127**, 11564.
100 Bojan, V.R., Fernández, E.J., Laguna, A., López-de-Luzuriaga, J.M., Monge, M., Olmos, M.E. and Silvestru, C. unpublished results.
101 Fenske, D., Rothenberger, A. and Wieber, S. (2007) *European Journal of Inorganic Chemistry*, 3469.
102 Schuster, O., Schier, A. and Schmidbaur, H. (2003) *Organometallics*, **22**, 4079.
103 Fernández, E.J., Laguna, A., López-de-Luzuriaga, J.M., Monge, M., Nema, M., Olmos, M.E., Pérez, J. and Silvestru, C. (2007) *Chemical Communications*, 571.
104 Kohler, K., Silverio, S.J., Hyla-Kryspin, I., Gleiter, R., Zsolnai, L., Driess, A., Huttner, G. and Lang, H. (1997) *Organometallics*, **16**, 4970.
105 Beckhaus, R., Oster, J., Wang, R. and Bohme, U. (1998) *Organometallics*, **17**, 2215.
106 Fischer, P.J., Young, V.G. Jr. and Ellis, J.E. (1997) *Chemical Communications*, 1249.
107 Drew, M.G.B. (1982) *Acta Crystallographica B*, **38**, 254.
108 Ellis, J.E. (1981) *Journal of the American Chemical Society*, **103**, 6106.
109 Wurst, K., Strahle, J., Beuter, G., Dell'Amico, D.B. and Calderazzo, F. (1991) *Acta Chemica Scandinavica*, **45**, 844.
110 Fajardo, M., Gomez-Sal, M.P., Royo, P., Carrera, S.M. and Blanco, S.G. (1986) *Journal of Organometallic Chemistry*, **312**, C44.
111 Antinolo, A., Burdett, J.K., Chaudret, B., Eisenstein, O., Fajardo, M., Jalon, F., Lahoz, F., Lopez, J.A. and Otero, A. (1990) *Chemical Communications*, 17.
112 Antinolo, A., Jalon, F.A., Otero, A., Fajardo, M., Chaudret, B., Lahoz, F. and Lopez, J.A. (1991) *Journal of the Chemical Society-Dalton Transactions*, 1861.
113 Fischer, P.J., Krohn, K.M., Mwenda, E.T. and Young, V.G. Jr. (2005) *Organometallics*, **24**, 5116.
114 Braunstein, P., Schubert, U. and Burgard, M. (1984) *Inorganic Chemistry*, **23**, 4057.
115 Green, M., Orpen, A.G., Salter, I.D. and Stone, F.G.A. (1982) *Chemical Communications*, 813.
116 Green, M., Orpen, A.G., Salter, I.D. and Stone, F.G.A. (1984) *Journal of the Chemical Society-Dalton Transactions*, 2497.
117 Edelmann, F., Tofke, S. and Behrens, U. (1986) *Journal of Organometallic Chemistry*, **309**, 87.
118 Esterhuysen, M.W. and Raubenheimer, H.G. (2003) *Acta Crystallographica C*, **59**, m286.
119 Vogel, U., Sekar, P., Ahlrichs, R., Huniar, U. and Scheer, M. (2003) *European Journal of Inorganic Chemistry*, 1518.
120 Strunin, B.N., Grandberg, K.I., Andrianov, V.G., Setkina, V.N., Perevalova, E.G., Struchkov, Yu.T. and Kursanov, D.N. (1985) *Doklady Akademii Nauk SSSR (Russ.) (Proceedings of the National Academy of Sciences of the USSR)* **281**, 599.
121 Balch, A.L., Noll, B.C., Olmstead, M.M. and Toronto, D.V. (1992) *Inorganic Chemistry*, **31**, 5226.

122 Pethe, J., Maichle-Mossmer, C. and Strahle, J. (1997) *Zeitschrift für Anorganische und Allgemeine Chemie*, **623**, 1413.

123 Brumas-Soula, B., Dahan, F. and Poilblanc, R. (1998) *New Journal of Chemistry*, **22**, 1067.

124 Braunstein, P., Cura, E. and Herberich, G.E. (2001) *Journal of the Chemical Society-Dalton Transactions*, 1754.

125 Du, S., Kautz, J.A., McGrath, T.D. and Stone, F.G.A. (2001) *Inorganic Chemistry*, **40**, 6563.

126 Carr, N., Gimeno, M.C. and Stone, F.G.A. (1990) *Journal of the Chemical Society-Dalton Transactions*, 2247.

127 Wang, M., Miguel, D., Lopez, E.M., Perez, J., Riera, V., Bois, C. and Jeannin, Y. (2003) *Dalton Transactions*, 961.

128 Ferrer, M., Reina, R., Rossell, O., Seco, M., Alvarez, S., Ruiz, E., Pellinghelli, M.A. and Tiripicchio, A. (1992) *Organometallics*, **11**, 3753.

129 Wilford, J.B. and Powell, H.M. (1969) *Journal of the Chemical Society A: Inorganic, Physical, Theoretical*, 8.

130 Carriedo, G.A., Hodgson, D., Howard, J.A.K., Marsden, K., Stone, F.G.A., Went, M.J. and Woodward, P. (1982) *Chemical Communications*, 1006.

131 Carriedo, G.A., Howard, J.A.K., Stone, F.G.A. and Went, M.J. (1984) *Journal of the Chemical Society-Dalton Transactions*, 2545.

132 Green, M., Howard, J.A.K., James, A.P., Nunn, C.M. and Stone, F.G.A. (1984) *Chemical Communications*, 1113.

133 Green, M., Howard, J.A.K., James, A.P., Nunn, C.M. and Stone, F.G.A. (1987) *Journal of the Chemical Society-Dalton Transactions*, 61.

134 Carriedo, G.A., Riera, V., Sanchez, G. and Solans, X. (1988) *Journal of the Chemical Society-Dalton Transactions*, 1957.

135 Jeffery, J.C., Jelliss, P.A. and Stone, F.G.A. (1994) *Organometallics*, **13**, 2651.

136 Carr, N., Gimeno, M.C., Goldberg, J.E., Pilotti, M.U., Stone, F.G.A. and Topaloglu, I. (1990) *Journal of the Chemical Society-Dalton Transactions*, 2253.

137 Goldberg, J.E., Mullica, D.F., Sappenfield, E.L. and Stone, F.G.A. (1992) *Journal of the Chemical Society-Dalton Transactions*, 2495.

138 Awang, M.R., Carriedo, G.A., Howard, J.A.K., Mead, K.A., Moore, I., Nunn, C.M. and Stone, F.G.A. (1983) *Chemical Communications*, 964.

139 Carriedo, G.A., Howard, J.A.K., Marsden, K., Stone, F.G.A. and Woodward, P. (1984) *Journal of the Chemical Society-Dalton Transactions*, 1589.

140 Carriedo, G.A., Riera, V., Sanchez, G., Solans, X. and Labrador, M. (1990) *Journal of Organometallic Chemistry*, **391**, 431.

141 Ipaktschi, J. and Munz, F. (2002) *Organometallics*, **21**, 977.

142 Albano, V.G., Busetto, L., Cassani, M.C., Sabatino, P., Schmitz, A. and Zanotti, V. (1995) *Journal of the Chemical Society-Dalton Transactions*, 2087.

143 Kim, H.P., Kim, S., Jacobson, R.A. and Angelici, R.J. (1986) *Journal of the American Chemical Society*, **108**, 5154.

144 Blum, T., Braunstein, P., Tiripicchio, A. and Tiripicchio Camellini, M. (1988) *New Journal of Chemistry*, **12**, 539.

145 Byers, P.K., Carr, N. and Stone, F.G.A. (1990) *Journal of the Chemical Society-Dalton Transactions*, 3701.

146 Carr, N., Fernandez, J.R. and Stone, F.G.A. (1991) *Organometallics*, **10**, 2718.

147 Kinsch, E.M. and Stephan, D.W. (1985) *Inorganica Chimica Acta*, **96**, L87.

148 Charnock, J.M., Bristow, S., Nicholson, J.R., Garner, C.D. and Clegg, W. (1987) *Journal of the Chemical Society-Dalton Transactions*, 303.

149 Canales, F., Gimeno, M.C., Jones, P.G. and Laguna, A. (1997) *Journal of the Chemical Society-Dalton Transactions*, 439.

150 Zheng, H.-g., Ji, W., Low, M.L.K., Sakane, G., Shibahara, T. and Xin, X.-q. (1997) *Journal of the Chemical Society-Dalton Transactions*, 2357.

151 Ma, M.-H., Zheng, H.-G., Tan, W.-L., Zhou, J.-L., Raj, S.S.S., Fun, H.-K. and Xin, X.-Q. (2003) *Inorganica Chimica Acta*, **342**, 151.

152 Zhou, J.-L., Song, Y.-L., Mo, H.-B., Li, Y.-Z., Zheng, H.-G. and Xin, X.-Q. (2005) *Zeitschrift für Anorganische und Allgemeine Chemie*, **631**, 182.

153 Pritchard, R.G., Moore, L.S., Parish, R.V., McAuliffe, C.A. and Beagley, B. (1988) *Acta Crystallographica C*, **44**, 2022.

154 Christuk, C.C., Ansari, M.A. and Ibers, J.A. (1992) *Inorganic Chemistry*, **31**, 4365.

155 Muller, A., Wienboker, U. and Penk, M. (1989) *Chimia*, **43**, 50.

156 Song, L., Xia, S.-Q., Hu, S.-M., Du, S.-W. and Wu, X.-T. (2005) *Polyhedron*, **24**, 831.

157 Lang, J.-P., Kawaguchi, H. and Tatsumi, K. (1998) *Journal of Organometallic Chemistry*, **569**, 109.

158 Hartung, H., Walther, B., Baumeister, U., Bottcher, H.-C., Krug, A., Rosche, F. and Jones, P.G. (1992) *Polyhedron*, **11**, 1563.

159 Braunstein, P., de Jesus, E., Tiripicchio, A. and Ugozzoli, F. (1992) *Inorganic Chemistry*, **31**, 411.

160 Reina, R., Rodriguez, L., Rossell, O., Seco, M., Font-Bardia, M. and Solans, X. (2001) *Organometallics*, **20**, 1575.

161 Pergola, R.D., Fumagalli, A., de Biani, F.F., Garlaschelli, L., Laschi, F., Malatesta, M.C., Manassero, M., Roda, E., Sansoni, M. and Zanello, P. (2004) *European Journal of Inorganic Chemistry*, 3901.

162 Galassi, R., Poli, R., Quadrelli, E.A. and Fettinger, J.C. (1997) *Inorganic Chemistry*, **36**, 3001.

163 Lin, J.T., Hsiao, Y.-M., Liu, L.-K. and Yeh, S.K. (1988) *Organometallics*, **7**, 2065.

164 Lin, J.T., Ch'ing, C., Lo, C.H., Wang, S.Y., Tsai, T.Y.R., Chen, M.M., Wen, Y.S. and Lin, K.J. (1996) *Organometallics*, **15**, 2132.

165 Berry, A., Green, M.L.H., Bandy, J.A. and Prout, K. (1991) *Journal of the Chemical Society-Dalton Transactions*, 2185.

166 Beuter, G., Brodbeck, A., Holzer, M., Maier, S. and Strahle, J. (1992) *Zeitschrift für Anorganische und Allgemeine Chemie*, **616**, 27.

167 Kappen, T.G.M.M., van den Broek, A.C.M., Schlebos, P.P.J., Bour, J.J., Bosman, W.P., Smits, J.M.M., Buerskens, P.T. and Steggerda, J.J. (1992) *Inorganic Chemistry*, **31**, 4075.

168 Beuter, G. and Strahle, J. (1988) *Angewandte Chemie (International Edition in English)*, **27**, 1094.

169 Beuter, G. and Strahle, J. (1989) *Journal of Organometallic Chemistry*, **372**, 67.

170 Pergola, R.D., Garlaschelli, L., Malatesta, M.C., Manassero, C. and Manassero, M. (2006) *Inorganic Chemistry*, **45**, 8465.

171 Mannan, K.A.I.F.M. (1967) *Acta Crystallographica*, **23**, 649.

172 Onaka, S., Katsukawa, Y. and Yamashita, M. (1998) *Journal of Organometallic Chemistry*, **564**, 249.

173 Nunokawa, K., Onaka, S., Yamaguchi, T., Yaguchi, M., Tatematsu, T., Watase, S., Nakamoto, M. and Ito, T. (2002) *Journal of Coordination Chemistry*, **55**, 1353.

174 Hata, M., Kautz, J.A., Lu, X.L., McGrath, T.D. and Stone, F.G.A. (2004) *Organometallics*, **23**, 3590.

175 Barrientos-Penna, C.F., Einstein, F.W.B., Jones, T. and Sutton, D. (1985) *Inorganic Chemistry*, **24**, 632.

176 Luo, S., Burns, C.J., Kubas, G.J., Bryan, J.C. and Crabtree, R.H. (2000) *Journal of Cluster Science*, **11**, 189.

177 Hodson, B.E., McGrath, T.D. and Stone, F.G.A. (2005) *Organometallics*, **24**, 3386.

178 Richter, M., Fenske, D. and Strahle, J. (2000) *Zeitschrift für Naturforschung*, **55b**, 907.

179 Iggo, J.A., Mays, M.J., Raithby, P.R. and Henrick, K. (1984) *Journal of the Chemical Society-Dalton Transactions*, 633.

180 Haupt, H.-J., Heinekamp, C., Florke, U. and Juptner, U. (1992) *Zeitschrift für Anorganische und Allgemeine Chemie*, **608**, 100.

181 Lee, K.H., Low, P.M.N., Hor, T.S.A., Wen, Y.-S. and Liu, L.K. (2001) *Organometallics*, **20**, 3250.

182 Low, P.M.N., Tan, A.L., Hor, T.S.A., Wen, Y.-S. and Liu, L.-K. (1996) *Organometallics*, **15**, 2595.

183 Haupt, H.-J., Schwefer, M. and Florke, U. (1995) *Inorganic Chemistry*, **34**, 292.

184 Florke, U. and Haupt, H.-J. (1995) *Zeitschrift für Kristallographie*, **210**, 471.

185 Haupt, H.-J., Schwefer, M., Egold, H. and Florke, U. (1995) *Inorganic Chemistry*, **34**, 5461.

186 Riera, V., Ruiz, M.A., Tiripicchio, A. and Tiripicchio Camellini, M. (1987) *Journal of the Chemical Society-Dalton Transactions*, 1551.

187 Sutherland, B.R., Ho, D.M., Huffman, J.C. and Caulton, K.G. (1987) *Angewandte Chemie (International Edition in English)*, **26**, 135.

188 Sutherland, B.R., Folting, K., Streib, W.E., Ho, D.M., Huffman, J.C. and Caulton, K.G. (1987) *Journal of the American Chemical Society*, **109**, 3489.

189 Moehring, G.A., Fanwick, P.E. and Walton, R.A. (1987) *Inorganic Chemistry*, **26**, 1861.

190 Haupt, H.-J., Heinekamp, C. and Florke, U. (1990) *Inorganic Chemistry*, **29**, 2955.

191 Haupt, H.-J., Schwefer, M. and Florke, U. (1995) *Zeitschrift für Anorganische und Allgemeine Chemie*, **621**, 1098.

192 Bruce, M.I., Low, P.J., Skelton, B.W. and White, A.H. (1996) *Journal of Organometallic Chemistry*, **515**, 65.

193 Egold, H., Schwarze, D. and Florke, U. (1999) *Journal of the Chemical Society-Dalton Transactions*, 3203.

194 Haupt, H.-J., Petters, D. and Florke, U. (1999) *Zeitschrift für Anorganische und Allgemeine Chemie*, **625**, 1652.

195 Koridze, A.A., Zdanovich, V.I., Sheloumov, A.M., Dolgushin, F.M., Ezernitskaya, M.G. and Petrovskii, P.V. (2001) *Izvestiia Akademii Nauk SSSR. Seriia Khimicheskaia*, 2331.

196 Carreno, R., Riera, V., Ruiz, M.A., Bois, C. and Jeannin, Y. (1992) *Organometallics*, **11**, 2923.

197 Riera, V., Ruiz, M.A., Tiripicchio, A. and Tiripicchio-Camellini, M. (1993) *Organometallics*, **12**, 2962.

198 Liu, X.-Y., Riera, V., Ruiz, M.A., Tiripicchio, A. and Tiripicchio-Camellini, M. (1996) *Organometallics*, **15**, 974.

199 Beringhelli, T., Ciani, G., D'Alfonso, G., De Maldé, V. and Freni, M. (1986) *Chemical Communications*, 735.

200 Haupt, H.-J., Florke, U. and Schnieder, H. (1991) *Acta Crystallographica C*, **47**, 2304.

201 Seewald, O., Florke, U., Egold, H., Haupt, H.-J. and Schwefer, M. (2006) *Zeitschrift für Anorganische und Allgemeine Chemie*, **632**, 204.

202 Haupt, H.-J., Seewald, O., Florke, U., Buss, V. and Weyhermuller, T. (2001) *Journal of the Chemical Society-Dalton Transactions*, 3329.

203 Haupt, H.-J., Heinekamp, C. and Florke, U. (1990) *Zeitschrift für Anorganische und Allgemeine Chemie*, **585**, 168.

204 Haupt, H.-J., Schwefer, M., Egold, H. and Florke, U. (1997) *Inorganic Chemistry*, **36**, 184.

205 Haupt, H.-J., Egold, H., Siefert, R. and Florke, U. (1998) *Zeitschrift für Anorganische und Allgemeine Chemie*, **624**, 1863.

206 Cabeza, J.A., Riera, V., Trivedi, R. and Grepioni, F. (2000) *Organometallics*, **19**, 2043.

207 Beringhelli, T., D'Alfonso, G., Garavaglia, M.G., Panigati, M., Mercandelli, P. and Sironi, A. (2002) *Organometallics*, **21**, 2705.

208 Katsukawa, Y., Onaka, S., Yamada, Y. and Yamashita, M. (1999) *Inorganica Chimica Acta*, **294**, 255.

209 Du, S., Kautz, J.A., McGrath, T.D. and Stone, F.G.A. (2003) *Angewandte Chemie (International Edition in English)*, **42**, 5728.

210 Holzer, M. and Strahle, J. (1994) *Zeitschrift für Anorganische und Allgemeine Chemie*, **620**, 786.

211 Pivoriunas, G., Richter, M. and Strahle, J. (2005) *Inorganica Chimica Acta*, **358**, 4301.

212 Haupt, H.-J., Petters, D. and Florke, U. (2000) *Zeitschrift für Anorganische und Allgemeine Chemie*, **626**, 2293.

213 Beuter, G. and Strahle, J. (1989) *Journal of the Less-Common Metals*, **156**, 387.

214 Nicholson, B.K., Bruce, M.I., bin Shawkataly, O. and Tiekink, E.R.T. (1992) *Journal of Organometallic Chemistry*, **440**, 411.

215 Alexander, B.D., Boyle, P.D., Johnson, B.J., Casalnuovo, J.A., Johnson, S.M., Mueting, A.M. and Pignolet, L.H. (1987) *Inorganic Chemistry*, **26**, 2547.

216 Boyle, P.D., Johnson, B.J., Buehler, A. and Pignolet, L.H. (1986) *Inorganic Chemistry*, **25**, 5.

217 Boyle, P.D., Johnson, B.J., Alexander, B.D., Casalnuovo, J.A., Gannon, P.R., Johnson, S.M., Larka, E.A., Mueting, A.M. and Pignolet, L.H. (1987) *Inorganic Chemistry*, **26**, 1346.

218 Mielcke, J. and Strahle, J. (1992) *Angewandte Chemie (International Edition in English)*, **31**, 464.

219 Latten, J.L., Hsu, G., Henly, T.J., Wilson, S.R. and Shapley, J.R. (1998) *Inorganic Chemistry*, **37**, 2520.

220 Henly, T.J., Shapley, J.R. and Rheingold, A.L. (1986) *Journal of Organometallic Chemistry*, **310**, 55.

221 Ma, L., Brand, U. and Shapley, J.R. (1998) *Inorganic Chemistry*, **37**, 3060.

222 Carreno, R., Riera, V., Ruiz, M.A., Tiripicchio, A. and Tiripicchio-Camellini, M. (1994) *Organometallics*, **13**, 993.

223 Burrell, A.K., Clark, D.L., Gordon, P.L., Sattelberger, A.P. and Bryan, J.C. (1994) *Journal of the American Chemical Society*, **116**, 3813.

224 Lewis, J. and Raithby, P.R. (1999) in *Metal Clusters in Chemistry* (eds P. Braunstein, L.A. Oro and P.R. Raithby), Wiley-VCH, Weinheim, pp. 348–380.

225 Simon, F.E. and Lauher, J.W. (1980) *Inorganic Chemistry*, **19**, 2338.

226 Schubert, U., Kunz, E., Knorr, M. and Muller, J. (1987) *Chemische Berichte*, **120**, 1079.

227 Braunstein, P., Knorr, M., Schubert, U., Lanfranchi, M. and Tiripicchio, A. (1991) *Journal of the Chemical Society-Dalton Transactions*, 1507.

228 Reinhard, G., Hirle, B. and Schubert, U. (1992) *Journal of Organometallic Chemistry*, **427**, 173.

229 Alexander, B.D., Johnson, B.J., Johnson, S.M., Boyle, P.D., Kann, N.C., Mueting, A.M. and Pignolet, L.H. (1987) *Inorganic Chemistry*, **26**, 3506.

230 Alexander, B.D., Gomez-Sal, M.P., Gannon, P.R., Blaine, C.A., Boyle, P.D., Mueting, A.M. and Pignolet, L.H. (1988) *Inorganic Chemistry*, **27**, 3301.

231 McLennan, A.J. and Welch, A.J. (1989) *Acta Crystallographica C*, **45**, 1721.

232 Ellis, D.D., Couchman, S.M., Jeffery, J.C., Malget, J.M. and Stone, F.G.A. (1999) *Inorganic Chemistry*, **38**, 2981.

233 Hill, A.F., Roper, W.R., Waters, J.M. and Wright, A.H. (1983) *Journal of the American Chemical Society*, **105**, 5939.

234 Arndt, L.W., Darensbourg, M.Y., Fackler, J.P., Lusk, R.J., Marler, D.O. and Youngdahl, K.A. (1985) *Journal of the American Chemical Society*, **107**, 7218.

235 Umland, H. and Behrens, U. (1985) *Journal of Organometallic Chemistry*, **287**, 109.

236 Rossell, O., Seco, M. and Jones, P.G. (1990) *Inorganic Chemistry*, **29**, 348.

237 Reina, R., Rossell, O., Seco, M., Ros, J., Yanez, R. and Perales, A. (1991) *Inorganic Chemistry*, **30**, 3973.

238 Delgado, E., Hernandez, E., Rossell, O., Seco, M., Gutierrez Puebla, E. and Ruiz, C. (1993) *Journal of Organometallic Chemistry*, **455**, 177.

239 Ferrer, M., Julia, A., Rossell, O., Seco, M., Pellinghelli, M.A. and Tiripicchio, A. (1997) *Organometallics*, **16**, 3715.

240 Andrianov, V.G., Struchkov, Yu.T. and Rossinskaya, E.R. (1974) *Zhurnal Strukturnoi Khimii*, **15**, 74.

241 Briant, C.E., Hall, K.P. and Mingos, D.M.P. (1983) *Chemical Communications*, 843.

242 Arndt, L.W., Ash, C.E., Darensbourg, M.Y., Hsiao, Y.M., Kim, C.M., Reibenspies, J. and Youngdahl, K.A. (1990) *Journal of Organometallic Chemistry*, **394**, 733.

243 Albano, V.G., Monari, M., Iapalucci, M.C. and Longoni, G. (1993) *Inorganica Chimica Acta*, **213**, 183.

244 Braunstein, P., Herberich, G.E., Neuschutz, M., Schmidt, M.U., Englert,

U., Lecante, P. and Mosset, A. (1998) *Organometallics*, **17**, 2177.

245 Engel, D.W., Haines, R.J., Horsfield, E.C. and Sundermeyer, J. (1989) *Chemical Communications*, 1457.

246 Engel, D.W., Field, J.S., Haines, R.J., Honrath, U., Horsfield, E.C., Sundermeyer, J. and Woollam, S.F. (1994) *Journal of the Chemical Society-Dalton Transactions*, 1131.

247 Alexander, B.D., Johnson, B.J., Johnson, S.M., Casalnuovo, A.L. and Pignolet, L.H. (1986) *Journal of the American Chemical Society*, **108**, 4409.

248 Johnson, B.F.G., Lewis, J., Raithby, P.R. and Sanders, A. (1984) *Journal of Organometallic Chemistry*, **260**, C29.

249 Bruce, M.I. and Nicholson, B.K. (1983) *Journal of Organometallic Chemistry*, **250**, 627.

250 Delgado, E., Hernandez, E., Rossell, O., Seco, M. and Solans, X. (1993) *Journal of the Chemical Society-Dalton Transactions*, 2191.

251 Rossell, O., Seco, M., Reina, R., Font-Bardia, M. and Solans, X. (1994) *Organometallics*, **13**, 2127.

252 Roof, L.C., Smith, D.M., Drake, G.W., Pennington, W.T. and Kolis, J.W. (1995) *Inorganic Chemistry*, **34**, 337.

253 Albano, V.G., Castellari, C., Femoni, C., Iapalucci, M.C., Longoni, G., Monari, M., Rauccio, M. and Zacchini, S. (1999) *Inorganica Chimica Acta*, **291**, 372.

254 Gubin, S.P., Polyakova, L.A., Churakov, A.V. and Kuz'mina, L.G. (1999) *Izvestiia Akademii Nauk SSSR. Seriia Khimicheskaia*, 1779.

255 Aitchison, A.A. and Farrugia, L.J. (1986) *Organometallics*, **5**, 1103.

256 Fischer, K., Muller, M. and Vahrenkamp, H. (1984) *Angewandte Chemie (International Edition in English)*, **23**, 140.

257 Ahlgren, M., Pakkanen, T.T. and Tahvanainen, I. (1987) *Journal of Organometallic Chemistry*, **323**, 91.

258 Braunstein, P., Knorr, M., Stahrfeldt, T., DeCian, A. and Fischer, J. (1993) *Journal of Organometallic Chemistry*, **459**, C1.

259 Albano, V.G., Castellari, C., Iapalucci, M.C., Longoni, G., Monari, M., Paselli, A. and Zacchini, S. (1999) *Journal of Organometallic Chemistry*, **573**, 261.

260 Alvarez, S., Rossell, O., Seco, M., Valls, J., Pellinghelli, M.A. and Tiripicchio, A. (1991) *Organometallics*, **10**, 2309.

261 Green, M., Mead, K.A., Mills, R.M., Salter, I.D., Stone, F.G.A. and Woodward, P. (1982) *Chemical Communications*, 51.

262 Bateman, L.W., Green, M., Mead, K.A., Mills, R.M., Salter, I.D., Stone, F.G.A. and Woodward, P. (1983) *Journal of the Chemical Society-Dalton Transactions*, 2599.

263 Mays, M.J., Raithby, P.R., Taylor, P.L. and Henrick, K. (1984) *Journal of the Chemical Society-Dalton Transactions*, 959.

264 Bateman, L.W., Green, M., Howard, J.A.K., Mead, K.A., Mills, R.M., Salter, I.D., Stone, F.G.A. and Woodward, P. (1982) *Chemical Communications*, 773.

265 Braunstein, P., Predieri, G., Tiripicchio, A. and Sappa, E. (1982) *Inorganica Chimica Acta*, **63**, 113.

266 Lavigne, G., Papageorgiou, F. and Bonnet, J.-J. (1984) *Inorganic Chemistry*, **23**, 609.

267 Bruce, M.I., Shawkataly, O.B. and Nicholson, B.K. (1985) *Journal of Organometallic Chemistry*, **286**, 427.

268 Bruce, M.I., Williams, M.L., Patrick, J.M., Skelton, B.W. and White, A.H. (1986) *Journal of the Chemical Society-Dalton Transactions*, 2557.

269 Jungbluth, H., Stoeckli-Evans, H. and Suss-Fink, G. (1990) *Journal of Organometallic Chemistry*, **391**, 109.

270 Bruce, M.I., Humphrey, P.A., Horn, E., Tiekink, E.R.T., Skelton, B.W. and White, A.H. (1992) *Journal of Organometallic Chemistry*, **429**, 207.

271 Mague, J.T. and Lloyd, C.L. (1992) *Organometallics*, **11**, 26.

272 Bruce, M.I., Zaitseva, N.N., Skelton, B.W. and White, A.H. (1999) *Journal of the Chemical Society-Dalton Transactions*, 2777.

273 Bruce, M.I., Skelton, B.W., White, A.H. and Zaitseva, N.N. (2000) *Journal of the Chemical Society-Dalton Transactions*, 881.

274 Braunstein, P., Oswald, B., Tiripicchio, A. and Ugozzoli, F. (2000) *Journal of the Chemical Society-Dalton Transactions*, 2195.

275 Bottcher, H.-C., Graf, M., Merzweiler, K. and Wagner, C. (2000) *Polyhedron*, **19**, 2593.

276 Sheloumov, A.M., Koridze, A.A., Doldushin, F.M., Starikova, Z.A., Ezernitskaya, M.G. and Petrovskii, P.V. (2000) *Izvestiia Akademii Nauk SSSR. Seriia Khimicheskaia*, 1295.

277 Khairul, W.M., Porres, L., Albesa-Jove, D., Senn, M.S., Jones, M., Lydon, D.P., Howard, J.A.K., Beeby, A., Marder, T.B. and Low, P.J. (2006) *Journal of Cluster Science*, **17**, 65.

278 Bruce, M.I., Zaitseva, N.N., Skelton, B.W. and White, A.H. (2005) *Journal of Organometallic Chemistry*, **690**, 3268.

279 Johnson, B.F.G., Kaner, D.A., Lewis, J. and Raithby, P.R. (1981) *Journal of Organometallic Chemistry*, **215**, C33.

280 Burgess, K., Johnson, B.F.G., Lewis, J. and Raithby, P.R. (1983) *Journal of the Chemical Society-Dalton Transactions*, 1661.

281 Bruce, M.I., Horn, E., Matisons, J.G. and Snow, M.R. (1985) *Journal of Organometallic Chemistry*, **286**, 271.

282 Deeming, A.J., Donovan-Mtunzi, S. and Hardcastle, K. (1986) *Journal of the Chemical Society-Dalton Transactions*, 543.

283 Farrugia, L.J. (1986) *Acta Crystallographica C*, **42**, 680.

284 Housecroft, C.E., Johnson, B.F.G., Lewis, J., Lunniss, J.A., Owen, S.M. and Raithby, P.R. (1991) *Journal of Organometallic Chemistry*, **409**, 271.

285 Johnson, B.F.G., Lahoz, F.J., Lewis, J., Prior, N.D., Raithby, P.R. and Wong, W.-T. (1992) *Journal of the Chemical Society-Dalton Transactions*, 1701.

286 Harding, M.M., Kariuki, B., Mathews, A.J., Smith, A.K. and Braunstein, P. (1994) *Journal of the Chemical Society-Dalton Transactions*, 33.

287 Dodd, I.M., Hao, Q., Harding, M.M. and Prince, S.M. (1994) *Acta Crystallographica B*, **50**, 441.

288 Hay, C.M., Leadbeater, N.E., Lewis, J., Raithby, P.R. and Burgess, K. (1998) *New Journal of Chemistry*, **22**, 787.

289 Ahrens, B., Cole, J.M., Hickey, J.P., Martin, J.N., Mays, M.J., Raithby, P.R., Teat, S.J. and Woods, A.D. (2003) *Dalton Transactions*, 1389.

290 Sanchez-Cabrera, G., Zuno-Cruz, F.J., Ordonez-Flores, B.A., Rosales-Hoz, M.J. and Leyva, M.A. (2007) *Journal of Organometallic Chemistry*, **692**, 2138.

291 Albinati, A., Venanzi, L.M. and Wang, G. (1993) *Inorganic Chemistry*, **32**, 3660.

292 Braunstein, P., Knorr, M., Tiripicchio, A. and Tiripicchio Camellini, M. (1992) *Inorganic Chemistry*, **31**, 3685.

293 Delgado, E., Donnadieu, B., Garcia, M.E., Garcia, S., Ruiz, M.A. and Zamora, F. (2002) *Organometallics*, **21**, 780.

294 Cabrera, A., Delgado, E., Pastor, C., Maestro, M.A. and Zamora, F. (2005) *Inorganica Chimica Acta*, **358**, 1521.

295 Ellis, D.D., Franken, A. and Stone, F.G.A. (1999) *Organometallics*, **18**, 2362.

296 Pergola, R.D., Garlaschelli, L., Demartin, F., Manassero, M., Masciocchi, N. and Sansoni, M. (1990) *Journal of the Chemical Society-Dalton Transactions*, 127.

297 Pergola, R.D., Fracchia, L., Garlaschelli, L., Manassero, M. and Sansoni, M. (1995) *Journal of the Chemical Society-Dalton Transactions*, 2763.

298 Roland, E., Fischer, K. and Vahrenkamp, H. (1983) *Angewandte Chemie (International Edition in English)*, **22**, 326.

299 Fischer, K., Deck, W., Schwarz, M. and Vahrenkamp, H. (1985) *Chemische Berichte*, **118**, 4946.

300 Poliakova, L.A., Gubin, S.P., Belyakova, O.A., Zubavichus, Y.V. and Slovokhotov, Y.L. (1997) *Organometallics*, **16**, 4527.

301 Horwitz, C.P., Holt, E.M. and Shriver, D.F. (1985) *Journal of the American Chemical Society*, **107**, 281.

302 Horwitz, C.P., Holt, E.M., Brock, C.P. and Shriver, D.F. (1985) *Journal of the American Chemical Society*, **107**, 8136.

303 Ferguson, G., Gallagher, J.F., Kelleher, A.-M., Spalding, T.R. and Deeney, F.T.

(2005) *Journal of Organometallic Chemistry*, **690**, 2888.

304 Draper, S.M., Housecroft, C.E., Rees, J.E., Shongwe, M.S., Haggerty, B.S. and Rheingold, A.L. (1992) *Organometallics*, **11**, 2356.

305 Johnson, B.F.G., Kaner, D.A., Lewis, J., Raithby, P.R. and Rosales, M.J. (1982) *Journal of Organometallic Chemistry*, **231**, C59.

306 Rossell, O., Seco, M., Segales, G., Alvarez, S., Pellinghelli, M.A., Tiripicchio, A. and de Montauzon, D. (1997) *Organometallics*, **16**, 236.

307 Thone, C. and Vahrenkamp, H. (1995) *Journal of Organometallic Chemistry*, **485**, 185.

308 Gubin, S.P., Galuzina, T.V., Golovaneva, I.F., Klyagina, A.P., Polyakova, L.A., Belyakova, O.A., Zubavichus, Ya.V. and Slovokhotov, Yu.L. (1997) *Journal of Organometallic Chemistry*, **549**, 55.

309 Sunick, D.L., White, P.S. and Schauer, C.K. (1993) *Inorganic Chemistry*, **32**, 5665.

310 Fumagalli, A., Italia, D., Malatesta, M.C., Ciani, G., Moret, M. and Sironi, A. (1996) *Inorganic Chemistry*, **35**, 1765.

311 Evans, J., Stroud, P.M. and Webster, M. (1991) *Journal of the Chemical Society-Dalton Transactions*, 1017.

312 Evans, J., Street, A.C. and Webster, M. (1987) *Organometallics*, **6**, 794.

313 Chen, C.-C., Chi, Y., Peng, S.-M. and Lee, G.-H. (1993) *Journal of the Chemical Society-Dalton Transactions*, 1823.

314 Farrugia, L.J., Freeman, M.J., Green, M., Orpen, A.G., Stone, F.G.A. and Salter, I.D. (1983) *Journal of Organometallic Chemistry*, **249**, 273.

315 Bruce, M.I. and Nicholson, B.K. (1984) *Organometallics*, **3**, 101.

316 Bruce, M.I., Horn, E., Shawkataly, O.B. and Snow, M.R. (1985) *Journal of Organometallic Chemistry*, **280**, 289.

317 Brown, S.S.D., Hudson, S., Salter, I.D. and McPartlin, M. (1987) *Journal of the Chemical Society-Dalton Transactions*, 1967.

318 Bruce, M.I., Horn, E., Humphrey, P.A. and Tiekink, E.R.T. (1996) *Journal of Organometallic Chemistry*, **518**, 121.

319 Bruce, M.I., Humphrey, P.A., Skelton, B.W. and White, A.H. (2004) *Journal of Organometallic Chemistry*, **689**, 2558.

320 Collins, C.A., Salter, I.D., Sik, V., Williams, S.A. and Adatia, T. (1998) *Journal of the Chemical Society-Dalton Transactions*, 1107.

321 Blohm, M.L. and Gladfelter, W.L. (1987) *Inorganic Chemistry*, **26**, 459.

322 Housecroft, C.E., Nixon, D.M. and Rheingold, A.L. (1999) *Polyhedron*, **18**, 2415.

323 Galsworthy, J.R., Housecroft, C.E. and Rheingold, A.L. (1996) *Journal of the Chemical Society-Dalton Transactions*, 2917.

324 Cowie, A.G., Johnson, B.F.G., Lewis, J. and Raithby, P.R. (1984) *Chemical Communications*, 1710.

325 Bruce, M.J., Gulbis, J.M., Humphrey, P.A., Surynt, R.J. and Tiekink, E.R.T. (1997) *Australian Journal of Chemistry*, **50**, 875.

326 Johnson, B.F.G., Kaner, D.A., Lewis, J., Raithby, P.R. and Taylor, M.J. (1982) *Chemical Communications*, 314.

327 Li, Y. and Wong, W.-T. (2003) *European Journal of Inorganic Chemistry*, 2651.

328 Amoroso, A.J., Edwards, A.J., Johnson, B.F.G., Lewis, J., Al-Mandhary, M.R., Raithby, P.R., Saharan, V.P. and Wong, W.T. (1993) *Journal of Organometallic Chemistry*, **443**, C11.

329 Amoroso, A.J., Johnson, B.F.G., Lewis, J., Massey, A.D., Raithby, P.R. and Wong, W.T. (1992) *Journal of Organometallic Chemistry*, **440**, 219.

330 Li, Y., Pan, W.-X. and Wong, W.-T. (2002) *Journal of Cluster Science*, **13**, 223.

331 Su, C.-J., Chi, Y., Peng, S.-M. and Lee, G.-H. (1996) *Journal of Cluster Science*, **7**, 85.

332 Braunstein, P., Rose, J., Lanfredi, A.M.M. and Tiripicchio, A. (1984) *Journal of the Chemical Society-Dalton Transactions*, 1843.

333 Burgess, K., Johnson, B.F.G., Kaner, D.A., Lewis, J., Raithby, P.R. and Syed-Mustaffa,

S.N.A.B. (1983) *Chemical Communications*, 455.
334 Puga, J., Sanchez-Delgado, R.A., Ascanio, J. and Braga, D. (1986) *Chemical Communications*, 1631.
335 Puga, J., Arce, A., Sanchez-Delgado, R.A., Ascanio, J., Andriollo, A., Braga, D. and Grepioni, F. (1988) *Journal of the Chemical Society-Dalton Transactions*, 913.
336 Rossell, O., Seco, M., Segales, G., Johnson, B.F.G., Dyson, P.J. and Ingham, S.L. (1996) *Organometallics*, **15**, 884.
337 Housecroft, C.E. and Rheingold, A.L. (1986) *Journal of the American Chemical Society*, **108**, 6420.
338 Housecroft, C.E. and Rheingold, A.L. (1987) *Organometallics*, **6**, 1332.
339 Housecroft, C.E., Shongwe, M.S. and Rheingold, A.L. (1989) *Organometallics*, **8**, 2651.
340 Housecroft, C.E., Shongwe, M.S., Rheingold, A.L. and Zanello, P. (1991) *Journal of Organometallic Chemistry*, **408**, 7.
341 Housecroft, C.E., Shongwe, M.S. and Rheingold, A.L. (1988) *Organometallics*, **7**, 1885.
342 Reina, R., Riba, O., Rossell, O., Seco, M., Font-Bardia, M. and Solans, X. (2002) *Organometallics*, **21**, 5307.
343 Beuter, G. and Strahle, J. (1989) *Zeitschrift für Naturforschung*, **44b**, 647.
344 Kakkonen, H.J., Tunkkari, L., Ahlgren, M., Pursiainen, J. and Pakkanen, T.A. (1995) *Journal of Organometallic Chemistry*, **496**, 93.
345 Brown, S.S.D., Salter, I.D. and Adatia, T. (1993) *Journal of the Chemical Society-Dalton Transactions*, 559.
346 Salter, I.D., Sik, V., Williams, S.A. and Adatia, T. (1996) *Journal of the Chemical Society-Dalton Transactions*, 643.
347 Cowie, A.G., Johnson, B.F.G., Lewis, J., Nicholls, J.N., Raithby, P.R. and Rosales, M.J. (1983) *Journal of the Chemical Society-Dalton Transactions*, 2311.
348 Johnson, B.F.G., Lewis, J., Nicholls, J.N., Puga, J. and Whitmire, K.H. (1983) *Journal of the Chemical Society-Dalton Transactions*, 787.
349 Cowie, A.G., Johnson, B.F.G., Lewis, J., Nicholls, J.N., Raithby, P.R. and Swanson, A.G. (1984) *Chemical Communications*, 637.
350 Henrick, K., Johnson, B.F.G., Lewis, J., Mace, J., McPartlin, M. and Morris, J. (1985) *Chemical Communications*, 1617.
351 Housecroft, C.E., Matthews, D.M. and Rheingold, A.L. (1992) *Organometallics*, **11**, 2959.
352 Hattersley, A.D., Housecroft, C.E. and Rheingold, A.L. (1996) *Journal of the Chemical Society-Dalton Transactions*, 603.
353 Adams, C.J., Bruce, M.I., Skelton, B.W. and White, A.H. (1992) *Journal of Cluster Science*, **3**, 219.
354 Freeman, M.J., Orpen, A.G. and Salter, I.D. (1987) *Journal of the Chemical Society-Dalton Transactions*, 379.
355 Brown, S.S.D., Salter, I.D., Dent, A.J., Kitchen, G.F.M., Orpen, A.G., Bates, P.A. and Hursthouse, M.B. (1989) *Journal of the Chemical Society-Dalton Transactions*, 1227.
356 Bates, P.A., Brown, S.S.D., Dent, A.J., Hursthouse, M.B., Kitchen, G.F.M., Orpen, A.G., Salter, I.D. and Sik, V. (1986) *Chemical Communications*, 600.
357 Brown, S.S.D., Salter, I.D., Dyson, D.B., Parish, R.V., Bates, P.A. and Hursthouse, M.B. (1988) *Journal of the Chemical Society-Dalton Transactions*, 1795.
358 Adatia, T. (1993) *Acta Crystallographica C*, **49**, 1926.
359 Ang, H.-G., Ang, S.-G. and Du, S. (1999) *Journal of Organometallic Chemistry*, **589**, 133.
360 Draper, S.M., Housecroft, C.E. and Rheingold, A.L. (1992) *Journal of Organometallic Chemistry*, **435**, 9.
361 Chipperfield, A.K., Housecroft, C.E. and Rheingold, A.L. (1990) *Organometallics*, **9**, 681.
362 Bruce, M.I., Skelton, B.W., White, A.H. and Zaitseva, N.N. (1999) *Inorganic Chemistry Communications*, **2**, 453.
363 Bruce, M.I., Shawkataly, O.B. and Nicholson, B.K. (1984) *Journal of Organometallic Chemistry*, **275**, 223.

364 Bruce, M.I., Humphrey, P.A., Skelton, B.W. and White, A.H. (1997) *Journal of Organometallic Chemistry*, **545**, 207.

365 Al-Mandhary, M.R.A., Buntem, R., Cathey, C., Lewis, J., de Arellano, M.C.R., Shields, G.P., Doherty, C.L. and Raithby, P.R. (2003) *Inorganica Chimica Acta*, **350**, 299.

366 Johnson, B.F.G., Khattar, R., Lewis, J. and Raithby, P.R. (1989) *Journal of the Chemical Society-Dalton Transactions*, 1421.

367 Johnson, B.F.G., Kaner, D.A., Lewis, J., Raithby, P.R. and Taylor, M.J. (1982) *Polyhedron*, **1**, 105.

368 Hay, C.M., Johnson, B.F.G., Lewis, J., McQueen, R.C.S., Raithby, P.R., Sorrell, R.M. and Taylor, M.J. (1985) *Organometallics*, **4**, 202.

369 Amoroso, A.J., Beswick, M.A., Li, C.-K., Lewis, J., Raithby, P.R. and de Arellano, M.C.R. (1999) *Journal of Organometallic Chemistry*, **573**, 247.

370 Johnson, B.F.G., Kaner, D.A., Lewis, J. and Rosales, M.J. (1982) *Journal of Organometallic Chemistry*, **238**, C73.

371 Rossell, O., Seco, M., Segales, G., Pellinghelli, M.A. and Tiripicchio, A. (1998) *Journal of Organometallic Chemistry*, **571**, 123.

372 Harpp, K.S., Housecroft, C.E., Rheingold, A.L. and Shongwe, M.S. (1988) *Chemical Communications*, 965.

373 Sunick, D.L., White, P.S. and Schauer, C.K. (1994) *Angewandte Chemie (International Edition in English)*, **33**, 75.

374 Johnson, B.F.G., Lewis, J., Nelson, W.J.H., Puga, J., Raithby, P.R., Braga, D., McPartlin, M. and Clegg, W. (1983) *Journal of Organometallic Chemistry*, **243**, C13.

375 Adatia, T., Curtis, H., Johnson, B.F.G., Lewis, J., McPartlin, M. and Morris, J. (1994) *Journal of the Chemical Society-Dalton Transactions*, 3069.

376 Galsworthy, J.R., Hattersley, A.D., Housecroft, C.E., Rheingold, A.L. and Waller, A. (1995) *Journal of the Chemical Society-Dalton Transactions*, 549.

377 Hattersley, A.D., Housecroft, C.E. and Rheingold, A.L. (1997) *Journal of Cluster Science*, **8**, 329.

378 Hattersley, A.D., Housecroft, C.E., Liable-Sands, L.M., Rheingold, A.L. and Waller, A. (1998) *Polyhedron*, **17**, 2957.

379 Hattersley, A.D., Housecroft, C.E. and Rheingold, A.L. (1999) *Inorganica Chimica Acta*, **289**, 149.

380 Housecroft, C.E., Matthews, D.M., Waller, A., Edwards, A.J. and Rheingold, A.L. (1993) *Journal of the Chemical Society-Dalton Transactions*, 3059.

381 Housecroft, C.E., Rheingold, A.L., Waller, A. and Yap, G.P.A. (1998) *Polyhedron*, **17**, 2921.

382 Chihara, T. and Yamazaki, H. (1992) *Journal of Organometallic Chemistry*, **428**, 169.

383 Cifuentes, M.P., Jeynes, T.P., Humphrey, M.G., Skelton, B.W. and White, A.H. (1994) *Journal of the Chemical Society-Dalton Transactions*, 925.

384 Bailey, P.J., Beswick, M.A., Lewis, J., Raithby, P.R. and de Arellano, M.C.R. (1993) *Journal of Organometallic Chemistry*, **459**, 293.

385 Bruce, M.I., Liddell, M.J., Williams, M.L. and Nicholson, B.K. (1990) *Organometallics*, **9**, 2903.

386 Galsworthy, J.R., Housecroft, C.E. and Rheingold, A.L. (1995) *Journal of the Chemical Society-Dalton Transactions*, 2639.

387 Bruce, M.I. and Nicholson, B.K. (1983) *Journal of Organometallic Chemistry*, **252**, 243.

388 Howard, J.A.K., Slater, I.D. and Stone, F.G.A. (1984) *Polyhedron*, **3**, 567.

389 Bruce, M.I. and Nicholson, B.K. (1982) *Chemical Communications*, 1141.

390 Adatia, T., McPartlin, M. and Salter, I.D. (1988) *Journal of the Chemical Society-Dalton Transactions*, 751.

391 Colbran, S.B., Hay, C.M., Johnson, B.F.G., Lahoz, F.J., Lewis, J. and Raithby, P.R. (1986) *Chemical Communications*, 1766.

392 Al-Mandhary, M.R.A., Buntem, R., Doherty, C.L., Edwards, A.J., Gallagher,

J.F., Lewis, J., Li, C.-K., Raithby, P.R., de Arellano, M.C.R. and Shields, G.P. (2005) *Journal of Cluster Science*, **16**, 127.

393 Johnson, B.F.G., Lewis, J., Nelson, W.J.H., Nicholls, J.N., Puga, J., Raithby, P.R., Rosales, M.J., Schroder, M. and Vargas, M.D. (1983) *Journal of the Chemical Society-Dalton Transactions*, 2447.

394 Camats, J., Reina, R., Riba, O., Rossell, O., Seco, M., Gomez-Sal, P., Martin, A. and de Montauzon, D. (2000) *Organometallics*, **19**, 3316.

395 Bunkhall, S.R., Holden, H.D., Johnson, B.F.G., Lewis, J., Pain, G.N., Raithby, P.R. and Taylor, M.J. (1984) *Chemical Communications*, 25.

396 Khimyak, T., Johnson, B.F.G., Hermans, S. and Bond, A.D. (2003) *Dalton Transactions*, 2651.

397 Feeder, N., Geng, J., Goh, P.G., Johnson, B.F.G., Martin, C.M., Shephard, D.S. and Zhou, W. (2000) *Angewandte Chemie (International Edition in English)*, **39**, 1661.

398 Housecroft, C.E., Nixon, D.M. and Rheingold, A.L. (2001) *Journal of Cluster Science*, **12**, 89.

399 Akhter, Z., Edwards, A.J., Gallagher, J.F., Lewis, J., Raithby, P.R. and Shields, G.P. (2000) *Journal of Organometallic Chemistry*, **596**, 204.

400 Al-Mandhary, M.R.A., Lewis, J. and Raithby, P.R. (1997) *Journal of Organometallic Chemistry*, **536**, 549.

401 Albano, V.G., Iapalucci, M.C., Longoni, G., Monari, M., Paselli, A. and Zacchini, S. (1998) *Organometallics*, **17**, 4438.

402 Albano, V.G., Castellari, C., Femoni, C., Iapalucci, M.C., Longoni, G., Monari, M. and Zacchini, S. (2001) *Journal of Cluster Science*, **12**, 75.

403 Adams, R.D., Barnard, T.S. and Cortopassi, J.E. (1995) *Organometallics*, **14**, 2232.

404 Hattersley, A.D., Housecroft, C.E. and Rheingold, A.L. (1999) *Collection of Czechoslovak Chemical Communications*, **64**, 959.

405 Johnson, B.F.G., Lewis, J., Nelson, W.J.H., Vargas, M.D., Braga, D. and McPartlin, M. (1983) *Journal of Organometallic Chemistry*, **246**, C69.

406 Johnson, B.F.G., Lewis, J., Nelson, W.J.H., Vargas, M.D., Braga, D., Henrick, K. and McPartlin, M. (1986) *Journal of the Chemical Society-Dalton Transactions*, 975.

407 Amoroso, A.J., Johnson, B.F.G., Lewis, J., Li, C.-K., Morewood, C.A., Raithby, P.R., Vargas, M.D. and Wong, W.-T. (1995) *Journal of Cluster Science*, **6**, 163.

408 Ahkter, Z., Edwards, A.J., Ingham, S.L., Lewis, J., Castro, A.M.M., Raithby, P.R. and Shields, G.P. (2000) *Journal of Cluster Science*, **11**, 217.

409 Johnson, B.F.G., Lewis, J., Nelson, W.J.H., Raithby, P.R. and Vargas, M.D. (1983) *Chemical Communications*, 608.

410 Akhter, Z., Ingham, S.L., Lewis, J. and Raithby, P.R. (1994) *Journal of Organometallic Chemistry*, **474**, 165.

411 Akhter, Z., Ingham, S.L., Lewis, J. and Raithby, P.R. (1998) *Journal of Organometallic Chemistry*, **550**, 131.

412 Akhter, Z., Gallagher, J.F., Lewis, J., Raithby, P.R. and Shields, G.P. (2000) *Journal of Organometallic Chemistry*, **614**, 231.

413 Dearing, V., Drake, S.R., Johnson, B.F.G., Lewis, J., McPartlin, M. and Powell, H.R. (1988) *Chemical Communications*, 1331.

414 Akhter, Z., Ingham, S.L., Lewis, J. and Raithby, P.R. (1996) *Angewandte Chemie (International Edition in English)*, **35**, 992.

415 Albano, V.G., Aureli, R., Iapalucci, M.C., Laschi, F., Longoni, G., Monari, M. and Zanello, P. (1993) *Chemical Communications*, 1501.

416 Albano, V.G., Monari, M., Demartin, F., Macchi, P., Femoni, C., Iapalucci, M.C. and Longoni, G. (1999) *Solid State Sciences*, **1**, 597.

417 Housecroft, C.E., Rheingold, A.L. and Shongwe, M.S. (1988) *Chemical Communications*, 1630.

418 Albano, V.G., Iapalucci, M.C., Longoni, G., Manzi, L. and Monari, M. (1997) *Organometallics*, **16**, 497.

419 Albano, V.G., Calderoni, F., Iapalucci, M.C., Longoni, G. and Monari, M. (1995) *Chemical Communications*, 433.

420 Cathey, C., Lewis, J., Raithby, P.R. and de Arellano, M.C.R. (1994) *Journal of the Chemical Society-Dalton Transactions*, 3331.

421 Johnson, B.F.G., Kaner, D.A., Lewis, J. and Raithby, P.R. (1981) *Chemical Communications*, 753.

422 Hay, C.M., Johnson, B.F.G., Lewis, J., Prior, N.D., Raithby, P.R. and Wong, W.T. (1991) *Journal of Organometallic Chemistry*, **401**, C20.

423 Blundell, T.L. and Powell, H.M. (1971) *Journal of the Chemical Society A*, 1685.

424 Bashkin, J., Briant, C.E., Mingos, D.M.P. and Wardle, R.W.M. (1985) *Transition Metal Chemistry*, **10**, 113.

425 Pons, A., Rossell, O., Seco, M. and Perales, A. (1995) *Organometallics*, **14**, 555.

426 Marsh, R.E. (2004) *Acta Crystallographica B*, **60**, 252.

427 Lauher, J.W. and Wald, K. (1981) *Journal of the American Chemical Society*, **103**, 7648.

428 Low, A.A. and Lauher, J.W. (1987) *Inorganic Chemistry*, **26**, 3863.

429 Braunstein, P., Rose, J., Dusausoy, Y. and Mangeot, J.-P. (1982) *Comptes Rendus de l'Academie des Sciences, SerieII*, **294**, 967.

430 Braunstein, P., Rose, J., Dedieu, A., Dusausoy, Y., Mangeot, J.-P., Tiripicchio, A. and Tiripicchio Camellini, M. (1986) *Journal of the Chemical Society-Dalton Transactions*, 225.

431 Kakkonen, H.J., Ahlgren, M., Pakkanen, T.A. and Pursiainen, J. (1994) *Acta Crystallographica C*, **50**, 528.

432 Pursiainen, J., Ahlgren, M. and Pakkanen, T.A. (1985) *Journal of Organometallic Chemistry*, **297**, 391.

433 Beuter, G., Mielcke, J. and Strahle, J. (1991) *Zeitschrift für Anorganische und Allgemeine Chemie*, **593**, 35.

434 Reina, R., Riba, O., Rossell, O., Seco, M., de Montauzon, D., Pellinghelli, M.A., Tiripicchio, A., Font-Bardia, M. and Solans, X. (2000) *Journal of the Chemical Society-Dalton Transactions*, 4464.

435 Holzer, M., Strahle, J., Baum, G. and Fenske, D. (1994) *Zeitschrift für Anorganische und Allgemeine Chemie*, **620**, 192.

436 van der Velden, J.W.A., Bour, J.J., Otterloo, B.F., Bosman, W.P. and Noordik, J.H. (1981) *Chemical Communications*, 583.

437 van der Velden, J.W.A., Bour, J.J., Bosman, W.P. and Noordik, J.H. (1983) *Inorganic Chemistry*, **22**, 1913.

438 Uson, R., Laguna, A., Laguna, M., Jones, P.G. and Sheldrick, G.M. (1981) *Journal of the Chemical Society-Dalton Transactions*, 366.

439 Bianchini, C., Elsevier, C.J., Ernsting, J.M., Peruzzini, M. and Zanobini, F. (1995) *Inorganic Chemistry*, **34**, 84.

440 McNair, R.J., Nilsson, P.V. and Pignolet, L.H. (1985) *Inorganic Chemistry*, **24**, 1935.

441 Esswein, A.J., Dempsey, J.L. and Nocera, D.G. (2007) *Inorganic Chemistry*, **46**, 2362.

442 Schiavo, S.L., Bruno, G., Nicolo, F., Piraino, P. and Faraone, F. (1985) *Organometallics*, **4**, 2091.

443 Fernandez, M.J., Modrego, J., Oro, L.A., Apreda, M.-C., Cano, F.H. and Foces-Foces, C. (1989) *Journal of the Chemical Society-Dalton Transactions*, 1249.

444 Bray, K.L., Drickamer, H.G., Mingos, D.M.P., Watson, M.J. and Shapley, J.R. (1991) *Inorganic Chemistry*, **30**, 864.

445 Balch, A.L., Fung, E.Y. and Olmstead, M.M. (1990) *Inorganic Chemistry*, **29**, 3203.

446 Howard, J.A.K., Jeffery, J.C., Jelliss, P.A., Sommerfeld, T. and Stone, F.G.A. (1991) *Chemical Communications*, 1664.

447 Jeffery, J.C., Jelliss, P.A. and Stone, F.G.A. (1993) *Journal of the Chemical Society-Dalton Transactions*, 1073.

448 Albinati, A., Demartin, F., Janser, P., Rhodes, L.F. and Venanzi, L.M. (1989) *Journal of the American Chemical Society*, **111**, 2115.

449 Kudinov, A.R., Muratov, D.V., Rybinskaya, M.I. and Turpeinen, U. (1993) *Mendeleev Communications*, 39.

450 Pursiainen, J., Pakkanen, T.A., Ahlgren, M. and Valkonen, J. (1993) *Acta Crystallographica C*, **49**, 1142.

451 Alexander, B.D., Mueting, A.M. and Pignolet, L.H. (1990) *Inorganic Chemistry*, **29**, 1313.

452 Casado, M.A., Perez-Torrente, J.J., Ciriano, M.A., Lahoz, F.J. and Oro, L.A. (2004) *Inorganic Chemistry*, **43**, 1558.

453 Shan, H. and Sharp, P.R. (1996) *Angewandte Chemie (International Edition in English)*, **35**, 635.

454 Bott, S.G., Fleischer, H., Leach, M., Mingos, D.M.P., Powell, H., Watkin, D.J. and Watson, M.J. (1991) *Journal of the Chemical Society-Dalton Transactions*, 2569.

455 Fumagalli, A., Martinengo, S., Albano, V.G. and Braga, D. (1988) *Journal of the Chemical Society-Dalton Transactions*, 1237.

456 Fumagalli, A., Martinengo, S., Albano, V.G., Braga, D. and Grepioni, F. (1989) *Journal of the Chemical Society-Dalton Transactions*, 2343.

457 Polyakova, L.A., Gubin, S.P., Churakov, A.V. and Kuzmina, L.G. (1999) *Koordinatsionnaya Khimiya (Russian Journal of Coordination Chemistry)*, **25**, 761.

458 Bott, S.G., Mingos, D.M.P. and Watson, M.J. (1989) *Chemical Communications*, 1192.

459 Fumagalli, A., Martinengo, S., Albano, V.G., Braga, D. and Grepioni, F. (1993) *Journal of the Chemical Society-Dalton Transactions*, 2047.

460 Albano, V.G., Fumagalli, A., Grepioni, F., Martinengo, S. and Monari, M. (1994) *Journal of the Chemical Society-Dalton Transactions*, 1777.

461 Lehner, H., Matt, D., Pregosin, P.S., Venanzi, L.M. and Albinati, A. (1982) *Journal of the American Chemical Society*, **104**, 6825.

462 Albinati, A., Anklin, C., Janser, P., Lehner, H., Matt, D., Pregosin, P.S. and Venanzi, L.M. (1989) *Inorganic Chemistry*, **28**, 1105.

463 Casalnuovo, A.L., Laska, T., Nilsson, P.V., Olofson, J., Pignolet, L.H., Bos, W., Bour, J.J. and Steggerda, J.J. (1985) *Inorganic Chemistry*, **24**, 182.

464 Casalnuovo, A.L., Laska, T., Nilsson, P.V., Olofson, J. and Pignolet, L.H. (1985) *Inorganic Chemistry*, **24**, 233.

465 Luke, M.A., Mingos, D.M.P., Sherman, D.J. and Wardle, R.W.M. (1987) *Transition Metal Chemistry*, **12**, 37.

466 Balch, A.L., Catalano, V.J. and Olmstead, M.M. (1990) *Inorganic Chemistry*, **29**, 585.

467 Balch, A.L. and Catalano, V.J. (1991) *Inorganic Chemistry*, **30**, 1302.

468 Jeffery, J.C., Jelliss, P.A. and Stone, F.G.A. (1993) *Journal of the Chemical Society-Dalton Transactions*, 1083.

469 Albinati, A., Chaloupka, S., Currao, A., Klooster, W.T., Koetzle, T.F., Nesper, R. and Venanzi, L.M. (2000) *Inorganica Chimica Acta*, **300**, 903.

470 Sykes, A.G. and Mann, K.R. (1990) *Journal of the American Chemical Society*, **112**, 7247.

471 Bianchini, C., Meli, A., Peruzzini, M., Vacca, A., Vizza, F. and Albinati, A. (1992) *Inorganic Chemistry*, **31**, 3841.

472 Balch, A.L., Catalano, V.J. and Olmstead, M.M. (1990) *Journal of the American Chemical Society*, **112**, 2010.

473 Balch, A.L., Oram, D.E. and Reedy, P.E. Jr. (1987) *Inorganic Chemistry*, **26**, 1836.

474 Balch, A.L., Nagle, J.K., Oram, D.E. and Reedy, P.E. Jr. (1988) *Journal of the American Chemical Society*, **110**, 454.

475 Balch, A.L., Davis, B.J. and Olmstead, M.M. (1989) *Inorganic Chemistry*, **28**, 3148.

476 Balch, A.L., Catalano, V.J., Noll, B.C. and Olmstead, M.M. (1990) *Journal of the American Chemical Society*, **112**, 7558.

477 Casalnuovo, A.L., Pignolet, L.H., van der Velden, J.W.A., Bour, J.J. and Steggerda, J.J. (1983) *Journal of the American Chemical Society*, **105**, 5957.

478 Comstock, M.C., Prussak-Wieckowska, T., Wilson, S.R. and Shapley, J.R. (1997) *Organometallics*, **16**, 4033.

479 Braga, D., Grepioni, F., Livotto, F.S. and Vargas, M.D. (1990) *Journal of Organometallic Chemistry*, **391**, C28.

480 Livotto, F.S., Vargas, M.D., Braga, D. and Grepioni, F. (1992) *Journal of the Chemical Society-Dalton Transactions*, 577.
481 Casalnuovo, A.L., Casalnuovo, J.A., Nilsson, P.V. and Pignolet, L.H. (1985) *Inorganic Chemistry*, **24**, 2554.
482 Bruce, M.I., Corbin, P.E., Humphrey, P.A., Koutsantonis, G.A., Liddell, M.J. and Tiekink, E.R.T. (1990) *Chemical Communications*, 674.
483 Bruce, M.I., Koutsantonis, G.A. and Tiekink, E.R.T. (1991) *Journal of Organometallic Chemistry*, **408**, 77.
484 Nicholls, J.N., Raithby, P.R. and Vargas, M.D. (1986) *Chemical Communications*, 1617.
485 Della Pergola, R., Demartin, F., Garlaschelli, L., Manassero, M., Martinengo, S., Masciocchi, N. and Sansoni, M. (1991) *Organometallics*, **10**, 2239.
486 Ceriotti, A., Della Pergola, R., Garlaschelli, L., Manassero, M. and Masciocchi, N. (1995) *Organometallics*, **14**, 186.
487 Chihara, T., Sato, M., Konomoto, H., Kamiguchi, S., Ogawa, H. and Wakatsuki, Y. (2000) *Journal of the Chemical Society-Dalton Transactions*, 2295.
488 Dempsey, J.L., Esswein, A.J., Manke, D.R., Rosenthal, J., Soper, J.D. and Nocera, D.G. (2005) *Inorganic Chemistry*, **44**, 6879.
489 Kim, H.P., Fanwick, P.E. and Kubiak, C.P. (1988) *Journal of Organometallic Chemistry*, **346**, C39.
490 Xia, B.-H., Zhang, H.-X., Che, C.-M., Leung, K.-H., Phillips, D.L., Zhu, N. and Zhou, Z.-Y. (2003) *Journal of the American Chemical Society*, **125**, 10362.
491 De Silva, N., Nichiporuk, R.V. and Dahl, L.F. (2006) *Dalton Transactions*, 2291.
492 Whoolery, A.J. and Dahl, L.F. (1991) *Journal of the American Chemical Society*, **113**, 6683.
493 Johnson, A.J.W., Spencer, B. and Dahl, L.F. (1994) *Inorganica Chimica Acta*, **227**, 269.
494 Teo, B.K., Zhang, H. and Shi, X. (1994) *Inorganic Chemistry*, **33**, 4086.
495 Crespo, O., Laguna, A., Fernández, E.J., López-de-Luzuriaga, J.M., Jones, P.G., Teichert, M., Monge, M., Pyykkö, P., Runeberg, N., Schutz, M. and Werner, H.-J. (2000) *Inorganic Chemistry*, **39**, 4786.
496 Burrows, A.D., Gosden, A.A., Hill, C.M. and Mingos, D.M.P. (1993) *Journal of Organometallic Chemistry*, **452**, 251.
497 Harvey, P.D., Eichhofer, A. and Fenske, D. (1998) *Journal of the Chemical Society-Dalton Transactions*, 3901.
498 Vicente, J., González-Herrero, P., Pérez-Cadenas, M., Jones, P.G. and Bautista, D. (2007) *Inorganic Chemistry*, **46**, 4718.
499 Copley, R.C.B., Hill, C.M. and Mingos, D.M.P. (1995) *Journal of Cluster Science*, **6**, 71.
500 Tran, N.T., Powell, D.R. and Dahl, L.F. (2004) *Dalton Transactions*, 209.
501 Tran, N.T., Powell, D.R. and Dahl, L.F. (2004) *Dalton Transactions*, 217.
502 Mednikov, E.G., Tran, N.T., Aschbrenner, N.L. and Dahl, L.F. (2007) *Journal of Cluster Science*, **18**, 253.
503 Mednikov, E.G. and Dahl, L.F. (2005) *Journal of Cluster Science*, **16**, 287.
504 Laupp, M. and Strahle, J. (1994) *Angewandte Chemie (International Edition in English)*, **33**, 207.
505 Laupp, M. and Strahle, J. (1995) *Zeitschrift für Naturforschung*, **50b**, 1369.
506 Takata, N.H., Felicissimo, A.M.P. and Young, V.G. Jr. (2001) *Inorganica Chimica Acta*, **325**, 79.
507 Sotelo, A.F., Felicissimo, A.M.P. and Gomez-Sal, P. (2003) *Inorganica Chimica Acta*, **348**, 63.
508 Ito, L.N., Felicissimo, A.M.P. and Pignolet, L.H. (1991) *Inorganic Chemistry*, **30**, 988.
509 Craighead, K.L., Felicissimo, A.M.P., Krogstad, D.A., Nelson, L.T.J. and Pignolet, L.H. (1993) *Inorganica Chimica Acta*, **212**, 31.
510 Ito, L.N., Johnson, B.J., Mueting, A.M. and Pignolet, L.H. (1989) *Inorganic Chemistry*, **28**, 2026.
511 Albinati, A., Lehner, H., Venanzi, L.M. and Wolfer, M. (1987) *Inorganic Chemistry*, **26**, 3933.

512 Crespo, M., Sales, J. and Solans, X. (1989) *Journal of the Chemical Society-Dalton Transactions*, 1089.

513 Shan, H., James, A. and Sharp, P.R. (1998) *Inorganic Chemistry*, **37**, 5727.

514 Batten, S.A., Jeffery, J.C., Jones, P.L., Mullica, D.F., Rudd, M.D., Sappenfield, E.L., Stone, F.G.A. and Wolf, A. (1997) *Inorganic Chemistry*, **36**, 2570.

515 Jeffery, J.C., Jelliss, P.A. and Stone, F.G.A. (1993) *Inorganic Chemistry*, **32**, 3943.

516 Douglas, G., Jennings, M.C., Manojlovic-Muir, L. and Puddephatt, R.J. (1988) *Inorganic Chemistry*, **27**, 4516.

517 Manojlović-Muir, L., Henderson, A.N., Treurnicht, I. and Puddephatt, R.J. (1989) *Organometallics*, **8**, 2055.

518 Yip, H.-K., Che, C.-M. and Peng, S.-M. (1991) *Chemical Communications*, 1626.

519 Yip, H.-K., Lin, H.-M., Wang, Y. and Che, C.-M. (1993) *Journal of the Chemical Society-Dalton Transactions*, 2939.

520 Xu, C., Anderson, G.K., Brammer, L., Braddock-Wilking, J. and Rath, N.P. (1996) *Organometallics*, **15**, 3972.

521 Yin, G.-Q., Wei, Q.-H., Zhang, L.-Y. and Chen, Z.-N. (2006) *Organometallics*, **25**, 580.

522 Cook, T.R., Esswein, A.J. and Nocera, D.G. (2007) *Journal of the American Chemical Society*, **129**, 10094.

523 Hayoun, R., Zhong, D.K., Rheingold, A.L. and Doerrer, L.H. (2006) *Inorganic Chemistry*, **45**, 6120.

524 King, C., Heinrich, D.D., Garzon, G., Wang, J.-C. and Fackler, J.P. Jr. (1989) *Journal of the American Chemical Society*, **111**, 2300.

525 Tanase, T., Toda, H. and Yamamoto, Y. (1997) *Inorganic Chemistry*, **36**, 1571.

526 Arsenault, G.J., Manojlovic-Muir, L., Muir, K.W., Puddephatt, R.J. and Treurnicht, I. (1987) *Angewandte Chemie (International Edition in English)*, **26**, 86.

527 Manojlovic-Muir, L., Muir, K.W., Treurnicht, I. and Puddephatt, R.J. (1987) *Inorganic Chemistry*, **26**, 2418.

528 Hallam, M.F., Luke, M.A., Mingos, D.M.P. and Williams, I.D. (1987) *Journal of Organometallic Chemistry*, **325**, 271.

529 Bennett, M.A., Berry, D.E. and Beveridge, K.A. (1990) *Inorganic Chemistry*, **29**, 4148.

530 Hill, C.M., Mingos, D.M.P., Powell, H. and Watson, M.J. (1992) *Journal of Organometallic Chemistry*, **441**, 499.

531 Hadj-Bagheri, N. and Puddephatt, R.J. (1993) *Inorganica Chimica Acta*, **213**, 29.

532 Toronto, D.V. and Balch, A.L. (1994) *Inorganic Chemistry*, **33**, 6132.

533 Yip, J.H.K., Wu, J., Wong, K.-Y., Ho, K.P., So-Ngan Pun, C. and Vittal, J.J. (2002) *Organometallics*, **21**, 5292.

534 Braunstein, P., Lehner, H., Matt, D., Tiripicchio, A. and Tiripicchio Camellini, M. (1984) *Angewandte Chemie (International Edition in English)*, **23**, 304.

535 Briant, C.E., Wardle, R.W.M. and Mingos, D.M.P. (1984) *Journal of Organometallic Chemistry*, **267**, C49.

536 Mingos, D.M.P. and Wardle, R.W.M. (1986) *Journal of the Chemical Society-Dalton Transactions*, 73.

537 Bour, J.J., Kanters, R.P.F., Schlebos, P.P.J., Bos, W., Bosman, W.P., Behm, H., Beurskens, P.T. and Steggerda, J.J. (1987) *Journal of Organometallic Chemistry*, **329**, 405.

538 Imhof, D., Burckhardt, U., Dahmen, K.-H., Ruegger, H., Gerfin, T. and Gramlich, V. (1993) *Inorganic Chemistry*, **32**, 5206.

539 Muir, K.W., Manojlovic-Muir, L. and Fullard, J. (1996) *Acta Crystallographica C*, **52**, 295.

540 Imhof, D., Burckhardt, U., Dahmen, K.-H., Joho, F. and Nesper, R. (1997) *Inorganic Chemistry*, **36**, 1813.

541 Peter, M., Wachtler, H., Ellmerer-Muller, E., Ongania, K.-H., Wurst, K. and Peringer, P. (1997) *Journal of Organometallic Chemistry*, **542**, 227.

542 Briant, C.E., Gilmour, D.I. and Mingos, D.M.P. (1984) *Journal of Organometallic Chemistry*, **267**, C52.

543 Briant, C.E., Gilmour, D.I. and Mingos, D.M.P. (1986) *Journal of the Chemical Society-Dalton Transactions*, 835.

544 Zhang, T., Drouin, M. and Harvey, P.D. (1999) *Inorganic Chemistry*, **38**, 4928.

545 Bender, R., Braunstein, P., Dedieu, A. and Dusausoy, Y. (1989) *Angewandte Chemie (International Edition in English)*, **28**, 923.

546 Mingos, D.M.P., Oster, P. and Sherman, D.J. (1987) *Journal of Organometallic Chemistry*, **320**, 257.

547 Payne, N.C., Ramachandran, R., Schoettel, G., Vittal, J.J. and Puddephatt, R.J. (1991) *Inorganic Chemistry*, **30**, 4048.

548 Breuer, M. and Strahle, J. (1993) *Zeitschrift für Anorganische und Allgemeine Chemie*, **619**, 1564.

549 Hallam, M.F., Mingos, D.M.P., Adatia, T. and McPartlin, M. (1988) *Journal of the Chemical Society-Dalton Transactions*, 335.

550 De Silva, N., Laufenberg, J.W. and Dahl, L.F. (2006) *Chemical Communications*, 4437.

551 Spivak, G.J., Vittal, J.J. and Puddephatt, R.J. (1998) *Inorganic Chemistry*, **37**, 5474.

552 Ivanov, S.A., de Silva, N., Kozee, M.A., Nichiporuk, R.V. and Dahl, L.F. (2004) *Journal of Cluster Science*, **15**, 233.

553 Smith, D.E., Welch, A.J., Treurnicht, I. and Puddephatt, R.J. (1986) *Inorganic Chemistry*, **25**, 4616.

554 Kanters, R.P.F., Bour, J.J., Schlebos, P.P.J., Bosman, W.P., Behm, H., Steggerda, J.J., Ito, L.N. and Pignolet, L.H. (1989) *Inorganic Chemistry*, **28**, 2591.

555 Ito, L.N., Sweet, J.D., Mueting, A.M., Pignolet, L.H., Schoondergang, M.F.J. and Steggerda, J.J. (1989) *Inorganic Chemistry*, **28**, 3696.

556 Krogstad, D.A., Young, V.G. Jr., Pignolet, L.H. (1997) *Inorganica Chimica Acta*, **264**, 19.

557 Bour, J.J., Kanters, R.P.F., Schlebos, P.P.J., Bosman, W.P., Behm, H., Beurskens, P.T. and Steggerda, J.J. (1987) *Recueil des Travaux Chimiques des Pays-Bas*, **106**, 157.

558 Kanters, R.P.F., Schlebos, P.P.J., Bour, J.J., Bosman, W.P., Behm, H.J. and Steggerda, J.J. (1988) *Inorganic Chemistry*, **27**, 4034.

559 Bour, J.J., Schlebos, P.P.J., Kanters, R.P.F., Bosman, W.P., Smits, J.M.M., Beurskens, P.T. and Steggerda, J.J. (1990) *Inorganica Chimica Acta*, **171**, 177.

560 Schoondergang, M.F.J., Bour, J.J., van Strijdonck, G.P.F., Schlebos, P.P.J., Bosman, W.P., Smits, J.M.M., Beurskens, P.T. and Steggerda, J.J. (1991) *Inorganic Chemistry*, **30**, 2048.

561 Rubinstein, L.I. and Pignolet, L.H. (1996) *Inorganic Chemistry*, **35**, 6755.

562 Kanters, R.P.F., Schlebos, P.P.J., Bour, J.J., Bosman, W.P., Smits, J.M.M., Beurskens, P.T. and Steggarda, J.J. (1990) *Inorganic Chemistry*, **29**, 324.

563 Kappen, T.G.M.M., Schlebos, P.P.J., Bour, J.J., Bosman, W.P., Beurskens, G., Smits, J.M.M., Beurskens, P.T. and Steggerda, J.J. (1995) *Inorganic Chemistry*, **34**, 2121.

564 Kappen, T.G.M.M., Schlebos, P.P.J., Bour, J.J., Bosman, W.P., Smits, J.M.M., Beurskens, P.T. and Steggerda, J.J. (1995) *Inorganic Chemistry*, **34**, 2133.

565 Kappen, T.G.M.M., Schlebos, P.P.J., Bour, J.J., Bosman, W.P., Smits, J.M.M., Buerskens, P.T. and Steggerda, J.J. (1995) *Journal of the American Chemical Society*, **117**, 8327.

566 Teo, B.K. and Zhang, H. (2000) *Journal of Organometallic Chemistry*, **614**, 66.

567 de Silva, N. and Dahl, L.F. (2005) *Inorganic Chemistry*, **44**, 9604.

568 Kappen, T.G.M.M., Schlebos, P.P.J., Bour, J.J., Bosman, W.P., Smits, J.M.M., Beurskens, P.T. and Steggerda, J.J. (1994) *Inorganic Chemistry*, **33**, 754.

569 Teo, B.K. and Zhang, H. (2001) *Journal of Cluster Science*, **12**, 349.

570 Teo, B.K., Zhang, H. and Shi, X. (1993) *Journal of the American Chemical Society*, **115**, 8489.

571 Schoondergang, M.F.J., Bour, J.J., Schlebos, P.P.J., Vermeer, A.W.P., Bosman, W.O., Smits, J.M.M., Beurskens, P.T. and Steggerda, J.J. (1991) *Inorganic Chemistry*, **30**, 4704.

572 Krogstad, D.A., Konze, W.V. and Pignolet, L.H. (1996) *Inorganic Chemistry*, **35**, 6763.

573 Pyykkö, P., Runeberg, N. and Mendizabal, F. (1997) *Chemistry – A European Journal*, **3**, 1451.

574 Pyykkö, P. and Mendizabal, F. (1998) *Inorganic Chemistry*, **37**, 3018.

575 Fernández, E.J., López-de-Luzuriaga, J.M., Monge, M., Rodríguez, M.A., Crespo, O., Gimeno, M.C., Laguna, A. and Jones, P.G. (1998) *Inorganic Chemistry*, **37**, 6002.

576 Fernandez, E.J., Lopez-de-Luzuriaga, J.M., Monge, M., Rodriguez, M.A., Crespo, O., Gimeno, M.C., Laguna, A. and Jones, P.G. (2000) *Chemistry – A European Journal*, **6**, 636.

577 Fernandez, E.J., Laguna, A., Lopez-de-Luzuriaga, J.M., Monge, M., Montiel, M. and Olmos, M.E. (2005) *Inorganic Chemistry*, **44**, 1163.

578 Fernandez, E.J., Laguna, A., Lopez-de-Luzuriaga, J.M., Monge, M., Montiel, M., Olmos, M.E. and Rodriguez-Castillo, M. (2006) *Organometallics*, **25**, 3639.

579 de la Riva, H., Nieuwhuyzen, M., Fierro, C.M., Raithby, P.R., Male, L. and Lagunas, M.C. (2006) *Inorganic Chemistry*, **45**, 1418.

580 Abu-Salah, O.M., Al-Ohaly, A.-R.A. and Knobler, C.B. (1985) *Chemical Communications*, 1502.

581 Yip, S.-K., Chan, C.-L., Lam, W.H., Cheung, K.-K. and Yam, V.W.-W. (2007) *Photochemical & Photobiological Sciences*, **6**, 365.

582 Hussain, M.S., Mazhar-Ul-Haque and Abu-Salah, O.M. (1996) *Journal of Cluster Science*, **7**, 167.

583 Hao, L., Mansour, M.A., Lachicotte, R.J., Gysling, H.J. and Eisenberg, R. (2000) *Inorganic Chemistry*, **39**, 5520.

584 Henkel, G., Krebs, B., Betz, P., Fietz, H. and Saatkamp, K. (1988) *Angewandte Chemie (International Edition in English)*, **27**, 1326.

585 Coucouvanis, D., Kanodia, S., Swenson, D., Chen, S.-J., Studemann, T., Baenziger, N.C., Pedelty, R. and Chu, M. (1993) *Journal of the American Chemical Society*, **115**, 11271.

586 Chen, Y.-D., Zhang, L.-Y., Qin, Y.-H. and Chen, Z.-N. (2005) *Inorganic Chemistry*, **44**, 6456.

587 Copley, R.C.B. and Mingos, D.M.P. (1996) *Journal of the Chemical Society-Dalton Transactions*, 491.

588 Olmos, M.E., Schier, A. and Schmidbaur, H. (1997) *Zeitschrift für Naturforschung*, **52b**, 203.

589 Rawashdeh-Omary, M.A., Omary, M.A. and Fackler, J.P. Jr. (2002) *Inorganica Chimica Acta*, **334**, 376.

590 Römbke, P., Schier, A., Schmidbaur, H., Cronje, S. and Raubenheimer, H. (2004) *Inorganica Chimica Acta*, **357**, 235.

591 Catalano, V.J., Malwitz, M.A. and Etogo, A.O. (2004) *Inorganic Chemistry*, **43**, 5714.

592 Catalano, V.J. and Moore, A.L. (2005) *Inorganic Chemistry*, **44**, 6558.

593 Catalano, V.J. and Etogo, A.O. (2005) *Journal of Organometallic Chemistry*, **690**, 6041.

594 Catalano, V.J. and Etogo, A.O. (2007) *Inorganic Chemistry*, **46**, 5608.

595 Vicente, J., Chicote, M.-T., Lagunas, M.C. and Jones, P.G. (1992) *Chemical Communications*, 1730.

596 Contel, M., Garrido, J., Gimeno, M.C., Jimenez, J., Jones, P.G., Laguna, A. and Laguna, M. (1997) *Inorganica Chimica Acta*, **254**, 157.

597 Mohammed, A.A., Burini, A. and Fackler, J.P. Jr. (2005) *Journal of the American Chemical Society*, **127**, 5012.

598 Ahmed, A.A., Galassi, R., Papa, F., Burini, A. and Fackler, J.P. Jr. (2006) *Inorganic Chemistry*, **45**, 7770.

599 Burini, A., Fackler, J.P. Jr., Galassi, R., Pietroni, B.R. and Staples, R.J. (1998) *Chemical Communications*, 95.

600 Zhang, K., Prabhavathy, J., Yip, J.H.K., Koh, L.L., Tan, G.K. and Vittal, J.J. (2003) *Journal of the American Chemical Society*, **125**, 8452.

601 Contel, M., Garrido, J., Gimeno, M.C. and Laguna, M. (1998) *Journal of the Chemical Society-Dalton Transactions*, 1083.

602 Cerrada, E., Contel, M., Valencia, A.D., Laguna, M., Gelbrich, T. and Hursthouse, M.B. (2000) *Angewandte Chemie (International Edition in English)*, **39**, 2353.

603 Uson, R., Laguna, A., Laguna, M., Uson, A., Jones, P.G. and Erdbrugger, C.F. (1987) *Organometallics*, **6**, 1778.

604 Lee, C.K., Lee, K.M. and Lin, I.J.B. (2002) *Organometallics*, **21**, 10.

605 Chen, J.-X., Zhang, W.-H., Tang, X.-Y., Ren, Z.-G., Li, H.-X., Zhang, Y. and Lang, J.-P. (2006) *Inorganic Chemistry*, **45**, 7671.

606 Schuster, O., Monkowius, U., Schmidbaur, H., Ray, R.S., Kruger, S. and Rosch, N. (2006) *Organometallics*, **25**, 1004.

607 Mazhar-Ul-Haque, Horne, W. and Abu-Salah, O.M. (1992) *Journal of Crystallographic and Spectroscopic Research*, **22**, 421.

608 Abu-Salah, O.M. and Knobler, C.B. (1986) *Journal of Organometallic Chemistry*, **302**, C10.

609 Vicente, J., Chicote, M.-T., Alvarez-Falcon, M.M. and Jones, P.G. (2005) *Organometallics*, **24**, 4666.

610 Contel, M., Jimenez, J., Jones, P.G., Laguna, A. and Laguna, M. (1994) *Journal of the Chemical Society-Dalton Transactions*, 2515.

611 Contel, M., Garrido, J., Gimeno, M.C., Jones, P.G., Laguna, A. and Laguna, M. (1996) *Organometallics*, **15**, 4939.

612 Fernandez, E.J., Laguna, A., Lopez-de-Luzuriaga, J.M., Montiel, M., Olmos, M.E., Perez, J. and Puelles, R.C. (2006) *Organometallics*, **25**, 4307.

613 Uson, R., Laguna, A., Laguna, M., Jones, P.G. and Sheldrick, G.M. (1981) *Chemical Communications*, 1097.

614 Uson, R., Laguna, A., Laguna, M., Manzano, B.R., Jones, P.G. and Sheldrick, G.M. (1984) *Journal of the Chemical Society-Dalton Transactions*, 285.

615 Fernandez, E.J., Gimeno, M.C., Laguna, A., Lopez-de-Luzuriaga, J.M., Monge, M., Pyykkö, P. and Sundholm, D. (2000) *Journal of the American Chemical Society*, **122**, 7287.

616 Fernandez, E.J., Laguna, A., Lopez-de-Luzuriaga, J.M., Monge, M., Montiel, M., Olmos, M.E., Perez, J., Puelles, R.C. and Saenz, J.C. (2005) *Dalton Transactions*, 1162.

617 Wei, Q.-H., Zhang, L.-Y., Yin, G.-Q., Shi, L.-X. and Chen, Z.-N. (2004) *Journal of the American Chemical Society*, **126**, 9940.

618 Wei, Q.-H., Zhang, L.-Y., Yin, G.-Q., Shi, L.-X. and Chen, Z.-N. (2005) *Organometallics*, **24**, 3818.

619 Wei, Q.-H., Yin, G.-Q., Zhang, L.-Y. and Chen, Z.-N. (2006) *Organometallics*, **25**, 4941.

620 Copley, R.C.B. and Mingos, D.M.P. (1992) *Journal of the Chemical Society-Dalton Transactions*, 1755.

621 Teo, B.K., Shi, X. and Zhang, H. (1991) *Journal of the American Chemical Society*, **113**, 4329.

622 Teo, B.K. and Zhang, H. (1991) *Inorganic Chemistry*, **30**, 3115.

623 Teo, B.K. and Zhang, H. (1992) *Angewandte Chemie (International Edition in English)*, **31**, 445.

624 Teo, B.K., Shi, X. and Zhang, H. (1993) *Journal of Cluster Science*, **4**, 471.

625 Teo, B.K., Dang, H., Campana, C.F. and Zhang, H. (1998) *Polyhedron*, **17**, 617.

626 Teo, B.K., Zhang, H. and Shi, X. (1990) *Journal of the American Chemical Society*, **112**, 8552.

627 Nunokawa, K., Ito, M., Sunahara, T., Onaka, S., Ozeki, T., Chiba, H., Funahashi, Y., Masuda, H., Yonezawa, T., Nishihara, H., Nakamoto, M. and Yamamoto, M. (2005) *Dalton Transactions*, 2726.

628 Toyota, A., Yamaguchi, T., Igashira-Kamiyama, A., Kawamoto, T. and Konno, T. (2005) *Angewandte Chemie (International Edition in English)*, **44**, 1088.

5
Supramolecular Architecture by Secondary Bonds
María Elena Olmos

5.1
Introduction

As described in previous chapters, many supramolecular gold materials are formed as a result of gold–gold or gold–heterometal interactions or bonds. Apart from these interactions between metal centers, amongst which aurophilic interactions are especially frequent, many supramolecular entities are formed from secondary bonds that appear between gold and a non-metal atom or between non-metal atoms of different ligands and gold.

The concept of the *secondary bond* was introduced in the seventies [1, 2] to describe *interactions characterized by interatomic distances longer than single covalent bonds but shorter than van der Waals interatomic distances*. Interatomic distances between the sum of covalent radii and the sum of van der Waals radii are quite frequently observed in crystal structure determinations and represent interactions weaker than covalent or dative bonds but strong enough to influence the coordination geometry of the atoms involved and to hold together pairs of atoms. Secondary interactions can be intra- or inter-molecular, and while intramolecular contacts lead to the formation of rings, intermolecular ones result in the formation of supramolecular associations, and are therefore of major importance in both inorganic [1] and organometallic compounds [3, 4]. This chapter will focus on intermolecular secondary bonds, which can be used to connect individual molecules into self-assembled supermolecules or supramolecular arrays. Secondary bonds are normally not strong enough to survive in solution, especially in coordinating solvents, but they can have major effects in the building of a crystal. This chapter will deal with supramolecular gold entities formed by means of secondary bonds in which gold may participate (such as Au$\cdots$NM) or may not participate (such as NM$\cdots$NM).

Hydrogen bonds (D—H$\cdots$A) [5, 6] are special types of secondary bonds. They are interactions that occur in compounds in which hydrogen is bonded to an electronegative element (such as halogens, oxygen, sulfur or nitrogen), and are interactions wherein a hydrogen atom is attracted to two atoms rather than just one. Hydrogen

Modern Supramolecular Gold Chemistry: Gold-Metal Interactions and Applications.
Edited by Antonio Laguna
Copyright © 2008 WILEY-VCH Verlag GmbH & Co. KGaA, Weinheim
ISBN: 978-3-527-32029-5

bonds are directional and of intermediate energy (0.5–40 kcal mol^{-1}), and can be formed easily, although they can also be easily broken in solution. Although hydrogen bonds fulfill the definition of a secondary bond (since the H$\cdots$A distance is longer than a covalent bond, but shorter than the expected van der Waals distance) and are indeed a class of secondary bond, here NM$\cdots$NM contacts will be referred to as secondary bonds and NM−H$\cdots$NM contacts as hydrogen bonds, and both will be treated in separate sections of this chapter.

Classical hydrogen bonds [7] are of the type D−H$\cdots$A, where D (the hydrogen bond donor) and A (the hydrogen bond acceptor) are both electronegative atoms. In the first fifty years of research, the term *hydrogen bond* was restricted to interactions stronger than 4 kcal mol^{-1}, but today other non-classical hydrogen bonds have emerged and this concept extends to encompass weaker interactions like C−H$\cdots$NM [8], D−H$\cdots\pi$ [9], M−H$\cdots$H−A [10, 11] or D−H$\cdots$M [11, 12]. The latter, in which a metal atom acts as the acceptor, are the most frequent in gold chemistry and most have a carbon atom as a hydrogen bond donor. Therefore, C−H$\cdots$Au and classical D−H$\cdots$A bonds in gold complexes will be treated in this chapter.

Finally, although much less frequent, there are also supramolecular gold materials formed as a result of the presence of different types of interactions in which π electron density is responsible for the formation of a supermolecule or a supramolecular array. Thus, Au$\cdots\pi$, $\cdots$C−H$\cdots\pi$ or $\pi\cdots\pi$ interactions will also be considered in this chapter.

5.2
Supramolecular Gold Entities Built with Gold–Non-Metal Secondary Bonds

This section will examine selected examples of supramolecular entities built by means of contacts in which gold is the acceptor of electron density, and which can have different classes of donors: non-metals different from hydrogen, which may be some of the atoms in Groups 17, 16, 15 or 14; a hydrogen atom of a covalent NM−H bond, where the non-metal is in most cases a carbon; or a cloud of π electron density.

5.2.1
Secondary Bonds: Au$\cdots$NM Interactions

Since the metal center in gold complexes usually displays acid character, Au$\cdots$NM contacts will be more easily formed as the basicity of the non-metal increases. Thus, there are many gold$\cdots$halogen weak interactions between cationic gold complexes and halides or anions of which the halogen forms part, but most are individual contacts between ions that do not generate supramolecular structures. For example, about 50% of Au$\cdots$Cl contacts are isolated cation$\cdots$anion interactions.

Nevertheless, in some cases, cation$\cdots$anion contacts give rise to extended structures, as in both the colorless and the yellow polymorph of [Au(C≡NC$_6$H$_{11}$)$_2$][PF$_6$] [13], where the gold cations aggregate via aurophilic contacts into infinite linear chains, which are further connected through Au$\cdots$F contacts to afford two-dimensional

5.2 Supramolecular Gold Entities Built with Gold–Non-Metal Secondary Bonds | 297

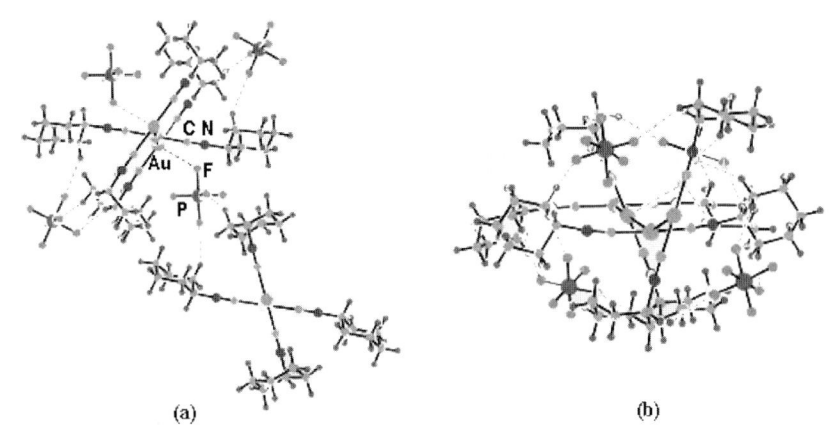

Figure 5.1 Chains of gold cations and PF_6^- anions in the colorless (a) and yellow (b) polymorphs of $[Au(C\equiv NC_6H_{11})_2][PF_6]$ showing the Au···F contacts.

arrays, as shown in Figure 5.1. The closest approach of a fluorine atom of a hexafluorophosphate ion to a gold center is 3.070 Å in the colorless polymorph and 3.152 Å in the yellow one. These distances are similar to the sum of the van der Waals radii for Au and F (3.13 Å).

In chlorine derivatives, where the presence of Au···X van der Waals interactions is much more frequent than in fluorine ones, different associations have been observed, the simplest being the formation of pairs of molecules. Close inspection of the crystal structure of the gold–germanium compound $[(o\text{-Tol}_3P)AuGeCl_3]$ [14] shows that there is evidence for a weak aggregation to give a centrosymmetrical dimer, resulting from an Au···Cl interaction less than the van der Waals radii of 3.299(3) Å that gives rise to a hexanuclear $Au_2Ge_2Cl_2$ ring (Figure 5.2). Curiously,

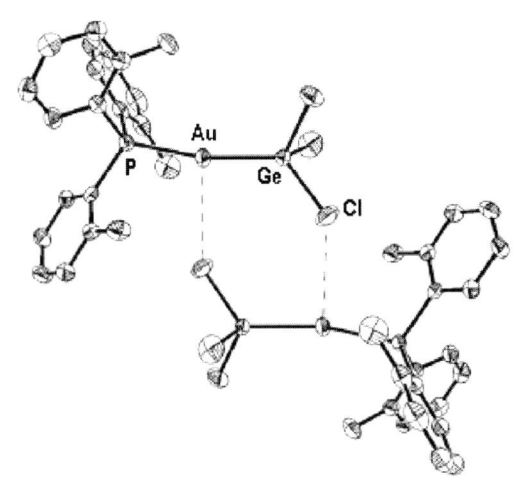

Figure 5.2 Dimers of $[(o\text{-Tol}_3P)AuGeCl_3]$ showing the Au···Cl contacts.

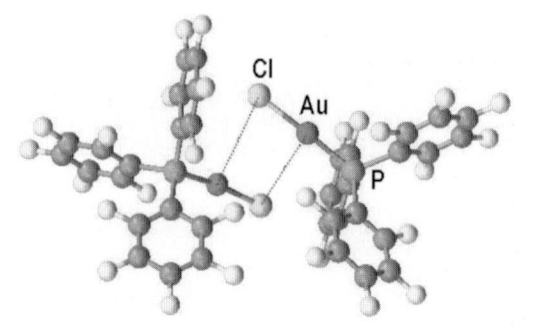

Figure 5.3 Dimers of [AuCl{(C$_6$F$_5$)$_3$P}] showing the Au···Cl contacts.

the variation of the substituents at the phosphine strongly affects the structure and the weak interactions present in the solid state. Thus, the dimethylphenylphosphino derivative displays an ionic structure in which the ions are arranged in tetramers {[(Me$_2$PhP)$_2$Au]$^+$[Au(GeCl$_3$)$_2$]$^-$}$_2$ via aurophilic contacts and Au–Cl distances shorter than the sum of the van der Waals radii are not observed in this case.

Another example of the association of neutral complexes of the type [AuClL] into pairs is found in [AuCl{(C$_6$F$_5$)$_3$P}] [15]. As highlighted in Figure 5.3, two independent molecules are in close proximity so as to form weak Au···Cl interactions of 3.5428(9) and 3.4111(9) Å, resulting in a dimer which in this case contains a tetranuclear Au$_2$Cl$_2$ ring. The Au···Au separation of 3.7828(3) Å is too long to consider any aurophilic interaction.

In the dinuclear complex [1,3-C$_6$H$_4$(OPPh$_2$AuCl)$_2$] [16] the molecules also form loosely associated dimers due to the presence of weak Au···Cl contacts of 3.407(2) and 3.616(2) Å, which are also reported to be aided by aurophilic attractions, as indicated in Figure 5.4, although the shortest gold–gold contact between

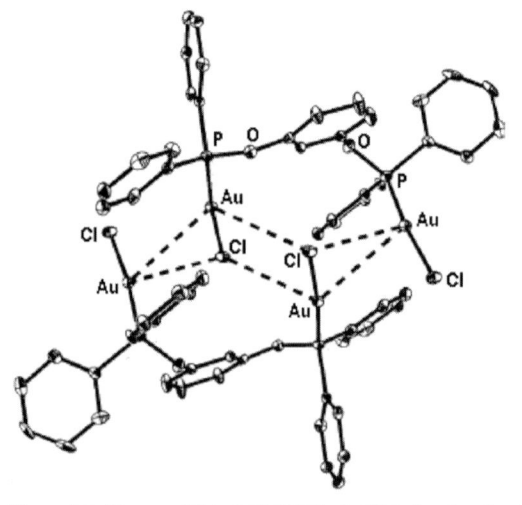

Figure 5.4 Dimers of [1,3-C$_6$H$_4$(OPPh$_2$AuCl)$_2$] showing the Au···Cl contacts.

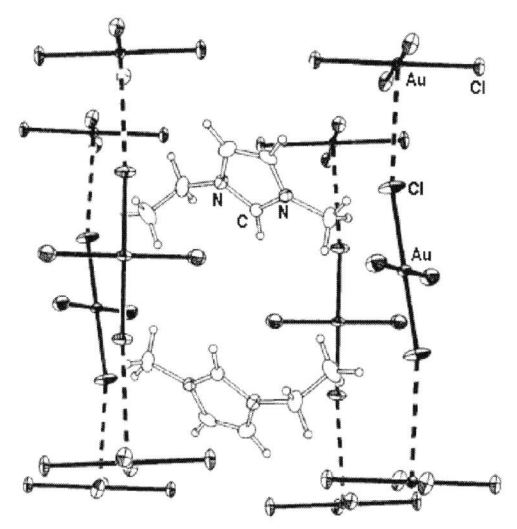

Figure 5.5 Chains of [AuCl$_4$]$^-$ in [EMIM][AuCl$_4$] showing the Au···Cl contacts.

adjacent molecules, 3.695(1) Å, is slightly longer than the sum of the van der Waals radii.

Apart from these types of associations observed in certain neutral gold(I) products, one-dimensional polymers can be formed by Au···Cl interactions between ions, which are more often found when the gold(III) anion [AuCl$_4$]$^-$ is present, as occurs, for example, in [EMIM][AuCl$_4$] and [BMIM][AuCl$_4$] ([EMIM]+ = 1-ethyl-3-methylimidazolium, [BMIM]+ = 1-(n-butyl)-3-methylimidazolium) [17]. Both salts are very similar in the solid state with each ion surrounded by eight counterions in a cubic fashion and containing two crystallographically independent [AuCl$_4$]$^-$ anions arranged perpendicularly to one another and connected by Au···Cl contacts. This results in infinite anionic chains of alternating corner-to-face arranged [AuCl$_4$]$^-$ units (Figure 5.5), a unique pattern in salts containing purely tetrachloroaurate(III) anions. The interionic Au···Cl distances in the EMIM and BMIM derivatives are 3.356(3) and 3.452(3) Å, respectively, the latter very close to the sum of the van der Waals radii of gold and chlorine (3.41 Å).

In the mixed-valent compound [Au(terpy)Cl]$_2$[AuCl$_2$]$_3$[AuCl$_4$] [18], an infinite chain of gold containing anions and cations is generated through two types of weak interactions: aurophilic contacts, which result in pentanuclear units, and Au···Cl interactions (Figure 5.6). Three [AuCl$_2$]$^-$ anions, arranged in a spiral, bridge two [Au(terpy)Cl]$^{2+}$ cations through long axial contacts. The remaining axial site of the [Au(terpy)Cl]$^{2+}$ cation is occupied by a chlorine atom from a square-planar [AuCl$_4$]$^-$ anion at distance of 3.401(4) Å, resulting in a polymeric ···Cl-Au^{III}-Cl···Au^{III}···Au^{I}···Au^{I}···Au^{I}···Au^{III}···Cl-Au^{III}-Cl··· chain.

A more complicated extended structure is present in [Au(HDMG)$_2$][AuCl$_4$] (H$_2$DMG = dimethylglyoxime, HON=C(Me)C(Me)=NOH) [19], where one-dimensional polymers are formed via interionic Au···Cl contacts of 3.323(4) and

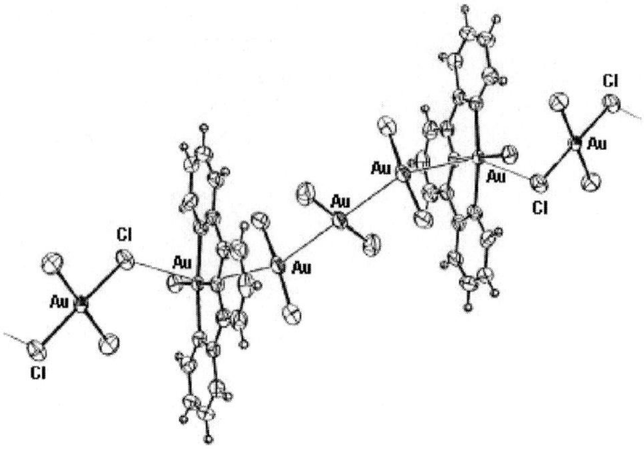

Figure 5.6 Extended chain structure of [Au(terpy)Cl]$_2$ [AuCl$_2$]$_3$[AuCl$_4$] showing the relationship between the [AuCl$_2$]$^-$, [AuCl$_4$]$^-$ and [Au(terpy)Cl]$^{2+}$ units.

3.401(4) Å (two crystallographically independent closed cation–anion pairs). There are two different mutual arrangements of cations and anions that are noteworthy in the closely packed pairs. In one of them, they approach orthogonally (dihedral angle between the AuN$_4$ and AuCl$_4$ fragments 91.4°), while in the other pair the same dihedral angle is equal to 8.1°. The pairs displaying nearly parallel cations and anions form chains along the *a* direction, and the [Au(HDMG)$_2$]$^+$ cations of the other pairs also expand along this direction as a consequence of O−H···O hydrogen bonds. Apart from these contacts, other weak interactions, such as Cl···Cl or weak C−H···Cl contacts, lead to the formation of layers, which are further combined through the Au···Cl van der Waals interactions between the cations and anions that approach orthogonally to give a three-dimensional network for the solid-state structure of [Au(HDMG)$_2$][AuCl$_4$] (Figure 5.7).

Although there are fewer examples of bromo-derivatives than chloro-derivatives of gold displaying intermolecular Au···X contacts, a certain degree of association through weak Au···Br interactions is detected in the X-ray diffraction analysis of the dinuclear gold(II) ylide [Au$_2${μ-(CH$_2$)$_2$PPh$_2$}(μ-C$_5$H$_4$NS)Br$_2$] [20]. The asymmetric unit contains three independent molecules, all displaying an eight-membered ring with a twisted conformation and a transannular metal–metal bond (Au–Au: 2.547(4), 2.548(4) and 2.564(4) Å), which are typical of dinuclear gold(II) species. There are also intermolecular Au···Au and Au···Br contacts, the latter displaying very different distances (3.461(6), 3.835(6) and 4.000(6) Å), which lead to a supramolecular structure.

The disproportionation of neutral five-coordinate [Au(phen)(CN)$_2$Br] (phen = 1,10-phenanthroline) in dimethylformamide gives a conducting solution from which two charged species, [Au(phen){(CN)$_{0.92}$Br$_{0.08}$}$_2$]Br and [Au(phen)(CN){(CN)$_{0.82}$Br$_{0.18}$}]·0.5*trans*-[Au(CN)$_2$Br$_2$]·0.5Br·phen, can be isolated [21]. The most striking feature of both solid-state structures is their ability to form secondary bonds

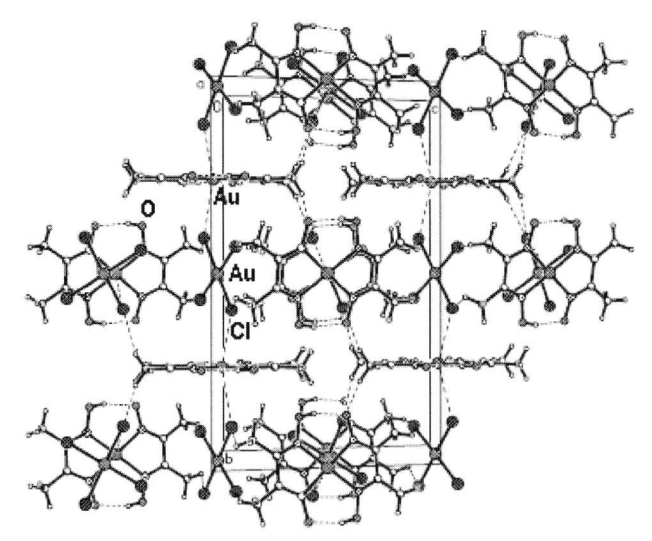

5.2 Supramolecular Gold Entities Built with Gold–Non-Metal Secondary Bonds | 301

Figure 5.7 Three-dimensional network of [Au(HDMG)$_2$][AuCl$_4$] viewed along the a-axis showing the Au$\cdots$Cl contacts and other weak interactions.

with bromide anions outside the square-planar coordination, although in different ways. Thus, while the latter displays individual Au$\cdots$Br contacts of 3.165(1) Å between the [Au(phen)(CN){(CN)$_{0.82}$Br$_{0.18}$}]$^+$ cations and the Br$^-$ (leading to a square pyramidal coordination), in the former two Br$^-$ anions are octahedrally coordinated at a distance of 3.277(1) Å. The Au$\cdots$Br line makes an angle of 6.20(3)° with the normal to the basal plane. Each bromide is shared by two gold(III) cations leading to a chain structure of octahedra (see Figure 5.8). The Au$\cdots$Br contacts in this complex represent stronger interactions than those commented above, since the de van der Waals distance is 3.51 Å, and are even shorter than the weak Au$\cdots$Cl interactions described above.

Regarding Au$\cdots$I contacts, which are also less frequent than Au$\cdots$Cl ones, they are responsible for the formation of two-dimensional networks in some examples, such as in the neutral complex [AuI(P^{i}Pr$_3$)]·1.5I$_2$ [22] or in the ionic species

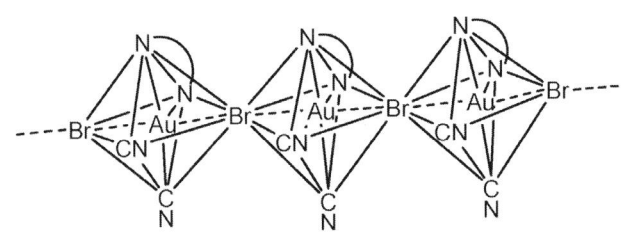

Figure 5.8 Octahedral environments of gold(III) in the one-dimensional structure of [Au(phen){(CN)$_{0.92}$Br$_{0.08}$}$_2$]Br by Au$\cdots$Br contacts.

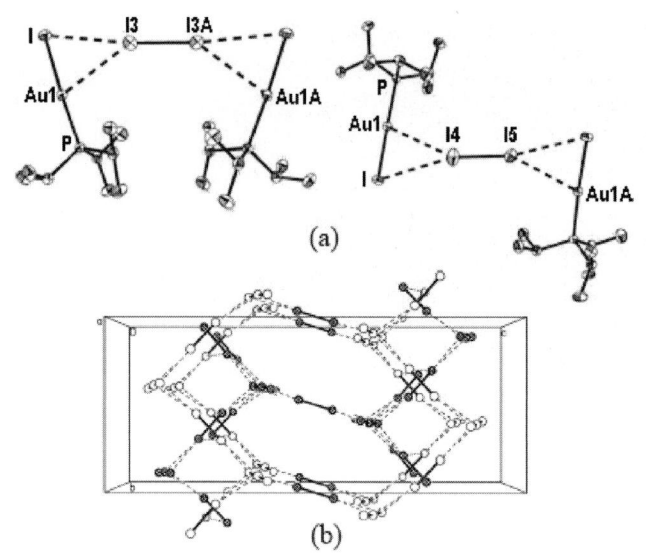

Figure 5.9 (a) Modes of supramolecular aggregation of the components of [AuI(P^{i}Pr$_3$)]·1.5I$_2$ (b) View of the cell along the a-axis.

(BEDT-TTF)$_6$[AuBr$_2$]$_6$Br(TIE)$_3$ (BEDT-TTF = bis(ethylenedithio)tetrathiafulvalene, TIE = tetraiodoethylene) [23] or Q$_2$[(Au^{I}I$_2$)(Au^{III}I$_4$)(I$_3$)$_2$] (Q=[NH$_3$(CH$_2$)$_8$NH$_3$], [NH$_3$(CH$_2$)$_7$NH$_3$]) [24]. In the former the iodine molecules are found to approach the [AuI(P^{i}Pr$_3$)] molecules end-on and midway between the Au−I bond, and this happens in two different ways: one produces a symmetrical I$_2$ bridge between two complexes with a twofold axis passing through the midpoint of the I$_2$ molecule, and the second is unsymmetrical and connects two complexes, which are inverted (head-to-tail) (Figure 5.9a). In both types of I$_2$ bridges, the Au···I and I···I distances are of comparable lengths (in the range 3.556–3.861 Å) and the I−I distances in the iodine molecules show no significant lengthening (2.7470(7) and 2.7078(5) Å). The aggregation of the components leads to the formation of two-dimensional arrays, in which zigzag chains ···Au1···I4−I5···Au1A···I4A−I5A···Au1B··· are linked to corrugated sheets by the ···Au1···I3−I3A···Au1A··· contacts (Figure 5.9b).

In the crystal structure of (BEDT-TTF)$_6$[AuBr$_2$]$_6$Br(TIE)$_3$ [AuBr$_2$]$^−$, Br$^−$, and TIE molecules construct two-dimensional sheets by means of Au···I and Br···I secondary bonds, as shown in Figure 5.10. There are two types of [AuBr$_2$]$^−$ anions and only one of them participates in the formation of Au···I interactions by coordination of four TIEs at the central Au atom, thus being incorporated in the sheet. The Au···I distances, of 3.271 and 3.369 Å, are shorter than in the previous example. The other type of [AuBr$_2$]$^−$ is located apart from the sheets and incorporated in the conduction layer. Both kinds of [AuBr$_2$]$^−$ anions are almost perpendicular to the two-dimensional sheets.

Finally, the crystal structure of [NH$_3$(CH$_2$)$_8$NH$_3$]$_2$[(Au^{I}I$_2$)(Au^{III}I$_4$)(I$_3$)$_2$] [24] features layers of corner-shared nominal AuI$_6$ octahedra stacked along the a axis, with

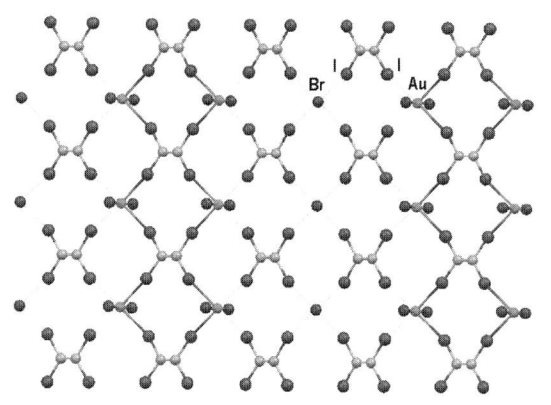

Figure 5.10 Association of [AuBr$_2$]$^-$ anions and TIE molecules in (BEDT-TTF)$_6$[AuBr$_2$]$_6$Br(TIE)$_3$ through Au···I and Br···I contacts.

interlayer distances of 15.5 Å (Figure 5.11). Adjacent layers are displaced by 5.8 Å along the c axis. Each inorganic sheet features two non-equivalent gold atoms corresponding to linear [AuI$_2$]$^-$ and square-planar [AuI$_4$]$^-$ units. A tetragonally compressed octahedron around each [Au$^I_2$]$^-$ unit is completed by four coplanar iodide ions (Au^I···I: 3.278(1) Å) from neighboring [AuI$_4$]$^-$ units. Similarly, an elongated octahedron around each [Au^{III}I$_4$]$^-$ unit is completed by two apical iodine atoms (Au^{III}···I: 3.752(7) Å) from I$_3^-$ ions sandwiched between the gold iodide layers. The [NH$_3$(CH$_2$)$_7$NH$_3$]$^+$ derivative displays a similar structure, the most significant differences being brought about by the different conformations of the organic cations.

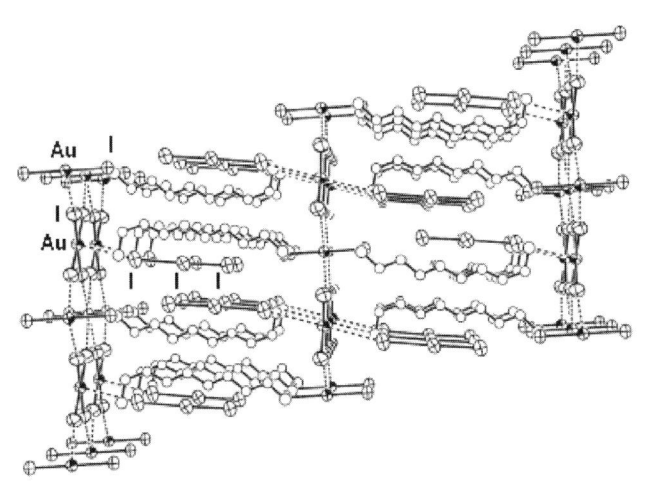

Figure 5.11 Crystal structure of [NH$_3$(CH$_2$)$_8$NH$_3$]$_2$[(Au^{I}I$_2$)(Au^{III}I$_4$)(I$_3$)$_2$] (viewed along the b-axis) showing the Au···I contacts.

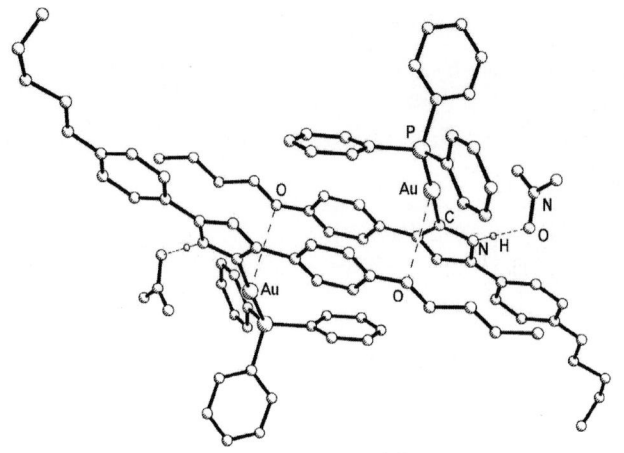

Figure 5.12 Molecular structure of [Au(HPz^{bp2})(PPh$_3$)](NO$_3$) showing the dimeric unit through Au···O weak interactions.

Apart from these van der Waals contacts of gold atoms with halogens, contacts with other nonmetals, such as calcogens, can also be present in some cases. The most frequent are those with oxygen or sulfur, but in the former about 50% are cation–anion individual interactions or contacts with oxygen atoms from solvent molecules, although in other cases Au···O interactions lead to supramolecular aggregations. For example, in [Au(HPz^{bp2})(PPh$_3$)](NO$_3$) (HPz^{bp2} = 3,5-bis(4-butoxyphenyl)pyrazole) [25] there is a very weak intermolecular coordinative interaction between the metal center and the oxygen atom of the C$_6$H$_4$OC$_4$H$_9$ substituent at the 5-position of the pyrazole ligand of a neighboring molecule (Au···O distance 3.540(6) Å), and this appears to be responsible for the formation of dimers in an antiparallel head-to-tail arrangement, as shown in Figure 5.12. The adjacent aromatic rings of opposite substituents of each dimer molecule are parallel, but are separated face-to-face by a distance of 4.4 Å. This distance suggests that π–π stacking interactions need not be considered, although they might be of help in adopting the observed dimeric arrangement.

In other cases, one-dimensional polymers are formed due to Au···O secondary bonds, as occurs in the gold(III) complexes [AuCl$_2$(o-NC$_5$H$_4$CH$_2$O)] [26] and [(H$_7$O$_3$)(15-crown-5)][AuCl$_4$] [27]. In the former the oxygen atom bonded to gold interacts with the metal center of an adjacent neutral molecule, thus forming a chain through Au···O contacts of 3.108 Å, considerably shorter than in the previous example (Figure 5.13). In the ionic compound, the Au(III) ions also exhibit very long axial interactions with two crown oxygen atoms of different cations, resulting in a cross-linked polymeric structure, as shown in Figure 5.14. The Au···O distances, 3.339(4) and 3.373(4) Å, are intermediate between those commented for the two previous derivatives. Moreover, adjacent chains are connected through hydrogen bonds to give rise to a two-dimensional network.

Although there are similar numbers of Au···S and Au···O intermolecular contacts, the number of Au···S weak bonds that lead to supramolecular structures is

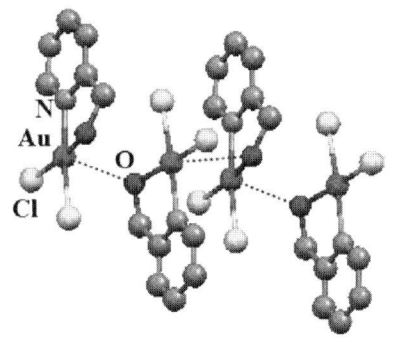

Figure 5.13 Extended chain structure of [AuCl$_2$(o-NC$_5$H$_4$CH$_2$O)] showing the Au···O contacts.

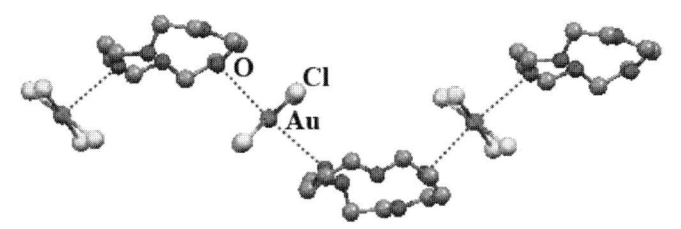

Figure 5.14 Extended chain structure of [(H$_7$O$_3$)(15-crown-5)][AuCl$_4$] showing the Au···O contacts.

higher due to the lower percentage of isolated cation–anion interactions. The most common structural motif in neutral gold(I) complexes with sulfur donor ligands is the formation of an Au$_2$S$_2$ parallelogram with Au···S edges and an Au···Au diagonal, such as that formed for example in the thiocyanate complex [Au(SCN){CN(2,6-C$_6$H$_3$Me$_2$)}] [28], shown in Figure 5.15. This mode of association, in which the molecules display a head-to-tail disposition around an inversion center, is very similar to that described above for [AuCl{(C$_6$F$_5$)$_3$P}] [15] and shown in Figure 5.3. The shape of such parallelograms can vary, depending on the interaction distances, from a more rectangular one (when the Au···Au distances are longer than the Au···S one) to a more rhomboidal shape (when the aurophilic interaction is stronger than the

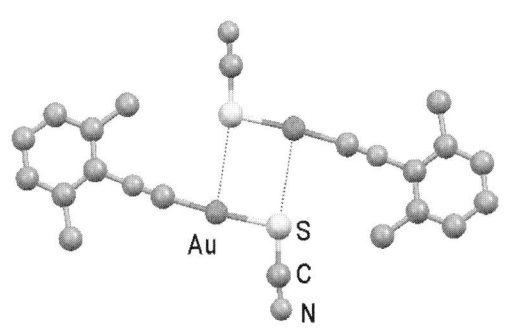

Figure 5.15 Dimers of [Au(SCN){CN(2,6-C$_6$H$_3$Me$_2$)}] showing the Au···S contacts.

Table 5.1 Gold(I) complexes containing [Au···S]$_2$ parallelograms as structural motifs.

Complex	Au···S (Å)	Au···Au (Å)	Ref.
[Au(SCN){CN(2,6-C$_6$H$_3$Me$_2$)}]	3.459(2)	3.983(1)	[28]
[Au(SCN){CN(2,4,6-C$_6$H$_2$Me$_3$)}]	3.938(2)	3.397(1)	[28]
	3.719(2)	5.125(1)	
[Au(SCN$_4$Me)(PPh$_3$)]	3.423	3.946	[29]
[Au{SC(=NPh)N(CH$_2$CH$_2$)$_2$O}(PPh$_3$)]	3.4211(7)		[30]
[Au(C$_6$F$_5$){S=$\overline{\text{CN(H)C(Me)=C(H)S}}$}]	3.529(4)	4.177(1)	[31]
[Au(SCH$_2$CO$_2$H)(PPh$_3$)]	3.131(2)		[32]
[Au(4-SC$_6$H$_4$CO$_2$H)(PEt$_3$)]	3.565(1)	3.629(1)	[33]
[Au(4-SC$_6$H$_4$CO$_2$H)(PPh$_2$py)]	3.212(1)	4.234(1)	[33]
	3.253(1)	4.232(1)	
[Au(2-Hmba){P(o-Tol)$_3$}]	3.4035(14)		[34]
[{Au{SC(=NPh)N(CH$_2$CH$_2$)$_2$O}$_2$(μ-dppe)]	3.8771(16)		[30]
[{Au(PPh$_3$)}$_2$(μ-SSS)]	3.85	4.872	[35]
	4.24		
[{Au(2-Hmba)}$_2$(μ-dppe)]	3.341(2)	4.194(2)	[36]
[{Au{SC(OMe)=NC$_6$H$_5$}}$_2${μ-Ph$_2$P(CH$_2$)$_4$PPh$_2$}]	3.4339(10)	3.7122(3)	[37]

gold···sulfur one). The Au···S and Au···Au separations within these motifs range from 3.131(2) to 4.24 Å and from 3.397(1) to distances longer than 5 Å, respectively, which implies that in most cases one of them is predominant.

Examples of molecules containing this type of association are shown in Table 5.1. In the mononuclear complexes [Au(SCN){CN(2,6-C$_6$H$_3$Me$_2$)}], [Au(SCN){CN(2,4,6-C$_6$H$_2$Me$_3$)}] [28], [Au(SCN$_4$Me)(PPh$_3$)] [29] and [Au{SC(=NPh)N(CH$_2$CH$_2$)$_2$O}(PPh$_3$)] [30] no additional supramolecular aggregation is observed, so the molecules are associated into dimers. In other cases, a second type of secondary interaction is also present, which leads to further aggregation. Thus, in [Au(C$_6$F$_5$){S=$\overline{\text{CN(H)C(Me)=C(H)S}}$}] [31] the molecules in the crystal are arranged in sheets as a result of N−H···F interactions.

A typical structural motif when carboxylic acid residues are present in the molecule is the formation of dimers from hydrogen bonding between pairs of carboxylate residues. Thus, in compounds [Au(SCH$_2$CO$_2$H)(PPh$_3$)] [32], [Au(4-SC$_6$H$_4$CO$_2$H)L] (L=PEt$_3$, PPh$_2$py) [33] and [Au(2-Hmba){P(o-Tol)$_3$}]·NCMe (Hmba = 2-mercaptobenzoate) [34] the carboxylate groups of neighboring molecules are also hydrogen bonded and these two categories of alternating intermolecular contacts produce a one-dimensional polymeric backbone, which dominates the supramolecular array (see Figure 5.16).

The [Au···S]$_2$ themselves can produce linear chains (Figure 5.17) instead of associating in pairs when the complex whose supramolecular structure is analyzed is a dinuclear species. This occurs, for example, in one of the two independent molecules of [{Au{SC(=NPh)N(CH$_2$CH$_2$)$_2$O}}$_2$(μ-dppe)] [30] or in complexes [{Au(PPh$_3$)}$_2$(μ-SSS)] (SSS = 1,3,4-thiadiazoledithiolate) [35], [{Au(2-Hmba)}$_2$(μ-dppe) [36]

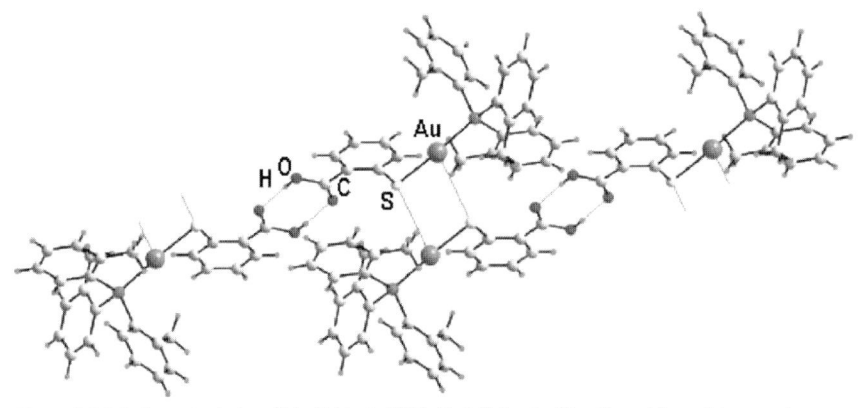

Figure 5.16 Polymeric chain of [Au(2-Hmba){P(o-Tol)$_3$}] by O−H···O and Au···S interactions.

or [{Au{SC(OMe)=NC$_6$H$_5$}}$_2${μ-Ph$_2$P(CH$_2$)$_4$PPh$_2$}] [37]. Furthermore, in the latter two cases, a close inspection of the structure reveals that centrosymmetrically-related pairs of molecules associated via Au···S interactions coupled with hydrogen bonding motifs result in two-dimensional or three-dimensional networks for [{Au(2-Hmba)}$_2$(μ-dppe) [36] or [{Au{SC(OMe)=NC$_6$H$_5$}}$_2${μ-Ph$_2$P(CH$_2$)$_4$PPh$_2$}] [37], respectively.

There are some other compounds in which the linkage of gold and sulfur does not occur through [Au···S]$_2$ dimers, as observed in the solid-state structure of [Au(SCN)(PMe$_3$)] [28]. In this case there are two independent monomers in the asymmetric unit, which are associated into dimers through short aurophilic contacts (Au···Au 3.0994(5) Å) and into tetramers mediated by Au···S interactions between gold and the sulfur atom of a S-bonded thiocyanate ligand of a neighboring dimer (Au···S 3.771(2) Å), as shown in Figure 5.18. In the resulting tetramer, the atoms are related by a crystallographic center of inversion.

Finally, in the neutral gold(III) complex [Au(SCN)(NCS)(ppy)] (ppy = 2-phenylpyridine) [38] and in the ionic mixed-valent product [Au^I(CN^{c}Hex)$_2$][Au^{III}(S$_2$C$_6$H$_4$)$_2$] [39] one-dimensional polymers are formed as a consequence of the Au···S van der Waals

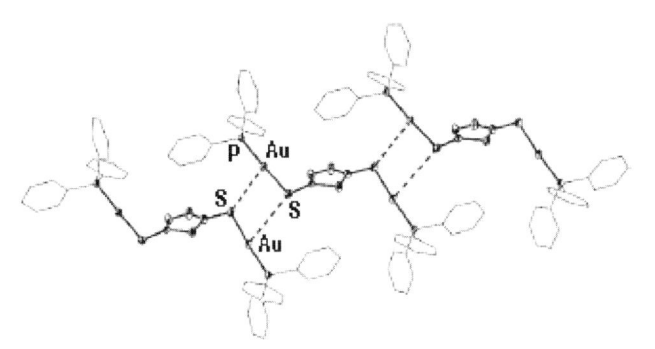

Figure 5.17 Extended chain structure of [{Au(PPh$_3$)}$_2$(μ-SSS)] showing the [Au···S]$_2$ motifs.

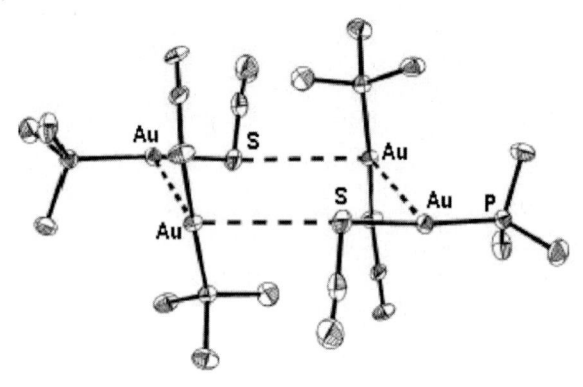

Figure 5.18 Tetrameric aggregates of [Au(SCN)(PMe$_3$)] by Au···Au and Au···S contacts.

contacts of 3.412 and 3.422 Å, respectively, which are shorter than in the previous example. In the neutral derivative, one thiocyanate ion binds to gold via nitrogen and the other through the sulfur. This binding mode leads to an interaction between gold and the terminal sulfur of thiocyanate from another molecule, which is still a good coordination center. This intermolecular interaction leads to the formation of a polymer structure (Figure 5.19), which may be attributed to the ability of thiocyanate to function as an ambidentate ligand.

In the mixed-valent [Au^I(CN^{c}Hex)$_2$][Au^{III}(S$_2$C$_6$H$_4$)$_2$] [39], the cations and anions are stacked alternately in columns parallel to the b axis of the crystal (Figure 5.20). Because of the tilt of the ions against this metal axis, the closest contact between the ions is between a sulfur of a dithiolate bonded to gold(III) and a gold(I) atom of a cation.

Although selenium derivatives of gold occur less frequently than sulfur ones, the presence of Au···Se intermolecular weak interactions has also been reported in a number of selenium complexes. For example, the cations of the tetranuclear complex [Se(AuPPh$_3$)$_4$](CF$_3$SO$_3$)$_2$ [40] are paired across symmetry centers to form loose dimers via an [Au···Se]$_2$ unit (Figure 5.21), which resembles the supramolecular motif usually found in sulfur compounds of gold. The intermolecular Au···Se

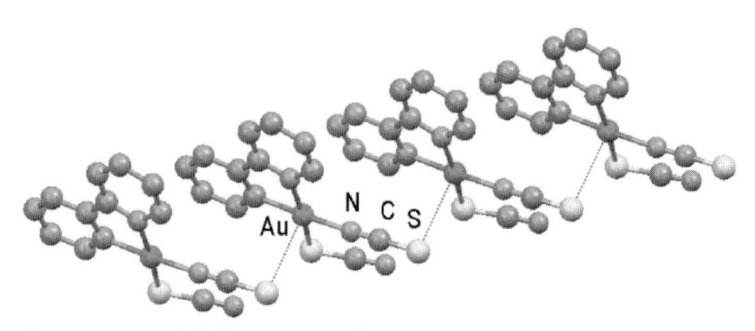

Figure 5.19 Extended chain structure of [Au(SCN)(NCS)(ppy)] showing the Au···S contacts.

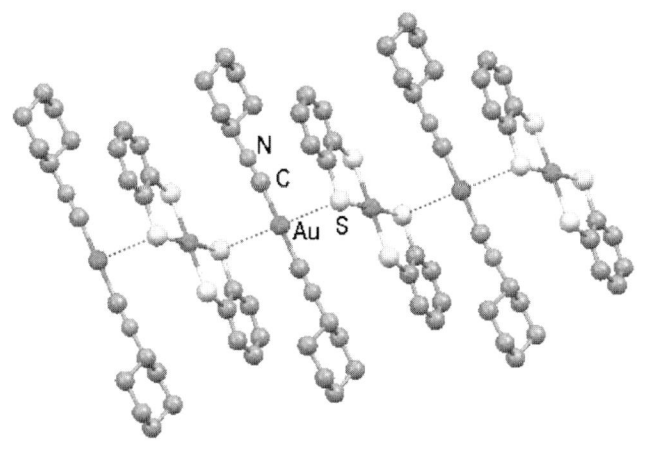

Figure 5.20 Extended chain structure of the mixed-valent complex [Au^{I}(CN^{c}Hex)$_2$][Au^{III}(S$_2$C$_6$H$_4$)$_2$] showing the Au···S contacts.

distance is 3.248 Å; therefore this interaction is stronger than most of the Au···S contacts commented above (see Table 5.1), since selenium is bigger than sulfur.

Association into pairs of molecules through Au···Se secondary bonds in a head-to-tail disposition is also observed in the ferrocene derivatives [AuX{Fc(SePh)PPh$_2$}] (X=Cl, C$_6$F$_5$; Fc=Fe(C$_5$H$_4$)$_2$) [41]. However, while in [Se(AuPPh$_3$)$_4$]$^+$ a four-membered ring is formed, in these two cases the gold and selenium atoms are

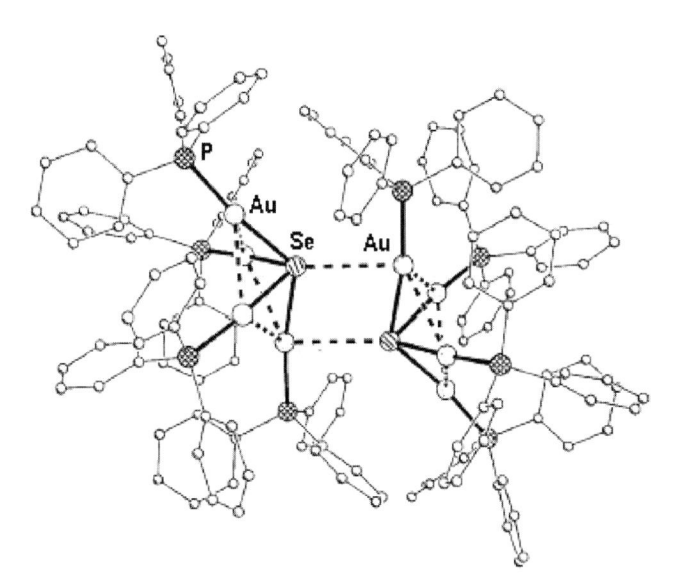

Figure 5.21 Association of two cations of [Se(AuPPh$_3$)$_4$](CF$_3$SO$_3$)$_2$ showing the Au···Se contacts.

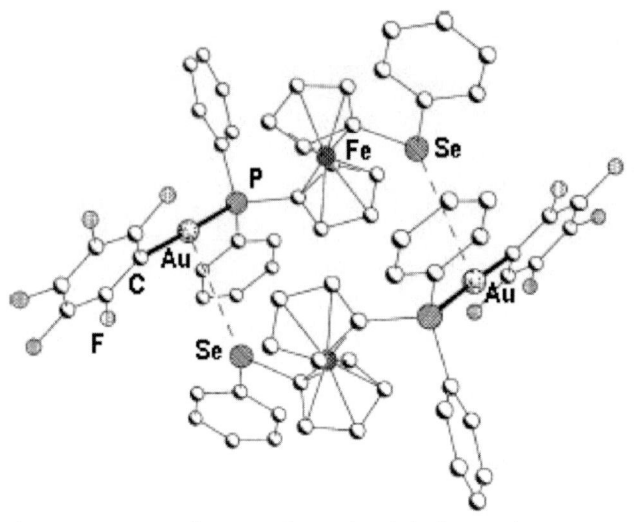

Figure 5.22 Dimers of [Au(C$_6$F$_5$){Fc(SePh)PPh$_2$}] showing the Au···Se contacts.

located far from each other in the monomer, which affords higher nuclearity cycles (see Figure 5.22). In addition, in the chloro-derivative, Au···Cl contacts between adjacent dimers are also present, which leads to the formation of a chain structure. The Au···Se distances in these species are longer than in the example commented above, being around 3.7 Å (similar to the Au···Cl interactions) in the chloro-complex and 3.643 Å in the pentafluorophenyl one. In these two complexes, other additional weak interactions such as Au···Ph, Au···H−C or X···H−C (X=Cl, F) hydrogen bonds are observed.

A different supramolecular synthon is found in the analysis of the solid-state structure of [Au$_2${(SePPh$_2$)$_2$N}(μ-dppm)](OSO$_2$CF$_3$) [42], where intramolecular aurophilic interactions are observed within the cations and the cations are arranged in pairs with sub-van der Waals Au···Se contacts of 3.331(1) and 3.447(1) Å (Figure 5.23). In the tetranuclear dimers of the cations, one of the selenium atoms of each monomer approaches the pair of gold atoms of the neighboring monomer, thus resulting in the formation of triangles related by inversion symmetry.

Finally, intermolecular formally non-bonding Au···Se contacts have also been detected in other gold(I) compounds, such as [{Au(PPh$_3$)$_2$}(SeC$_{10}$H$_7$)]SbF$_6$·[Au(PPh$_3$)(SeC$_{10}$H$_7$)] (3.226(3)–3.357(2) Å) [43], [Au(PPh$_3$)(SePh)] (3.381 and 3.460 Å) [44], [Au(PPh$_3$)(SePMe$_2$Ph)]$^+$ (3.353 Å) [45] or [AuCl{Se(dppm)}] (3.277 Å) [46].

Secondary bonding interactions between gold and Group 15 elements are limited only to nitrogen atoms. Intermolecular Au···N contacts have been recently described in several complexes, such as those observed between Au^I and N≡C substituents of tris(cyanoethyl)phosphine in [{Au(SSNH$_2$)}{P(CH$_2$CH$_2$CN)$_3$}] and [{Au(SSCNH$_2$)}{P(CH$_2$CH$_2$CN)$_3$}] (SSNH$_2$ = 2-amino-5-mercapto-1,3,4-thiadiazolate; SSCNH$_2$ = 6-amino-2-mercaptobenzothiazolate) [47], which are further associated into

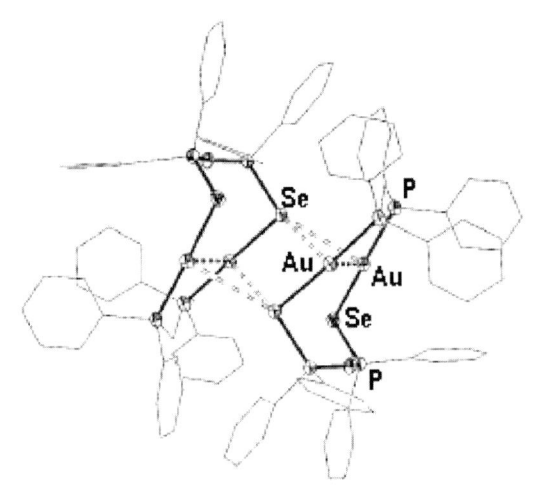

Figure 5.23 Dimers of [Au$_2${(SePPh$_2$)$_2$N}(μ-dppm)](OSO$_2$CF$_3$) showing the Au$_2$Se triangles.

polymers via hydrogen bonds. The long Au···N distances (3.71 and 3.52 Å, respectively) indicate that these contacts are very weak.

This type of interaction is most commonly present in complexes containing tetracyanoaurate(III) anions, which can be present in the solid state in different association degrees. The simplest one is found in [Cu(bipy)(H$_2$O)$_2${Au(CN)$_4$}$_{0.5}$][Au(CN)$_4$]$_{1.5}$ (bipy = 2,2′-bipyridine) [48] or in [{Cu(tmeda)(μ-OH)}$_2$Au(CN)$_4$][Au(CN)$_4$] (tmeda = N,N,N′,N′-tetramethylethylenediamine) [49], where part of the [Au(CN)$_4$]$^-$ anions form a linear trimeric anionic cluster via Au···N van der Waals interactions of the nitrogen atoms of two cyano groups *trans* to the gold(III) centers of two adjacent anions (Figure 5.24a). The interacting distances, 3.052(9) or 3.062 Å, respectively, are

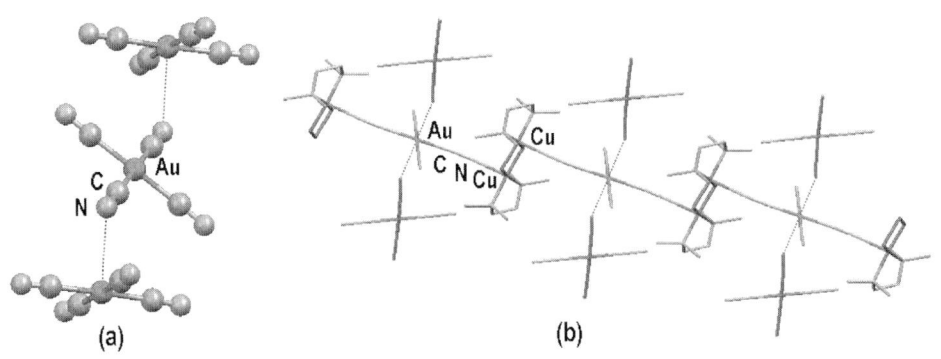

Figure 5.24 (a) Trimeric anionic {[Au(CN)$_4$]$^-$}$_3$ formed by Au···N contacts. (b) Extended chain structure of [{Cu(tmeda)(μ-OH)}$_2$Au(CN)$_4$][Au(CN)$_4$].

considerably shorter than those mentioned above. However, there is a remarkable difference between these two structures: in the former the interaction does not propagate, but the copper coordinated water and bipy molecules serve to increase dimensionality through weak hydrogen bonds and π-interactions, respectively, affording a three-dimensional array. In contrast, in [{Cu(tmeda)(μ-OH)}$_2$Au(CN)$_4$][Au(CN)$_4$] the central [Au(CN)$_4$]$^-$ of the trinuclear anions forms part of the one-dimensional chain resulting from the coordination of two of its four nitrogen atoms to the copper center of two dimer cations {Cu(tmeda)(μ-OH)}$_2$$^+$, thus resulting in a chain polymer with pendant [Au(CN)$_4$]$^-$ units, as shown in Figure 5.24b.

In [Cu(dmeda)$_2$Au(CN)$_4$][Au(CN)$_4$] (dmeda = N,N- dimethylethylenediamine) [48] both the cation and the anion display independent extended chains, the former arising from the coordination of nitrogen atoms of [Au(CN)$_4$]$^-$ units to copper centers, in a similar way to that observed in [{Cu(tmeda)(μ-OH)}$_2$Au(CN)$_4$][Au(CN)$_4$]. Secondary Au$\cdots$N bonds between adjacent [Au(CN)$_4$]$^-$ anions generate an anionic zigzag chain in which, considering the Au$\cdots$N contacts, all the metal centers display a square pyramidal environment (Figure 5.25). The Au$\cdots$N distance of 2.963(13) Å is comparable in length to those commented above.

The same type of Au$\cdots$N interaction between adjacent [Au(CN)$_4$]$^-$ anions is found in [Cu(dien)][Au(CN)$_4$]$_2$ and [Cu(en)$_2$][Au(CN)$_4$]$_2$ (dien = diethylenetriamine; en = ethylenediamine) [48], although in these examples the copper cations act as bridges between neighboring chains generating 2-D sheets, as shown in Figure 5.26. The Au$\cdots$N separations, of 3.001 and 3.137 Å in the triamine complex and 3.035(8) Å in the diamine one are similar to those in other tetracyanoaurate(III) derivatives. Additional weak hydrogen bonds between amino hydrogen and N(cyano) atoms increase the dimensionality of the supramolecular structure.

Finally, secondary Au$\cdots$C bonds have been found in some adducts of the triangular gold(I) complex [Au$_3$(MeN=COMe)$_3$], where they function as electron donors, with nitro-9-fluorenones as acceptors. The solid-state structures of [Au$_3$(MeN=COMe)$_3$]·[2,4,7-trinitro-9-fluorenone] and [Au$_3$(MeN=COMe)$_3$]·[2,4,5,7-tetranitro-9-fluorenone] [50] consist of one-dimensional arrays in which the planar gold(I) trimers and the nearly planar organic molecules are interleaved with the gold trimers, making face-to-face contact with the nitro-aromatic portion of

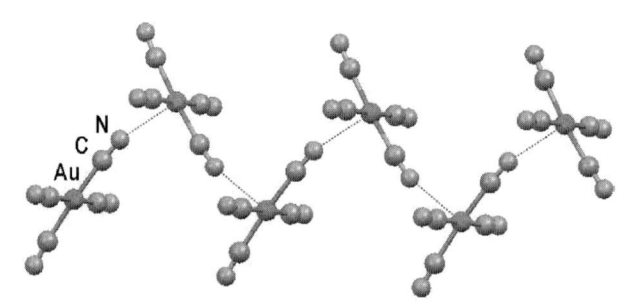

Figure 5.25 Extended structure of the [Au(CN)$_4$]- chain in [Cu(dmeda)$_2$Au(CN)$_4$][Au(CN)$_4$] through Au$\cdots$N contacts.

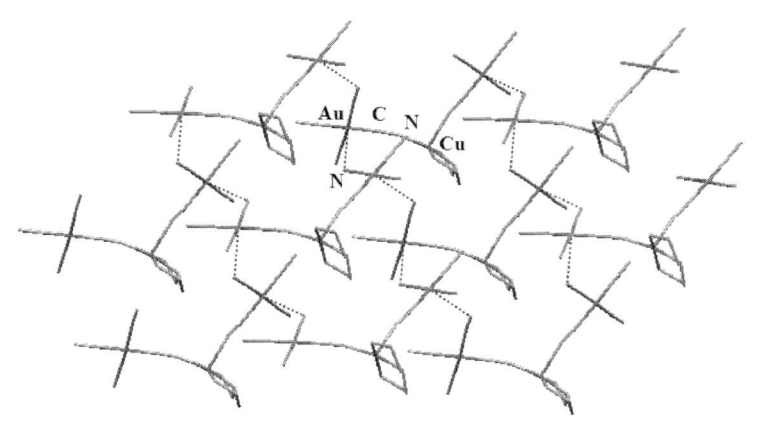

5.2 Supramolecular Gold Entities Built with Gold–Non-Metal Secondary Bonds | 313

Figure 5.26 Two-dimensional structure of [Cu(dien)][Au(CN)$_4$]$_2$ showing the Au···N contacts.

the electron acceptor. Thus, the organic acceptors disrupt the aurophilic interactions present in crystalline [Au$_3$(MeN=COMe)$_3$] itself. The association of the individual molecules within the crystal occurs via weak Au···C van der Waals contacts in the range 3.326–3.589 Å in the former and 3.314–3.572 Å in the latter. The main difference between both structures is that while the latter displays a stepwise arrangement of molecules (Figure 5.27a), in the former they arrange in linear columns (Figure 5.27b).

The same column arrangement is also present in the crystal structure of the chloroform solvate of [Au{C(OMe)NMeH}$_2$][C$_7$Cl$_2$NO$_3$] [51], where the cation is

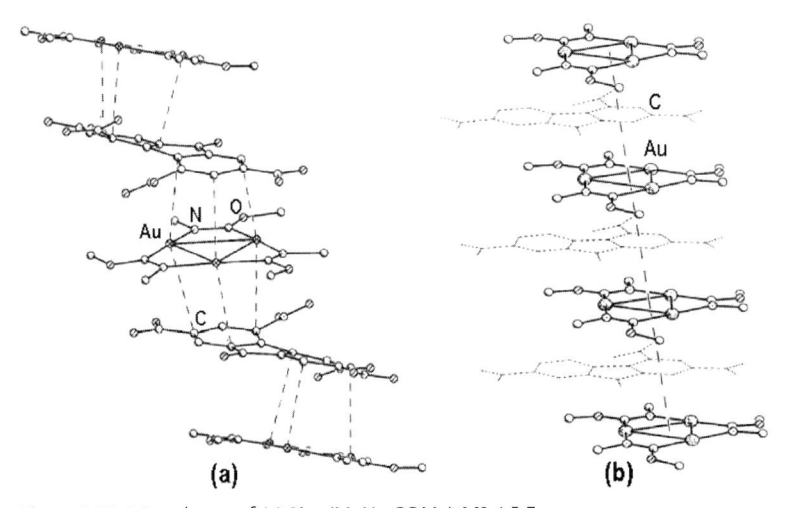

Figure 5.27 1-D polymer of (a) [Au$_3$(MeN=COMe)$_3$]·[2,4,5,7-tetranitro-9-fluorenone]; (b) [Au$_3$(MeN=COMe)$_3$]·[2,4,7-trinitro-9-fluorenone] showing the Au···C interactions.

formed by protonation of [Au$_3$(MeN=COMe)$_3$] and the anion is obtained by hydrolysis of 2,3-dichloro-4,5-dicyano-1,4-benzoquinone. The gold ion is 3.320 Å away from the nearest carbon atom of the anions.

5.2.2
Hydrogen Bonds: Au···H—NM Interactions

This section will look at "non-classical" hydrogen bonds in which a gold atom acts as the acceptor of the hydrogen bond. The distinction between hydrogen bonds and van der Waals interactions is mainly based on the directionality of the former versus the isotropic character of the latter. X—H···M hydrogen bonds (M being a metal and X, in most cases, O, N or C) have received little attention until quite recently. These interactions are favored for electron-rich metals, such as late transition metals in low oxidation states, and thus, some of them involve gold(I). These bonds are different from the X—H···M agostic interactions, in which the metal center acts as a Lewis acid with respect to the X—H bond density. The main structural differences between both types of interactions are the X—H···M angle, which is narrower for agostic (<100–130°) than for hydrogen bonds (>130°), and the X···M distance, which is shorter for agostic than for hydrogen bond interactions.

Although many structures of gold(I) complexes show contacts interpretable as C—H···Au hydrogen bonds, only a few have been discussed explicitly. This could be due to the presence of other classical hydrogen bonds or Au^I···Au^I aurophilic interactions or to the fact that the search for C—H···M interactions is not well integrated into common program systems; correspondingly, it may be assumed that many such interactions fail to be reported. They could explain the packing architecture in many complexes, particularly when no other stronger intermolecular interactions are present.

This section will examine some of the hydrogen bonds involving gold that have been described. The vast majority of them have a C—H bond as hydrogen bond donor; however, an example with an intermolecular B—H bond as donor has been described recently. In the crystal structure of [NBu$_3$Me][Au(PPh$_2$CH$_2$SnB$_{11}$H$_{11}$)$_2$] [52], a weak interaction (H···Au: 2.68 Å) between a BH unit and the gold center, which results in the formation of a dimer, can be detected (Figure 5.28). Much shorter B—H···Au interatomic distances in the range 190–210 pm have been described by Stone [53], although these are intra- instead of inter-molecular contacts.

Recent examples of intermolecular C—H···Au hydrogen bonds are collected in Table 5.2. The Au···H distances range from approximately 2.7 to 3.18 Å, the latter corresponding to the weak interaction present between the molecules of [AuCl{PPh$_2$C(S)N(H)Me}] [54], and the angles vary from 120 to 169°, which appears in the macrocycle [Ph$_2$C(4-C$_6$H$_4$OCH$_2$C≡CAu)$_2${μ-PPh$_2$(CH$_2$)$_4$PPh$_2$}] [56], and is similar to the value of 167° found in the related organometallic ring [O(4-C$_6$H$_4$OCH$_2$C≡CAu)$_2${μ-PPh$_2$(CH$_2$)$_5$PPh$_2$}] [56]. Curiously, the C—H···Au contacts in the [2]catenane [(CF$_3$)$_2$C(4-C$_6$H$_4$OCH$_2$C≡CAu)$_2${μ-PPh$_2$(CH$_2$)$_4$PPh$_2$}]

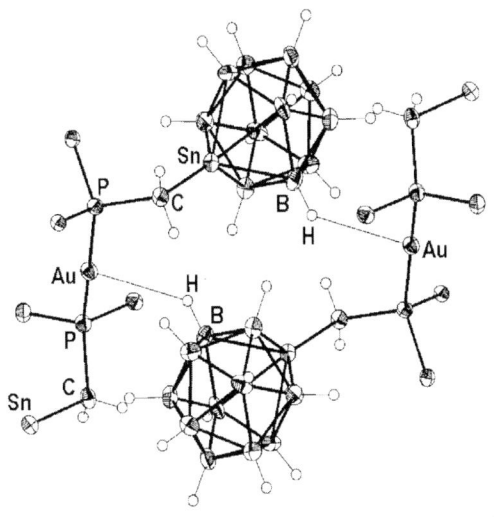

Figure 5.28 Dimer of [NBu₃Me][Au(PPh₂CH₂SnB₁₁H₁₁)₂] formed via B-H···Au bonds. Phenyl substituents and part of the boron cluster cages have been omitted for clarity.

[56], together with aryl···aryl attraction, could stabilize one isomer with respect to another.

The degree of association through C–H···Au bonds can vary depending on the complex analyzed and, thus, we can find one-dimensional arrays, such as in the

Table 5.2 Gold complexes containing C–H···Au hydrogen bonds.

Complex	Au···H (Å)	C–H···Au (Å)	Ref.
[AuCl{PPh₂C(S)N(H)Me}]	3.18	139	[54]
[{Au(C₆F₅)}₂(μ-PPh₂CH₂P^{i}Pr₂)]	2.93	163	[55]
[Ph₂C(4-C₆H₄OCH₂C≡CAu)₂{μ-PPh₂(CH₂)₄PPh₂}]	2.92–3.03	145–169	[56]
[O(4-C₆H₄OCH₂C≡CAu)₂{μ-PPh₂(CH₂)₅PPh₂}]	2.75–3.11	130–167	[56]
[(CF₃)₂C(4-C₆H₄OCH₂C≡CAu)₂{μ-PPh₂(CH₂)₄PPh₂}]	2.89–3.09	137–165	[56]
[{Au(C≡N^{t}Bu)}₂{μ-1,2-C₆Me₄(C≡C)₂}]	2.95–3.06	136–161	[57]
[{AuC(NH^{t}Bu)NEt₂}₂{μ-1,2-C₆Me₄(C≡C)₂}]	2.75–2.99	146–157	[57]
PPN[Au{C≡C-1,3,5-C₆Me₃(C≡CH)₂}]	3.07–3.10	120–135	[58]
PPN[Au(C≡C-4-C₆H₄Me)₂]	2.85	147	[58]
	2.76	136	
[AuCl{Fc(SPh)PPh₂}]	2.9	117	[41]
[AuCl{Fc(SePh)PPh₂}]	3.1	143	[41]
[Au(C₆F₅){Fc(SPh)PPh₂}]	2.9	120	[41]
[AuCl(PPh₂CH₂Fc)]	2.73	158	[59]
	2.76	151	
[Au(C₆F₅)(pzCH₂Fc)]	2.93	123	[60]
[Au(C₆F₅)₃(η³-Fcterpy)]	2.74	166	[61]

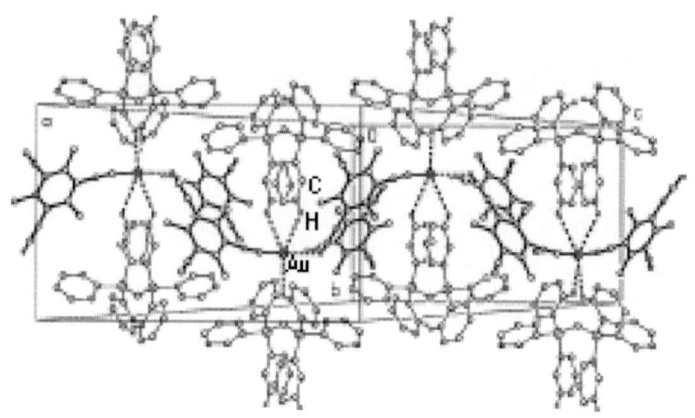

Figure 5.29 Extended chain structure of the the (arenetriethynyl)gold(I) complex PPN[Au{C≡C-1,3,5-C$_6$Me$_3$(C≡CH)$_2$}] through C—H···Au hydrogen bonds.

(areneethynyl)gold(I) complexes [{AuC(NH^{t}Bu)NEt$_2$}$_2${μ-1,2-C$_6$Me$_4$(C≡C)$_2$}] [57], PPN[Au{C≡C-1,3,5-C$_6$Me$_3$(C≡CH)$_2$}] or PPN[Au(C≡C-4-C$_6$H$_4$Me)$_2$] [58], two-dimensional polymers, such as in the ferrocenyl derivative [AuCl(PPh$_2$CH$_2$Fc)] (PPh$_2$CH$_2$Fc = (ferrocenylmethyl)phosphine) [59], or three-dimensional networks, such as that observed in the (arenediethynyl)gold compound [{Au(C≡N^{t}Bu)}$_2${μ-1,2-C$_6$Me$_4$(C≡C)$_2$}] [57].

Figure 5.29 shows the corrugated ribbons parallel to the z axis formed in PPN[Au{C≡C-1,3,5-C$_6$Me$_3$(C≡CH)$_2$}] [58], where each gold is involved in a short contact to three independent hydrogen atoms, two of them representing the coordination of the anionic chain by the cations.

Some C—H···Au contacts have also been described in gold complexes with ferrocenyl derivatives as ligands, such as those included in Table 5.2 containing 1-(diphenylphosphino)-1′-(phenylsulfanyl)ferrocene, 1-(diphenylphosphino)-1′-(phenylselenyl)ferrocene [41], (ferrocenylmethyl)phosphine [59], (ferrocenylmethyl)pyrazole [60] or 4′-ferrocenyl-2,2′:6′,2″-terpyridine [61]. Figure 5.30 displays the secondary contacts (C—H···Au and C—H···Cl) in [AuCl(PPh$_2$CH$_2$Fc)] [59], which give rise to layers in the solid state.

In the case of [{Au(C≡N^{t}Bu)}$_2${μ-1,2-C$_6$Me$_4$(C≡C)$_2$}] [57] the molecules pack in sheets (Figure 5.31a), which, in turn, stack to give a three-dimensional supramolecular structure (Figure 5.31b); this packing seems to be attributable to intermolecular van der Waals C—H···Au interactions.

In other cases, such as in [Au(C$_6$F$_5$)(pzCH$_2$Fc)] [60], different types of van der Waals contacts contribute to the supramolecular three-dimensional structure. In the lattice, the molecules are associated into pairs via an intermolecular Au···Au interaction, and the pairs form chains held together by C—H···Au and C—H···F hydrogen bonds. The chains are further linked to form a three-dimensional structure through additional C—H···F interactions, as shown in Figure 5.32.

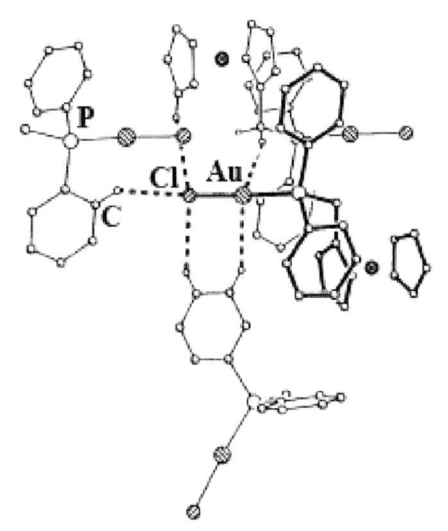

5.2 Supramolecular Gold Entities Built with Gold–Non-Metal Secondary Bonds | 317

Figure 5.30 Environment of the molecule of [AuCl(PPh$_2$CH$_2$Fc)] showing the C–H···Au and C–H···Cl secondary bonds.

Regarding the hydrogen bond donors, different organic substituents, such as CH$_3$, CH$_2$, C≡CH or CH groups from aromatic rings, have been shown to act as donors. Most of the examples shown in Table 5.2 contain a gold(I) atom as hydrogen bond acceptor, but the C–H···Au hydrogen bond in [Au(C$_6$F$_5$)$_3$(η^3-Fcterpy)] [61] demonstrates that even a metal ion with less electron density, such as a gold(III) center, can also act as acceptor of the hydrogen bond.

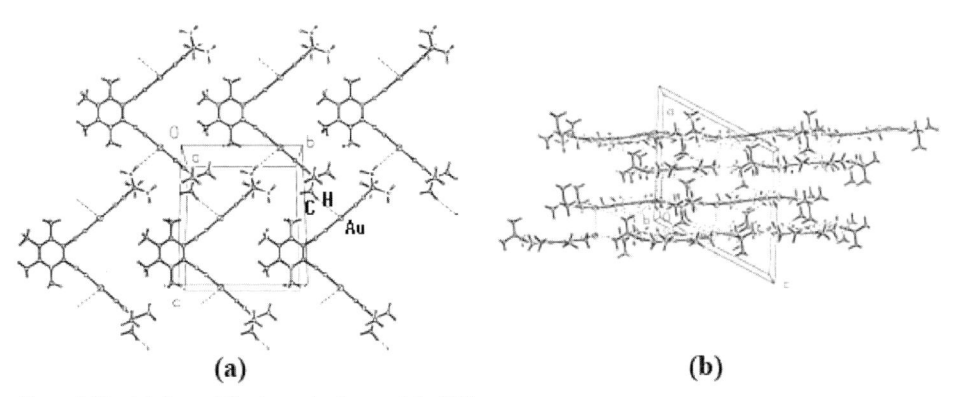

Figure 5.31 (a) One of the layers in the crystal of [{Au(C≡N^{t}Bu)}$_2${μ-1,2-C$_6$Me$_4$(C≡C)$_2$}]. (b) Three-dimensional polymeric structure viewed parallel to the layers built by means of C–H···Au contacts.

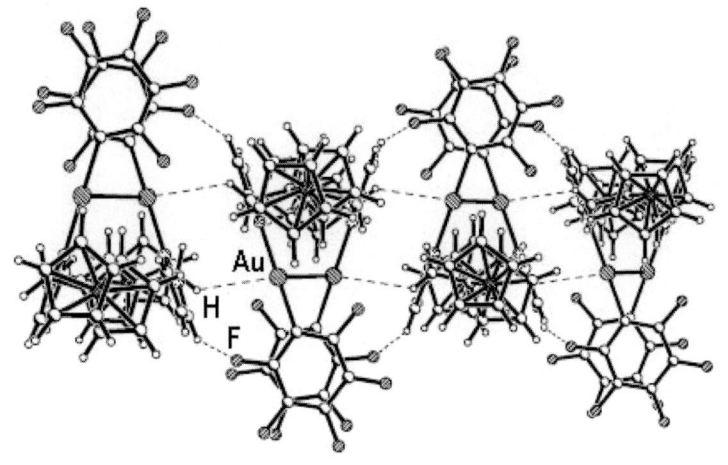

Figure 5.32 Part of the three-dimensional structure of [Au(C$_6$F$_5$)(pzCH$_2$Fc)] showing the C–H···Au hydrogen bonds.

5.2.3
Au···π Interactions

As far as the Group 11 metal···arene π complex is concerned, there are numerous η^2-bonding complexes between Ag$^+$ and aromatic compounds and a few analogous Cu$^+$ derivatives. However, no Au$^+$ η^2-arene complex had been reported until very recently [62] and only a few gold(I) η^6-arene compounds have been described, the first being an *endocyclic* metal-η^6-arene bonding reported in 2003 [63]. In the phosphino derivative [Au(py)L] (py = pyridine) (L = 9-{[N-n-propyl-N-(diphenylphosphino)amino]methyl}anthracene [62], the linear two-coordinated P–Au–N unit folds toward the anthracene plane, resulting in an Au···η^2-arene interaction of 3.233 Å. However, apart from this intramolecular contact, the molecules stack forming a one-dimensional array generated by an intermolecular Au···η^2-arene interaction, as shown in Figure 5.33, with a separation between the gold atom and the centroid of the aromatic carbon–carbon bond of 3.439 Å.

A couple of examples of complexes with intermolecular gold(I)···η^6-arene interactions have been reported recently [64, 65]. They were found when studying the nucleophilic trinuclear gold(I) ring complex [Au$_3$(p-tolN=COEt)$_3$]. Trinuclear Au(I) compounds with aromatic-substituted imidazolate and carbeniate bridging ligands can form sandwich adducts with a variety of electrophiles to produce supramolecular chains. Most of them show 2:1 assemblies, although [Au$_3$(p-tolN=COEt)$_3$] forms sandwich adducts with hexafluorobenzene (C$_6$F$_6$) [64] or octafluoronaphthalene (C$_{10}$F$_8$) [65], which are assembled in a 1:1 ratio forming supramolecular stacks. The stack corresponding to the octafluoronaphthalene derivative is shown in Figure 5.34, displaying a repeated pattern ···(Au$_3$)(μ-C$_{10}$F$_8$)(Au$_3$)(μ-C$_{10}$F$_8$)···. The distance between the centroid of the organic molecule and the centroid of [Au$_3$(p-tolN=COEt)$_3$]

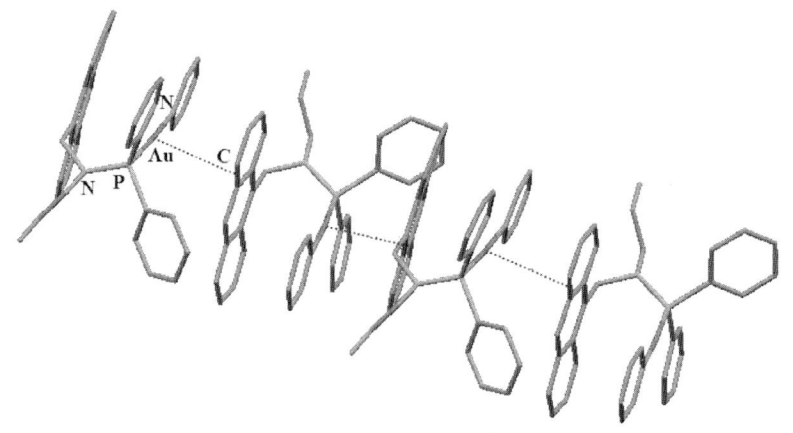

Figure 5.33 Extended chain of [Au(py)L] through Au···η^2-arene interactions.

is 3.565 Å in the hexafluorobenzene derivative and it is slightly shorter (3.509 Å) in the octafluoronaphthalene one.

Finally, a metal···η^6-arene interaction that generates a supramolecular one-dimensional array has been reported in the solid-state structure of the (2,6-diphenylpyridine)gold(III) derivative [Au(CN){2,6-NC$_5$H$_3$(C$_6$H$_4$)$_2$-C,C',N}] [66]. Its crystal structure shows that the planes of the planar molecules are parallel, but slip-stacked via intermolecular Au···η^6-arene contacts in which the distance between the gold atom and the centroid of the arene is 3.428 Å. A representation of this structure is shown in Figure 5.35.

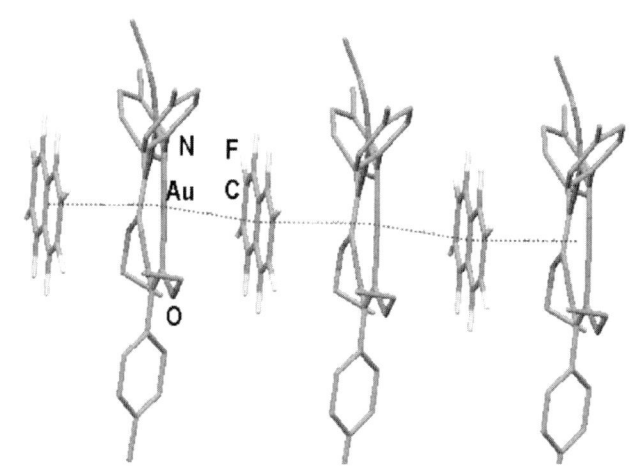

Figure 5.34 Extended chain of [Au$_3$(p-tolN=COEt)$_3$]·C$_{10}$F$_8$ formed through Au···η^6-arene interactions.

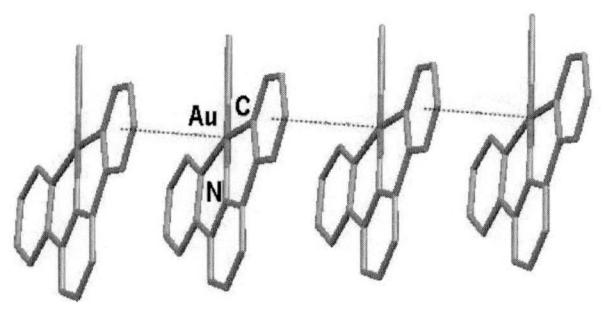

Figure 5.35 Extended chain of [Au(CN){2,6-NC$_5$H$_3$(C$_6$H$_4$)$_2$-C,C', N}] formed through Au···η^6-arene contacts.

5.3
Supramolecular Gold Entities Built with Non-Metal–Non-Metal Secondary Bonds

In this section I will comment on some examples of gold complexes that display supramolecular entities in the solid state, although in these cases the metal atom is not directly involved in the van der Waals interactions. As in the previous section, the electron density donor can be a non-metal different from hydrogen, which can be a halogen or a chalcogen; a hydrogen atom of a covalent NM−H bond; or a cloud of π electron density.

5.3.1
Secondary Bonds: NM···NM Interactions

This type of interatomic contact is sometimes found in halogen or chalcogen gold derivatives, and the vast majority of the examples described in the literature are of the type X···X, that is, they contain the same element in both extremes of the interaction. Of these, the most frequent are I···I or S···S interactions, although there are other non-metal···non-metal contacts, such as F···F, Br···Br, Se···Se or Te···Te.

Section 2.2 presented the two-dimensional network resulting from secondary C−H···Au and C−H···Cl bonds in the (ferrocenylmethyl)phosphine derivative [AuCl(PPh$_2$CH$_2$Fc)] [59] (Figure 5.30). The solid-state structure of its pentafluorophenyl analog [Au(C$_6$F$_5$)(PPh$_2$CH$_2$Fc)] [59] again does not display aurophilic interactions, but in the lattice there are F···F contacts of 2.858 Å, which give rise to the association of the molecules in dimers, one of which is shown in Figure 5.36. The presence of additional C−H···F hydrogen bonds between different pairs affords a supramolecular structure in the form of layers.

Although the most usual intermolecular halogen···halogen contacts observed in gold chemistry appear in adducts with X$_2$, iodine-containing gold complexes can also associate though I···I secondary bonds. For example, in the solid structure of the methanide derivative [Ru(C≡N^{t}Bu)$_4${(PPh$_2$)$_2$C(AuCl)I}]I [67] every three cations interestingly form unusual supramolecular triangular aggregates with central I···I

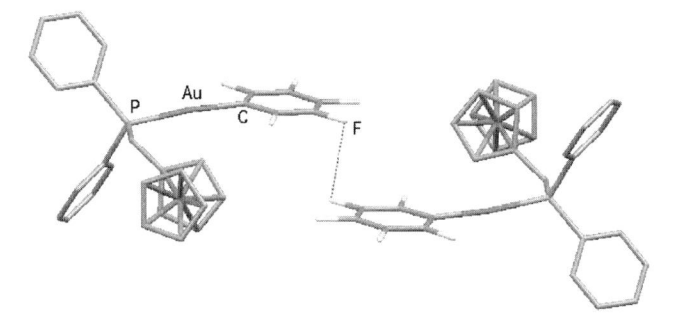

Figure 5.36 Dimers of [Au(C_6F_5)(PPh_2CH_2Fc)] showing the F···F contact.

contacts and peripheral C—H···Cl hydrogen bonds, as shown in Figure 5.37. Intramolecular Au···I interactions of 3.35 Å are also present within each molecule.

In adducts of gold(III) compounds of the type $[AuX_3(PR_3)]\cdot X_2$ (X = halogen, R = alkyl group) the substitution of the halogen or of the alkyl group strongly influences its supramolecular structure in the solid state [68]. Thus, in the PMe_3 adduct $[AuI_3(PMe_3)]_2\cdot I_2$ a pair of complex molecules is connected by two I_2 molecules forming a supermolecule containing a 10-membered ring (Figure 5.38a), which is formed by four I···I contacts of 3.570(1) and 3.360(1) Å. In contrast, when bromide is employed instead of iodine, the components of the adduct also aggregate through halogen···halogen secondary bonds, but the molecules are now linked, via bromine molecules, into chains (Figure 5.38b). The two bromine atoms located *cis* to the phosphine in the complex are mainly engaged in these aggregations (Br···Br

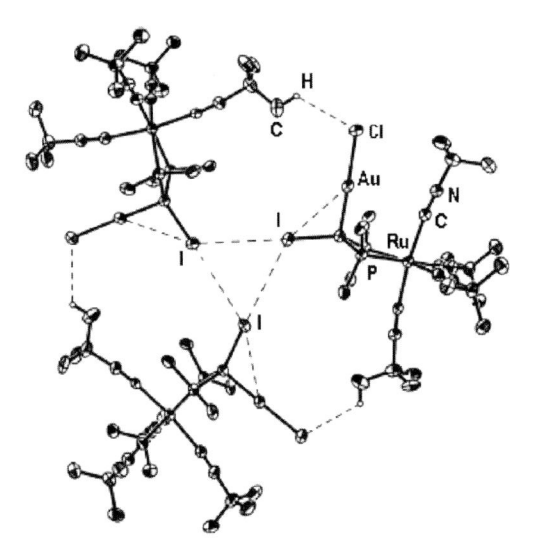

Figure 5.37 Aggregation of three cations of [Ru(C≡N^tBu)_4{(PPh_2)_2C(AuCl)I}]I showing the I···I contacts.

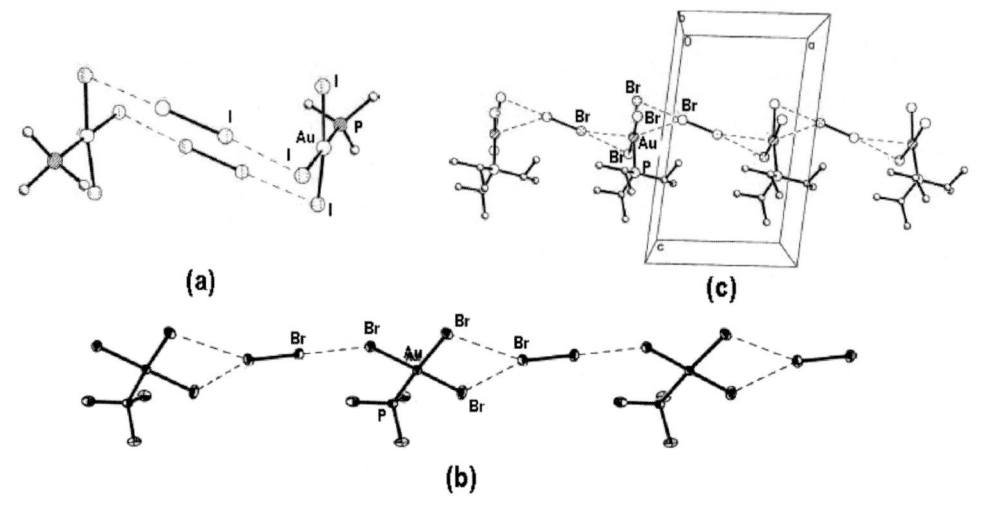

Figure 5.38 Association of the components of the adducts (a) [AuI₃(PMe₃)]₂·I₂, (b) [AuBr₃(PMe₃)]·Br₂ and (c) [AuBr₃(PⁱPr₃)]·Br₂ showing the I···I, Br···Br and Au···Br contacts.

distances of 3.219 and 3.291 Å), but there is also a sub-van der Waals contact to the *trans*-bromine atom. Moreover, when the methyl groups in the phosphine are replaced by isopropyl substituents in the bromine adduct, the coordination compounds are also aggregated into chains via the bromine molecules but the connectivity pattern is significantly different. As shown in Figure 5.38c, the bromine atom *trans* to phosphorus and one of the *cis*-bromine atoms become involved in the Br···Br contacts, and the approach of the Br$_2$ molecules is not strictly to the terminal bromine atoms but about midway between a gold and a bromine atom. Thus, Au···Br (3.534 Å) and Br···Br (3.099 and 3.309 Å) are present in the crystal structure of [AuBr$_3$(PⁱPr$_3$)]·Br$_2$.

Similarly, different supramolecular arrangements have been observed in the crystal structures of gold(I) adducts of the type [AuI(PR$_3$)]·nI$_2$ [22]. The mode of supramolecular aggregation in the isopropyl derivative [AuI(PⁱPr$_3$)]·1.5I$_2$ was commented in Section 2.1, since Au···I, in addition to I···I contacts, appear between the components. In this case, the conjunction of both types of secondary bonds results in a two-dimensional array in which zigzag chains are linked into corrugated sheets via Au···I contacts (Figure 5.9b). In contrast, in the structure of the phenyl derivative [AuI(PPh$_3$)]·I$_2$ the iodine atoms form zigzag chains, as shown in Figure 5.39a, in which the iodide atom attached to the gold center occupies a bridging position between two iodine molecules with almost symmetrical contacts, with I···I distances of 3.4361(5) and 3.5782(5) Å. Lastly, in [AuI{P(o-Tol)$_3$}]$_2$·I$_2$ the components are arranged in groups consisting of one iodine molecule flanked by two complex molecules (with a center of inversion located at the mid-point of the I$_2$ molecule) and joined through long I···I contacts (3.441(1) Å). These trimolecular units are

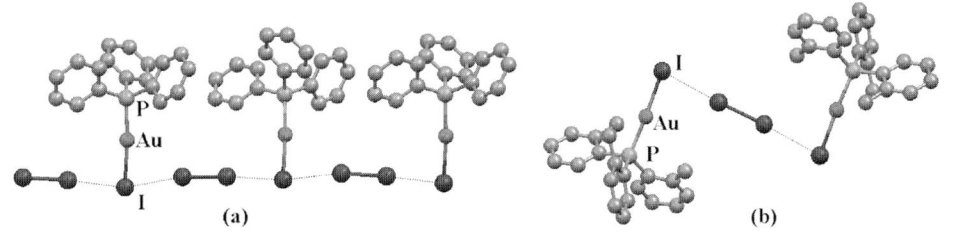

5.3 Supramolecular Gold Entities Built with Non-Metal–Non-Metal Secondary Bonds | 323

Figure 5.39 Extended chains in the adducts (a) [AuI(PPh$_3$)]·I$_2$ or (b) [AuI{P(o-Tol)$_3$}]$_2$·I$_2$ formed via I···I contacts.

further aggregated through even longer I···I contacts (4.189(1) Å) to give zigzag chains along the a axis of the crystal (Figure 5.39b).

The compound of composition Au$_2$I$_4$(C≡N^{t}Bu)$_2$ [69], obtained from [AuI(C≡N^{t}Bu)] and iodine, has been shown to feature ionic components with intercalated iodine molecules: [Au(C≡N^{t}Bu)$_2$][AuI$_2$]·I$_2$, with linear strings of gold atoms formed by aurophilic bonding between alternating cations and anions running parallel to the b-axis. The intercalated iodine molecules reside on centers of inversion and have I···I contacts with a distance of 3.424(1) Å, typical for metal polyiodide compounds, which give rise to a layer polymer (Figure 5.40a). Crystals of the composition AuI$_5$(C≡N^{c}Hex)$_2$ [69], obtained from [AuI(C≡N^{c}Hex)] and excess iodine, have been shown to contain [AuI(C≡N^{c}Hex)$_2$]$^+$ cations and corrugated polyiodide sheets formed by I···I secondary bonds between I$^-$ anions and I$_2$ molecules (Figure 5.40b). In the corrugated polyiodide layers, the iodide anion is surrounded by four iodine molecules with I···I distances of 3.331(1) and 3.334(1) Å.

Apart from these I···I interactions, S···S is one of the intermolecular contacts most frequently observed in gold complexes. Sulfur-rich dithiolate ligands having a planar geometry, such as C$_3$S$_5$$^{2-}$ [4,5-disulfanyl-1,3-dithiole-2-thionate] or C$_8$H$_4$S$_8$$^{2-}$ [2-{(4,5-ethylenedithio)-1,3-dithiole-2-ylidene}-1,3-dithiole-4,5-dithionate], have been

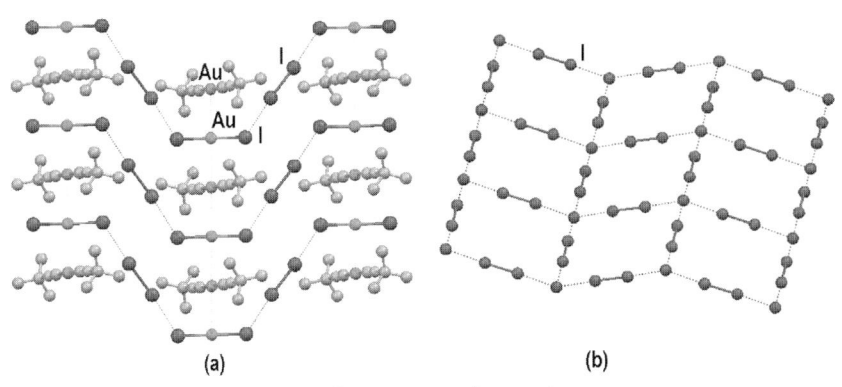

Figure 5.40 Two-dimensional arrays of (a) Au$_2$I$_4$(C≡N^{t}Bu)$_2$ and (b) AuI$_5$(C≡N^{c}Hex)$_2$ showing the Au···I and I···I contacts.

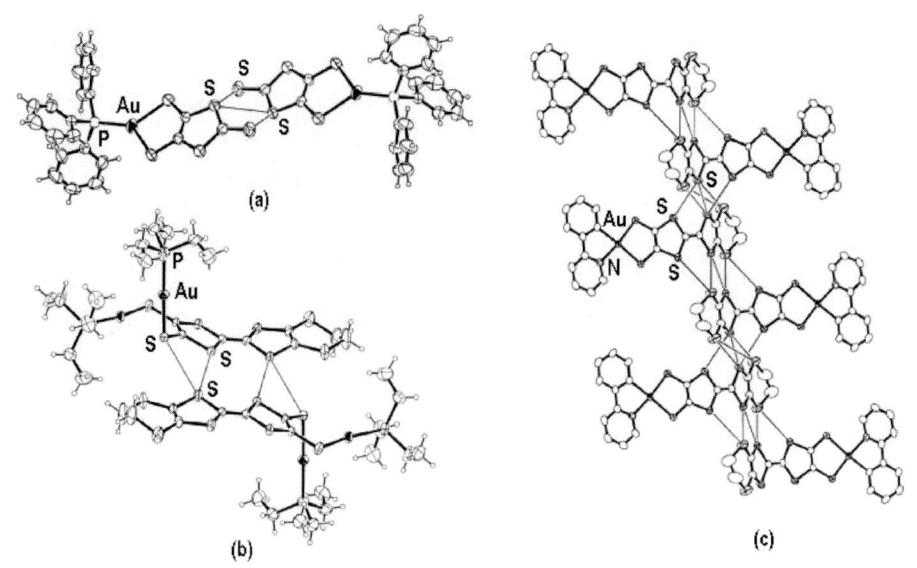

Figure 5.41 Association in dimers in [NMe$_4$][Au(PPh$_3$)(C$_3$S$_5$)] (a) or [{Au(PEt$_3$)}$_2$(C$_8$H$_4$S$_8$)] (b), or in sheets in [Au(ppy)(C$_8$H$_4$S$_8$)]·0.5DMF (c) through S···S interactions.

employed in the synthesis of gold conducting materials which display S···S contacts in the solid state. For example, in [NMe$_4$][Au(PPh$_3$)(C$_3$S$_5$)] [70] two anion moieties form a dimer with S···S contacts of 3.520(2) and 3.698(2) Å (see Figure 5.41a), which contrasts with the isolated anion moiety of [N^{n}Bu$_4$][Au(PPh$_3$)(C$_3$S$_5$)] [71]. Similarly, [{Au(PEt$_3$)}$_2$(C$_8$H$_4$S$_8$)] has some slightly longer S···S contacts of 3.791 and 3.733 Å between the molecules to form a dimeric unit (Figure 5.41b). This complex, however, does not form any further molecular interaction through S···S contacts in the solid state, which was observed for several C$_8$H$_4$S$_8$–metal complexes, such as [Au(ppy)(C$_8$H$_4$S$_8$)]·0.5DMF (ppy$^-$ = C-deprotonated-2-phenylpyridine(1-), DMF = dimethylformamide) [72]. In this last case, there are several S···S contacts between 3.36 and 3.64 Å between the molecules, which form a two-dimensional molecular interaction through these S···S contacts almost parallel to the ab plane. Figure 5.41c shows a projection of the packing diagram of [Au(ppy)(C$_8$H$_4$S$_8$)]·0.5DMF along the bc plane.

Apart from thiolate ligands, S, Se or Te-containing fulvalene derivatives have been employed as electron donors in the synthesis of gold complexes. The combination of tetrathiafulvalene, TTF, with the gold complex bis(pyrazine-2,3-diselenolate)aurate (III), [Au(pds)$_2$]$^-$, affords three different compounds depending on the preparative conditions: (TTF)$_2$[Au(pds)$_2$]$_2$, (TTF)$_3$[Au(pds)$_2$]$_2$ and (TTF)$_3$[Au(pds)$_2$]$_3$ [73]. The crystal structure of the former contains two types of TTF motifs: dimers and isolated units. There are several short distances established between Se atoms of [Au(pds)$_2$]$^-$ anions and S atoms of TTF (S···Se: 3.297(6)–3.840(6) Å) and between S atoms belonging to the two TTF units within a dimer (S···S: 3.429(8) and 3.510(8) Å) forming the network shown in Figure 5.42a. These distances are much smaller than

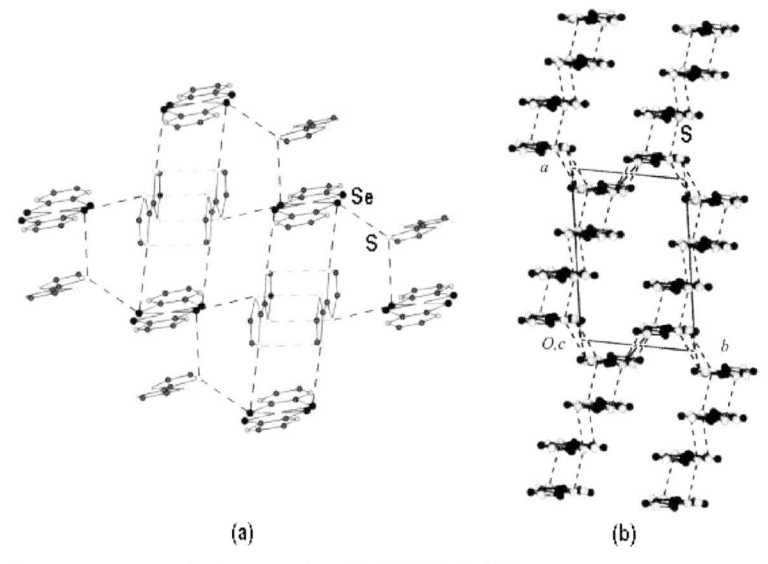

Figure 5.42 Supramolecular networks in (a) (TTF)$_2$[Au(pds)$_2$]$_2$ or (b) (DT-TTF)$_4$[Au(pds)$_2$]$_3$ through S···S and S···Se contacts. Part of the [Au(pds)$_2$]$^-$ units is omitted for clarity.

the sum of the van der Waals radii, thus denoting strong interactions. In the structures of the other two compounds, the TTF units are arranged in trimers, the closest distances between sulfur atoms of neighboring TTF units within a trimer being 3.502(5) and 3.511(5) Å in (TTF)$_3$[Au(pds)$_2$]$_2$ or 3.567(9), 3.532(8), 3.554(8) and 3.631(8) Å in (TTF)$_3$[Au(pds)$_2$]$_3$.

Moreover, the crystal structure of the related charge transfer salt (DT-TTF)$_4$[Au(pds)$_2$]$_3$ (DT-TTF, $\Delta^{2,2'}$-bithieno[3,4-d]-1,3-dithiol) [74] contains a two-dimensional network of DT-TTF tetramers (Figure 5.42b) and orthogonal anions. The DT-TTF units within each tetramer are stacked with a similar slipping along the short axis of these units, with average interplanar distances of 3.9332 and 3.5243 Å, and there are several S···S contacts (3.457(7)–3.682(8) Å) between DT-TTF units within these tetramers which are below the sum of the van der Waals radii. The end units of the donor tetrads are connected to two other tetrads by longer S···S contacts between 3.826(7) and 4.133(8) Å, that is, slightly above the sum of the van der Waals radii, and therefore there is a two-dimensional network of short interdonor S···S contacts. There are also several short contacts between sulfur atoms of the DT-TTF units and gold or selenium atoms of the [Au(pds)$_2$]$^-$ anionic complexes in the range 3.401(6)–3.810(5) or 3.535(6)–3.691(5) Å, respectively.

Electrocrystallization of the donor bis(trimethylenedithio)-bridged TSF-phane (TSFphane = tetraselenafulvalenophane) with the [Au(CN)$_2$]$^-$ as counterion gives a radical cation salt that shows a very high conductivity at room temperature [75]. X-ray crystallographic analysis reveals that its crystal structure contains many intra- and inter-molecular Se···Se contacts. The donor gives rise to a stacking column with

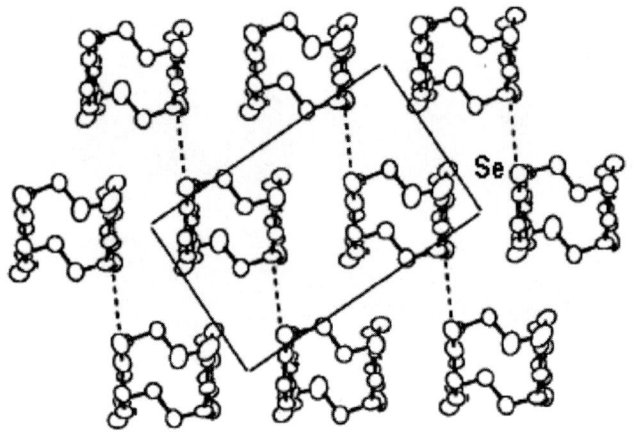

Figure 5.43 Supramolecular structure of (TSF-phane)[Au(CN)$_2$] showing the Se···Se contacts.

other donors, with slight sliding along the long axis direction of the molecule. The shortest intermolecular Se···Se contacts along the columnar direction are 3.84 Å. In addition, there are close side-by-side Se···Se contacts (3.68 Å) between the neighboring donor columns, forming a sheet-like interaction (Figure 5.43). This sheet-like interaction array must contribute to the high conductivity of this salt.

Finally, a supramolecular three-dimensional network formed through Te···Te secondary bonds has been described for the related tellurium compound (TMTTeN)$_2$[Au(CN)$_2$] (TMTTeN = (2,3,6,7-tetramethylnaphtho[1,8-cd:4,5-$c'd'$]bis[1,2]ditellurole), a salt which is a quasi-three-dimensional conductor [76]. The donor molecules are stacked along the c-axis and take the same orientation with a short interplanar distance of about 3.4 Å and a large slip distance of approximately 4 Å. There are several intermolecular Te···Te contacts, and a three-dimensional network through tellurium atoms is developed between the intracolumns and intercolumns.

5.3.2
Hydrogen Bonds: NM···H−NM Interactions

As commented in the introduction, hydrogen bonds are special secondary interactions that have traditionally referred to strong interactions of the type D−H···A, where D and A are both electronegative atoms, such as halogens, oxygen, sulfur or nitrogen. Non-classical hydrogen bonds, that is, hydrogen bonds with C−H as donor group, were once the subject of controversy but nowadays they are accepted as valid and the subject of great interest. In Section 2.2 we looked at a series of examples of supramolecular entities built by means of non-classical hydrogen bonds containing gold as acceptor, mainly of the type C−H···Au. In this section, supramolecular arrays built with hydrogen bonds without the participation of gold will be considered. Classical hydrogen bonds with O−H or N−H groups as hydrogen bond donors will be treated first, and hydrogen bonds containing C−H as donor will be discussed later.

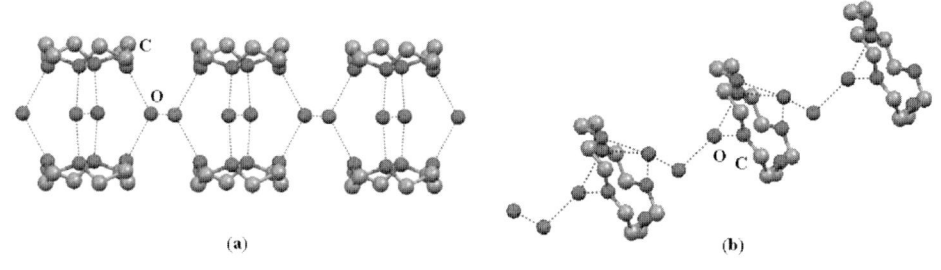

Figure 5.44 One-dimensional polymer in (a) $[(H_5O_2)_2(12\text{-crown-}4)_2][AuCl_4]_2$ or (b) $[(H_7O_3)(15\text{-crown-}5)][AuCl_4]$ formed through O−H···O bonds.

One of the best-known hydrogen bonds is the O−H···O bond found in a variety of oxonium ions. Apart from existing as isolated supermolecules, some of these oxonium ions can act as hydrogen bond donors and thus interact with, for example, crown ethers, to form bigger supermolecules or even polymeric networks. In the X-ray crystal structure of $[(H_5O_2)_2(12\text{-crown-}4)_2][AuCl_4]_2$ [27] (shown in Figure 5.44a), a 12-crown-4 acceptor is too small to accommodate even H_3O^+ and, as a consequence, a sandwich structure is adopted in which an $H_5O_2^+$ cation is situated between facing pairs of 12-crown-4 molecules, hydrogen bonding to two ether oxygen atoms on each crown. The remaining ether oxygen atoms interact with a second $H_5O_2^+$ cation, which bridges between two "$H_5O_2^+ \cdot (12\text{-crown-}4)_2$" pairs, thus resulting in a hydrogen-bonded chain polymer. Although hydrogen atoms could not be located, the O···O distances within the $H_5O_2^+$ cations of 2.419 and 2.452 Å are typical of the strong hydrogen bonding in $H_5O_2^+$, and hydrogen bonds to ether oxygen atoms are longer, although still short at 2.576 and 2.674 Å.

Similarly, complex $[(H_7O_3)(15\text{-crown-}5)][AuCl_4]$ [27] also displays a polymeric one-dimensional structure built via O−H···O hydrogen bonds of 2.432(7) and 2.563(7) Å within $H_7O_3^+$ ions and longer O−H···O hydrogen bonds (O···O 2.649(7)–2.775(8) Å) to the crown ether oxygen atoms (see Figure 5.44b). All of these are steric requirements. Interestingly, the Au(III) ions also exhibit very long axial interactions with crown oxygen atoms, resulting in a cross-linked polymeric structure (not shown in the figure).

The importance of hydrogen bonding interactions in crystal engineering is undisputed for both organic systems and coordination chemistry. Carboxylic acid groups are one of the most frequently employed in the design of crystal structures, and thus, the familiar carboxylic dimer structural motif is also present in the solid-state structure of certain gold derivatives. Dimers are formed from hydrogen bonding between pairs of carboxylate residues in $[Au(PPh_3)(SCMe_2CO_2H)]$ [32] (Figure 5.45a), with an O···O separation of 2.605(6) Å. In this case, the bulkier thiolate ligand prevents further association, as observed in the related compound $[Au(PPh_3)(SCH_2CO_2H)]$ [32], where alternating intermolecular Au···S and O−H···O contacts (each type of interaction forming dimers) produce a one-dimensional polymeric backbone which dominates the supramolecular array. As

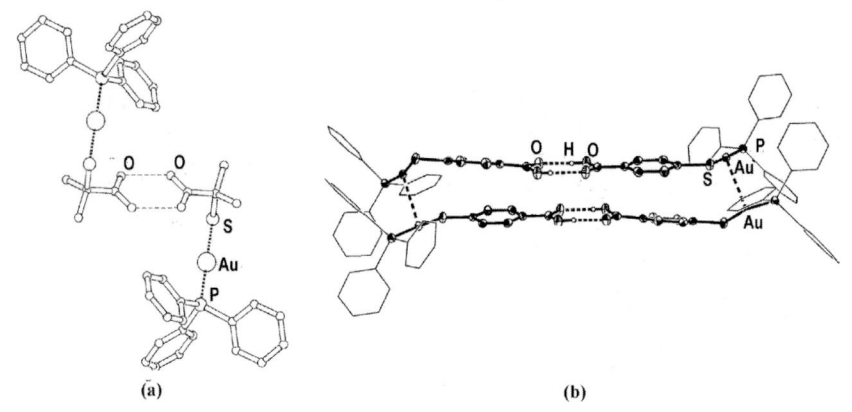

Figure 5.45 (a) Dimer and (b) tetramer formed via O—H···O bonds between carboxylate residues aided by Au···Au contacts in the latter.

commented in Section 5.2.1, in [Au(4-SC$_6$H$_4$CO$_2$H)L] (L=PEt$_3$, PPh$_2$py) [33] and [Au(2-Hmba){P(o-Tol)$_3$}]·NCMe (Hmba = 2-mercaptobenzoate) [34] the same arrangement in chains via O—H···O hydrogen bonds aided by Au···S interactions is also observed (see Figure 5.16). A higher dimensionality is found in the solid-state structure of [{Au(2-Hmba)}$_2$(μ-dppe) [36], where O—H···O hydrogen bonds between carboxylic acids coupled with Au···S interactions result in a two-dimensional network. Other secondary bonds, such as the rather frequent intermolecular aurophilic interactions, can be present in other cases, such as in the triphenylphosphine derivative [Au(4-SC$_6$H$_4$CO$_2$H)(PPh$_3$)] [33]. In this complex, two molecules form dimers via a short intermolecular aurophilic contact (Au···Au 3.0756(2) Å) between crossed P—Au—S moieties, and the carboxylic acid end groups are able to engage in parallel paired hydrogen bonding with another dinuclear unit. In total, the compound forms *tetramers* through a large elongated macrocycle comprising two pairs of gold atoms and two pairs of thiolato benzoic acid groups (Figure 5.45b).

Less common is the form of association found in [{Au$_2$(μ-SPh)(PPh$_2$O)(PPh$_2$OH)}$_2$] [16], which can be considered to comprise two Au$_2$(μ-SPh)(PPh$_2$O)(PPh$_2$OH) units linked by both Au···Au and O—H···O secondary bonds. The four gold atoms are arranged in a distorted square, and the Au$_4$S$_2$ core gives a six-membered ring in a chair conformation. The tetranuclear entity bears two neutral diphenylphosphinous acid groups, Ph$_2$POH, and two anionic diphenylphosphinite ligands, Ph$_2$PO$^-$, and these groups combine intermolecularly by P—O···H—O—P bonding, thus forming two peripheral seven-membered rings (Figure 5.46).

Apart from OH groups, NH or NH$_2$ functions are well-established hydrogen bond donors. In the particular case of gold(I) complexes, where aurophilic interactions are attractive forces of major importance, N—H···X (X=halogen, O, S, N) hydrogen bonds have been seen to frequently compete against Au···Au contacts, both being of similar strength. [Au(C$_6$F$_5$){N(H)=CPh$_2$}] [77] displays discrete dimers in an antiparallel conformation in which both Au···Au interactions and N—H···F$_{ortho}$

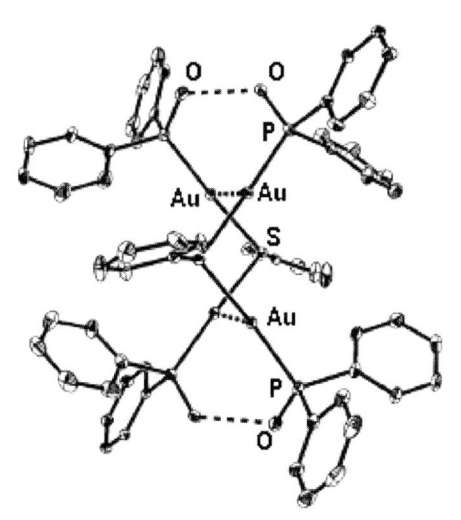

Figure 5.46 Dimer of [{Au$_2$(μ-SPh)(PPh$_2$O)(PPh$_2$OH)}$_2$] via Au···Au and O−H···O secondary bonds.

hydrogen bonds appear within the dimeric units (Figure 5.47). In contrast, the silver derivative shows a ladder-type structure in which two [Ag(C$_6$F$_5$){N(H)=CPh$_2$}] units are linked by an Ag···Ag interaction and the dimeric units are further associated through hydrogen bonds of the type N−H···F$_{ortho}$. These coexisting interactions have been studied theoretically studied separately by *ab initio* calculations, concluding that the aurophilic interaction is stronger than the argentophilic one, and that N−H···F hydrogen bonding and Au···Au contacts have a similar strength in the same molecule, which enables competition between these two structural motifs, giving rise to different structural arrangements.

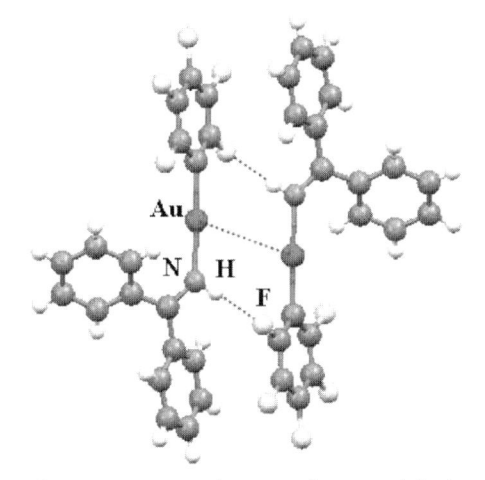

Figure 5.47 Dimers of [Au(C$_6$F$_5$){N(H)=CPh$_2$}] through N−H···F and Au···Au interactions.

Table 5.3 Some gold(I) complexes containing N−H···X hydrogen bonds.

Complex	X	N···X (Å)	Au···Au (Å)	Ref.
[Au(C$_6$F$_5$){N(H)=CPPh$_2$}]	F	3.225(10)	3.5884(7)	[77]
[Au$_3$Cl$_3$(pyrr)$_4$]	Cl	3.179(6)	3.2041(7)	[78]
		3.284(6)	3.5834(4)	
[AuCl(pip)]	Cl	3.346, 3.358	3.301(5)	[79]
[AuCl(Me$_4$pip)]	Cl	3.417(3)	3.2792(4)	[80]
[AuCl(NHCy$_2$)]	Cl	3.391(8)	3.2676(14)	[81]
[Au(pip)$_2$]Cl	Cl	3.108(6),3.122(7)	4.085(2)	[81]
[Au(pyrr)$_2$]Cl	Cl	3.149(7)–3.311(5)	3.2790(7)	[80]
[Au(NH$_2$Cy)$_2$]Cl	Cl	3.230(9)–3.343(9)	[4.740(1)]	[80]
[Au(NH$_3$)$_2$]Br	Br	3.38–3.58	3.414(1)	[82]
[Au(SSNH$_2$)(PMe$_3$)]	N	2.950,3.161	3.0581(6)	[83]
	S	3.505		
[{Au(SSNH$_2$)}$_2$(μ-dppe)]·0.5 MeOH	N	2.957–3.136	3.0676(1)	[83]
[Au{SC(=NCN)N(H)Me}(PPh$_3$)]	N	2.995(8)		[30]
[Au(SSNH$_2$)(PPh$_3$)]·0.5Et$_2$O	N	2.943		[83]
[Au(SSNH$_2$)(PPh$_3$)]·MeOH	N	3.169		[83]
	O	2.816,2.855		
[{Au(SSNH$_2$)}$_2$(μ-dppm)]·2DMF	N	2.921		[83]
	O	2.878,2.958		
[Au(SSNH$_2$){P(CH$_2$CH$_2$CN)$_3$}]	N	3.019		[47]
[Au(SSNH$_2$){P(CH$_2$CH$_2$CN)$_3$}]·DMF	N	2.891		[47]
	O	3.021		
[{Au(SSNH$_2$)}$_2$(μ-dpph)]	N	2.972–3.024		[47]
[Au(SSCNH$_2$){P(CH$_2$CH$_2$CN)$_3$}]	N	3.068–3.130		[47]
[{Au(SSCNH$_2$)}$_2$(μ-dpph)]	S	3.530		[47]
[Au(SC$_6$H$_3$N$_2$O$_2$)(PPh$_3$)]	O	2.71,2.75		[84]
[{Au(SC$_6$H$_3$N$_2$O$_2$)}$_2${μ-PPh$_2$(CH$_2$)$_3$PPh$_2$}]	O	2.76,2.66		[84]
[{Au(SC$_6$H$_3$N$_2$O$_2$)}$_2$(μ-PPh$_2$CH=CHPPh$_2$)]	O	2.73,2.70		[84]
[Au$_2$(μ-SC$_4$H$_2$N$_2$O$_2$)(μ-dppe)]	O	2.737		[84]
{[Au$_2${1,2-C$_6$H$_4$(NCO-4-C$_5$H$_4$N)$_2$}{μ-PPh$_2$(CH$_2$)$_4$PPh$_2$}]$^{2+}$}$_n$	O	2.73(1)		[85]

On other occasions, both types of secondary interactions are present in the same compound and can afford supramolecular entities via cooperative interactions. Table 5.3 shows examples of gold complexes containing N−H···X bonds in addition to or instead of Au···Au contacts.

Amine complexes of gold(I) in the absence of stabilizing ligands such as phosphines are generally regarded as relatively unstable. This can be rationalized in terms of incompatibility of the soft metal center with the hard nitrogen donor. Additional stabilisation may, however, be provided by secondary interactions, such as aurophilic interactions or N−H···X hydrogen bonds. A series of chlorogold(I) aliphatic amines of the types [AuClL] (L = piperidine [79], 2,2,6,6,-tetramethylpiperidine [80], dicyclohexylamine [81]) or [AuL$_2$]Cl (L = piperidine [81], pyrrolidine [80], cyclohexylamine [80]) have been structurally characterized, most of them crystallizing as

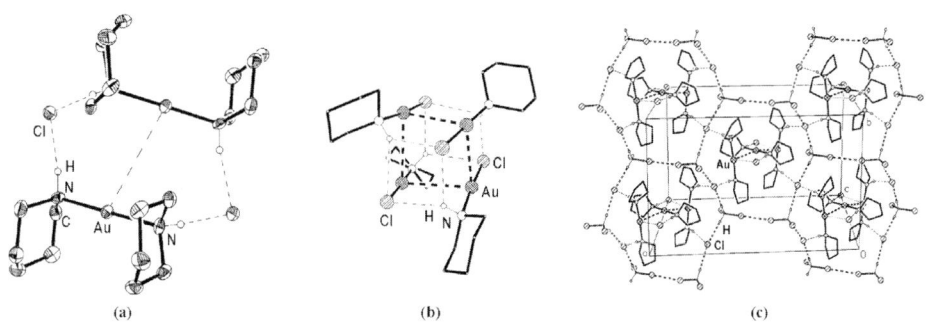

5.3 Supramolecular Gold Entities Built with Non-Metal–Non-Metal Secondary Bonds | 331

Figure 5.48 (a) Dimer, (b) tetramer and (c) layer structures of some [AuClL] or [AuL$_2$]Cl via N—H$\cdots$Cl and Au$\cdots$Au interactions.

macromolecules formed by the aggregation of two, three or four molecules or cations and anions through both N—H$\cdots$Cl and Au$\cdots$Au interactions. Analysis of the secondary interactions in these compounds suggests an inverse correlation between the lengths of aurophilic interactions and hydrogen bonds. For example, [Au(pip)$_2$]Cl [81] displays antiparallel dimers (Figure 5.48a) with short N—H$\cdots$Cl hydrogen bonds (N$\cdots$Cl 3.108(6) and 3.122(7) Å) and a long Au$\cdots$Au distance of 4.085(2) Å, while in the neutral dimer of [AuCl(Me$_4$pip)] [80] weaker N—H$\cdots$Cl bonds (N$\cdots$Cl 3.417(3) Å) and stronger aurophilic contacts (Au$\cdots$Au 3.2792(4) Å) are observed. In [AuCl(pip)] [79] the molecules are associated into tetramers (Figure 5.48b) through Au$\cdots$Au contacts (3.301(5) Å) and bifurcated hydrogen bonds (N$\cdots$Cl 3.346, 3.358 Å). The formation of trimers has been observed in the ionic [Au(pyrr)$_2$]Cl and [Au(NH$_2$Cy)$_2$]Cl [80], the latter not displaying any aurophilic contact, with additional N—H$\cdots$Cl interactions which result in layer structures, as shown in Figure 5.48c.

The thiolate complexes [Au(SSNH$_2$)(PMe$_3$)] and [{Au(SSNH$_2$)}$_2$(μ-dppe)] (SSNH$_2$ = 2-amino-5-mercapto-1,3,4-thiadiazolate) [83] are rare examples of two-dimensional frameworks built on cooperative intermolecular aurophilic and hydrogen-bonding interactions in the solid state. The resulting hydrogen-bonding pattern is similar in both complexes, and can be illustrated as dinuclear units based on aurophilic interactions and further associated into a two-dimensional hydrogen-bonded framework through multiple hydrogen-bonding. The extended two-dimensional sheet structure of the former is shown in Figure 5.49.

In contrast, in other related thiolate derivatives no intermolecular aurophilic contacts are present, and supramolecular entities are built only by means of hydrogen bonds. The presence or absence, as well as the nature, of solvent in the lattice has a marked influence on the solid-state structure. For example, the triphenylphosphine complex [Au(SSNH$_2$)(PPh$_3$)]·0.5Et$_2$O [83] forms a dimer solely via bifurcated hydrogen bonds (Figure 5.50a), while in the presence of methanol it gives rise to a one-dimensional ladder structure via intermolecular hydrogen-bonding interactions with molecules of methanol as bridges (Figure 5.50b).

On other occasions, the change in one of the ligands bonded to gold leads to different supramolecular arrays, such as in [{Au(SSNH$_2$)}$_2$(μ-dpph)] and [{Au(SSCNH$_2$)}$_2$(μ-dpph)] (SSCNH$_2$ = 2-amino-5-mercapto-1,3,4-thiadiazolate; SSNH$_2$ =

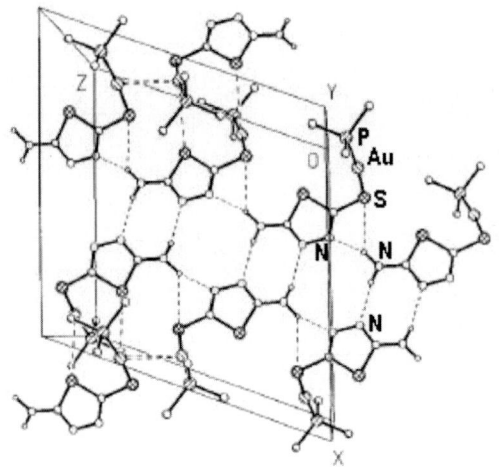

Figure 5.49 Two-dimensional structure of [Au(SSNH$_2$)(PMe$_3$)] built via N—H···N and N—H···S interactions.

6-amino-2-mercaptobenzothiazolate; dpph = 1,6-bis(diphenylphosphino)hexane) [47], where dramatic structural differences between the complexes seem only to be due to the variation in structural moieties of SSNH$_2$ and SSCNH$_2$. Thus, while the former displays a pseudo-double-helical structure, which is further assembled into a two-dimensional sheet via N—H···N hydrogen bonds, the latter features a dinuclear S-shaped structure, which is further associated into a one-dimensional chain via a weak N—H···S interaction.

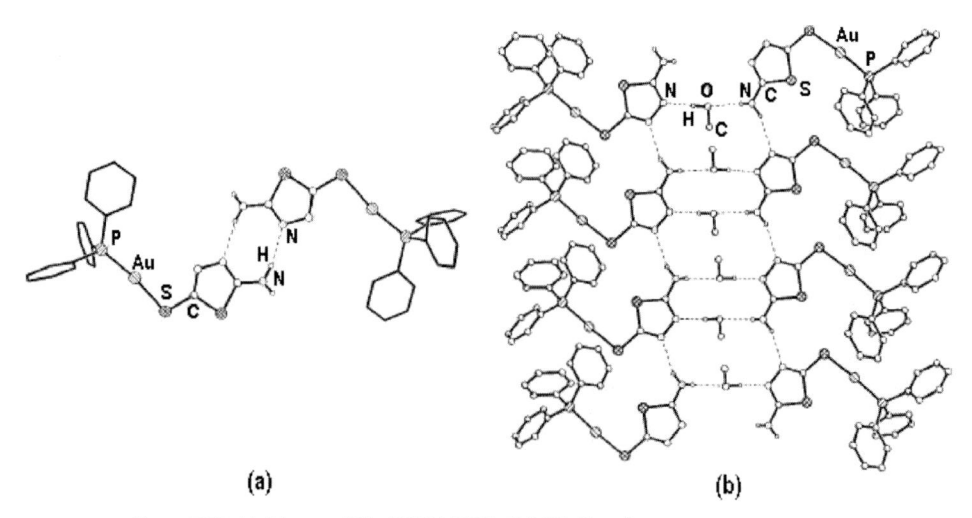

Figure 5.50 (a) Dimers of [Au(SSNH$_2$)(PPh$_3$)]·0.5Et$_2$O and (b) ladder structure of [Au(SSNH$_2$)(PPh$_3$)]·MeOH formed through hydrogen bonds.

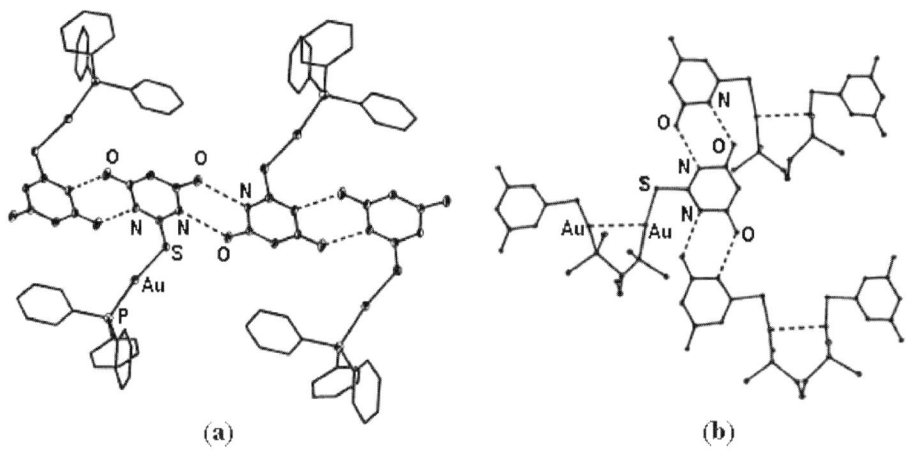

Figure 5.51 (a) Polymeric tape structure of [Au(SC$_6$H$_3$N$_2$O$_2$)(PPh$_3$)] and (b) sheet structure of [{Au(SC$_6$H$_3$N$_2$O$_2$)}$_2${μ-PPh$_2$(CH$_2$)$_3$PPh$_2$}] formed via N—H···O hydrogen bonds.

Classical N—H···O hydrogen bonds have been frequently identified in complexes containing amide functions, and, resembling the carboxylic acid dimer, they usually associate in head-to-tail pairs. This type of dimer is present in the crystal structures of some gold(I) derivatives of the ligands thiobarbiturate (SC$_6$H$_3$N$_2$O$_2$) [84] or 1,2-bis(amidopyridine)benzene (1,2-C$_6$H$_4$(NCO-4-C$_5$H$_4$N)$_2$) [85], which contain the functional group -N(H)C(=O)-. Depending on the nature of the ancillary ligands at gold, and on the presence or absence of additional secondary bonds, chains, sheets or three-dimensional polymers can be observed in the solid state. For example, the simplest one is found in the triphenylphosphine derivative [Au(SC$_6$H$_3$N$_2$O$_2$)(PPh$_3$)] [84], in which the association of pyrimidine groups in a pair-wise fashion (each pyrimidine unit is involved in four strong N—H···O contacts) results in a polymeric tape structure, as shown in Figure 5.51a. In contrast, in the bis(diphenylphosphino)propane derivative [{Au(SC$_6$H$_3$N$_2$O$_2$)}$_2${μ-PPh$_2$(CH$_2$)$_3$PPh$_2$}] [84], the presence of the bidentate phosphine leads to an overall sheet structure (Figure 5.51b), in which each digold(I) unit is involved in eight strong intermolecular hydrogen bonds, so the network is strongly bonded and the compound is insoluble in most organic solvents. Finally, in [Au{1,2-C$_6$H$_4$(NCO-4-C$_5$H$_4$N)$_2$}{μ-PPh$_2$(CH$_2$)$_4$PPh$_2$}](CF$_3$CO$_2$)$_2$ [85] both ligands at gold act as bridges between different metal atoms, which is why the complex exists as polymeric cations with repeated units [-NN-Au-PP-Au-]$^{2+}$, that are further associated through interchain amide N—H···O hydrogen bonds to give a three-dimensional network.

As commented above, gold(I) complexes with aliphatic amines are relatively unstable, but the use of disulfonylamides as counterions allows the synthesis of stable bis(amine)gold(I) cations [86]. Specifically, compounds with cyclohexylamine, benzylamine, 3-iodobenzylamine, morpholine, pyrrolidine, and piperidine as the amine ligands and bis(methanesulfonyl)amide, bis(p-chlorobenzenesulfonyl)amide, bis(p-iodobenzenesulfonyl)amide and o-benzenedisulfonylamide as the

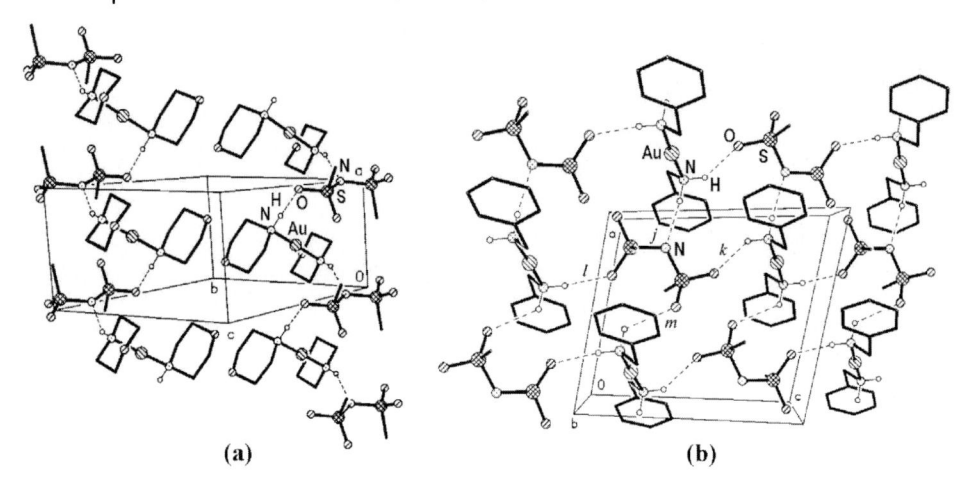

Figure 5.52 (a) Chain structure of [Au{NH(CH$_2$CH$_2$)$_2$O}$_2$](NMs$_2$) and (b) layer structure of [Au(NH$_2$Bz)$_2$](NMs$_2$) formed via N−H···O and N−H···N hydrogen bonds.

anions have been reported. Extensive systems of N−H···O hydrogen bonds are present in their crystal structures and hydrogen bonds seem to be particularly important in stabilizing such complexes. Amines with NH functions (morpholine and pyrrolidine) tend to form hydrogen-bonded chain motifs (Figure 5.52a), while those with NH$_2$ as donor group (with four amine hydrogen atoms per formula unit) form layers of linked ring patterns (Figure 5.52b). Thus, it seems that the packing can more easily be classified in terms of the influence of the cation than that of the anion. In some of the complexes, the oxygen atoms not involved in classical hydrogen bonding showed hydrogen bonds of the type C−H···O, but these played a less important role. These hydrogen bonds can link neighboring chains, such as in [Au{NH(CH$_2$CH$_2$)$_2$O}$_2$](NMs$_2$) or layers, such as in [Au(NH$_2$Cy)$_2$](NMs$_2$), [Au(NH$_2$Bz)$_2$](NMs$_2$) or [Au{NH$_2$(3-CH$_2$C$_6$H$_4$I)}$_2$](NMs$_2$) (NMs$_2$ = bis(methanesulfonyl)amide); others connect molecules within the layer or chain.

In other gold complexes with the same type of anion, such as the bis(thione)gold(I) bis(4-halobenzenesulfonyl)amide derivatives [Au(tzt)$_2$][N(SO$_2$-4-C$_6$H$_4$Cl)$_2$] (tzt = bis(1,3-thiazolidine-2-thione-κS^2), C$_3$H$_5$NS$_2$) or [Au(Me-etu)$_2$][N(SO$_2$-4-C$_6$H$_4$I)$_2$] (Me-etu = bis(1-methylimidazolidine-2-thione-κS^2), C$_4$H$_8$N$_2$S) [87] (see Figure 5.53a),

(a) Z = S, X = Cl
Z = NMe, X = I

Figure 5.53 Schematic drawing of (a) [AuL$_2$][N(SO$_2$-4-C$_6$H$_4$X)$_2$], (b) bis(tetrahydrothiophene)gold(I) benzene-1,2-disulfonimidate.

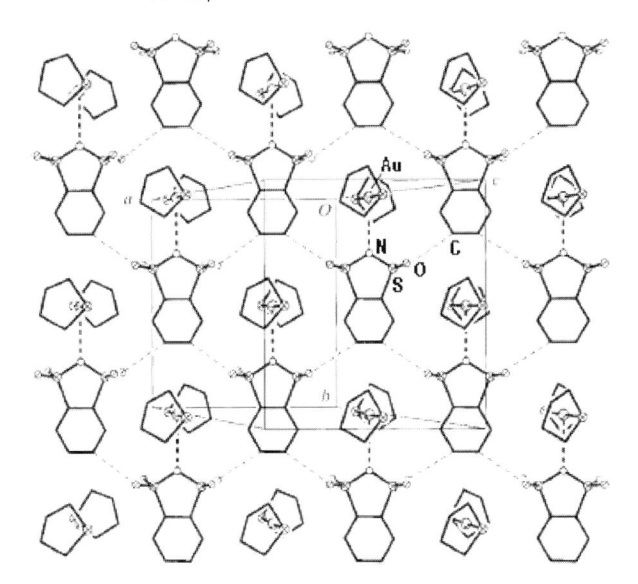

Figure 5.54 Layer structure of [Au(SC$_4$H$_8$)$_2$][C$_6$H$_4$(SO$_2$)$_2$N] showing the C—H···O and Au···N contacts.

classical N—H···N and/or N—H···O hydrogen bonds are also present between anions and cations. These two compounds display a lot of additional secondary bonds, such as C—H···X (X=O, Cl, S) non-classical hydrogen bonds, X···X (X=Cl, I) or Au···I contacts, which afford extended systems in the solid state.

In contrast, the crystal packing of the related complex bis(tetrahydrothiophene)gold(I) benzene-1,2-disulfonimidate [88] (Figure 5.53b), where the only possible hydrogen bond donors are the C—H bonds, consists of layers of anions connected by short C—H···O hydrogen bonds (C···O 3.165(7) Å, C—H···O 138°). The cations occupy cavities in these layers and are linked to the anions by Au···N contacts of 3.009(7) Å (see Figure 5.54). Further C—H···O interactions perpendicular to these layers (C···O 3.416(8) Å, C—H···O 156°) connect the layers linking alternate anions and cations to form chains, and therefore neighboring layers are staggered.

The same kind of secondary bonds, C—H···O interactions, now involving nitro-oxygen atoms, have been found to quite clearly dominate the crystal packing of the phosphinegold(I) thiolates [Au$_2${SC(OR) = NC$_6$H$_4$Y-4}$_2${μ-Ph$_2$P(CH$_2$)$_4$PPh$_2$}] (R=Me, Et, ^{i}Pr; Y=NO$_2$) [37]. While in the compounds with Y=H or Me, synthons of the type [Au···S]$_2$ (R=Me; Y=H) or C—H···S non-conventional hydrogen bonds (usually leading to chains) are the dominant interactions, in the nitro-derivative chains C—H···O interactions are found in the crystal structures of the methyl and isopropyl complexes and, in the case of R=Et, a layer motif mediated by C—H···O interactions is observed. Thus, the influence of Y upon crystal packing is quite plainly apparent in the case of Y=NO$_2$, where the nitro-oxygen atoms directly participate in C—H···O interactions. This indicates that C—H···O interactions are more important than C—H···S interactions, which in turn are more important than

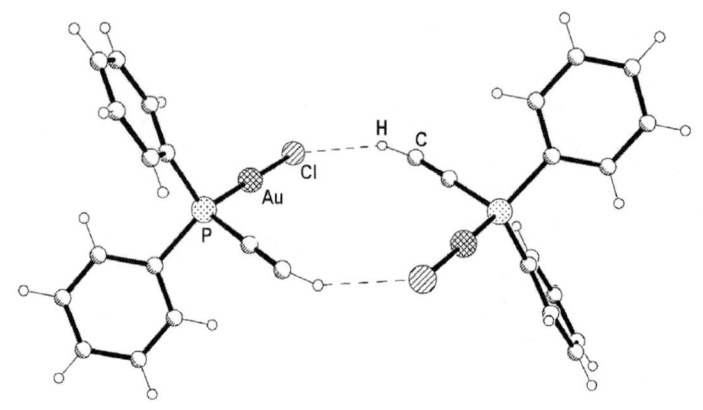

Figure 5.55 Dimer of [AuCl(PPh$_2$C≡CH)] formed by C−H···Cl hydrogen bonds.

the other types of intermolecular interactions, such as C−H···π, π···π or C−H···N interactions, also observed in these structures.

Apart from oxygen, sulfur or nitrogen, halogen atoms can also act as acceptors of non-classical hydrogen bonds. For example, in the solid-state structure of [AuCl(PPh$_2$C≡CH)] [89] the molecules form dimers through two symmetry-equivalent hydrogen bonds (Figure 5.55) and no short Au···Au contacts are found. The acidic alkynyl proton acts as donor to the chloro ligand of a second molecule related by inversion, with a strikingly short C−H···Cl distance of 2.45 Å and a corresponding wide angle of 156°. Nevertheless, the main feature in the structure of [AuI(PPh$_2$C≡CH)] is the presence of aurophilic interactions, and, although there are several C−H···I contacts of 3.2–3.3 Å, most of them are very far from linear, with the exception of one of them with an H···I distance of 3.23 Å and angle of 144°. Curiously, the molecular packing of the pentafluorophenyl derivative [Au(C$_6$F$_5$)(PPh$_2$C≡CH)] differs from the two previous examples and involves some C−H···Au contacts that link the molecules by translation parallel to the *a* axis. There are also several borderline C−H···F interactions, but no aurophilic contacts are detected.

It is reasonable to think that the formation of hydrogen bonds will be more favored in ionic compounds. In accordance with this, the extended structure of [Au(3-NC$_5$H$_4$Br)$_2$][AuCl$_2$] [90] contains tetranuclear dimers formed via aurophilic interactions and essentially linear C−H···Cl contacts (see Figure 5.56a), which are charge-assisted by the opposite charges of the ions. An analysis of the packing of the dimeric units reveals further secondary interactions that afford a layer structure by additional, slightly longer and less linear C−H···Cl contacts, leading to five-membered rings C−H···Cl···H−C.

Finally, in the crystal structures of the chloroform and ethanol solvates of [(dppmSe)$_2$Au]Cl [91] the ions are connected into chains by C−H···Cl hydrogen bonds, whereby the chloride accepts two hydrogen bonds. Additionally, there is a C−H···Cl contact from a solvent molecule in the chloroform solvate (Figure 5.56b) and an O···Cl contact involving the solvent in the ethanol solvate.

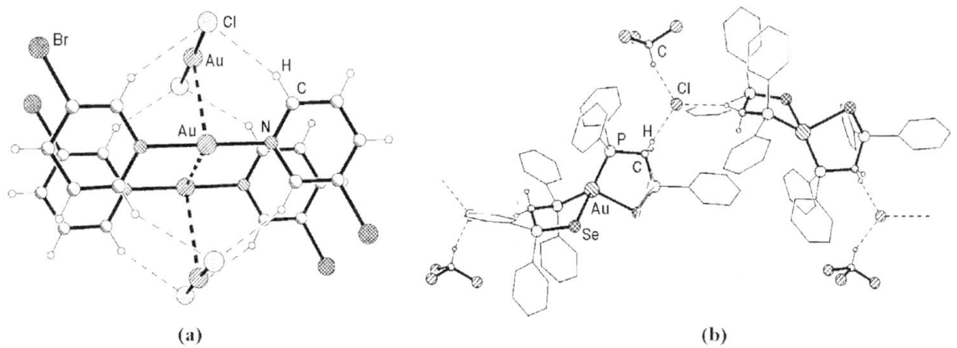

Figure 5.56 (a) Dimeric unit of [Au(3-NC$_5$H$_4$Br)$_2$][AuCl$_2$] and (b) chain structure of [(dppmSe)$_2$Au]Cl·2CHCl$_3$ formed via C−H···Cl contacts.

Therefore, it may be concluded that non-conventional hydrogen bonds can play an important role in determining the supramolecular structure of gold complexes.

5.3.3
C−H···π Interactions and π···π Stacking

In this last section of the chapter other van der Waals interactions in which C−H bonds and/or carbon atoms or, more correctly, π electron density, are involved will be discussed. C−H···π interactions are considered to be present when the hydrogen atom of a C−H bond interacts with an aromatic ring in a nearly normal fashion and with H···centroid distances shorter than about 3 Å. Interactions between clouds of π electron density of neighboring nearly parallel aromatic rings that lead to intercentroid distances below 3.5 Å are considered as effective π···π interactions. Although these contacts are not reported very often, they may play an important role in the packing forces of metal complexes.

Self-assembly through C−H···π interactions does occur in the solid-state structure of a number of gold compounds. For example, examination of the crystal structures of the metallamacrocycles [Au$_3$(CNC)(μ-dppm)$_2$]Cl·8.5CHCl$_3$ and [Au$_3$(CNC)(μ-dppm)$_2$]PF$_6$·CH$_2$Cl$_2$·5CHCl$_3$ (CNC = pyridyl-2,6-diphenyl^{2-}) [92] reveals the existence of dimers, each of which consists of a pair of enantiomers linked by two intermolecular C−H···π interactions between a C−H bond of a dppm phenyl group and the nearest phenyl ring of the CNC ligand. The closest H···C distances are 2.826 Å in the former and 2.848 Å in the latter. For each of the dimers in the chloride compound there are also four intermolecular C−H···π interactions that arise from the C−H bonds of four molecules and the nearest CNC phenyl groups.

On most occasions C−H···π interactions are not the only secondary bonds present in the supramolecular structure, but can afford higher dimensionality. In the ethanol and methanol solvates of the thiolate complex [Au(2-Hmba){P(o-Tol)$_3$}] (Hmba = 2-mercaptobenzoate) [34], two molecules are hydrogen bonded via two alcohol molecules about a center of inversion and centrosymmetrically related pairs aggregate via

C−H···π interactions. The structure of the acetonitrile solvate, commented above in Section 5.2.1, features a one-dimensional chain aligned along the *a* axis formed via O−H···O from the carboxylate groups and Au···S contacts. Reinforcement within the chain is found in C−H···O interactions involving the carbonyl oxygen atom. In the dimethylsulfoxide solvate, each molecule is associated with the DMSO oxygen atom through an O−H···O interaction and the carbonyl oxygen atom forms C−H···O non-conventional hydrogen bonds. The lattice of these two last solvates displays further associations of the type C−H···π with C−H bonds at 3.17 Å from the centroid of a phosphine phenyl ring (with an angle of 131°) in the former or at 2.75 Å from the ring centroid of the thiolate ligand (with an angle of 167°). In the acetonitrile complex, links between the chains are afforded by these C−H···π contacts and by additional π···π interactions (3.798(3) Å between the centroids of the interacting rings).

Similarly, other thiolate derivatives of gold(I) with diphosphines as ancillary ligands feature C−H···π and/or π···π interactions in addition to Au···S contacts or hydrogen bonds. For example, a close inspection of the structure of [{Au(2-Hmba)}$_2$(μ-dppe)] [36] (commented in Section 5.2.1) reveals a two-dimensional network resulting from the association of molecules via Au···S interactions and hydrogen bonds. Cohesion between the layers is afforded by a combination of π···π and C−H···π contacts, as shown in Figure 5.57. Thiolate rings approach each other at distances indicative of π···π interactions (separation between the rings 3.35 Å) and

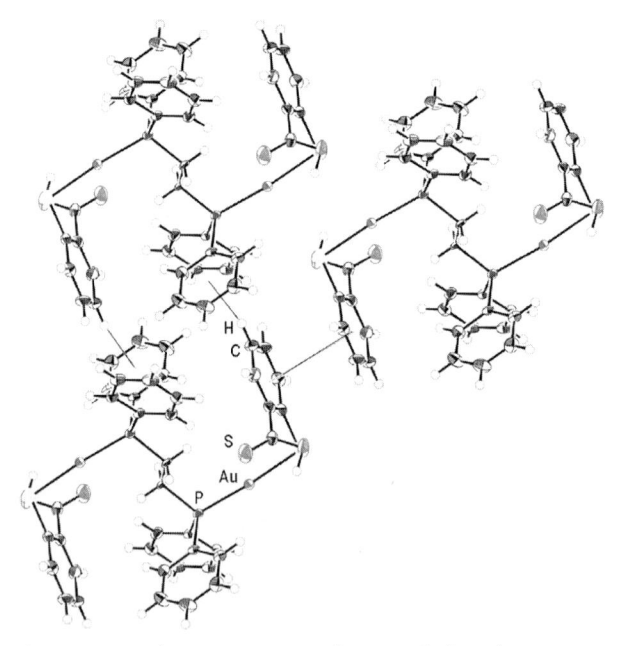

Figure 5.57 Interlayer association in [{Au(2-Hmba)}$_2$(μ-dppe)] formed via π···π and C−H···π contacts.

C−H···π interactions between thiolate C−H bonds and phosphine phenyl rings of different layers (H···ring centroid distance 2.75 Å; C−H···centroid angle 156°). Similar observations are made in its dimethylsulfoxide solvate, where, between the layers, centrosymmetrically related pairs of phenyl rings have an average separation of 3.67 Å, and several C−H···π contacts are formed between the layers, involving both phenyl- and DMSO-bound hydrogen atoms. C−H···π interactions are also responsible for the formation of layers in $[Au_2\{SC(OR)=NC_6H_4Y\text{-}4\}_2\{\mu\text{-}Ph_2P(CH_2)_4PPh_2\}]$ (Y=H, R=Et; Y=Me, R=Et, ^{i}Pr) or three-dimensional arrays in $[Au_2\{SC(OMe)=NC_6H_5\}_2\{\mu\text{-}Ph_2P(CH_2)_4PPh_2\}]$ [37] and serve to reinforce the supramolecular networks in these species and in related complexes.

π···π interactions themselves can lead to different degrees of association in a number of gold complexes. The dimeric complex $[Au(4,6\text{-}Me_2pym\text{-}2\text{-}S)]_2$ (4,6-Me$_2$-pym-2-S = 4,6-dimethylpyrimidinethiolate) [93] and the aminoquinoline derivatives [Au(Quingly)Cl]Cl and [Au(Quinala)Cl]Cl (Quingly = N-(8-quinolyl)glycinecarboxamide, Quinala = N-(8-quinolyl)-L-alanine-carboxamide) [94] (represented in Figure 5.58) are examples of solid-state structures in which π stacking clearly dominates intermolecular Au···Au bonding.

The most striking feature of their crystal structures is the packing of pairs of molecules by overlapping of aromatic rings in adjacent molecules (Figure 5.59) with interplanar distances of 3.416(1) and 3.474(1) Å in the former, and 3.1906 Å or 3.2119 Å in the latter two, which show stronger π···π interaction than the former.

π···π stacking can also lead to the formation of chains, as observed in the crystal structures of the heterometallic acetylacetonate complexes *trans*-$[Cr(acac)_2\{(S\text{-}4\text{-}py)AuPPh_3\}_2](ClO_4)$ (acac = acetylacetonate) [95] or $[\{Cu(acac)(phen)\}_2\{Au(CN)_2\}](ClO_4)$ [96]. The crystal packing diagram of the former shows that the partial π···π stacking between phenyl rings, each of which comes from another PPh$_3$ ligand, forms a quasi-one-dimensional structure in which two phenyl rings are almost parallel and the closest C−C distance is 3.44 Å. In the copper derivative every trinuclear cation interacts through the chelating ligands with two other trinuclear units generating supramolecular chains running along the crystallographic *b*-axis

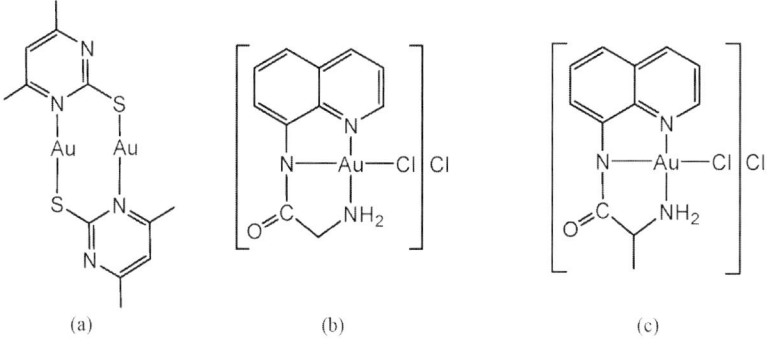

Figure 5.58 Schematic drawing of (a) $[Au(4,6\text{-}Me_2pym\text{-}2\text{-}S)]_2$, (b) [Au(Quingly)Cl]Cl and (c) [Au(Quinala)Cl]Cl.

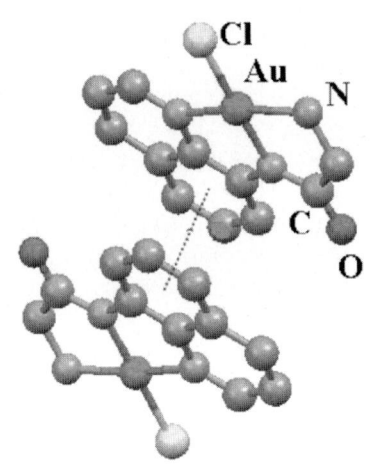

Figure 5.59 Association of pairs of molecules in [Au(4,6-Me$_2$pym-2-S)]$_2$ via π···π interactions.

(Figure 5.60). The distances associated with the π···π contacts ranging between 3.30 and 3.56 Å. Similarly, the analysis of the solid-state structure of the di- or tetranuclear compounds [Au$_2$(PPh$_3$)$_2$L](ClO$_4$)$_2$ or [Au$_2$(μ-dppm)L]$_2$(ClO$_4$)$_4$ (L = N,N'-bis-4-methylpyridil oxalamide, (4-NC$_5$H$_4$CH$_2$NHC=O)$_2$ [97] shows that there are weak intermolecular π···π stacking interactions between parallel pyridyl rings leading to the formation of a one-dimensional chain structure. These contacts are, however, weaker than in the previous cases, since the centroid-to-centroid distances are 3.954 or 3.715 Å, respectively.

On other occasions, π···π interactions are not unique secondary bonds and sometimes they appear coupled with the more common aurophilic contacts, resulting in the formation of a supramolecular array, such as the tetramer found in [Au(SPh)(PPh$_3$)] [98]. Its molecular packing shows a dimer-of-dimer structure comprising Au···Au contacts between two molecules and π···π interactions between a phenyl group of triphenylphosphine and the phenyl group of the SPh ligand of an adjacent dimmer (Figure 5.61a). The phenyl groups overlap with closest interatomic distance among twelve carbon atoms of 3.22(4) Å, which is good evidence of π···π interaction between these two phenyl groups.

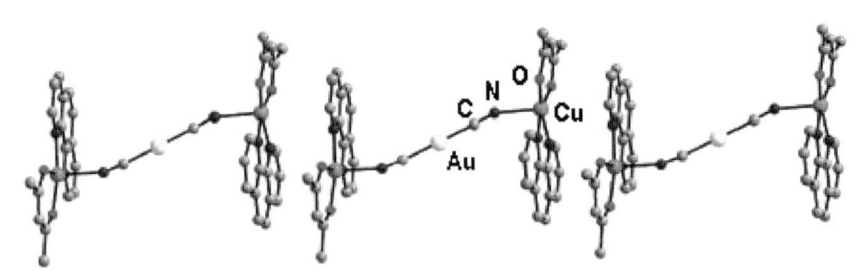

Figure 5.60 Extended chain of [{Cu(acac)(phen)}$_2${Au(CN)$_2$}](ClO$_4$) through π···π stacking interactions.

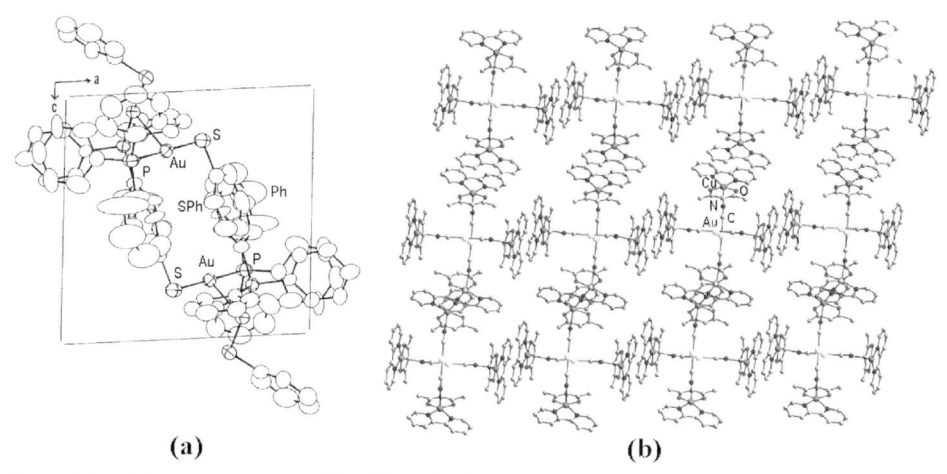

Figure 5.61 (a) Tetramer structure of [Au(SPh)(PPh₃)] and (b) gridlike layer of [{Cu(acac)(bipy)}₂{Au(CN)₂}](ClO₄)·0.5CH₃CN via Au···Au and π···π interactions.

Finally, combinations of the same types of van der Waals interactions can afford two-dimensional networks. The substitution of the nitrogen donor ligand phenanthroline in [{Cu(acac)(phen)}$_2${Au(CN)$_2$}](ClO$_4$) by 2,2′-bipyridine reveals a different form of packing. The analysis of the packing diagram of [{Cu(acac)(bipy)}$_2${Au(CN)$_2$}](ClO$_4$)·0.5CH$_3$CN [96] reveals the formation of gridlike supramolecular layers (Figure 5.61b) formed via aurophilic contacts of 3.2946(6) Å (absent in the phen derivative) associated with π···π contacts, in which distances range between 3.37 and 3.62 Å. A two-dimensional layered structure, also formed by aurophilicity and π···π interactions, is found for [(AuSPh)$_2$(μ-dpen)] (dpen = *trans*-1,2-bis(diphenylphosphino)ethylene) [98], where neighboring chains formed through Au···Au contacts are connected with partial π···π stacking between the phenyl rings of the SPh groups. These latter interactions are stronger than in the previously described complexes, since the closest contact between carbon atoms in these phenyl rings is 3.0(1) Å.

5.4
Concluding Remarks

In spite of the marked tendency of gold compounds to form aggregates through aurophilic contacts, there are other types of intermolecular van der Waals interactions that, alone or in cooperation with other types of intermolecular interactions, can serve as a versatile bonding motif for the self-assembly of gold complexes. Thus, secondary bonds with non-metal atoms, hydrogen bonding or π···π stacking may play an important role in the supramolecular chemistry of gold and a wide range of compositions and structures can be expected.

References

1. Alcock, N.W. (1972) *Advances in Inorganic Chemistry and Radiochemistry*, **15**, 1.
2. Alcock, N.W. and Countryman, R.M. (1977) *Journal of The Chemical Society-Dalton Transactions*, 217.
3. Haiduc, I. and Edelmann, F.T. (1999) *Supramolecular Organometallic Chemistry*, Wiley-VCH, Weinheim. Chapters 1 and 4.
4. Haiduc, I. (2004) Secondary Bonding, in *Encyclopedia of Supramolecular Chemistry* (eds J.L. Atwood and J.W. Steed), Dekker, New York, pp. 1215–1224.
5. Desiraju, G.R. (2004) Hydrogen Bonding, in *Encyclopedia of Supramolecular Chemistry* (eds J.L. Atwood and J.W. Steed), Dekker, New York, pp. 658–664.
6. Haiduc, I. and Edelmann, F.T. (1999) *Supramolecular Organometallic Chemistry*, Wiley-VCH, Weinheim. Chapters 1 and 5.
7. Jeffrey, G.A. and Saenger, W. (1994) *Hydrogen Bonding in Biological Structures*, Springer, Berlin.
8. Desiraju, G.R. (1996) *Accounts of Chemical Research*, **29**, 565.
9. Nishio, M., Hirota, M. and Umezawa, Y. (1998) *The CH/π Interaction. Evidence, Nature and Consequences*, Wiley-VCH, New York.
10. Crabtree, R.H. Siegbahn, P.E.M., Einstein, O. and Reingold, A.L. (1996) *Accounts of Chemical Research*, **29**, 348.
11. Crabtree, R.H. (2004) Hydrogen Bonds to Metals and Metal Hydrides, in *Encyclopedia of Supramolecular Chemistry* (eds J.L. Atwood and J.W., Steed), Dekker, New York, pp. 666–672.
12. Brammer, L., Zhao, D., Ladipo, F.T. and Braddock-Wilking, J. (1995) *Acta Crystallographica Section B-Structural Science*, **51**, 632.
13. White-Morris, R.L., Olmstead, M.M. and Balch, A.L. (2003) *Journal of the American Chemical Society*, **125**, 1033.
14. Bauer, A. and Schmidbaur, H. (1996) *Journal of the American Chemical Society*, **118**, 5324.
15. Chen, H.W. and Tiekink, E.R.T. (2003) *Acta Crystallographica Section E*, **59**, m50.
16. Hunks, W.J., Jennings, M.C. and Puddephatt, R.J. (2000) *Inorganic Chemistry*, **39**, 2699.
17. Hasan, M., Kozhevnikov, I.V., Siddiqui, M.R.H., Steiner, A. and Winterton, N. (1999) *Inorganic Chemistry*, **18**, 5637.
18. Hollis, L.S. and Lippard, S.J. (1983) *Journal of the American Chemical Society*, **105**, 4293.
19. Simonov, Y., Bologa, O., Bourosh, P., Gerbeleu, N., Lipkowski, J. and Gdaniec, M. (2006) *Inorganica Chimica Acta*, **359**, 721.
20. Bardají, M., Connelly, N.G., Gimeno, M.C., Jones, P.G., Laguna, A. and Laguna, M. (1995) *Journal of The Chemical Society-Dalton Transactions*, 2245.
21. Marangoni, G., Pitteri, B., Bertolasi, V., Ferretti, V. and Gilli, G. (1987) *Journal of The Chemical Society-Dalton Transactions*, 2235.
22. Schneider, D., Schier, A. and Schmidbaur, H. (2004) *Dalton Transactions*, 1995.
23. Yamamoto, H.M., Yamaura, J.-I. and Kato, R. (1998) *Journal of the American Chemical Society*, **120**, 5905.
24. Castro-Castro, L.M. and Guloy, A.M. (2003) *Angewandte Chemie-International Edition*, **42**, 2771.
25. Claramunt, R.M., Cornago, P., Cano, M., Heras, J.V., Gallego, M.L., Pinilla, E. and Torres, M.R. (2003) *European Journal of Inorganic Chemistry*, 2693.
26. Hashmi, A.S.K., Rudolph, M., Weyrauch, J.P., Wölfle, M., Frey, W. and Bats, J.W. (2005) *Angewandte Chemie-International Edition*, **44**, 2798.
27. Calleja, M., Johnson, K., Belcher, W.J. and Steed, J.W. (2001) *Inorganic Chemistry*, **40**, 4978.
28. Mathieson, T., Schier, A. and Schmidbaur, H. (2001) *Journal of The Chemical Society-Dalton Transactions*, 1196.
29. Nöth, H., Beck, W. and Burger, K. (1998) *European Journal of Inorganic Chemistry*, 93.

30 Henderson, W., Nicholson, B.K. and Tiekink, E.R.T. (2006) *Inorganica Chimica Acta*, **359**, 204.

31 Cronje, S., Raubenheimer, H.G., Spies, H.S.C., Esterhuysen, C., Schmidbaur, H., Schier, A. and Kruger, G.J. (2003) *Dalton Transactions*, 2859.

32 Bishop, P., Marsh, P., Brisdon, A.K., Brisdon, B.J. and Mahon, M.F. (1998) *Journal of The Chemical Society-Dalton Transactions*, 675.

33 Wilton-Ely, J.D.E.T., Schier, A., Mitzel, N.W. and Schmidbaur, H. (2001) *Journal of The Chemical Society-Dalton Transactions*, 1058.

34 Yun, S.S., Kim, J.K., Jung, J.S., Park, C., Kang, J.G., Smyth, D.R. and Tiekink, E.R.T. (2006) *Crystal Growth and Design*, **6**, 899.

35 Wilton-Ely, J.D.E.T., Schier, A., Mitzel, N.W. and Schmidbaur, H. (2001) *Inorganic Chemistry*, **40**, 6266.

36 Smyth, D.R., Vincent, B.R. and Tiekink, E.R.T. (2000) *CrystEngComm*, **2**, 115.

37 Ho, S.Y. and Tiekink, E.R.T. (2007) *CrystEngComm*, **9**, 368.

38 Fan, D., Yang, C.T., Ranford, J.D., Vittal, J.J. and Lee, P.F. (2003) *Dalton Transactions*, 3376.

39 Ehlich, H., Schier, A. and Schmidbaur, H. (2002) *Zeitschrift für Naturforschung*, **57b**, 890.

40 Canales, S., Crespo, O., Gimeno, M.C., Jones, P.G. and Laguna, A. (1999) *Chemical Communications*, 679.

41 Aguado, J.E., Canales, S., Gimeno, M.C., Jones, P.G., Laguna, A. and Villacampa, M.D. (2005) *Dalton Transactions*, 3005.

42 Wilton-Ely, J.D.E.T., Schier, A. and Schmidbaur, H. (2001) *Inorganic Chemistry*, **40**, 4656.

43 Eikens, W., Kienitz, C., Jones, P.G. and Thöne, C. (1994) *Journal of The Chemical Society-Dalton Transactions*, 83.

44 Jones, P.G. and Thöne, C. (1990) *Chemische Berichte*, **123**, 1975.

45 Jones, P.G. and Thöne, C. (1991) *Inorganica Chimica Acta*, **181**, 291.

46 Thöne, C. and Jones, P.G. (1992) *Acta Crystallographica Section C-Crystal Structure Communications*, **48**, 2114.

47 Tzeng, B.C., Huang, Y.C., Wu, W.M., Lee, S.Y., Lee, G.H. and Peng, S.M. (2004) *Crystal Growth and Design*, **4**, 63.

48 Shorrock, C.J., Jong, H., Batchelor, R.J. and Leznoff, D.B. (2003) *Inorganic Chemistry*, **42**, 3917.

49 Katz, M.J., Shorrock, C.J., Batchelor, R.J. and Leznoff, D.B. (2006) *Inorganic Chemistry*, **45**, 1757.

50 Olmstead, M.M., Jiang, F., Attar, S. and Balch, A.L. (2001) *Journal of the American Chemical Society*, **123**, 3260.

51 Jiang, F., Olmstead, M.M. and Balch, A.L. (2000) *Journal of The Chemical Society-Dalton Transactions*, 4098.

52 Ronig, B., Schulze, H., Pantenburg, I. and Wesemann, L. (2005) *European Journal of Inorganic Chemistry*, 314.

53 Jeffrey, J.C., Jelliss, P.A. and Stone, F.G.A. (1994) *Organometallics*, **13**, 2651.

54 Crespo, O., Fernández, E.J., Jones, P.G., Laguna, A., López-de-Luzuriaga, J.M., Monge, M., Olmos, M.E. and Pérez, J. (2003) *Dalton Transactions*, 1076.

55 Bardají, M., Jones, P.G., Laguna, A., Villacampa, M.D. and Villaverde, N. (2003) *Dalton Transactions*, 4529.

56 Habermehl, N.C., Jennings, M.C., McArdle, C.P., Mohr, F. and Puddephatt, R.J. (2005) *Organometallics*, **24**, 5004.

57 Vicente, J., Chicote, M.-T., Alvarez-Falcón, M.M. and Jones, P.G. (2005) *Organometallics*, **24**, 4666.

58 Vicente, J., Chicote, M.-T., Alvarez-Falcón, M.M. and Jones, P.G. (2005) *Organometallics*, **24**, 5956.

59 Barranco, E.M., Crespo, O., Gimeno, M.C., Laguna, A., Jones, P.G. and Ahrens, B. (2000) *Inorganic Chemistry*, **39**, 680.

60 Barranco, E.M., Gimeno, M.C., Laguna, A. and Villacampa, M.D. (2005) *Inorganica Chimica Acta*, **358**, 4177.

61 Aguado, J.E., Calhorda, M.J., Gimeno, M.C. and Laguna, A. (2005) *Chemical Communications*, 3355.

62 Li, Q.S., Wan, C.Q., Zou, R.Y., Xu, F.B., Song, H.B., Wan, X.J. and Zhang, Z.Z. (2006) *Inorganic Chemistry*, **45**, 1888.

63 Xu, F.B., Li, Q.S., Wu, L.Z., Leng, X.B., Li, Z.C., Zeng, X.S., Chow, Y.L. and Zhang, Z.Z. (2003) *Organometallics*, **22**, 633.

64 Rawashdeh-Omary, M.A., Omary, M.A., Fackler, J.P., Jr, Galassi, R., Pietroni, B.R. and Burini, A. (2001) *Journal of the American Chemical Society*, **123**, 9689.

65 Mohamed, A.A., Rawashdeh-Omary, M.A., Omary, M.A. and Fackler, J.P. Jr., (2005) *Dalton Transactions*, 2597.

66 Crowley, J.D., Steele, I.M. and Bosnich, B. (2005) *Inorganic Chemistry*, **44**, 2989.

67 Ruiz, J., Mosquera, M.E.G., García, G., Patrón, E., Riera, V., García-Granda, S. and Van der Maelen, F. (2003) *Angewandte Chemie-International Edition*, **42**, 4767.

68 Schneider, D., Schuster, O. and Schmidbaur, H. (2005) *Dalton Transactions*, 1940.

69 Schneider, D., Schuster, O. and Schmidbaur, H. (2005) *Organometallics*, **24**, 3547.

70 Ryowa, T., Nakano, M., Tamura, H. and Matsubayashi, G. (2004) *Inorganica Chimica Acta*, **357**, 3532.

71 Cerrada, E., Jones, P.G., Laguna, A. and Laguna, M. (1996) *Inorganic Chemistry*, **35**, 2995.

72 Kubo, K., Nakano, M., Tamura, H., Matsubayashi, G. and Nakamoto, M. (2003) *Journal of Organometallic Chemistry*, **669**, 141.

73 Morgado, J., Santos, I.C., Veiros, L.F., Rodrigues, C., Henriques, R.T., Duarte, M.T., Alcácer, L. and Almeida, M. (2001) *Journal of Materials Chemistry*, **11**, 2108.

74 Dias, J.C., Morgado, J., Alves, H., Lopes, E.B., Santos, I.C., Duarte, M.T., Henriques, R.T., Almeida, M., Ribas, X., Rovira, C. and Veciana, J. (2003) *Polyhedron*, **22**, 2447.

75 Takimiya, K., Oharuda, A., Morikami, A., Aso, Y. and Otsubo, T. (2000) *European Journal of Organic Chemistry*, 3013.

76 Fujiwara, E., Fujiwara, H., Narymbetov, B. Zh., Kobayashi, H., Nakata, M., Torii, H., Kobayashi, A., Takimiya, K., Otsubo, T. and Ogura, F. 2005, *European Journal of Inorganic Chemistry* 3435.

77 Codina, A., Fernández, E.J., Jones, P.G., Laguna, A., López-de-Luzuriaga, J.M., Monge, M., Olmos, M.E., Pérez, J. and Rodríguez, M.A. (2002) *Journal of the American Chemical Society*, **124**, 6781.

78 Jones, P.G. and Ahrens, B. (1997) *Chemische Berichte*, **130**, 1813.

79 Guy, J.J., Jones, P.G., Mays, M.J. and Sheldrick, G.M. (1977) *Journal of The Chemical Society-Dalton Transactions*, 8.

80 Ahrens, B., Jones, P.G. and Fischer, A.K. (1999) *European Journal of Inorganic Chemistry*, 1103.

81 Jones, P.G. and Ahrens, B. (1998) *New Journal of Chemistry*, 1041.

82 Mingos, D.M.P., Yau, J., Mentzer, S. and Williams, D.J. (1995) *Journal of The Chemical Society-Dalton Transactions*, 319.

83 Tzeng, B.C., Schier, A. and Schmidbaur, H. (1999) *Inorganic Chemistry*, **38**, 3978.

84 Hunks, W.J., Jennings, M.C. and Puddephatt, R.J. (2002) *Inorganic Chemistry*, **41**, 4590.

85 Burchell, T.J., Eisler, D.J., Jennings, M.C. and Puddephatt, R.J. (2003) *Chemical Communications*, 2228.

86 Ahrens, B., Friedrichs, S., Herbst-Irmer, R. and Jones, P.G. (2000) *European Journal of Inorganic Chemistry*, 2017.

87 Jones, P.G. and Friedrichs, S. (2006) *Acta Crystallographica Section C-Crystal Structure Communications*, **62**, m623.

88 Jones, P.G. and Friedrichs, S. (2000) *Acta Crystallographica Section C-Crystal Structure Communications*, **56**, 56.

89 Bardají, M., Jones, P.G. and Laguna, A. (2002) *Journal of The Chemical Society-Dalton Transactions*, 3624.

90 Freytag, M. and Jones, P.G. (2000) *Chemical Communications*, 277.

91 Jones, P.G. and Ahrens, B. (1998) *Chemical Communications*, 2307.

92 Kui, S.C.F., Huang, J.S., Sun, R.W.-Y., Zhu, N. and Che, C.M. (2006) *Angewandte Chemie-International Edition*, **45**, 4663.

93 Hao, L., Lachicotte, R.J., Gysling, H.J. and Eisenberg, R. (1999) *Inorganic Chemistry*, **38**, 4616.

94 Yang, T., Tu, C., Zhang, J., Lin, L., Zhang, X., Liu, Q., Ding, J., Xu, Q. and Guo, Z. (2003) *Dalton Transactions*, 3419.

95 Nunokawa, K., Onaka, S., Mizuno, Y., Okazaki, K., Sunahara, T., Ito, M., Yaguchi, M., Imai, H., Inoue, K., Ozeki, T., Chiba, H. and Yosida, T. (2005) *Journal of Organometallic Chemistry*, **690**, 48.

96 Madalan, A.M., Avarvari, N. and Andruh, M. (2006) *Crystal Growth and Design*, **6**, 1671.

97 Tzeng, B.C., Yeh, H.T., Wu, Y.L., Kuo, J.H., Lee, G.H. and Peng, S.M. (2006) *Inorganic Chemistry*, **45**, 591.

98 Onaka, S., Katsukawa, Y., Shiotsuka, M., Kanegawa, O. and Yamashita, M. (2001) *Inorganica Chimica Acta*, **312**, 100.

6
Luminescence of Supramolecular Gold-Containing Materials
José María López-de-Luzuriaga

6.1
Introduction. Conditions for Luminescence in Gold Complexes

In addition to the characteristics that make gold a unique element, such as the lowest electrochemical potential of all metals, its high electronegativity, or the fact that it may have a mononegative oxidation state, perhaps the most intriguing characteristic is the tendency of many gold complexes to aggregate into oligomers or supramolecular assemblies through metal–metal interactions. This fact, far from being a chemical curiosity, has become an object of study because of the associated properties. One of the most studied properties and that which involves many groups all over the world is the luminescence displayed by many gold-containing complexes. In fact, after the discovery of the luminescence of the three-coordinate complex [AuCl(PPh$_3$)$_2$] by Dori and coworkers in 1970 [1], the number of groups that have studied this topic has increased rapidly in recent years, together with the number of synthesized luminescent gold molecules. At the same time, the large number of studies performed in this area has provided knowledge of the conditions that gold complexes require to display luminescence. For instance, luminescence can originate from *ligands*, assigned to *certain geometries* around the gold atom or from the presence of *metal–metal interactions* in the complexes. More specifically, it can be produced by transitions between orbitals of the metal center exclusively, in orbitals of the ligands, usually among π orbitals, or in transitions involving both metal and ligands, where these can act as donors or acceptors of electronic density (charge transfer transitions). In gold (I), with a d^{10} closed shell configuration, the ground state is $^1S_0$, while the excited states are $^3D_2$, $^3D_1$, $^3D_0$ and $^1D_0$ (see Figure 6.1) [2].

Of all these, the only permitted electronic transition is $^1D_0 \leftarrow {}^1S_0$, while the rest are forbidden according to the spin rule, leading to phosphorescent emissions. However, these are precisely the ones responsible for the emissions found in many luminescent gold complexes, appearing in a range between 500 and 700 nm (665 nm in the gaseous ion).

6 Luminescence of Supramolecular Gold-Containing Materials

▬▬▬▬▬	$^1D_0$	29620 cm^{-1} (338 nm)
▬▬▬▬▬	$^3D_1$	27764 cm^{-1} (360 nm)
▬▬▬▬▬	$^3D_2$	17639 cm^{-1} (567 nm)
▬▬▬▬▬	$^3D_3$	15039 cm^{-1} (665 nm)
▬▬▬▬▬	$^1S_0$	0 cm^{-1} (0 nm)

Figure 6.1 Atomic energy levels for Au(I).

Another very important factor that should be considered is the expected large spin–orbit effect promoted by a heavy atom such as gold. This effect produces a relaxation of the spin rule and makes orbital forbiddenness instead of the spin rule the key factor for determining whether a gold complex will show luminescence. For instance, phosphorescence is generally not observed for linear molecules (where the ligands are not involved in the transitions), since the mixing of s, p_z and d_{z^2} of gold form a HOMO of Σ^+ symmetry, while the LUMO orbital is formed mainly from p_x and p_y leading to a Π symmetry, consequently, there is no orbital forbiddenness in the transition and phosphorescence is not observed [3]. Another possible coordination environment at gold is the tetrahedral one. In that case, the HOMO and the LUMO are $^1T_2$ in symmetry and the transition is allowed according to Laporte's rule; therefore, phosphorescence is not observed in these complexes either. In contrast, when the environment around gold is planar-three-coordinate (placing the z-axis perpendicular to the molecular plane), the LUMO is composed mainly of the p_z orbital and the HOMO of the $d_{x^2-y^2}$ and d_{xy} orbitals, in the perpendicular plane; thus, the transition is orbitally forbidden, giving rise to phosphorescence (Figure 6.2).

In general, according to the previous comments, we can assume that a change in the linear geometry around the gold(I) center from linear to trigonal-planar leads to a higher probability of metal-centered phosphorescent emission and that a subsequent change to tetrahedral leads to its loss. Nevertheless, the three-coordination is not a necessary condition to produce luminescent gold complexes because a moderate deviation from linearity produces a shortening of the HOMO–LUMO gap. For example, molecular orbital calculations reveal an energy variation in the frontier orbitals when a linear gold(I) molecule bends its L–Au–L angle to 120°. Thus, in the figure we can observe a qualitative molecular orbital diagram of a linear gold(I) fragment with phosphorus donor ligands (the molecular axis coincides with the z-axis) and the decrease in the energy of the HOMO–LUMO gap when the ligands are placed at 120° to each other (Figure 6.3)[3].

6.1 Introduction. Conditions for Luminescence in Gold Complexes

Figure 6.2 Qualitative molecular orbital diagram for $[Au(PR_3)_3]^+$.

Additionally, Figure 6.4 shows the variation in the energy of the occupied 5d orbitals when the P–Au–P angle varies from 180° (linear coordination) to 120°. Thus, while the d_{z^2} orbital stabilizes slightly, the d_{xz} destabilizes as a consequence of the interaction of this orbital with the $3p_x$ and $3p_z$ orbitals of phosphorus. The main consequence is that, below 168°, the former HOMO (d_{z^2} in linear molecules) is replaced by the d_{xz} orbital, which displays a higher energy and, consequently, a lower energy is needed to reach the excited state.

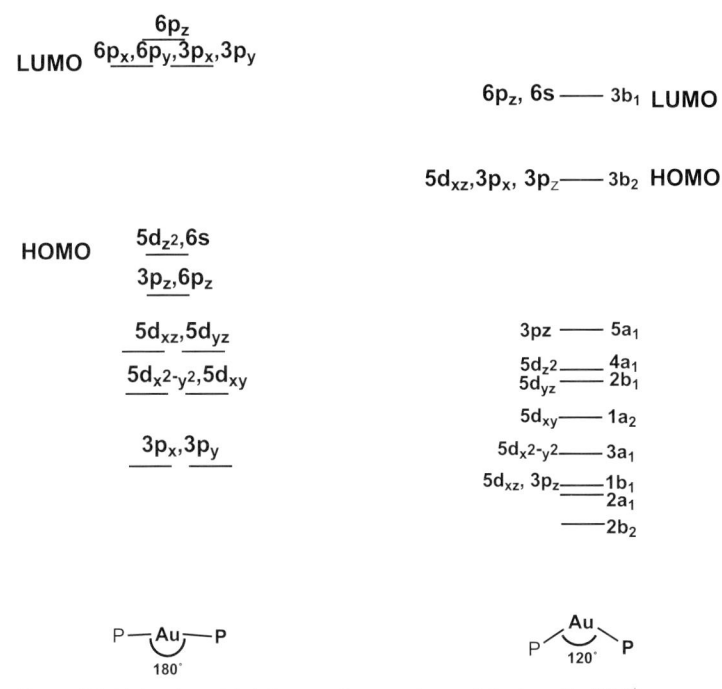

Figure 6.3 Molecular orbital diagram for complexes of the type $[Au(P)_2]^+$.

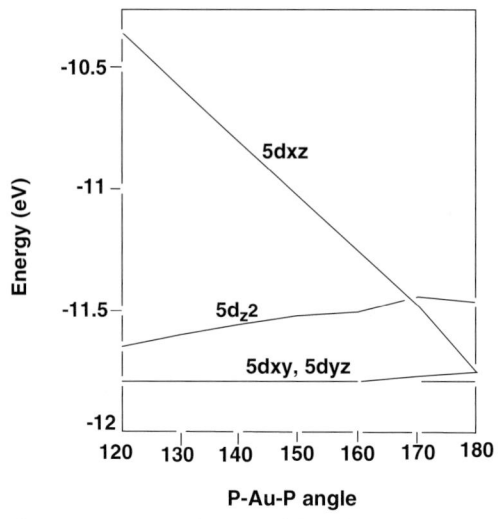

Figure 6.4 Variation of energy of the 5d orbitals with the P–Au–P angle.

In addition, while the π → σ transition in the linear configuration is orbitally allowed, the corresponding $b_1 \to b_2$ is forbidden, leading to phosphorescence. At this point, we can assume that any factor that leads to a deviation of the expected linear environment of gold(I) would contribute to the gap shortening and therefore to a higher probability of luminescence. Thus, it is foreseeable that steric effects produced by bulky ligands or strong gold–gold interactions, both affecting the linearity of polynuclear gold molecules, are very important factors that can promote luminescence in these materials.

Although the origin is different, the HOMO–LUMO gap shortening is also observed in extended systems built by gold–gold interactions. In this case, in addition to the expected more or less pronounced loss of linearity produced by the metal–metal interactions or ligands, the interactions themselves lead to the same phenomenon due to overlapping of the orbitals of each interacting gold atom in the system. In this case, the degree of shortening is dependent on the number of interacting gold atoms. For instance, Figure 6.5 shows a comparison of simplified molecular orbital diagrams of an isolated molecule or interacting $[Au(phosphine)_2]^+$ cations (z-axis along the gold–gold interactions).

In this case the interaction between $[Au(P)_2]^+$ fragments produces a decrease in the HOMO–LUMO gap giving rise to a $(d\delta^*)^1(p\sigma)^1$ excited state by combination of the filled $d_{x^2-y^2}$ and empty p_z orbitals [4]. Nevertheless, the assignation of this state is not unambiguous because by defining the Au–Au axis as the z-axis, the d_{z^2} would be the most affected by the interaction and could become the HOMO, giving rise to a $(d\sigma)^1(p\sigma)^1$ excited state [5]. The latest studies on binuclear gold complexes with bridging bis(phosphine) ligands seem to demonstrate that the last assignment is the most plausible. However, at the same time, these studies suggest that the excited states responsible for the visible emission are solvent or anion exciplexes [6–8].

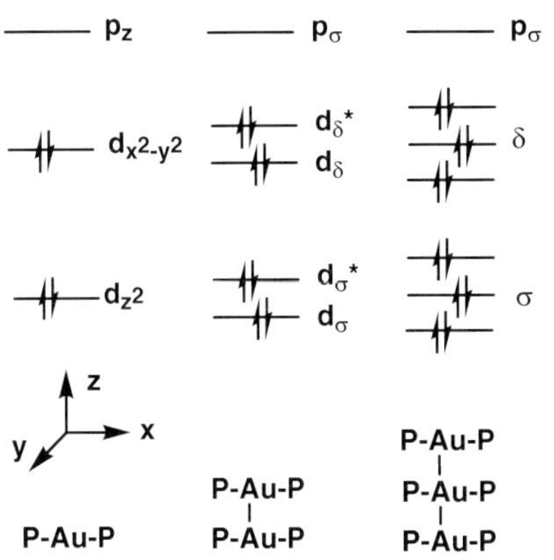

Figure 6.5 Qualitative molecular orbital diagram of [Au(phosphine)$_2$]$^+$ species.

Taking into account the above-mentioned comments and once the different possibilities giving luminescence in gold complexes have been analyzed, the following sections will examine selected examples of luminescent supramolecular gold architectures and the factors that affect the observed emissions. At this point I would like to comment that, in spite of the great number of luminescent gold complexes reported, the chosen examples correspond to those forming supramolecular entities preferably built making use of the ability of gold to interact with itself or with other metals, since those gold-containing complexes built using ancillary ligands or other secondary interactions are not usually different in their behavior from the simple isolated molecules. Additionally, in some selected cases I will comment on examples of other supramolecular entities in which the optical behavior depends on the whole network, even when gold–gold interactions do not exist. Finally, I will only consider those structures built by "weak" (sometimes not so weak!) gold–gold or gold–heterometal interactions, but in which the metal centers have a non-zero formal oxidation state, and not those molecules named clusters that are described in another chapter of this book. Thus, I will divide this chapter into two main parts: luminescent supramolecular gold entities and luminescent supramolecular gold–heterometal entities.

6.2
Luminescent Supramolecular Gold Entities

In this section, I will comment on selected examples of luminescent supramolecular architectures built through Au–Au interactions, both in the solid state and/or, when there is enough evidence, in solution. This topic will be divided into four parts based on the unit that by repetition gives rise to the supramolecular network, that is mononuclear, binuclear, trinuclear and higher nuclearity systems.

6.2.1
Networks from Mononuclear Units

One of the most studied mononuclear systems that usually leads to supramolecular networks and that also exhibits very rich photophysics and photochemistry is the [Au(CN)$_2$]$^-$ anion. This complex is among the most stable two-coordinate complexes of the transition ions, with a stability constant of 10^{37} [9], being reasonably stable to air, moisture, temperature and light, which could make it appropriate for practical applications.

Patterson et al. performed a thorough photophysical study of the solid state [10] and solution ground and excited states properties [11, 12] of the potassium salt, suggesting that in the solid state the strength of the metal–metal interactions has a great effect on luminescence.

Crystals of the complex display a layered structure of Au(CN)$_2$$^-$ linear ions, alternating with layers of K$^+$ ions. The gold centers within a layer are separated by 3.64 Å by the potassium ions at room temperature. However, this distance is temperature dependent, X-ray measurements on K[Au(CN)$_2$] single crystals between room temperature and 78 K indicate that the Au–Au separation decreases from 3.64 Å at 278 K to 3.58 Å at 78 K. At the same time, decreasing temperature leads to an increase of 0.06 Å in the separation between adjacent layers. These variations lead to a change in the luminescence spectra. For instance, when excited with a pulsed nitrogen laser at 337 nm, the 295 K emission spectrum shows a high energy emission band at 390 nm and a low energy band at 630 nm. As the temperature of the crystal is lowered, the low energy band decreases in intensity and disappears below 120 K. At 8 K, the high energy band shows a vibronic structure with peaks at 363, 388, 411 and 430 nm. In addition, both bands show a shift in position. The assignment for the high energy emission is based on extended Hückel molecular orbital calculations and these indicate that the HOMO and the LUMO comprise mainly 6s and 6p$_x$, 6p$_y$ atomic orbitals of gold that overlap with neighboring Au(CN)$_2$$^-$ ions. They also indicate that as the Au–Au separation is decreased, the energies between the HOMO and LUMO also decrease. The vibronic structure at low temperature is assigned to a symmetric stretch mode progression. In contrast, the low energy emission is assigned to the presence of AuCN groups in sites of the structure, obtained as products of the dissociation of the Au(CN)$_2$$^-$ ions.

The same group studied the ground and excited states properties of the same material in solution. These studies demonstrated that increases in the concentration of the solutions lead to an oligomerization process. This was identified in the ground state because the absorption or excitation bands shifted to lower energies and new bands assigned to oligomers of higher nuclearity also appeared. As before, this behavior is attributed to the formation of gold–gold interactions between neighboring Au(CN)$_2$$^-$ ions, giving rise to a supramolecular assembly in solution. Very interestingly, the comparison of the excitation spectrum of the solid K[Au(CN)$_2$] with those of the solutions revealed that, while the solid has an excitation of 332 nm, the concentrated aqueous solutions exceeded that value. This result was interpreted as follows: the two-dimensional layers of the Au(CN)$_2$$^-$ ions in the crystals are separated by layers of K$^+$ ions, whereas in solution the motion is random and can lead to three-dimensional interactions between Au(CN)$_2^-$ anions. Therefore, the supramolecular entity built by gold–gold interactions in solution can be even larger than that produced in the solid state. Similar observations can be made with other solvents, for instance methanol. Consequently, oligomerization in solution is not dependent on the solvent. The excitation spectra of frozen solutions contain peaks at much longer wavelengths than the absorption peaks of the same solution, indicating that oligomerization occurs much more easily at 77 K.

As regards the excited state properties of this material in solution, the emission spectra also give evidence of an oligomerization process. Thus, the emission spectra at different concentrations give two emission bands in the ranges 400–410 and 430–470 nm. An increase in concentration produces an increase in the intensity of the low energy band, when compared with the high energy band, and a red shift of the former, both with the same excitation spectra. This result is interpreted in terms of the presence of different oligomers in the excited state and the shift of the low energy band with the formation of larger oligomers as the concentration increases. In fact, the plot of the emission energy versus molar concentration shows a quadratic relationship, which is consistent with the two-dimensional layered structure observed in the solid state. Nevertheless, the geometry of the excited state relative to the ground state in these oligomers should be very distorted because the emission bands are structureless, have large Stokes shifts and are red shifted from the absorption bands. Furthermore, theoretical studies on these systems indicate that for a given [Au(CN)$_2^-$]$_n$ oligomer, the first excited state has a deeper potential well (higher binding energy) and a shorter Au–Au equilibrium distance (2.664 Å) than the ground state (2.960 Å). Consequently, taking into account the foregoing, the supramolecular architectures found in the solid state are also observed in solution in both the ground and the excited states, producing *different luminescent materials depending on the aggregation state, concentration, temperature and solvent.*

The importance of gold–gold interactions in the emissions and the solvent dependence is illustrated in a curious example reported by Balch and coworkers [13]. As before, the behavior differed depending on the aggregation state and, in both the solid state and in frozen solutions, the emissions were attributed to supramolecular species built by means of aurophilic interactions. Thus, the carbene complex [Au{C(NHMe)$_2$}$_2$]PF$_6$·0.5Me$_2$C=O displayed an extended chain structure with the anions hydrogen bonded to the cations (see Figure 6.6).

The Au–Au distances are 3.1882(1) Å and they are considered to be responsible for the emission band that appears at 460 nm at room temperature. When the anion in the carbene complex is BF$_4^-$, the structure is similar although the Au–Au distances are substantially longer (3.4615(2) Å). The different distance leads to a different emission band that is blue-shifted. This indicates a greater orbital interaction in the former, consistent with its shorter Au–Au distance. Nevertheless, the behavior in solution is similar for both. Thus, at room temperature in solution they lose their emissive properties but they recover them in frozen solutions at 77 K. Interestingly, the emission differs in color, depending on the solvent, ranging from orange (acetone) to blue (pyridine), which would seem to result from the self-association

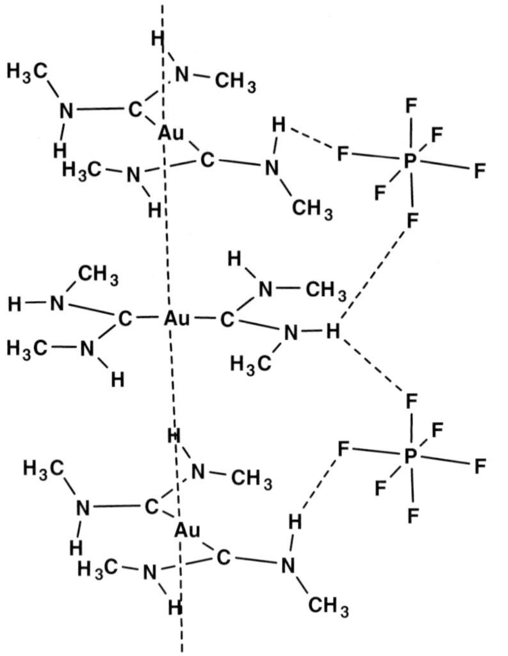

Figure 6.6 Schematic diagram of the extended chain of [Au{C(NHMe)}$_2$]PF$_6$·0.5 acetone.

of the cations through aurophilic interactions. Situations in which no luminescence is observed are assigned to monomeric species. A particularly important factor that should be considered in these systems is the role of *hydrogen bonding* in building the polymer. In this regard, it is well known that gold–gold interactions are comparable in strength to certain hydrogen bonds ($\sim$40 kJ mol^{-1}), but in systems in which gold–gold interactions and hydrogen bonding are both possible it has been demonstrated that both contribute to the stability of the molecular architecture [14]. An example of this is the homologous carbene complex [Au{C(NMe$_2$)(NHMe)}$_2$]PF$_6$ which is not luminescent at room temperature and at 77 K in the solid state but exhibits luminescence in frozen solutions. The structure does not show aurophilic interactions between the cations (shortest Au–Au distance = 7.1109 Å) and it is proposed that adding the methyl group to the carbene ligand inhibits hydrogen bonding and does not promote aurophilic interaction between cations.

Despite the foregoing and as commented in Section 6.1, gold–gold interactions are not the only source of luminescence, even when they are present in the structures. For example, the complexes [(o-xylylNC)AuX] (X = Cl, Br, I, CN) [15] show quite similar molecular structures, consisting of nearly linear dispositions around gold and in the Au–C–N portions. Nevertheless, there are considerable variations in intermolecular organization in the solid state. Thus, while the chloride complex is a dimer with a single close Au–Au contact of 3.3570(11) Å; the iodine complex organizes into chains (d(Au–Au) = 3.4602(3) Å); the bromine crystallizes forming slightly kinked chains (d(Au–Au) = 3.3480(5) Å); and the cyanide complex forms a grid (Au–Au average distances 3.4 Å). Far from exhibiting structural diversity, all the complexes display similar emission bands around 420 and 500 nm. In this case, the authors assigned the emissions as arising from xylyl based $\pi\pi^*$ states, since the isocyanide ligand showed a band at 415 nm in its emission spectrum.

A ligand-based emission is also proposed in the complex [PhC≡CAu]$_n$, which displays a honeycomb network with PhC≡C pillars [16]. In this case, the Au–Au interactions are even shorter than in the previous cases (2.98(1)–3.26(1) Å) and the emission spectrum displays a very complicated pattern with peak maxima at 413, 468, 550, 592 nm and shoulders at 621, 636, 653, 670 and 694 nm. The analysis of these bands indicates an intraligand $^3(\pi\pi^*)$ transition influenced by Au–Au interactions. In these two cases, it is likely that the presence of ligands with π-electronic density in the complexes favors the intraligand transitions.

However, a small variation in the complexes such as, for example, in the isocyanide group of the related (RNC)AuX complexes (R = Cy; X = Cl, Br, I, CN)(R = ^{n}Bu, ^{i}Pr, Cy, Me, ^{t}Bu; X = CN), can lead to a different origin of the optical behavior [17, 18]. For example, in the case of the halogen derivatives, as in the previous case, the complexes show different supramolecular organization: chains, dimers and isolated molecules. All show similar optical behavior with orange luminescence and a strikingly large Stokes' shift (~21 000 cm^{-1}). In this case, the similarity in the absorption, excitation and emission spectra suggests that differences in supramolecular organization do not influence the optical properties. An excited state arising from a ^{3}(MLCT) transition is proposed in this case. This assignment is in accordance with the extremely large Stokes' shift, since a geometric distortion of the Au–C–N–C unit could be a common feature in each of the three molecules and that distortion would produce a significant change in the excited state structure without affecting the bulky cyclohexyl groups or the Au–Au interactions.

In contrast, in the case of the cyanide derivatives, the variation in the alkyl groups produces solids with different patterns of association, such as chains or two-dimensional sheets. This leads to different absorption bands and different emissions that are justified by the different supramolecular organization of the molecules in the solids and in which the aurophilic interactions are considered to be a crucial factor in the optical properties. In the specific case of the (CyNC)AuCN complex, which can be compared with the previous halogen derivatives, this does not exhibit the above-mentioned large Stokes shift and exhibits emissions that depend on the excitation wavelength. As shown in these examples, changing from a π-acceptor anionic ligand to a π-donor anionic ligand alters the electronic properties of the complexes. Therefore, a small variation in the components actually gives rise to dramatic differences in the optical properties.

These comments prompt the next question, namely whether the structure itself can determine the optical properties while keeping the components of the complex unchanged; in other words, *is polymorphism another possible factor that should be taken into account?*

We will now comment on the case of two recent examples reported by Balch and coworkers and by Che and coworkers. In the first case, [Au(CyNC)$_2$]PF$_6$ [19], a complex closely related to those described previously crystallizes into two forms: a colorless form and a yellow form. Neither involves the inclusion of any solvent molecules into the solid and they are therefore genuine polymorphs. The crystal structure of the colorless form consists of a semi-staggered array and the structure corresponding to the yellow form consists of a helical ribbon. Both exhibit short gold–gold distances of 3.1822(3) and 2.9643(6)–2.9803(6) Å, respectively, and both emit in the solid state at 424 nm (exc. 353 nm) and 480 nm (exc. 394 nm) with lifetimes in the range of microseconds, indicative of phosphorescent processes. Both also lose their emissive properties in solution. Taking into account the previous data, the authors suggest that luminescence must result from the extended Au–Au interactions between cations in their solid state supramolecular structures, assigning the transitions as MC (metal centered), from the filled $5d_{z^2}$ to the empty $6p_z$ (z-axis collinear with the Au–Au stacking direction). In this case, the less energetic emission corresponded to the shortest gold–gold distance, as expected. As before, the behavior in frozen solvents was exactly as expected, since different emissions were obtained when the solvents changed; this result was interpreted as a variation in the Au–Au separations, the Au–Au–Au angles or, even, the relative orientation of the isocyanide ligands. Therefore, in this example, the polymorphism determined the optical properties of the material, but based only on the well-known factor, namely the distance between adjacent gold centers.

In contrast, in the case of the complex [Au(C≡C–C$_6$H$_4$–4–NO$_2$)(PCy$_3$)] [20], the material appeared as two crystalline forms: a rod-like form, and a plate-like form. The major structural differences between them were the orientations of the molecular dipoles, the dihedral angles between neighboring 4-nitrophenyl moieties and the polymeric nature of only one of them (see Figure 6.7).

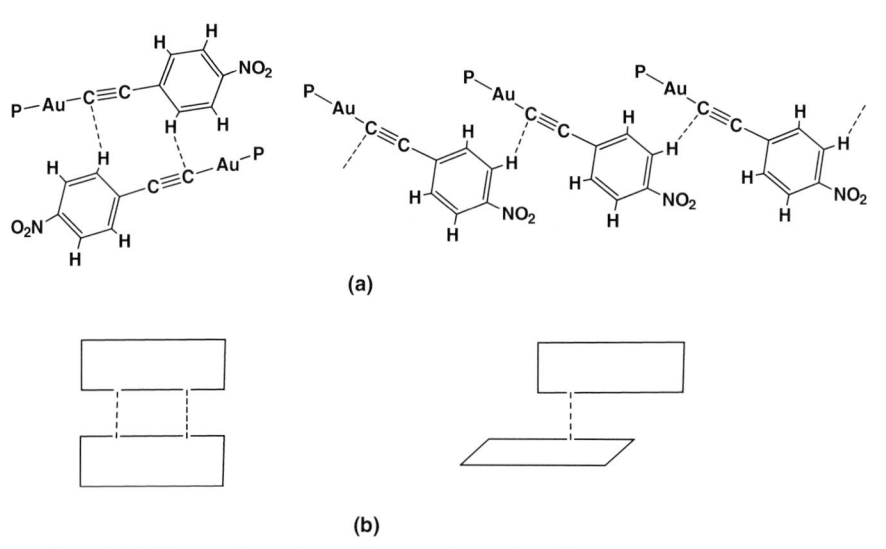

Figure 6.7 (a) Schematic structure of the two polymorphs of [Au (C≡C–C$_6$H$_4$–4–NO$_2$)(PCy$_3$)] showing the dimer (E-form) (left) or polymer (N-form) (right) arrangements. (b) Representation of the geometry of adjacent molecules in the crystal.

The optical behavior of both forms differs; thus, while the dimer (E-form) is strongly luminescent in the solid state at room temperature (max. em. 504 nm), the polymer (N-form) is luminescent only at 77 K (max. em. 486 nm). Both emissions, within the microseconds range, are considered to originate from triplet intraligand excited states. However, the authors also suggest in this case that the angle between the two molecular planes of the interacting lumophores is related to the emissive properties of these solids (Figure 6.7b). In fact, the structural differences between both polymorphs lead to a different orientation of the dipoles, head-to-tail and slipped, for the dimer and polymer, respectively. Therefore, the interaction between the neighboring chromophores depends on these angles. Thus, while the dihedral angle leads to a significant excitonic coupling in the case of the N-form, in the E-form with a 180° angle the excitonic coupling is negligible. The former results in the transfer of excitonic energy to non-radiative energy traps that are located at the lattice flaws. Although this is a plausible assignment, the authors suggest that this is not completely clear, since this process is suppressed at low temperature and excitonic splitting is much larger than in other conventional examples. Nevertheless, this example seems to show that *the relative order of the molecules in the structure can also affect the optical properties*, even when there are no interactions between the gold centers.

The importance of the structure can also be seen in the next example, reported by Fackler and coworkers [21]. It describes the optical properties of the complexes [Au(TPA)$_2$][Au(CN)$_2$] and [AuCl(TPA)] (TPA = 1,3,5-triaza-7-phosphaadamantane). Both form supramolecular structures through gold–gold interactions giving rise to an extended linear chain and a helical chain, respectively. The metal–metal interactions are of the same magnitude for both complexes, 3.457(1) and 3.394 Å, respectively, but the optical properties are considerably different. Thus, the chloro-complex emits weakly at 78 K, exhibiting an emission at 580 nm, which is interpreted as a metal-centered transition. The dicyanide complex behaves differently depending on the size of the sample. For instance, the bulk powder of this complex exhibits two emissions, whose relative intensities depend on the excitation wavelength and temperature (luminescence thermochromism). Thus, at wavelengths below 360 nm, the powder exhibits a blue emission at 425 nm, while excitation with longer wavelengths leads to a green emission near 500 nm. The green emission dominates at ambient temperature but cooling at cryogenic temperatures leads to the dominance of the blue emission. Interestingly, single crystals of this complex do not luminesce visibly, but grinding the crystals finely initiates a strong green emission under UV irradiation at room temperature. In this case, the authors suggest that changes occurring near the surface of the material and/or formation of defect sites with little communication between them may be the origin of this behavior. In fact, the chains in the crystal run parallel to the long crystal axis and the grinding process leads to more chain ends than in the single crystal, and in these parts the Au–Au distances are supposed to be shorter, leading to lower energy emission. In contrast, in the single crystal the metal-based emission is too weak to be observed.

As observed in these last examples and in spite of the above-mentioned foreseeable simplicity of the conditions that a gold complex requires to display luminescence,

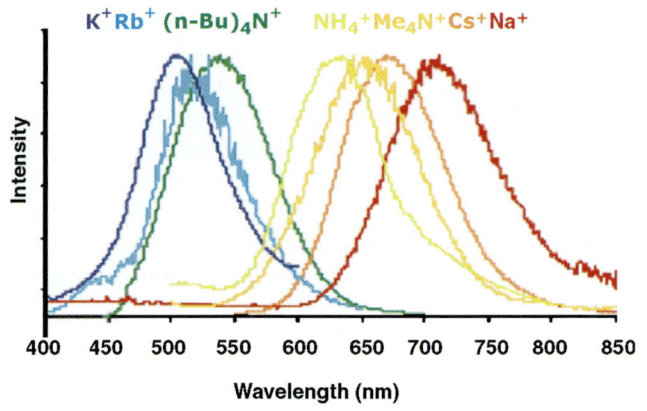

Figure 6.8 Emission spectra of oligomeric salts of [Au(SCN)$_2$]$^-$ at 77 K (Reprinted with permission from Reference [22]. Copyright 2004 American Chemical Society).

many factors should be taken into account. A common objective of research groups devoted to this topic has been to identify a simple relationship between gold–gold distances and the energy of the emissions. However, as we can observe, many factors are potential sources of variations in these conditions.

In this regard, perhaps the most direct relationship was reported by Elder et al. recently [22]. This result is particularly striking since the structures of the Au(SCN)$_2$ anions are different depending on the counterion, and there are examples of infinite linear chains, kinked chain of trimers joined at a shared gold atom such as the kink, or isolated dimers, as in the case of the tetrabutylammonium salt. In all cases the shorter the gold–gold distance, the lower the energy of the emission and cation size is not a determining factor (see Figure 6.8).

Based on the analysis of the data, the authors concluded that a single gold–gold pair serves as the emissive source in infinite chains, since in the dimeric tetrabutylammonium salt a single gold–gold pair is the source of emission. Similarly, a trimeric group of equidistant atoms behaves as though a single pair is the emissive source. Obviously there is a striking simplicity in the result compared with the observations reported by other authors in related examples (see above).

Finally, in contrast to Au(I), Au(III) complexes have rarely been observed to emit luminescence and in most cases the observed luminescence is due to intraligand transitions [23]. Furthermore, extended systems built through Au(III)–Au(III) interactions are even more unusual. Thus, the conjunction of both characteristics is very rare. An example is the gold (III) complex [Au(dbbpy)Cl$_2$]PF$_6$ (dbbpy = 4,4′-di-tert-butyl-2,2′-bipyridine) [24], which is luminescent in a glass matrix with a vibrational spacing consistent with ring breathing modes in the aromatic diimine ligands and whose luminescence originates from a lower intraligand (π–π^*) excited state. Although the crystal structure of this complex has not been determined, the substitution of both chlorides by a dithiolate ligand such as tdt (tdt = 3,4-toluene-dithiolate) leads to a linear stacking complex with gold(III)–gold(III) interactions of

3.60 and 3.75 Å alternating between neighboring molecules. This contrasts with what occurs with the precursor; in this case, the complex was not luminescent, a result that is not fully comprehended and is probably related to a quenching process originated by the dithiolate ligand, which probably affects the d-orbital energies.

As commented in Section 6.1, in some cases the role of the ligands is critical in explaining the optical properties of the systems. In fact, the existence of gold(I) chains in the structure does not guarantee that the complex will exhibit luminescent properties due to these atoms either. For instance, and among many other examples, that is the case for the complex $(C_{12}H_{14}N_2)[Au_2I_4]$ [25]. The crystal structure shows a stacking of layers of methyl viologen and $[AuI_2]^-$ ions. The anions form a single one-dimensional chain with gold–gold interactions of 3.3767 (3) Å, which is a reasonably short distance. Nevertheless, the solid is not luminescent, a fact that the authors assign to a quenching of the luminescence promoted by methyl viologens.

6.2.2
Networks from Dinuclear Units

A very common strategy in the synthesis of complexes that display gold–gold interactions is the use of polydentate ligands that can act as bridges between the metal centers. Of these, perhaps the most accessible and commonly used ligands are bidentates and, in the case of gold, soft donor ligands such as phosphorus or sulfur and to a lesser extent hard donor ligands such as nitrogen. Among the former, diphosphines of the type (bis(diphenylphosphino)methane), dithiocarbamates or dithiolates are very common ligands in dinuclear complexes. Interestingly, in some cases, in addition to the gold–gold interactions that appear as a consequence (or not) of the bridging ligands, interactions between adjacent binuclear units lead to extended systems. In principle, the same conditions considered here to describe luminescence in supramolecular systems built with mononuclear units are valid in these molecules. Nevertheless, as selected examples will show, the optical properties are also influenced by the intrinsic characteristics of the dinuclear building blocks.

In the case of the complex $[\{Au_2L(C\equiv CPh)_2\}_\infty]$ (L = 2,6-bis(diphenylphosphino)pyridine) [26], the structure is a one-dimensional polymer with the repeating unit $Au_2L(C\equiv CPh)_2$ held together by gold–gold interactions of 3.252(1) Å between neighboring dinuclear units in a zig-zag fashion (see Figure 6.9). Interestingly, the gold centers of the dinuclear unit do not interact.

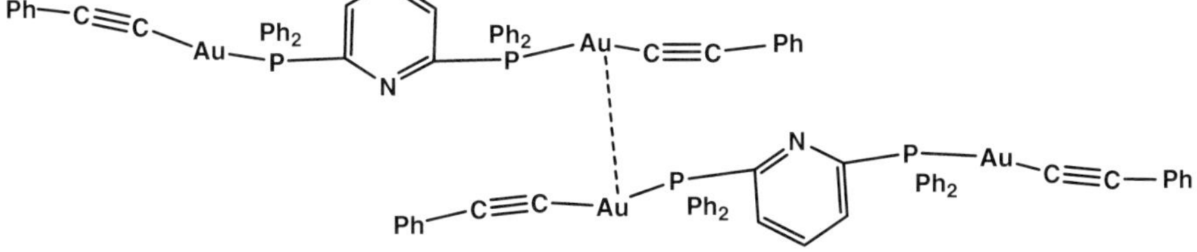

Figure 6.9 Schematic structure of $[\{Au_2L(C\equiv CPh)_2\}_\infty]$.

This product is strongly emissive in the solid state at room temperature, exhibiting an emission at 500 nm, lower in energy than the intraligand phosphorescence attributed to the acetylide ligand but higher than in the related [Au$_2$dppe(C≡CPh)$_2$] (550 nm) [27]. Considering that the emission arises from MC excited states of triplet parentage ($^3[d\delta^*p\sigma]$ or $^3[d\sigma^*p\sigma^*]$) (see Section 6.1), the lower emission of the related [Au$_2$dppe(C≡CPh)$_2$] complex could be rationalized by the shorter gold–gold distance (3.153(2) Å) in this complex.

What is more common is to find metal–metal interactions between the gold atoms belonging to the same binuclear unit and that the system polymerizes through interactions between adjacent units. In that case, two types of gold–gold interactions can be considered: intramolecular interactions; and intermolecular interactions. For example, in the case of the complexes of stoichiometry [Au$_2$(L-L)(dtc)]Cl (L-L = dmpm (bis(dimethylphosphino)methane); dppm (bis(diphenylphosphino)methane); dppe (bis(diphenylphosphino)ethane; dtc = diethyldithiocarbamate) [28, 29] only in the first case, [Au$_2$(dmpm)(dtc)]Cl, does the crystal structure display an extended linear chain through intermolecular contacts of 3.061(2)–3.135(2) Å. The intramolecular gold–gold distances are very short, 2.882(2)–2.892(2) Å. Surprisingly, the homologous complexes with the dppm or dppe ligands are a tetramer and a monomer, respectively. Although the latter is not an extended system, information can be obtained by comparing it with the former. Thus, in the tetramer the intra-gold distances are 2.877(2)–2.894(3) and the intermolecular ones between 3.086(3) and 3.222(3) Å. In the dppe monomer, the Au–Au distance is 2.943(1) Å. The three complexes emit in the solid state at 77 K at 541, 535 and 520 nm, respectively. Since the intramolecular distances for polymer, tetramer and monomer are almost the same, an assignment involving isolated dinuclear units is not likely. Instead, in similar phosphine-gold-thiolate dinuclear complexes a triplet ligand-to-metal charge transfer (^{3}LMCT) excited state from the sulfur ligand to gold, influenced by the gold–gold interactions, is proposed [30]. Nevertheless, in this case the authors add another possible assignment of the excited state, an MC transition. According to the authors, the effect that gold–gold interactions produce in the excited state when two S–Au–P units come together is minimal in the HOMO dithiocarbamate orbitals, since they are almost perpendicular and the interaction between the 6p$_z$ orbitals giving rise to the LUMO has a greater influence. The influence on the energy of the emission in an LMCT is less than in an MC transition. This, together with the expected slight blue shift in the dmpm complex due to the electronic effects originated by methyl substitution at the phosphorus atoms instead of the phenyl groups, agrees with the alternative assignment. In contrast, other authors consider that while the substituents on the thiolates have direct effects on the S → Au transition energy, the substituents on the phosphines only have indirect effects [31]. Nevertheless, whatever the assignment, in both cases the higher the number of interactions, the lower the energy of the emission, as observed experimentally. As can be seen, the debate continues.

The same authors also studied the related complexes [Au$_2$(L-L)(i-mnt)] (L-L = dmpm, dppe; i-mnt = S$_2$C$_2$(CN)$_2^{2-}$) [29], the first (dmpm complex) being a helical extended polymer and the second (dppe) a monomer. In the former, the intramolecular distances are 2.925(3)–2.893(3) Å, and the intermolecular distances 3.171

(3)–3.095(3) Å. The latter has an intra-annular gold–gold distance of 2.850(1), shorter than those of the dmpm complex. Surprisingly, taking into account the previous comments, the dmpm complex displayed an emission at 558 nm, while the dppe monomer one emitted at lower energy (576 nm). It seems that in this case the length of the interactions is more important than their number. In fact, the dmpm complex that displayed the largest distances also showed the most energetic emission.

These results seem to suggest that ligands have a strong influence on the structure and orbitals involved in electronic transitions. For instance, the dpmp ligand, probably because of its lower steric demand, influences the polymeric arrangement of the [Au$_2$(dpmp)(dtc)]Cl or [Au$_2$(dpmp)(i-mnt)] complexes, in contrast to what occurs with the homologous dppm or dppe complexes. Similarly, the exchange of dtc by i-mnt between these two complexes leads to different assignments of the excited states responsible for the emissions.

In this case, and as described in the previous section in some of the systems built with mononuclear units, these systems also display luminescence in solution with emissions depending on the concentration, although, according to the authors, the length of the species is not more than a few gold atoms. Thus, for instance, concentration studies of the absorption and emission spectra reveal concentration-dependent absorption shoulders and emissions. These appear in the range 400–440 nm and are assigned to spin-allowed metal-centered transitions from simple dinuclear units at lower concentrations and dimeric units at higher concentrations. This result is also confirmed by variable-temperature NMR spectroscopy.

In contrast, in the related ylide complexes of stoichiometry [Au$_2$(L-L){(CH$_2$)$_2$S(O)NMe$_2$}] (L-L = dpmp, dppm, dppe) [32], the same group describes oligomerization processes in solution, even when these molecules are not polymers in the solid state. In this case, the oligomerization process is detected by absorption measurements because the spectra of the dppm and dpmp complexes are concentration dependent. For instance, the dppm complex has a shoulder at 230, a band at 280 and a low energy band at about 340 nm. However, in contrast to the mononuclear molecules described in the previous section, oligomerization in solution does not produce a shift of the band to lower energies but, rather, abnormal growth of the band. Thus, the band at 340 nm grows faster with increasing concentration than expected from the Lambert–Beer law. This growing band was assigned to an MC transition. However, in the solid-state precursor, which is also luminescent, the emission is assigned to an LMCT. In the case of the complex [Au$_2$(dpmp){(CH$_2$)$_2$S(O)NMe$_2$}], a larger association constant appears, probably due to the sterically less crowded dmpm ligand that favors oligomerization.

At this point, it seems that in these binuclear units, which act as building blocks of supramolecular systems, the existence of different gold–gold interactions does not affect the excited states from which the emissions are produced. In other words, the emissions do not depend on the type of interactions in the complexes but only on their number or length. This is surprising since one would expect different excited states to arise from the interactions that appear as a consequence of intramolecular and intermolecular interaction, above all when the distances are quite different. This problem was first considered by Fackler and coworkers in complexes of the type

Table 6.1 Solid-state luminescence of dinuclear gold–sulfur complexes.

Complex	298 K emission (nm)	77 K emission (nm)	d(Au–Au)
$\{[AuS_2PPh(OC_3H_5)]_2\}_n$	443	445, 491	3.10, 3.12
$\{[AuS_2PPh_2]_2\}_n$	461	451, 495	2.96, 3.09
$\{[AuS_2PPh(OEt)]_2\}_n$	447	453, 496	3.10, 3.12
$[AuS_2P(4\text{-}C_6H_4OMe)(OEt)]_2$	467	453, 494	structure not determined
$\{[AuS_2PPh(OC_5H_9)]_2\}_n$	487	491, 530	2.93, 2.95
$\{[AuS_2P(O^iPr)_2]_2\}_n$	450	468, 524	3.05, 3.10, 2.91
$[AuS_2P(4\text{-}C_6H_4OMe)(OSiPh_3)]_2$		417	3.14
$[AuS_2P(4\text{-}C_6H_4OMe)(OMenthyl)]_2$		447	3.04
$[AuS_2PEt_2]_2$		423	3.18
$[AuS_2PMe_2]_2$		421	3.19
$[AuS_2P(4\text{-}C_6H_4OMe)_2]_2$		418	structure not determined
$[Au_2\{S_2PPh_2\}\{(CH_2)_2PMe_2\}]$		451	3.10
$[Au_2\{S_2PEt_2\}\{(CH_2)_2PMe_2\}]$		437	structure not determined
$[NBu_4]_2[Au_2\{S_2C=C(CN)_2\}]$		495, 527	2.80
$K_2[S_2C=C(CN)_2]$		553	

$[Au_2(S_2PRR')_2]$ or $[Au_2(S_2PR(OR'))_2]$ [33, 34] (see Table 6.1) to explain the optical properties that these complexes display in the solid state at different temperatures. The crystal structures of the dithiophosphate and dithiophosphinate complexes appear in two forms: as dinuclear discrete molecules or as polymeric extended systems, similar to those observed in the examples given previously. All of them had neutral eight-member metallacyclic rings in an elongated chair conformation with short transannular gold–gold interactions.

Interestingly, some of the complexes display emissive properties at room temperature and at 77 K in the solid state, but others are only luminescent at low temperature. None of them are luminescent in solution at ambient temperature. The comparison between optical properties and structures reveals a clear relationship. While the extended systems are luminescent at room temperature, the discrete molecules are not. Furthermore, the former display two bands at 77 K, whereas the latter only exhibit one at the same temperature (see Table 6.1). Previous studies of dinuclear gold(I) dithiolate systems indicate that the emissions arise from an S–Au charge transfer transition with contribution from the metal–metal bond formed in the excited state ligand-to-metal–metal charge transfer (LMMCT) [35]. Since these complexes do not show ligand-based emissions, the authors propose an LMMCT as the dominant feature in the extended chain systems, with the metal–metal bond formed in the excited state strongly perturbed by the intermolecular gold–gold interactions. In an S → Au···Au charge transfer, the interaction between the gold centers destabilizes the filled d_{z^2} orbital (z-axis along the metal–metal interaction), but also stabilizes the empty p_z orbital, which is the LUMO. The net effect is a lowering of the energy of the transition. Consequently, as the number of gold–gold interactions increases, the gap between the HOMO and the LUMO is reduced. Also, when the gold–gold interactions are larger the emissions are blue-shifted. In the case of discrete dinuclear units with only intramolecular interactions, as in the case of the complexes [AuS$_2$P(4-C$_6$H$_4$OMe)(OSiPh$_3$)]$_2$ or [AuS$_2$(4-C$_6$H$_4$OMe)(O-menthyl)]$_2$, the emissions are blue-shifted, and even disappear when the HOMO–LUMO gap is larger.

Based on lifetime measurements, the authors also propose that the high energy emission is fluorescent and the low energy one is phosphorescent, the latter being strongly influenced by the close proximity of the gold centers of different molecules because it disappears completely when there are no intermolecular interactions.

The significance of these results is that, in addition to the assignment of the excited states responsible for the emissions, they are useful for predicting the presence of weak intermolecular Au–Au interactions in other dinuclear gold–sulfur complexes. For instance, as shown in Table 6.1, for complexes [AuS$_2$P(4-C$_6$H$_4$OMe)(OEt)]$_2$, [AuS$_2$P(4-C$_6$H$_4$Me)$_2$]$_2$ and [Au$_2${S$_2$PEt$_2$}{(CH$_2$)$_2$PMe$_2$}] the crystal structures have not been determined, but analysis of the emission data allows the structures to be predicted. Thus, in the case of the former, the existence of luminescence at both room temperature and 77 K indicates that the structure will probably consist of an extended linear chain of dinuclear units with gold–gold interactions between them. Additionally, the comparison of the energy of the emission with that of known structures suggests that the gold–gold interactions will be in the range 2.9–3.0 Å. In contrast, the other two examples are only luminescent at low temperature, displaying only one band; consequently, the proposal is that the structures will exhibit discrete dinuclear complexes with intramolecular distances of around 3.2 and 3.1 Å, respectively. A further probe of the validity of these conclusions arises from the study of the luminescence of the complex [NBu$_4$][Au$_2${S$_2$C=C(CN)$_2$}], whose structure is known [36], showing a discrete diauracycle. In this case, the solid was not luminescent at room temperature, but displayed two bands at 77 K. This apparent contradiction is also attributed by the authors to the fact that the ligand itself is also luminescent at low temperature. In summary, *the emission profile alone is a useful predictor of the presence of intermolecular linear chain Au–Au interactions for the dinuclear gold(I)–sulfur compounds.*

A next step in the assignment of the excited states responsible for emissions in gold-dithiophosphate dimers is the contribution of Eisemberg et al. [37]. They analyzed the optical properties of the complexes [Au$_2${S$_2$P(OR)$_2$}$_2$] (R = Me, Et, n-Pr, n-Bu), for which the crystal structure determinations of the complexes with R = Me and R = Et revealed that these are extended linear chain polymers formed by gold interactions between dinuclear units of about 3 Å, of the same type as those described previously.

These materials were luminescent in the solid state, exhibiting white luminescence that became brilliant at 77 K. For example, the room temperature emission spectrum of the methyl derivative revealed, as expected from the previous comments, one emission band at 422 nm, but at 77 K three emission bands were detected at 415, 456 and 560 nm, with the first two being much more intense. The time-resolved measurements revealed that the first band had a lifetime of 20 ns and the second a lifetime of 2.2 μs, in accordance with fluorescence and phosphorescence, respectively.

As before, one of the main factors contributing to emission was molecular aggregation. For example, the measurement of the emission spectra of solid samples prepared from solvent evaporation of solutions with concentrations greater than 1×10^{-3} M showed all the three bands and a white luminescence, but for solids obtained from dilute solutions the higher energy emissions were greatly attenuated and an orange luminescence appeared. This result seems to suggest that the higher energy bands had a concentration dependence consistent with excited states formed as a consequence of molecular aggregation, while the lowest energy band was probably localized in a dimeric unit.

Taking into account that a simplified molecular orbital diagram of an S—Au—S unit consists of $d_z 2\sigma^*$ and $p_z\sigma$ orbitals of gold, filled and empty, respectively, and a nonbonding filled S(p) orbital between them, a transition S(p) → $p_z\sigma$, which is a charge transfer, should appear at lower energy than an MC transition between orbitals of gold ($d_z 2\sigma^*$ → $p_z\sigma$), even with the existence of molecular aggregation. Consequently, the authors assigned the bands of the emission spectrum as arising from ^{1}MC, ^{3}MC and ^{3}LMCT excited states for the 415, 456 and 560 nm bands, respectively. Very interestingly, this case is one of the few in which a singlet gold-centered transition was observed in a spectrum. In fact, there has been only one report of ^{1}MC transitions in gold [29], and this was the first case in which the two counterparts (^{1}MC and ^{3}MC) of this transition were observed in the same molecule.

Finally, the behavior in solution was striking since the emissions were different for different solvents and also exhibited thermochromism. Thus, the complexes were not luminescent in solution at room temperature, but luminesced when the solutions were cooled. In frozen glasses, only a single MC emission was observed around 440 nm (singlet) and the LMCT emission appeared at 600 nm. The relative intensity of these bands was different for different solvents and this leads to different colors. In addition, thermochromism was evident when the frozen samples were allowed to warm slowly, showing a progression of color. For instance, in the case of the methyl complex, the color changed from orange to green and disappeared at room temperature. These results were justified based on the size of the linked oligomers and, therefore, the relative intensity of the emission bands. It is worth remembering that the metal-centered transition had a concentration dependence and appeared as a consequence of molecular aggregation; consequently, a different number of units led to different excited states.

However, extensive research on these systems has also led to the identification of interesting properties that can even have future practical applications or to the discovery of exciting new phenomena.

As an example of the former, Eisemberg et al. reported the interesting properties of the complex $[Au(S_2CN(C_5H_{11})_2)]_2$ [38]. This complex was prepared by reaction of the salt $K[S_2CN(C_5H_{11})_2]$ and $[AuCl_2]^-$ in water and recrystallized from acetone, obtaining an orange microcrystalline solid which crystallizes as a solvate. The structure of the DMSO solvate showed discrete dimers stacked along an axis and forming an infinite chain of gold atoms. The intra- and inter-molecular contacts were short

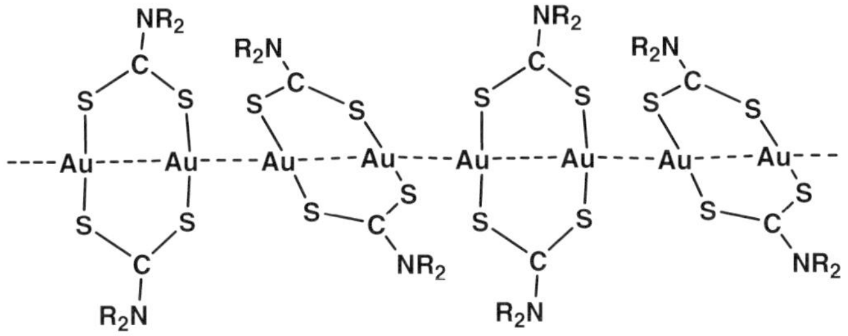

Figure 6.10 Schematic structure of the [{Au(S$_2$CN(C$_5$H$_{11}$)$_2$)}$_2$]$_n$ chain.

(2.7690(7) and 2.9617 (7) Å, respectively) and the dimers were rotated by approximately 90°, resulting in a staggered arrangement (see Figure 6.10).

This complex was strongly luminescent at room temperature, showing a band centered at 631 nm that shifted to 604 nm at 77 K. Interestingly, when this solid was dried in vacuo, it became colorless and non-emissive, but exposure to vapors of aprotic solvents such as acetonitrile, chloroform or dichloromethane, prompted it to revert to its original properties. This effect did not occur in protic solvents such as alcohols. The structure of the non-emissive solid consisted of a discrete dinuclear unit with an intramolecular Au–Au distance of 2.7653(3) Å, that did not interact with adjacent units (shortest intermolecular distance of 8.135 Å). The authors suggest that the formation of linear chains of gold centers promoted by the solvents was the main factor responsible for the orange color and the luminescence. Significantly, there were no interactions between solvents and gold centers in the chains.

Aggregation as the cause of luminescence was also probed in fluid solutions, since a dilute colorless solution was not emissive, but a concentrated (2×10^{-2} M) solution was weakly emissive; a result that is consistent with the aggregation of these units in solution. Undoubtedly, this complex has potential for practical applications as a luminescent sensor for the detection of volatile organic compounds (VOCs).

The same laboratory describes a group of complexes with a unique property among metal complexes, referred to as *luminescence tribochromism* [39], which is a substantial change in the emission of the solid upon application of pressure. This phenomenon contrasts with the more common triboluminescence, which refers to the transient emission seen upon sample grinding or crushing. In this case, the effect was observed in a set of complexes of formula [Au$_2$(μ-TU)(μ-dppm)]Y and [Au$_2$(μ-MeTU)(μ-dppm)]Y (TU = 2-thiouracyl; MeTU = 6-methyl-2-thiouracyl; Y = CF$_3$COO$^-$, NO$_3^-$, ClO$_4^-$, Au(CN)$_2^-$).

The crystal structures of the trifloroacetate derivatives have been determined and both consist of a dinuclear gold complex with a thiouracilate and a dppm ligand bridging both metal centers. For example, in the case of the thiouracyl complex the gold–gold distance is 2.8797(4) Å. These dinuclear units bind through intermolecular gold–gold interactions giving rise to a helical arrangement with gold–gold interactions of 3.3321(5) Å (see Figure 6.11)

The solid samples of these complexes are non- or weakly emissive at room temperature, but when the samples are gently crushed a dramatic change occurs

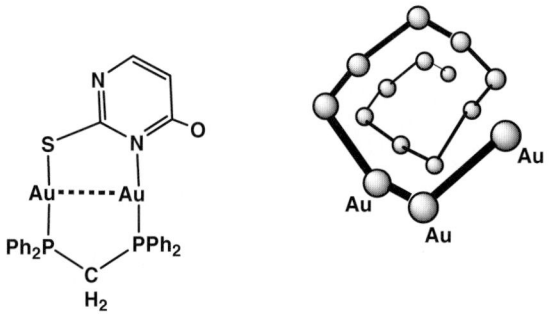

Figure 6.11 Schematic structure of [Au$_2$(μ-MeTU)(μ-dppm)]+ and helical arrangement of the gold atoms.

to give samples exhibiting bright blue or cyan luminescence. Powder X-ray diffraction experiments on both samples, crystalline and powder, indicated that there was no phase change when the sample was crushed and, therefore, the change in the optical properties cannot be attributed to this effect. In addition, derivatives with different anions exhibited similar behavior and heating the samples or sonication led to the same emissive form. Interestingly, in the process of conversion from non-emissive to emissive a release of acid was detected. Therefore, the emissive complex can be prepared chemically by stirring a solution of the initial complex with a base. The resulting complex is a neutral dinuclear compound in which the thioauracyl ligand is doubly deprotonated. The structure also shows dimers in a head-to-head arrangement held together by a stronger aurophilic interaction of 2.9235(4) Å. Starting from this complex, the process can be reversed by adding trifluoroacetic acid. If the deprotonated complex is exposed to vapors of this acid, conversion is also possible but slower. The addition of triethylamine to the former also leads to the luminescent complex.

In contrast, dichloromethane solutions of the nondeprotonated complex are blue emitters (489 nm) but the deprotonated form is non-emissive. The addition of triethylamine to the former makes it non-emissive and, similarly, addition to the latter of trifluroacetic acid forms the emissive complex.

Taking into account these facts, what is proposed is first that the probable site of protonation/deprotonation is the uncoordinated pyrimidine nitrogen atom, which is in accordance with the H-bonding found in the structures. Secondly, the arrangement of the gold centers has a dramatic effect on the optical properties of the complexes. Thus, the non-emissive complex has a helical structure and the intermolecular gold–gold interactions are weak and kinked. In the emissive form, the intermolecular interaction is strong and the arrangement of the gold centers is more linear. Consequently, in addition to the strength, *the geometry of the gold–gold interactions is very important for the luminescence*. Finally, the application of pressure by grinding appears to induce cleavage of the weakest bonds of the Au$_n$ helix, the release of volatile acid and rearrangement into dimers.

The next contribution, from Puddephatt's laboratory, can be considered as an example of the fusion of the two types of building blocks analyzed separately here, i.e.

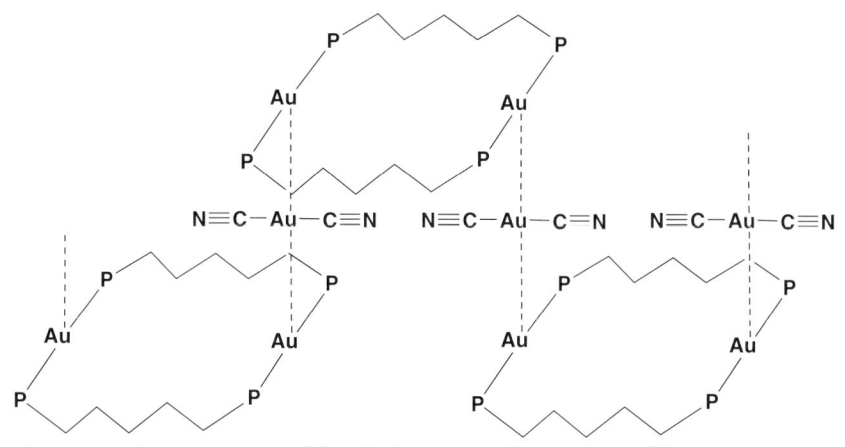

Figure 6.12 Schematic structure of $\{[Au_2\{PPh_2(CH_2)_5PPh_2\}_2][Au(CN)_2]_2\}_n^{2n+}$.

mononuclear and dinuclear units [40]. In this case, the complex $[Au(CN)_2]^-$ acts as a linker of gold rings.

As shown in Figure 6.12, in the case of the cationic gold ring $[Au_2\{PPh_2\text{-}(CH_2)_5PPh_2\}_2]^{2+}$, the reaction with the dicyanoaurate anion leads to a pleated chain polymeric structure.

Similar reactions with other gold rings with a smaller number of CH_2 groups between the donor phosphorus atoms did not lead to polymeric materials and, for example, when the number of CH_2 groups is three, a pentanuclear complex formed by two gold rings linked by $Au(CN)_2$ unit appeared. This result was explained by the fact that the polymeric structures need longer transannular gold–gold distances since the $Au(CN)_2$ linkers are approximately collinear (see Figure 6.12).

In the case of the polymeric complex, an emission at 411 nm appeared. This energy was similar to that in non-extended structures and, consequently, the extended structure did not affect the photophysical property of the complex. Although the authors did not assign the origin of the emissions, the similarity in energy for all the complexes (polymeric or not) and the high energy values obtained seem to suggest π–π^* transitions as the origin of these emissions.

6.2.3
Networks from Trinuclear Units

Cyclic trinuclear gold(I) complexes provide a novel and productive strategy for achieving supramolecular structures. While molecules of this type have been known for more than twenty years, some of their remarkable properties have only been recognized recently. Some can form liquid crystals at room temperature [41], while others lead to luminescent materials with surprising properties. We will now summarize some selected examples to illustrate the behavior of these trinuclear systems.

In 1997 Balch *et al.* published an outstanding study describing the synthesis, structural characterization and unexpected optical properties of the columnar

Figure 6.13 Schematic representation of an individual molecule of [Au$_3$(CH$_3$N=COCH$_3$)$_3$].

complex [Au$_3$(CH$_3$N=COCH$_3$)$_3$]$_n$ [42, 43]. The structure of this complex is a trinuclear molecule formed from three gold centers and three ligands bonded to them through a carbon atom and a nitrogen center forming a triangle (see Figure 6.13). The intramolecular gold–gold distance is 3.308(2) Å.

The trinuclear molecules aggregate along a crystallographic axis to form columnar stacks. The intermolecular gold–gold distance was 3.346(1) Å. In addition to these ordered stacks, there are other parallel columns with two sets of positions for the gold triangles. These appear as a result of the two positions for the methoxy-methyl groups giving rise to two orientations (see Figure 6.14).

The solid is luminescent, showing dual emission, at 446 nm, structured and short-lived (1 ms), and at 552 nm, broad and long-lived (1.4, 4.4 and 31 s). Solutions of the complex in chloroform showed an emission band at 422 nm. Very interestingly, the solid displayed very unusual behavior, described by the authors as *solvent-stimulated luminescence*, consisting of light emission that was triggered by contact with a liquid after the solid had been previously irradiated with a conventional hand-held UV lamp. This phenomenon is related to the already known *lyoluminescence*, consisting of light emission from dissolution of solid samples irradiated with ionizing radiation. Nevertheless, the phenomenon described here was different because it was the result of sample excitation with near-visible light.

Thus, the addition of a few drops of acetone to a previously irradiated solid led to an intense yellow emission, visible to the human eye for a few seconds. This emission was the same as the lower energy emission detected in the solid state. Several other liquids such as chloroform, dichloromethane, toluene, methanol, hexane or even water produced the same phenomenon and the intensity of the emission depended

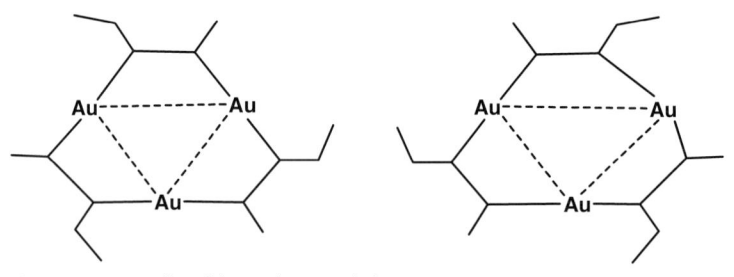

Figure 6.14 Disorder of the methoxy-methyl groups giving rise to two orientations in the trinuclear [Au$_3$(CH$_3$N=COCH$_3$)$_3$].

on the solubility of the solid in the solvent. Moreover, the solvoluminescence was not dependent on the atmosphere in which the reaction occurred. No chemical transformation was detected in the emission process, since the recovered solid produced the same emission after several cycles of irradiation and liquid contact.

The authors suggested that the gold–gold interactions that gave rise to the supramolecular aggregate seemed to be important for the energy storage. Therefore, the energy had to be stored somehow in the columnar structures and released with the addition of solvent. The authors proposed that the presence of a considerable degree of disorder within the solid could have been responsible for the creation of sites where electrons and holes could be trapped. Energy storage was produced by charge or electron separation, which was facilitated by conduction of electrons along the columnar structures. The emission resulted from the recombination that occurred when the addition of solvents altered those sites in the stacks of molecules and the migration of charge through the stacks facilitated energy transfer from the bulk to the surface.

At this point, a question arises: is the structural arrangement so important in the storage of energy?

The same laboratory attempted to answer this question by performing an in-depth study. Thus, if we analyze the optical properties of the already commented solvoluminescent material $[Au_3(CH_3N=COCH_3)_3]$ and the related $[Au_3(PhCH_2N=COCH_3)_3]$ in relation to their structures [44], further knowledge can be obtained.

Unlike the results obtained with solvoluminescent $[Au_3(CH_3N=COCH_3)_3]$, the benzyl complex packed in a stair-step fashion (see Figure 6.15). The trinuclear units that formed the stack had an average intramolecular gold–gold separation of 3.316 Å, while in $[Au_3(CH_3N=COCH_3)_3]$ it was 3.308 Å. The intermolecular distances ranged from 3.662 to 4.100 Å. Therefore, this complex could be considered as an association of single molecules since the distances were within the limit to be considered interactions because most of them were larger than the sum of the van der Waals radii of gold. It seems likely that the substituents on the periphery of the individual molecules affected the extent of the aurophilic interactions between these trinuclear complexes. Nevertheless, in spite of the larger distances, the molecules maintained an ordered arrangement that was repeated along the crystal, giving rise to a supramolecular architecture.

As regards the optical properties of this complex, the absorption spectrum in chloroform solution revealed absorption maxima at 290 ($\varepsilon = 2.5 \times 10^3$) and 204 nm ($\varepsilon = 7.7 \times 10^3$). These values were similar to those found in $[Au_3(CH_3N=COCH_3)_3]$ and were likely due to metal-to-ligand charge transfer. It was luminescent at room temperature in the solid state showing an intense band at 404 nm and shoulders at 525 and 793 nm, the third with a very low intensity. The complexity of this pattern was

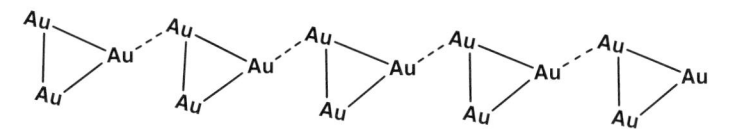

Figure 6.15 View of molecular packing in $[Au_3(PhCH_2N=COCH_3)_3]$.

not unexpected and was likely due to the complexity of the molecular structure in the solid, since there were eight different trinuclear molecules in the asymmetric unit. In solution, the complex exhibited two closely spaced bands at 482 and 508 nm that the authors correlated with the shoulders in the solid-state emission spectrum. Interestingly, this complex did not show any liquid-triggered solvoluminescence.

In other studies, changes in the structural arrangement of the trinuclear units was not only due to the change in substituents. In fact, another unusual structural finding, that also led to a difference in the optical properties of the complex [$Au_3(CH_3N=COCH_3)_3$], appeared as a consequence of the discovery of polymorphic structures in the crystallization process [45].

Thus, pure samples of the complex formed crystals with varying luminescence. Most of the crystals exhibited the described yellow luminescence, but others displayed a bluish-white luminescence and a few of the crystals showed a pink luminescence. The two new forms of trinuclear complex can be obtained by evaporation of dichloromethane solutions of the complex. They consisted of one triclinic form and one monoclinic form and both crystallized as colorless blocks, in contrast to the results with the original solvoluminescent hexagonal form that crystallized as colorless needles. These polymorphs differed in the packing of the nearly planar molecules and in the nature of the aurophilic interactions between the trinuclear units (see Figure 6.16).

Thus, the triclinic polymorph formed a chain of pairs of trinuclear units. The trinuclear complexes associated through interactions at 3.2201(9) Å and the interactions between pairs of molecules are longer, 3.583(12).

The monoclinic form contained 13.5 gold atoms in the asymmetric unit; nine of them formed a prismatic stack (Au···Au interactions of 3.28(3) Å), and each of these stacks bound to another stack with a gold–gold distance of 3.65(3) Å to form infinite chains. The remaining 4.5 gold centers in the asymmetric units formed a second nine-prism stack by a two-fold operation leading to an intramolecular distance of 3.28(3) Å and an intermolecular distance of 3.48(3) Å.

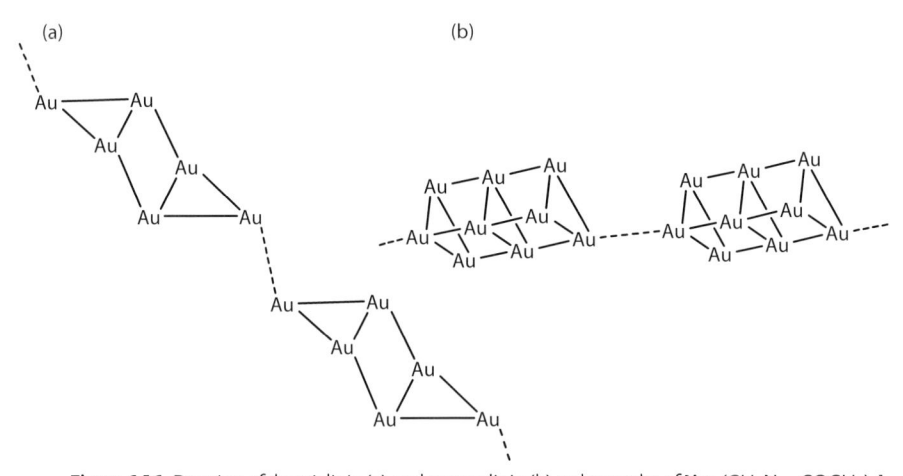

Figure 6.16 Drawing of the triclinic (a) and monoclinic (b) polymorphs of [$Au_3(CH_3N=COCH_3)_3$].

6.2 Luminescent Supramolecular Gold Entities | 371

Each of the three polymorphs of [Au$_3$(CH$_3$N=COCH$_3$)$_3$] exhibited a distinctive emission spectrum and only the hexagonal polymorph displayed solvoluminescence. Thus, while the hexagonal polymorph showed two emission bands at 450 and 520 nm, the triclinic one displayed only one emission at 444 nm by excitation at 390 nm and there was no emission counterpart for the low energy long-lived emission observed in the hexagonal polymorph, which was responsible for the solvoluminescent emission. The same situation was found for the monoclinic form, which exhibited a single emission at 431 nm (exc. 349 nm).

Polymorphism also appeared in the related complex [Au$_3$(n-PentN=COCH$_3$)$_3$] [45]. In this case, by evaporation of ether solutions of the complex, two crystalline forms were obtained: an orthorhombic form and a triclinic form. The orthorhombic form contained one trinuclear unit in the asymmetric unit with intramolecular gold–gold distances in the range 3.260(2) to 3.332(2) Å. The trinuclear units interacted with others in a stair-step arrangement with distances of 3.618(2) Å, shorter than in the similar structure found in the previously commented [Au$_3$(PhCH$_2$N=COCH$_3$)$_3$]. The triclinic polymorph exhibited four independent molecules in the asymmetric unit. Each of the four molecules interacts with two adjacent molecules through gold–gold interactions in the range 3.3458(6) to 3.8125(6) Å (see Figure 6.17).

In contrast to the previous case, only the triclinic polymorph was luminescent at room temperature, displaying a single emission at 654 nm (exc. 326 nm). The authors assigned this luminescence to the existence of shorter gold–gold contacts between adjacent trinuclear molecules. This result is also striking since it displayed a red shift

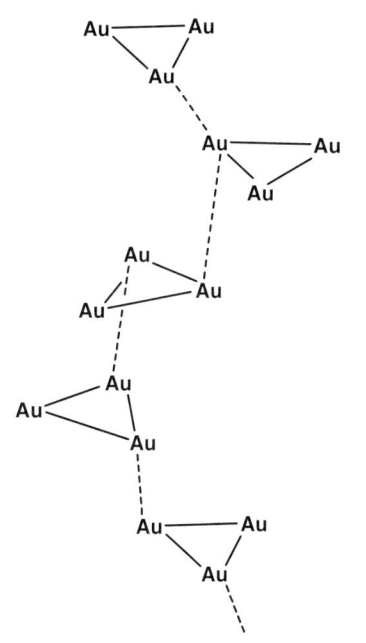

Figure 6.17 Drawing of the triclinic polymorph of [Au$_3$(n-PentN=COCH$_3$)$_3$].

compared to any of the polymorphs of $[Au_3(CH_3N=COCH_3)_3]$. The large Stokes shift is indicative of a large distortion in the excited state.

Another interesting characteristic of the trinuclear $[Au_3(CH_3N=COR)_3]$ (R = Me, Et) complexes is that they can act as electron donors with organic acceptor molecules such as nitro-9-fluorenones [46]. Thus, they form adducts with 2,4,7-trinitro-9-fluorenone (deep yellow and red; R = Me, Et, respectively), 2,4,5,7-tetranitro-9-fluorenone (red; R = Me), 2,7-dinitro-9-fluorenone (red; R = Et). The solid state structures of the complexes formed with $[Au_3(CH_3N=COMe)_3]$ and 2,4,7-trinitro-9-fluorenone or 2,4,5,7-tetranitro-9-fluorenone consist of planar gold(I) trimers interleaved with the nitro-9-fluorenones to form columns in the crystal. The distances between the faces of both portions indicate that interactions occur between the gold atoms, rich in electronic density, and the nitroaromatic portion of the electron acceptor. The difference between them stems from the greater complexity of the 2,4,5,7-tetranitro-9-fluorenone adduct because of the presence of four independent molecules in the stack and the distortion from planarity of the fluorenones.

In the case of the other two complexes, two trinuclear molecules interact through the gold atoms forming hexanuclear systems. These prismatic $[Au_3(MeN=COEt)_3]_2$ units have intermolecular distances of 3.2328(10)–3.3545(10) Å, in the adduct with 2,7-dinitro-9-fluorenone, and 3.2200(11) and 3.3175(11) Å, in the adduct with the 2,4,7-trinitro-9-fluorenone. In both cases, the hexanuclear prismatic units are interleaved with the organic acceptors forming columns in the solid state.

In all these examples, the color changes observed with respect to the original trinuclear colorless species and the pale yellow fluorenones to the deep yellow or red of the products of the reactions are indicative of the formation of charge transfer adducts. These charge transfer interactions were also observed in the UV–vis spectra, since the spectra of the mixtures showed enhanced absorption at longer wavelengths than in the starting products. For instance, in the case of the complex $[Au_3(CH_3N COMe)_3]$ with 2,4,5,7-tetranitro-9-fluorenone two new shoulders at 580 and 470 nm were visible in the spectrum. For the mixture of $[Au_3(CH_3N=COEt)_3]$ with 2,4,7-trinitro-9-fluorenone, the enhanced absorption appeared in the range 400–600 nm with a shoulder at 460 nm.

In all of these cases, the products of the reactions did not exhibit any luminescence when irradiated with UV light. This result was not surprising since the rupture of the columnar structures of the starting materials with the interleaved fluorenones probably led to a loss of luminescence. In fact, *luminescence in these systems seems to be associated with the supramolecular organization of the solid.*

Therefore, taking into account the previous comments on these and related systems, we may conclude that the absence of a columnar structure in the solid, as well as the absence of different orientations of the substituents that lead to disordered stacks, are essential in energy storage and the solvoluminescent behavior observed only for the hexagonal form of $[Au_3(CH_3N=COCH_3)_3]$.

The use of other potential bidentate ligands also permits the synthesis of cyclic trinuclear gold complexes that form, by stacking, supramolecular assemblies. This is the case for the trinuclear complex $[Au_3(NC_5H_4)_3]$ [47] that crystallizes in two

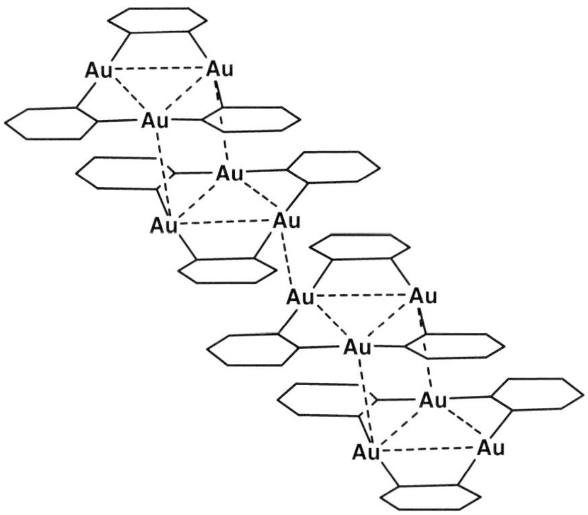

Figure 6.18 Schematic structure of $[Au_3(NC_5H_4)_3]_n$.

different forms, an extended chain of molecules formed by interactions between two pairs of gold atoms (Au–Au distance of 3.146(3) Å) and individual gold–gold contacts (3.077(2) Å) (see Figure 6.18), and discrete dimers linked by two gold–gold interactions (3.105(2) Å).

The complex is luminescent in the solid state and in solution in pyridine. The absorption spectrum shows a maximum at 340 nm and the emission spectrum displays a band at 425 nm. The crystals are also luminescent, but the excitation and emission spectra appear at different wavelengths. Thus, the complex emits at 490 nm and the excitation spectrum is complicated, with bands in a wide range from 300 to 450 nm. This result suggests that different species are responsible for luminescence in the solid state and in solution, with the isolated trinuclear complex being the emitting species in solution, while the luminescent properties of the crystal are the result of the extended supramolecular aggregation in the solid. As before, the complex did not exhibit solvoluminescence.

In the case of the trinuclear $[\mu-N^1,C^2\text{-bzimAu}]_3$ (bzim = benzylimidazolate), in addition to the extended structures that form with other metals (see Section 6.3), it also forms supramolecular networks, acting as an electron donor with small organic acids [48]. For example, it reacts with TCNQ (tetracyanoquinodimethane) giving rise to a columnar structure in which each TCNQ molecule is sandwiched between two units of the trinuclear complex in a face-to-face manner. Thus, the repetition of this pattern leads to a stacking of the type $(Au_3)(Au_3)(\mu\text{-TCNQ})(Au_3)$ $(Au_3)(\mu\text{-TCNQ})\ldots$. The complex contains two very short intermolecular gold–gold distances of 3.152 Å, shorter than the intramolecular distances, which are in the range 3.471–3.534 Å. The reason for this shortening of the intermolecular distances is ascribed to a partial oxidation of the gold centers in the trinuclear molecules because of the charge transfer from these atoms to the TCNQ molecules.

In fact partial oxidation of the Au(I) atoms leads to a shortening of the Au–Au distances, having as the limiting case the oxidation to Au(II), where a formal metal–metal bond is formed.

The adduct is not luminescent, a result that is interpreted as due to quenching of the luminescence by the rupture of the columnar stack.

The same effect was observed by these authors for the adduct formed by reaction of the carbeniate [AuC(OEt) = $NC_6H_4CH_3$]$_3$ and C_6F_6. In this case, the crystal structure showed a columnar stack consisting of alternating C_6F_6 and gold-trimer molecules. While the original gold material displayed a blue luminescence, the column did not exhibit any emission. This result was also obtained when the gold complex was exposed to C_6F_6 vapors at room temperature and atmospheric pressure.

Lastly, a study of the excited states in trinuclear pyrazolate coinage complexes has been reported by Omary et al. [49]. They studied the photophysical properties of the cyclic complexes of formula [M(3,5-$(CF_3)_2$Pz)]$_3$ (M = Cu, Ag, Au; Pz = pyrazolato). All of them packed as infinite chains of trimers but, with the exception of silver, the intertrimer distances were long (3.885 and 3.956 Å for the gold complex). These long distances did not explain the striking luminescent thermochromism observed for the three complexes. Specifically, in the case of the gold complex by excitation at short wavelengths ($\leq$290 nm), the emissions varied from the UV region (365 nm) to visible green (535 nm) or even orange (660 nm) upon heating from 4 K to room temperature. These emissions were phosphorescent and unstructured but solutions of the complex had high molar absorption bands and a weak intensity shoulder at longer wavelengths. The former were assigned as charge transfer transitions. DFT calculations suggested a transition from a filled orbital with strong ligand characteristics to a vacant orbital with strong metal–metal bonding characteristics. Therefore, these absorptions were assigned to an LMMCT. The corresponding weak band at longer wavelengths corresponded to the crystal' phosphorescence excitation spectrum.

As regards phosphorescence, its first characteristic was a large Stokes shift, indicative of a large excited-state distortion. Secondly, the complex showed multiple emissions as a function of temperature, but excitation profiles independent of temperature. This circumstance ruled out the possibility of a structural phase transition being responsible for these properties, since a phase transition is a ground-state behavior that should lead to changes in the absorption or excitation energies. Thus, regarding the multiple emissions as a function of temperature, one would expect internal contractions upon cooling and, thus, as the major consequence of such thermal compressions, a gradual red shift in emission. However, the opposite trend was observed. The authors suggested that these emissions were likely to arise from different Au–Au bound excited states instead of changes in the ground state or in a given excited state. In fact, the multiple emissions observed at different temperatures were assigned to different phosphorescent excimeric states that exhibited enhanced gold–gold bonding relative to the ground state. Therefore, and interestingly, although the molecules did not show interactions in the ground state, these interactions appeared with UV light, leading even to metal–metal bonding in the excited states.

6.2.4
Networks from Higher Nuclearity Systems

As commented in Section 6.1, among the consequences that may derive from the special situation of gold in a hypothetical relativistic effects ranking, perhaps the most well known is the tendency of gold to interact with other centers and, even in mononuclear complexes, to aggregate into oligomers, as shown in the different examples presented in the previous sections. However, in selected examples where the ligands had several donor centers, or when these could act as donors to more than one gold center, the molecules displayed a larger number of gold atoms and, usually, the aggregation occurred inside the molecules. Nevertheless, what is more unusual is that these large molecules interact with one another, giving rise to supramolecular assemblies. For instance, Fackler et al. and Fenske et al. recently reported the structural characterization and luminescence of molecules with arsenic or phosphorus donor ligands containing between nine and nineteen gold centers [50–52]. However, the molecules were organized in different polyhedral arrangements of gold atoms with no connections between the different molecules.

An example of a luminescent gold(I) supermolecule that interacts with others around it through gold–gold interactions leading to a two-dimensional structure is that reported by Che et al. [53]. It has the empirical formula [(LAu)(AuPPhMe$_2$)$_2$]$_2$ (L = trithiocyanurate) and the structure consists of a hexanuclear gold(I) molecule, in which four gold(I) centers are arranged in the form of a parallelogram with gold–gold distances of 2.964(2) and 2.987(2) Å, while the other two lie outside the molecule core and connect different molecules through interactions of 3.130(2) Å (see Figure 6.19), resulting in a two-dimensional structure in which the hexanuclear molecule is the repeating unit.

The absorption spectrum of this complex displayed an intense absorption at 320 nm ($\varepsilon = 44870$ M^{-1} cm^{-1}) assigned to a 5d(σ^*) → 6p(σ) transition influenced by gold–gold interactions. This complex showed a broad emission at 520 nm by excitation in a range between 300 and 400 nm, with a lifetime of 11.6 µs in the solid state. This emission was assigned to the S → Au interaction.

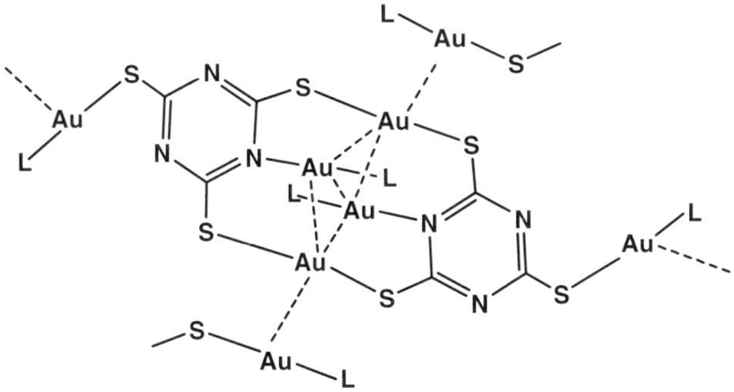

Figure 6.19 Schematic structure of {[(C$_3$S$_3$N$_3$Au)(AuL)$_2$]$_2$}$_n$ (L = PPhMe$_2$).

As shown in this example, although the structure displayed a complexity not found in the previous examples, the assignment of the origin of the luminescence did not differ much from simpler sulfur–gold complexes in less aggregated systems.

In contrast, with different donor ligands, complexity in the structure can result in a more complex assignment, and multistate photoluminescence can even appear in the solid state. This happens with the complexes of general formula [(AuX)$_4$L] (L = 1,4,8,11-tetraazacyclotetradecane; X = Cl, Br, I), reported by the same group [54]. The chlorine-derivative has a two-dimensional structure in which the four Ph$_2$P–Au–Cl arms of each (AuCl)$_4$L unit maintain intermolecular contact with other molecules through gold–gold interactions of 3.104(1) Å.

The three complexes are luminescent in the solid state at 77 K. The chlorine derivative shows an emission at 470 nm and a shoulder at 600 nm; the bromine and the iodine derivatives exhibit dual emissions at 530 and 600 nm and at 530 and 700 nm, respectively. The authors attributed the high energy emissions to a MLCT from the gold atoms to the phosphorus or, alternatively, to intraligand transitions. In the case of the low energy emissions, these were attributed to MC transitions (5d/6s → 6p) mixed with LMCT (X → Au) modified by gold–gold interactions. The fact that the iodine derivative appears at lower energy than the bromine derivative was attributed to the foreseeable stronger interactions between the gold centers in the former if both complexes had similar crystal structures. This effect was previously predicted theoretically by Pyykkö et al. [55].

6.3
Luminescent Supramolecular Gold–Heterometal Entities

Another class of supramolecular gold-containing complexes is that in which gold interacts with other metal centers. In these complexes, whose number is increasing every day, the metallophilicity promoted by gold seems to be a key factor. In fact, the well-known aurophilicity that explains aggregation in gold complexes does not seem to be an isolated phenomenon; instead, the presence of gold in the complexes seems to increase the metallophilicity of the other metals and, consequently, the presence of metal–metal interactions. Similar to the complexes described in the previous section, the supramolecular entities include linear chains, two-dimensional sheets or three-dimensional networks. Of these, perhaps the most common structural arrangement is that in which the metals form extended linear systems and these can be built using several building blocks, mononuclear units, dinuclear or more. Following the same approach used previously, this section will focus on structures built by gold–heterometal interactions but in which the metal centers have a non-zero formal oxidation state. In this way, it is possible to determine the influence of the interaction between gold and other metals in terms of the optical properties. I will not consider formal clusters or molecules built by other secondary interactions since these are described elsewhere in this book.

As regards the heterometals that, together with gold, form supramolecular entities, there are not very many. It would be tempting to think that the group congeners silver and copper are electronically the most favorable candidates for such gold–heterometal interactions but, as we will see, although luminescent (and non-luminescent) supramolecular gold–silver entities are known, they are not a large family and the gold–copper examples are much less numerous. Of the rest, some examples of interactions with other late transition and post-transition metals will be described.

This topic will be divided into two parts according to the heterometal interacting with gold: the first will focus on the luminescent supramolecular structures that contain gold and other coinage elements; the second will examine those containing gold and other heterometals.

6.3.1
Supramolecular Gold–Group 11 Metal Complexes

A common strategy in the synthesis of heteronuclear materials is the use of polydentate donor ligands. If they have different donor centers, different metal centers can be coordinated selectively and in consecutive steps. Among the ligands, perhaps the most used in the coordination chemistry that gives rise to interactions between the different donor atoms are bidentate ligands. Interactions between both metals are normally intramolecular but sometimes, albeit not very often, the bidentate units bind to one another, leading to extended structures through metallophilic interactions.

Thus, for example, Fackler et al. reported the synthesis of a polymeric Au–Ag complex using the gold precursor complex [PPN][Au(MTP)$_2$] (PPN = bis(triphenylphosphoranylidene)ammonium; MTP = diphenylmethylenethiophosphinate), by reaction with AgNO$_3$ [56]. The structure of the derivative, of formula [AuAg(MTP)$_2$]$_n$, in the solid state consists of a one-dimensional chain in which the MTP ligands bridge the gold and silver centers and the units are bound together through gold–silver interactions between different units. In each unit, both carbon atoms are bonded to gold, while both sulfur atoms are bonded to silver, with the S–Ag–S angle exhibiting a deviation from linearity since the Ag atom is directed toward the Au atom. This indicates that intramolecular interaction is not forced by the ligand architecture. Thus the intramolecular Au–Ag distance is short (2.9124(13) Å), although it displays an appreciably longer intermolecular distance of 3.635 Å (see Figure 6.20).

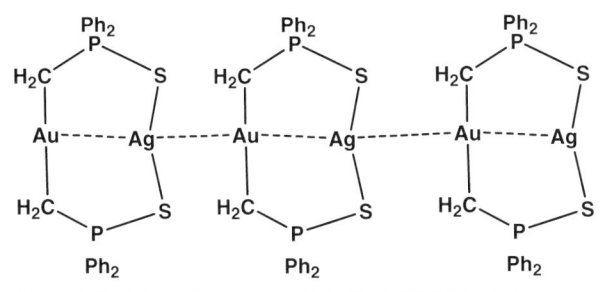

Figure 6.20 Schematic structure of the [AuAg(MTP)$_2$]$_n$ chain.

This complex had an absorption spectrum that was dependent on concentration. Thus, it exhibited a band with a large extinction coefficient ($\sim 10^4$ M^{-1} cm^{-1}) at 275 nm that was consequently assigned to a permitted transition. This also appeared in the homologous homonuclear gold and silver species and was therefore assigned to an intraligand π–π^* transition in the MTP ligand. The same happened to the extended chains built with mononuclear entities; the edge of the band shifted to the red when the concentration increased, suggesting molecular aggregation. This result was consistent with the extended-chain structure observed for the solid. Nevertheless, the maximum of the excitation spectrum appeared at lower energy ($\sim$320 nm) than the band edge of the absorption spectrum in solution, which is indicative of the presence of discrete oligomers formed by a few complexes in concentrated solutions. The emission of the complex appeared at 424 nm and the DFT calculations suggested an excited state which resulted from *argento–aurophilic* bonding.

Trinuclear donor ligands can also lead to supramolecular extended systems. For instance, Catalano et al. reported the reaction between the gold-imadazolium precursor with silver salts. For example, the simple reaction of the gold monomer [Au(py$_2$im)$_2$][BF$_4$] (py$_2$im = 1,3-bis(2-pyridinyl)imidazolium) with Ag[BF$_4$] in acetonitrile led to the helical polymer {[AuAg(py$_2$im)$_2$(CH$_3$CN)][BF$_4$]}$_n$. The chiral nature of this material was confirmed by the authors with the resolution of the crystal structures of both enantiomers. The extended structures displayed very short alternating Au–Ag separations of 2.8359(4)–2.9042(4) Å [57] (Figure 6.21).

This complex was luminescent in solution and in the solid state. In acetonitrile solution, it displayed an emission at 345 nm, by excitation at 284 nm, which is similar to that of the ligand precursor and is attributed to a fluorescent process. In the solid state, it emitted at 515 nm and the ligand precursor, [H(py)$_2$im]BF$_4$, at 514 nm. Consequently, the remarkable similarity between the emissions of the ligand and the metal salt in both the solid state and solution suggested a ligand-centered emission.

Starting from the complex [Au(Mepyim)$_2$]PF$_6$, the reactions with equimolecular amounts of silver salts such as AgBF$_4$ or AgNO$_3$ in acetonitrile, benzonitrile or benzylnitrile gave rise to the synthesis of the polymeric species {[AuAg(Mepyim)$_2$L]X}$_n$ (X = PF$_6$, L = CH$_3$CN, C$_6$H$_5$CN, C$_6$H$_5$CH$_2$CN; X = NO$_3$, L = NO$_3$ (CH$_3$CN solvate)) [58]. All exhibited alternate sequences of gold and silver atoms that had Au–Ag interactions in a range between 2.8125(2) and 2.9428(2) Å. The silver atoms, in addition to two interactions with two gold atoms, were bonded to two pyridine

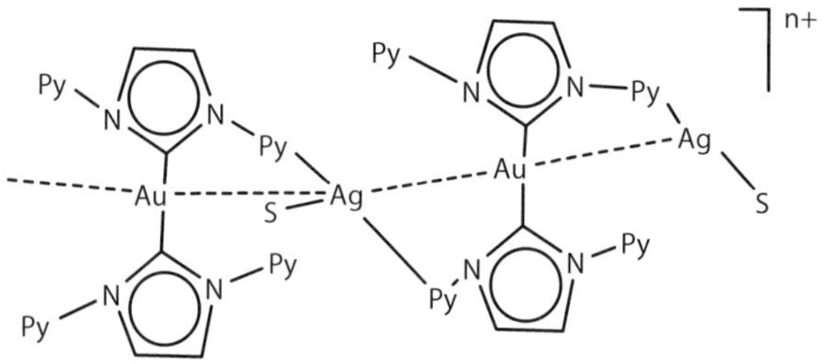

Figure 6.21 Schematic representation of the helical polymer {[AuAg(py$_2$im)$_2$(CH$_3$CN)][BF$_4$]}$_n$.

groups of different Mepyim ligands and to one L (nitrile or NO_3^-) showing a distorted trigonal planar environment.

The absorption spectra of the complexes were similar to those of the precursor [HMepyim]PF_6 that exhibited bands between 260 and 270 nm, attributed to π–π^* transitions. The solid-state emission spectra of the nitrile-containing polymers displayed intense bands at 480 (acetonitrile), 474 (benzonitrile) and 522 (benzylnitrile) nm. The nitrate-containing polymer showed a band at 469 nm. All exhibited emissions independent of the excitation wavelengths. The effect of cooling the samples prompted the bands to sharpen and shift to the blue, 450, 480, 453 and 466 nm, respectively. In solution, all showed the same emissions, which were also the same as those of the precursor gold complex. Taking into account that the identified order of the increasing energy of the emissions for the complexes was benzylnitrile < acetonitrile < benzonitrile and that the emission energies did not correlate with the Au–Ag separation or the donor ability of the nitrile ligand, the authors proposed that the excited states responsible for the emissions were ligand-based although strongly influenced by metal chain geometry, ancillary ligands and steric properties.

As explained at the beginning of the previous section, the $Au(CN)_2^-$ anion can auto-aggregate itself and give rise to supramolecular networks. However, with this anion, in addition to the possibility of using both donor atoms of the cyanide group, the gold center can itself act as a donor, since the negative charge of each cyanide ligand increases the electronic density on this metal, making it a potential donor center. To a certain degree it is a potential tridentate complex. Nevertheless, in the case described here, the complex with the potassium cation consisted of $Au(CN)_2^-$ linear ions, alternating with layers of K^+ ions; the gold centers and the potassium cations were situated at distances larger than the sum of their van der Waals radii. Nevertheless, the presence of another heavier element in the cyanide unit or acting as counter-cation should facilitate the presence of interactions between the metals. This was the case for the silver–gold mixed metal systems of stoichiometry $M[Ag_xAu_{1-x}(CN)_2]_3 \cdot 3H_2O$ (M = La [59], Eu [60]) obtained and studied by the same group, and which exhibited tunable photoluminescence depending on the Au/Ag stoichiometric ratio in the complexes.

All the determined crystal structures exhibited hexagonal arrays of metal–metal interactions with diagonal interactions in the layer. All the lanthanum salts and the europium one were isostructural.

In spite of the potential basic properties of the $Au(CN)_2^-$ anion, these complexes were not obtained in an acid–base reaction with silver salts. Instead, they were prepared by slow crystallization of pure gold and silver dicyanide complexes in different molar ratios ($x = 0.25, 0.50, 0.75, 0.90$). In the case of the two lanthanum derivatives, the exact compositions established by X-ray diffraction studies gave empirical formula in which $x = 0.33$ and 0.78. For the europium complexes, only one structural determination was performed with $x = 0.14$.

These complexes were strongly luminescent at room temperature, in contrast to the precursor homonuclear derivatives, and at an energy that was tunable and depended on the Au : Ag molar ratio, lying between the emission bands of the pure Ag and Au compounds. For instance, while the homonuclear derivative $La[Ag(CN)_2]_3$

emitted at 77 K at 345 and 470 nm (exc. 310 nm) and the gold complex La[Au(CN)$_2$]$_3$ at 431 and 493 nm (exc. 310 nm), the mixed metal systems tended towards Ag or Au peak positions, depending on loading. Therefore, the excited states responsible for the emissions were located on the gold and silver atoms in each complex. The theoretical TD-DFT calculations coincided with these results, with higher oscillator strengths being obtained for the mixed systems, possibly making them better candidates than the pure metal systems for photoluminescence applications.

The basicity of other gold(I) complexes can be used in the synthesis of heteropolynuclear networks in their reactions with different silver salts. Of these, perhaps the best-known Lewis basic precursor is the electron-rich bis(pentafluorophenyl)aurate (I) complex. For example, the reaction between NBu$_4$[Au(C$_6$F$_5$)$_2$] and AgClO$_4$ in acetone leads to a heterometallic extended linear chain of formula [Au$_2$Ag$_2$(C$_6$F$_5$)$_4$-(Me$_2$CO)$_2$]$_n$ [61]. The crystal structure consists of a polymeric chain built by repetition of the Au$_2$Ag$_2$ core through short gold–gold contacts of 3.1674(11) Å. The gold–silver distances are 2.7903(9) and 2.7829(9) Å and both silver atoms have an interaction of 3.1810(13) Å. Therefore, the same complex contains the three types of metal–metal interactions: gold–gold, gold–silver and silver–silver (see Figure 6.22).

In spite of the foreseeable complexity of the emission due to the different metal–metal interactions, this complex is luminescent in the solid state displaying a single band at 546 nm, which shifts to 554 nm at 77 K. In dilute solutions (5×10^{-4} M) it is also luminescent, although at a very different wavelength (em = 405 nm by excitation at 332 nm). Very interestingly, in this state, changes in the concentration of the sample give rise to changes in emission wavelengths and, therefore, a deviation from the Lambert–Beer law. The authors explain the result by considering that isolated tetranuclear units are responsible for the luminescence behavior observed in dilute solutions. Therefore the increase in concentration would produce oligomerization of tetranuclear [Au$_2$Ag$_2$(C$_6$F$_5$)$_4$(Me$_2$CO)$_2$] units through gold–gold contacts. In that case, as commented in Section 6.1, the increase in gold–gold interactions should shift the emissions to lower energies. Indeed, a linear fit is obtained when emission wavelengths are plotted versus the inverse of concentration. The value on the y axis represents the solid state because it can be considered to be the value corresponding to an infinite concentration. The value obtained (547.04 nm) matches the experimental value obtained in the solid state (546 nm at room temperature). In this case, the theoretical TD-DFT calculations mainly suggest

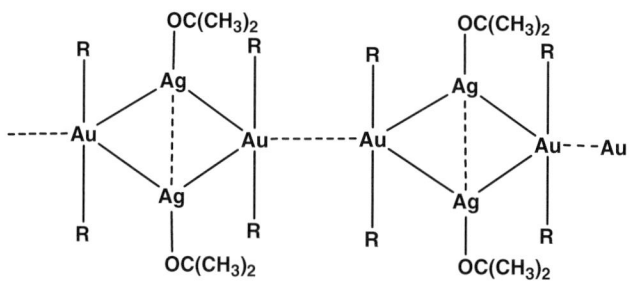

Figure 6.22 Schematic representation of two units of [Au$_2$Ag$_2$(C$_6$F$_5$)$_4$(Me$_2$CO)$_2$] (R = C$_6$F$_5$).

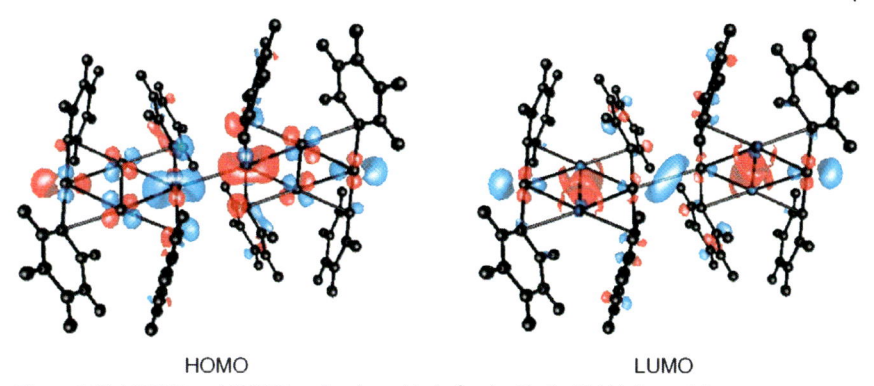

Figure 6.23 HOMO and LUMO molecular orbitals for the [AuAg(C$_6$H$_5$)$_2$]$_4$ model system.

that metal centered $(d\sigma^*)^1(p\sigma)^1$ or $(d\delta^*)^1(p\sigma)^1$ gold–gold and/or silver–silver excited states are responsible for the emission observed (see Figure 6.23). Nevertheless, the role of the perhalophenyl groups cannot be excluded.

For many years, the anionic nature of this gold precursor has been considered the key to its donor characteristics in its reactions with Lewis acid salts. Nevertheless, very recently, our research group has found that even neutral gold starting materials have similar potential in their reactions with silver salts. This is the case for the well-known gold starting materials of formula [AuR(tht)] (R = C$_6$F$_5$, 3,5-C$_6$Cl$_2$F$_3$, C$_6$Cl$_5$). Their reaction with silver trifluoroacetate leads to complexes of stoichiometry [AgAu(C$_6$F$_5$)(CF$_3$CO$_2$)(tht)]$_n$ and [Ag$_2$Au(3,5-C$_6$Cl$_2$F$_3$)(CF$_3$CO$_2$)(tht)]$_n$, [AgAu(C$_6$F$_5$)(CF$_3$CO$_2$)(tht)]$_n$, respectively [62]. Although the three structures are different, all of them are polymers, the first is two-dimensional and the other two are mono-dimensional, formed by association of linear [AuR(tht)] units and eight-membered [Ag$_2$(CF$_3$CO$_2$)$_2$] rings joined through Au–Ag interactions. In the pentafluorophenyl complex, in addition to the Au–Ag interactions that build the polymer and the intramolecular Ag–Ag of the carboxylate dimers, Au–Au interactions also appear. In all the complexes, the metal–metal interactions are shorter than the sum of the van der Waals radii and have approximately the same magnitude.

The three complexes show luminescence in the solid state at room temperature and at 77 K, or in frozen solutions. The different optical behavior in the solid state is related to the different number and types of metal–metal interactions present in each complex and, in agreement with this assumption, the complexes do not show luminescence in fluid solutions, where the metal–metal interactions are not present. Thus, for instance, at 77 K the pentafluorphenyl derivative shows two emissions at 430 and 480 nm, while in the other perhalophenyl complexes the emissions appear at 590 and 495 nm, respectively. All the emissions show lifetimes in the microsecond range, suggesting phosphorescent processes. The DFT calculations carried out using two different models representing the structural situation of the three complexes (see Figure 6.24) suggest that the highest occupied molecular orbitals are mostly placed at the perhalophenyl groups with some contribution from the gold centers, while the lowest unoccupied molecular orbitals are mainly located at gold and silver centers.

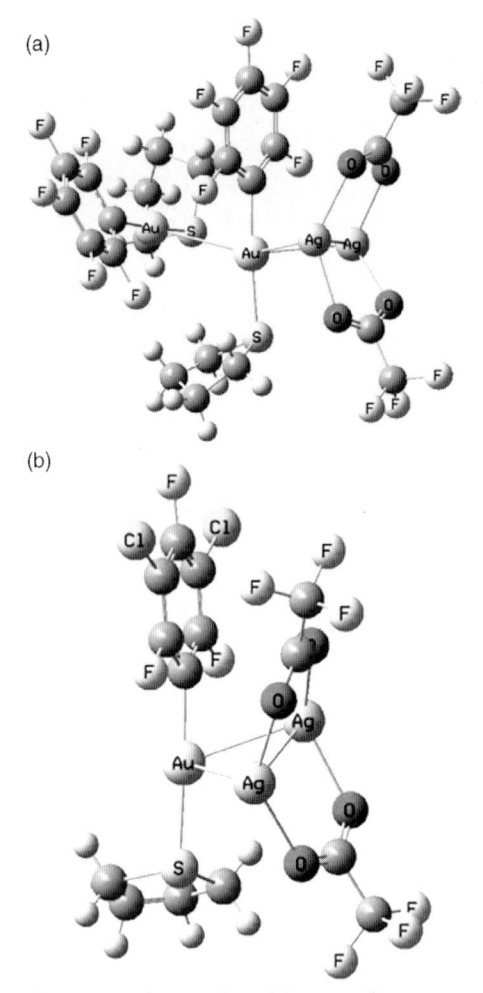

Figure 6.24 Theoretical model systems for [Ag$_2$Au$_2$(C$_6$F$_5$)$_2$(CF$_3$CO$_2$)$_2$(tht)$_2$] (a) and [Ag$_2$Au(C$_6$Cl$_2$F$_3$)$_2$(CF$_3$CO$_2$)$_2$(tht)] (b).

The two emissions observed for the pentafluorophenyl complex seem to be related to the two different types of metal–metal interactions present in the complex, gold–gold and gold–silver. In contrast, the other two complexes with only gold–silver interactions displayed one emission. Therefore, based on theoretical calculations, it appears that in the three complexes the transitions that lead to the excited states responsible for the emissions can be considered as ^{3}LMMCT (ligand(perhalophenyl) metal(gold)-to-metal(silver) charge transfer) in which the basic part (perhalophenyl-gold) interacts with the acid silver atom. Additionally, in the case of the pentafluorophenyl part, which exhibits an additional emission, it is likely to originate from a cooperative contribution of both gold centers ((C$_6$F$_5$)Au···Au(C$_6$F$_5$) → Ag) to the acidic silver, leading to a different triplet emission.

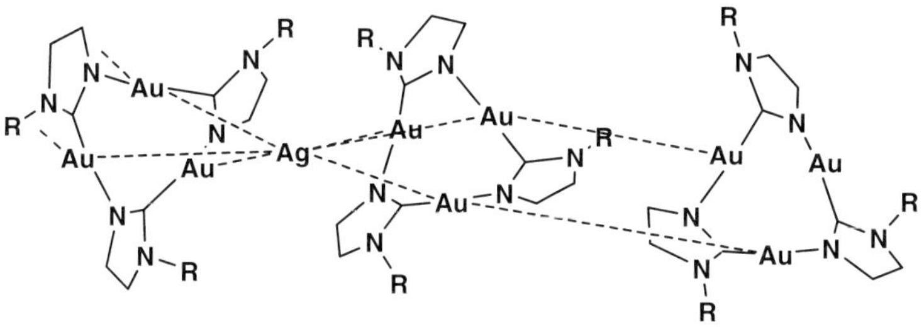

Figure 6.25 Schematic representation of $\{Ag([Au((\mu\text{-}C^2,N^3\text{-}bzim)]_3)_2\}_n$.

R = Bz

Another electron-rich gold complex that is able to react with acid silver salts is the trinuclear pyrazolate complex $[Au(\mu\text{-}C^2,N^3\text{-}bzim)]_3$ (bzim = benzylimidazolate) [63, 64]. Thus, the reaction between this complex and $AgBF_4$ leads to a luminescent extended chain, in which each unit is formed by a naked silver ion center bonded to six gold atoms that form Ag(I) centered trigonal prisms. These Au_3AgAu_3 prisms interact through gold–gold interactions between molecular units (see Figure 6.25). The Au–Ag distances range from 2.731(2) to 2.922(2) Å, indicating relatively strong metal–metal bonding. The shortest intermolecular Au–Au interactions are 3.2678(12) and 3.1157(11) Å, and the average intramolecular Au–Au distance is 3.19 Å.

The complex shows low energy visible emission at room temperature (535 nm) that shifts to a lower energy at 77 K (570 nm), a phenomenon referred to as *luminescence thermochromism*. The lifetime measurements are in the range of hundreds of nanoseconds (115 and 550 ns at 298 K), attributed by the authors to phosphorescence. They propose that these emissions are probably associated with excited states that are delocalized along the crystallographic axis of the chain. The thermal contraction that occurs when the material is cooled leads to a reduction in the metal–metal distances and, consequently, to a reduction in the band-gap energy.

As regards the gold–copper supramolecular systems, the complex $[Au_2Cu_2(C_6F_5)_4(MeCN)_2]_n$ [65] consists of an extended linear chain formed by tetranuclear gold–copper units through gold–gold interactions, a structure that exhibits a similar arrangement of metals to that of the $[Au_2Ag_2(C_6F_5)_4(Me_2CO)_2]_n$ complex described previously [61]. The starting material used to achieve this complex was the homologous gold–silver compound $[Au_2Ag_2(C_6F_5)_4(MeCN)_2]_n$, whose reaction with CuCl in acetonitrile leads to $[Au_2Cu_2(C_6F_5)_4(MeCN)_2]_n$ in a transmetallation reaction. Interestingly, both complexes – the copper and the silver derivatives – are isostructural ($C2/c$ space group), allowing an interesting comparison of the *influence of the heterometal in the optical properties*.

The gold–gold distance between tetranuclear units was 2.8807(4) Å in the gold–silver derivative and 2.9129(3) Å in the gold–copper derivative, suggesting a substantial bonding interaction between the gold centers. The gold–silver interactions were 2.7577(5) and 2.7267(5) Å and the gold–copper ones 2.5741(6) and 2.5876(5) Å, which is indicative of stronger bonding interaction in the silver derivative than in the copper one. In addition, while the silver–silver distance was 3.1084(10) Å, shorter than twice the van der Waals radius of silver, the copper–copper distance was 3.0197(11), which is longer than twice the corresponding van der Waals radius.

Both complexes emitted in the solid state at room temperature (547 nm for the silver complex and 570 nm for the copper one). These emissions shifted to lower energies when the temperature was lowered to 77 K (567 nm and 594 nm, respectively). They also had lifetimes in the range of nanoseconds, suggesting fluorescence. In dilute acetonitrile solutions (ca. 5×10^{-4} M), both complexes displayed absorptions assigned to π–π^* transitions in the pentafluorphenyl rings and lost their emissive properties, which can be interpreted in terms of the rupture of the metal–metal interactions provoked by the solvent. Nevertheless, in the solid state, the absorption spectra displayed bands at lower energies – 493 and 484 nm – for the gold–silver and gold–copper derivatives, respectively, which were not present in the precursor gold complexes and that were assigned to transitions influenced by the gold–gold or gold–heterometal interactions. Interestingly, both derivatives behaved differently in glassy solutions (EtOH/MeOH/CH$_2$Cl$_2$) at 77 K. Thus, whereas the gold–silver complex shows three concentration-dependent emission bands at 470, 501 and 540 nm, the gold–copper complex is not luminescent. In the case of the gold–silver complex, the result is explained by the presence of oligomers of different length in solution as a function of the concentration, while in the case of the gold–copper complex, the presence of solvents with donor characteristics quenches the emission by formation of exciplexes. In fact, when the measurement is carried out in CH$_2$Cl$_2$ or toluene at 77 K, there is no quenching and the complex shows luminescence at 514 or 481 nm, respectively.

Based on the previously described silver complex [Au$_2$Ag$_2$(C$_6$F$_5$)$_4$(MeCN)$_2$]$_n$, the same reaction with CuCl in acetonitrile but with addition of an equivalent of pyrimidine leads to the polymeric [CuAu(C$_6$F$_5$)$_2$(MeCN)(μ_2-C$_4$H$_4$N$_2$)]$_n$ [66]. The polymerization of this complex is produced by covalent copper–pyrimidine bonds. The environment of the copper centers also comprises unsupported gold–copper interactions of 2.8216(6) Å and one molecule of acetonitrile, leading to a distorted tetrahedral arrangement (see Figure 6.26).

This complex is luminescent in the solid state, exhibiting an emission of 525 nm (exc. 390 nm) at room temperature that shifts slightly to 529 nm (exc. 371 nm) at 77 K. The lifetime of 10 μs suggests a phosphorescent process. In solution, it displays very interesting solvent dependence. It emits at 355 (exc. 281 nm) in MeOH, 365 (exc.

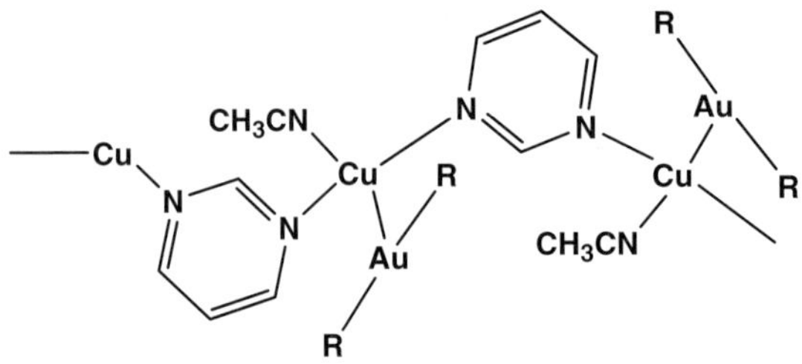

Figure 6.26 Schematic representation of [CuAu(C$_6$F$_5$)$_2$(MeCN)(μ_2-C$_4$H$_4$N$_2$)]$_n$.

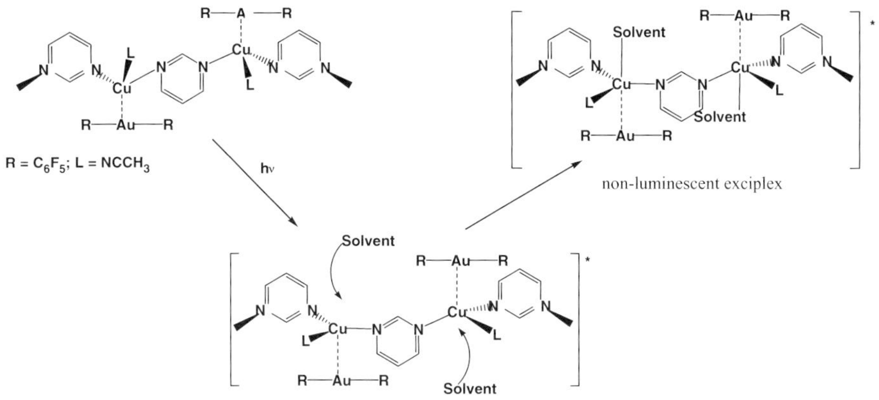

Scheme 6.1 Quenching mechanism of the complex [CuAu$(C_6F_5)_2$(MeCN)(μ_2-$C_4H_4N_2$)]$_n$ in donor solvents.

290 nm) in CH_3CN and exhibits a dual emission at 394 (exc. 263 nm) and 530 (exc. 369 nm) in CH_2Cl_2. The high-energy emissions are assigned to intraligand transitions in the pyrimidine ligand. In contrast, the low-energy emission that appeared only in CH_2Cl_2 is assigned to arise from a ^{3}(MLCT) excited state. It is noteworthy that this emission is similar to that observed in the solid state. Therefore, a similar origin is proposed. The fact that this emission is not observed in donor solvents is interpreted, as before, as an exciplex quenching mechanism where the formally d^9 copper, which is formed in the excitation process, is able to coordinate nucleophiles leading to exciplexes. In particular, the formation of a Cu(II) character in the excited state promotes a flattening distortion of the copper environment and allows an associative attack by the solvent. This results in stabilization of the excited state and destabilization of the ground state, inevitably giving rise to quenching (Scheme 6.1).

6.3.2
Other Supramolecular Gold–Heterometal Complexes

Of the metals that form luminescent supramolecular entities with gold, that for which most complexes are known is thallium in its +1 oxidation state. As described below, in recent years the contributions of several laboratories have been reported. Nevertheless, in some cases, the papers also report similar reactions with other metals, leading to similar structures. In order to maintain a congruent synthetic description, those examples will be discussed together as they appear in the original work.

The first descriptions of heteronuclear luminescent supramolecular complexes were given by Fackler *et al.* in 1988 and 1989. In these studies, one gold–thallium and one gold–lead complex were reported. As in the case of the gold–silver dinuclear systems, the extended systems appeared as a result of the unidirectional polymerization of dinuclear or trinuclear units through metal–metal interactions. These were prepared by reaction of the gold precursor [PPN][Au(MTP)$_2$] (PPN = N(PPh$_3$)$_2$;

MTP = Ph$_2$P(CH$_2$)S with Tl$_2$SO$_4$ (1:1) or Pb(NO$_3$)$_2$ (2:1) leading to [AuTl(MTP)$_2$] [67, 68] or [Au$_2$Pb(MTP)$_4$] [68], respectively. The gold–thallium complex in the solid state formed a one-dimensional polymer in which dinuclear units bound to each other as a result of the short relativistic Tl–Au bonds of about 3 Å. The gold–lead derivative consisted of an extended linear chain formed by the interactions between trinuclear units through the gold centers (see Figure 6.27). The gold–lead distances within the trinuclear unit were 2.896(1) and 2.963(2) Å. The intermolecular distance between the gold centers was 3.149(2) Å.

The absorption spectrum of the gold–thallium complex showed one band at 320 nm ($\varepsilon = 2900$ M^{-1} cm^{-1}), assigned to the $\sigma_1^* \rightarrow \sigma_2$ transition localized in the Au–Tl moiety. The gold–lead complex exhibited two bands at 290 nm ($\varepsilon = 28598$ M^{-1} cm^{-1}) and 385 nm ($\varepsilon = 7626$ M^{-1} cm^{-1}). The emission spectrum in the solid state at 77 K for the former showed a band at 602 nm, and the corresponding spectrum of the gold–lead complex showed a band at 752 nm, at room temperature. In both cases, they were attributable to transitions between orbitals that appeared as a result of the heterometal interaction. Fenske–Hall molecular orbital calculations indicated that in the case of the gold–thallium complex the ground state was the result of the mixing of the empty 6s and 6pz orbitals of gold(I) with the filled 6s and the empty 6pz orbitals of thallium(I); while the calculations for the gold–lead compound indicated that the HOMO was formed by the 6p$_z$ orbital of gold and the 6s orbital of lead and the LUMO was entirely formed by the 6p$_z$ orbitals of these atoms.

The complexes exhibited different behavior in solution. The gold–thallium derivative showed a shift of the emission to 536 nm when the measurement was carried out in frozen solution. This was explained by a higher aggregation of [AuTl(MTP)$_2$] units in the solid state compared to the situation in solution. In the case of the Au–Pb compound, the emission spectrum showed a strong dependence on the aggregation state and temperature. Thus, the emission band in THF solution, which appeared at 555 nm (298 K) ($\tau = 57$ ns), was shifted to 480 nm in frozen solution ($\tau = 2.3$ μs) or appeared at 752 nm in solid state ($\tau = 22$ ns). As with the thallium complex, the shift to high energy in solution may have been related to the polymeric structure of the complex in the solid state that was not reproduced in solution.

As in the case of the gold–silver analogous systems, perhaps the most productive method for preparing extended systems through metallophilic interactions is the acid–base process, in which basic gold(I) precursors react with metallic Lewis acids forming supramolecular networks via acid–base stacking. The cation–anion interactions assist the formation of extended chains.

An example of this is the gold–thallium complex Tl[Au(CN)$_2$] [69], prepared by reaction of TlNO$_3$ and K[Au(CN)$_2$]·2H$_2$O in the 90s by Patterson *et al.* This group carried out a pioneering study of the luminescence properties of this complex and their theoretical interpretation. The structure showed a complexity that made the analyses difficult. Thus, neutron diffraction studies showed thallium–gold interactions of 3.446 and 3.463 Å, shorter than the sum of their van der Waals radii (3.62 Å) in one of the three crystallographically-distinct Au sites in the crystal.

Interestingly, the emission of the complex was thermochromic and, thus, the temperature influenced emission energies and lifetimes. For instance, two bands at

575 and 518 nm appeared at 5 K, but at 300 K the former disappeared. This was attributed to luminescent traps. In contrast, the most energetic band shifted to the blue under the same conditions. It appeared at 509 at 40 K and at 483 nm at 360 K.

Similarly, the dependence of lifetime on temperature was also very pronounced, changing from 176 μs at 1.7 K to about 50 ns at 400 K. The theoretical calculations indicated that the composition of the HOMO remained the same, independently of the Au–Tl distance, since it had 60% gold and 25% CN^- character. The LUMO had mostly CN^- character with a small contribution of $Tl(6p_z)$ and gold characters, with a greater contribution of the thallium orbital when the distance diminished. The authors concluded that the absorptions and the emissions resulted from the gold–thallium interactions, influencing the energies, intensities and rates of deactivation, since the homologous $Cs[Au(CN)_2]$, which contained only Au–Au interactions, did not exhibit similar optical behavior.

Years later, the same precursor gold complex was used by Balch et al. in the synthesis of the complex $[(NH_3)_4Pt][Au(CN)_2]·1.5(H_2O)$ [70]. Its structure displayed two $[Au(CN)_2]^-$ ions above and below the plane of one $[Pt(NH_3)_4]^{2+}$ with Pt–Au interactions of 3.2804(4) and 3.2794(4) Å. These units were arranged into extended chains through short gold–gold interactions of 3.1902(4) Å. Additional cross-linking of these chains occurred through Au–Au contacts of similar lengths.

The complex was non-luminescent at room temperature, but it displayed luminescence in the solid state at 77 K, showing an unstructured emission at 443 nm. This emission was attributed to the gold centers or to the interaction between the gold and platinum ions. In fact, the isostructural $[(NH_3)_4Pt][Ag(CN)_2]·1.4(H_2O)$ was not luminescent and, similarly, the corresponding $[Ni(NH_3)_2][Au(CN)_2]_2$, whose structure only showed gold–gold interactions (3.0830(5) Å), did not display luminescence at either room temperature or at 77 K.

Nevertheless, with different gold precursors, our group followed the acid–base strategy and contributed a number of papers, many of them dealing with supramolecular systems. The most-studied metal was thallium. We focused mainly on this metal because of its very special characteristics: it displays an astonishing complexity in coordination numbers and geometries in complexes; it represents the lowest extreme of metallophilicity. Note that gold is the metal that exhibits the highest level of this characteristic. Therefore, the combination of both metals in heterometal complexes with metal–metal interactions, is a challenge from the synthetic and

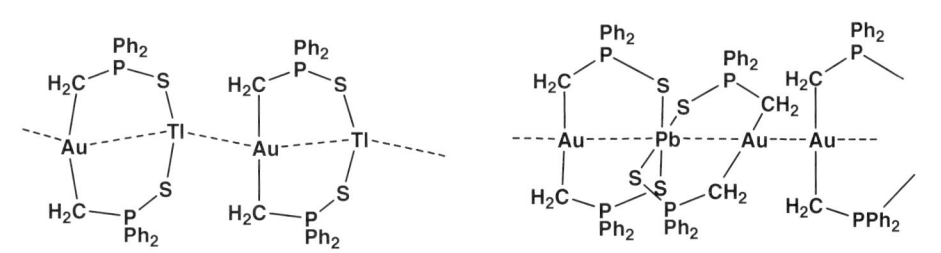

Figure 6.27 Schematic representation of portions of the structures of complexes $[AuTl(MTP)_2]_n$ and $[Au_2Pb(MTP)_4]_n$.

theoretical viewpoints. As described below, the synthesized complexes display a very intense photoluminescence that depends on metal–metal interactions, synthetic pathways, temperature, solvents, precursors and so on. This makes the study of all these aspects all the more interesting and may even help to exemplify similar situations that occur in future syntheses with other metals.

Our first study of these systems was the synthesis in 1998 of the polymeric complex $[AuTl(C_6F_5)_2(OPPh_3)_2]_n$ through the reaction between triphenylphosphine oxide, thallium nitrate and lithium bis(pentafluorophenyl)aurate(I) [71]. This complex consisted of an extended unsupported linear chain of alternate gold and thallium centers. These atoms displayed Au–Tl interactions of 3.0358(8) and 3.0862(8) Å, and the thallium atoms showed a distorted pseudo-trigonal-bipyramidal environment, taking into account the stereochemically active inert pair of this atom. As described below, *the environment around thallium* is one of the main factors that affects the optical properties of these mixed systems.

The complex exhibited luminescence in the solid state at 293 K (494 nm) and at 77 K (494 and 530 nm), but was not emissive in solution, suggesting that the emission was the result of interactions between the metals. This complex also displayed very unusual behavior that may even make it suitable for practical applications. Thus, when this complex was saturated with halogenated solvents and irradiated with UV light, the emission was quenched and the evaporation of the solvent did not regenerate the original complex; instead, a gray solid appeared that exhibited an emission at 476 nm. When the process was performed in the dark, in the absence of radiation, the original solid was recovered without any change. This result was indicative of an excited state reaction with halocarbons (see Figure 6.28).

Similar complexes of stoichiometry $\{[Tl(OPPh_3)][Tl(OPPh_3)L][Au(C_6Cl_5)_2]_2\}_n$ (L=THF, Me$_2$CO) [72] were obtained by reaction of the precursor NBu$_4$[Au(C$_6$Cl$_5$)$_2$] with TlPF$_6$ and OPPh$_3$, using acetone or tetrahydrofuran as solvent. The structures were also extended linear chains of alternating metals, but in this case the environment around thallium was different, alternating between pseudo-tetrahedral and distorted trigonal-bipyramidal, taking into account the stereochemically active lone pair (see Figure 6.29). The metal–metal distances ranged

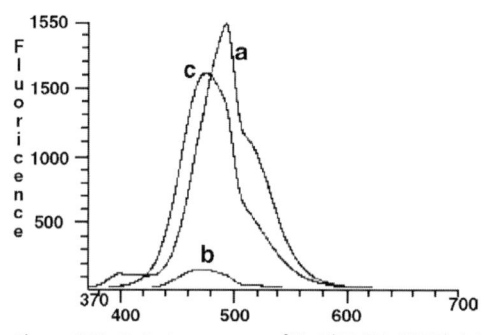

Figure 6.28 Emission spectra of $[AuTl(C_6F_5)_2(OPPh_3)_2]_n$: (a) in the solid state (green solid); (b) saturated with dichloromethane (quenched); (c) after evaporation of dichloromethane (gray solid).

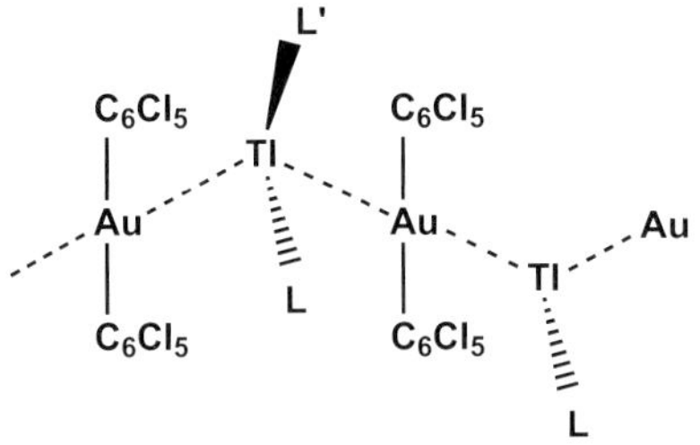

```
          L'
          |
C₆Cl₅   Tl      C₆Cl₅
  |    ⋰ ⋮        |
  Au ⋯     ⋯ Au ⋯       Au
  |         |    ⋯ Tl ⋯
C₆Cl₅     C₆Cl₅    ⋮
                   L
```

L = OPPh$_3$, L' = Me$_2$CO, THF

Figure 6.29 Schematic representation of the structure of the complexes {[Tl(OPPh$_3$)][Tl(OPPh$_3$)L'][Au(C$_6$Cl$_5$)$_2$]$_2$}$_n$ (L' = THF, Me$_2$CO).

between 3.0529(3) and 3.3205(3) Å for the tetrahydrofuran derivative and between 3.0937(3) and 3.2705(4) for the acetone derivative. In both cases, four different metal–metal distances were found.

Neither complex was luminescent in solution but they were strongly luminescent in the solid state, displaying a single band at room temperature (497 (exc. 400 nm) (THF complex) and 501 (exc. 390 nm) (Me$_2$CO complex). At 77 K, each displayed two independent bands at 461 (exc. 330 nm) and 510 (exc. 412 nm) in the case of the former and 465 (exc. 335 nm) and 526 (exc. 410 nm) in the case of the latter. With these data, a number of conclusions were drawn: first, that the absence of luminescence in solution at room temperature suggested that the metal–metal interactions were no longer present in these conditions; secondly, that the four different metal–metal distances in each complex did not seem to be responsible for the different emissions found in the solid state at room temperature and at 77 K, since a more complicated pattern would be expected in that case; finally, that the similarity displayed by both complexes in the energies of the emissions seemed to suggest that the coordinated solvents did not influence the excited states.

Theoretical calculations helped to establish the nature of the metal–metal interactions, as well as the interpretation of the origin of the excited states responsible for the emissions. As regards the former, *ab initio* calculations revealed that the energy associated with the Au–Tl interaction reached an impressive value of 275.7 kJ mol^{-1}. Of this, almost 20% was due to van der Waals interactions and 80% to ionic interactions. This result, far from being theoretically unusual, indicated a very strong metallophilic interaction between both extremes of metallophilicity. In terms of the unusual optical behavior, the TD-DFT calculations revealed that, as observed experimentally, the excited states were not determined by the length of the gold–thallium interactions. Instead, what seemed likely was that those states were related to a charge transfer transition between the electron-rich {Au(C$_6$Cl$_5$)} units and the thallium atoms in different coordination modes, existing simultaneously in each complex. Thus, the tetrahedral thallium atoms accounted for the low energy emissions and the trigonal-bipyramidal thallium centers, with very different gold thallium interactions (3.05 and 3.32 Å) accounted for the high energy ones. Therefore, the main conclusion that can be drawn from these studies is that *the importance of the environment was greater than the metal–metal distances in gold–thallium extended systems*. Similar conclusions were reached separately by Che et al. for d^{10}–d^{10} metal complexes [73–75].

The importance of the environment was also evident in the 1D, 2D or 3D polymers $\{[Tl(4,4'-bipy)THF][Au(C_6F_5)_2]\}_n$, $\{[Tl(1,10-phen)][Au(C_6F_5)_2]\}_n$, $\{[Tl(py)_2][Au(C_6F_5)_2]\}_n$ and $\{[Tl(2,2'-bipy)][Au(C_6F_5)_2]\}_n$ [76]. In all cases, the structures displayed an alternate arrangement of gold and thallium centers, which exhibited a wide range of metal–metal distances, even within the same complex. For instance, the 1,10-phenatroline derivative displayed very different distances ranging from 3.0120(6) to 3.4899(6) Å. In spite of this situation and as in the previous case, a single emission appeared in the solid state for each complex, at both room temperature and 77 K, since all the thallium centers in each complex had a similar environment.

Nevertheless, the environment around the metals and the metal–metal distances were not the only factors that could have influenced the optical properties in such complicated systems. For example, on many occasions, *the experimental conditions in which the reaction was carried out conditioned the structure and therefore the luminescence of the systems*. For instance, the reaction between $[AuTlR_2]_n$ and bipyridine led to different 2D and 3D supramolecular assemblies, depending on the solvent employed in the reaction or the crystallization process. This was the case for the complexes of stoichiometry $[AuTlR_2(bipy)_{0.5}]_n$, $[AuTlR_2(bipy)]_n$, $\{[Tl(bipy)][Tl(bipy)_{0.5}(THF)][AuR_2]_2\}_n$, $\{[Tl(bipy)][Tl(bipy)_{0.5}(THF)][AuR_2]_2 \cdot THF\}_n$ and $\{[AuTlR_2(bipy)] \cdot 0.5\text{toluene}\}_n$ (bipy = 4,4' bipyridine; R = C_6Cl_5) [77]. As can be seen, in all cases the ligands around the thallium were 4,4'-bipyridine and solvent molecules in different ratios. Even within the same crystallization tube, crystals with different ligand ratios appeared. Therefore, different compositions led to different emission energies and, as a result, emitted at 646, 609, 620, 606 and 683 nm by excitation at ca. 550 nm, respectively. There was no correspondence between the gold–thallium distances and the emissions since all of them ranged from 2.9 to 3.1 Å. Similarly, there was also no correspondence with the environment around the thallium centers. The proposed explanation in this case was that all the 2D or 3D networks could probably lead to different excited states and the formation of these networks might be influenced by the presence or absence of coordinating solvents in their structures.

All these examples share a common feature, the synthesis follows an acid–base strategy. Nevertheless, a variation in the basic characteristics of the gold precursor should affect the number of ancillary ligands bonded to thallium and, hence, the structure of the complex. This circumstance was exemplified in the reaction between $NBu_4[Au(C_6F_5)_2]$, $TlPF_6$ and 4,4'-bipyridine, a reaction which was similar to that described previously. In this case, the reaction led to the synthesis of the complex of stoichiometry $\{[Tl(bipy)]_2[Au(C_6F_5)_2]_2\}_n$ [78]. This crystallized in a 2D structure, in which tetramer Au_2Tl_2 units are bound to each other with bridging 2,2'-bipyridine ligands (see Figure 6.30). Interestingly, the tetramer unit, which was the repeating motif, exhibited a non-alternating sequence of ions (+ − − +), which was at variance with the classical Coulomb rules that would give a more stable (+ − + −) arrangement. The metal–metal distances were for gold–thallium 3.0161(2) and for gold–gold 3.4092(3) Å.

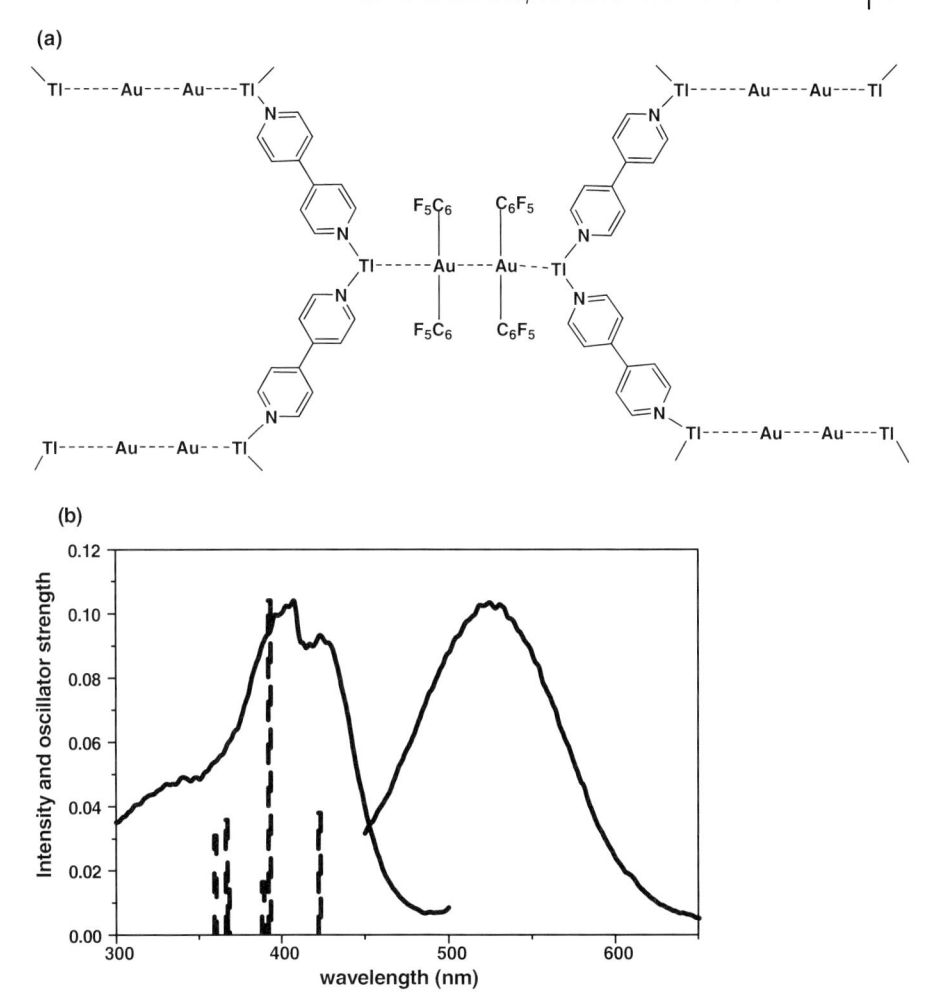

6.3 Luminescent Supramolecular Gold–Heterometal Entities

Figure 6.30 (a) Schematic representation of the structure of complex $\{[Tl(bipy)]_2[Au(C_6F_5)_2]_2\}_n$. (b) Experimental excitation and emission spectra (solid curves) and theoretical excitations (dashed lines).

As before, the complex did not display emissions in solution at room temperature and the absorption spectrum showed only the absorptions due to the bis(perhalophenyl) group. In spite of the two different metal–metal distances, the complex displayed only a single emission at 525 nm in the solid state at room temperature, with a lifetime in agreement with a fluorescent transition ($\tau_1 = 79$ ns; $\tau_2 = 256$ ns). Interestingly, this emission shifted to the blue (507 nm) at 77 K. The shift of the emission to the blue with decreasing temperature has been observed in certain other luminescent complexes [68, 79] and has been explained as a substantial dependence of the emission maxima on environmental rigidity. The TD-DFT calculations showed a good agreement between theoretical and experimental excitations (see Figure 6.30

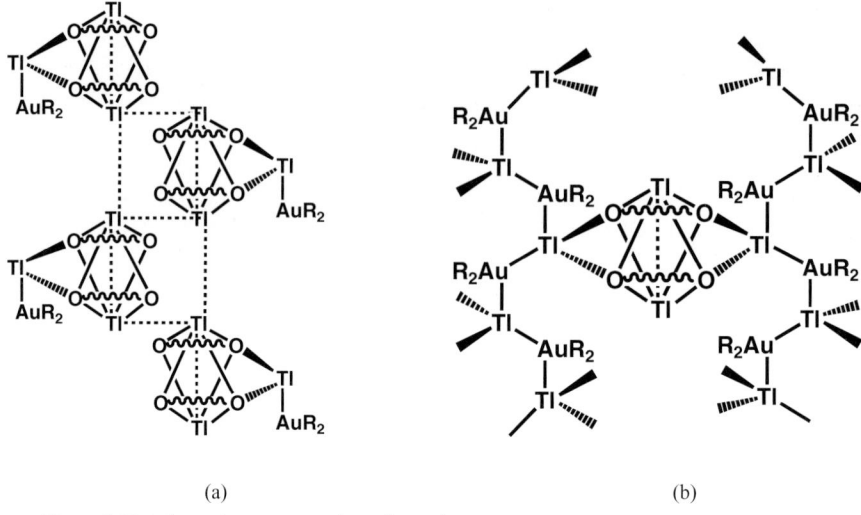

Figure 6.31 Schematic representation of complexes [AuTl$_3$(acac)$_2$(C$_6$F$_5$)$_2$]$_n$ (a) and [AuTl$_2$(acac)(C$_6$Cl$_5$)$_2$]$_n$ (b).

(below)), and the analyses of these theoretical transitions indicated an LMMCT, from an anti-bonding orbital of the [Au(C$_6$F$_5$)$_2$]$^-$ groups to an anti-bonding orbital with the contribution of the four metallic centers.

Another example of different structures that have their origin in the different basicity of the precursor gold complexes are the complexes from the reactions between the linear chain [AuTlR$_2$]$_n$ (R = C$_6$F$_5$, C$_6$Cl$_5$) and [Tl(acac)] with stoichiometry [AuTl$_3$(acac)$_2$(C$_6$F$_5$)$_2$]$_n$ or [AuTl$_2$(acac)(C$_6$Cl$_5$)$_2$]$_n$ [80]. In this case, one complex consists of a chain, while the other consists of a 2D network. In both cases Tl$_2$(acac)$_2$ units act as connections between [AuTl(C$_6$F$_5$)$_2$] units, or between linear chains of [AuTl(C$_6$Cl$_5$)$_2$]$_n$, respectively (see Figure 6.31). In the Tl$_2$(acac)$_2$ units, each Tl(I) linked the four oxygen atoms of the two acetylacetonate groups. Very interestingly, in addition to the expected gold–thallium interactions of 3.0963(7) and 3.2468(7) Å, for the pentachlorophenyl derivative, or 3.0653(4) Å for the pentafluorophenyl derivative, both displayed thallium–thallium interactions of 3.6774(11) in the former and between 3.6688(4) and 3.7607(4) in the latter, respectively.

Therefore, in these complexes, a new structural characteristic appeared that had consequences for their optical properties. Thus, as expected, both complexes displayed luminescence in the solid state at room temperature, showing a single band, but at 77 K they displayed two independent emissions (463 and 588 nm for pentachlorophenyl-derivatives and 427 and 507 nm for pentafluorophenyl-derivatives, respectively) (see Figure 6.32), a result that was repeated in glassy solutions at the same temperature. In addition, and unlike the results with the other polymeric gold–thallium systems, the products exhibited luminescence in dilute solution, showing similar profiles and energies (390 nm). What is proposed is that these high energy emissions, which appeared in solution and in the solid state at low temperature, originate in the Tl$_2$(acac)$_2$ units, which even remain in solution. In fact,

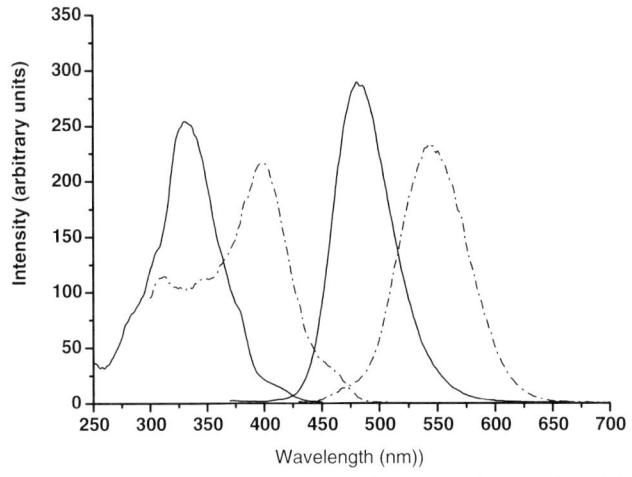

Figure 6.32 Luminescence spectra of [AuTl$_2$(acac)(C$_6$Cl$_5$)$_2$]$_n$ in the solid state at 77 K.

Tl(acac) showed similar behavior. In this case, TD-DFT calculations showed that the transitions arose from ligand-based (acac) orbitals to Tl-based molecular orbitals.

This result was further confirmed by the reactions of the same precursor gold complexes [AuTlR$_2$]$_n$ (R = C$_6$F$_5$, C$_6$Cl$_5$) with an oxygen donor ligand such as dimethylsulfoxide. Thus, the reaction of the pentafluorophenyl complex with DMSO in toluene in a 1:1, 1:1,5 or 1:2 molar ratio led to the product of stoichiometry [Tl$_2${Au(C$_6$F$_5$)$_2$}$_2${μ-DMSO}$_3$]$_n$, while when the starting material was the pentachlorophenyl derivative, the reactions with the same ligand in the same molar ratios led to [Tl$_2${Au(C$_6$Cl$_5$)$_2$}$_2${μ-DMSO}$_2$]$_n$ [81]. The first complex was a monodimensional polymer formed by repetition of the [Au···Tl(μ-O = SMe$_2$)$_3$Tl] units, with gold–thallium interactions of 3.2225(6)–3.5182(8), while the second had two bridging (μ-O = SMe$_2$) units and an additional Au(C$_6$Cl$_5$)$_2$ one. In addition, in this complex a thallium–thallium interaction of 3.7562(6) Å appeared.

Similarly to the results observed in the previous case, the structural differences affected the optical properties of the complexes. While in the former a single emission was observed in the solid state, both at room temperature (440, exc. 390 nm) and at 77 K (460, exc. 360 nm); the latter showed one band (510, exc. 450 nm) and one shoulder (560 nm) at room temperature, and two independent emissions (510, exc. 370 nm; 550, exc. 480 nm) at 77 K. This pentachlorophenyl derivative was also luminescent in solution, displaying a band at 530 nm (exc. 345 nm), which was not present in the precursor complexes or in the pentafluorophenyl derivative. Therefore, in this case, as in the previous one, it is proposed that *the Tl–Tl interaction in the solid state remained in solution and was responsible for the luminescence behavior observed in this state.*

The previous reactions highlighted the role of the bis(perhalophenyl)aurates as donors of electronic density to the thallium atoms. These reactions can be considered to be Lewis neutralizations and lead to polymeric systems. In addition, the thallium centers incorporated in the reactions additional ligands (sometimes the solvent) into

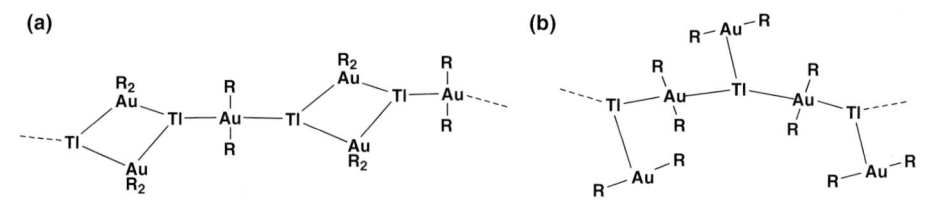

Figure 6.33 Schematic representation of the metal arrangements in $\{NBu_4[Tl_2\{Au(C_6Cl_5)_2\}\{\mu\text{-}Au(C_6Cl_5)_2\}_2]\}_n$ (a) and $\{NBu_4[Tl\{Au(3,5\text{-}C_6Cl_2F_3)_2\}\{\mu\text{-}Au(C_6Cl_2F_3)_2\}]\}_n$ and $\{NBu_4[Tl\{Au(C_6Cl_5)_2\}\{\mu\text{-}Au(3,5\text{-}C_6Cl_2F_3)_2\}]\}_n$ (b).

their coordination sphere, which also gave electronic density to them. Therefore, a plausible possibility is that the bis(perhalophenyl)aurates can act as the only source of electrons, that is they can act as metalloligands. In that case, the resulting complexes would have an anionic nature.

Indeed, the reaction between the metallic gold precursors and thallium(I) in different molar ratios led to the heteropolynuclear complexes $\{NBu_4[Tl_2\{Au(C_6Cl_5)_2\}\{\mu\text{-}Au(C_6Cl_5)_2\}_2]\}_n$, $\{NBu_4[Tl\{Au(3,5\text{-}C_6Cl_2F_3)_2\}\{\mu\text{-}Au(3,5\text{-}C_6Cl_2F_3)_2\}]\}_n$ or $\{NBu_4[Tl\{Au(C_6Cl_5)_2\}\{\mu\text{-}Au(3,5\text{-}C_6Cl_2F_3)_2\}]\}_n$, a result that is proposed to depend on the basicity of the bis(perhalophenyl)aurates. The structures consisted of linear chains formed by alternating gold and thallium centers that had incorporated one half (Figure 6.33a) or one (Figure 6.33b) additional AuR_2 unit per thallium atom [82]. In all of these anionic complexes, the gold–thallium interactions had the same strength; thus, there was no difference in the interactions between the angular bridging AuR_2 units and the linear units (from 3.0559(4) to 3.1678(4) (angular) and 3.0940(3) and 3.1001(3) (linear)) in the first complex, and those of the other complexes with linear bridging AuR_2 units and terminal units (from 2.9704(7) to 3.1196 (5) Å).

Since the metal–metal interactions were similar in the three complexes, their lengths were not considered to be responsible for the different optical properties observed. Instead, the emissions were probably influenced by the number and type of Au–Tl interactions, as well as by the aryl groups bonded to gold. In each case, the emission spectra in the solid state at low temperature displayed two emissions: 560 and 612 nm for $\{NBu_4[Tl_2\{Au(C_6Cl_5)_2\}\{\mu\text{-}Au(C_6Cl_5)_2\}_2]\}_n$, 470 and 520 nm for $\{NBu_4[Tl\{Au(3,5\text{-}C_6Cl_2F_3)_2\}\{\mu\text{-}Au(3,5\text{-}C_6Cl_2F_3)_2\}]\}_n$ and 580 and 620 nm for $\{NBu_4[Tl\{Au(C_6Cl_5)_2\}\{\mu\text{-}Au(3,5\text{-}C_6Cl_2F_3)_2\}]\}_n$. These emissions were probably related to the presence of a different type of Au–Tl interaction.

As regards the other examples reported previously, the high energy bands can be assigned to excited states arising from the new Au–Tl interactions formed by the addition of $[AuR_2]^-$ groups to the extended chains, while the low-energy bands are likely assigned to delocalized excitons along the main axis. The difference between the two isostructural complexes could be related to the different aryl groups present in each complex. As expected, none of the complexes were luminescent in solution at room temperature, indicating that the metal–metal interactions were the origin of the emissive behavior. Based on these data, it is proposed that these emissions resulted from an admixture of MMCT and LMCT.

As we have seen and as occurred with the heteropolynuclear bis(perhalophenyl) gold–silver chains, the acid–base synthetic strategy enabled the synthesis of many supramolecular materials by reaction of the metallic precursors in the presence of several donor ligands. Nevertheless, the reaction in the absence of these ligands led to a potential starting material that could react with a multitude of molecules bearing donor centers. This was the complex [AuTl(C_6Cl_5)$_2$]$_n$ [83], prepared by reaction of the metallic precursors NBu$_4$[Au(C_6Cl_5)$_2$] and TlPF$_6$ in tetrahydrofuran. Surprisingly, the material did not contain THF molecules bonded to the thallium centers but consisted of a perfectly linear chain (180°) built via unsupported Au–Tl interactions of 3.0045(5) and 2.9726(5) Å. The unsaturated thallium atoms maintained contacts with the chlorine atoms of the adjacent linear chains and this led to a 3D structure containing holes parallel to the z-axis of about 10 Å. The weak thallium–chlorine contacts between different chains were easily replaced by stronger bonds with donor molecules, and this took place even with such molecules in the gas phase. Consequently, it behaved like a VOCs sensor. In fact, it reacted with molecules of THF, tetrahydrothiophene (THT), acacH, 2-fluorpyridine (2-FPy), NEt$_3$, py, CH$_3$CN or acetone, producing very evident changes in color that were reversible when the sample was heated for a few minutes. The change in color was even greater under UV radiation and, for example, the complex that exhibited an emission at 531 nm, when exposed to the different vapors, shifted emission to 507 (THF), 567 (THT), 650 (acacH), 627 (2-Fpy), 511 (NEt$_3$), 646 (py), 513 (CH$_3$CN) or 532 (acetone) nm, respectively, which is easily detectable by the human eye (see Figure 6.34). The mechanism responsible for this unusual behavior was interpreted based on the information obtained from the products through conventional reactions between the starting materials in solution. Thus, when we compared the structure and the optical properties of the materials formed with the vapors, with the products obtained in solution in the reaction of the gold–thallium precursor with those ligands, we concluded that the vapor molecules penetrated through the tunnels into the structure, bound to some of the thallium centers and produced a variation of the Au–Tl distances leading to different bandgaps that resulted in different emissions. Reversibility was achieved because the molecules interacted with only some of the thallium centers, thus preventing the

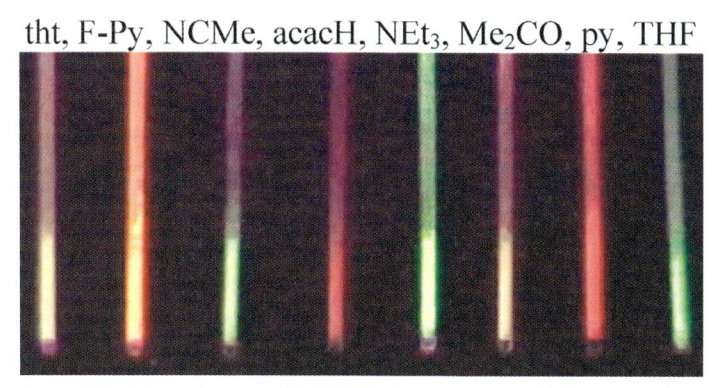

Figure 6.34 Colour changes of [AuTl(C_6Cl_5)$_2$]$_n$ with vapors of donor molecules under UV-light.

supramolecular structure from collapsing. It is important to note that if all the thallium atoms were involved in the bonds with the vapor molecules, the contacts between different chains would have disappeared, together with the holes, and the reaction would have been irreversible.

Another possible practical application of this heteropolynuclear system is described in a very recent contribution by this laboratory that reports on the study of the gold–thallium interaction as an assistant in the synthesis of organic derivatives such as imines. Starting with the ethylenediamine complexes $[AuTlR_2(en)]_n$ (R = C_6Cl_5, C_6F_5), the reaction with ketones, even in the gas phase, led to new imine or diimine complexes. Thus, the reactions of these derivatives with acetone or phenylmethylketone led to complexes with the formulae $[AuTlR_2\{(CH_3)_2C=N(CH_2)_2NH_2\}]_n$, $[AuTlR_2\{((CH_3)_2C=N)_2(CH_2)_2\}]_n$ and $[AuTlR_2\{((PhCH_3)C=N)_2(CH_2)_2\}]_n$ [84].

To test the importance of the gold–thallium interaction in these reactions, when these were performed in the absence of one of the metals, gold or thallium, the imine or diimine products were not isolated and instead the starting materials were recovered unchanged. At this point, it is interesting to note that the reported methods for the synthesis of imines require high temperatures, prolonged reaction periods and the presence of external or *in situ* dehydrating agents. In this case the reaction between the starting gold–thallium product and the ketones took place even in the gas phase. For example, Figure 6.35 shows the step-by-step transformation of the derivative $[AuTl(C_6F_5)(en)]$ subjected to acetone vapor.

As shown in the figure, both the starting material and the imine products were strongly luminescent under UV light. The starting ethylenediamine products displayed emissions at 515 and 505 nm for the pentafluoro- and pentachlorophenyl complexes, respectively. The result of their reactions with both ketones shifted the emissions to 670 and 625 nm for $[AuTlR_2\{(CH_3)_2C=N(CH_2)_2NH_2\}]_n$, 640 and 675 nm for $[AuTlR_2\{((CH_3)_2C=N)_2(CH_2)_2\}]_n$ and 560 and 575 nm for

Figure 6.35 Conversion of the derivative $[AuTl(C_6F_5)_2(en)]_n$ with vapors of acetone under UV-light.

[AuTlR$_2$ {((PhCH$_3$)C=N)$_2$(CH$_2$)$_2$}]$_n$. As expected, none of the complexes exhibited emissions in solution and, consequently, the assignment of the excited states was similar to that of other chains with similar donor centers, that is delocalized excitons along the heterometallic chains influenced by the coordination environment of the thallium centers. Finally, the fact that the starting ethylenediamine precursors displayed high energy emissions when compared with the imine ones was in accordance with this assignment, since the only structurally characterized ethylenediamine complex, the pentachlorophenyl one, consisted of discrete dinuclear molecules [AuTl(C$_6$Cl$_5$)$_2$(en)] with an intramolecular distance of 3.1 Å and an intermolecular distance of 4.04 Å. Consequently, the emission was shifted to higher energies.

Lastly, as occurred with the trinuclear pyrazolate complex [Au(μ-C^2,N^3-bzim)]$_3$ (bzim = benzylimidazolate), the carbeniate [AuC(O−R)=NR′]$_3$ (R = Et; R′ = C$_6$H$_4$CH$_3$) or the pyrazolate [Au(μ-C^2,N^3-Rim)]$_3$ (Rim = benzylimidazolate; methylimidazolate) complexes can also lead to luminescent-extended chains reacting with thallium(I) salts or even with the trimer complex [Hg(μ-C,C-C$_6$F$_4$)]$_3$ [63, 64, 85]. Thus, the resultant {Tl([AuC(O-R)=NR′]$_3$)$_2$[PF$_6$]}$_n$ and {Tl([Au(μ-C^2,N^3-Rim)]$_3$)$_2$[PF$_6$]}$_n$ consisted of extended chains by repetition of (Au$_3$TlAu$_3$ ...) through aurophilic interactions between Au$_3$TlAu$_3$ molecular units. In the case of the hexanuclear {[Hg(μ-C,C-C$_6$F$_4$)]$_3$([AuC(O-R)=NR′]$_3$)$_2$}$_n$ and {[Hg(μ-C,C-C$_6$F$_4$)]$_3$([Au(μ-C^2,N^3-Rim)]$_3$)$_2$}$_n$ complexes, the repeating motif was the Au$_3$Hg$_3$Au$_3$ unit.

All these complexes displayed a strong visible luminescence under UV excitation, which was sensitive to temperature as well as to the heterometal interacting with the gold centers. In the case of the gold–thallium complexes, as in the case of the gold–silver pyrazolate described previously, the authors attributed the emissions to phosphorescence from excited states formed by the metal–metal interactions that were delocalized along the crystallographic axis of the chain. Thus, for instance, the gold–thallium carbeniate complex exhibited an emission at 525 nm at ambient temperature that became orange (580 nm) at 77 K and the benzylimidazolate gold–thallium complex showed a blue emission at room temperature and a green one at 77 K. In the latter, the emission profile was dependent on the excitation wavelength.

The gold–mercury complexes {[Hg(μ-C,C-C$_6$F$_4$)]$_3$([AuC(O−R)=NR′]$_3$)$_2$}$_n$ and {[Hg(μ-C,C-C$_6$F$_4$)]$_3$([Au(μ-C^2,N^3-Rim)]$_3$)$_2$}$_n$ also showed strong luminescence. The spectrum of the former at 77 K exhibited a fluorescent blue emission with a vibronic structure. This had an average spacing of $(1.4 \pm 0.2) \times 10^3$ cm^{-1}, corresponding to a progression in the μ(C=N) vibrational mode of the carbeniate ligand ($\sim$1500 cm^{-1} in the infrared spectrum). This vibronic pattern indicated that the HOMO had strong C–N ligand characteristics, in contrast to what occurred in the phosphorescent gold–thallium and gold–silver complexes described. In contrast, in the case of the imidazolate gold–mercury complex, a strong green luminescence appeared at ambient temperature with vibronic progression, but with increased metallophilic bonding in the ground and excited states, as occurred in the homologous complexes with thallium and silver.

6.4
Concluding Remarks

To conclude this chapter, I would like to emphasize that these supramolecular entities built through metallophilic interactions provide a promising area in which interaction between different metals determines not only the structural motifs but also the optical properties. As observed in many examples throughout this chapter, the gold is not the only factor influencing the exhibiting of luminescence by a supramolecular entity; factors such as ligand type and number, the coordination environments around the metal centers, temperature, heterometal or metal–metal distances have an enormous effect on the energy and type of the emissions. Of course, research on this subject will be fascinating and will open up promising areas of future application, including imaging technology, vapor sensors and light-emitting devices, where the demand for optoelectronic devices is increasing every day.

References

1 Ziolo, R.F., Lipton, S. and Dori, Z. (1970) *Chemical Communications*, 1124.
2 Moore, C.E. (1948) *Atomic Energy Levels*, National Bureau of Standards, Washington, DC, p. 186.
3 Assefa, Z., Staples, R.J. and Fackler, J.P., Jr (1994) *Inorganic Chemistry*, **33**, 2790.
4 Che, C.M., Kwong, H.L., Poon, C.K. and Yam, V.W.W. (1990) *Journal of The Chemical Society Dalton Transactions*, 3215.
5 Jaw, H.R.C., Savas, M.M., Rodgers, R.D. and Mason, W.R. (1989) *Inorganic Chemistry*, **28**, 1028.
6 Fu, W.F., Chan, K.C., Miskowski, V.M. and Che, C.M. (1999) *Angewandte Chemie-International Edition*, **38**, 2783.
7 Leung, K.H., Phillips, D.L., Tse, M.C., Che, C.M. and Miskowski, V.M. (1999) *Journal of the American Chemical Society*, **121**, 4799.
8 Fu, W.F., Chan, K.C., Cheung, K.K. and Che, C.M. (2001) *Chemistry – A European Journal*, **7**, 4656.
9 Sharpe, A.G. (1976) *The Chemistry of Cyano Complexes of the Transition Metals*, Academic Press, London.
10 Nagasundaram, N., Roper, G., Biscoe, J., Chai, J.W., Patterson, H.H., Blom, N. and Ludi, A. (1986) *Inorganic Chemistry*, **25**, 2947.
11 Rawashdeh-Omary, M.A., Omary, M.A. and Patterson, H. 2000, *Journal of the American Chemical Society*, **122**, 10371.
12 Rawashdeh-Omary, M.A., Omary, M.A., Patterson, H. and Fackler, J.P., Jr (2001) *Journal of the American Chemical Society*, **123**, 11237.
13 White-Morris, R.L., Olmstead, M.M., Jiang, F., Tinti, D.S. and Balch, A.L. (2002) *Journal of the American Chemical Society*, **124**, 2327.
14 Codina, A., Fernández, E.J., Jones, P.G., Laguna, A., López-de-Luzuriaga, J.M., Monge, M., Olmos, M.E., Pérez, J. and Rodríguez, M.A. (2002) *Journal of the American Chemical Society*, **124**, 6781.
15 Ecken, H., Olmstead, M.M., Noll, B.C., Attar, S., Schlyer, B. and Balch, A.L. (1998) *Journal of The Chemical Society, Dalton Transactions*, 3715.
16 Chui, S.Y., Ng, M.F.Y. and Che, C.M. (2005) *Chemistry – A European Journal*, **11**, 1739.
17 White-Morris, R.L., Olmstead, M.M., Balch, A.L., Elbjeirami, O. and Omary, M. (2003) *Inorganic Chemistry*, **42**, 6741.
18 White-Morris, R.L., Stender, M., Tinti, D.S., Balch, A.L., Rios, D. and Attar, S. (2003) *Inorganic Chemistry*, **42**, 3237.

19 White-Morris, R.L., Olmstead, M.M. and Balch, A.L. (2003) *Journal of the American Chemical Society*, **125**, 1033.
20 Lu, W., Zhu, N. and Che, C.M. (2003) *Journal of the American Chemical Society*, **125**, 16081.
21 Assefa, Z., Omary, M.A., McBurnett, B.G., Mohamed, A.A., Patterson, H.H., Staples, R.J. and Fackler, J.P., Jr (2002) *Inorganic Chemistry*, **41**, 6274.
22 Coker, N.L., Krause Bauer, J.A. and Elder, R.C. (2004) *Journal of the American Chemical Society*, **126**, 12.
23 Vogler, A. and Kunkely, H. (2001) *Coordination Chemistry Reviews*, **219–221**, 489.
24 Mansour, A., Lachicotte, R.J., Gysling, H.J. and Eisemberg, R. (1998) *Inorganic Chemistry*, **37**, 4625.
25 Tang, Z., Litvinchuk, A.P., Lee, H.G. and Guloy, A.M. (1998) *Inorganic Chemistry*, **37**, 4752.
26 Shieh, S.J., Hong, X., Peng, S.M. and Che, C.M. (1994) *Journal of The Chemical Society, Dalton Transactions*, 3067.
27 Li, D., Hong, X., Che, C.M., Lo, W.C. and Peng, S.M. (1993) *Journal of The Chemical Society, Dalton Transactions*, 2929.
28 Tang, S.S., Lin, I.J.B., Liu, L.S. and Wang, J.C. (1996) *Journal of the Chinese Chemical Society*, **43**, 327.
29 Tang, S.S., Chang, C.P., Lin, I.J.B., Liou, L.S. and Wang, J.C. (1997) *Inorganic Chemistry*, **36**, 2294.
30 Forward, J.M., Bohmann, D., Fackler, J.P., Jr and Staples, R.J. (1995) *Inorganic Chemistry*, **34**, 6330.
31 Tzeng, B.C., Huang, Y.C., Wu, W.M., Lee, S.Y., Lee, G.H. and Peng, S.M. (2004) *Crystal Growth and Design*, **4**, 63.
32 Feng, D.F., Tang, S.S., Liu, C.W. and Lin, I.J.B. (1997) *Organometallics*, **16**, 901.
33 Van Zyl, W.E., López-de-Luzuriaga, J.M. and Fackler, J.P., Jr (2000) *Journal of Molecular Structure*, **516**, 99.
34 Van. Zyl, W.E., López-de-Luzuriaga, J.M., Mohamed, A.A., Staples, R.J. and Fackler, J.P., Jr. (2002) *Inorganic Chemistry*, **41**, 4579.
35 Yam, V.W.W., Chan, C.L., Li, C.K. and Wong, K.M.C. (2001) *Coordination Chemistry Reviews*, **216–217**, 173.
36 Khan, M.N.I., Wang, S. and Fackler, J.P., Jr (1989) *Inorganic Chemistry*, **28**, 3579.
37 Lee, Y.A., McGarrah, J.E., Lachicotte, R.J. and Eisemberg, R. (2002) *Journal of the American Chemical Society*, **124**, 10662.
38 Mansour, M.A., Connick, W.B., Lachicotte, R.J., Gysling, H.J. and Eisemberg, R. (1998) *Journal of the American Chemical Society*, **120**, 1329.
39 Lee, Y.A. and Eisemberg, R. (2003) *Journal of the American Chemical Society*, **125**, 7778.
40 Brandys, M.C. and Puddephatt, R.J. (2001) *Chemical Communications*, 1280.
41 Barberá, J., Elduque, A., Giménez, R., Oro, L.A. and Serrano, J.L. (1996) *Angewandte Chemie-International Edition*, **35**, 2832.
42 Vickery, J.C., Olmstead, M.M., Fung, E.Y. and Balch, A.L. (1997) *Angewandte Chemie-International Edition*, **36**, 1179.
43 Fung, E.Y., Olmstead, M.M., Vickery, J.C. and Balch, A.L. (1998) *Coordination Chemistry Reviews*, **171**, 151.
44 Balch, A.L., Olmstead, M.M. and Vickery, J.C. (1999) *Inorganic Chemistry*, **38**, 3494.
45 White-Morris, R.L., Olmstead, M.M., Attar, S. and Balch, A.L. (2005) *Inorganic Chemistry*, **44**, 5021.
46 Olmstead, M.M., Jiang, F., Attar, S. and Balch, A.L. (2001) *Journal of the American Chemical Society*, **123**, 3260.
47 Hayashi, A., Olmstead, M.M., Attar, S. and Balch, A.L. (2002) *Journal of the American Chemical Society*, **124**, 5791.
48 Rawashdeh-Omary, M.A., Omary, M.A. and Fackler, J.P., Jr (2001) *Journal of the American Chemical Society*, **123**, 9689.
49 Omary, M.A., Rawashdeh-Omary, M.A., Gonser, M.W.A., Elbjeirami, O., Grimes, T., Coundari, T.R., Diyabalanage, H.V.K., Gamage, C.S.P. and Dias, H.V.R. (2005) *Inorganic Chemistry*, **44**, 8200.
50 Chen, J., Mohamed, A.A., Abdou, H.E., Krause Bauer, J.A., Fackler, J.P., Jr, Bruce, A.E. and Bruce, M.R.M. (2005) *Chemical Communications*, 1575.

51 Sevillano, P., Fuhr, O., Kattannek, M., Nava, P., Hampe, O., Lebedkin, S., Ahlrichs, R., Fenske, D. and Kappes, M.M. (2006) *Angewandte Chemie-International Edition*, **45**, 3702.

52 Sevillano, P., Fuhr, O., Hampe, O., Lebedkin, S., Matern, E., Fenske, D. and Kappes, M.M. (2007) *Inorganic Chemistry*, **46**, 7294.

53 Tzeng, B.C., Che, C.M. and Peng, S.M. (1997) *Chemical Communications*, 1771.

54 Tzeng, B.C., Cheung, K.K. and Che, C.M. (1996) *Chemical Communications*, 1681.

55 Pyykkö, P., Li, J. and Runeberg, N. (1994) *Chemical Physics Letters*, **218**, 133.

56 Rawashdeh-Omary, M.A., Omary, M.A. and Fackler, J.P., Jr. (2002) *Inorganica Chimica Acta*, **334**, 376.

57 Catalano, V.J., Malwitz, M.A. and Etogo, A.O. (2004) *Inorganic Chemistry*, **43**, 5714.

58 Catalano, V.J. and Etogo, A.O. (2005) *Journal of Organometallic Chemistry*, **690**, 6041.

59 Colis, J.C.F., Larochelle, C., Fernández, E.J., López-de-Luzuriaga, J.M., Monge, M., Laguna, A., Tripp, C. and Patterson, H. (2005) *The Journal of Physical Chemistry B*, **109**, 4317; Colis, J.C.F., Larochelle, C., Staples, R., Irmer, R.H. and Patterson, H. (2005) *Dalton Transactions*, 675; Larochelle, C. and Patterson, H. (2006) *Chemical Physics Letters*, **429**, 440.

60 Colis, J.C.F., Staples, R., Tripp, C., Labrecque, D. and Patterson, H. (2005) *The Journal of Physical Chemistry B*, **109**, 102.

61 Fernández, E.J., Gimeno, M.C., Laguna, A., López-de-Luzuriaga, J.M., Monge, M., Pyykkö, P. and Sundholm, D. (2000) *Journal of the American Chemical Society*, **122**, 7287.

62 Fernández, E.J., Jones, P.G., Laguna, A., López-de-Luzuriaga, J.M., Monge, M., Olmos, M.E. and Puelles, R.C. (2007) *Organometallics*, **26**, 5931.

63 Burini, A., Fackler, J.P., Jr, Galassi, R., Pietroni, B.R. and Staples, R.J. (1998) *Journal of the Chemical Society, Chemical Communications*, 95.

64 Burini, A., Bravi, R., Fackler, J.P., Jr Galassi, R., Grant, T.A., Omary, M.A., Pietroni, B.R. and Staples, R.J. (2000) *Inorganic Chemistry*, **39**, 3158.

65 Fernández, E.J., Laguna, A., López-de-Luzuriaga, J.M., Monge, M., Montiel, M., Olmos, M.E. and Rodríguez-Castillo, M. (2006) *Organometallics*, **25**, 3639.

66 Fernández, E.J., Laguna, A., López-de-Luzuriaga, J.M., Monge, M., Montiel, M. and Olmos, M.E. (2005) *Inorganic Chemistry*, **44**, 1163.

67 Wang, S., Fackler, J.P., Jr, King, C. and Wang, J.C. (1988) *Journal of the American Chemical Society*, **110**, 3308.

68 Wang, S., Garzón, G., King, C., Wang, J.C. and Fackler, J.P., Jr (1989) *Inorganic Chemistry*, **28**, 4623.

69 Assefa, Z., DeStefano, F., Garepapaghi, M.A., LaCasce, J.H., Jr, Oulette, S., Corson, M.R., Nagle, J.K. and Patterson, H.H. (1991) *Inorganic Chemistry*, **30**, 2868; Fischer, P., Mesot, J., Lucas, B., Ludi, A., Patterson, H. and Hewat, A. (1997) *Inorganic Chemistry*, **36**, 2791.Blom, N., Ludi, A., Bürgi, H.B. and Tichy, K. (1984) *Acta Crystallographica, Part C*, **40**, 1767.

70 Stender, M., White-Morris, R.L., Olmstead, M.M. and Balch, A. (2003) *Inorganic Chemistry*, **42**, 4504.

71 Crespo, O., Fernández, E.J., Jones, P.G., Laguna, A., López-de-Luzuriaga, J.M., Mendía, A., Monge, M. and Olmos, E. (1998) *Journal of the Chemical Society, Chemical Communications*, 2233.

72 Fernández, E.J., Laguna, A., López-de-Luzuriaga, J.M., Mendizábal, F., Monge, M., Olmos, M.E. and Pérez, J. (2003) *Chemistry – A European Journal*, **9**, 456.

73 Fu, W.F., Chan, K.C., Miskowsky, V.M. and Che, C.M. (1999) *Angewandte Chemie-International Edition*, **38**, 2783.

74 Fu, W.F., Chan, K.C., Cheung, K.K. and Che, C.M. (2001) *Chemistry – A European Journal*, **7**, 4656.

75 Zhang, H.X. and Che, C.M. (2001) *Chemistry – A European Journal*, **7**, 4887.

76 Fernández, E.J., Jones, P.G., Laguna, A., López-de-Luzuriaga, J.M., Monge, M.,

Montiel, M., Olmos, M.E. and Pérez, J. (2004) *Zeitschrift fur Naturforschung*, **59b**, 1379.

77 Fernández, E.J., Laguna, A., López-de-Luzuriaga, J.M., Olmos, M.E. and Pérez, J. (2004) *Dalton Transactions*, 1801.

78 Fernández, E.J., Jones, P.G., Laguna, A., López-de-Luzuriaga, J.M., Monge, M., Olmos, M.E. and Pérez, J. (2002) *Inorganic Chemistry*, **41**, 1056.

79 Lees, A. (1987) *Journal of Chemical Reviews*, **87**, 711. Itokazu, M.K., Polo, A.S. and Iha, N.Y.M. (2003) *Journal of Photochemistry and Photobiology A: Chemistry*, **160**, 27.

80 Fernández, E.J., Laguna, A., López-de-Luzuriaga, J.M., Monge, M., Montiel, M., Olmos, M.E. and Pérez, J. (2004) *Organometallics*, **23**, 774.

81 Fernández, E.J., Laguna, A., López-de-Luzuriaga, J.M., Montiel, M., Olmos, M.E. and Pérez, J. (2005) *Inorganica Chimica Acta*, **358**, 4293.

82 Fernández, E.J., Laguna, A., López-de-Luzuriaga, J.M., Montiel, M., Olmos, M.E. and Pérez, J. (2005) *Organometallics*, **24**, 1631.

83 Fernández, E.J., López-de-Luzuriaga, J.M., Monge, M., Olmos, M.E., Pérez, J., Laguna, A., Mohamed, A.A. and Fackler, J.P., Jr. (2003) *Journal of the American Chemical Society*, **125**, 2022; Fernández, E.J., López-de-Luzuriaga, J.M., Monge, M., Montiel, M., Olmos, M.E., Pérez, J., Laguna, A., Mendizábal, F., Mohamed, A.A. and Fackler, J.P., Jr (2004) *Inorganic Chemistry*, **43**, 3573.

84 Fernández, E.J., Laguna, A., López-de-Luzuriaga, J.M., Montiel, M., Olmos, M.E. and Pérez, J. (2006) *Organometallics*, **25**, 1689.

85 Burini, A., Fackler, J.P., Jr, Galassi, R., Grant, T.A., Omary, M.A., Rawashdeh-Omary, M.A., Pietroni, B.R. and Staples, R.J. (2000) *Journal of the American Chemical Society*, **122**, 11264.

7
Liquid Crystals
Manuel Bardají

7.1
Introduction

7.1.1
What Are Liquid Crystals?

The history of liquid crystals started with the pioneer works of Reinitzer and Lehmann (the latter constructed a heating stage for his microscope) at the end of the nineteenth century. Reinitzer was studying cholesteryl benzoate and found that this compound has two different melting points and undergoes some unexpected color changes when it passes from one "phase" to another [1]. In fact, he was observing a chiral nematic liquid crystal.

As its name suggests, a liquid crystal is a fluid (liquid) with some long-range order (crystal) and therefore has properties of both states: mobility as a liquid, self-assembly, anisotropism (refractive index, electric permittivity, magnetic susceptibility, mechanical properties, depend on the direction in which they are measured) as a solid crystal. Therefore, the liquid crystalline phase is an intermediate phase between solid and liquid. In other words, macroscopically the liquid crystalline phase behaves as a liquid, but, microscopically, it resembles the solid phase. Sometimes it may be helpful to see it as an ordered liquid or a disordered solid. The liquid crystal behavior depends on the intermolecular forces, that is, if the latter are too strong or too weak the mesophase is lost. Driving forces for the formation of a mesophase are dipole–dipole, van der Waals interactions, π–π stacking and so on.

Liquid crystals are classified into two broad groups depending on the method used to get this state:

1. *Thermotropic* liquid crystals: the temperature is the variable that determines which phase of matter exists.
2. *Lyotropic* liquid crystals: they display liquid crystalline behavior when mixed with another material in the right concentration (typically a solvent). They can also be a mixture of more than two components (e.g., cell membranes).

Modern Supramolecular Gold Chemistry: Gold-Metal Interactions and Applications.
Edited by Antonio Laguna
Copyright © 2008 WILEY-VCH Verlag GmbH & Co. KGaA, Weinheim
ISBN: 978-3-527-32029-5

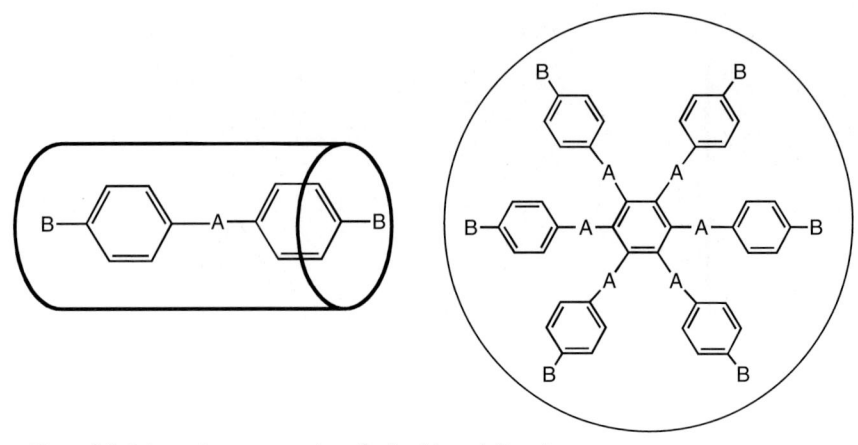

Figure 7.1 Schematic representation of calamitic and discotic liquid crystals. A = alkynyl, CH_2, $C(O)O$, $N=N$, ...; B = OC_nH_{2n+1}, C_nH_{2n+1}.

Moreover, there are amphiphilic molecules that can behave as both thermotropic and lyotropic liquid crystals.

Some basic *terms* are:

Mesogen: a substance that shows liquid crystal behavior (exhibits mesomorphism).
Isotropic state: conventional liquid state.
Mesophase: liquid crystalline phase.
Melting point: the temperature at which a thermotropic liquid crystal passes from the solid to the mesophase (or to an isotropic liquid).
Clearing point: the temperature at which the mesophase transforms into an isotropic fluid.

Thermotropic liquid crystals are divided into two principal types, by considering the shape of the units forming the mesophase (Figure 7.1):

1. *Calamitic*: rod-like. A typical structure consists of a rigid core formed by aromatic rings connected by a bridge (A) and flexible wings (B), in order to extend the conjugated system but maintain the linearity and generate permanent dipole moments. A higher number of aromatic rings implies a higher melting point. However, the presence of alkyl chains acts as a thermal stabilizer. The axial component is longer than the radial component.

2. *Discotic*: disc-like. A similar structure but the radial component is more important. The core is often aromatic and surrounded by 6–8 alkyl chains.

Nowadays, rods and discs are considered as model structures because some other unconventional shapes have been recently described such as bananas (bent-core), cones and rings.

Calamitic mesophases (Figure 7.2) are classified into:

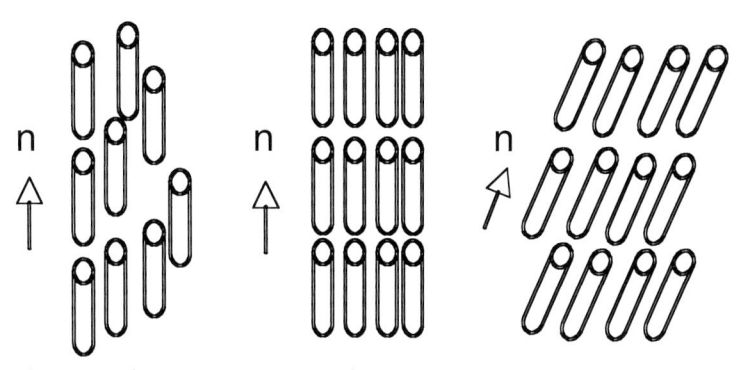

Figure 7.2 Schematic representation of N, SmA and SmC calamitic mesophases with the corresponding director vector **n**.

1. *Nematic* phase: this is the simplest structure. It is the most disordered mesophase and therefore very fluid. It is called N. In the nematic phase, the molecules are ordered mainly in one dimension with their long axes parallel, and they are free to move parallel to this axis (there is no long-range order). Nematic liquid crystal mixtures, containing various amounts of different liquid crystal compounds, are used in electro-optic display systems such as flat-panel displays.

2. *Smectic* phases: they are more highly ordered than the nematic. They are characterized by some positional correlation of the molecules into layers, in addition to orientational correlations. The simplest smectic phase is the smectic A (SmA). The long axes of the molecules are oriented on average in the same direction but, additionally, the molecules are loosely associated into layers, with the orientational direction perpendicular to the layer normal. If the SmA is modified by tilting the molecules within the layer plane, the smectic C (SmC) is obtained. The SmA may be modified by retaining the orthogonality of the molecules with respect to the layer normal and introducing hexagonal symmetry into the layer. Then, the smectic B (SmB) is obtained. Other more ordered smectic phases are denoted by the letters E, G, H, J and K.

As we will see below, N, SmA and SmC are the most typical mesophases found for gold mesogens.

There are also *chiral phases* that exhibit chirality. The simplest is the chiral nematic phase N*, which is often called the cholesteric phase because it was first observed for cholesterol compounds (Reinitzer). In the chiral nematic phase the director has a helical shape. This phase exhibits a twisting of the molecules along the director. The finite twist angle between adjacent molecules is due to their asymmetric packing, which results in longer-range chiral order. The chiral pitch refers to the distance (along the director) over which the mesogens undergo a full 360° twist. The pitch may be varied by adjusting the temperature or adding other molecules to the liquid crystal fluid. For many types of liquid crystals, the pitch is of the same order as the wavelength of visible light. This causes these systems to exhibit unique optical properties, such as selective reflection. These properties are

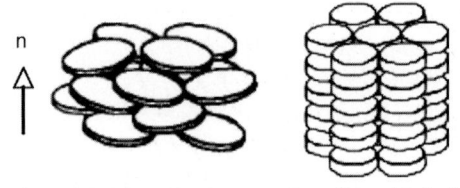

Figure 7.3 Schematic representation of N and Col$_h$ discotic mesophases with the director vector **n**.

exploited in a number of optical applications. The chiral smectic C (SmC*) phase gives rise to ferro-, ferri- or antiferroelectric properties. N* and SmC* mesophases have been found for gold mesogens.

Discotic mesophases (Figure 7.3): nematic discotic N is similar to the calamitic nematic, being very fluid and possessing only orientational order. More common are the columnar phases, such as the hexagonal columnar phase Col$_h$ with the molecules arranged in columns in a hexagonal array. In addition, rectangular, oblique and tilted modifications are known. Col$_h$ mesophases have been found for gold mesogens.

Transitions follow an order–disorder sequence on heating: C (crystal) – Sm – N – I.

7.1.2
What Are Metallomesogens?

When a liquid crystal is a metal-containing material, it is called a metallomesogen. It is commonly accepted that the first report on metallomesogens was by Vorländer in 1923, who reported a series of mercury mesomorphic complexes (Figure 7.4) [2]. The subject was not very popular until the end of the 1970s, then the number of reports increased regularly until the 1990s and then stabilized.

The introduction of a metal results in the modification of physical properties such as conductivity, color and magnetism, which increases the scope for the preparation of new materials with new applications. As shown for organic liquid crystals there is a relationship between the shape and the type of a complex, and the mesomorphism. Actually, it is often found that metals can induce mesomorphism in non-mesomorphic ligands, the ligand being known as a promesogen (it possesses a suitable structure for a mesogen although it is not). In general, higher transition temperatures than for the ligands are observed and therefore easier decomposition. Therefore, it is necessary to reduce the transition temperatures and to increase the chemical stability of the complexes.

The main differences between metallomesogens and organic liquid crystals are: (i) an increase in electric anisotropy because of the high polarizability of metals;

Figure 7.4 First metallomesogens.

(ii) color, because of d–d electronic transitions; (iii) paramagnetism, by using paramagnetic metal centers; (iv) metallic complexes allow a greater variety of geometries than carbon-based derivatives; (v) apart from metals, auxiliary ligands can be changed by inducing different molecular shapes and physical properties; (vi) changes in viscoelastic properties, and (vii) low symmetry molecules, which can cause liquid crystal behavior.

There has been a series of previous reviews on metallomesogens [3, 4] and even a book [5]. There is a also specific review of gold mesogens, published in 1999 [6].

7.1.3
Liquid Crystals Characterization

First, liquid crystals are characterized in the way Lehmann did at the beginning of liquid crystals history, by Polarized Optical Microscopy (POM). This technique consists of an optical microscope with a heating stage and two polarizing filters (only an anisotropic material appears bright between the crossed polarizers because it displays birefringence or double refraction whereas an isotropic material appears dark) in order to determine the intermediate transitions from solid to isotropic liquid and to characterize the different mesophases. Every mesophase appears to have a distinct texture resulting from the different domains, although within a domain the molecules are well ordered. Therefore this technique allows us to decide whether a sample behaves as a liquid crystal, and we can often assign the type of mesophase by the characteristic texture. Additionally, for thermotropic liquid crystals, we measure the temperature range of every mesophase, on heating and on cooling. The liquid crystal is called *enantiotropic* if the mesophase is observed on heating and on cooling and *monotropic* when it is observed only on cooling.

Secondly, a liquid crystal is always characterized by Differential Scanning Calorimetry (DSC). This technique allows us to fix more precisely the transition temperatures and the energy involved in the transition which, in turn, is useful to determine the temperature ranges, the type of mesophase (logically large order–disorder transitions imply more energy) and the thermal stability (reproducibility of the heating–cooling cycles).

7.1.4
Liquid Crystals Applications

The first application described was as temperature sensors by using a chiral nematic liquid crystal, which displays different colors at different temperatures. It is also worth noting that many common fluids are in fact liquid crystals. Soap, for instance, is a liquid crystal, and forms a variety of liquid crystal phases depending on its concentration in water.

Afterwards there appeared what has become the main application: liquid crystal displays (LCDs) based on the twisted nematic (TN) mode. These are commonly used for flat panel displays (e.g., desk calculators). Thin film transistor (TFT) LCDs enabled a large number of segments (e.g., 640×1024) to be used and they had advantages like low weight, low space requirement and low power consumption. That is why they were used firstly in notebooks and later for desktop computers and television monitors. The display performance and size has been dramatically improved, mainly by the introduction of fast switching, better viewing angle dependence, high brightness and contrast, good color quality, to such an extent that LCDs have almost eliminated the traditional cathode ray tube (CRT) monitors. It is important to note that LCDs are not a single liquid crystal but mixtures of several liquid crystals in order to get the proper temperature range, the appropriate response to an electric field, the required stability and suitable viscosity.

These applications are always for organic liquid crystals. Metallomesogens continue to be promising compounds in materials science on account of their optical, mechanical and electronic properties. A critical review of the possible applications was published in 2002 [7]. At that time, the only true application was the use of some lyotropic metallomesogens as templates for mesoporous materials. Since then, much effort has been made to prepare polyfunctional materials by combining the self-organizational ability of metallomesogens with the properties of metal complexes. For instance, those with luminescence have been proposed for applications in light emitting diodes (LEDs), information storage and sensors.

7.1.5
The Advantages of Gold

Gold(I) displays a quite simple chemistry in terms of its shape (linear two-coordinated compounds) that hinders some ways of improving metallomesogens (reduced symmetry or high coordination number). However, it is evident that the formation of linear compounds [X–Au–L] will facilitate the formation of calamitic liquid crystals. In fact, non-mesomorphic ligands behave as liquid crystals after coordination to gold centers. Moreover, gold(I) has a strong affinity for forming metal–metal interactions with gold(I) or other metallic centers, which produces a rich supramolecular chemistry. These interactions could also promote the formation of liquid crystals and some examples will be shown.

Gold(III) usually displays a square planar geometry, typically observed in d^8 metallic complexes such as palladium(II), platinum(II), rhodium(I) and iridium (I), for which an enormous number of liquid crystals have been described [3–5], mainly as orthometallated compounds. However, only a gold(III) metallomesogen has been published. Since the first gold mesogen was reported in 1986, many other compounds have been described.

7.2
Gold Mesogens

As stated above, gold(I) has a strong tendency to give linear two-coordinated compounds, which means that a maximum of two different ligands can be used. This chapter is organized by the mesogen or promesogen ligand involved, in

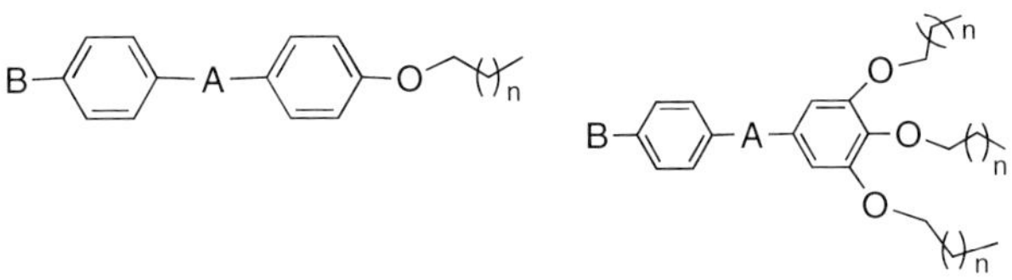

Figure 7.5 Scheme of a mesogen or promesogen ligand.
A = nothing, O(O)C, C(O)O, CH=CH; B: ethynyl, isocyanide, Li$^+$,
CS$_2^-$, N (included in ring); n = 2–10.

alphabetical order. These ligands are very often arranged as shown in Figure 7.5: one, two or three alkyl or alkoxy chains around a phenyl ring which, in turn, is bonded through a bridge A to a second phenyl (or perhalophenyl) ring functionalized by B, a functional group able to coordinate to the gold center. In some simple cases, there is neither a bridge nor two phenyl rings. Alkoxy chains are preferred to alkyl chains due to the introduction of a new dipole via the electronegative oxygen atom.

7.2.1
Alkynyl Ligands

Rod-like mononuclear derivatives have been prepared by the reaction of a polymeric alkynyl gold(I) complex with isocyanides. The mesomorphic properties of three types of gold(I) acetylide complexes, namely, [Au(CC-Ar)(CN-Ar)], [Au(CC-R)(CN-Ar)] and [Au(CC-Ar)(CN-R)] (Ar = aryl with aliphatic chain, R = alkyl chain), have been systematically examined.

Derivatives shown in Figure 7.6, where both ligands are promesogens, behave mostly as liquid crystals [8]. Actually, 20 of 24 reported complexes display smectic A phases with melting points in the range 100–170 °C and clearing points from 150 to 170 °C accompanied by some decomposition (gold mirrors). Melting points decrease regularly for equal total length of the molecule (same $m + n$) as m increases and n decreases, which means that the alkynyl substituent chain favors this behavior more than the isocyanide chain.

The close compounds bearing three aromatic rings (two of them related by an ester bridge as part of an isocyanide promesogen, Figure 7.7) also show an SmA phase

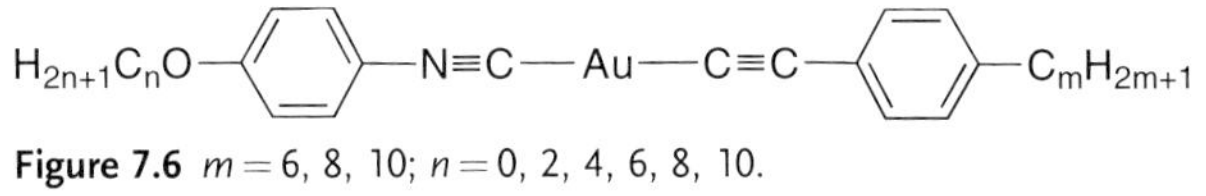

Figure 7.6 $m = 6, 8, 10$; $n = 0, 2, 4, 6, 8, 10$.

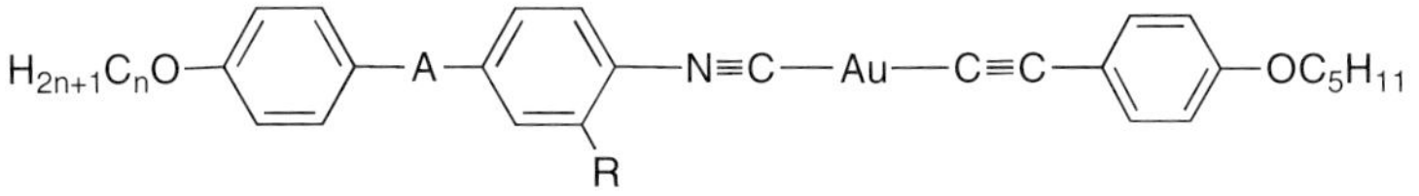

Figure 7.7 A = C(O)O, $n = 9$, R = H; A = O(O)C, $n = 10$, R = OC$_2$H$_5$, OC$_6$H$_{13}$.

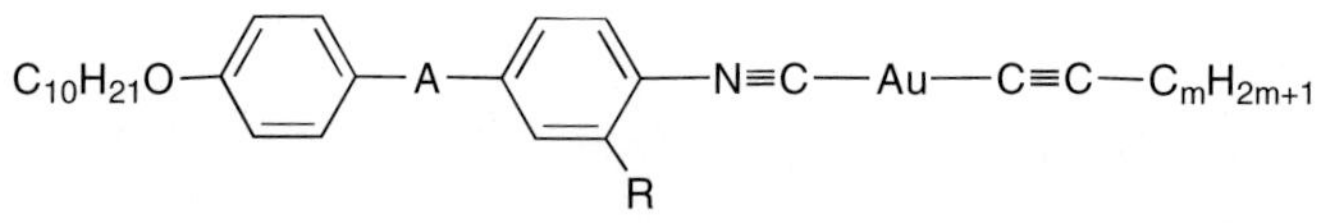

Figure 7.8 $C_{10}H_{21}O-\text{Ar}-A-\text{Ar}(R)-N\equiv C-Au-C\equiv C-C_mH_{2m+1}$

Figure 7.8 A = O(O)C, m = 4, 5, 6, R = OCH$_3$, OC$_2$H$_5$, OC$_3$H$_7$; A = C(O)O, m = 4, R = Et, OMe, OEt.

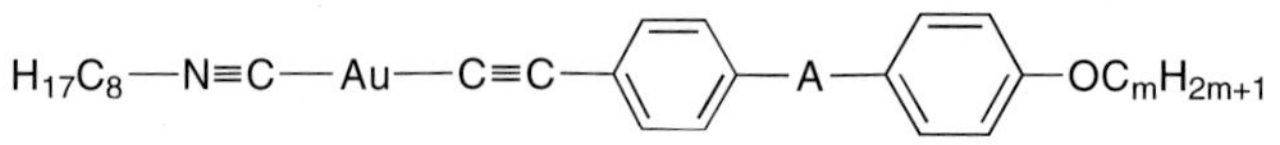

$H_{17}C_8-N\equiv C-Au-C\equiv C-\text{Ar}-A-\text{Ar}-OC_mH_{2m+1}$

Figure 7.9 A = O(O)C, C(O)O; m = 10, 12.

around 165 °C but with extensive decomposition [9]. The analogs alkynyl phosphano gold (I) complexes are not mesomorphic. Introduction of a lateral group *ortho* to the isocyanide results in the lowering of the melting points to 142 and 87.5 °C, respectively, for ethyloxy or hexyloxy. They show SmA or N phases and are more thermally stable (low or no decomposition). Introduction of a lateral group increases the width of the rod, the interactions diminish and the transition temperatures are lowered. Consequently there is less decomposition. However, the order is also reduced from SmA to N and the compound can even become non-mesomorphic.

Complexes resulting from an aromatic isonitrile and an aliphatic acetylide (Figure 7.8) are less stable, but the use of lateral substituents allows the formation of SmA and N phases at low temperatures [9].

Compounds with an alkyl chain in the isocyanide and two aromatic rings in the acetylide gave the best results in this study (Figure 7.9). They display N phases over a wide range. Transition temperatures are lower if the octyl isocyanide is branched, in fact, when the branch is in the α position the nematic phase does not appear on heating, only on cooling to 69 °C [9]. This illustrates the possibility of tuning the transition temperatures by suitable choice of the width of the same metallomesogen.

7.2.2
Azobenzene Ligands

4,4′-Disubstituted azobenzene compounds are liquid crystals that have been used as ligands to prepare metallomesogens because of their photosensitivity. This optical property can be exported to the new derivatives. The chemistry of gold is very scarce and only some gold(III) derivatives have been synthesized (Figure 7.10) with the ligand acting as mono- or bi-dentate (orthometallated). Surprisingly, none of the complexes shows mesomorphic behavior [10].

7.2.3
Carbene Ligands

The nucleophilic addition of alcohols (with alkyl chains) to gold (I) isocyanide complexes [AuCl{CNC$_6$H$_4$A(C$_6$H$_4$)$_x$OC$_n$H$_{2n+1}$}], which will be described below,

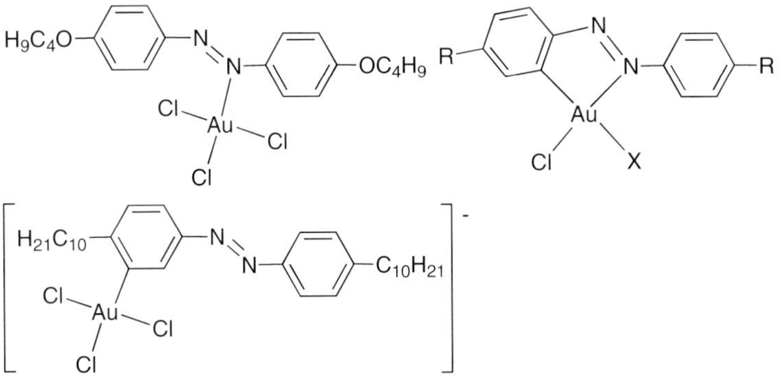

Figure 7.10 R = C$_4$H$_9$, OC$_4$H$_9$; X = Cl, CH$_2$C(O)Me.

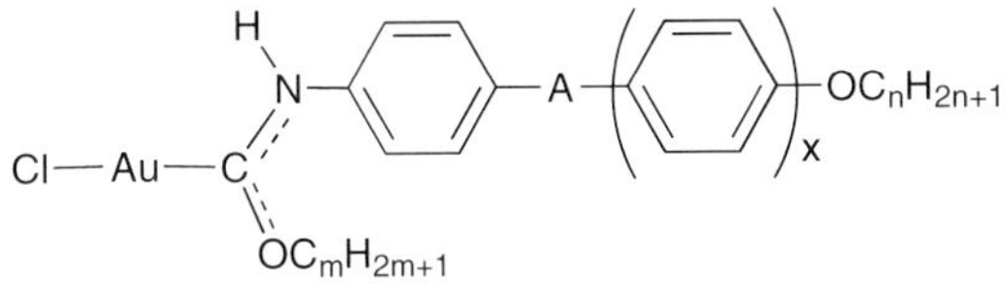

x = 1: A = OC(O), n = 9-12, m = 2-5; A = C(O)O, n = 10-12, m = 2, 3;
A = CC, n = 8-10, m = 2, 3; A = CH$_2$-CH$_2$, n = 8, m = 2.
x = 2: A = C(O)O, n = 8, m = 2

Figure 7.11 N, O-carbene gold mesogens.

affords calamitic liquid crystals with a variety of bridges (Figure 7.11) [11]. The complexes show enantiotropic smectic A or smectic C phases, in general more stable than the gold isocyanide starting materials. The partial double-bond character of the C−N bond in the carbene complex gives rise to two geometrical isomers E:Z (about 3:1 in favour of isomer E), which are separated by means of fractional crystallization. The liquid crystal behavior of both isomers is almost the same.

The same reaction carried out with amines leads to diaminocarbenes (Figure 7.12) [11]. Only complexes prepared from primary amines show smectic A mesophases (melting points from 113 to 137 °C), although they gradually decompose before the clearing points.

A dinuclear gold-carbene complex is obtained when the reaction is carried out with the diamine HNCH$_3$C$_6$H$_{12}$NHCH$_3$ (Figure 7.13) [11]. This dinuclear compound

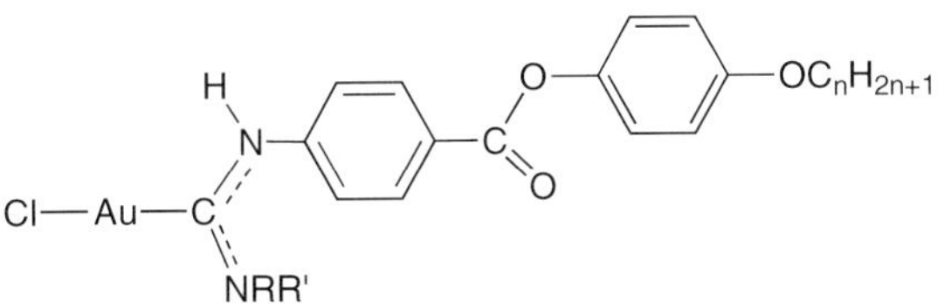

Figure 7.12 N, N-carbene gold mesogens. R = H, R′ = n-Bu, n-Pr, Et; R = Me, Et, R′ = n-Bu, Et.

Figure 7.13 N, N-carbene dinuclear mesogen.

generates a smectic mesophase at 211 °C, although it decomposes before transition to the isotropic state.

Treatment of [AuCl(SMe$_2$)] with 1,3-dialkylbenzimidalozolium cation under phase-transfer catalysis conditions leads to the quantitative formation of a cationic biscarbene derivative (Figure 7.14) [12]. X-ray diffraction studies in the mesophase and POM show a *lamellar β phase*, rather unusual in gold mesogens. Melting points are in the range 92–102 °C and clearing points range from 152 to 162 °C. These

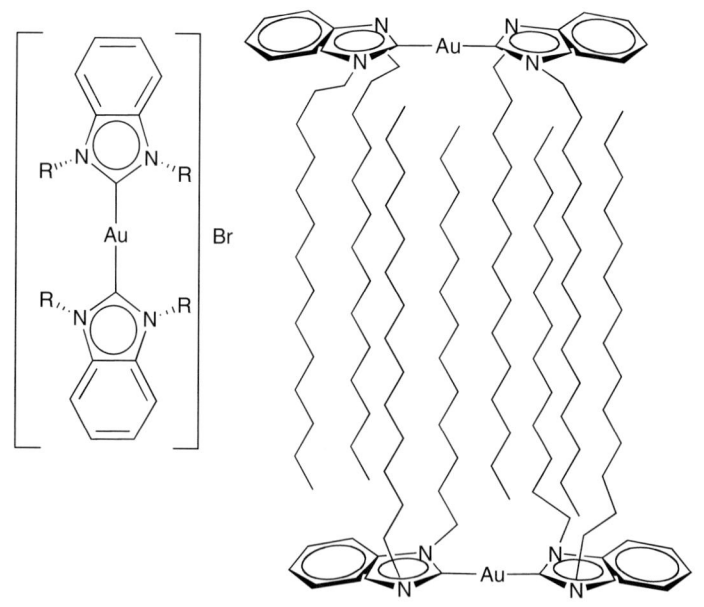

Figure 7.14 N, N-heterocyclic bis(carbene) gold mesogen and schematic representation of the bilayer. R = C_nH_{2n+1}; n = 12, 14, 16.

ordered liquids can be considered as *ionic liquids* due to their low melting point (below 100 °C). The structure of the solid is solved by monocrystal X-ray diffraction. It reveals that the four-alkyl chains are perpendicular to the aromatic gold plane like a tabletop with four legs. Interdigitation of two planes with interpenetrating alkyl chains leads to a *bilayer* with a thickness of 22 Å (Figure 7.14). Stacking of bilayers with Au–Au distances of 3.66 Å and Br–O bridges (anion with water of crystallization) produces the lamellar structure. A lamellar spacing of about 26 Å is measured, which is consistent with the repeating layer distance found by X-ray diffraction studies in the mesophase (27.9 Å). A careful inspection of the X-ray data and the fact that the clearing enthalpies are quite high (11.4–31.1 kJ mol^{-1}; although the melting enthalpies 48–118.2 kJ mol^{-1} are also high) have led to other authors claiming that these materials would be better described as crystals [4].

7.2.4
Dithiobenzoate Ligands

Amongst the first ligands used in metallomesogen chemistry were the alkoxydithiobenzoates. The reaction with [AuCl(tht)] gives the dinuclear homoleptic gold(I) derivatives [Au$_2$(μ-S$_2$C-C$_6$H$_4$OC$_n$H$_{2n+1}$)$_2$] [13]. Oxidative addition with halogens affords the mononuclear gold(III) complexes from which the methyl derivative can be obtained by a substitution reaction of chloride (Figure 7.15). All the gold(III) complexes are liquid crystals, although the ligands are non-mesomorphic. It is worth remarking that these are the first and the *only gold (III) mesogens* reported so far. The halo complexes show SmA mesophases with melting points in the range 145–166 °C and clearing points from 190 to 200 °C, although with a lot of decomposition. The methyl compound displays a lower melting point of 60 °C, but the clearing point at 130 °C is also accompanied by decomposition.

7.2.5
Isocyanide Ligands

Isocyanide or isonitrile derivatives are thermally unstable and are considered pestilent substances (not true for those bearing aromatic substituents, some of which are even fragrant [14]), which have delayed their use in coordination chemistry. Anyway, these ligands are the most commonly found in gold metallomesogens and therefore the number of reported examples is, comparatively, quite high. Isocyanide is strongly attached to the gold centers, which is not the case for other mesogen or promesogen fragments with N- or O-donor atom ligands. This leads to suitable thermal stability of the complexes and therefore they melt without decomposition. As

Figure 7.15 X = Cl, Br; n = 8, 10; X = Me, n = 8.

Figure 7.16 R = C_5H_4N, C_6H_4CN, $C_6H_4CCC_5H_4N$.

stated in Section 7.1 this is critical for obtaining liquid crystals. Moreover, these ligands maintain an appropriate molecular shape due to their linear geometry. On the other hand, phosphano derivatives are even more stable but their tetrahedral geometry makes liquid crystal behavior difficult. Isocyanide gold(I) derivatives are well known to give short gold–gold contacts, which can perturb the mesogen behavior by increasing the melting point. Isocyanides also possess a considerable dipole moment (3.44 D for CNPh).

7.2.5.1 Combined with an Alkynyl Ligand

In Section 7.2.1 we dealt with derivatives obtained by mixing alkynyl and isocyanide promesogens. In a similar way, rod-like organogold(I) complexes can be prepared by combining an isocyanide (already a mesogen) with standard alkynyl ligands (Figure 7.16) [15]. They exhibit a SmA mesophase (except the non-mesogen for R = $C_6H_4CCC_5H_4N$) with melting points in the range 87–144 °C and clearing points from 101 to 215 °C.

7.2.5.2 Combined with Halides or Pseudohalides

The *first isocyanide gold(I) mesogens* were reported in 1994. They show SmA and SmC mesophases (existence in the range 170–270 °C), while the ligands display nematic and SmA mesophases at lower temperatures and over a much smaller range (70–85 °C) (Figure 7.17) [16]. Complexes are prepared by displacement of a weakly coordinated ligand in [AuCl(SMe$_2$)]. This is the standard procedure in isocyanide gold(I) chemistry. Derivatives decompose around the clearing point due to the high transition temperature (270 °C). The introduction of lateral substituents has been carried out to weaken the interactions and therefore to decrease the transition temperatures (Figure 7.18). Actually, the isocyanide ligands become

Figure 7.17 $n = 9, 10$.

Figure 7.18 $n = 10$, $m = 1–6$; $n = 8, 9, 11, 12$, $m = 2$.

X—Au—C≡N—⟨C6H4⟩—OC$_n$H$_{2n+1}$

Figure 7.19 X = Cl, Br, I; n = 2, 4, 6, 8, 10, 12.

non-mesomorphic. All the gold complexes are mesogens, even with a long lateral chain such as hexyloxy. They exhibit N and/or SmA mesophases at lower temperature and no decomposition is observed at the clearing points. This shows that the longer the lateral alkoxy chains, the more the transition temperatures are lowered. Moreover, the less ordered mesophase N and monotropic transitions (only on cooling) are favored, as expected after decreasing the intermolecular interactions. Preliminary X-ray diffraction studies on a monocrystal of the methoxy laterally substituted compound show a short gold–gold distance of 3.40 Å, which means that a weak gold–gold interaction still occurs.

Complexes with the simplest alkoxyphenylisocyanide and several halides are prepared by metathetical reactions of [AuCl(CNR)] with KX salts (Figure 7.19) [17]. The chloro-derivatives ($n \geq 4$) and the bromo-complexes ($n \geq 6$) display SmA phases. However, the ligands and the iodo-complexes are not liquid crystals. The transition temperatures decrease in the order Cl > Br > I, according to the decrease in polarity of the Au—X bond. It is important to note that the coordination of a very simple non-mesomorphic isocyanide (only one alkoxy chain and one aromatic ring) to Au—Cl allows the formation of a quite ordered and stable smectic mesophase.

By the same procedure are obtained the corresponding biphenyl isocyanide derivatives (Figure 7.20) [18]. Now, the free isonitriles are already liquid crystals displaying nematic and SmA phases with a short range of existence at moderate temperatures (40–85 °C), while the complexes show a marked increase in the melting points and also an expansion of the range of existence of the mesophase (up to 140 °C; N and SmA phases). The exception is the shortest iodo-derivative, which is not a mesogen. Most of the complexes decompose into the isotropic state (above 220 °C). The biphenyl moiety increases the polarizability anisotropy compared to the phenyl and hence facilitates liquid crystal behavior.

Several modifications of the linear isocyanide CN-C$_6$H$_4$-OC$_n$H$_{2n+1}$-p have been carried out in order to study the structure–mesomorphism relationship [19]. None of the free isocyanides is a liquid crystal. The introduction of two new alkoxy chains in [AuX(CN-C$_6$H$_4$-OC$_n$H$_{2n+1}$-p)] leads to a dramatic change in the molecular shape, which cannot be considered rod-like but rather semidiscotic (Figure 7.21). Therefore, the trialkoxy derivatives exhibit *hexagonal columnar mesophases* at room temperature from $n = 6$, through a disc-like arrangement of alkoxy chains. This can be achieved simply if the molecules are considered by pairs in an antiparallel arrangement and probably short gold–gold distances, as is often observed in [AuXL] molecules

X—Au—C≡N—⟨C6H4⟩—⟨C6H4⟩—OC$_n$H$_{2n+1}$

Figure 7.20 X = Cl, Br, I; n = 4, 6, 8, 10, 12.

Figure 7.21 X=Cl, n=4, 6, 8, 10; X=Br, I, n=10 and schematic representation of the disc-like arrangement.

(Figure 7.21). Actually, [AuCl(CNR)] (R = Me, Et, i-Pr, Cy) compounds typically show infinite zigzag chains formed by antiparallel molecules with gold–gold distances in the range 3.388–3.637 Å [20, 21]. X-ray diffraction studies in the mesophase show an intra-columnar separation of only 4 Å, which supports this explanation. This is a clear example of the so-called *complementary molecular shape approach*: supramolecular correlations between non-discotic metal complexes produce a resulting structure further organized to form a columnar mesophase. The transition temperatures follow the expected sequence: Cl > Br > I.

Lateral fluorination affords derivatives (Figure 7.22) which display smectic A or nematic mesophases (except the iodo-2-fluoro derivative). The melting points decrease in the order: 3-F ≫ > H > 2-F while the clearing points follow the trend 3-F > H > 2-F, with no changes if the chain length n varies. It is evident that lateral substituents diminish the liquid crystal ranges that follow the sequence H > 3-F ≫ 2-F, as expected due to the broadening of the molecule which reduces the intermolecular attractions. However, it should lead to lower transition temperatures, which is not the case with 3-F derivatives. We must consider the electronic effect because higher polarization will increase the intermolecular attractions and so the transition temperatures. The presence of the electronegative F substituent in the *ortho*-position (3-F) of the alkoxy chain increases the polarization effect. As stated above, the transition temperatures of 3-F complexes decrease in the order Cl > Br > I, according to the decrease in polarity of the Au—X bond.

Finally, it is worth introducing the case of the derivatives containing alkylisocyanides (Figure 7.23) [21]. These derivatives have been studied by DSC and POM, which

Figure 7.22 X=Cl, n=6, 8, 10, 12; X=Br, I; n=12.

Figure 7.23 n=7–11.

has revealed some intermediate phases. These phases are described as crystalline but mechanically soft and easy to deform. The transition temperatures are in the range 44.5–52.2 °C for the melting points and 51–58.7 °C for the clearing points. It has been suggested that these phases should be described as *rotator phases* by taking into account the physical characteristics and the cylindrical shape of the molecule. Rotator phases have previously been described for primary alcohols as a result of the supramolecular order created by hydrogen bonds. These gold derivatives are geometrically similar to the alcohols and form aurophilic bonds of similar energy to the hydrogen bonds. Therefore, it has been suggested that rotator phases are induced by aurophilic bonding, similarly to those induced by hydrogen bonding. X-ray diffraction studies performed at variable temperature have been unable to determine the precise structure of the layers.

An analogous structure–mesomorphism relationship study has been carried out, but using pseudohalo ligands instead of halo ligands [22]. The corresponding cyano and thiocyano derivatives were prepared by reaction of [AuCl(CNR)] with KSCN or AgCN (Figure 7.24). All the gold complexes are liquid crystals, except the SCN derivative ($n = 4$), and exhibit smectic A (one alkoxy chain) or hexagonal columnar (three alkoxy chains) mesophases. Variation in transition temperatures CN > SCN can be explained because the SCN complexes are a mixture (thiocyano as S- and N-donor) whereas the CN complexes are pure. Further, the SCN ligand increases the molecular breadth considered as a cylinder, which diminishes interactions and transition temperatures. X-ray diffraction studies in the mesophase of the thiocyano compound confirm the hexagonal columnar mesophase with an intra-columnar separation of only 3.3 Å, shorter than found for the chloro complex. Again, disc-like arrangements of antiparallel molecules could explain the mesophase. As found in other series of compounds, lateral fluorination leads to a variation in the transition temperatures following the order 3-F > H > 2-F, while the biphenyl linker increases the mesophase range.

7.2.5.3 Combined with a Perhaloaryl Ligand

There has been an increasing interest in liquid crystals containing fluorine atoms because fluorination produces important changes in the melting temperatures, viscosity, birefringence, dielectric anisotropy and mesophase stability. This why structure–mesomorphism relationship studies were carried out in mono- and

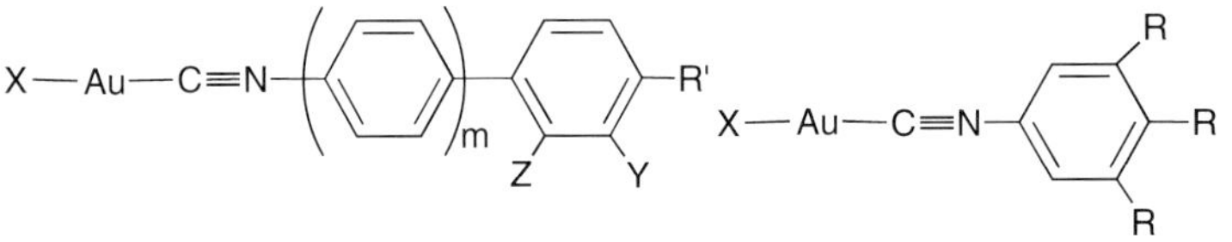

Figure 7.24 (a) One alkoxy chain: R′ = OC$_n$H$_{2n+1}$, n = 4, 6, 8, 10, 12; X = CN, SCN; Y = Z = H, m = 0, 1. R′ = OC$_{10}$H$_{21}$; X = CN, SCN, m = 0; Y = F, Z = H and Y = H, Z = F; (b) Three alkoxy chains: R = OC$_{10}$H$_{21}$, X = CN, SCN.

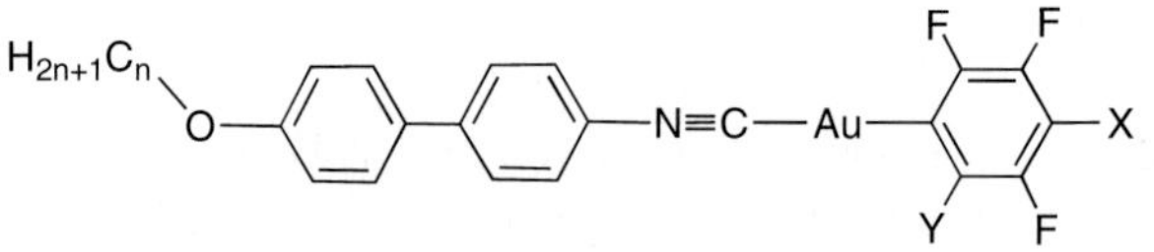

Figure 7.25 $n=4, 6, 8, 10, 12$; $X=Y=F$, $Y=Br$, $X=F$; $X=Br$, $Y=F$.

dinuclear gold(I) mesogens containing phenyl or biphenylisocyanide as promesogens and perhaloaryl as auxiliary ligands (Figure 7.25) [23].

The mononuclear derivatives show nematic and/or SmA mesophases, depending on the chain length: as usual, the longer the terminal chains, the lower the melting points and the more ordered the mesophases. The variation in transition temperatures depends on the alkoxy chain length and the substituents. Br is more polarizable than F, which increases interactions and transition temperatures, but Br in the *ortho* position leads to an increase in the molecular width, which reduces interactions and transition temperatures. The experimental results for the transition temperatures are as follows: C_6F_4Br-$p \geq C_6F_5 > C_6F_4Br$-o ($n=4, 6$) and C_6F_4Br-$p > C_6F_4Br$-$o > C_6F_5$ ($n=8, 10, 12$).

The dinuclear complexes display only N mesophases, whose transition temperatures decrease in the order biphenyl > phenyl and with longer alkoxy chains (Figure 7.26) [23]. The analogous mononuclear complex with phenylisocyanide [$Au(C_6F_5)(CNC_6H_4OC_{10}H_{21}$-$p)$] is not mesomorphic, although the chlorocompounds show SmA phases [17]. The corresponding gold(III) complexes obtained by oxidative addition of Br_2 or I_2 are not liquid crystals, since they are thermally unstable and decompose to give halo isocyanide gold(I) complexes by cleavage of the Au−C bonds.

Similar compounds to those shown in Figure 7.25 have been prepared with a different mesomorphic isocyanide (Figure 7.27) [15]. All of them are mesogens and exhibit short-range SmA mesophases, except the perfluorophenylpyridine derivative, which shows an SmA phase in the range 87.4–215 °C. This much longer

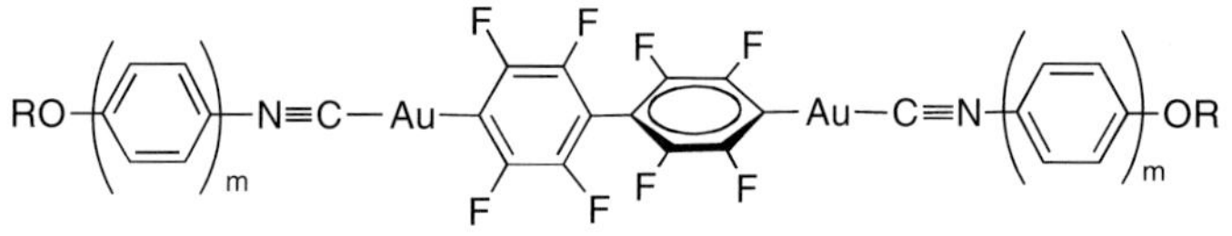

Figure 7.26 $m=1, 2$; $R=C_nH_{2n+1}$, $n=4, 6, 8, 10, 12$.

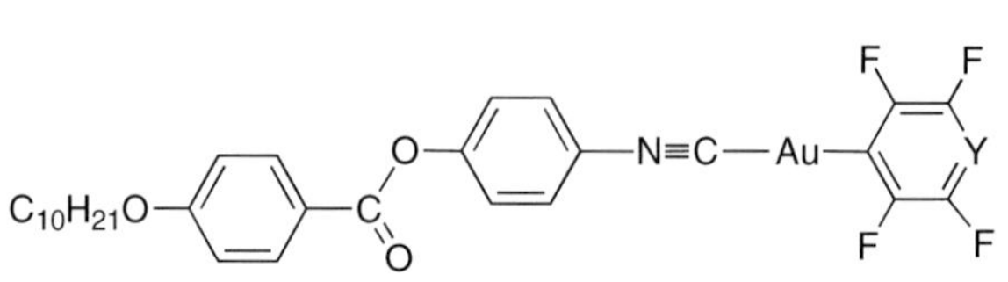

Figure 7.27 $Y=N$, C-C_5H_4N, CF.

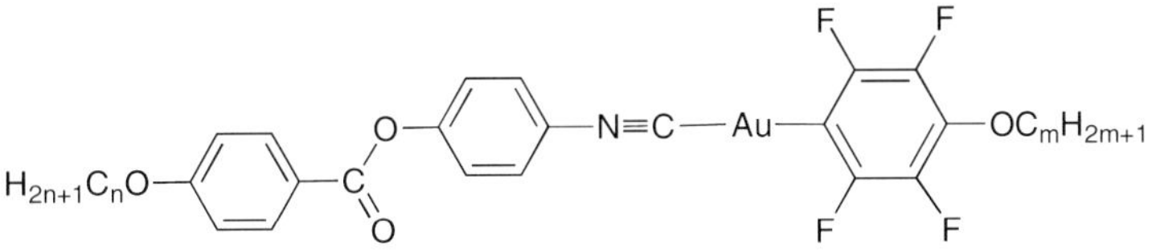

Figure 7.28 $m = 4, 8; n = 4, 6, 8, 10.$

range seems to be related to the non-planar structure because, as is well known, the perfluorophenyl ring is significantly twisted with respect to the plane of the pyridine ring.

An interesting structure–mesomorphism relationship study has been carried out on mononuclear compounds by combining two aromatic ring isocyanides (mesogens themselves) and fluoroaryl ligands with an alkoxy chain (Figure 7.28) [24]. Therefore, both ligands are going to promote liquid crystal behavior. All the derivatives are liquid crystals and their thermal stability is high, even in the isotropic state. Complexes with a short alkoxy chain ($m = 4$) in the fluoroaryl ligand display both smectic A and nematic phases for shorter chains in the isocyanide ($n = 4, 6$), whereas only smectic A phases are observed for longer chains ($n = 8, 10$). When the alkoxy substituent in the tetrafluorophenyl group is longer ($m = 8$), both SmA and SmC phases are obtained, irrespective of n.

The analogous series of compounds that uses biphenylisocyanides (also mesogens themselves) instead of the two aromatic ring isocyanides also exhibit N, SmA and SmC mesophases (Figure 7.29) [25]. In addition, they show photoluminescence in the mesophase, as well as in the solid state and in solution. There are only a few metallomesogens based on lanthanides [26], copper [27] or zinc [28] that are luminescent in the mesophase. Therefore, this is the first case of *luminescent gold mesogens* where both properties are displayed at the same time. The emission spectra show maxima at around 390, 485 and 520 nm in the solid and liquid crystal states, while only the maxima at about 390 nm is kept in solution (although all the emissions are recovered in solution at 77 K). X-ray diffraction studies confirm its rod-like structure and the absence of gold–gold interactions. However, short F–F intermolecular distances of 2.66 Å were reported.

7.2.5.4 Cationic bis(Isocyanide) Compounds

Substitution reactions on cationic gold(I) derivatives afford the corresponding complexes containing two isocyanides (Figures 7.30 and 7.31) [29]. A structure—

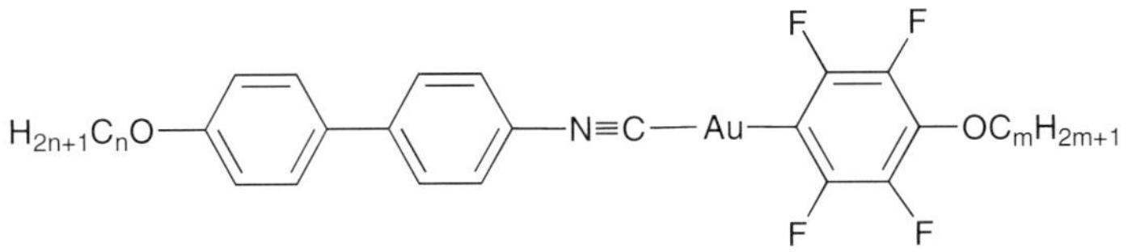

Figure 7.29 $m = 2, n = 4, 10; m = 6, n = 10; m = 10, n = 6, 10.$

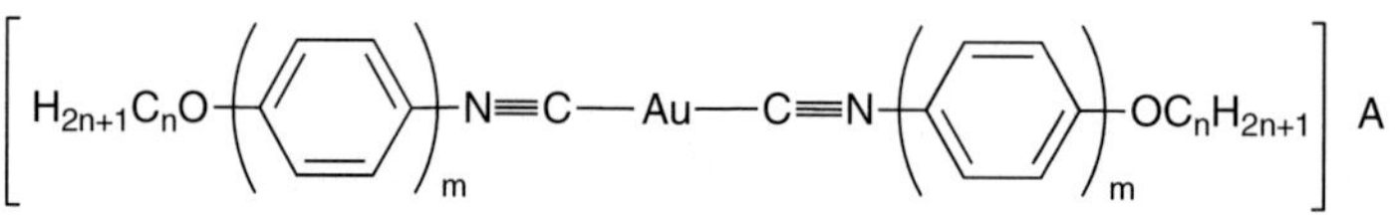

Figure 7.30 One alkoxy chain cationic mesogens. A = PF_6, BF_4; $m = 1, 2$; $n = 4, 8, 12$.

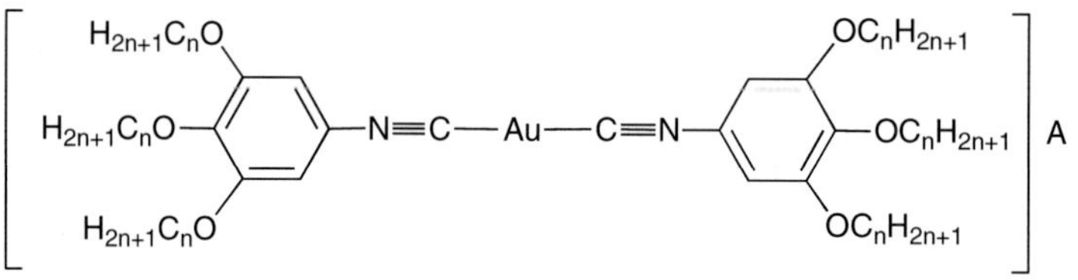

Figure 7.31 Three alkoxy chains cationic mesogens. A = PF_6, BF_4; $n = 4, 8, 12$.

mesomorphism study was carried out by taking into account the number of phenyl rings (one or two), the length and number of alkoxy chains (one or three) and the counter-ions (BF_4, PF_6, NO_3). All the new compounds are liquid crystals (except PF_6 complexes with $n = 4$), although the analogues with nitrate decompose on melting. These complexes possess double the number of chains per molecule and an additional ionic interaction compared to the corresponding [AuCl(CNR)] complexes, however, there are few changes: (i) for alkoxyphenyl isocyanide: they display the same SmA mesophase and similar mesophase ranges (the mesogenic range is greater for neutral complexes when the chains are short, and becomes wider for ionic complexes for the longest chain); (ii) for biphenyl isocyanide: they display similar SmA and SmC mesophases with a small increase in the mesophase range, although they decompose near the clearing point; (iii) for trialkoxyphenyl isocyanide: they display the same hexagonal columnar mesophases, for some of them at room temperature. X-ray diffraction studies in the mesophase confirm the hexagonal columnar mesophase, although in this case the molecules could stack rotated by 90° instead of 180°, to fill the column space more efficiently. The counter-ion makes a difference in the case of nitrate: all the gold nitrate derivatives are non-mesomorphic and undergo extensive decomposition at relatively low temperatures, which is likely related to the easy decomposition of the anion in the presence of many heavy metal cations. Some of these compounds can be described as *ionic liquids* due to their low melting point (below 100 °C).

7.2.5.5 Chiral Isocyanide

The *first helical SmC** mesophase for a gold(I) mesogen has been prepared by means of complexation to an enantiomerically pure chiral isocyanide (Figure 7.32) [30]. This

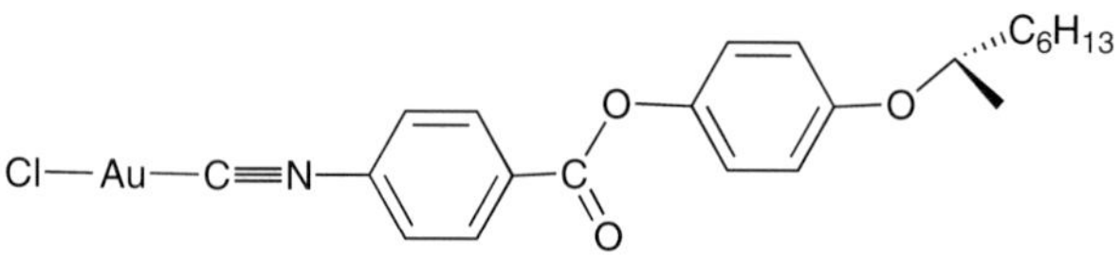

Figure 7.32 First chiral gold mesogen.

Figure 7.33 $R^1 = $ (R)-2-butyl, $R^2 = C_nH_{2n+1}$ ($n = 2$, 10); $R^1 = C_mH_{2m+1}$ ($m = 2$, 10), $R^2 = $ (R)-2-butyl; $R^1 = R^2 = $ (R)-2-butyl.

ligand is non-mesomorphic and unstable in air, but after coordination to the Au–Cl fragment it becomes stabilized and produces a mesogen like the non-chiral analogs. In addition, the chirality of the isocyanide is transferred to the mesophase. Over the chiral SmC* mesophase there is a non-chiral SmA phase. This has been related to the strong intermolecular interactions in the mesophase due to the terminal strong dipole associated with the Cl–Au–CN unit. It decomposes at the clearing point (285 °C). The ferroelectric behavior characteristic of SmC* phases has been studied although, due to experimental problems, the complex was used as a chiral dopant in a binary mixture and the maximum P_s (spontaneous polarization) value was estimated to be very small: 5 nC cm^{-2}.

The effect of introducing one or two chiral centers in gold(I) perfluoroaryl isocyanide complexes has been studied (Figure 7.33) [31]. This approach provides examples of metallomesogens presenting *cholesteric phases* (N*), as well as rare *twist-grain boundary* (TGBA*) and *blue phases* (BP), two types of frustrated phases, which are found exclusively in optically active systems. These are the first gold mesogens exhibiting these mesophases.

The complexes bearing one chiral substituent display a smectic A mesophase when the non-chiral chain is long, or an enantiotropic cholesteric and a monotropic SmA phase for shorter alkoxy chains. A TGBA* phase is observed for the derivative which contains the chiral isocyanide combined with the diethyloxy, when the SmA to cholesteric transition is studied. The compound with two chiral ligands shows a monotropic chiral nematic transition. When this compound is cooled very slowly from the isotropic liquid it exhibits blue phases BP-III, BP-II, and BP-I.

7.2.5.6 Binary Mixtures of Isocyanide Derivatives

As described in Section 7.1, applications for liquid crystals are very demanding with respect to temperature range, adequate response to an electric field, viscosity, stability, and so on. The desired properties do not occur in a unique liquid crystal and consequently all the industrial devices use mixtures. In the field of metallomesogens, these are usually limited to binary mixtures.

The thermotropic behavior and phase diagrams of binary mixtures of copper and gold complexes of the type [MX(CNR)] (X = anionic ligand, R = *p*-alkoxyaryl group) have been studied [32] and the main results are:

1. The mixture of the phenyl and biphenylisocyanide gold(I) complexes [AuCl(CN-C_6H_4-$OC_{12}H_{25}$-*p*)] and [AuCl(CN-C_6H_4-C_6H_4-$OC_{12}H_{25}$-*p*)], produces an enhance-

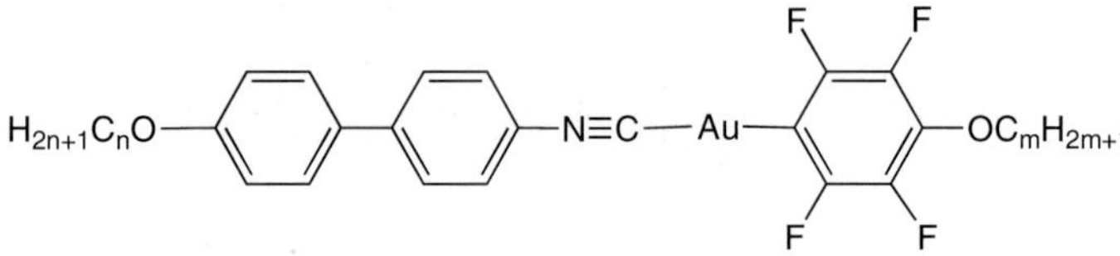

Figure 7.34 $n = 10$ and $m = 6$; $n = 6$ and $m = 10$.

ment in the range of the SmA mesophase mainly at the eutectic composition, with respect to the pure components.

2. The mixture of [AuCl(CN-C$_6$H$_4$-C$_6$H$_4$-OC$_{12}$H$_{25}$-p)] and [AuCl(CN-C$_6$H$_4$-C$_6$H$_4$-OC$_4$H$_9$-p)], the only difference being the alkoxy length, gives rise to a wider range of SmA mesophases with lower melting points and to SmC mesophases for mixtures richer in the dodecyloxy compound.

3. The mixture of two derivatives (Figure 7.34), which contain two promesogen or mesogen ligands, the only difference being where the alkoxy chain is bonded, shows melting temperatures lower than those of pure complexes, leading to room-temperature SmC mesophases, and to a higher temperature range of mesophases than for the pure components. Actually, by tuning the mixture, liquid crystal behavior can be obtained from room temperature to about 200 °C. This is the only mixture in these studies that gives a *solid solution*, whereas the others separated into their components after crystallization.

4. Finally, the mixture of two compounds which differ only in the metal (Figure 7.35), exhibits liquid crystal behavior for a large range of concentrations (at least 30 % mol of gold complex), although the starting copper compound is non-mesomorphic. Moreover, a *single fluid material* with an ordered homogeneous distribution of the two metal complexes is obtained.

7.2.6
Perhaloaryl Ligands

Most gold mesogens with perhaloaryl ligands as promesogen also contain isocyanide (mesogen or promesogen) ligands. Therefore, they were introduced in Section 7.2.5.3.

Dinuclear gold(I) complexes containing two equal promesogen alkoxytetrafluoroaryl fragments and a non-mesogenic biphenyl diisocyanide as bridging ligand have been reported (Figure 7.36) [33]. All of them are liquid crystals and display N mesophases. The transition temperatures decrease in the order 4-4′-biphenylene > 2,2′-dichloro-4-4′-biphenylene > 2,2′dimethyl-4-4′-biphenylene, which has been

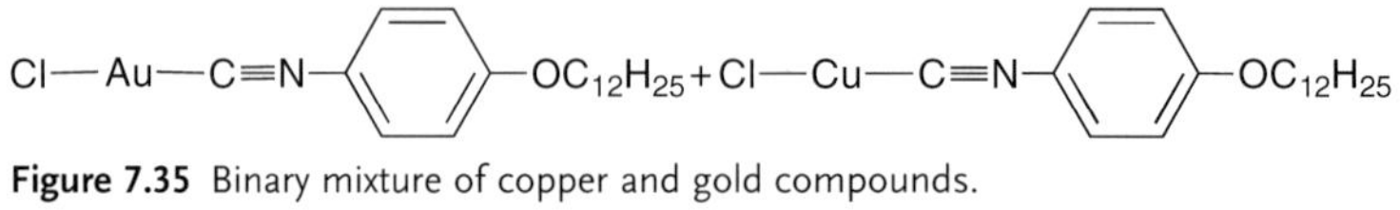

Figure 7.35 Binary mixture of copper and gold compounds.

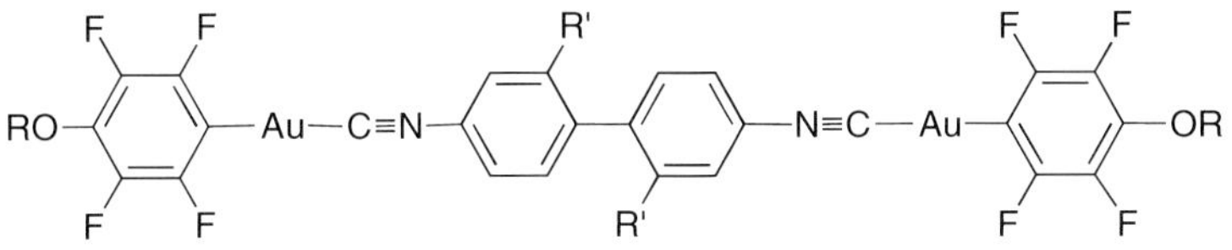

Figure 7.36 R = C$_n$H$_{2n+1}$, n = 4, 6, 8, 10, R′ = H, Cl, Me.

related to the planarity of the molecule: the twist angle of the biphenyl follows the trend Me > Cl ≫ H and more planar means stronger interactions and therefore higher transition temperatures. All compounds show photoluminescence at room temperature in the solid state (maxima from 480 to 532 nm) and in solution (maxima from 452 to 524 nm).

7.2.7
Pyrazolate Ligands

We have explained that gold facilitates the formation of molecular rods and therefore of calamitic liquid crystals. We have also shown that the use of a semidiscotic promesogen attached to a gold center affords discotic liquid crystals by the adequate supramolecular rearrangement. The mesogen or promesogen ligand has always been a monodentate ligand with a rigid core (one or two aromatic rings) and one, two or three alkyl or alkoxy flexible chains (Figure 7.5). The synthesis of an anionic bidentate pyrazolate with three aromatic rings and four to six alkoxy chains opened the possibility of making trinuclear gold(I) derivatives [{Au(pz)}$_3$]. These compounds are well known in the gold coordination chemistry of this bidentate ligand [34]. These kinds of derivatives have been called metallocrowns as structural analogs of organic crown ethers. The reaction of the potassium pyrazolate salts (Figure 7.37) with [AuCl(tht)] in a 1 : 1 molar ratio affords the expected trinuclear compounds containing 12, 15 and 18 alkoxy chains (Figure 7.38) [35]. Although the ligands are not mesomorphic, all the corresponding gold complexes show hexagonal columnar mesophases. X-ray diffraction measurements in the mesophase confirm the supramolecular columnar arrangement. The melting points are in the range 35–55 °C (one derivative is already a liquid crystal) and the clearing points are in the range 22–64 °C. After cooling, these complexes exhibit a hysteresis phenomenon and the mesophase

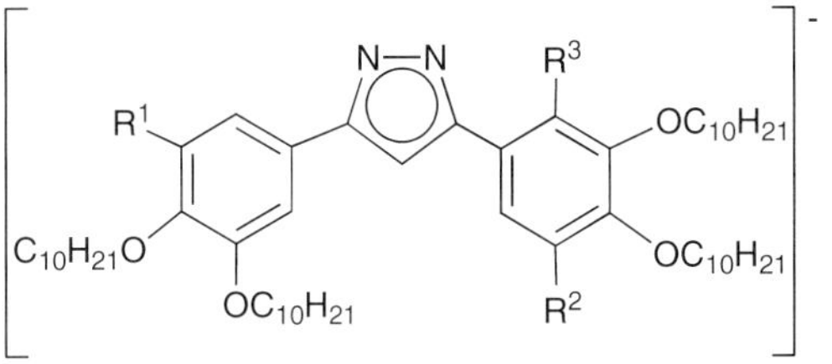

Figure 7.37 4, 5, 6 alkoxy chain pyrazolate ligands.
R^1 = R^2 = R^3 = H; R^1 = OC$_{10}$H$_{21}$, R^2 = R^3 = H;
R^1 = R^3 = OC$_{10}$H$_{21}$, R^2 = H; R^1 = R^2 = OC$_{10}$H$_{21}$, R^3 = H.

Figure 7.38 Schematic representation of the trinuclear discotic mesogens.

remains metastable at room temperature for long periods. X-ray diffraction studies carried out in the analogous compound without chains [{Au(3,5-(MeOPh)$_2$pz)}$_3$] show that the molecular structure is based on a roughly triangular planar nine-membered metallacycle core, which favors the columnar arrangements of these discotic-shaped complexes. It is worth recalling that these are the *first discotic gold mesogens*, reported in 1996.

Related complexes containing only three side chains from 4-substituted pyrazolate groups have also been found to behave as discotic metallomesogens (Figure 7.39) [36]. It is noticeable that they form hexagonal columnar mesophases with only three aliphatic side chains per molecule, although this was already reported for semi-discotic-shaped gold compounds [19]. The mesophase only appears on cooling to 112 or 85 °C, crystallisation occurring at 99 or 61 °C, respectively, for $n=7$ or 8.

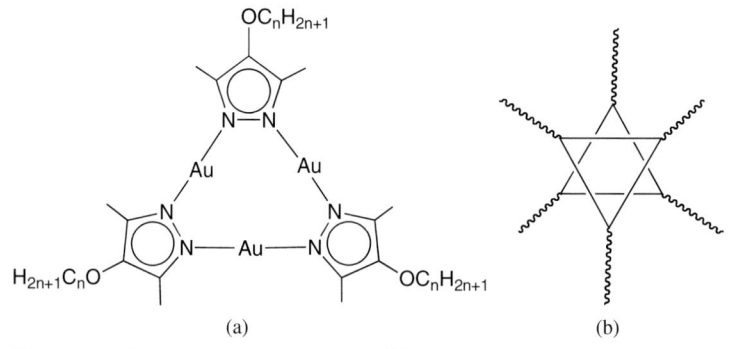

Figure 7.39 (a) Schematic representation of discotic mesogens with only three alkoxy chains ($n=7, 8$); (b) schematic representation of the star-shaped dimer.

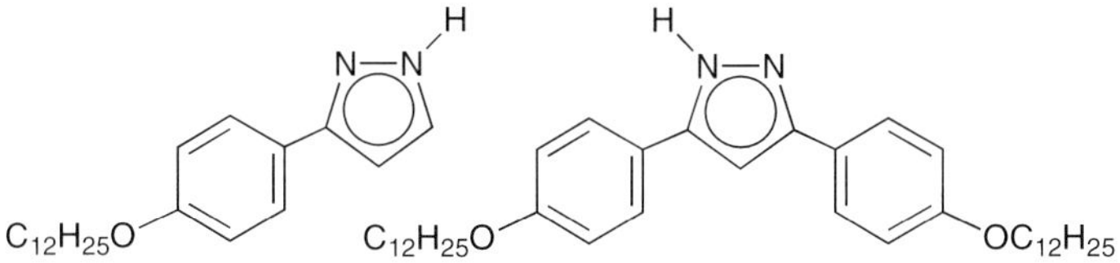

Figure 7.40 Mono- and di-substituted pyrazole ligands.

Surprisingly, the same complexes with $n = 9$ and 11 are not liquid crystals. The X-ray diffraction crystal structure of the octyloxy compound reveals an interesting feature: the trinuclear gold complex forms a *star-shaped dimer* from two triangular monomers stacked in a tilted fashion, through intermolecular Au–Au contacts of 3.255(2) Å. The mean distance between the planar cores of two dimers is 3.238(2) A and therefore each dimer behaves as a planar disc with six side chains (Figure 7.39). This is a new example of the complementary molecular shape approach taking advantage of the gold–gold interactions.

The related mono- and di-substituted pyrazole ligands have been synthesized (Figure 7.40). The free disubstituted pyrazole ligands are calamitic liquid crystals, while the monosubstituted are non-mesomorphic materials. However, only the metal derivative [{Au(pz)}$_3$] containing the asymmetric 3-[4-(dodecyloxy)phenyl]pyrazolate ligand is found to have liquid-crystal properties [37]. On the other hand, the trinuclear derivative containing double the amount of aliphatic side chains is not a mesogen. Moreover, if the dodecyloxy chain is changed to a shorter butyloxy or a longer tetradecyloxy chain, the mesomorphism is also lost. The compound exhibits a columnar mesophase, which is only observed on cooling very slowly in a narrow temperature range (76.4–83.5 °C). These facts illustrate that the mesomorphism in this kind of complex is related to the disc-like shape of the molecular core and its supramolecular rearrangement, which is affected by the length, number and position of the alkyloxy chains. Very subtle changes can inhibit this property.

The corresponding pyrazole mononuclear derivatives [AuCl(pzH)] have been prepared with the same ligands (Figure 7.41) [38]. Only the monosubstituted non-mesomorphic pyrazole leads to a liquid crystal, not the disubstituted mesomorphic ligand. The complex displays a SmA mesophase on cooling to 69.2 °C, although this is mixed with some crystallization. Moreover, this derivative is luminescent at 77 K in the solid state. In addition, the corresponding trinuclear and mononuclear derivatives, namely [{Au(pz)}$_3$] and [Au(Hpz)$_2$]$^+$, which could be expected to be liquid crystals are non-mesomorphic.

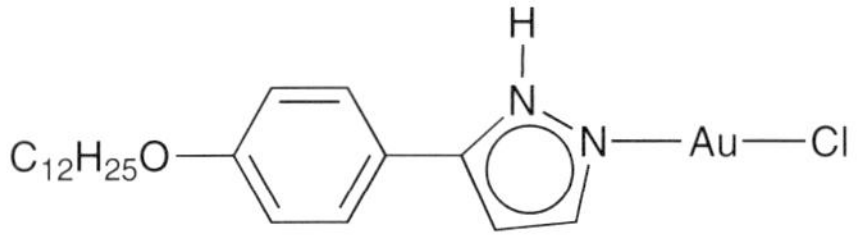

Figure 7.41 [AuCl(pzH)] derivative.

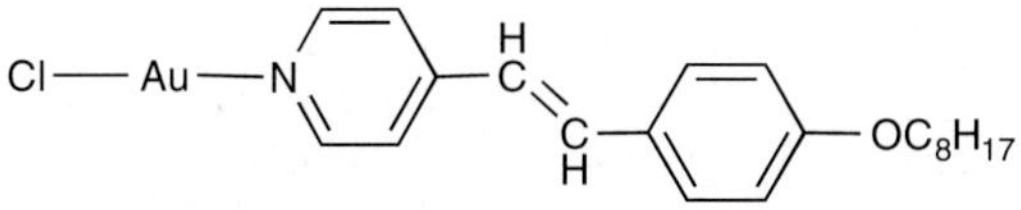

Figure 7.42 Stilbazol gold complex.

7.2.8
Stilbazol Ligands

Stilbazol ligands, or more properly 4-alkyloxy-4'-stilbazoles, are *p*-substituted pyridines with liquid crystal behavior. Only the compound [AuClL] has been reported. It exhibits an unidentified mesophase in the range 120–200 °C before decomposing (Figure 7.42) [39]. This is the *first gold metallomesogen*, reported in 1986, although it is the last of this chapter dedicated to gold mesogens.

7.3
Concluding Remarks

The literature of gold liquid crystals is somewhat limited and is dominated by organometallic gold(I) complexes of isocyanide ligands. Gold(III) chemistry remains unexplored. Clear relationships between the molecular or supramolecular structures and the physical properties can be deduced by taking advantage of the relatively simple chemistry of gold. The role of the metal is more relevant than in other metallomesogens. All kinds of mesophases can be produced with linearly coordinated gold(I) compounds (rod-like), even discotic columnar mesophases by using specific ligands and/or the rich supramolecular chemistry of gold. These mesogens provide a fluid processable source of gold, in which the molecules can be easily oriented. The synthesis of polyfunctional materials, such as luminescent liquid crystals increases the chances of future applications.

References

1 Reinitzer, F. (1888) *Monatshefte fur Chemie,* **9**, 421.
2 Vorländer, D. (1910) *Berichte der Deutschen Chemischen Gesellschaft,* **43**, 3120.
3 Donnio, B. and Bruce, D.W. (1999) *Structure and Bonding,* **95**, 193.
4 Donnio, B., Guillon, D., Bruce, D.W. and Deschenaux, R. (2006) Metallomesogens, in *Comprehensive Organometallic Chemistry III: From Fundamentals to Applications* (eds R.H. Crabtree and D.M.P. Mingos), Elsevier, Oxford, UK, Volume 12 (ed. D. O'Hare) *Applications III: Functional Materials Environmental and Biological Applications,* Chapter 12. 05, p. 195.
5 Serrano, J.L. (ed.) (1996) *Metallomesogens: Synthesis, Properties and Applications,* Wiley-VCH, Weinheim.
6 Espinet, P. (1999) *Gold Bulletin,* **32**, 127.
7 Giménez, R., Lydon, D.P. and Serrano, J.L. (2002) *Current Opinion in Solid State & Materials Science,* **6**, 527.

8 Alejos, P., Coco, S. and Espinet, P. (1995) *New Journal of Chemistry*, **19**, 799.
9 Kaharu, T., Ishii, R., Adachi, T., Yoshida, T. and Takahashi, S. (1995) *Journal of Materials Chemistry*, **5**, 687.
10 Vicente, J., Bermúdez, M.D., Carrión, F.J. and Martínez-Nicolás, G. (1994) *Journal of Organometallic Chemistry*, **480**, 103.
11 Ishii, R., Kaharu, T., Pirio, N., Zhang, S.W. and Takahashi, S. (1995) *Journal of the Chemical Society, Chemical Communications*, 1215; Zhang, S.W., Ishii, R. and Takahashi, S. (1997) *Organometallics*, **16**, 20.
12 Lee, K.M., Lee, C.K. and Lin, I.J.B. (1997) *Angewandte Chemie-International Edition*, **36**, 1850.
13 Adams, H., Bailey, N.A., Bruce, D.W., Dhillon, R., Dunmur, D.A., Hunt, S.E., Lalinde, E., Maggs, A.A., Orr, R., Styring, P., Wragg, M.S. and Maitlis, P.M. (1988) *Polyhedron*, **7**, 1861; Adams, H., Albéniz, A.C., Bailey, N.A., Bruce, D.W., Cherodian, A.S., Dhillon, R., Dunmur, D.A., Espinet, P., Feijoo, J.L., Lalinde, E., Maitlis, P.M., Richardson, R.M. and Ungar, G. 1991 *Journal of Materials Chemistry*, **1**, 843.
14 Pirrung, M.C. and Ghorai, S. (2006) *Journal of the American Chemical Society*, **128**, 11772.
15 Ferrer, M., Mounir, M., Rodríguez, L., Rossell, O., Coco, S., Gómez-Sal, P. and Martín, A. (2005) *Journal of Organometallic Chemistry*, **690**, 2200.
16 Kaharu, T., Ishii, R. and Takahashi, S. (1994) *Journal of the Chemical Society, Chemical Communications*, 1349.
17 Coco, S., Espinet, P., Falagán, S. and Martín-Alvarez, J.M. (1995) *New Journal of Chemistry*, **19**, 959.
18 Benouazzane, M., Coco, S., Espinet, P. and Martín-Alvarez, J.M. (1995) *Journal of Materials Chemistry*, **5**, 441.
19 Coco, S., Espinet, P., Martín-Alvarez, J.M. and Levelut, A.-M. (1997) *Journal of Materials Chemistry*, **7**, 19.
20 Schneider, W., Angermaier, K., Sladek, A. and Schmidbaur, H. (1996) *Zeitschrift fur Naturforschung Section B: Journal of Chemical Sciences*, **51**, 790; White-Morris, R.L., Olmstead, M.M., Balch, A.L., Elbjeirami, O. and Omary, M.A. (2003) *Inorganic Chemistry*, **42**, 6741.
21 Bachman, R.E., Fioritto, M.S., Fetics, S.K. and Cocker, T.M. (2001) *Journal of the American Chemical Society*, **123**, 5376.
22 Benouazzane, M., Coco, S., Espinet, P. and Martín-Alvarez, J.M. (1999) *Journal of Materials Chemistry*, **9**, 2327.
23 Bayón, R., Coco, S., Espinet, P., Fernández-Mayordomo, C. and Martín-Alvarez, J.M. (1997) *Inorganic Chemistry*, **36**, 2329.
24 Coco, S., Falagán, S., Fernández-Mayordomo, C. and Espinet, P. (2003) *Inorganica Chimica Acta*, **350**, 366.
25 Bayón, R., Coco, S. and Espinet, P. (2005) *Chemistry – A European Journal*, **11**, 1079.
26 Suárez, S., Mamula, O., Imbert, D., Piguet, C. and Bünzli, J.-C.G. (2003) *Chemical Communications*, 1226; Suárez, S., Imbert, D., Gumy, F., Piguet, C. and Bünzli, J.-C.G. (2004) *Chemistry of Materials*, **16**, 3257.
27 Kishimura, A., Yamashita, T., Yamaguchi, K. and Aida, T. (2005) *Nature Materials*, **4**, 546.
28 Cavero, E., Uriel, S., Romero, P., Serrano, J.L. and Giménez, R. (2007) *Journal of the American Chemical Society*, **129**, 11608.
29 Benouazzane, M., Coco, S., Espinet, P., Martín Álvarez, J.M. and Barberá, J. (2002) *Journal of Materials Chemistry*, **12**, 691.
30 Omenat, A., Serrano, J.L., Sierra, T., Amabilino, D.B., Minguet, M., Ramos, E. and Veciana, J. (1999) *Journal of Materials Chemistry*, **9**, 2301.
31 Bayón, R., Coco, S. and Espinet, P. (2002) *Chemistry of Materials*, **14**, 3515.
32 Ballesteros, B., Coco, S. and Espinet, P. (2004) *Chemistry of Materials*, **16**, 2062.
33 Coco, S., Cordovilla, C., Espinet, P., Martín-Alvarez, J.M. and Muñoz, P. (2006) *Inorganic Chemistry*, **45**, 10180.
34 Minghetti, G., Banditelli, G. and Bonati, F. (1979) *Inorganic Chemistry*, **18**, 658; Murray, H.H., Raptis, R.G. and Fackler, J.P. (1988) *Inorganic Chemistry*, **27**, 26.
35 Barberá, J., Elduque, A., Giménez, R. Oro, L.A. and Serrano, J.L. (1996) *Angewandte Chemie-International Edition*, **35**, 2832;

Barberá, J., Elduque, A., Giménez, R., Lahoz, F.J., López, J.A., Oro, L.A. and Serrano, J.L. (1998) *Inorganic Chemistry*, **37**, 2960.

36 Kim, S.J., Kang, S.H., Park, K.M., Kim, H., Zin, W.C., Choi, M.G. and Kim, K. (1998) *Chemistry of Materials*, **10**, 1889

37 Torralba, M.C., Ovejero, P., Mayoral, M.J., Cano, M., Campo, J.A., Heras, J.V., Pinilla, E. and Torres, M.R. (2004) *Helvetica Chimica Acta*, **87**, 250.

38 Ovejero, P., Mayoral, M.J., Cano, M. and Lagunas, M.C. (2007) *Journal of Organometallic Chemistry*, **692**, 1690.

39 Bruce, D.W., Lalinde, E., Styring, P., Dunmur, D.A. and Maitlis, P.M. (1986) *Journal of the Chemical Society. Chemical Communications*, 581.

8
Catalysis
M. Carmen Blanco Ortiz

8.1
Introduction

Although the latent principle of catalysis was not recognized at the time, catalysis was already used in antiquity. For example, enzymes (biocatalysts) catalyze the malting procedure in beer brewing (6000 bc) or the preparation of bread and other leavened bakery products by carbon dioxide and alcohol (2000 bc). However, the scientific method for catalysis development only began about 200 years ago, and its importance has continued to grow until the present day.

Catalysts have been used in the chemical industry for hundreds of years and many large-scale industrial processes can only be carried out thanks to the presence of catalysts. However, it is only since the 1970s that catalysis has become familiar to the general public, mainly because of developments in environmental protection, such as the well-known and widely used catalytic converter for automobiles [1].

The modern industrialized world would be inconceivable without catalysts. Catalysis is a multidisciplinary area of chemistry, particularly industrial chemistry where around 85% of all products pass through at least one catalytic stage. Anyone who is involved with chemical reactions will eventually have something to do with catalysts. For example, the contact process for the production of sulfuric acid was introduced as early as 1880. After World War II, some catalysts for crude oil processing appeared on the US and European markets and, from an environmental standpoint, they became crucial from 1970 onwards because of their contribution to the protection of the environment and thus to a generally higher standard of living.

Berzelius introduced the term "catalysis" as early as 1836 to explain various decomposition and transformation reactions. He later referred to the "special power" that some substances (catalysts) have for influencing the affinity of chemical substances. According to the Ostwald definition of catalyst (1895), it was assumed that the catalyst remained unchanged in the course of the reaction but now it is known that it is involved in chemical bonding with the reactants during the catalytic cycle. Thus, catalysis is a process in which the rate of a reaction is enhanced under

Modern Supramolecular Gold Chemistry: Gold-Metal Interactions and Applications.
Edited by Antonio Laguna
Copyright © 2008 WILEY-VCH Verlag GmbH & Co. KGaA, Weinheim
ISBN: 978-3-527-32029-5

milder reaction conditions (lower temperatures, lower pressure, etc.) by a relatively small amount of a different substance (catalyst) that does not undergo any permanent change itself.

In theory, an ideal catalyst would not be consumed but this is not always the case in practice. Owing to competing reactions, the catalyst undergoes chemical changes and its activity becomes lower (catalyst deactivation). Thus catalysts must be regenerated or eventually replaced [2].

Chemical reactions may be performed for two reasons:

1. To convert chemicals into environmentally-acceptable compounds (waste management).
2. To prepare a desired chemical compound (synthesis of bulk or fine chemicals).

Catalysts do not only accelerate reactions, they can also influence the selectivity of chemical reactions.

This means that completely different products can be obtained from a given starting material by using different catalyst systems. Industrially, this targeted reaction control is often even more important than catalytic activity [3].

8.1.1
Transition Metal Catalysis

Many transformations that were inaccessible by "traditional" organic chemistry are now possible and thus significantly increase synthesis efficiency. In just one step, catalysts enable certain transformations that increase the complexity of the molecules in such a way that it is difficult to recognize the relationship between the product and the starting material.

Important achievements in industrial chemistry have been based on the development of organometallic catalysts. The number of organometallic complexes has vastly increased in the last few decades, and their potential applications as industrial catalysts have fuelled rapid development in organic chemistry.

Progress has also been made in asymmetric synthesis through the use of a chiral metal catalyst, so-called catalytic asymmetric synthesis, which is one of the most promising methods for obtaining optically active compounds, since a small amount of chiral material can produce a large amount of chiral product [4].

It is clear that in the future new transition metal catalysts with new ligands, newly discovered reactions and improvements to existing processes will be introduced in industry.

8.1.2
Gold Catalysis

The use of gold in catalysis has been undervalued for many years due to the preconceived opinion that gold is an expensive and extremely chemically inert metal [5, 6].

8.1 Introduction

Standard textbooks normally paid little attention to gold chemistry compared with that of other metals, even other noble metals. This tendency has changed in the last two decades, with impressive development in its stoichiometric coordination and organometallic chemistry [2]. However, while platinum and palladium have been extensively used as catalysts for a long time, and copper and silver (partners of gold in the periodic table) are used in many large-scale processes, gold was not considered for these types of transformations [7].

Although isolated reports on "gold catalysis" were published before the nineteen seventies, none of them reported a better efficiency or selectivity with respect to other known catalysts [8]. It was not until 1973 that a process was developed in which gold catalytic activity was good enough compared with other metals, when Bond *et al.* reported the hydrogenation of olefins over supported gold catalysts [9]. In 1985, Hutchings predicted the potential of gold as a catalyst and proved this by studying the hydrochlorination of ethyne to vinyl chloride [10]. Almost simultaneously, but independently, Haruta published the low-temperature oxidation of CO by gold heterogeneous catalysis [11]. These studies were the first examples in which gold was proposed as the best catalyst for a process, in contrast to the poor activity of gold reported previously.

Until then, only heterogeneous catalyst had been successful. However, in the mid-1980s, the work of Ito *et al.* led to an outstanding discovery in a catalytic asymmetric aldol reaction. In this case, enantioselectivity was given by a chiral ferrocene diphosphine ligand, with a carbon nucleophile addition to a carbonyl group [12].

In chronological order, the next milestones in research were the studies by Fukuda and Utimoto on the addition of nucleophiles (water, alcohols and amines) to alkynes [13]. A decade later, Teles obtained notable turnover numbers (TONs) and turnover frequencies (TOFs) in the addition of alcohols to alkynes [14].

The oxidation of propene to propene oxide is considered an essential practice in industrial chemistry [1]. Haruta *et al.* showed that this process can be led by heterogeneous catalysis with gold supported over titania [15, 16]. Another goal in the gold catalysis sequence is the selective oxidation of some alcohols and carbohydrates with molecular oxygen, as studied by Prati and Rossi [17].

At the beginning of the new millennium, Hashmi *et al.* presented a broad research study on both intramolecular and intermolecular nucleophilic addition to alkynes and olefins [18]. One of the areas covered by these authors was the isomerization of ω-alkynylfuran to phenols [19]. After that, Echavarren and coworkers identified the involvement of gold-carbene species in this type of process, thus opening a new branch in gold chemistry [20]. And subsequently, Yang and He demonstrated the initial activation of aryl C−H bonds in the intermolecular reaction of electron-rich arenes with O-nucleophiles [21, 22].

In 2002, Hutchings began working on the direct synthesis of hydrogen peroxide, a process with well-known problems of selectivity, where the efficiency of gold made it possible to manage safe conditions [23].

This chapter offers an overview of the most relevant and most studied processes catalyzed by gold. Homogeneous and heterogeneous catalysis are presented in different sections, each organized by chemical transformation in order to facilitate comprehension.

8.2
Homogeneous Catalysis

Nowadays many reports based on homogeneous gold catalysis are published every year, even every day. Despite this, our understanding of the mechanisms involved is of great current interest and many authors try to find a rational explanation for this reactivity. An integrating theoretical (relativistic effects must be kept in mind) and synthetic study of gold catalysts will provide a deeper understanding of the fundamental properties of gold, and of course its catalytic activity [24].

8.2.1
Nucleophilic Additions to C—C Multiple Bonds

The coordination of gold complexes to the C—C Π-system activates them very efficiently in order to attack a nucleophile. The carbonylic double bond can also be activated for nucleophilic addition.

Most gold-catalyzed reactions involve this pattern of reactivity that has been studied since the eighties. The first examples were found with allenes as substrates and subsequently alkenes and alkynes started to be used, the latter being the most popular in the last five years.

The general mechanism proposed for these additions starts with the interaction of the gold catalyst with the Π-system of the substrate (Scheme 8.1). This forms an intermediate where the double or triple bond is activated for nucleophilic attack. In most examples, addition is *anti* to gold delivering vinyl gold species. The final step of the reaction involves the release of the gold catalyst by protodemetallation and the addition product.

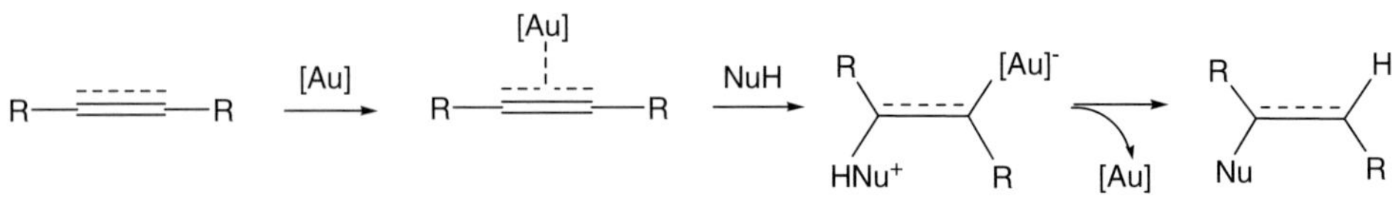

Scheme 8.1

However, in the case of norbornenes as substrates (Equation 8.1) a *syn* addition can be provided by the preferred *exo* addition at the strained cycle [25, 26].

$$(8.1)$$

8.2.1.1 Allenes as Substrates

Allenes are versatile synthetic precursors in organic chemistry and the synthesis of many natural products involves the use of allenic compounds [27]. However, they have received much less attention than alkenes or alkynes for transition metal catalyzed reactions. The explanation lies in the problems of selectivity that these substrates display as a result of their reactivity and their inherent chirality [28].

Hydroalkoxylation of Allenes In the year 2000, during their investigation of transition metal catalyzed reactions of allenyl ketones [29], Hashmi et al. discovered that gold(III) salts were able to lead the cycloisomerization and dimerization of these substrates (Equation 8.2) with a considerable improvement related to other assays with Ag (I) or Pd (II) catalysts [18].

$$(8.2)$$

Krause et al. worked on the conversions of 2-hydroxy-3,4-dienoates in the corresponding tri- and tetrasubstituted 2,5-dihydrofurans by treatment with HCl gas in chloroform. Since this reaction was not accessible with acid-labile substrates [30, 31], these conversions were tested through gold catalysis, obtaining better reaction rates and transformations in more difficult substrates compared to the well-established Ag(I)-promoted method [32].

$$(8.3)$$

8 Catalysis

Some years later, this activation was also applied in the synthesis of new 2,5-dihydrofuranes from allenamide. This reaction was achieved without loss of stereochemistry and with good yields [33].

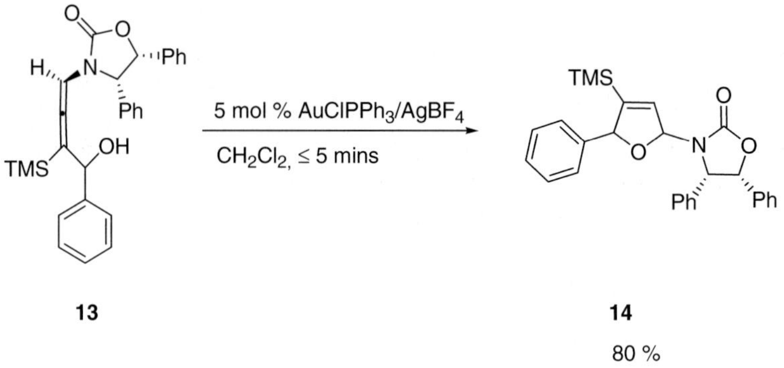

$$\text{13} \xrightarrow[\text{CH}_2\text{Cl}_2, \leq 5 \text{ mins}]{5 \text{ mol \% AuClPPh}_3/\text{AgBF}_4} \text{14} \quad (8.4)$$

80 %

An extension to the synthesis of dihydropyrans was reported by Gockel and Krause [34]. The use of the higher homologs β-hydroxyallenes afforded efficient access to the six-membered rings with axis-to-center chirality transfer, which can be explained using the mechanistic model shown in Scheme 8.2, and which once again suggests that reaction proceeds via a vinyl gold species.

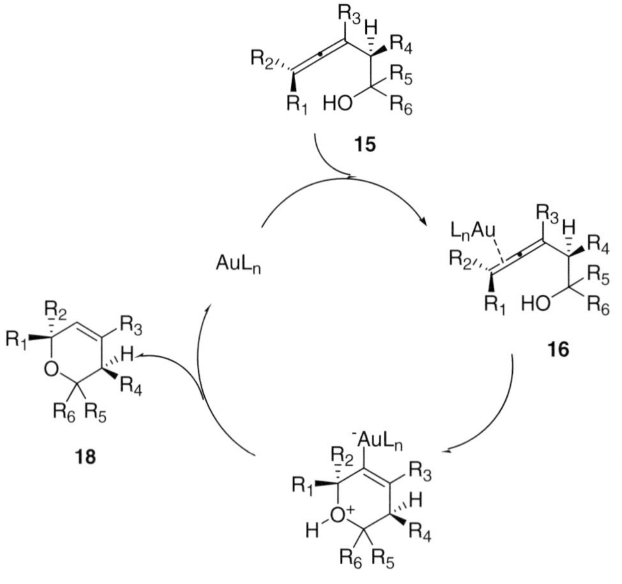

Scheme 8.2

This method was also tested with the β-aminoallene **19** and a slow but clean conversion into the tetrahydropyridine **20** was observed.

$$\text{Bu}\diagup\diagdown\text{(Me)}\diagdown\text{NH}_2 \quad \xrightarrow[\text{CH}_2\text{Cl}_2,\ \text{rt,\ 6\ d}]{5\ \text{mol\ \%\ AuCl,\ 5\ mol\ \%\ pyridine}} \quad \text{Bu-(Me)-HN-cycle} \qquad (8.5)$$

19 → **20** (76 %)

At the beginning of 2007, Zhang and Widenhoefer published a very interesting study of the enantioselective hydroalkoxylation of allenes using certain dinuclear chlorodiarylphosphino complexes and AgOTs as cocatalyst [35].

$$\textbf{21} \xrightarrow[\text{toluene,\ }-20°\text{C}]{[\text{Au}_2\{(\text{P-P})\}\text{Cl}_2]/\ \text{AgOTs}} \textbf{22} \qquad (8.6)$$

Yield 96 %
E/Z 88 %

It is worth mentioning that in some of these reactions in which gold (III) catalysts were used, some evidence for the *in situ* reduction of gold was obtained, delivering oxidative processes under exclusion of other oxidants [36].

Widenhoefer *et al.* carried out an extensive study of intramolecular exo-hydrofunctionalization with different nucleophiles [37].

Hydrothiolation of Allenes The first example of a gold-catalyzed carbon–sulfur bond formation was published by Kraus *et al.* who synthesized 2,5-dihydrothiophenes by allenyl thiocarbinols [38]. The best results were obtained with AuCl in CH_2Cl_2, providing an impressive diastereoselectivity. In the same study, other coin-metal precatalysts were tested but only gold afforded the cyclization product.

$$\textbf{23} \xrightarrow[\text{CH}_2\text{Cl}_2,\ 20\,°\text{C}]{5\ \text{mol\ \%\ AuCl}} \textbf{24} \qquad (8.7)$$

Hydroamination of Allenes Different related amines can also be cyclized. The use of free amino groups led to long reaction times (several days), but sulfonamides, acetyl or BOc as protecting group led to fast conversion (in the latter case, problems of diastereoselectivity were observed). Optimization studies showed that, although cationic gold (I) complexes were not effective for these conversions, AuCl was a very good catalyst for these reactions.

Morita and Krause first showed this with the conversion type of Equation 8.8 [39, 40].

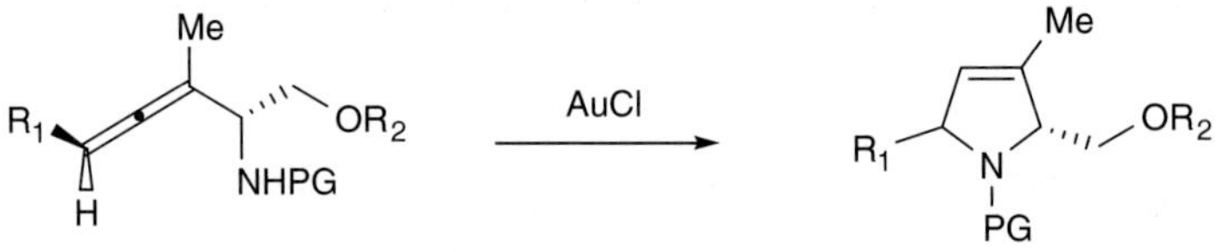

$$R_1\text{-CH=CH-C(Me)(NHPG)-CH(OR}_2\text{)} \xrightarrow{\text{AuCl}} \text{pyrroline}\quad(8.8)$$

25 → **26**

Once again mechanistic studies suggested that a gold(I) compound was the catalytically active species, even if the reaction was started with a gold (III) precatalyst.

During the studies of Lee et al. with several organometallic reagents for the development of new β-lactam antibiotics a gold-catalyzed synthesis of bicyclic β-lactams **28** was described. In this case, the suggested reaction pathway also involved a vinyl gold intermediate [41].

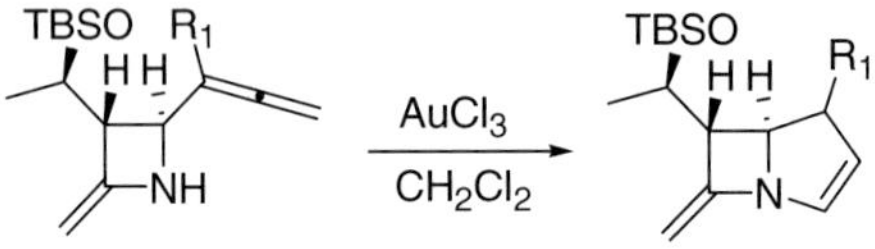

$$\textbf{27} \xrightarrow[\text{CH}_2\text{Cl}_2]{\text{AuCl}_3} \textbf{28} \quad(8.9)$$

Initial studies by Yamamoto et al. developed a highly efficient gold-catalyzed intramolecular hydroamination of allenes under very mild conditions [42].

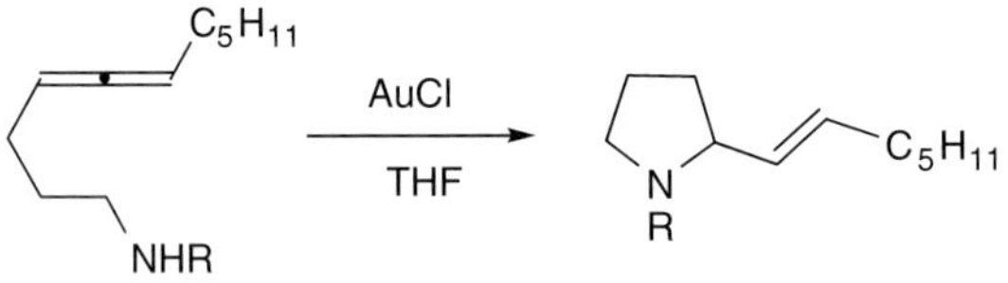

$$\textbf{29} \xrightarrow[\text{THF}]{\text{AuCl}} \textbf{30} \quad(8.10)$$

29 96 % ee

30 99 % 94% ee

Subsequently it was found that this method can also be applied to intermolecular reactions, opening further applications towards the synthesis of natural products [43].

Another application was described by Reissig et al. with the cyclization of alkoxyallenes. The most relevant finding reported in this paper was the obtainment of aromatic pyrrole, with the absence of dihydropyrrole product [44].

Despite the significance of these reactions, care must be taken when interpreting the results because some can also be obtained by silver (I) catalysts or even, as shown in a recent study by Ohno and Tanaka, in the case of sulfonamides, cyclization readily occurs under DMF reflux and basic conditions without any metal catalysts [45].

At the end of 2007, Widenhoefer *et al.* reported the first examples of the dynamic kinetic enantioselective hydroamination of axially chiral allenes, catalyzed by a dinuclear complex of gold (Figure 8.1) and silver perchlorate [46, 47].

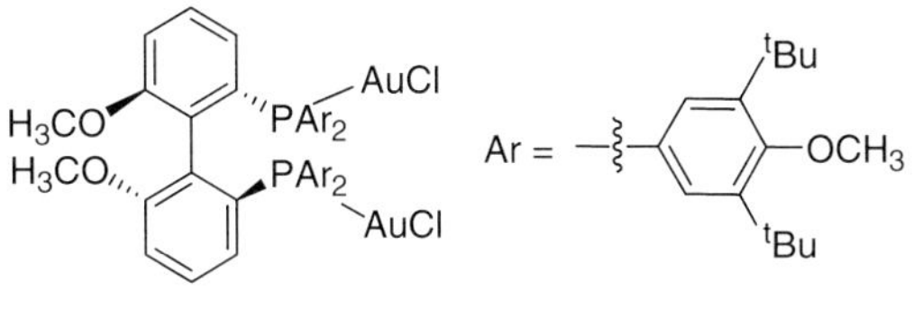

Figure 8.1 Gold complex used as catalyst for the hydroamination of axially chiral allenes.

Hydroacylation of Allenes The use of *tert*-butyl esters allowed Shin to lactonize allenoates using $AuCl_3$ as a catalyst [48]. This paper also studies the effect of different phosphine ligands used as co-catalysts.

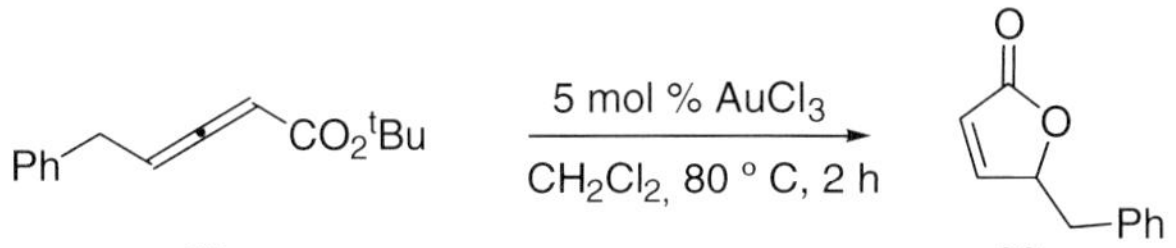

(8.11)

Reactions by Arenes The work of Hashmi *et al.* to form α,β-unsaturated ketones by reacting allenyl ketones and furans presented the first results in these types of gold-catalyzed processes [28] (Scheme 8.3).

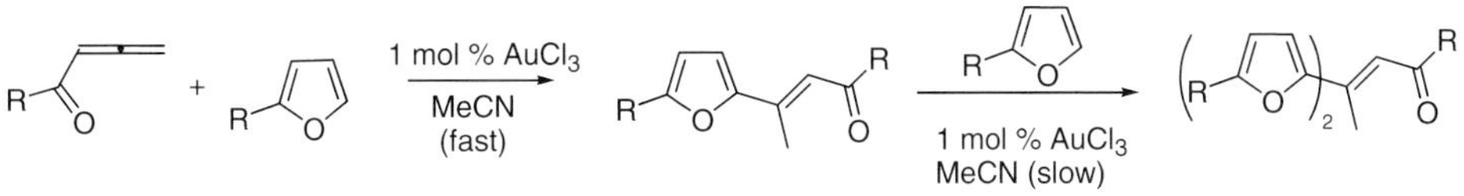

Scheme 8.3

Based on this precedent, Nelson *et al.* used an intramolecular reaction in the total synthesis of (–)-rhazinilam. In this context, trisubstituted allenes are excellent precursors for a diastereoselective heterocyclic annulation that highlights the usefulness of this reaction in target-oriented synthesis [49]. In this case, the aforementioned catalyst that yielded the best results was [AuPPh$_3$OTf], affording a higher yield and diastereoselectivity than Pd(II).

[Scheme/Equation (8.12): Compound **38** → **39**, 5 mol % [AuOTfPPh$_3$], 92 %, d.r. 97:3]

Very recently, Fujii and Ohno developed a route for the synthesis of dihydroquinoline and chromene derivatives under mild reaction conditions. Hydroarylation leads to a highly selective formation of six-membered rings, depending on the carbon (terminal or central allenic) that reacts with the aryl moiety [50].

[Scheme/Equation (8.13): Compound **40** → **41**, 1 mol % (AuClP{(^{t}Bu)$_2$(o-biphenyl)} /AgOTf), dioxane, 100 °C, 1 h, 90 %]

[Scheme/Equation (8.14): Compound **42** → **43**, 1 mol % (AuClP{(^{t}Bu)$_2$(o-biphenyl)} /AgOTf), dioxane, 60 °C, 1 h, 98 %]

Reactions by Other Nucleophiles As in the case of the formal cycloadditions of alkenes to allyl cations, the addition of alkenes to gold(I)-activated allenes generates intermediates that determine which cycloaduct formed. Based on this hypothesis, Toste et al. recently developed enantiorich bicycle-[3.2.0] structures by [2+2]-cycloaddition reaction catalyzed by chiral biarylphosphinegold(I) complexes [51].

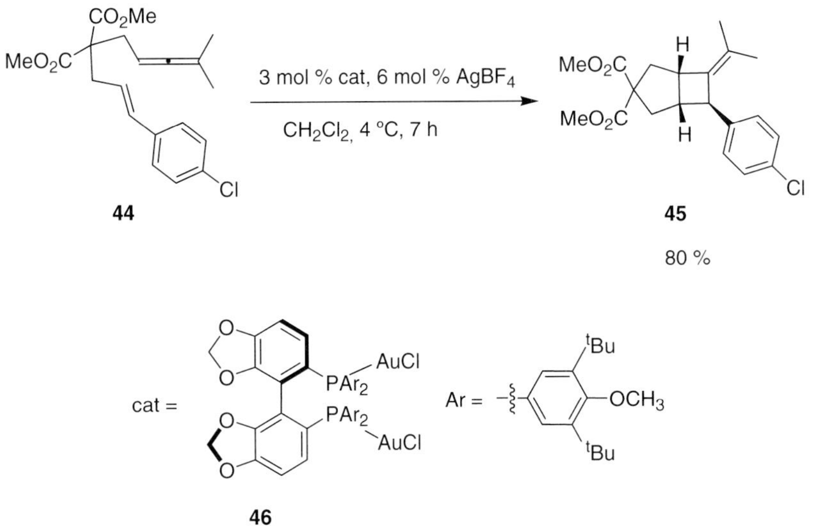

8.2.1.2 Alkenes as Substrates

The first experiments in this area were carried out by Thomas et al. in 1974 [52], inspired by their earlier work on similar reactions from alkynes and cyclopropanes [53]. However, in the case of alkenes a stoichiometric amount of gold was indeed needed for the process: gold was an oxidant and not a catalyst.

Hydroalkoxylation of Alkenes The first example was reported by Hashmi et al. when an intramolecular addition of a hydroxy group to an alkene was proposed as part of a tandem reaction [28] (Scheme 8.4).

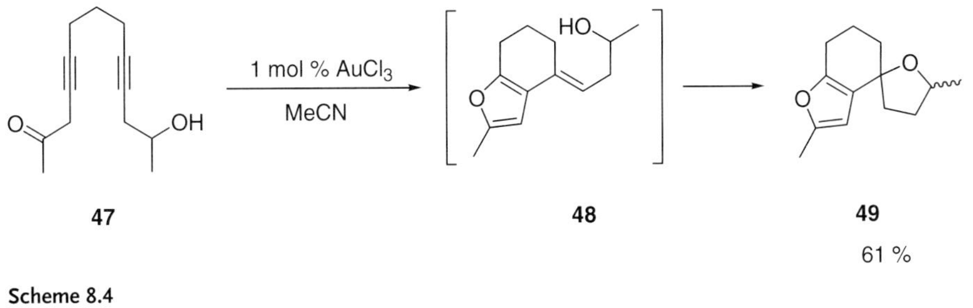

Scheme 8.4

Using *in situ* prepared [AuOTfPPh$_3$] as catalyst and relatively mild conditions, He and Yang obtained the first intramolecular addition to terminal alkenes [54]. Phenols and carboxylates were the nucleophiles chosen and several types of olefins, even unactivated ones, worked well in the reaction. In this study, the aforementioned authors also discovered gold's capacity for the constitutional isomerization of olefins, as demonstrated by the obtainment of the two isomers shown in Equation 8.16.

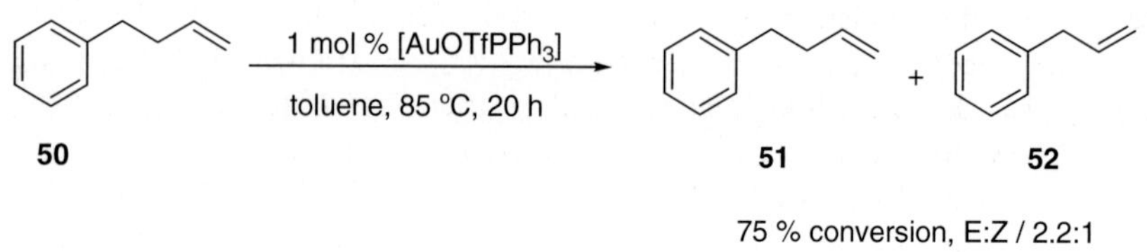
1 mol % [AuOTfPPh3]
toluene, 85 °C, 20 h
50
51
52
75 % conversion, E:Z / 2.2:1

(8.16)

Moreover, intramolecular addition of alcohol to olefin was shown to give comparable yields to the platinum-based system [55].

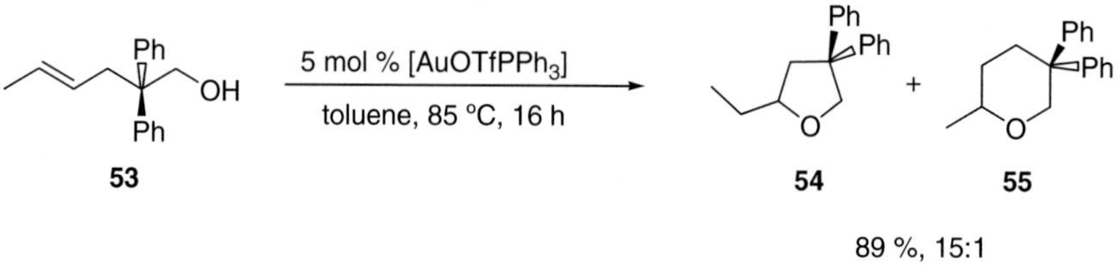
Ph
OH
Ph
5 mol % [AuOTfPPh3]
toluene, 85 °C, 16 h
Ph
Ph
O
Ph
Ph
O
53
54
55
89 %, 15:1

(8.17)

From the encouraging results obtained in the reactions of a series of gold(III) oxo complexes with olefins [56], Cinellu *et al.* tried to achieve the supposed oxametallacyclic intermediate, which had never been isolated before [25]. In the reaction of **8** and norbornene **56**, if the μ-oxo atoms were considered to be equivalents of coordinated water, and it was therefore possible to talk about the gold-catalyzed addition of water to an alkene. The metallaoxetane **58** was separated from the gold-alkene complex **57** and characterized by X-ray crystal structure analysis. The subsequent stoichiometric reaction yielded epoxide **59** (Scheme 8.5).

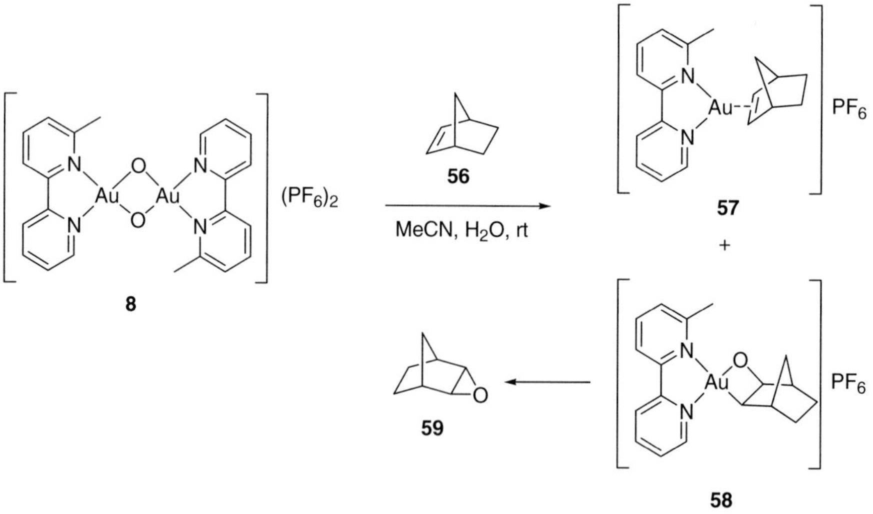
(PF6)2
56
MeCN, H2O, rt
PF6
57
PF6
8
59
58

Scheme 8.5

A tandem reaction was also tried by Floreancig *et al.*, who reported an example of intramolecular addition of alcohol to olefin [57]. In this case, reaction started with the gold-catalyzed hydration of the terminal alkyne **60** (the dichloromethane used was water-saturated). The subsequent elimination reaction and the metal-promoted conjugated addition delivered the cyclized product. It is worth noting the high diastereoselectivity obtained in these transformations, predicted on the basis of product stability. For this study, the comparison between gold(III) (NaAuCl$_4$) and gold(I) catalysts (AuClPPh$_3$/AgSbF$_6$) concluded that the gold(I) system produced a more active catalyst that was preferable for slower reactions, although gold(III) was also suitable for many reactions (Scheme 8.6).

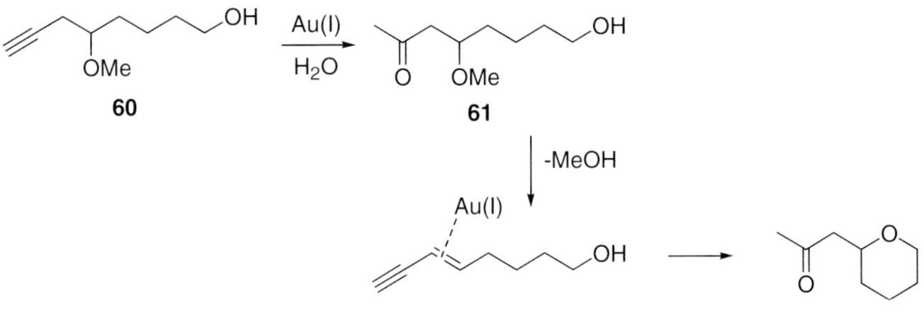

Scheme 8.6

Research in this area was developed further by Li et al. using 1,3-dienes **64** for the gold-catalyzed annulation of phenols and naphthols [58, 59]. These generated various dihydrobenzofuran derivatives. The best yields were achieved when the catalytic system included enough AuCl$_3$ and silver salt to remove halogen atoms and deliver cationic gold.

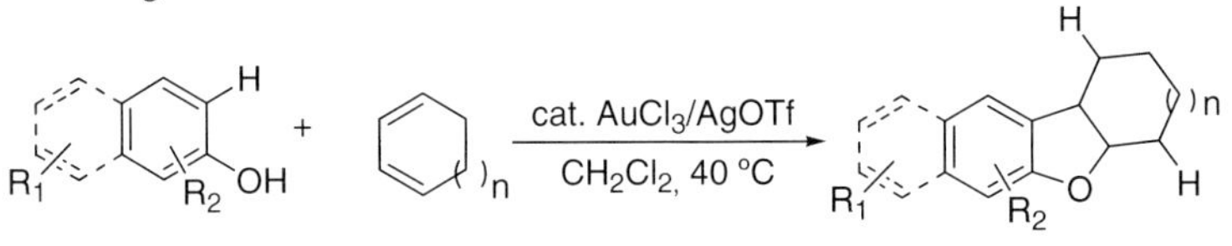

$$(8.18)$$

Hydroamination of Alkenes Kobayashi et al. found that several transition metal salts displayed high catalytic activity in aza-Michael reactions of enones with carbamates, while conventional Lewis acids (BF$_3$·OEt$_2$, AlCl$_3$, TiCl$_4$...) were much less active. Both Au(I) and Au(III) showed excellent results [60].

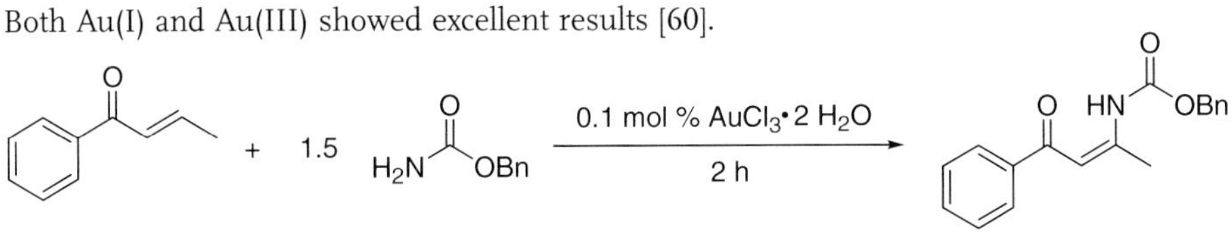

Quantitative

$$(8.19)$$

Some gold catalyst species proved to be better than platinum in the intramolecular reactions of unactivated alkenes, as studied by Widenhoefer et al. [61–63]. Gold was allowed to work under mild conditions and the scope of the reaction was also broader than with other late-transition-metal catalyst systems, leading to the formation of five- and six-membered rings.

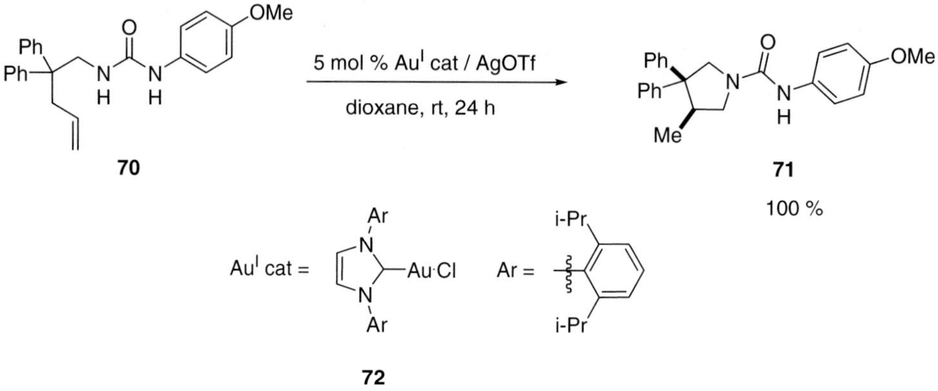

$$\text{70} \xrightarrow{\text{5 mol \% Au}^{I} \text{ cat / AgOTf, dioxane, rt, 24 h}} \text{71 (100 \%)}$$

Au^I cat = 72, Ar = 2,6-(i-Pr)$_2$C$_6$H$_3$

(8.20)

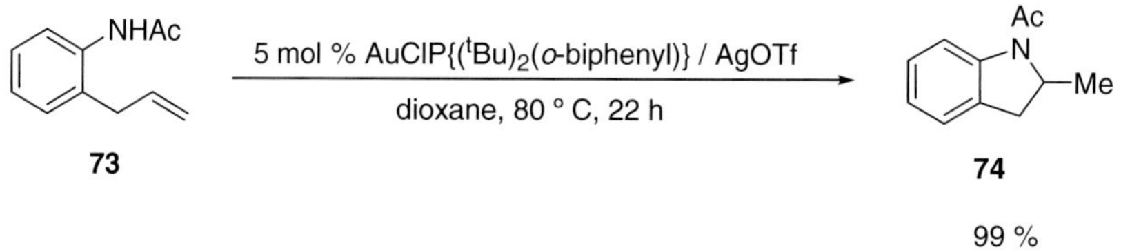

$$\text{73} \xrightarrow{\text{5 mol \% AuClP\{(}^{t}\text{Bu)}_2\text{(o-biphenyl)\} / AgOTf, dioxane, 80 °C, 22 h}} \text{74 (99 \%)}$$

(8.21)

$$\text{75} \xrightarrow{\text{5 mol \% AuClP\{(}^{t}\text{Bu)}_2\text{(o-biphenyl)\} / AgOTf, dioxane, 60 °C, 23 h}} \text{76}$$

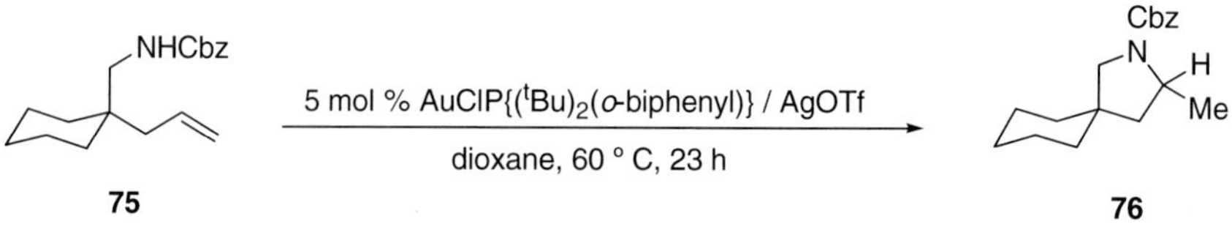

(8.22)

If the amine is protected by tosyl groups, intermolecular reaction can also be achieved [64]. In this study by He et al., a d$_2$-labeled substrate was followed by NMR. This experiment suggested that the amine attacks from the opposite face of a gold(I)-bound olefin to give the trans-addition product.

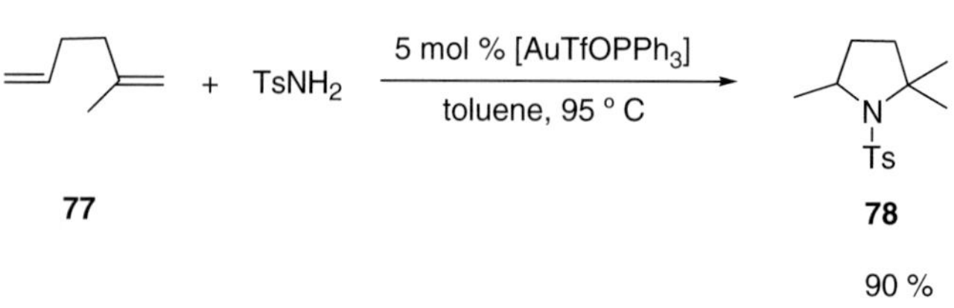

$$\text{77} + \text{TsNH}_2 \xrightarrow{\text{5 mol \% [AuTfOPPh}_3\text{], toluene, 95 °C}} \text{78 (90 \%)}$$

(8.23)

For the synthesis of protected allylic amines, a variety of synthetically useful carbamates and sulfonamides can be used, added to conjugated dienes [65]. Scheme 8.7 shows the proposed mechanism for these reactions.

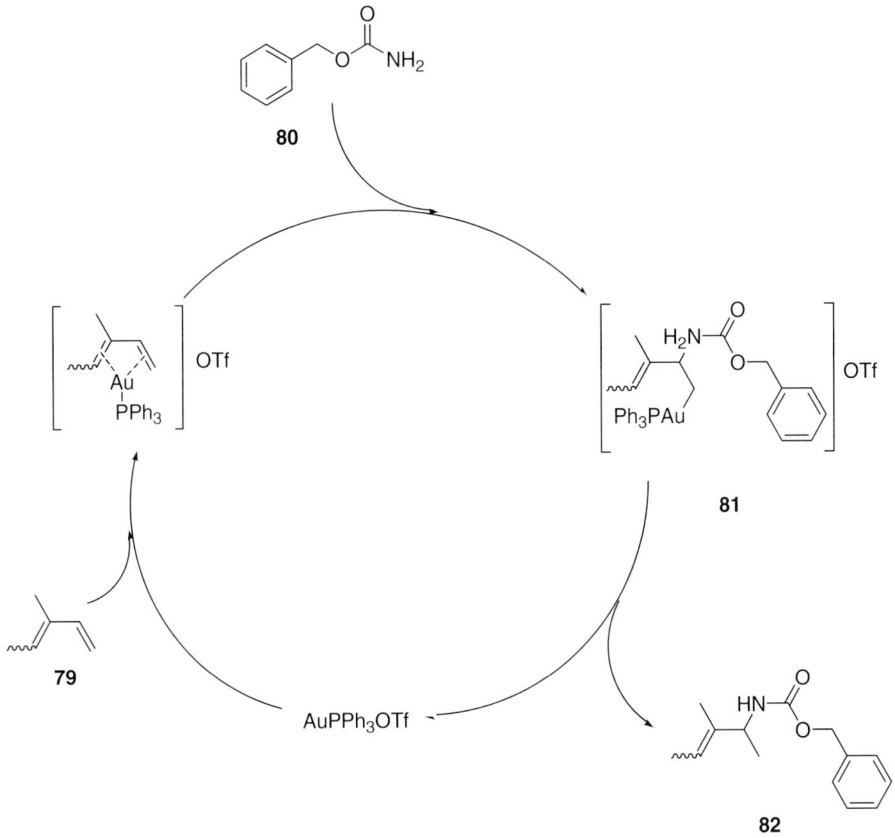

Scheme 8.7

In order to accelerate these types of reactions, Che et al. applied microwave conditions, a variety of phosphine gold(I) complexes as catalysts and carboxamides as nucleophiles, in both intra- and inter-molecular transformations [66].

Despite the relevance of these results, they must be considered with caution because some authors have described similar processes catalyzed by Brønsted acid systems [67].

Reactions by Arenes In 2000, the first evidence of a gold-catalyzed hydroarylation of α,β-unsaturated alkenes was reported by Hashmi et al. in their research with furans. In this publication, they proposed two possible mechanisms for the process:

- Direct auration of the furan followed by 1,4- addition to the enone.

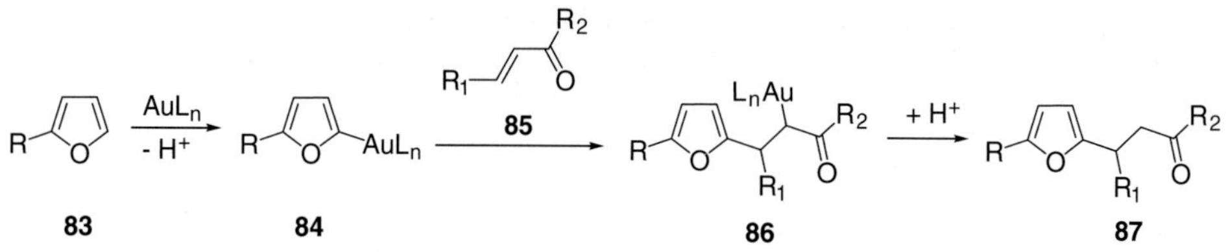

Scheme 8.8

83 → 84 (AuL$_n$, -H$^+$) → with 85 → 86 → (+ H$^+$) → 87

- Activation of the alkene and electrophilic substitution, where the furan (electron-rich) would act as a nucleophile.

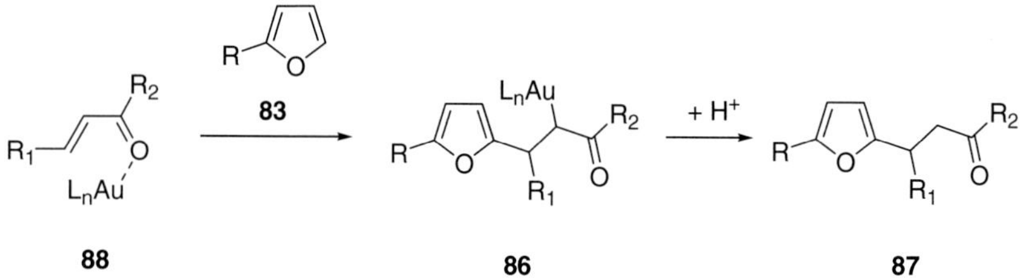

Scheme 8.9

88 + 83 → 86 → (+ H$^+$) → 87

Auration proposed in the first mechanism (Scheme 8.8) is possible in similar species, as Kharasch [68] and Fuchita [69] observed in stoichiometric reactions. In both pathways, the same intermediate was formed and no β-hydrogen elimination was observed and not only furan but other electron-rich arenes could react [70]. Although AuCl$_3$ was the precatalyst used, it was not possible to determine whether Au(III) or Au(I) were the real catalytic species (Scheme 8.9).

Further preliminary studies of direct auration of furane and thiophene derivatives were performed by Schmidbaur et al., who synthesised several organogold compounds similar to **89** or **90** [71] and the research of He and Shi on the gold(III)-catalyzed inter- or intramolecular functionalization of aromatic C−H bonds confirmed the possibility of such processes [21] (Figure 8.2).

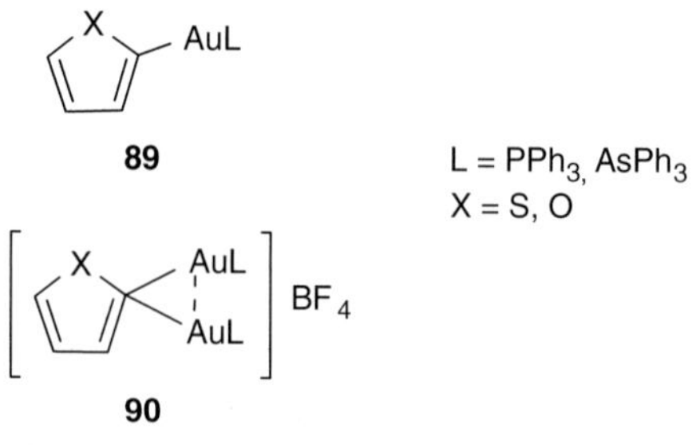

89

L = PPh$_3$, AsPh$_3$
X = S, O

90 (BF$_4$)

Figure 8.2 Gold complexes showing furane auration.

Related studies were recently reported by Contel, Urriolabeitia et al. using organogold(III) iminophosphorane complexes as catalysts for the addition of 2-methylfuran and electron-rich arenes to methyl vinyl ketone [72].

In the case of pyrroles with α,β-unsaturated carbonyl compounds, the reactions have low selectivity because these substrates are excellent nucleophiles, thus twofold addition products are mainly obtained [73, 74].

$$\mathbf{91} \xrightarrow{\text{5 mol \% Na[AuCl}_4\text{]} \cdot 2\text{ H}_2\text{O}}_{\text{MeCN, 20 °C}} \mathbf{92} \quad (8.24)$$

However selectivity increases if the substrates are indoles, 7-aza-indoles, benzofuranes or benzopyrroles [75–77].

$$\mathbf{93} + 1.5\ \mathbf{94} \xrightarrow{\text{0.05 mol \% AuCl}_3} \mathbf{95}\ (95\%) \quad (8.25)$$

Addition of Active Methylenes In the previous decade Cinellu et al. studied certain 1,3-dicarbonyl gold complexes and some applications in organic synthesis were subsequently proposed [78]. Intermolecular addition of activated methylene compounds to alkenes was developed by using $AuCl_3/AgOTf$ [79].

$$\mathbf{96} + \mathbf{97} \xrightarrow[\text{CH}_2\text{Cl}_2]{\text{5 mol \% AuCl}_3,\ 15\text{ mol \% AgOTf}} \mathbf{98}\ (74\%) \quad (8.26)$$

This methodology was later successfully applied to cyclic alkenes, dienes and trienes [59]. The tentative mechanism proposed by Li et al. involves the activation of the C–H bond by gold(I) species (from in situ reduction of gold(III)). Then, the reaction is followed by the alkene attack on the alkylgold hydride intermediate.

However, deuterated labelling experiments performed by Toste et al. in carbocyclizations of acetylenic dicarbonyl compounds suggest the involvement of nucleophile for enol, making the reaction more straightforward [80–82].

Che et al. recently applied gold catalysis to the synthesis of lactams by the intramolecular addition of β-ketoamide to unactivated alkenes, as shown in Equation 8.27 [83].

$$\mathbf{99} \xrightarrow[\text{toluene, 95 °C, 48 h}]{20 \text{ mol \% [AuClP}(^t\text{Bu})_2(o\text{-biphenyl})] / \text{AgOTf}} \mathbf{100} \quad 67\% \quad (8.27)$$

8.2.1.3 Alkynes as Substrates

Alkynes are by far the most popular and studied functional group in gold catalyzed reactions since Thomas et al. chose them for their research in 1976 [53, 84].

Hydrochlorination of Alkynes When Thomas and coworkers treated different alkynes in aqueous methanol with $HAuCl_4$ and observed the corresponding ketones as major products (Equation 8.28), with less than 5% of methyl vinyl ethers and vinyl chlorides, they were unaware of the fascinating treasure that was in front of them. Some of the most important types of products for gold catalysis were reported in the aforementioned study, but unfortunately at that time this process was believed to be a gold(III) oxidation process, despite the fact that the reaction achieved almost six turnovers.

$$\mathbf{101} \xrightarrow[\text{MeOH/H}_2\text{O, reflux}]{6.5 \text{ mol \% HAuCl}_4} \mathbf{102} + \mathbf{103} + \mathbf{104} \quad (8.28)$$
$$\quad \quad \quad \quad 38\% \quad <5\% \quad <5\%$$

Gold(III) was identified as the most active catalyst for that process in 1985, when Hutchings recognized that the efficiency in catalyzing the hydrochlorination of ethyne to vinyl chloride (a very important industrial process that previously used mercury salts as catalysts) correlated with the standard reduction potential of the supported metal cation. That meant that the metal could be found as a transient species in the reaction [10].

$$\mathbf{105} \xrightarrow[\text{HCl}]{\text{Au/C catalyst}} \mathbf{106} \quad (8.29)$$

This was the first time that anyone stated that gold catalysis is an area worthy of further research [85].

Hydration and Hydroalkoxylation of Alkynes Gold compounds were first applied to catalyze these types of reactions by Utimoto et al. in 1991, when they studied the use of Au(III) catalysts for the effective activation of alkynes. Previously, these reactions were only catalyzed by palladium or platinum(II) salts or mercury(II) salts under strongly acidic conditions. Utimoto et al. reported the use of Na[AuCl$_4$] in aqueous methanol for the hydration of alkynes to ketones [13].

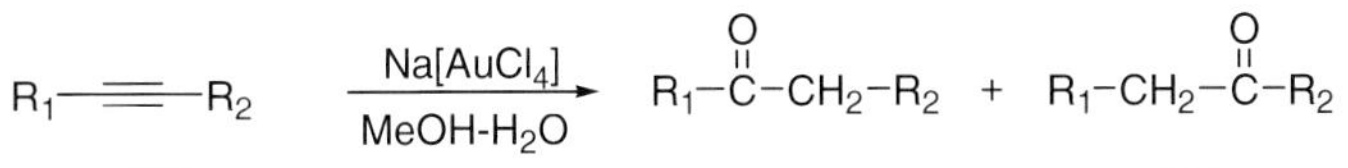

$$R_1 \equiv\!\!\!\equiv R_2 \xrightarrow[\text{MeOH-H}_2\text{O}]{\text{Na[AuCl}_4]} R_1\text{-}\overset{\text{O}}{\overset{\|}{\text{C}}}\text{-CH}_2\text{-R}_2 + R_1\text{-CH}_2\text{-}\overset{\text{O}}{\overset{\|}{\text{C}}}\text{-R}_2 \quad (8.30)$$

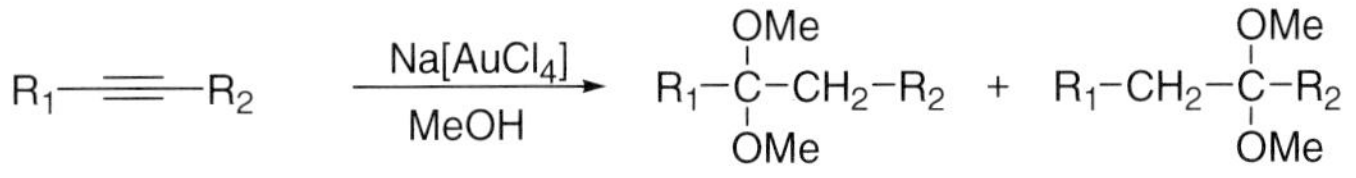

107 → 108 + 109

$$R_1 \equiv\!\!\!\equiv R_2 \xrightarrow[\text{MeOH}]{\text{Na[AuCl}_4]} R_1\text{-}\underset{\text{OMe}}{\overset{\text{OMe}}{\text{C}}}\text{-CH}_2\text{-R}_2 + R_1\text{-CH}_2\text{-}\underset{\text{OMe}}{\overset{\text{OMe}}{\text{C}}}\text{-R}_2 \quad (8.31)$$

107 → 110 + 111

In the case of internal symmetric or terminal alkynes, reaction takes place according to Markovnikov selectivity, unlike the problem of regioselectivity that appears when internal asymmetric alkynes are used. Unfortunately, at that time only the gold(I) compound K[Au(CN)$_2$] was tested, a compound that is now known not to be effective as a catalyst, unlike many other gold(I) compounds.

A sequential propargylation/Au(III) catalyzed hydration was obtained by Arcadi et al. in their attempts to develop more efficient synthetic strategies [86].

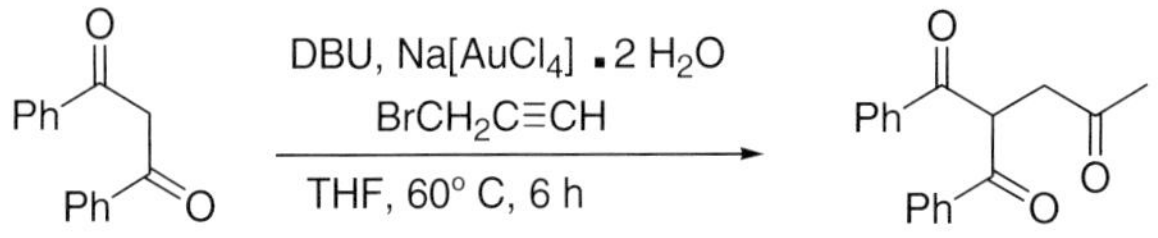

$$\text{112} \xrightarrow[\text{THF, 60° C, 6 h}]{\text{DBU, Na[AuCl}_4]\cdot 2\text{H}_2\text{O}, \text{BrCH}_2\text{C}\equiv\text{CH}} \text{113} \quad (8.32)$$

40 %

Following these results, Utimoto and Fukuda applied the same gold(III) catalyst for the conversion of propargyl alcohols into α,β-unsaturated ketones in good yields under mild and neutral conditions [87] (Scheme 8.10).

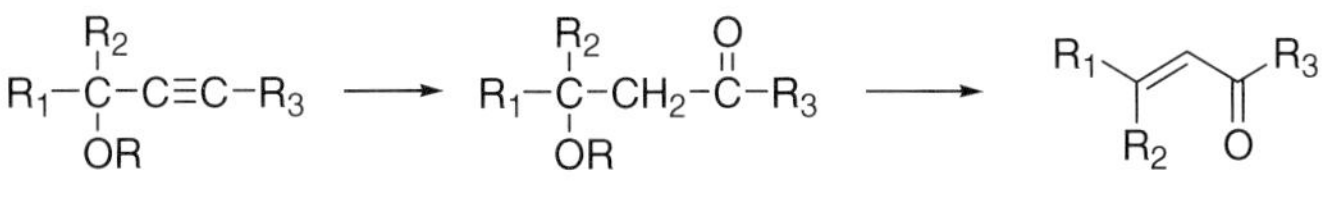

114 → 115 → 116

Scheme 8.10

The stereochemistry around the double bond was exclusively *trans* in the case of disubstituted olefins and it was also noteworthy that the hydration of methyl propargyl ether alkynes was regioselective, which was not the case with other alkynes.

In the case of terminal alkynes, the propargyl position promoted the normal Markovnikov product.

$$R_1-\underset{OR}{\underset{|}{\overset{\overset{R_2}{|}}{C}}}-C\equiv C-H \longrightarrow R_1-\underset{OR}{\underset{|}{\overset{\overset{R_2}{|}}{C}}}-\overset{\overset{O}{\|}}{C}-CH_3 \qquad (8.33)$$

117 118

Dudley et al. applied $AuCl_3$ to the rearrangement of alkoxy-alkynes for the formation of α,β-unsaturated esters (Equation 8.34), a procedure which is very difficult with other strategies when dealing with sterically-demanding groups as substitutes [88]. The use of $AuCl_3$ was crucial because propargylic alcohols did not react in the case of $Na[AuCl_4]$, as studied by Utimoto.

$$\underset{119}{R_1 \underset{OH}{\overset{R_2}{\diagup}} {-}OEt} \xrightarrow[CH_2Cl_2,\ rt,\ 5\ min]{5\ mol\ \%\ AuCl_3,\ 5\ equiv.\ EtOH} \underset{\underset{>95\ \%}{120}}{R_1 \overset{R_2\ \ O}{\diagup\!\!\diagup\!\!\diagdown} OEt} \qquad (8.34)$$

Until 1998, only gold(III) was believed to be effective for catalyzing these processes because, as mentioned previously, only the gold(I) compound $K[Au(CN)_2]$ was tested and it was inert to catalysis. Fortunately, Teles et al. reported very strong activity in the addition of alcohols to alkynes when they used cationic gold(I)-phosphane complexes [14]. In this study, the aforementioned authors tested for the first time the suitability of nucleophilic carbenes that displayed even greater activity than other gold complexes, but they were unable to synthesize the subsequent cationic derivatives.

The system $[AuClPR_3]$ is still the most used type of homogeneous gold catalyst although other gold cationic systems were synthesized by Schmidbaur et al. and were extremely effective as aurating agents or as precursors for gold deposition processes [89]. The catalysts used by Teles achieved TON of up to 10^5 with TOFs of up to $5400\,h^{-1}$. This high activity was strongly dependent on the ligand (initial TOF [h^{-1}] in parentheses): $AsPh_3$ (430) < PEt_3 (550) < PPh_3 (610) < $P(4\text{-F-}C_6H_4)_3$ (640) < $P(OMe)_3$ (1200) < $P(OPh)_3$ (1500). This series showed that electro-poor ligands led to increased activity, in the same direction as the catalyst stability decreased. For these systems, acidic co-catalysts were required to release the cationic fragment $[AuL]^+$, which is the agent that activates the alkynes, as shown in the proposed mechanism for the addition of methanol to propyne (Scheme 8.11).

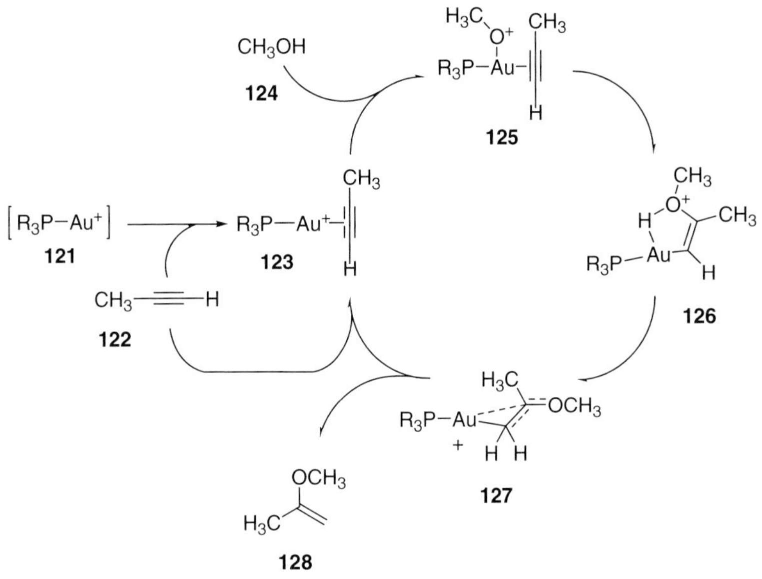

8.2 Homogeneous Catalysis

[Scheme 8.11 reaction cycle showing species 121, 122, 123, 124, 125, 126, 127, 128]

Scheme 8.11

From this perspective, Teles used [AuMePPh$_3$] as a catalyst and methanesulfonic acid as a co-catalyst when adding alcohols to alkynes, and achieved excellent results [14]. The scope of this reaction covered both internal and terminal alkynes. In the case of symmetrical alkynes, the only product formed in the presence of excess alcohol came from dialkoxylation.

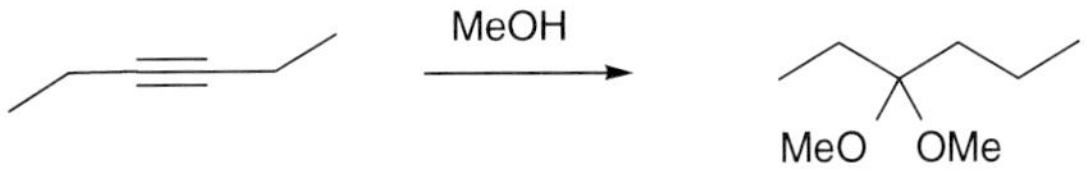

$$129 \xrightarrow{\text{MeOH}} 130 \quad (8.35)$$

With unsymmetrical alkynes, acetal forms as a result of addition to the less sterically hindered position, but a small amount of enol ether also forms. The latter is the only species obtained in the case of diphenylacetylene.

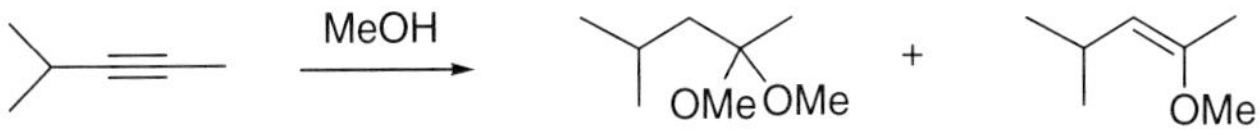

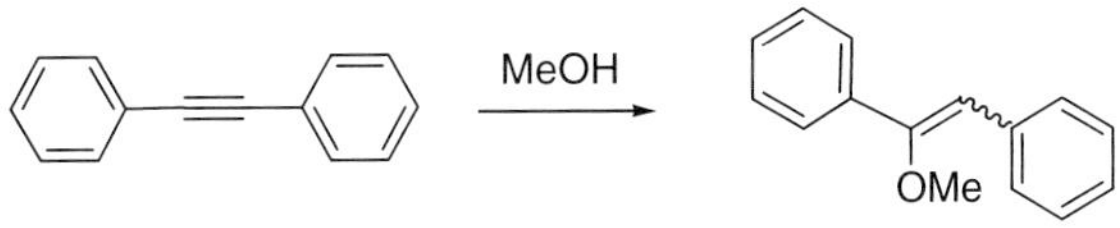

$$131 \xrightarrow{\text{MeOH}} 132 + 133 \quad (8.36)$$

$$134 \xrightarrow{\text{MeOH}} 135 \quad (8.37)$$

The Z isomer predominates (Z:E ≈ 8:1) in the initial reaction cycles but at the end of the reaction the ratio is about 2:1 due to partial isomerization. Computational studies of this reaction suggest that the alkynes coordinate more easily to AuL$^+$ fragments than to methanol.

In the study by Schwarz et al. on this transformation in the gas phase, the authors observed that the reaction did not occur when the solvent was absent, so hydrogen migration assisted by solvent seems to be essential [90].

Gold intramolecular processes are also possible. Using AuCl and K_2CO_3 as catalysts, Harkat et al. achieved γ- and δ-alkylidene lactonizations from ω-acetylenic acids by efficient and stereoselective reactions [91].

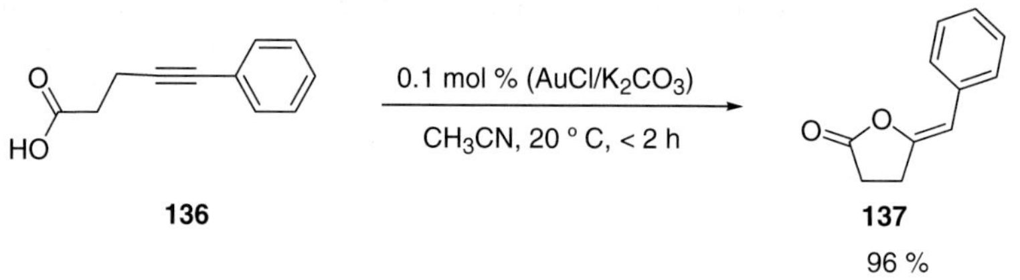

$$\text{136} \xrightarrow[\text{CH}_3\text{CN, 20 °C, < 2 h}]{0.1 \text{ mol \% (AuCl/K}_2\text{CO}_3\text{)}} \text{137} \quad 96\%$$

(8.38)

The impressive activity achieved by Teles' catalyst was improved some years later by the use of CO as an additive [92]. In this study, Hayashi and Tanaka reported a TOF of $15600\,h^{-1}$, at least two orders of magnitude higher than [cis-PtCl$_2$(tppts)$_2$], for the hydration of alkynes, providing an alternative synthetic route to the Wacker oxidation. Although several solvents were tested, the best results were obtained with aqueous methanol, and sulfuric acid or HTfO as acidic promoters. Unlike Utimoto's observation, in this case terminal propargylic alcohols partially (17–20%) delivered anti-Markovnikov product, in addition to the Markovnikov species. Some years before, Wakatsuki et al. had already reported the anti-Markovnikov hydration of terminal alkynes catalyzed by ruthenium(II) [93].

From these results, many groups have tested the activity of different gold complexes when adding alcohols to alkynes, in order to identify a "key-compound" that could be applied in many other transformations. Moreover, the great progress achieved in gold organometallic chemistry has opened up a wide range of possibilities.

A nucleophilic addition to the triple bond in alkynylphosphine derivatives was observed by Laguna and Bardají and although there was no evidence for a gold catalyzed cycle, reaction conditions were extremely milder than in the classic Reppe vinylation [94] (Scheme 8.12).

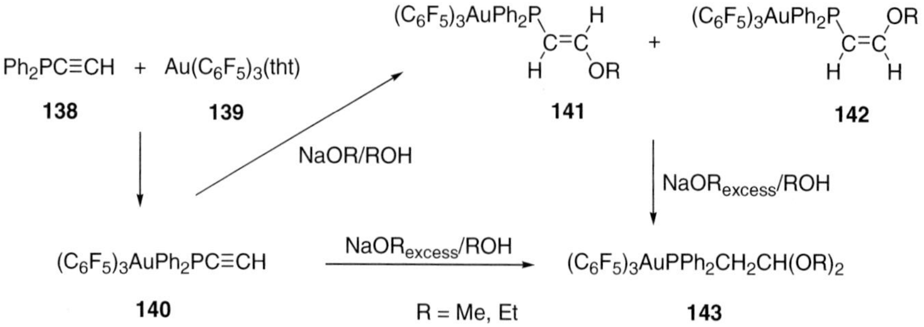

$$\text{Ph}_2\text{PC≡CH} + \text{Au(C}_6\text{F}_5)_3(\text{tht)}$$
138   139

140: $(C_6F_5)_3\text{AuPh}_2\text{PC≡CH}$

141: $(C_6F_5)_3\text{AuPh}_2\text{P}\diagdown_{H}\text{C=C}\diagup^{H}_{OR}$

142: $(C_6F_5)_3\text{AuPh}_2\text{P}\diagdown_{H}\text{C=C}\diagup^{OR}_{H}$

143: $(C_6F_5)_3\text{AuPPh}_2\text{CH}_2\text{CH(OR)}_2$

NaOR/ROH; NaOR$_{\text{excess}}$/ROH

R = Me, Et

Scheme 8.12

Two tetrahalogen-gold(III) compounds ($AuCl_4^-$, $AuBrCl_3^-$) in ionic liquids were studied by Raubenheimer et al. for the hydration of phenylacetylene. Although the activity was lower than that reported in previous studies, this strategy described the re-use of the catalyst achieved by recycling the ionic liquid phase [95].

The addition of water and methanol to terminal alkynes has also been studied by Laguna et al. by pentafluorophenyl and mesityl gold derivatives. Both acidic and non-acidic conditions led to high activity, even in the presence of as little as 0.5 mol% of catalyst. The use of pentafluorophenyl compounds allowed them to obtain additional spectroscopic information in the stoichiometric reaction of the complex [Au$(C_6F_5)_2Cl]_2$ and phenylacetylene, which showed that gold(III) was the active species in the catalytic process. The reaction followed the Markovnikov rule, as shown in the proposed mechanism (Scheme 8.13), delivering the corresponding ketones or diacetal products [96].

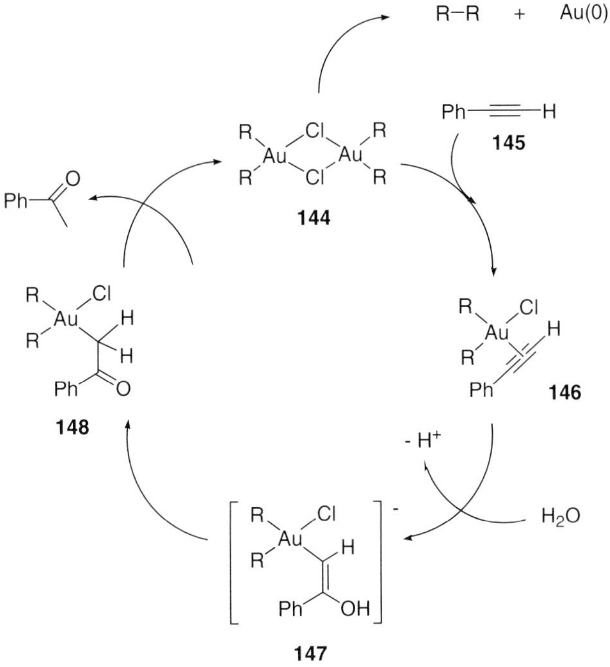

Scheme 8.13

In this case, the presence of at least one chloride ligand bonded to gold(III) seemed to be essential, unlike the inhibition produced by neutral coordinating ligands such as tetrahydrothiophene or triphenyl phosphine. In subsequent studies, water-soluble alkynyl complexes were used to catalyze the hydration of phenylacetylene, even in water alone. The main advantage of this method is that the catalyst contained in the aqueous phase can be recycled at least three times, giving the highest TON reported for this process (up to $1000\,h^{-1}$) [97].

The next study in this field was carried out by Herrmann et al. on the application of N-heterocyclic gold carbenes to the hydration of 3-hexynes in the presence of a Lewis acid as co-catalyst. Furthermore, the complex chosen as catalyst (**152**) was the first example of a carbene complex with a gold–oxygen bond, whose synthesis is shown in Scheme 8.14 [98].

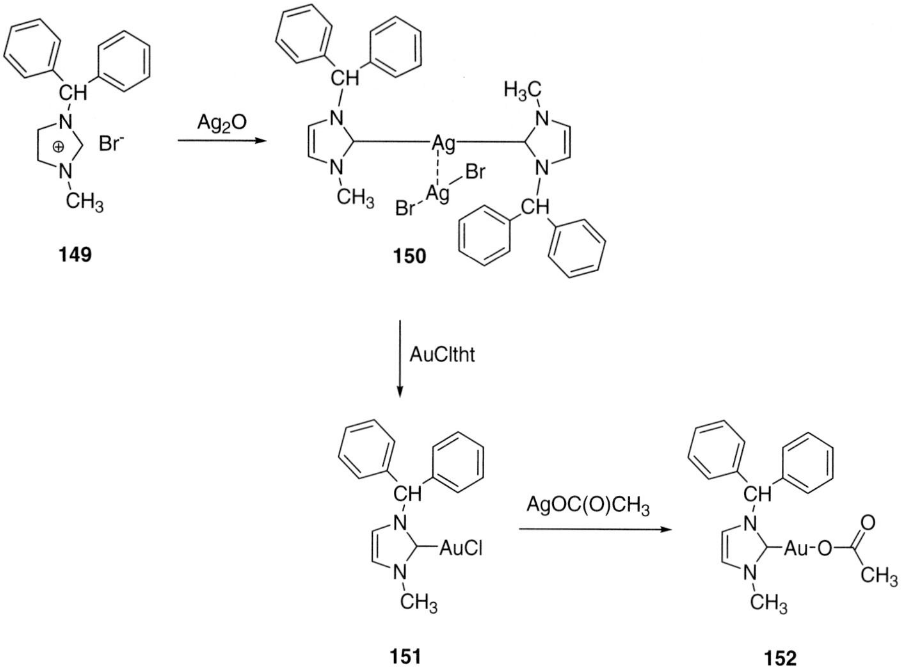

Scheme 8.14

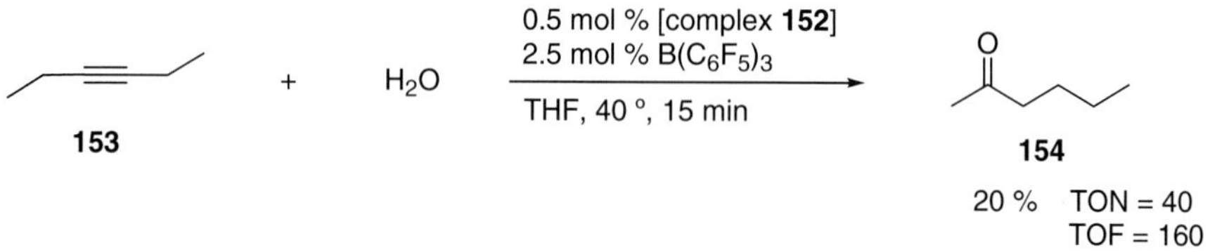

In a joint study by Schmidbaur and Raubenheimer, several phosphine carboxylates and sulfonates of gold and silver were tested as catalysts for the hydration of non-active alkynes [99]. While the gold complexes showed high activity for these reactions, analogous silver (I) complexes were not active in them. This different behavior was due to the fact that gold cations are weaker acceptors for their ligands and counterions than silver (I) cations (Figure 8.3).

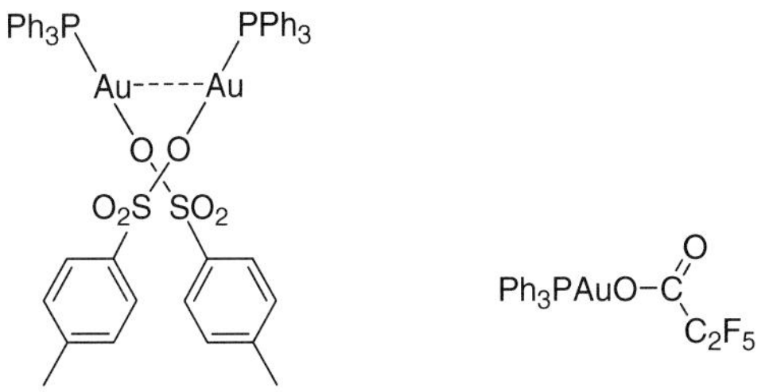

Figure 8.3 Gold complexes tested for the hydration of non-active alkynes.

Alcohol addition was also studied by Hashmi *et al.* in intramolecular processes [28]. Through gold catalyzed cyclizations of (Z)-3-ethynylallyl alcohols **157**, these authors were able to obtain furans **159**. Reaction occurred via intermediate **158**, which tautomerized the heteroaromatic furan, which is thermodynamically more stable (Scheme 8.15).

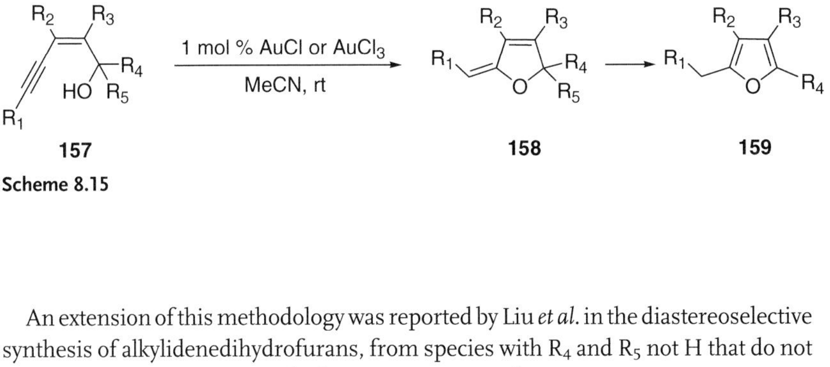

Scheme 8.15

An extension of this methodology was reported by Liu *et al.* in the diastereoselective synthesis of alkylidenedihydrofurans, from species with R_4 and R_5 not H that do not permit the tautomerization of a heteroaromatic product [100].

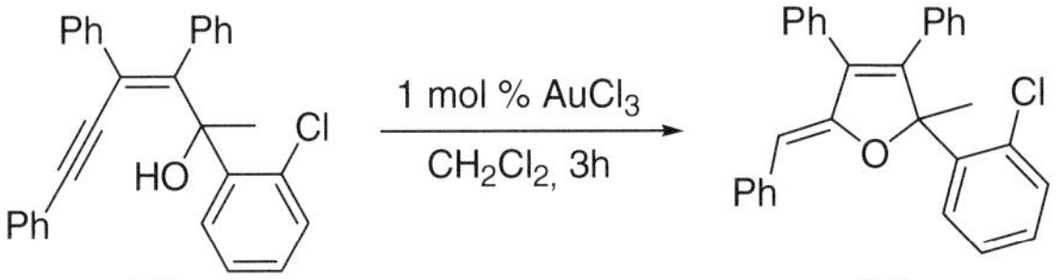

(8.40)

A reaction mechanism was proposed to explain the formation of tetrasubstituted furans when R_4 or R_5 are H (Scheme 8.16).

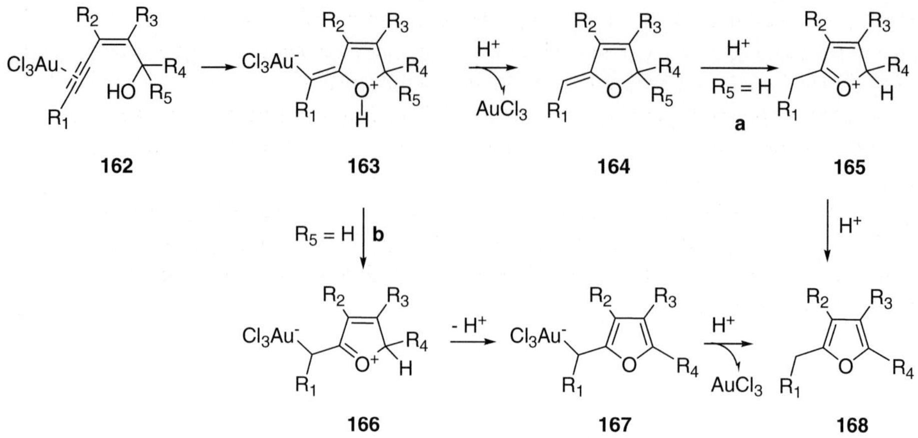

Scheme 8.16

Recently, Gagosz et al. extended the methodology to other substrates for the synthesis of functionalized furans via Claisen-type rearrangement. The future asymmetric version of the process could be a very useful tool for the synthesis of natural products [101].

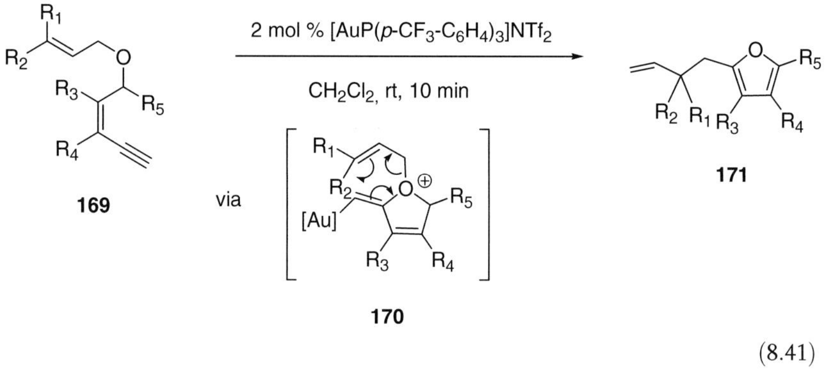

$$(8.41)$$

Genêt et al. used AuCl and $AuCl_3$ as catalysts and two intramolecular hydroxy groups as nucleophiles to obtain bicyclical ketals **173**. Chemoselectivity of this reaction was very high, showing the preference of the hydroxy groups for the alkyne fragment even, in the presence of styrene-like olefins [102].

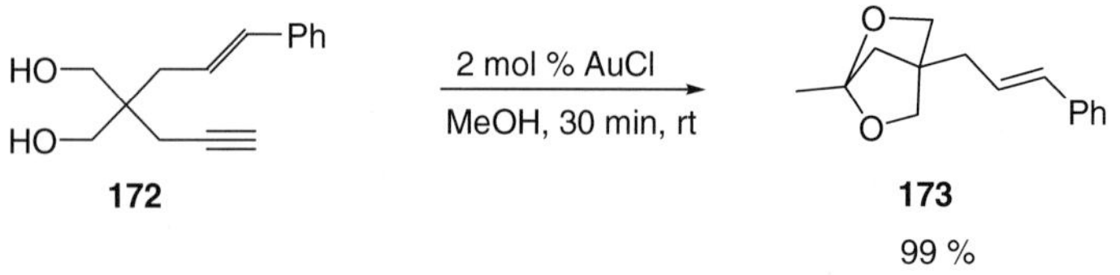

$$(8.42)$$

It was determined that these processes can be applied to synthesize trioxadispiroketal containing A-D rings of azaspiracid (a marine toxin) via a two-fold cyclization.

In the designed synthesis, gold first catalyzed an exo-addition of an OH group across an alkyne resulting in an enol ether. This transient enol ether could engage a ketal oxygen under protic conditions to form the bis-spiroketal [103].

In the development of domino reactions as efficient and environmentally suitable tools, Barluenga et al. reported the diastereoselective synthesis of eight-membered carbocycles using a tandem platinum or gold-catalyzed cycloisomerization/Prins type cyclization reaction. Although reaction was catalyzed not only by $AuCl_3$ but also by $PtCl_2(COD)$, the reaction conditions were milder in the case of gold (platinum requires a reaction temperature of 65 °C while gold reaction works at room temperature) [104] (Scheme 8.17).

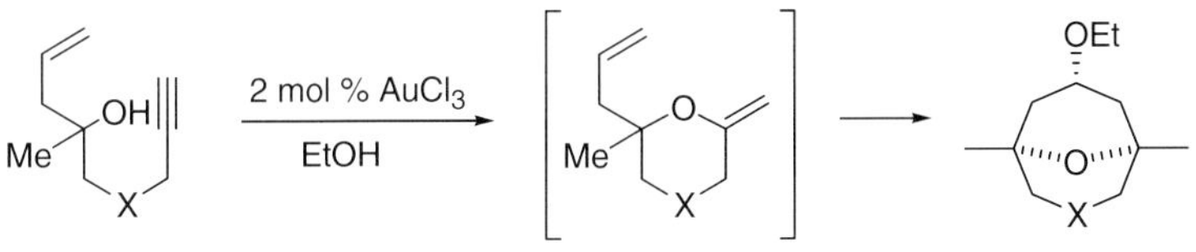

Scheme 8.17

Krause and Belting studied a tandem catalyzed reaction, in this case intramolecular cyclization and intermolecular hydroalkoxylation. The substrates were various homopropargylic alcohols in the presence of non-tertiary alcohols and a dual catalyst system consisting of Brønsted acids and a gold precatalyst (Equation 8.43).

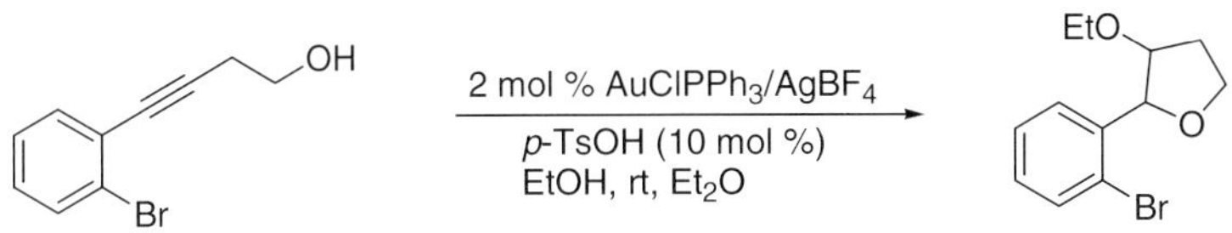

$$(8.43)$$

The reaction worked with both internal and terminal alkynes (except silylated alkynes) and in many solvents, even in the neat alcohol added [105]. The mechanism proposed involved two catalytic cycles: first, gold catalysis would lead to dihydrofuran by a fast intramolecular reaction; then, the subsequent slower intermolecular reaction would be produced by the addition of alcohol to the enol ether to deliver a ketal (Scheme 8.18).

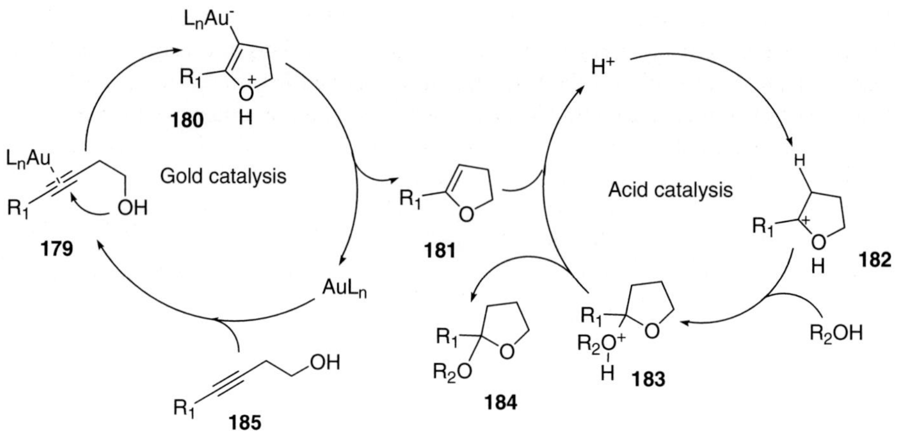

8 Catalysis

Scheme 8.18

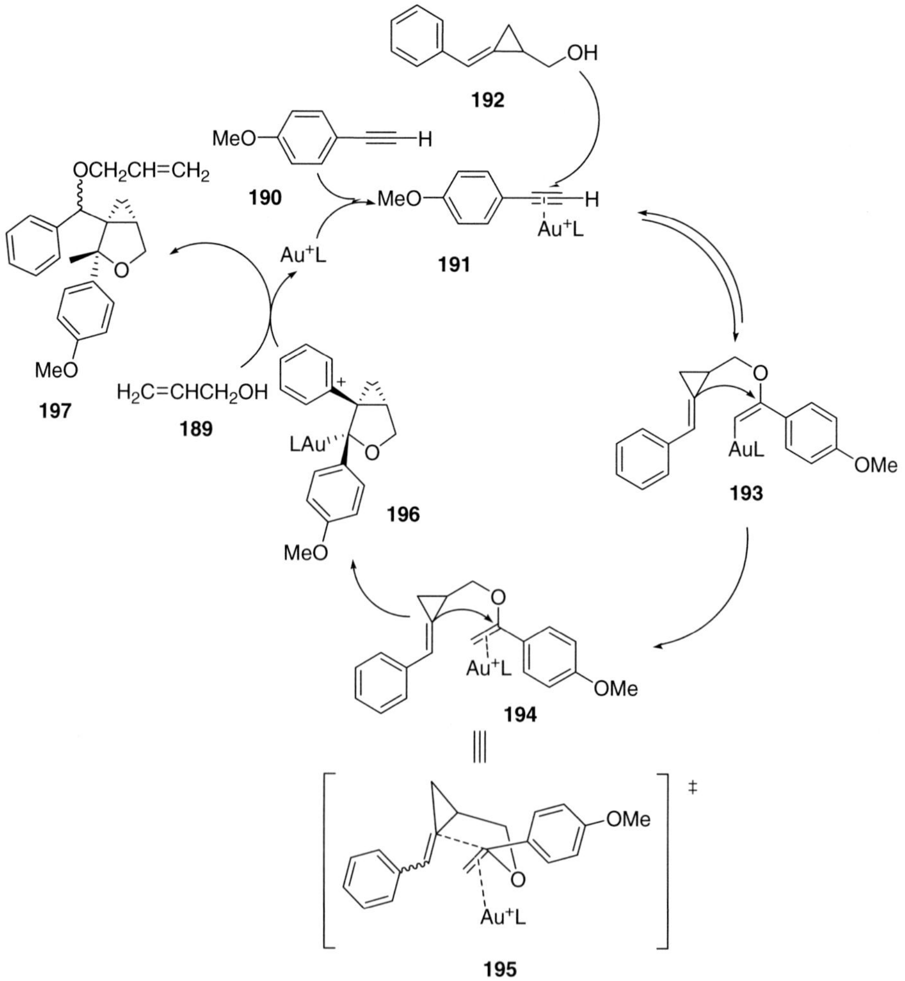

Scheme 8.19

A novel gold catalyzed example of three-component addition was recently reported by Shi et al. (Equation 8.44) [106]. Terminal aryl alkynes, alcohols and 2-(arylmethylene) cyclopropylcarbinols provided an intermolecular tandem hydroalkoxylation/Prins-type reaction to form 3-oxabicyclo[3.1.0]hexanes from simple materials and under mild conditions, catalyzed by the system AuClPPh$_3$/AgOTf. The proposed mechanism for this reaction is shown in Scheme 8.19.

$$(8.44)$$

Indenyl ethers were synthesized via intramolecular carboalkoxylation of alkynes. In this process, a benzylic ether group played a nucleophile role to capture a vinyl gold intermediate obtained by alkyne activation. The first catalytic system tested by Toste and Dubé in this study was a mixture of [AuClPPh$_3$] and AgBF$_4$. However, the moderate yield prompted them to research the use of more electrophilic gold(I) complexes such as [AuP(p-CF$_3$-C$_6$H$_4$)$_3$]BF$_4$, which increased the yield of cyclized products by 70% [107].

$$(8.45)$$

Related studies have recently been reported by Nakamura et al. In this case, the reaction, named silyldemetalation, required the capture of the vinyl gold intermediate by a silicon electrophile [108].

$$(8.46)$$

In a recent report, Shi et al. developed a valuable tool for the synthesis of 2,6-trans substituted morpholines by addition of water and alcohol to epoxy alkynes [109]. The procedure involved a domino three-membered ring opening, 6-exo-cycloisomerization, and subsequent intra-or intermolecular nucleophilic addition or a double-bond sequence.

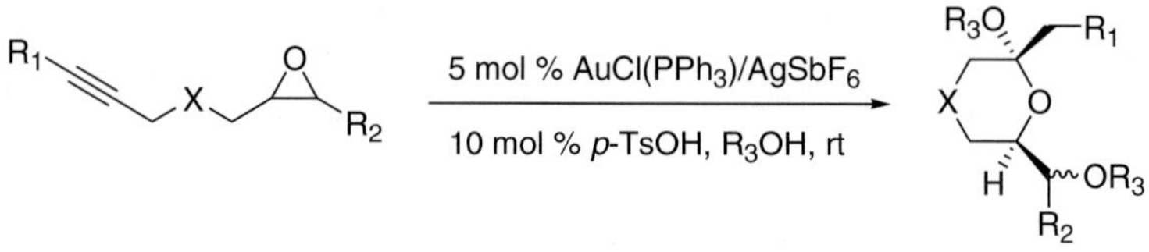

$$203 \xrightarrow[\text{10 mol \% } p\text{-TsOH, } R_3\text{OH, rt}]{\text{5 mol \% AuCl(PPh}_3)/\text{AgSbF}_6} 204 \quad 44\text{-}80\% \qquad (8.47)$$

Hashmi et al. also applied gold catalysis to the isochromene derivatives. This paper also reported a benzylic C–H activation that provided unprecedented dimerization from the formation of eight new bonds [110].

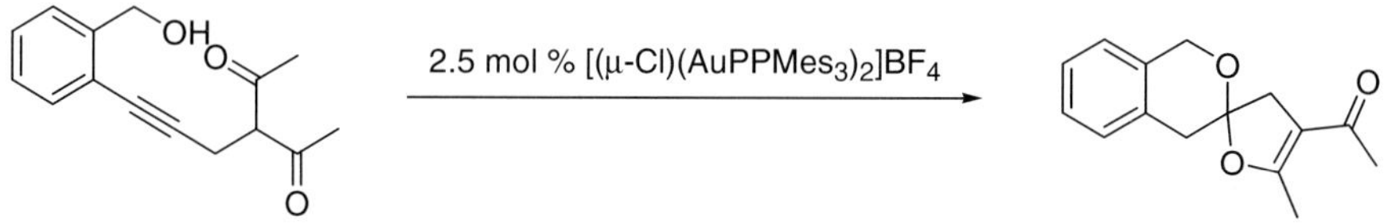

$$205 \xrightarrow{2.5 \text{ mol \% } [(\mu\text{-Cl})(\text{AuPPMes}_3)_2]\text{BF}_4} 206 \quad 18\% \qquad (8.48)$$

Hydroamination of Alkynes The discovery of palladium-catalyzed intramolecular addition of amines to acetylene coupled with the spectacular contribution of Hutchings opened the door for the synthesis of several nitrogen heterocycles. The first study in this field was performed by Utimoto et al., who researched gold catalyzed intramolecular 6-exo-dig hydroamination. Tautomerization of the initial enamines allowed them to obtain imines, which were thermodynamically more stable [111] (Scheme 8.20).

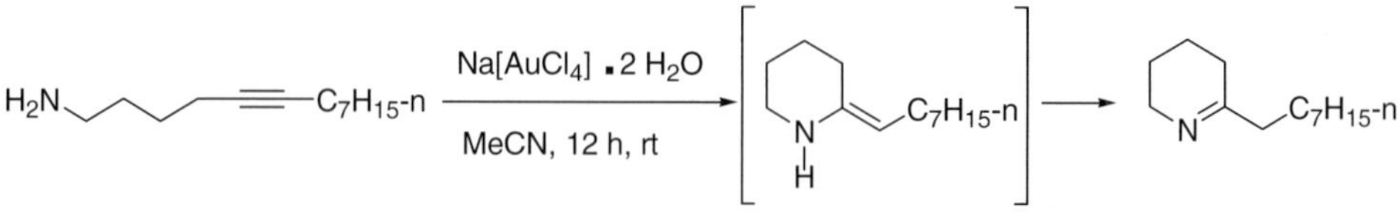

$$207 \xrightarrow[\text{MeCN, 12 h, rt}]{\text{Na[AuCl}_4] \cdot 2\text{ H}_2\text{O}} [208] \longrightarrow 209 \text{ quantitative}$$

Scheme 8.20

Subsequent studies led to 5-exo-dig cyclizations, showing even better results [112].

$$H_2N\text{-}CH_2CH_2CH_2\text{-}C{\equiv}C\text{-}C_7H_{15}\text{-}n \xrightarrow[\text{MeCN}]{Na[AuCl_4] \cdot 2 H_2O} \text{pyrrolidine-}C_7H_{15}\text{-}n \quad (8.49)$$

210 → **211** quantitative

Müller compared the activity of several metal catalysts for this reaction. He later reported that AuCl$_3$ was not at all effective for the conversion [113]. However, later studies showed that the complex [AuCl(triphos)](NO$_3$)$_2$ (1 mol%) could provide much better results: TOF = 212 h^{-1} and quantitative conversions [114]. Lok *et al.* showed that gold can catalyze the rearrangements of alkynylamino heterocycles, albeit with concomitant gold mirror formation (Equation 8.50) [115].

$$\text{benzoxazole-NHCH}_2C{\equiv}CCH_3 \xrightarrow{AuClPPh_3} \text{fused heterocycle} \quad (8.50)$$

212 → **213**

Solvent-free conditions were used by Tanaka *et al.* in their assays to obtain intermolecular reactions from alkynes and anilines. The chosen catalyst was [AuMe(PPh$_3$)] with an acidic promoter [92]. Reaction, whose effectiveness was greater in the case of aromatic amines, proceeded via Markovnikov by amine electrophilic attack of the alkyne in a similar way to the methanol addition proposed by Teles (see Section 2.1.3.2) and provided high yields and TONs.

$$PhC{\equiv}CH + 4\text{-NC-C}_6H_4\text{-NH}_2 \xrightarrow[\text{20 h}]{0.01 \text{ mol \% [AuMePPh}_3\text{]},\ 1.05 \text{ mol \% H}_3PW_{12}O_{40}} \text{ketimine product} \quad (8.51)$$

214 + **215** → **216**
90 %, TON : 8600

Dyker *et al.* reported an effective combination of an Ugi four-component reaction and gold catalysis to build highly functionalized isoindoles and dihydroisoquinolines with relatively good stereoselectivity [116] (Scheme 8.21).

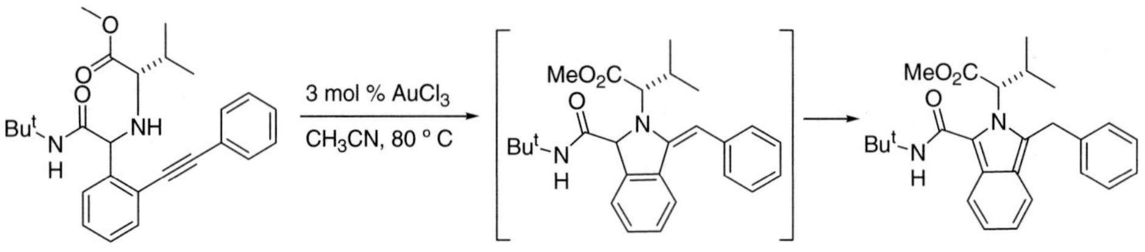

217 218 219

35 %

Scheme 8.21

A gold(III) catalyzed multicomponent synthesis of aminoindolizines was recently reported by Liu and Yan. This reaction of heteroaryl aldehydes, amines and alkynes took place under solvent-free conditions or in water and represented a high atom economic process. Especially noteworthy was the obtainment of *N*-indolizine-incorporated amino acid derivatives by enantiomerically-enriched amino acid substrates without loss of enantiomeric purity [117].

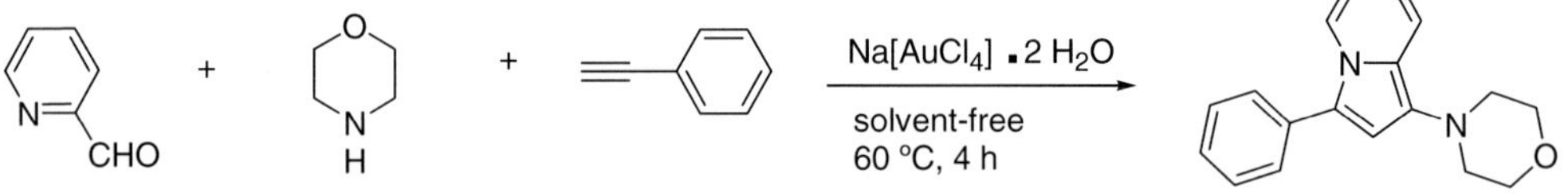

220 221 222 223

96 %

(8.52)

Another application in the construction of biological components was the formation of a hexacyclic substructure of Communesin B by Crawley and Funk, via intramolecular hydroamination [118].

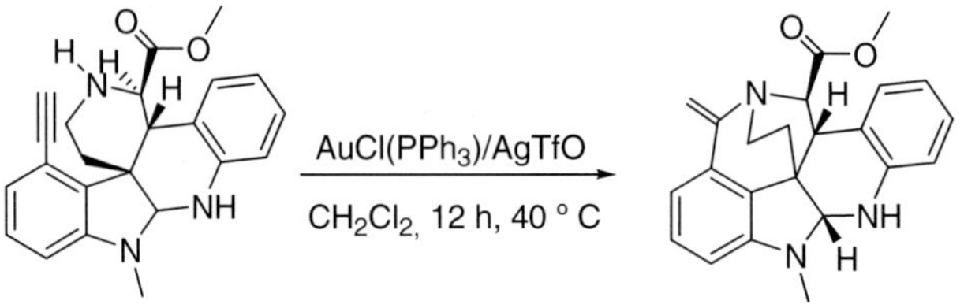

(8.53)

224 225

89 %

Floreancig and Hood recently incorporated gold-catalyzed heterocycle formation in the total synthesis of (+)-Andrachcinidine, a natural component extracted from the plant *Andrachne aspera,* which has medicinal properties. Gold catalysis was required in the last step of the synthesis, as shown in Scheme 8.22 [119].

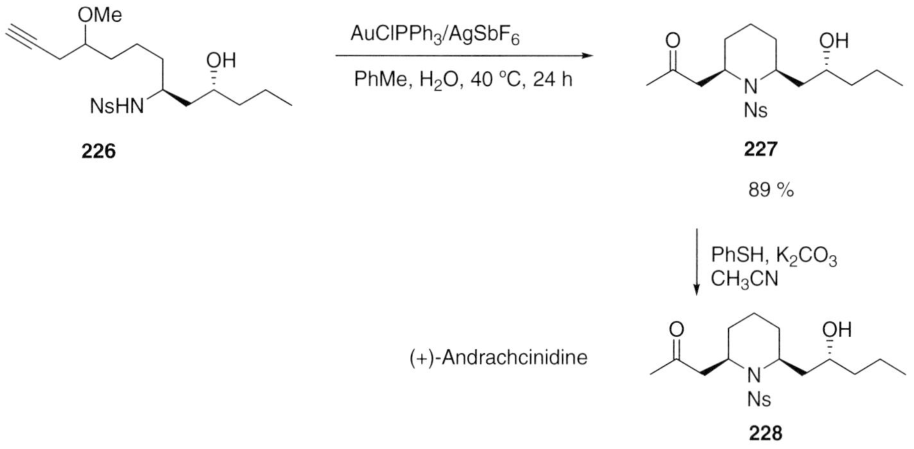

Scheme 8.22

A related work by Nakamura *et al.* was recently reported to show the gold-catalyzed process of aminosulfonylation, the formal addition of a nitrogen–sulfur bond to an alkyne moiety, and environmentally benign synthesis of a wide variety of 3- and 6-sulfonylindoles, present in many biologically active compounds [120].

Hydrocarboxylation of Alkynes Intramolecular addition of carboxylic acids (weak nucleophiles) to alkynes led to lactones, which were first reported by Schmidbaur *et al.* in the reaction of acetic acid with 3-hexynes to obtain, in addition to enol ester, 3-hexanone. Traces of water were probably present in the solvent to enable the process to be carried out [99].

Tert-butyl carbonates **229** can be used instead of free acids to obtain cyclic carbonates **230**. The influence of the electronic and steric properties of phosphine ligand in the catalytic system was studied and optimal behavior was observed for electron-deficient phosphines such as $P(C_6F_5)_3$ [48].

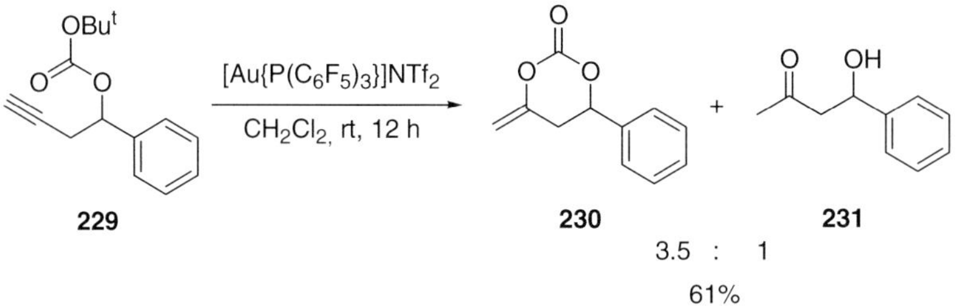

(8.54)

Genêt *et al.* used carboxylic acids again to obtain exo-methylene lactones. An important consideration is the fact that AuCl smoothly catalyzes this reaction under a 5-exo mode, without reaction of the alkenyl moieties present in the molecule [121].

$$\underset{\textbf{232}}{\text{MeO}_2\text{C}\diagdown\diagup\diagdown\diagup\diagdown\text{Ph}\text{, HO-C(=O)}}\quad\xrightarrow[\text{CH}_3\text{CN, 2h, rt}]{\text{5 mol \% AuCl}}\quad\underset{\substack{\textbf{233}\\90\%}}{\text{MeO}_2\text{C-lactone-CH=CHPh}}\quad(8.55)$$

In the case of substituted alkynyl derivatives, the 5-exo process was still effective and the stereochemistry of the resulting alkenes was proven to be exclusively Z.

$$\underset{\textbf{234}}{\text{EtO}_2\text{C, Bu}^n\text{, HO}_2\text{C}\diagdown\equiv\diagdown\text{Ph}}\quad\xrightarrow[\text{CH}_3\text{CN, 4 h, rt}]{\text{5 mol \% AuCl}}\quad\underset{\substack{\textbf{235}\\73\%}}{\text{EtO}_2\text{C, Bu}^n\text{-cyclopentane=CHPh}}\quad(8.56)$$

Even six or seven-membered ring lactones were reported by AuCl catalyzed reactions. Pale et al. prepared γ- and δ-alkylidene lactones by intramolecular cyclizations of ω-acetylenic acids. Once again, the process was stereospecific and only Z products were obtained in the case of substituted alkynes [91].

$$\underset{\textbf{236}}{\text{Br}\diagdown\equiv\diagdown\diagdown\text{C(=O)OH}}\quad\xrightarrow[\text{CH}_3\text{CN, 12 h, rt}]{0.1\text{ mol \% AuCl/K}_2\text{CO}_3}\quad\underset{\substack{\textbf{237}\\98\%}}{\text{Br-lactone}}\quad(8.57)$$

Enol or Arene Double Bonds as Nucleophiles Enolate alkylation has been a powerful tool for the formation of C–C bonds, despite the strong conditions needed. Gold catalysis allowed Toste et al. to perform a Conia-ene reaction of β-ketoesters with alkynes under mild conditions [81]. Two possible pathways could explain the reaction (Scheme 8.23). However, deuterated labelled experiments supported mechanism A, which involved enol addition to an Au-alkyne complex.

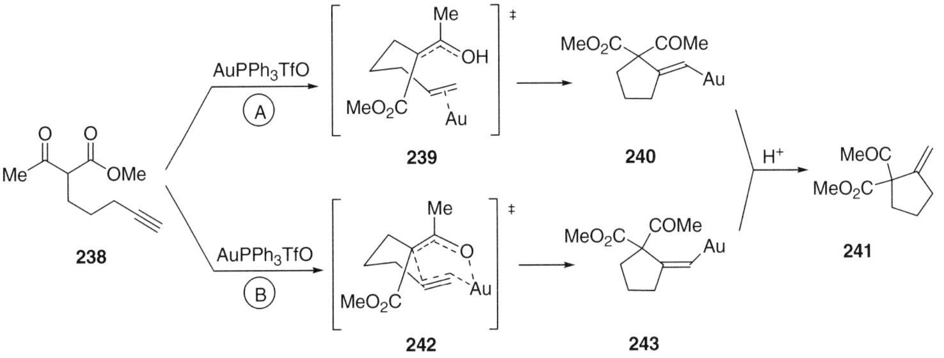

239 **240** **241** **238** **242** **243**

Scheme 8.23

The method was subsequently extended to non-terminal alkynes [80].

Sames *et al.* studied the cyclization of phenyl/propargyl ethers catalyzed by different metal salts since this Friedel–Crafts alkenylation had previously been reported to be catalyzed by metals such as Pd, Zr In and Sc [122–124] or even zeolites by hereogeneous catalysis [125]. In this preliminary research, the maximum yield of the desired product achieved by gold was only 6%, the best results being obtained with $PtCl_2$ as catalyst [126].

Reetz and Sommer then studied the intramolecular hydroarylation of alkynes when they were looking for carbon triple bond coupling reactions. Depending on the substrate, the choice of gold(I) or gold(III) species was crucial [127].

From an atom economy perspective, He and Shi achieved efficient "solventless" hydroarylation of alkynes. Reaction took place efficiently under air atmosphere at ambient temperature and different functional groups could be tolerated. The catalytic system employed was $AuCl_3$/AgTfO (1:3) and both inter-molecular and intra-molecular processes were described [128].

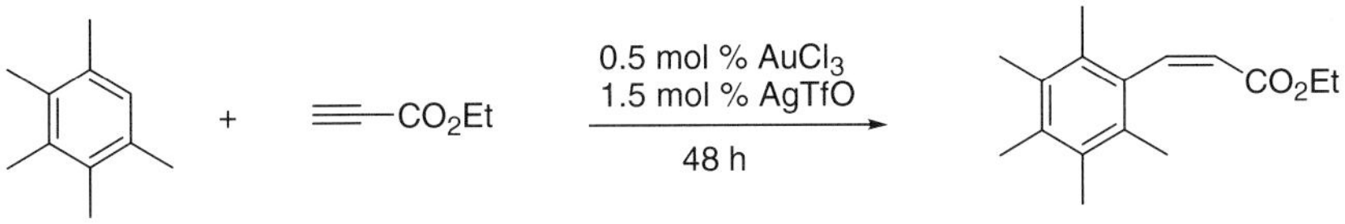

244 **245** **246**

99 %

(8.58)

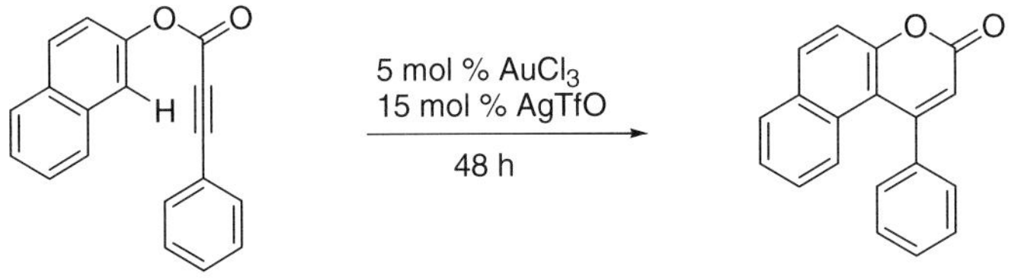

(8.59)

247 **248**

99 %

This method was used by Li et al. to construct heterocycle-based structures. $AuCl_3$ was once again the catalyst used [77].

$$\text{249} + \text{245} \xrightarrow[\text{CH}_3\text{CN, rt}]{5 \text{ mol \% AuCl}_3} \text{250} \quad 74\% \tag{8.60}$$

In this case, double addition products were obtained. More recently, Hashmi and Blanco described the twofold intermolecular hydroarylation of unactivated C–C triple bonds, using the chloride-bridged gold (I) salt [(μ-Cl)(AuPMes$_3$)$_2$]BF$_4$ [129, 130].

$$\text{251} + \text{252} \xrightarrow[50\,°\text{C, 7 d}]{5 \text{ mol \% [μ-Cl)(AuPMes}_3)_2]\text{BF}_4} \text{253} \quad 58\% \tag{8.61}$$

Reports of the intramolecular versions of these processes were published by Nevado and Echavarren [131] and Hashmi et al. [132, 133].

A notable eight-membered ring annulation was reported by Echavarren et al. in their study of indoles as substrates for the intramolecular reaction [134].

$$\text{254} \xrightarrow{\text{AuCl}_3} \text{255} \quad 75\% \tag{8.62}$$

Related studies have recently been reported by the same author on propargyl steres reactions with dicarbonyl compounds or electron-rich arenes [135], to provide an atom-economical functionalization of carbon nucleophiles under catalytic conditions, using a very different method of addition catalyzed by Lewis acids [136].

Imines and Ketones as Nucleophiles Research by Gevorgyan and Seregin on alkyne–vinylidene isomerization prompted them to develop a mild cycloisomerization of propargyl N-containing heterocycles into pyrroloheterocycles by gold catalysis. Deuterium-labeling experiments suggested that reaction took place via alkyne–vinylidene isomerization with concomitant 1,2-migration [137] (Scheme 8.24).

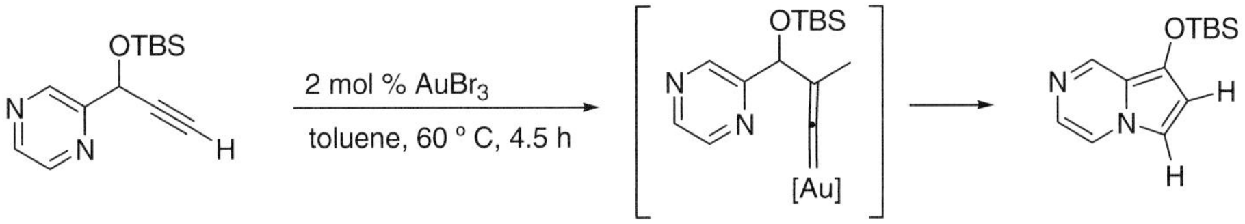

256 257 258
72 %

Scheme 8.24

Similar results were obtained by Dake *et al.* with silver catalyzed reactions, showing that the thermal stability of gold was higher than silver but silver catalyzed reactions were faster [138].

In research on ketones as nucleophiles, Hashmi *et al.* studied the cycloisomerization of propargyl ketones. Reaction took place by nucleophilic attack of the ketone followed by aromatization to furan [28] (Scheme 8.25).

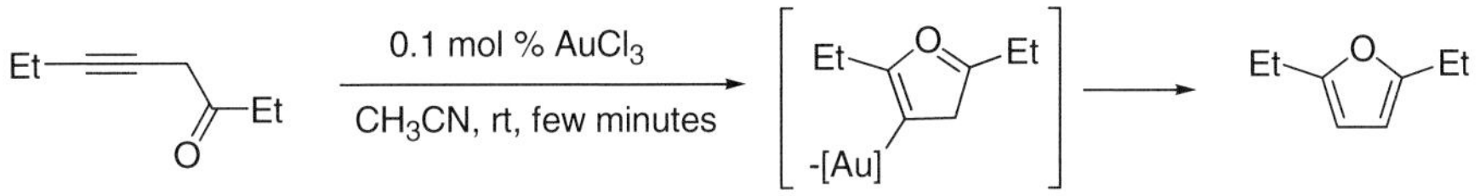

259 260 261
quantitative

Scheme 8.25

For the synthesis of furanones, Kirsch *et al.* studied the gold-catalyzed heterocyclization/1,2-migration cascade reaction of α-hydroxy propargyl ketones [139].

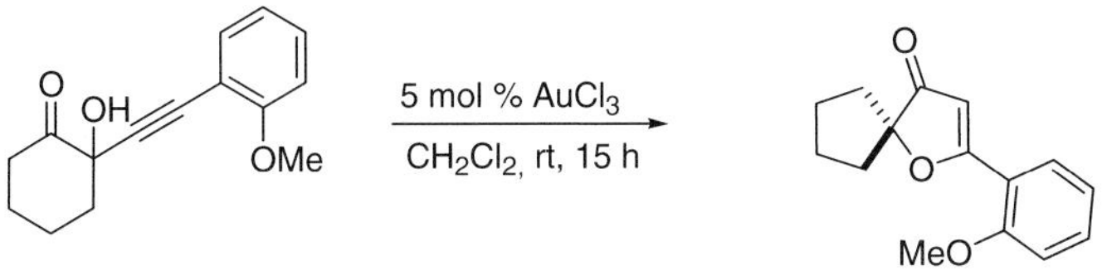

(8.63)

262 263
81 %

A related synthesis of highly substituted furans was reported by Schmalz and Zhang from cyclopropyl alkynyl ketones. The reaction scope included a great variety of nucleophiles such as several alcohols, indole or even acetic acid [140].

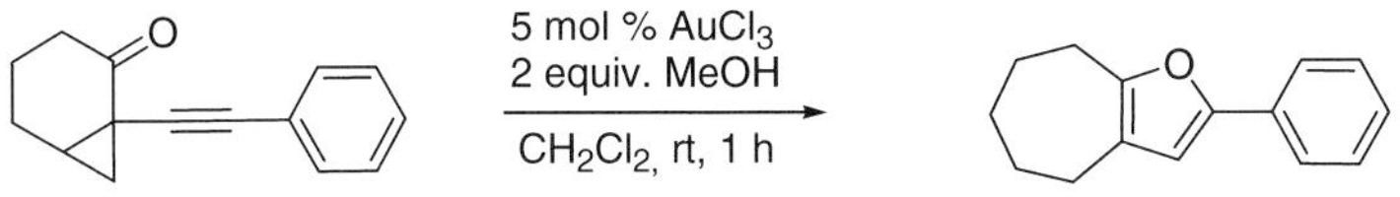

(8.64)

264 265
81 %

A recent study by Yamamoto *et al.* on the intramolecular carbocyclization of alkynyl ketones provided a synthetic route to highly substituted cyclic enones [141].

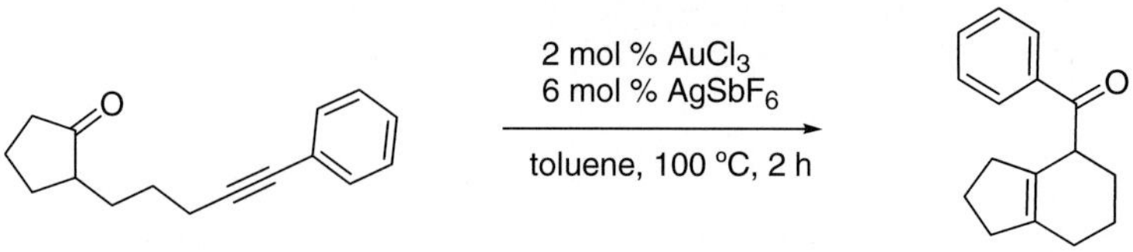

$$\text{266} \xrightarrow[\text{toluene, 100 °C, 2 h}]{\text{2 mol \% AuCl}_3 \text{, 6 mol \% AgSbF}_6} \text{267} \quad (8.65)$$

55 %

Oxidative Rearrangements Toste *et al.* recently developed various oxidative rearrangements of alkynes using sulfoxides as stoichiometric oxidants through carbenoid intermediates. These reactions could provide an entry into products that contain a carbonyl group susceptible to further functionalization [142] (Scheme 8.26).

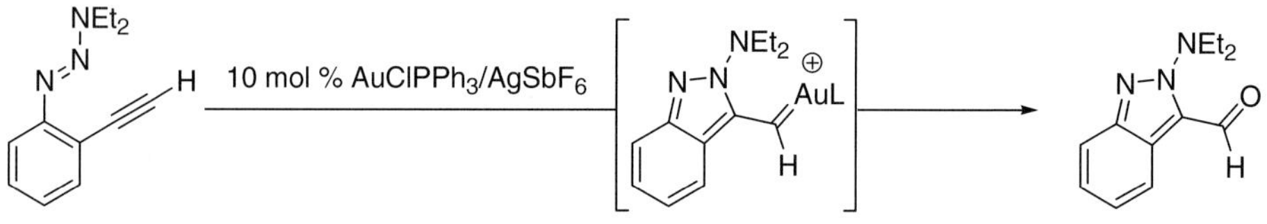

$$\text{268} \xrightarrow{10 \text{ mol \% AuClPPh}_3/\text{AgSbF}_6} [\text{269}] \rightarrow \text{270}$$

66 %

Scheme 8.26

8.2.1.4 Enynes as Substrates

Enyne Cycloisomerization and Enyne Metathesis These reactions are very interesting because they do not require additional reactants and produce few by-products, making this procedure desirable in terms of synthesis efficiency and the minimization of atomic waste [143, 144]. Transition metal catalysis has been used for these processes for many years and appears to be a highly promising approach for ring construction under very mild conditions [145, 146]. However, nowadays, gold catalysis contributes new product types for these processes. This reveals the fascinating nature of gold catalysis research. The wide scope and high efficiency of gold observed in catalyzed enyne reactions have been recently highlighted and summarized [145, 146]. One of the main reasons for the significant interest in this field is the mild conditions needed to access complex or hitherto inaccessible products.

A very common point for most of these processes is the presence of triarylphosphine bonded to gold in the catalytic system employed. This seems to be a crucial factor for achieving high selectivity in enyne metathesis. Thus, gold is the most active metal for catalyzing these processes using the mildest reaction conditions [147].

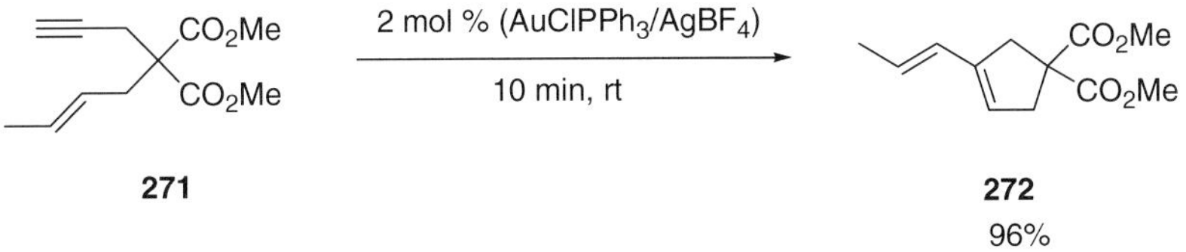

8.2 Homogeneous Catalysis

(8.66)

Methylene cyclohexenes are other products obtained by cyclopropanation of the carbene gold intermediate over a second olefin unit [148].

If the substrate contains additional double bonds, bi- or even tri-cyclic products can be obtained [149]. Furthermore, Echavarren et al. synthesized several gold(I) cationic derivatives to avoid having to use silver salts that could lead to unwanted side reactions [150] or the hygroscopic character of $AuCl_3$ [151].

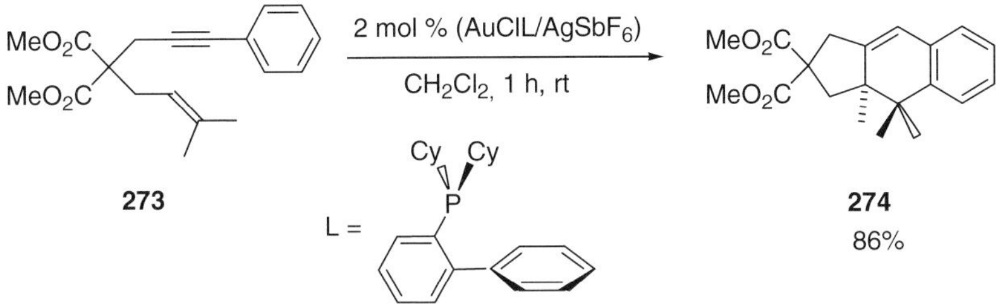

(8.67)

Further research on this subject was recently reported, in relation to the use of dienynes as substrates for intramolecular cycloaddition. While thermal intramolecular [4+2] cycloadditions of enynes with alkenes only took place at high temperatures, the gold(I) catalyzed transformations provided bi- or tri-cyclic ring systems under mild conditions [152].

Gagosz studied the gold(I)-catalyzed isomerization of 3-hydroxylated 1,5-enynes and observed that reaction could follow different reaction pathways depending on the substituents and relative configuration of the substrate [153] (Scheme 8.27).

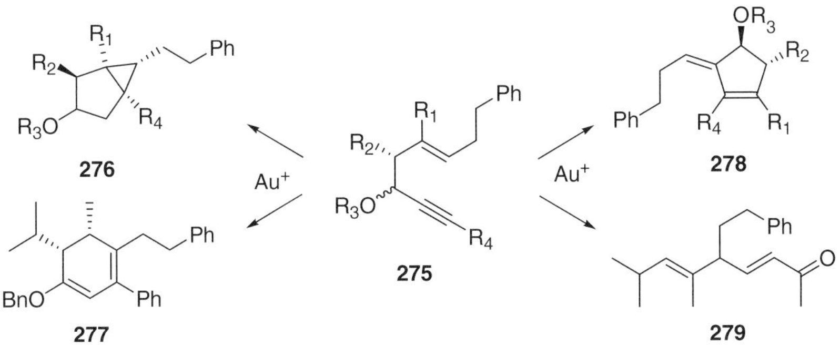

Scheme 8.27

Double cyclization was observed with siloxy enynes when a new cycloisomerization mechanism was used that involved a cascade of 1,2- alkyl shifts [154].

Related to these studies, Gagosz et al. synthesized new phosphine gold (I) complexes using the bis(trifluoromethanesulfonyl)imidate moiety as a weakly coordinating counteranion suitable for stabilizing cationic gold (I). The main advantage of these compounds was that they avoided the addition of silver derivatives to remove the halogen atoms coordinated to gold, a step that sometimes favored the appearance of side products. These imidate species were also air-stable, easy to prepare, store and handle, and their efficiency as catalysts had been demonstrated [155].

$$Cl-Au-L + AgNTf_2 \xrightarrow{CH_2Cl_2,\ rt} AgCl + [Au-L]NTf_2 \quad \mathbf{280} \tag{8.68}$$

The authors then prepared N-heterocyclic carbenes with the same counteranion and their effectiveness was observed in certain previously reported gold(I)-catalyzed transformations [156].

The use of 1,5-enynes instead of 1,6-enynes delivered bicycle derivatives such as **282** [157], and a hydroxy group in the tether enabled the formation of rings (**284**), as reported by Grisé and Barriault (Equation 8.71) [158].

$$\mathbf{281} \xrightarrow[CH_2Cl_2,\ rt]{[AuPPh_3]PF_6} \mathbf{282} \quad 99\ \% \tag{8.69}$$

$$\mathbf{283} \xrightarrow[CH_2Cl_2,\ rt,\ 18\ h]{2\ mol\ \%\ (AuClPPh_3/AgTfO)} \mathbf{284}\ \text{good yields} \tag{8.70}$$

The use of propargyl vinyl ethers prompted Toste et al. to develop a stereoselective preparation of 2-hydroxy-3,6-dihydropyrans, suitable for the synthesis of spirocyclic compounds. The reaction was catalyzed by a small amount (1 mol%) of [O(AuPPh$_3$)$_3$] BF$_4$ under mild conditions [159].

In a recent report, Toste and Shen developed a gold(I)-catalyzed cyclization of alkynes using silyl ketene amides that, by means of prior hydrolysis, provided 1,6-enyne (**285**) or 1,5-enyne systems (**287**) activated for the intramolecular cycloisomerization [160].

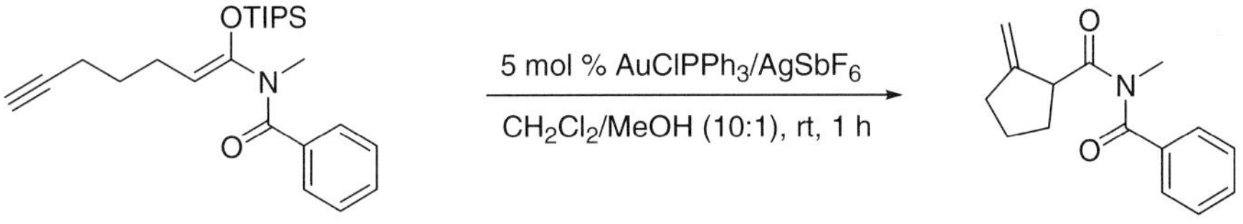

$$\text{285} \xrightarrow[\text{CH}_2\text{Cl}_2/\text{MeOH (10:1), rt, 1 h}]{\text{5 mol \% AuClPPh}_3/\text{AgSbF}_6} \text{286}$$

68 %

(8.71)

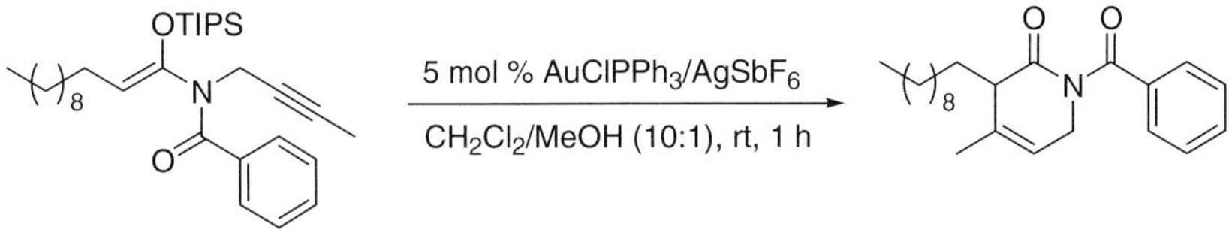

$$\text{287} \xrightarrow[\text{CH}_2\text{Cl}_2/\text{MeOH (10:1), rt, 1 h}]{\text{5 mol \% AuClPPh}_3/\text{AgSbF}_6} \text{288}$$

77 %

(8.72)

A mechanism proposed for the skeletal rearrangement of enynes involved the presence of gold carbenes [161]. This proposed mechanism was supported by the capture of intermediate gold carbenoids trapped by reactive alkenes in intermolecular cyclopropanation reactions [162].

The assembly of highly functionalized cyclohexene derivatives was achieved by Lee and Shin using a highly stereoselective tandem reaction. In this process, the nucleophilic participation of the O-Boc group appeared to intercept a carbocationic (or cyclopropyl carbene) gold intermediate [163].

Very recently, Genêt and Michelet developed a diastereoselective reaction that took place by hydroamination of an unactivated alkene followed by a cyclization process under very mild conditions. This 1,6-enyne reaction was compatible with electron-poor aromatic amines, including amines bearing chloride atoms (Equation 8.73), which is very useful for further functionalization of the substituted aniline ring [164].

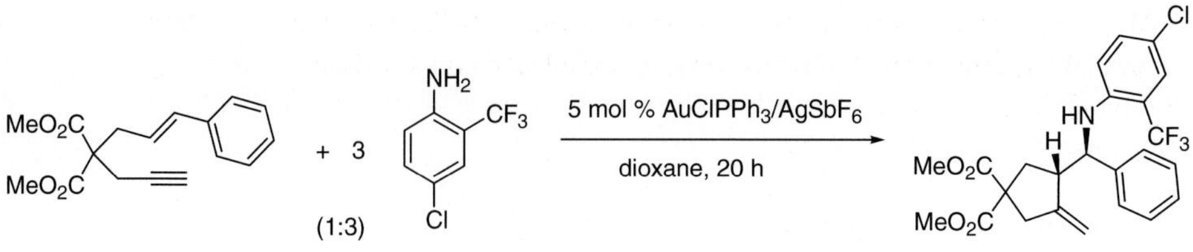

$$\text{MeO}_2\text{C}\diagup\diagdown\diagup\text{Ph} \quad + \quad 3 \quad \underset{(1:3)}{\underset{\text{Cl}}{\text{2-NH}_2\text{-4-Cl-C}_6\text{H}_3\text{-CF}_3}} \quad \xrightarrow[\text{dioxane, 20 h}]{5\ \text{mol}\ \%\ \text{AuClPPh}_3/\text{AgSbF}_6} \quad \mathbf{291}$$

289 **290** **291** 86%

(8.73)

Phenol Synthesis A new method for obtaining arenes from easily available furans was reported by Hashmi *et al.* [19]. In this first paper, AuCl$_3$ was used to produce a highly substituted phenol without side products.

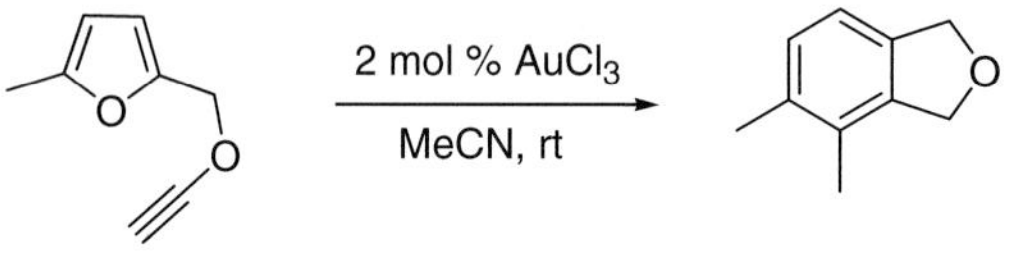

$$\mathbf{292} \xrightarrow[\text{MeCN, rt}]{2\ \text{mol}\ \%\ \text{AuCl}_3} \mathbf{293}$$

292 **293** 69 %

(8.74)

Although Echavarren *et al.* reported similar conversions by platinum catalysis, lower selectivity was observed [165–167]. Computational studies then suggested the initial formation of a cyclopropyl carbenoid intermediate **213**, as shown in Scheme 8.28.

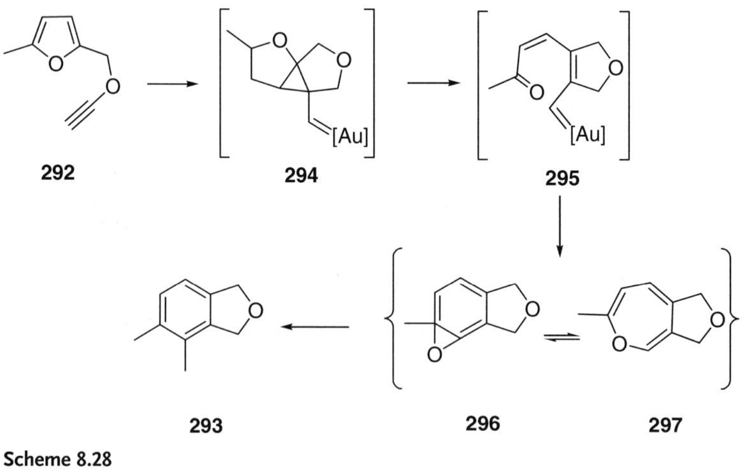

292 → **294** → **295** → { **296** ⇌ **297** } → **293**

Scheme 8.28

Hashmi *et al.* then reported the first direct experimental evidence for the formation of type **298** species by their capture via dienophile **299** addition [168].

$$\text{(8.75)}$$

298 **299** **300**

A phenol synthesis reaction induced by gold catalysts without steric limitations for the substituents was also reported [169]. These results provided a very helpful tool for organic synthesis of a large variety of derivatives such as biaryls, isochromanes, benzofurans, tetrahydroisoquinolines and other natural products [133, 170, 171].

The application of this principle to the intermolecular process was first achieved by Hashmi *et al.* using the dinuclear gold compound [(μ-Cl){AuPMes$_3$}$_2$]BF$_4$ as catalyst. Although neat substrates were used at 60 °C, reaction took a long time and it also led to a hydroarylation product [132]. The formation of the tetrasubstituted arene **303** was confirmed by crystal structure analysis.

301 **302** **303** 38 % **304** 30 %

$$\text{(8.76)}$$

In a joint study by Corma and Hashmi, heterogeneous gold catalysts based on nanogold on nanocerium oxide support were employed for phenol synthesis [172].

Alkoxy Cyclization After the studies of the 1,6-enynes with furans, Echavarren *et al.* studied 1,6-enynes which did not contain furan rings in their structure as they had previously studied these with PtCl$_2$ catalysis [173], obtaining promising results in the case of substrates, which displayed enol ether substructures [174]. DFT calculations showed that *exo-dig* and *endo-dig* cyclization transition states differed little in energy, suggesting that the strategic position of the substituents could have a strong influence on the final pathway. This methodology allowed them to obtain useful results, such as the diastereoselective synthesis of α-glucosides [175] or other complex substrates with high enantioselectivity [168].

8.2.2
Activation of Carbonyl Groups and Alcohols

The gold catalyst has provided some very important achievements in chemistry in general, such as the asymmetric aldol reaction of aldehydes with isocyanoacetates reported by Ito, Sawamura and Hayashi [12, 176]. The use of chiral ferrocenylphosphine gold (I) complexes allowed them to obtain enantiomerically-pure oxazolines.

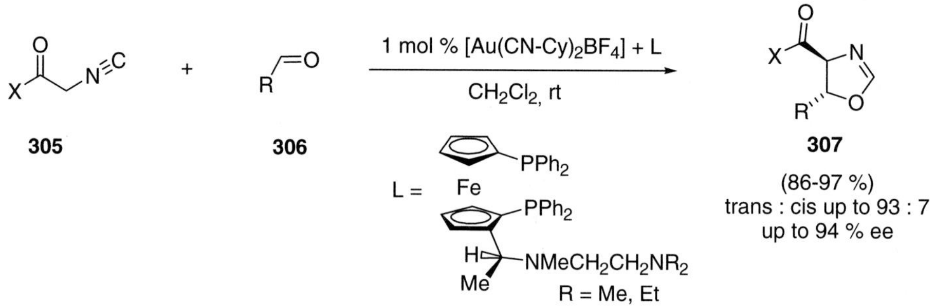

$$\underset{305}{X\overset{O}{\underset{}{\parallel}}\text{—}CH_2\text{—}N\!\equiv\!C} + \underset{306}{R\text{—}CHO} \xrightarrow[\text{CH}_2\text{Cl}_2,\ rt]{1\ mol\ \%\ [Au(CN\text{-}Cy)_2BF_4]\ +\ L} \underset{307}{X\overset{O}{\underset{}{\parallel}}\text{—}\underset{R'}{\overset{}{\text{CH}}}\text{—}\underset{O}{\overset{N}{\diagup\!\!\diagdown}}}$$

L = ferrocenyl-PPh$_2$/PPh$_2$ with –CH(Me)–NMeCH$_2$CH$_2$NR$_2$, R = Me, Et

(86-97 %)
trans : cis up to 93 : 7
up to 94 % ee

(8.77)

The postulated transition state **308** explained the stereoselectivity found for the process, showing that the gold catalyst activated both the carbonyl group and the nucleophile [176] (Figure 8.4).

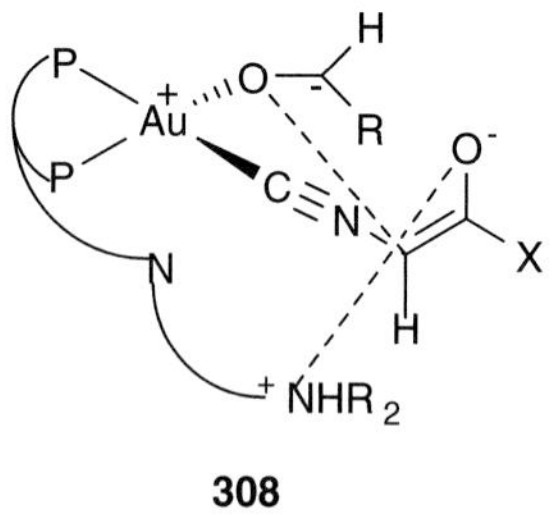

308

Figure 8.4 Proposed transition state to explain the double activation of the carbonyl group and the nucleophile by the gold catalyst.

The substrate was cyclized to the oxazoline after aldol addition. This reaction generated such interest that it has been repeatedly reviewed since the 1990s [4, 177, 178].

The next milestone came with the description of the condensation reactions between 1,3- dicarbonyl compounds and alcohols, amines or thiol derivatives. This reaction was a great success in the search for simplicity as an efficient tool for green chemical processes, thanks to the reduction of the number of synthetic steps and auxiliaries present [179]. The main catalysts employed in these reactions were NaAuCl$_4$ or AuCl$_3$.

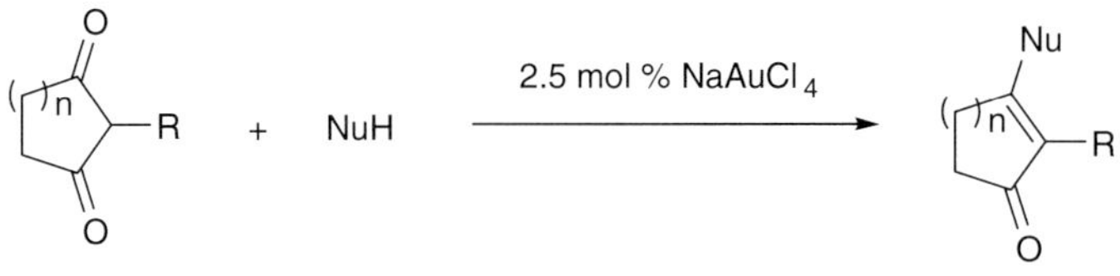

$$\text{309} + \text{NuH} \xrightarrow{2.5 \text{ mol \% NaAuCl}_4} \text{310} \quad (8.78)$$

Since then, similar reactions have been reported by different authors; examples include the synthesis of quinolines [180] or even pyrroles [181] or enamines from chiral amines [182]. Indoles can also participate in these reactions [183].

In some cases, it was observed that Lewis acids could also activate the processes. However, AuCl$_3$ displayed higher yields and higher selectivity [184].

Stradiotto et al. prepared some Au(I) complexes supported by donor-functionalized indene ligands that showed catalytic properties for aldehyde hydrosilylation [185].

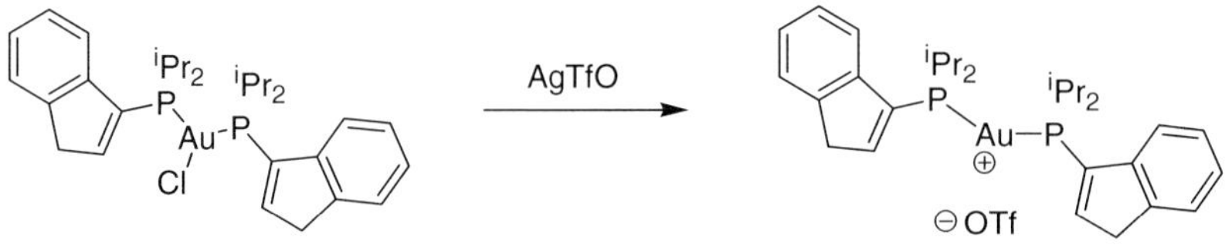

$$\text{311} \xrightarrow{\text{AgTfO}} \text{312} \quad (8.79)$$

Gold has even shown its ability as a nucleophile activator in three-component reactions of terminal alkynes, aldehydes and amines [186]. In the case of chiral amines, excellent diastereoselectivities were obtained [187] (Scheme 8.29).

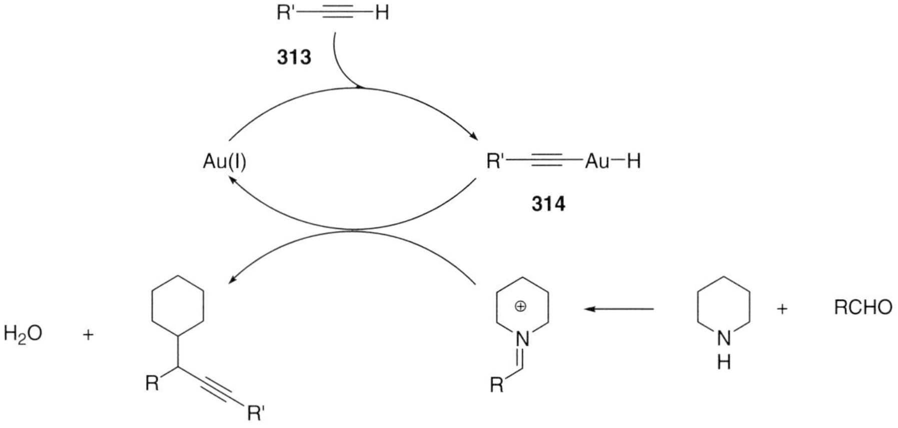

Scheme 8.29

8.2.3
Hydrogenation Reactions

Muller was the first to describe homogeneous gold catalyzed hydrogenation using $HAuCl_4/Sb(S(C_6F_5))_3$ as a catalytic system to hydrogenate ethene to ethane. Reaction took place in ethanol at 0 °C and 1 atm total hydrogen pressure, but no more details were provided [188].

No further research was performed in this field until three decades later, when Arcadi et al. developed a one-pot entry into functionalized pyridines. Reaction required a catalyst to dehydrogenate a dihydropyridine intermediate to pyridine. At that time, the liberated hydrogen was believed to be a consequence of aromatization [189].

Hosomi et al. reported an unprecedented hydrosilylation of ketones and imines displaying remarkable chemoselectivity that enabled ketones to be differentiated from aldehydes. Although the active species were not clear, they accepted that monomeric gold complexes stabilized by excess of tributylphosphine played a crucial role in controlling the reaction [190] (Scheme 8.30).

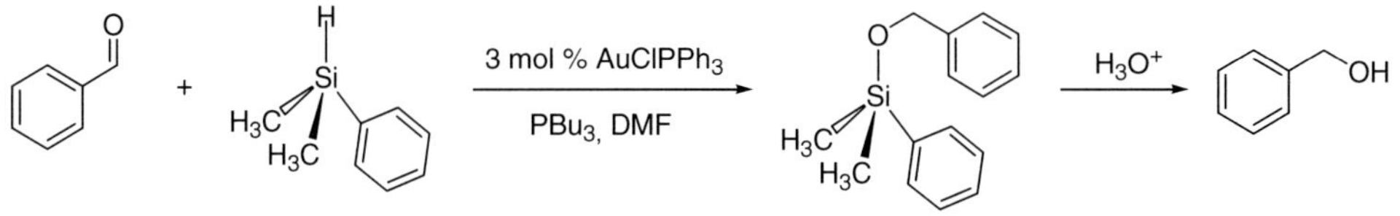

Scheme 8.30

Similarly, the hydroboration of imines was described, showing that $AuClPR_3$ can act as an efficient catalysts for the process [191].

Through a combination of experimental and theoretical calculations, Corma and coworkers proposed a mechanism for the hydrogenation of olefinic molecules by gold catalysis. In this study, Au(III)-Schiff base complexes proved to be as active as the corresponding Pd complexes. Some gold(III) intermediate species were proposed but were not detected [192].

A gold monohydride species was also suggested in the report by Ito and Sawamura et al. on the dehydrogenative silylation of alcohols by $HSiEt_3$ and a diphosphine gold(I) complex. Reaction was selective for the silylation of hydroxy groups in the presence of alkyl halides, ketones, aldehydes, alkenes, alkynes and other functional groups [193].

(8.80)

The gold(I) catalyzed alkylation of alcohols and aromatic compounds was described by Asao et al. very recently. Reaction produced corresponding Friedel–Crafts (Equation 8.81) or ether alkylation derivatives (Equation 8.82) under mild conditions [194].

(8.81)

(8.82)

Only one paper has reported on catalytic asymmetric hydrogenation. In this study by Corma et al., the neutral dimeric duphos-gold(I)complex **332** was used to catalyze the asymmetric hydrogenation of alkenes and imines. The use of the gold complex increased the enantioselectivity achieved with other platinum or iridium catalysts and activity was very high in the reaction tested [195] (Figure 8.5).

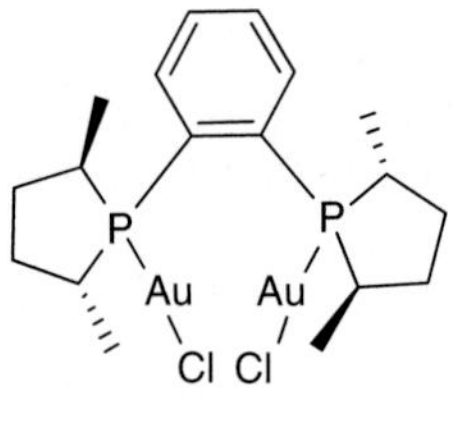

332

Figure 8.5 Proposed structure for the gold(I) complex used in catalytic asymmetric hydrogenation.

8.3
Heterogeneous Catalysis

The special nature of gold chemistry and gold catalysis is now known and many applications can be based on the low-temperature activity of supported gold compared to that of other metals [196].

In the same way as the synthesis of new organometallic derivatives or new applications of catalytic systems are sought in the field of homogeneous catalysis, in heterogeneous catalysis there are several aspects to bear in mind and sometimes the significance of these aspects becomes apparent when gold is concerned [197–199]:

- The activity of a gold catalyst depends largely on the conditions used for calcination or on the preparation method.
- The support plays a crucial role, for example in increasing the activity of gold by supplying enough oxygen.
- The interface between the gold particles and the support or even their size seems to be very important also.
- The different oxidation states that gold can present can of course have a strong influence.

There is also growing interest today in the development of gold nanoparticle-based systems that are providing new sources of substrates for study in the field of catalysis.

8.3.1
Hydrogenation Reactions

8.3.1.1 Alkenes Hydrogenation
This is the oldest gold catalyzed reaction and was first reported in the studies by Bone and Wheeler in 1906 on the capture of hydrogen by gold in the presence of oxygen at 600 °C. Despite the interpretation of the experiment, results were poor since the only observable parameter was pressure, although some conclusions of this experiment became very relevant some time later. It was then observed that hydrogen pressure directly influenced the reaction rate and oxygen had a minor retarding effect, which

could have meant that the interaction of the hydrogen with the gold catalyst was the rate-limiting step [200]. This hypothesis was confirmed half a century later [201–204].

A number of later studies on hydrogen exchange reactions showed that hydrogen could be activated by gold surfaces [205–207].

First, an example of alkene hydrogenation was reported by Erkelens, Kemball and Galwey using cyclohexene as hydrogen donor and substrate for hydrogenation. Reaction needed a temperature range of 196–342 °C to deliver both benzene and cyclohexane [208]. Chambers and Boudart changed the gold films used in this study for gold powder and obtained similar results. The ratio of the compounds obtained depended on the temperature and hydrogen pressure, as expected from thermodynamic principles [209].

Gas phase hydrogenation was also studied for alkene or alkyne substrates. Supported gold or Pt/Au alloys delivered interesting results [210, 211].

In 1973, Bond et al. performed a groundbreaking experiment when they achieved efficient alkene hydrogenation with several gold supports and at only 100–217 °C. Reaction took place with high chemoselectivity to monohydrogenation for alkenes and dialkenes; however, the process was only diastereoselective in the case of monoalkenes [9] (Scheme 8.31).

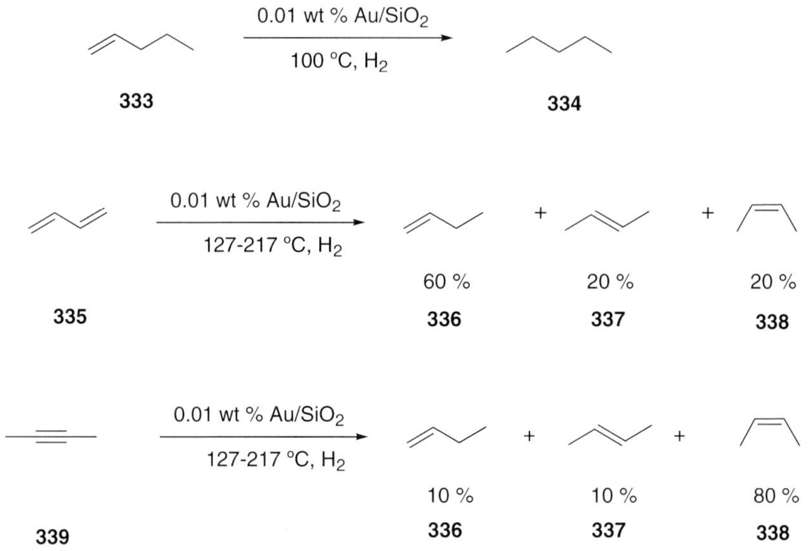

Scheme 8.31

Subsequent studies showed that the process depended enormously on the choice of catalyst, wt.% of gold and even gold particle size [212]. In terms of kinetic information, the breaking of the H–H bond was suggested as the rate-determining step [213]. As regards the oxidation state, Guzman and Gates described the hydrogenation of ethane, with mononuclear gold complexes on MgO powder at 80 °C and atmospheric pressure. IR spectroscopy suggested ethyl gold species as reactive intermediates, although EXAFS and XANES data seemed to evidence gold(III) as the active species [214].

After compiling many results obtained in similar studies of different substrates (alkenes, dienes, alkynes and so on), the results cannot be correlated to draw definitive conclusions due to the wide variety of parameters that can influence the reaction (substrates, catalyst precursors, supports, pressure, temperature and so on) [9, 208–214]. This is maybe the main reason why there are no clear mechanistic explanations for this "simple" reaction, unlike homogeneous gold-catalyzed processes.

8.3.1.2 Hydrogenation of α,β-Unsaturated Aldehydes

Claus recently reviewed the hydrogenation of α,β-unsaturated aldehydes, a process that displays very high selectivity by supported gold catalysts [215].

Bailie *et al.* were the first to mention alcohol formation from aldehydes by supported gold-catalyzed selective hydrogenation. The reaction of the formation of crotyl alcohol from crotonaldehyde showed high selectivity (up to 81%) at conversions of 5–10%, with preferential hydrogenation of C=O rather than the C=C bond [216]. The addition of thiophene promoted this selective hydrogenation. This promotional effect was also observed in similar situations for Cu and Ag, but it was not very common for gold.

Claus *et al.* applied gold-catalyzed hydrogenation to acrolein, using several Au/metal oxide supports [217, 218]; this method was subsequently used for more complex substrates [219]. They focused mainly on synthesing gold nanoparticles and studying their morphologies. A more rounded morphology (for example, in Au/TiO_2 catalysts) seems to be responsible for a higher relative amount of low-coordinate surface sites [218, 219].

One more step was proposed after it was found that the addition of a second metal to the supported gold catalyst could refine the selectivity of the process.

An example of the usefulness of the methodology was the new industrial route to cyclohexanone oxime developed by Corma and Serna, a process which is based on the selective hydrogenation of 1-nitro-1-cyclohexene by gold nanoparticles supported on TiO_2 or Fe_2O_3 [220].

8.3.2
Oxidation Reactions

8.3.2.1 C–H Bond Activation

This is one of the most interesting commercial objectives of such research. An industrial application consists in the oxidation of cyclohexene to cyclohexanol and cyclohexanone, a central step in the production of certain polymers such as nylon-6 or nylon-6,6, whose production exceeds 10^6 tonnes per annum. The established industrial process only reached 4% conversion and 70–85% selectivity [221]. Subsequent studies have attempted to improve the process by using different catalytic systems [222–224].

Zhao *et al.* achieved selectivities of about 90% in a process to activate cyclohexane. In this first study with gold-supported catalysts, temperatures of 140–160 °C were needed and although the catalysts activated C–H bonds, activity and selectivity decreased after a number of cycles [225].

In a recent study, Xu *et al.* managed to lower temperatures to 100 °C by using gold catalyst and oxygen as oxidant. The authors compared the Au/C catalyst and supported Pd or Pt catalysts, and concluded that these systems offered similar performance and that selectivity generally depended on cyclohexene conversion [226].

8.3.2.2 Oxidation of Alcohols and Aldehydes

The main problem of catalyzed processes based on platinum or palladium nanoparticles for such a needed target as the oxidation of alcohols and polyols is the scarce selectivity achieved with a complex substrate.

Rossi, Pratti *et al.* showed the effectiveness of supported gold nanoparticles for the oxidation of alcohols and diols [17, 227–229]. One of the most notable differences from Pd or Pt catalysts was the essential presence of base, required for initial hydrogen abstraction. However, this requirement was no longer necessary in Au/C systems applied to gas-phase reactants [230]. The studies were extended to the oxidation of sugars with high catalytic efficiency [231]. A synergistic effect of the addition of Pd or Pt to the Au/C catalyst was even observed in the selective oxidation of D-sorbitol to gluconic and gulonic acids.

Total selectivity was achieved by Carretin *et al.* when studying glycerol oxidation by Au supported on graphite using dioxygen as oxidant under relatively mild conditions [232], and Basheer *et al.* showed that the selective oxidation of glucose was also possible using a supported gold catalyst in a capillary flow reactor [233].

A recent contribution reported by Aoshima and Tsukuda showed the aerobic oxidation of alcohols such as benzyl alcohol catalyzed by gold nanoclusters. These stable and durable clusters of less than 4 nm were prepared using thermosensitive vinyl ether star polymers previously obtained by living cationic polymerization. One advantage of this method is that the clusters can be easily separated from the reaction mixture due to their thermosensitive nature, allowing for repeated reuse [234].

$$\text{PhCH}_2\text{OH} \xrightarrow[\text{H}_2\text{O, 27 °C}]{\text{gold nanoclusters, KOH}} \text{PhCOOH} \quad (8.83)$$

340 → **341**

99% (even after the sixth reuse)

8.3.2.3 Epoxidation

The oxidation of propene to propene oxide, a strategic compound in the manufacture of polyurethane and polyols, displays very low selectivity with many catalysts, unlike the epoxidation of ethane, whose selectivity may be as high as 90% when a supported Ag catalyst is used [235]. Lambert *et al.* recently showed that selectivities of about 50% can be achieved at 0.25% conversion by supported catalysts, although selectivity declines when the conversion increases [236].

The first evidence of the capacity of supported gold catalysts to epoxidate propene was described by Haruta et al. using dioxygen in the presence of H_2 as reductant, which allows O_2 activation at low temperatures [237]. Although initial selectivities by Au/TiO_2 were low, promising improvements were achieved with different supports [230–232, 238–246]. One of them, TS-1, is known to be suitable for the selective epoxidation of propene with H_2O_2 [246]. For that reason, many early studies focused on its use.

It has recently been found that NEt_3 is a gas-phase promoter for propene epoxidation by supported gold catalysts [245]. In more recent studies, Hughes et al. reported that catalytic amounts of peroxides could initiate the oxidation of alkenes with O_2, without the need for sacrificial H_2 [243]. The process worked for a range of substrates (cyclohexene, cis-stilbene, styrene and so on) and even in the absence of solvent; hence, we may refer to this as "green" technology.

Gao and Friend showed that chlorine is an effective promoter for enhancing the selectivity of styrene epoxidation on Au(1 1 1) by inhibiting secondary oxidation [247].

8.3.2.4 Direct Synthesis of Hydrogen Peroxide

Hydrogen peroxide is a large-scale manufactured chemical compound due to its application as a disinfectant and in bleaching, although it also has an important use as a reactant in many minor-scale fine-chemical productions.

At present, its synthesis requires a sequential hydrogenation and subsequent oxidation of alkyl anthraquinone [248]. One of the main problems of this process is the high cost of the quinine solvent and the need for continuous anthraquinone replacement. The manufacturing process is also hindered by frequent storage and transport problems. In view of these obstacles, the development of a new synthetic method is of great interest from a commercial standpoint.

Although much research over the last century has focused on a direct synthetic process, at present no commercial route is known. For many years, the catalysts used were based on Pd [249–251].

Hutchings et al. initially showed that Au/Al_2O_3 catalysts could promote direct reactions and results were subsequently drastically improved with the use of Au/Pd alloys supported on alumina [23, 252].

However, selectivity continued to be a problem in the process due to the hard conditions required by the method. Thus, in parallel research, Haruta et al. [253] and Ishihara and coworkers [254] achieved the reaction at only 10 °C. These studies compared the activity of Au/SiO_2 and Au–Pd/SiO_2 catalysts and the authors concluded that the enhancement observed when Pd was added to Au was directly related to the activation of hydrogen. However, excess Pd also induced rapid decomposition of H_2O_2.

A recent improvement was reported by Hutchings et al. who showed that the use of Fe_2O_3 and TiO_2 as supports increased the selectivity for H_2 utilization [255, 256]. Selectivity up to 95% was achieved for the reaction of H_2/O_2 mixtures (1:1, 5 vol%) diluted with CO_2 (95% vol).

8.3.3
Reactions Involving Carbon Monoxide

8.3.3.1 Carbon Monoxide Oxidation

Haruta *et al.* recognized the high effectiveness of supported gold nanocrystals for the oxidation of CO at very low temperatures, unlike the low activity displayed by other metals [11, 257].

It is worth mentioning the importance of the preparation method and the excellent behavior of the α-Fe_2O_3 support. Later studies showed that Au/TiO_2 was equally effective [257].

Electron microscopy studies showed that the most active catalysts contained gold nanoparticles of 2–4 nm in diameter.

In recent studies, Lahr and Ceyer achieved high activity even at $-203\,°C$, by using an Au/Ni(III) surface; this questions the real role of oxide supports [258].

The high activity that supported gold catalysts have shown for CO oxidation at ambient temperature makes them ideal candidates for use as respiratory protectors. A copper manganese oxide, Hopcalite, has been used for many years to remove CO in toxic environments. Thus, supported gold catalysts may be chosen in the future.

However, one application where gold could be most applied is in fuel cells used in electric vehicles, with operating temperatures of 80–100 °C. Another field of application for supported gold catalysts is the production of hydrogen by steam reforming of methanol.

Despite the efforts of many research groups due to the high level applicability of this field, many questions remain unanswered, for example, the reaction mechanism, the nature of the active site and so on. One of the main problems to finding answers is the correlation of the results obtained after using such a variety of parameters: different catalysts, supports, preparation methods and so on. Therefore, a great deal of research is still required.

8.3.3.2 Water Gas Shift Reaction

After showing the high activity of gold for the activation of carbon monoxide oxidation, the water gas shift reaction is another of the most relevant reactions to be studied.

For the production of hydrogen on an industrial scale, CO oxidation is considered to be a participant in the water gas shift process, acting as an intermediate to react with adsorbed oxygen atoms [259]. Reactions in the water gas shift process are shown in Scheme 8.32.

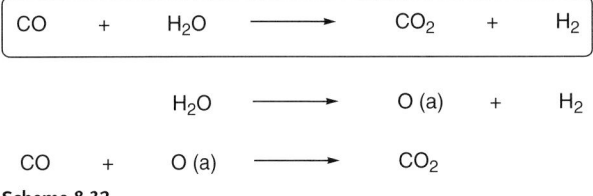

$$CO + H_2O \longrightarrow CO_2 + H_2$$

$$H_2O \longrightarrow O\,(a) + H_2$$

$$CO + O\,(a) \longrightarrow CO_2$$

Scheme 8.32

The catalytic activity of gold for this reaction at low temperature was reported by Andreeva et al., who used Au/Fe$_2$O$_3$ catalysts [260, 261]. Venugopal et al. also studied this system and showed that the combination Au–Ru/Fe$_2$O$_3$ or even gold supported on hydroxyapatite were more useful for the transformation [262].

The next improvement was reported by Hua et al. by means of the addition of metal oxides to Au/Fe$_2$O$_3$ [263].

Traces of water can enhance the rate of CO oxidation at low temperature but if the water concentration is too high, much higher temperatures are required in order to avoid the reduction of cationic gold to metallic gold [264]. Haruta et al. [265] and Idakiev et al. [266] studied TiO$_2$ and mesoporous TiO$_2$ and both demonstrated their effectiveness.

Subsequent studies focused on Au/CeO$_2$ catalysts [267]. One study showed that the addition of CN$^-$ ions increased the catalytic activity of Au/CeO$_2$ by removing most of the gold from the catalyst, and it was proposed that the resulting cationic gold was the active species [268].

A subsequent combined experimental and theoretical study by Tibiletti et al. suggested that gold was present in a partially oxidized form at a Ce^{4+} vacancy [269].

8.4
Concluding Remarks

While there was a preconception that gold was a noble immutable element, nowadays it is clear that this metal offers many opportunities in catalysis, and its increasing appreciation is revealed with tens of papers appearing weekly. One of the main advantages that gold offers is high activity in very mild reaction conditions compared to other catalytic systems.

Transformations achieved by gold catalysis are more and more complex, with a higher number of bonds formed and many tandem reactions.

C,H activation seems to be one of the most promising fields combined with the search for stereoselectivity that could lead to asymmetric catalysis.

The gold catalysis field is expanding so quickly that future reviews will probably cover only the most relevant developments or the advances in some certain subfields.

Clearly, gold has just opened a New Age in catalysis.

References

1 Hagen, J. (1999) *Industrial Catalysis: A Practical Aproach*, Wiley-VCH, Weinheim.
2 Hashmi, A.S.K. (2004) *Gold Bulletin*, **37**, 51.
3 Hagen, J. (1992) *Chemische Reaktionstechnik: eine Einführung mit Übungen*, Wiley-VCH, Weinheim.
4 Sawamura, M. and Ito, Y. (1992) *Chemical Reviews*, **92**, 857.
5 Schmidbaur, H. (1995) *Naturwissenschaftliche Rundschau*, **48**, 443.
6 Hashmi, A.S.K. and Hutchings, G.J. (2006) *Angewandte Chemie International Edition*, **45**, 7896.

7 Hutchings, G.J. (2005) *Catalysis Today*, **100**, 55.
8 Bond, G.C. (1972) *Gold Bulletin*, **5**, 11.
9 Bond, G.C., Sermon, P.A., Webb, G., Buchanan, D.A. and Wells, P.B. (1973) *Journal of the Chemical Society-Chemical Communications*, 444.
10 Hutchings, G.J. (1985) *Journal of Catalysis*, **96**, 292.
11 Haruta, M., Kobayashi, T., Sano, H., Yamada, N. (1987) *Chemistry Letters*, **16**, 405.
12 Ito, Y., Sawamura, M. and Hayashi, T. (1986) *Journal of the American Chemical Society*, **108**, 6405.
13 Fukuda, Y. and Utimoto, K. (1991) *The Journal of Organic Chemistry*, **56**, 3729.
14 Teles, J.H., Brode, S. and Chabanas, M. (1998) *Angewandte Chemie-International Edition*, **37**, 1415.
15 Haruta, M., Ueda, A., Sanchez, R.M.T. and Tanaka, K. (1996) Prp. Pet. Div. ACS Symposium, New Orleans.
16 Hayashi, T., Tanaka, K. and Haruta, M. (1996) Prp. Pet. Div. ACS Symposium, New Orleans.
17 Prati, L. and Rossi, M. (1998) *Journal of Catalysis*, **176**, 552.
18 Hashmi, A.S.K., Schwarz, L., Choi, J.-H. and Frost, T.M. (2000) *Angewandte Chemie*, **112**, 2382.
19 Hashmi, A.S.K., Frost, T.M. and Bats, J.W. (2000) *Journal of the American Chemical Society*, **122**, 11553.
20 Martin-Matute, B., Nevado, C., Cárdenas, D.J. and Echavarren, A.M. (2003) *Journal of the American Chemical Society*, **125**, 5757.
21 Shi, Z. and He, C. (2004) *Journal of the American Chemical Society*, **126**, 13596.
22 Yang, C.-G. and He, C. (2005) *Journal of the American Chemical Society*, **127**, 6966.
23 Landon, P., Collier, P.J., Papworth, A.J., Kiely, C.J. and Hutchings, G.J. (2002) *Chemical Communications*, 2058.
24 Gorin, D.J. and Toste, F.D. (2007) *Nature*, **446**, 395.
25 Cinellu, M.A., Minghetti, G., Cocco, T., Stoccoro, S., Zucca, A. and Manassero, M. (2005) *Angewandte Chemie-International Edition*, **44**, 6892.
26 Hashmi, A.S.K. (2005) *Angewandte Chemie-International Edition*, **44**, 6990.
27 Schuster, H.F. and Coppola, G.M. (1984) *Allenes in Organic Synthesis*, Wiley-VCH, New York.
28 Hashmi, A.S.K., Schwarz, L., Choi, J.-H. and Frost, T.M. (2000) *Angewandte Chemie-International Edition*, **39**, 2285.
29 Hashmi, A.S.K., Schwarz, L. and Bats, J.W. (2000) *Journal für Praktische Chemie*, **342**, 40.
30 Hoffmann-Röder, A. and Krause, N. (2001) *Organic Letters*, **3**, 2537.
31 Krause, N., Hoffmann-Roder, A. and Canisius, J. (2002) *Synthesis*, 1759.
32 Marshall, J.A. and Bartley, G.S. (1994) *The Journal of Organic Chemistry*, **59**, 7169.
33 Hyland, C.J.T. and Hegedus, L.S. (2006) *The Journal of Organic Chemistry*, **71**, 8658.
34 Gockel, B. and Krause, N. (2006) *Organic Letters*, **8**, 4485.
35 Zhang, Z.B. and Widenhoefer, R.A. (2007) *Angewandte Chemie-International Edition*, **46**, 283.
36 Hashmi, A.S.K., Blanco, M.C., Fischer, D. and Bats, J.W. (2006) *European Journal of Organic Chemistry*, 1387.
37 Zhang, Z., Liu, C., Kinder, R.E., Han, X., Qian, H. and Widenhoefer, R.A. (2006) *Journal of the American Chemical Society*, **128**, 9066.
38 Morita, N. and Krause, N. (2006) *Angewandte Chemie-International Edition*, **45**, 1897.
39 Morita, N. and Krause, N. (2004) *Organic Letters*, **6**, 4121.
40 Morita, N. and Krause, N. (2006) *European Journal of Organic Chemistry*, 4634.
41 Lee, P.H., Kim, H., Lee, K., Kim, M., Noh, K. and Seomoon, D. (2005) *Angewandte Chemie-International Edition*, **44**, 1840.
42 Patil, N.T., Lutete, L.M., Nishina, N. and Yamamoto, Y. (2006) *Tetrahedron Letters*, **47**, 4749.
43 Nishina, N. and Yamamoto, Y. (2006) *Angewandte Chemie-International Edition*, **45**, 3314.

44 Kaden, S., Reissig, H.U., Brudgam, I. and Hard, H. (2006) *Synthesis*, 1351.
45 Ohno, H., Kadoh, Y., Fujii, N. and Tanaka, T. (2006) *Organic Letters*, **8**, 947.
46 Zhang, Z., Bender, C.F. and Widenhoefer, R.A. (2007) *Journal of the American Chemical Society*, **129**, 14148.
47 Zhang, Z., Bender, C.F. and Widenhoefer, R.A. (2007) *Organic Letters*, **9**, 2887.
48 Shin, S. (2005) *Bulletin of the Korean Chemical Society*, **26**, 1925.
49 Liu, Z., Wasmuth, A.S. and Nelson, S.G. (2006) *Journal of the American Chemical Society*, **128**, 10352.
50 Watanabe, T., Oishi, S., Fujii, N. and Ohno, H. (2007) *Organic Letters*, **9**, 4821.
51 Luzung, M.R., Mauleón, P. and Toste, F.D. (2007) *Journal of the American Chemical Society*, **129**, 12402.
52 Norman, R.O.C., Parr, W.J.E. and Thomas, C.B. (1976) *Journal of the Chemical Society-Perkin Transactions 1*, 811.
53 Norman, R.O.C., Parr, W.J.E. and Thomas, C.B. (1976) *Journal of the Chemical Society-Perkin Transactions 1*, 1983.
54 Yang, C.G. and He, C. (2005) *Journal of the American Chemical Society*, **127**, 6966.
55 Qian, H., Han, X. and Widenhoefer, R.A. (2004) *Journal of the American Chemical Society*, **126**, 9536.
56 Cinellu, M.A., Minghetti, G., Stoccoro, S., Zucca, A. and Manassero, M. (2004) *Chemical Communications*, 1618.
57 Jung, H.H. and Floreancig, P. (2006) *Organic Letters*, **8**, 1949.
58 Nguyen, R., Yao, X. and Li, C. (2006) *Organic Letters*, **8**, 2397.
59 Nguyen, R.V., Yao, X.Q., Bohle, D.S. and Li, C.J. (2005) *Organic Letters*, **7**, 673.
60 Kobayashi, A., Kakumoto, K. and Sugiura, M. (2002) *Organic Letters*, **4**, 1319.
61 Bender, C.F. and Widenhoefer, R.A. (2006) *Chemical Communications*, 4143.
62 Bender, C.F. and Widenhoefer, R.A. (2006) *Organic Letters*, **8**, 5303.
63 Han, X.Q. and Widenhoefer, R.A. (2006) *Angewandte Chemie-International Edition*, **45**, 1747.
64 Zhang, J.L., Yang, C.G. and He, C. (2006) *Journal of the American Chemical Society*, **128**, 1798.
65 Brouwer, C. and He, C. (2006) *Angewandte Chemie-International Edition*, **45**, 1744.
66 Liu, X.Y., Li, C.H. and Che, C.M. (2006) *Organic Letters*, **8**, 2707.
67 Anderson, L.L., Arnold, J. and Bergman, R.G. (2005) *Journal of the American Chemical Society*, **127**, 14542.
68 Kharasch, M.S. and Beck, T.M. (1934) *Journal of the American Chemical Society*, **56**, 2057.
69 Fuchita, Y., Utsunomiya, Y. and Yasutake, M. (2001) *Journal of the Chemical Society, Dalton Transactions*, 2330.
70 Dyker, G., Muth, E., Hashmi, A.S.K. and Ding, L. (2003) *Advanced Synthesis and Catalysis*, **345**, 1247.
71 Porter, K.A., Schier, A. and Schmidbaur, H. (2003) *Organometallics*, **22**, 4922.
72 Aguilar, D., Contel, M., Navarro, R. and Urriolabeitia, E.P. (2007) *Organometallics*, **26**, 4604.
73 Hashmi, A.S.K., Salathe, R., Frost, T.M., Schwarz, L. and Choi, J.H. (2005) *Applied Catalysis A: General*, **291**, 238.
74 Hashmi, A.S.K., Salathe, R. and Frey, W. (2007) *European Journal of Organic Chemistry*, 1648.
75 Arcadi, A., Bianchi, G., Chiarini, M., D'Anniballe, G. and Marinelli, F. (2004) *Synlett*, 944.
76 Alfonsi, M., Arcadi, A., Bianchi, G., Marinelli, F. and Nardini, A. (2006) *European Journal of Organic Chemistry*, 2393.
77 Li, Z., Shi, Z. and He, C. (2005) *Journal of Organometallic Chemistry*, **690**, 5049.
78 Cinellu, M.A., Minghetti, G., Pinna, M.V., Stoccoro, S., Zucca, A. and Manassero, M. (1999) *Journal of The Chemical Society, Dalton Transactions*, 2823.
79 Yao, X. and Li, C.J. (2004) *Journal of the American Chemical Society*, **126**, 6884.
80 Staben, S.T., Kennedy-Smith, J.J. and Toste, F.D. (2004) *Angewandte Chemie-International Edition*, **43**, 5350.

81 Kennedy-Smith, J.J., Staben, S.T. and Toste, F.D. (2004) *Journal of the American Chemical Society*, **126**, 4526.
82 Corkey, B.K. and Toste, F.D. (2005) *Journal of the American Chemical Society*, **127**, 17168.
83 Zhou, C.Y. and Che, C.M. (2007) *Journal of the American Chemical Society*, **129**, 5828.
84 Hashmi, A.S.K. (2007) *Chemical Reviews*, **107**, 3180.
85 Hutchings, G. (1996) *Gold Bulletin*, **29**, 123.
86 Arcadi, A., Cerichelli, G., Chiarini, M., Di Giuseppe, S. and Marinelli, F. (2000) *Tetrahedron Letters*, **41**, 9195.
87 Fukuda, Y. and Utimoto, K. (1991) *Bulletin of the Chemical Society of Japan*, **64**, 2013.
88 Engel, D.A. and Dudley, G.B. (2006) *Organic Letters*, **8**, 4027.
89 Preisenberger, M., Schier, A. and Schmidbaur, H. (1999) *Journal of The Chemical Society, Dalton Transactions*, 1645.
90 Roithova, J., Hrusak, J., Schroder, D. and Schwarz, H. (2005) *Inorganica Chimica Acta*, **358**, 4287.
91 Harkat, H., Weibel, J.M. and Pale, P. (2006) *Tetrahedron Letters*, **47**, 6273.
92 Mizushima, E., Sato, K., Hayashi, T. and Tanaka, M. (2002) *Angewandte Chemie-International Edition*, **41**, 4563.
93 Tokunaga, M. and Wakatsuki, Y. (1998) *Angewandte Chemie-International Edition*, **37**, 2867.
94 Bardají, M. and Laguna, A. (2001) *Organometallics*, **20**, 3906.
95 Deetlefs, M., Raubenheimer, H.G. and Esterhuysen, M.W. (2002) *Catalysis Today*, **72**, 29.
96 Casado, R., Contel, M., Laguna, M., Romero, P. and Sanz, S. (2003) *Journal of the American Chemical Society*, **125**, 11925.
97 Sanz, S., Jones, L.A., Mohr, F. and Laguna, M. (2007) *Organometallics*, **26**, 952.
98 Schneider, S.K., Herrmann, W.A. and Herdtweck, E. (2003) *Zeitschrift Fur Anorganische Und Allgemeine Chemie*, **629**, 2363.
99 Roembke, P., Schmidbaur, H., Cronje, S. and Raubenheimer, H. (2004) *Journal of Molecular Catalysis A-Chemical*, **212**, 35.
100 Liu, Y., Song, F., Song, Z., Liu, M. and Yan, B. (2005) *Organic Letters*, **7**, 5409.
101 Istrate, F.M. and Gagosz, F.L. (2008) *The Journal of Organic Chemistry*, **73**, 730.
102 Antoniotti, S., Genin, E., Michelet, V. and Genet, J.P. (2005) *Journal of the American Chemical Society*, **127**, 9976.
103 Li, Y., Zhou, F. and Forsyth, C. (2007) *Angewandte Chemie-International Edition*, **46**, 279.
104 Barluenga, J., Dieguez, A., Fernandez, A., Rodriguez, F. and Fananas, F.J. (2006) *Angewandte Chemie-International Edition*, **45**, 2091.
105 Belting, V. and Krause, N. (2006) *Organic Letters*, **8**, 4489.
106 Tian, G.Q. and Shi, M. (2007) *Organic Letters*, **9**, 4917.
107 Dubé, P. and Toste, F.D. (2006) *Journal of the American Chemical Society*, **128**, 12062.
108 Nakamura, I., Sato, T., Terada, M. and Yamamoto, Y. (2007) *Organic Letters*, **9**, 4081.
109 Dai, L.Z., Qi, M.J., Shi, Y.L., Liu, X.G. and Shi, M. (2007) *Organic Letters*, **9**, 3191.
110 Hashmi, A.S.K., Schäfer, S., Wölfle, M., Diez, C., Fischer, P., Laguna, A., Blanco, M.C. and Gimeno, M.C. (2007) *Angewandte Chemie-International Edition*, **46**, 6184.
111 Fukuda, Y., Utimoto, K. and Nozaki, H. (1987) *Heterocycles*, **25**, 297.
112 Fukuda, Y. and Utimoto, K. (1991) *Synthesis*, 975.
113 Mueller, T.E. (1998) *Tetrahedron Letters*, **39**, 5961.
114 Mueller, T.E., Grosche, M., Herdtweck, E., Pleier, A.-K., Walter, E. and Yan, Y.-K. (2000) *Organometallics*, **19**, 170.
115 Lok, R., Leone, R.E. and Williams, A. (1996) *The Journal of Organic Chemistry*, **61**, 3289.
116 Kadzimirsz, D., Hildebrandt, D., Merz, K. and Dyker, G. (2006) *Chemical Communications*, 661.

117 Yan, B. and Liu, Y. (2007) *Organic Letters*, **9**, 4323.
118 Crawley, S.L. and Funk, R.L. (2006) *Organic Letters*, **8**, 3995.
119 Jung, H.H. and Floreancig, P.E. (2007) *The Journal of Organic Chemistry*, **72**, 7359.
120 Nakamura, I., Yamagishi, U., Song, D., Konta, S. and Yamamoto, Y. (2007) *Angewandte Chemie-International Edition*, **46**, 2284.
121 Genin, E., Toullec, P.Y., Antoniotti, S., Brancour, C., Genet, J.P. and Michelet, W. (2006) *Journal of the American Chemical Society*, **128**, 3112.
122 Jia, C., Lu, W., Oyamada, J., Kitamura, T., Matsuda, K., Irie, M. and Fujiwara, Y. (2000) *Journal of the American Chemical Society*, **122**, 7252.
123 Jia, C., Piao, D., Oyamada, J., Lu, W., Kitamura, T. and Fujiwara, Y. (2000) *Science*, **287**, 1992.
124 Tsuchimoto, T., Maeda, T., Shirakawa, E. and Kawakami, Y. (2000) *Chemical Communications*, 1573.
125 Sartori, G., Bigi, F., Pastorio, A., Porta, C., Arienti, A., Maggi, R., Moretti, N. and Gnappi, G. (1995) *Tetrahedron Letters*, **36**, 9177.
126 Pastine, S.J., Youn, S.W. and Sames, D. (2003) *Organic Letters*, **5**, 1055.
127 Reetz, M.T. and Sommer, K. (2003) *European Journal of Organic Chemistry*, 3485.
128 Shi, Z. and He, C. (2004) *The Journal of Organic Chemistry*, **69**, 3669.
129 Hashmi, A.S.K. and Blanco, M.C. (2006) *European Journal of Organic Chemistry*, 4340.
130 Beakley, L.W., Yost, S.E., Cheng, R. and Chandler, B.D. (2005) *Applied Catalysis A: General*, **292**, 124.
131 Nevado, C. and Echavarren, A.M. (2005) *Chemistry – A European Journal*, **11**, 3155.
132 Hashmi, A.S.K., Blanco, M.C., Kurpejovic, E., Frey, W. and Bats, J.W. (2006) *Advanced Synthesis and Catalysis*, **348**, 709.
133 Hashmi, A.S.K., Haufe, P., Schmid, C., Nass, A.R. and Frey, W. (2006) *Chemistry – A European Journal*, **12**, 5376.
134 Ferrer, C. and Echavarren, A.M. (2006) *Angewandte Chemie-International Edition*, **45**, 1105.
135 Amijs, C.H.M., López-Carrillo, V. and Echavarren, A.M. (2007) *Organic Letters*, **9**, 4021.
136 Endo, K., Hatakeyama, T., Nakamura, M. and Nakamura, E. (2007) *Journal of the American Chemical Society*, **129**, 5264.
137 Sergeev, G.B. (2001) *Uspekhi Khimii*, **70**, 915.
138 Harrison, T.J., Kozak, J.A., Corbella-Pané, M. and Dake, G.R. (2006) *The Journal of Organic Chemistry*, **71**, 4525.
139 Kirsch, S.F., Binder, J.T., Liebert, C. and Menz, H. (2006) *Angewandte Chemie-International Edition*, **45**, 5878.
140 Zhang, J. and Schmalz, H.G. (2006) *Angewandte Chemie-International Edition*, **45**, 6704.
141 Jin, T. and Yamamoto, Y. (2007) *Organic Letters*, **9**, 5259.
142 Witham, C.A., Mauleón, P., Shapiro, N.D., Sherry, B.D. and Toste, F.D. (2007) *Journal of the American Chemical Society*, **129**, 5838.
143 Trost, B.M. (1990) *Accounts of Chemical Research*, **23**, 34.
144 Trost, B.M. and Krische, M.J. (1998) *Synlett*, 1.
145 Bruneau, C. (2005) *Angewandte Chemie-International Edition*, **44**, 2328.
146 Ma, S.M., Yu, S.C. and Gu, Z.H. (2006) *Angewandte Chemie-International Edition*, **45**, 200.
147 Nieto-Oberhuber, C., Muñoz, M.P., Buñuel, E., Nevado, C., Cárdenas, D.J. and Echavarren, A.M. (2004) *Angewandte Chemie-International Edition*, **43**, 2402.
148 Diaz-Requejo, M.M. and Perez, P.J. (2005) *Journal of Organometallic Chemistry*, **690**, 5441.
149 Nieto-Oberhuber, C., López, S. and Echavarren, A.M. (2005) *Journal of the American Chemical Society*, **127**, 6178.
150 Herrero-Gómez, E., Nieto-Oberhuber, C., López, S., Benet-Buchholz, J. and Echavarren, A.M. (2006) *Angewandte Chemie-International Edition*, **45**, 5455.

151 Méndez, M., Muñoz, M.P., Nevado, C., Cárdenas, D.J. and Echavarren, A.M. (2001) *Journal of the American Chemical Society*, **123**, 10511.

152 Nieto-Oberhuber, C., Pérez-Galán, P., Herrero-Gómez, E., Lauterbach, T., Rodríguez, C., López, S., Bour, C., Rosellón, A., Cárdenas, D.J. and Echavarren, A.M. (2008) *Journal of the American Chemical Society*, **130**, 269.

153 Gabor, S.A. (1992) *Magyar Kemiai Folyoirat*, **98**, 1.

154 Zhang, L. and Kozmin, S.A. (2004) *Journal of the American Chemical Society*, **126**, 11806.

155 Mezailles, N., Ricard, L. and Gagosz, F. (2005) *Organic Letters*, **7**, 4133.

156 Ricard, L. and Gagosz, F. (2007) *Organometallics*, **26**, 4704.

157 Zhang, L. and Kozmin, S.A. (2005) *Journal of the American Chemical Society*, **127**, 6962.

158 Grisé, C.M. and Barriault, L. (2006) *Organic Letters*, **8**, 5905.

159 Sherry, B.D., Maus, L., Laforteza, B.N. and Toste, F.D. (2006) *Journal of the American Chemical Society*, **128**, 8132.

160 Minnihan, E.C., Colletti, S.L., Toste, F.D. and Shen, H.C. (2007) *The Journal of Organic Chemistry*, **72**, 6287.

161 Nieto-Oberhuber, C., López, S., Muñoz, M.P., Cárdenas, D.J., Buñuel, E., Nevado, C. and Echavarren, A.M. (2005) *Angewandte Chemie-International Edition*, **44**, 6146.

162 López, S., Herrero-Gómez, E., Pérez-Galán, P., Nieto-Oberhuber, C. and Echavarren, A.M. (2006) *Angewandte Chemie-International Edition*, **45**, 6029.

163 Lim, C., Kang, J.E., Lee, J.E. and Shin, S. (2007) *Organic Letters*, **9**, 3539.

164 Leseurre, L., Toullec, P.Y., Genêt, J.P. and Michelet, V. (2007) *Organic Letters*, **9**, 4049.

165 Martin-Matute, B., Cardenas, D.J. and Echavarren, A.M. (2001) *Angewandte Chemie-International Edition*, **40**, 4754.

166 Méndez, M., Muñoz, M.P., Nevado, C., Cárdenas, D.J. and Echavarren, A.M. (2001) *Journal of the American Chemical Society*, **123**, 10511.

167 Echavarren, A.M., Méndez, M., Muñoz, M.P., Nevado, C., Martín-Matute, B., Nieto-Oberhuber, C. and Cárdenas, D.J. (2004) *Pure and Applied Chemistry*, **76**, 453.

168 Hashmi, A.S.K., Rudolph, M., Weyrauch, J.P., Woelfle, M., Frey, W. and Bats, J.W. (2005) *Angewandte Chemie-International Edition*, **44**, 2798.

169 Hashmi, A.S.K., Salathe, R. and Frey, W. (2006) *Chemistry – A European Journal*, **12**, 6991.

170 Hashmi, A.S.K., Weyrauch, J.P., Rudolph, M. and Kurpejovic, E. (2004) *Angewandte Chemie-International Edition*, **43**, 6545.

171 Hashmi, A.S.K., Ata, F., Bats, J.W., Blanco, M.C., Frey, W., Hamzic, M., Rudolph, M., Salathé, R., Schäfer, S. and Wölfle, M. (2007) *Gold Bulletin*, **40**, 31.

172 Carrettin, S., Blanco, M.C., Corma, A. and Hashmi, A.S.K. (2006) *Advanced Synthesis and Catalysis*, **348**, 1283.

173 Mendez, M., Muñoz, M.P. and Echavarren, A.M. (2000) *Journal of the American Chemical Society*, **122**, 11549.

174 Nevado, C., Cárdenas, D.J. and Echavarren, A.M. (2003) *Chemistry – A European Journal*, **9**, 2627.

175 Kashyap, S. and Hotha, S. (2006) *Tetrahedron Letters*, **47**, 2021.

176 Sawamura, M., Nakayama, T., Kato, T. and Ito, Y. (1995) *The Journal of Organic Chemistry*, **60**, 1727.

177 Machajewski, T.D. and Wong, C.H. (2000) *Angewandte Chemie-International Edition*, **39**, 1352.

178 Sawamura, M. and Ito, Y. (2000) in *Catalytic Asymmetric Synthesis*, 2nd Edn (ed. I. Ojima), Wiley-VCH, New York.

179 Arcadi, A., Bianchi, G., Di Giuseppe, S. and Marinelli, F. (2003) *Green Chemistry*, **5**, 64.

180 Arcadi, A., Chiarini, M., Di Giuseppe, S. and Marinelli, F. (2003) *Synlett*, 203.

181 Arcadi, A., Di Giuseppe, S., Marinelli, F. and Rossi, E. (2001) *Tetrahedron: Asymmetry*, **12**, 2715.

182 Arcadi, A., Di Giuseppe, S., Marinelli, F. and Rossi, E. (2001) *Advanced Synthesis and Catalysis*, **343**, 443.

183 Hashmi, A.S.K., Schwarz, L., Rubenbauer, P. and Blanco, M.C. (2006) *Advanced Synthesis and Catalysis*, **348**, 705.
184 Nair, V., Vidya, N. and Abhilash, K.G. (2006) *Synthesis*, 3647.
185 Wile, B.M., McDonald, R., Ferguson, M.J. and Stradiotto, M. (2007) *Journal of the American Chemical Society*, **26**, 1069.
186 Wei, C. and Li, C.J. (2003) *Journal of the American Chemical Society*, **125**, 9584.
187 Lo, V.K.Y., Liu, Y.G., Wong, M.K. and Che, C.M. (2006) *Organic Letters*, **8**, 1529.
188 Muller, M.C. (1974) *Gold Bulletin*, **7**, 39.
189 Abbiati, G., Arcadi, A., Bianchi, G., Di Giuseppe, S., Marinelli, F. and Rossi, E. (2003) *The Journal of Organic Chemistry*, **68**, 6959.
190 Ito, H., Yajima, T., Tateiwa, J.-i. and Hosomi, A. (2000) *Chemical Communications*, 981.
191 Baker, R.T., Calabrese, J.C. and Westcott, S.A. (1995) *Journal of Organometallic Chemistry*, **498**, 109.
192 Comas-Vives, A., González-Arellano, C., Corma, A., Iglesias, M., Sánchez, F. and Ujaque, G. (2006) *Journal of the American Chemical Society*, **128**, 4756.
193 Ito, H., Takagi, K., Miyahara, T. and Sawamura, M. (2005) *Organic Letters*, **7**, 3001.
194 Asao, N., Aikawa, H., Tago, S. and Umetsu, K. (2007) *Organic Letters*, **9**, 4299.
195 Gonzalez-Arellano, C., Corma, A., Iglesias, M. and Sanchez, F. (2005) *Chemical Communications*, 3451.
196 Min, B.K. and Friend, C.M. (2007) *Chemical Reviews*, **107**, 2709.
197 Haruta, M. (2004) *Gold Bulletin*, **37**, 27.
198 Chen, M.S. and Goodman, D.W. (2004) *Science*, **306**, 252.
199 Straub, B.F. (2004) *Chemical Communications*, 1726.
200 Bone, W.A. and Wheeler, R.V. (1906) *Philosophical Transactions of the Royal Society of London. Series A, Mathematical and Physical Sciences*, **206**, 45.
201 Trapnell, B.M.W. (1953) *Proceedings of the Royal Society of London. Series A*, **218**, 566.
202 Pritchard, J. (1963) *Transactions of the Faraday Society*, **59**, 437.
203 Attard, G.A. and King, D.A. (1989) *Surface Science*, **223**, 1.
204 Stobinski, L. and Dus, R. (1993) *Czechoslovak Journal of Physics*, **43**, 1035.
205 Couper, A. and Eley, D.D. (1950) *Discussions of the Faraday Society*, **8**, 172.
206 Bond, G.C. and Sermon, P.A. (1973) *Gold Bulletin*, **6**, 102.
207 Saito, A. and Tanimoto, M. (1988) *Journal of the Chemical Society-Chemical Communications*, 832.
208 Erkelens, J., Kemball, C. and Galway, A.K. (1963) *Transactions of the Faraday Society*, **59**, 1181.
209 Chambers, R.P. and Boudart, M. (1966) *Journal of Catalysis*, **5**, 517.
210 Wood, B.J. and Wise, H. (1966) *Journal of Catalysis*, **5**, 135.
211 Dessing, R.P., Ponec, V. and Sachtler, W.M.H. (1972) *Journal of the Chemical Society-Chemical Communications*, 880.
212 Sermon, P.A., Bond, G.C. and Wells, P.B. (1979) *Journal of the Chemical Society, Faraday Transactions 1*, **75**, 385.
213 Saito, A. and Tanimoto, M. (1988) *Journal of the Chemical Society, Faraday Transactions 1*, **84**, 4115.
214 Guzman, J. and Gates, B.C. (2004) *Journal of Catalysis*, **226**, 111.
215 Claus, P. (2005) *Applied Catalysis A: General*, **291**, 222.
216 Bailie, J.E., Abdullah, H.A., Anderson, J.A., Rochester, C.H., Richardson, N.V., Hodge, N., Zhang, J.-G., Burrows, A., Kiely, C.J. and Hutchings, G.J. (2001) *Physical Chemistry Chemical Physics*, **3**, 4113.
217 Claus, P., Brueckner, A., Mohr, C. and Hofmeister, H. (2000) *Journal of the American Chemical Society*, **122**, 11430.
218 Mohr, C., Hofmeister, H. and Claus, P. (2003) *Journal of Catalysis*, **213**, 86.
219 Fordham, P., Besson, M. and Gallezot, P. (1997) *Studies in Surface Science and Catalysis*, **108**, 429.
220 Corma, A. and Serna, P. (2006) *Science*, **313**, 332.

221 Hill, C.L. and Weinstock, I.A. (1997) *Nature*, **388**, 332.
222 Raja, R., Sankar, G. and Thomas, J.M. (1999) *Journal of the American Chemical Society*, **121**, 11926.
223 Chavan, S.A., Srinivas, D. and Ratnasamy, P. (2002) *Journal of Catalysis*, **212**, 39.
224 Nowotny, M., Pedersen, L.N., Hanefeld, U. and Maschmeyer, T. (2002) *Chemistry – A European Journal*, **8**, 3724.
225 Zhao, R., Ji, D., Lu, G., Qian, G., Yan, L., Wang, X.C. and Suo, J. (2004) *Chemical Communications*, 904.
226 Xu, Y.J., Landon, P., Enache, D., Carley, A.F., Roberts, M.W. and Hutchings, G.J. (2005) *Catalysis Letters*, **101**, 175.
227 Prati, L. and Martra, G. (1999) *Gold Bulletin (London)*, **32**, 96.
228 Prati, L., Porta, F., Biella, S. and Rossi, M. (2003) *Catalysis Letters*, **90**, 23.
229 Prati, L. and Porta, F. (2005) *Applied Catalysis A: General*, **291**, 199.
230 Biella, S. and Rossi, M. (2003) *Chemical Communications*, 378.
231 Beltrame, P., Comotti, M., Della Pina, C. and Rossi, M. (2006) *Applied Catalysis A: General*, **297**, 1.
232 Carrettin, S., McMorn, P., Johnston, P., Griffin, K., Kiely, C.J., Attard, G.A. and Hutchings, G.J. (2004) *Topics in Catalysis*, **27**, 131.
233 Basheer, C., Swaminathan, S., Lee, H.K. and Valiyaveettil, S. (2005) *Chemical Communications*, 409.
234 Kanaoka, S., Yagi, N., Fukuyama, Y., Aoshima, S., Tsunoyama, H., Tsukuda, T. and Sakurai, H. (2007) *Journal of the American Society*, **129**, 12060.
235 G Boxhoorn, Shell Internationale Research Maatschappij B. V., Netherlands EP 255975 1988, 8.
236 Vaughan, O.P.H., Kyriakou, G., Macleod, N., Tikhov, M. and Lambert, R.M. (2005) *Journal of Catalysis*, **236**, 401.
237 Haruta, M. (2005) *Nature*, **437**, 1098.
238 Biella, S., Castiglioni, G.L., Fumagalli, C., Prati, L. and Rossi, M. (2002) *Catalysis Today*, **72**, 43.
239 Comotti, M., Della Pina, C., Matarrese, R., Rossi, M. and Siani, A. (2005) *Applied Catalysis A: General*, **291**, 204.
240 Davis, R.J. (2003) *Science*, **301**, 926.
241 Comotti, M., Della Pina, C., Matarrese, R. and Rossi, M. (2004) *Angewandte Chemie-International Edition*, **43**, 5812.
242 Nijhuis, T.A., Huizinga, B.J., Makkee, M. and Moulijn, J.A. (1999) *Industrial and Engineering Chemistry Research*, **38**, 884.
243 Hughes, M.D., Xu, Y.J., Jenkins, P., McMorn, P., Landon, P., Enache, D.I., Carley, A.F., Attard, G.A., Hutchings, G.J., King, F., Stitt, E.H., Johnston, P., Griffin, K. and Kiely, C.J. 2005 *Nature*, **437**, 1132.
244 Dimitratos, N., Porta, F., Prati, L. and Villa, A. (2005) *Catalysis Letters*, **99**, 181.
245 Chowdhury, B., Bravo-Suarez, J.J., Date, M., Tsubota, S. and Haruta, M. (2006) *Angewandte Chemie-International Edition*, **45**, 412.
246 Nijhuis, T.A., Visser, T. and Weckhuysen, B.M. (2005) *The Journal of Physical Chemistry B*, **109**, 19309.
247 Pinnaduwage, D.S., Zhou, L., Gao, W. and Friend, C.M. (2007) *Journal of the American Chemical Society*, **129**, 1872.
248 Hess, H.T., Kroschwitz, I. and Howe-Grant, M. (1995) *Kirk-Othmer Encyclopedia of Chemical Engineering*, Wiley, New York.
249 Henkel, H. and Weber, W. (1914). *Chemical Abstracts*, **8**, 2463.
250 Cook, G.A. (1945) Carbide & Carbon Chemical Corp, US 2368640.
251 Zhou, B. and Lee, L.K. (2001) Hydrocarbon Technologies, Inc. USA, US 6168775, 11.
252 Landon, P., Collier, P.J., Carley, A.F., Chadwick, D., Papworth, A.J., Burrows, A., Kiely, C.J. and Hutchings, G.J. (2003) *Physical Chemistry Chemical Physics*, **5**, 1917.
253 Okumura, M., Kitagawa, Y., Yamagcuhi, K., Akita, T., Tsubota, S. and Haruta, M. (2003) *Chemistry Letters*, **32**, 822.
254 Ishihara, T., Ohura, Y., Yoshida, S., Hata, Y., Nishiguchi, H. and Takita, Y. (2005) *Applied Catalysis A: General*, **291**, 215.

255 Edwards, J.K., Solsona, B., Landon, P., Carley, A.F., Herzing, A., Kiely, C.J. and Hutchings, G.J. (2005) *Journal of Catalysis*, **236**, 69.

256 Edwards, J.K., Solsona, B., Landon, P., Carley, A.F., Herzing, A., Watanabe, M., Kiely, C.J. and Hutchings, G.J. (2005) *Journal of Materials Chemistry*, **15**, 4595.

257 Haruta, M., Yamada, N., Kobayashi, T. and Iijima, S. (1989) *Journal of Catalysis*, **115**, 301.

258 Lahr, D.L. and Ceyer, S.T. (2006) *Journal of the American Chemical Society*, **128**, 1800.

259 Rhodes, C., Hutchings, G.J. and Ward, A.M. (1995) *Catalysis Today*, **23**, 43.

260 Andreeva, D., Idakiev, V., Tabakova, T. and Andreev, A. (1996) *Journal of Catalysis*, **158**, 354.

261 Andreeva, D. (2002) *Gold Bulletin*, **35**, 82.

262 Venugopal, A., Aluha, J., Mogano, D. and Scurrell, M.S. (2003) *Applied Catalysis A: General*, **245**, 149.

263 Hua, J., Zheng, Q., Zheng, Y., Wei, K. and Lin, X. (2005) *Catalysis Letters*, **102**, 99.

264 Daniells, S.T., Makkee, M. and Moulijn, J.A. (2005) *Catalysis Letters*, **100**, 39.

265 Sakurai, H., Ueda, A., Kobayashi, T. and Haruta, M. (1997) *Chemical Communications*, 271.

266 Idakiev, V., Tabakova, T., Yuan, Z.Y. and Su, B.L. (2004) *Applied Catalysis A: General*, **270**, 135.

267 Andreeva, D., Ivanov, I., Ilieva, L. and Abrashev, M.V. (2006) *Applied Catalysis A: General*, **203**, 127.

268 Fu, Q., Saltsburg, H. and Flytzani-Stephanopoulos, M. (2003) *Science*, **301**, 935.

269 Tibiletti, D., Amiero-Fonseca, A., Burch, R., Chen, Y., Fischer, J.M., Goguet, A., Hardacre, C., Hu, P. and Thompsett, D. (2005) *The Journal of Physical Chemistry B*, **109**, 22553.

9
Concluding Remarks

M. Concepción Gimeno and Antonio Laguna

The chemistry of gold was once termed by Schmidbaur as the "Sleeping Beauty" [1]. That was true for many decades when the chemistry of many transition metals was developing rapidly. The new age of gold chemistry started with the discovery, using the more readily-available X-ray technique, of gold–gold interactions in simple or more complicated gold complexes. The development of gold chemistry since then has been spectacular. Interesting complexes from the chemical, structural, theoretical or practical standpoints have been prepared. It would have been very difficult previously to predict the existence of many of these species; some of them are against all the classical rules of bonding and structure and it is the Au–Au interactions which give them unexpected stability [2]. So, the compounds involved in the chemistry of gold may still hold many surprises, waiting for a synthetic chemist to discover them, and this promises a bright future for gold chemistry, where undeveloped areas can emerge simply through efforts to find new preparation methods, a new way to stabilize unstable compounds or new structural complexes. A clear example of this fact is the enormous and growing body of work that has been performed in recent years on gold complexes in catalysis or in metal-assisted organic synthesis, particularly striking bearing in mind that only a few years ago gold was considered an almost inactive metal in catalytic reactions.

In the development of gold chemistry, new species with interesting properties have emerged or the potential applications of new or known compounds have been discovered. In recent years, a great deal of attention has focused on the potential uses of gold compounds, but "bulk" gold has been used in the industry for multiple applications such as: jewellery, the primary use of gold; currency; electronics, because it is a highly efficient conductor and is not affected by corrosion; the transmission of digital information in computers; in aerospace in gold-coated polyester films to reflect infrared radiation and help maintain temperature; and in medical devices including wires for pacemakers, gold-plated stents used to inflate and support arteries in the treatment of heart disease and implants that are a high risk of infection, such as in the inner ear, and so on.

Modern Supramolecular Gold Chemistry: Gold-Metal Interactions and Applications.
Edited by Antonio Laguna
Copyright © 2008 WILEY-VCH Verlag GmbH & Co. KGaA, Weinheim
ISBN: 978-3-527-32029-5

The main type of compounds and areas in which the chemistry of gold has practical or potential applications have been discussed. The first is the field of gold clusters or nanoparticles. Important research has been carried out in this field in recent decades by chemists and physicists, as exemplified by the many articles published in the bibliography dealing with the synthesis, properties and uses of gold clusters and nanoparticles [3]. Gold presents unique properties at the nanoscale level, such as its optical properties. Gold nanoparticles have a color varying from red to purple, depending on particle size, a property that can be successfully exploited in a range of applications. Gold colloids with the ruby color "Purple of Cassius" have been known for centuries and used for decorating ceramics. These gold inks were essentially metallic gold dispersed in gold-base resinous materials. The discovery of thiol-stabilized gold nanoparticles, which dissolve in organic solvents and could therefore be used for printing, has helped to improve the production of these colored ceramics by eliminating specific quality control issues [4]. Other optical properties of gold nanoparticles are intense surface plasmon resonance, strong optical absorption due to the collective electronic response of metal to light, which made them attractive candidates as colorimetric sensors for the detection of cations such as K^+ or biological molecules such as oligonucleotides, polynucleotides with single base imperfections, antigens, and so on. These interesting optical properties and their biocompatibility mean that gold nanoparticles can also be used for rapid testing or biomedical assays, for example, disposable membrane-based assays that, through visual evidence, confirm or refute the presence of an analyte in a liquid sample. These include clinical applications (fertility, allergies, forensic, toxicology, and so on), agricultural (food safety, plant disease, and so on) and environmental contamination [5]. Gold nanoshells are optically-tunable nanoparticles composed of a dielectric core, for example, silica, coated with an ultra-thin metallic layer [6]. The optical response of the gold nanoshells depends dramatically on the relative sizes of the nanoparticle core and the thickness of the gold shell. By modifying these two variables, the color of gold nanoshells can be varied across a broad range of energies that span the visible and near-infrared spectral regions. Human blood and tissue minimally absorb certain near-infrared wavelengths of light, enabling an external laser to be used to deliver light to nanoshells either in a tumor (for thermal destruction or imaging), a wound (for wound closure or tissue repair) or whole blood (diagnosis). The commercialization of this type of technology, based on gold nanoshells, is being pursued by companies such as Nanospectra Biosciences Inc., Houston, USA.

In the chemistry of gold in the oxidation state +1 there has been a huge increase in the research into supramolecular materials obtained through the formation of aurophilic or metallophilic bonding, and also through the presence of other secondary bonds such as hydrogen bonding or gold–donor atom interactions. These complexes present fascinating structural motifs and although, in principle, no practical applications for these complexes have been discovered to date, in the near future some of the properties shown by these materials, such as the interesting luminescence properties arising in many cases in aurophilic bonding, may yield practical uses as sensors for volatile organic compounds, light emitting diodes, biological tracers, and so on. These luminescence properties can be easily modulated by several factors such as aurophilic or metallophilic interactions, the nature of the ligand, the coordination geometry around the metal centers and even the types of metals present in the supramolecular entities or the difference in the metal–metal distances. These all have an enormous effect on the energy and origin of the emissions. Other types of supramolecular complexes are metallomesogens, which are promising compounds in materials science due to their optical, mechanical and electronic properties. Gold is a good metal for improving these properties because its tendency to produce aurophilic bonding and the special optical properties of its compounds can lead to luminescent liquid crystals with possible applications in light emitting diodes, information storage and sensors.

The importance of gold metal and gold nanoparticles in medicine has been described but many simple gold complexes display biological activity. The uses of gold in medicine date back many of thousand of years to ancient cultures (China, 2500 BC). The modern use of gold thiolates in the treatment of rheumatoid arthritis is well known, but many gold(I) and gold(III) complexes have been reported to be active against HIV, bacteria, malaria, Chaga disease or cancer cells. Several thiolate, phosphine or carbene gold(I) complexes of different nature have been reported to have antitumoral activity and also gold(III) derivatives with chelated nitrogen or orthometallated ligands [7]. A new class of enantiomerically-pure phosphine-supported gold complexes have been recently patented with promising activity. They work on a dual-action mechanism; the phosphorus zooms in on the sulfur-content of cancer cells while the gold kills off the cells. Two drugs in particular have shown great promise as cures for lymphoma, leukemia and liver cancer and clinical trials are likely to begin in the near future. Furthermore, bronchial asthma has been treated with gold thioglucose (oil-soluble) or gold sodium thiomalate (water-soluble) with good results.

Gold has been considered to be too inert and noble to provide active surfaces for catalyzing chemical reactions. For that reason, its chemical reactivity has not yet been researched in as much depth as that of the platinum group metals and its remarkable catalytic properties have been ignored until recently. However, Haruta *et al.* have shown that if the metal is sufficiently dispersed (less than 5 nm particles) on an appropriate oxide support material, it has potent catalytic properties, easily converting carbon monoxide into carbon dioxide, even at temperatures as low as $-70\,^\circ$C [8]. Therefore, gold nanoparticles supported on carbon have potential applications in the field of fuel cells. Gold catalysts have already been used commercially in Japan for removing odors from toilets. The gold on iron oxide catalyst is used to oxidize nitrogen-containing odor compounds. Haruta *et al.* showed the oxidation of propene to propene oxide, which is essential in the chemical industry, can be carried out with gold supported on titania [9].

Homogeneous catalysis by gold has experienced amazing growth in recent years and many papers dealing with this subject have been published in the most prestigious chemical journals. Gold has a unique ability to activate double and triple carbon–carbon bonds as a soft nucleophile. Gold compounds in their oxidation states I and III are highly efficient catalysts for the formation of C–C, C–O, C–N and C–S bonds, often in cyclization reactions. Gold also activates C–H bonds in aromatic compounds or other substrates and this fact has enormous potential in the search for the functionalization of different substrates [10]. This began with the pioneering work of Ito *et al.* who made an outstanding discovery, the catalytic asymmetric aldol reaction [11]. In this case enantioselectivity was given by a chiral ferrocene diphosphine ligand, with a carbon nucleophile addition to a carbonyl group. Since then many reactions have been catalyzed by gold compounds, including different processes such as hydrogenation–oxidation or many tandem reactions.

In conclusion, it seems that the greatest prospects in the field of gold chemistry for practical or industrial applications lie in the fields of nanoparticles, catalysis or optoelectronic devices. However, this is very difficult to predict because new compounds and new areas of research always appear in surprising ways in gold chemistry. Thirty years ago no one could have imagined the tremendous development of gold chemistry that has taken place in both basic and applied research.

References

1 Schmidbaur, H. (ed.) (1999) *Gold, Progress in Chemistry, Biochemistry and Technology*, John Wiley & Sons.
2 Schmidbaur, H. (1995) *Chemical Society Reviews*, **24**, 391.
3 Daniel, M.C. and Astruc, D. (2004) *Chemical Reviews*, **104**, 293.
4 Bishop, P.T. (2002) *Gold Bulletin*, **35**, 89.
5 Wilson, M., Kannangara, K., Smith, G., Simmons, M. and Raguse, B. (2002) *Nanotechnology – Basic Science and Emerging Technologies*, Chapman Hall.
6 West, J.L., Halas, N.J. and Hirsch, L.R. (2002) Optically Active Nanoparticles for Use in Therapeutic and Diagnostic Methods, United States Patent Application, US 2002/0103517.
7 Tiekink, E.R.T. (2003) *Gold Bulletin*, **36**, 117.
8 Haruta, M., Kobayashi, T., Sano, H. and Yamda, N. (1987) *Chemistry Letters*, 405.
9 Hashmi, A.S.K. and Hutchings, G.J. (2006) *Angewandte Chemie (International Edition in English)*, **45**, 7896.
10 Hashmi, A.S.K. (2007) *Chemical Reviews*, **107**, 3180.
11 Ito, Y., Sawamura, M. and Hayashi, T. (1986) *Journal of the American Chemical Society*, **108**, 6405.

Index

a

ab initio calculations   72, 329, 389
acetonitrile complex   338
acetophenone ligands   205
acetylacetonate gold(I) metal   81
– derivatives   81
– ligands   205
acid-base reactions   142
acidic alkynyl proton   336
addition-fragmentation chain-transfer polymerization   150
AFM microscopy   143
Ag-Ag interactions   270
alkali metal aurides   5, 183
– CsAu   5
– RbAu   5
alkali metal salts   217
alkenes   439, 441, 462, 474
– hydroalkoxylation   439
– hydroamination   441
– hydrogenation   475
– stereochemistry   462
alkoxy cyclization   471
alkyl gold(III) complexes   41
alkynes   447, 458, 461, 463
– hydration   447
– hydroalkoxylation   447
– hydroamination   458
– hydroarylation   463
– hydrocarboxylation   461
alkynyl gold complexes   33
– derivatives   44
alkynyl ketones   466
– carbocyclization   466
alkynyl ligands   30, 409
alkynyl phosphano gold(I) complexes   410
alkynyl phosphine derivatives   450
amine ligands   147

amine-stabilized gold nanoparticles   147
– synthesis   147
amino hydrogen atoms   312
amino indolizines   460
– gold(III) catalyzed multicomponent synthesis   460
aminopropyltriethoxylsilane (APS) monolayer   167
amphiphilic molecules   404
analogous hexaaurated complexes   78
ancillary ligands   390
anionic bidentate pyrazolate   423
– synthesis   423
anionic heterometallic chains   211
anionic ligands   22, 92
– halogens/pseudohalogens   22
anions   300
– mutual arrangements   300
antiarthritic drug   13, 87
– Auranofin   13, 87
– gold thiomalate (myocrysine)   13
– thiolate-goldphosphine derivatives   87
aprotic solvents   365
– acetonitrile   365
– chloroform   365
– dichloromethane   365
arborescent monodisperse nanometer sized molecules   157
– dendrimers   157
aromatic ring isocyanides   419
aromatic-substituted imidazolate   318
arsenic   80
– donor ligands   89
– tetrahedral geometry   80
arsine, *see* phosphine ligand
aryl gold(I) complexes   31
aryl gold(III) complexes   42
aryl ligands   41–42

Modern Supramolecular Gold Chemistry: Gold-Metal Interactions and Applications.
Edited by Antonio Laguna
Copyright © 2008 WILEY-VCH Verlag GmbH & Co. KGaA, Weinheim
ISBN: 978-3-527-32029-5

ascorbic acid 161
Au-Ag-Au system 268
Au-Ag distance 377
Au-Ag interactions 33, 260, 264, 266, 269, 378, 381
– heterometallophilic 260
Au-Ag supraclusters 7
Au-Au-Au angles 356
Au-Au bond lengths 139
Au-Au interactions 15–16, 18, 25, 107, 223, 356, 363, 381, 491
– interatomic distance 229
Au-Au single bonds 4
Au-centered icosahedral species 263
Au-centered pentagonal antiprisms 272
Au-Cu interactions 260
– heterometallophilic 260
Au-Hg bonds/interactions 186
Au-metal bonds 251
– interactions 251
Au-Si bonds/interactions 219
Au-Tl interactions 196, 212
Au-transition metal bonds/interactions 235
Au nanorods 161–163
– mechanism 162–163
Au NP 160
– DNA hybrid method 172
– dodecanethiol-stabilized 160
– peptide chain hybrids 164
Au NP-biomolecule interactions, *see* specific affinity interactions
Au(I)-Au(I) interactions 48
Au(III) ions 327
Au/Al$_2$O$_3$ catalysts 479
Au55 clusters 142
– 2D cubic lattices 142
– 2D hexagonal 142
AuCl catalyzed reactions 461
AuPTA fragment 18
aurophilic attraction 4, 113
aurophilic bonding 11, 296, 492
aurophilic interactions 10, 25, 28, 33, 65–66, 70, 78, 80, 85, 91–92, 94, 97, 109, 117, 313, 331, 397, 493
Au-Xe bonding energy 37
auxiliary ligands 28, 34
azobenzene ligands 410

b

benzimidazolate 112
– derivatives 112
– luminescent properties 112
bi-cyclic ring systems 466
bidentate diphosphines 96
– bis(diphenylphosphino) methane (dppm) 96
bidentate ligands 16, 23, 47, 90, 95, 101, 263, 372
– bis(ylide) 16
– dithiocarbamate(s) 16, 47
– dithiolate 16
– dithiophosphates 16, 47
– phosphoniodithioformate 16
– pyridine-2-thiolate 16
– xantate 16
bifunctional linkers 164
– disulfides 164
– phosphines 164
– thiols 164
biocatalytic electrodes 172
biological molecules 492
– oligonucleotides 492
– polynucleotides 492
biomolecule-nanoparticle interaction 164
biotin-streptavidin interaction 165
biotinylated proteins 164
– immunoglobulins 164
– oligonucleotides 164
– serum albumins 164
bipyridine ligands 216
bluish-white luminescence 370
bridging diphosphine ligands 91
bridging ligands 27, 38, 46, 263
bromine atoms 321
bronchial asthma 493
– gold sodium thiomalate 493
– gold thioglucose 493
Brust–Schiffrin method 145–146, 148, 151
bulky monophosphines 81
bulky phosphines 26
butterfly metallic cores 245, 253
tert-butyl esters 437

c

carbene ligands 30
calamitic liquid crystals 408, 411, 425
– pyrazole ligands 425
calamitic mesophases 404
– nematic phase 405
– smectic phases 405
Cambridge Structure Database 182, 186
capping ligands 148
capping-reducing agent 154
carbene compound [Au{C(NHMe)NEt$_2$}(SCN)] 71
carbene gold complexes 32
carbeniate ligands 24
– [Au$_3$(MeN=COMe)$_3$] 24

carbon-carbon bonds   493
carbon-halogen bond dissociation
   energies   39
carbon monoxide oxidation   481
carbonyl gold(I) complexes   30
carbonyl groups   472
carbonyldicyclopentadienyl-niobium
   fragments   237
carborane-dithiolate ligand   106
carboxylic acid   327, 461
– acid residues   306
– dimer   333
– end groups   328
catalysis   429
– principle   429
– scientific method   429
catalytic system   461
cathode ray tube (CRT) monitors   408
cation-anion interactions   305, 386
cation-anion pairs   300
cationic bis(isocyanide) compounds   419
cationic gold(I) complexes   435
cations   300
– mutual arrangements   300
cetyltrimethylammonium bromide
   (CTAB)   161
Chaga disease   493
chalcogen atoms   83
chalcogen-centered gold(I) complexes   27
– chemistry   28
chalcogen ligands   46
– donor ligands   76
– gold(III) complexes   46
chalcogenolate gold(I) complexes   29
charge transfer interactions   372
chelating amine-imine   209
chelating diphosphines   22
chelating ligands   41
chiral ferrocene diphosphine ligand   494
chiral isocyanide   420
Chitosan-stabilized Au NPs   155
chloride ligands   451
chlorogold(I) aliphatic amines   330
cholesteryl benzoate   403
citrate-capped Au NPs   157
citrate-stabilized gold nanoparticles   151, 161
citrate-stabilized seed particles   162
closed-shell interactions   65
closed-shell metal atoms   4
condensation reactions   472
Conia-ene reaction   462
core-shell particles   152
Coulomb repulsion   226
Coulomb rules   390

crystal phosphorescence excitation
   spectrum   374
Cu(II)-pyrazine units   77
cyclic hexamers   23
cyclic trinuclear gold(I) complexes   367
cyclometallated gold(III) complexes   42
cyclometallated ligands   44
cyclopropyl carbenoid intermediate   470

d

d-orbital energies   359
DDT-protected gold nanoparticles   144
decomposition reactions   429
dendrimer-Au NPs composites   160
dendrimer encapsulated nanoparticles
   (DENs)   157
dendrimer-nanoparticle composites
   (DNCs)   157
dendrimer sequesters gold ions   157
dendrimer stabilized nanoparticles
   (DSNs)   157, 158
dendrimer terminal nitrogen ligands/gold
   atoms   159
dendrimers   157
dentritic polymers   17
– phosphorus atoms   17
DFT calculations   10, 36, 91, 374, 381
di-substituted pyrazole ligands   425
dialquinyl groups   95
diamagnetic species   38
diastereoselective reaction   469
dicoordinated gold(I) mononuclear
   species   68
dimercaptosuccinic acid (DMSA)   146
dimeric xenon cation   49
dimerization process   81
dinuclear aggregates   68
dinuclear derivatives   263
dinuclear diphosphine gold(I) thiolate   30
dinuclear gold(I) complexes   16, 422
dinuclear gold(I) mesogens   418
– biphenylisocyanide   418
– phenylisocyanide   418
dinuclear gold(II) complexes   38
dinuclear gold(II) derivatives   39
dinuclear homoleptic gold(II) complexes   38
dinuclear species   102
dinuclear units   359
diphenylphosphinite acid ligands   70
diphenylphosphinous acid ligands   70
diphosphine ligands   19, 22, 34, 83, 240
– 2-thioxo-1,3-dithiole-4,5-dithiolate   19
dipole–dipole interactions   403
discotic mesogens   424

– schematic representation 424
discotic mesophases 406
dispersive interactions 230
distorted trigonal bipyramidal geometry 245
distorted trigonal prismatic geometry 240
dithiobenzoate ligands 413
dithiolato ligands 93, 262
dithiolene or dithiolate systems 49
dithiophosphate complexes 362
– crystal structures 362
dithiophosphinate complexes 362
– crystal structures 362
DMSO-bound hydrogen atoms 339
DMSO molecules 209–210
DMSO oxygen atom 338
DMSO solvate 364
DNA–DNA hybridization 164
DNA-functionalized Au NPs 164
DNA-nanoparticle conjugates 164
DNA double-helix 46, 172
– antiproliferative effects 46
– denaturation-rehybridization 172
DNA hybridization 172
donor-acceptor complex 42
donor ligands 23, 41, 91, 183, 305, 375, 378
– arsenic 375
– carbon 23, 41
– chalcogen 183
– phosphorus 24, 91, 375
– sulfur 305
– tetrahydrothiophene 39
– trinuclear 378
double-bond sequence 457
doubly-bridging chalcogenolate complexes 30

e

Egyptian tombs 1
eight-membered ring annulation 463
Einstein's theories of relativity 3
electro-optic display systems 405
– flat-panel displays 405
electron-deficient phosphines 461
– $P(C_6F_5)_3$ 461
electron-rich arenes 444, 464
– addition of 444
electronegative atoms 326
– halogens 326
– nitrogen 326
– oxygen 326
– sulfur 326
electronegative element 295
– halogens 295
– nitrogen 295

– oxygen 295
– sulfur 295
electronegative ligands 49
– fluorine 49
electrostatic interactions 163
emission bands 353
emission spectra 355
endo-dig cyclization transition states 470
enolate alkylation 462
environmentally-acceptable compounds, see waste management
enyne metathesis 465
EPR simulations 36
ESI mass spectrometry 137
ESI-MS spectra 145
ethanol solvates 336
ethylene glycol dimethacrylate (EGDMA) 155
EXAFS measurements 140
exo-dig cyclization transition states 470
exo-methylene lactones 461

f

Fenske–Hall molecular orbital calculations 386
ferrocenyl derivatives 316
fluorine atom 297
fluoroaryl ligands 419
four-coordinate gold(I) complexes 21, 23
four-coordinate homoleptic species 23
Frechet-type dendrons 160
free electron models 6
frozen solutions 353
– excitation spectra 353
furane derivatives 444

g

gas phase hydrogenation 476
Ge-atom bridges 226
gel electrophoresis 164
gluconic acids 478
α-glucosides 471
– diastereoselective synthesis 471
gold 1, 408
– advantages 408
– auxiliary ligands 33
– based metallocryptate 184
– chemical properties 2
– chemistry 4
– edge-sharing bitetrahedral arrangement 133
– electronic configuration 2
– history 1
– nanotechnology 131
– physical properties 2

gold(I)-activated allenes 438
gold(I)-catalyzed transformations 467
gold(I)-gold(I) interactions 66
gold(I)-gold(III) distances 113, 116
gold(I) mesogen 420
gold(I)-sulfur compounds 363
gold(II) derivatives 39
gold(III) catalysts 435
gold(III) complexes 41, 44–45, 47
gold(III) cyclometallated complexes 42
gold(III) porphyrins 45
gold-antimony compounds 232–234
gold atom(s) 136, 247, 252
– icosahedron 136
– T-shaped 252
gold-bismuth bonds/interactions 231
gold-bismuth compounds 235
gold-carbene complex 411
gold catalysis 430–431, 460
gold-catalyzed reactions 432, 475
– carbon-sulfur bond 435
– cycloisomerization 454
– field 481
gold chloride isocyanide complexes 11
gold-chlorine interactions 225
gold-chromium compounds 238
gold-coated polyester films 491
gold-containing complexes 347, 351
gold-containing materials 347
gold-containing pentanuclear clusters 253
– heterometallic framework 253
gold-copper supramolecular systems 383
gold clusters 139
gold-donor atom interactions 492
gold fullerenes 10
gold-gallium compounds 192–196
gold-germanium bonds 222
gold-germanium clusters 226
gold-germanium compound 222–227, 297
gold-gold bonds 35, 131, 295
gold-gold contacts 10, 18, 74, 216, 371, 380
gold-gold distance 69–70, 100, 115
gold-gold interactions 26, 44, 66, 69–70, 78, 83, 97, 101, 110, 117, 189, 295, 350–354, 357, 359–366, 369, 371, 383, 387, 415, 491
– importance 353
gold-halogen bonds 222
gold-heterometal bonds 181, 295
gold-heterometal complexes 385
gold-heterometal compounds 272
gold-heterometal contacts 66
gold-heterometal interactions 181, 295, 351, 376, 384
gold-indium compounds 194–196

gold inks 492
gold-lead compounds 231
gold-manganese compounds 241
gold-mercury bonds 189
gold-mercury compounds 186–192
gold mesogens 408–409, 422
gold metal 493
– importance 493
gold-molybdenum compounds 239
gold nanoclusters 8–10
gold nanoparticle(s) 9, 131, 144, 146–147, 152, 154, 169, 492–493
– applications 169–172
– cored dendrimers 160
– importance 493
– properties 169–172
– synthesis 144, 147
gold nanorods 166
– self-assemblies 166
gold nanotubes 10
gold-phosphine cluster 137
– thiol-modified 137
gold-rhenium compounds 241
gold-silicon compounds 221
gold-silver analogous systems 386
gold-silver compounds 263
gold-silver entities 377
gold-silver interactions 382, 383
gold-stibine complexes 234
gold sulfur ligands 239
– affinity 239
gold-technetium compounds 243
gold-thallium complexes 397
gold-thallium compounds 196–217
gold-thallium interactions 392, 396
gold-tin compounds 227–231
gold-titanium compounds 237
gold-tungsten compounds 239
gold vanadium compounds 238
gulonic acids 478

h
halide ligands 13
haloalkanes 38
halogen(s) 22, 38, 92
– oxidative addition reaction 38
heteroatom bridged gold(I) complexes 26
heteroatom-centered complexes 26
N-heterocyclic carbene coordination
    species 264
N-heterocyclic gold carbenes 43, 451, 467
– ligands 33
heterodimetallic/trimetallic cores 245
heterogeneous catalysis 431–432, 475

heteroleptic gold clusters 135–136
– structure 135
heterometal atoms 85
heterometal entities 376
– luminescent supramolecular gold 376
heterometallic acetylacetonate
  complexes 339
heterometallic Au-Ag compounds 260
heterometallic Au-Cu compounds 260
heterometallic chains 397
heterometallic complexes 110
– gold 110
– iridium 110
– ruthenium 110
heterometallic cores 246
heterometallic gold-metal bonds 181, 268
heteronuclear arrays 77
heteronuclear derivatives 77
heteronuclear materials 377
– synthesis 377
heteropolymetallic compounds 85
heteropolynuclear system 396
hexafluorobenzene derivative 319
hexafluorophosphate ion 297
hexagonal columnar mesophases 415
hexanuclear prismatic units 372
hexanuclear systems 246
high-angle annular dark-field scanning
  transmission electron microscopy (HAADF-
  STEM) 137
high-resolution transmission electron
  microscopy 150
higher nuclearity systems networks 375
higher oxidation states 49
HIV-associated glycoprotein 165
homo-bridged ligand 38
homo/hetero-nuclear compounds 6
HOMO dithiocarbamate orbitals 360
HOMO-LUMO gap 348, 350
HOMO molecular orbitals 381
homobridged diauracycles 16
homobridged gold(II) complexes 38
homogeneous gold catalysis 432, 493
homoleptic dithiocarboxilates 18
homoleptic gold(III) derivatives 47
homoleptic isomeric forms 11
homoleptic species 78
homometallic dimers 260
homonuclear derivatives 379
– La[Ag(CN)$_2$]$_3$ 379
homonuclear gold cluster compounds 131
– small-size 131
HPLC analysis 141
HRTEM image 140

Hückel molecular orbital 352
– calculations 4, 352
hydrocarbons 42
– benzene 42
– toluene 42
hydrogen-bonded chain motifs 334
hydrogen-bonded chain polymer 327
hydrogen bonds 11, 33, 73, 295, 311, 320, 326
– donors 317, 327
– interactions 331
– systems 5
hydrogen peroxide 480
– direct synthesis 480
hydrogenation reactions 474–476
α-hydroxy propargyl ketones 465

i

in situ dehydrating agents 396
intercalated iodine molecules 323
intermetallic interactions 206
intermolecular aurophilic interactions 66, 88, 93, 95, 112, 331
intermolecular metal-metal interactions 14, 255
intermolecular reactions 436
intra/inter-molecular Au-Au interactions 14
intra/intermolecular nucleophilic
  addition 458
intraligand transitions (ILT) 65
intramolecular gold-gold distances 114
ionic compounds 224
iridium catalysts 475
IR spectroscopy 141, 477
isocyanide derivatives 413, 421–423
– binary mixtures 421–423
isocyanide gold(I) derivatives 414
isocyanide gold(I) mesogens 414
isocyanide ligand(s) 355–356, 413–414
– orientation of 356
isonitrile derivatives 413

l

lactones 462
– seven-membered ring 462
– six-membered ring 462
ladder-type structure 329
Lambert–Beer law 361, 380
Langmuir–Blodgett (LB) technique 143
lanthanum salts 379
Laporte's rule 348
large-scale icosahedral gold nanocrystals 153
large-scale industrial processes 429
large-size gold clusters 139
– Au$_{55}$ 139

Lewis acid
– metals 31
– salts 381
Lewis acid-base interactions 181
ligand-based emission 355
ligand exchange reaction 137, 160, 165
ligand-protected nanoparticles 147
– synthesis 147
ligand-radical species 36
ligand stabilizers 136, 157
– dendrons 157
– gold clusters 136
ligand-to-metal-metal charge transfer processes (LMMCT) 28, 70, 74, 94, 362
ligand-to-metal charge transfer transitions (LMCT) 70
– emission 364
light emitting diodes (LEDs) 408, 492
liquid crystal displays (LCDs) 407
liquid crystalline phase 403
liquid crystals 403
– applications 407
– characterization 407
– history 403, 407
– lyotropic 403
– thermotropic 403
living radical polymerization (LRP) 149
luminescence thermochromism 383
luminescent complex [Au(Tab)$_2$][Au(CN)$_2$] 73
luminescent emissions 76, 99, 104
luminescent gold mesogens 419
luminescent supramolecular gold 351
– entities 351, 385
– gold architectures 351
– heterometal entities 351
luminescent supramolecular complexes 375, 385
– heteronuclear 385
luminescent trigonal gold(I) metallocryptates 22
LUMO molecular orbitals 348, 381
lyoluminescence process 368
lyotropic liquid crystals 404

m
M-centered square-antiprism 191
Mössbauer isomer 5
Mössbauer spectroscopy 113, 140
macrocycles 6
– silicalix[n]phosphinines 6
– synthesis 6
macroreticular ion exchange resin 5
MALDI-MS spectra 145
membrane-based assays 492

mercaptosuccinic acid (MSA) 166
– water-soluble 166
mesityl gold derivatives 451
mesogen 404
– ligand 408, 423
mesomorphic isocyanide 418
metal-assisted organic synthesis 491
metal-based emission 357
metal-centered phosphorescent emission 348
– transitions 65
metal-containing material 406
metal-gold compounds 183
metal-metal bond 102, 200, 362
metal-metal distance 21
metal-metal interactions 200–201, 262, 266, 347, 352, 360, 362, 376, 381–382, 384, 387, 389, 394, 408
metal-to-ligand charge transfer 369
metal-xenon compounds 36
metallic gold precursors 394
metallomesogens 406
metallophilic bonding 492
metallophilic interactions 398, 493
methane sulfonic acid 449
methanide ligands 43
methanide-type complexes 45
methoxy-methyl groups 368
methyl methacrylate (MMA) 150
methyl propargyl ether alkynes 447
methylenethiophosphinate ligand 186
Michael-type additions 45
microcrystalline solid 364
microcrystals 140
– HRTEM micrographs 140
micrometer-scale mirrors 167
Mie theory 155
mixed phosphine-arsine ligands 17
mixed-valent compound 38
molecular gold clusters 131, 138
molecular orbitals 7
– spectra 7
monoanionic tridentate or dianionic tetradentate ligands 32
monodentate ligands [Au(L)$_4$]$^+$ 21, 86–90
– complexes 86
monodentate phosphines 20
monodimensional polymer 209
monolayer protected clusters (MPCs) 144
– water soluble 144
mononuclear derivatives 425
mononuclear dicoordinated gold units 66–67
– aggregation 67
mononuclear gold units 67

mononuclear gold(II) complexes 35
mononuclear precursor 135
multifunctional ligands 165
myocrysine 13

n

nano-bio hybrids synthesis 144
nanoparticle assembly 165–169
nanoparticle-biomolecule hybrid 163–165
nanoparticle cored dendrimers (NCDs) 157
nematic liquid crystal mixtures 405
neutral ligands 11
– amine 11
– arsine 11
– carbine 11
– isocyanide 11
– phosphine 11
– ylide 11
neutral/ionic clusters 241
nido-carborane ligands 95
nitrate-containing polymer 379
nitrate-isocyanide species 76
– [Au(ONO$_2$)(CNC$_6$H$_{11}$)] 76
– [Au(ONO$_2$)(CNiPr)] 76
nitrogen atom 80
nitrogen-sulfur bond 461
nuclear magnetic resonance (NMR) 91
– experiments 91
– spectroscopy 141
– temperature-dependent 66
– time scale 20
non-discotic metal complexes 416
non-mesogenic biphenyl diisocyanide 422
non-mesomorphic ligands 406
nonanuclear gold clusters 134
– synthesis 134
notable turnover numbers (TONs) 431
NP-biomolecule hybrids 163
nucleophilic addition 450
nucleophilic carbenes 448

o

octadecylamine 146
octahedral geometry 46
octanuclear gold cluster structures 134
octyloxy compound 425
– crystal structure 425
– X-ray diffraction 425
oligomeric gold(I) complexes 23
oligomeric rings 24
oligomerization process 361
oligopeptide capping-agents 163
one-dimensional arrays 65, 315

one-pot method 147
optically detected magnetic resonance (ODMR) 69
optically-tunable nanoparticles 492
organogold(III) iminophosphorane complexes 445
organometallic gold(I)-alkynyl complexes 147, 261
organometallic gold(I) complexes 30, 41
organometallic ligands 30
organoruthenium moiety 249
ortho-tolyl groups 222
oxidants, see halogens
oxidation reactions 478
– C-H bond activation 478

p

palladium-catalyzed intramolecular amines 458
palladium nanoparticles 479
PAMAM dendrimer 159–160
– gold colloid nanocomposites 158
PDMS films 155
pentachloro phenyl derivative 216, 396, 451
pentafluoro phenyl complexes 382, 396
pentanuclear cycle 89
perhaloalkyl group 31
perhaloaryl ligands 417, 422–423
peripheral gold atoms 137
peripheral metal atoms 134
peripheral tolyl groups 227
π-π stacking interactions 304
phase transfer reagent tetraoctylammonium bromide (TOAB) 148
phenylene-dithiolate ligands 18
phenol synthesis method 470–471
phenyl/propargyl ethers 463
phenyl-transfer reaction 21
phenylthiolate/pyridilthiolate species 93
phosphine-based systems 136
phosphine gold(I) complexes 443
phosphine gold(I) organosulfonates 71
phosphine ligands 81, 132, 135, 247
phosphine/polyphosphine ligands 10
phosphine-stabilized gold clusters 131–132, 139
– structural characterization 132
– synthesis 132
phosphine-stabilized hexanuclear gold clusters 133
phosphine thiolate complexes 69
phosphino derivative 318
phosphino-thiolate ligands 48

phosphinocarborane ligands 132, 139
phosphonium/sulfonium ylides 31
phosphonium salts 31
phosphorescent processes 381
phosphorus atoms
– based ligands 231, 348
planar-three-coordinate 348
plasma-assisted physical vapor deposition (PAPVD) 143
platinum catalysts 475
platinum nanoparticles 479
PNIPAM polymers 152
PNIPAM samples 152
polarized optical microscopy (POM) 407
polydentate ligands 17, 19, 66, 86, 359
– use of 359
polydentate nitrogen ligands 20, 45
– ferrocenyl-terpyridine ligand 20
polydentate phosphines 34
polyethylene glycol 154
polymer-stabilized Au NPs 149
polymeric micelles 155
polynuclear derivatives 29
– [RP(AuPPh$_3$)]BF$_4$ 29
polynuclear gold(II) complex 38, 40
polynuclear species 240
polypyridines 44
– phenanthroline 44
– pyrazolate ligands 44
– terpyridine 44
porphyrin diads and triads 45
post-lanthanide elements 3
– protons 3
Prins type cyclization reaction 455
promesogen ligand 408, 423
propargyl vinyl ethers 469
pseudo-pentacoordinate gold(III) derivatives 44
pseudo-trigonal bipyramidal geometry 208
pyrazolate ligands 70, 423
pyrazole mononuclear derivatives 425

q

quasi-relativistic pseudopotentials 83
quasi-three-dimensional conductor 326

r

redox enzymes/proteins 172
– co-deposition of 172
reversed hybrid polymeric micelles (RHPMs) 155
reversible vapochromic behavior 31
Rh–Rh interaction 253
rock-salt-like lattice structure 272

s

SCN complexes 417
SCN ligand 417
S-donor ligands 36
– dithiocarbamates 36
– dithiolates 36
– dithiolenes 36
secondary bonds 295
selenide ligands 47
selenium donor ligands 47, 48
– dithiocarbamates 47
– dithiolates 47
– selenolates 47
– thiolates 47
selenolate/telurolate ligands 88
– examples 88
self-assembled supermolecules 295
scanning electron microscopy (SEM) 143
semi-staggered array 356
shining yellow nuggets, *see* gold
silicalix[n]phosphinines 6
– synthesis 6
silver-silver interactions 72
– excited states 381
single-crystal X-ray technique 181
– diffraction 49, 131, 183, 201, 217
single-electron transfer process 48
singlet ligand (halide) 68
singly-occupied molecular orbital (SOMO) 36
sleeping beauty, *see* gold
small-angle neutron scattering (SANS) 159
small-angle X-ray scattering (SAXS) 159
– measurements 145
Sn-Au bonding interaction 229
sol–gel methods 168
sol–gel monomers 169
solid-state materials 183
solvent-stimulated luminescence 368
specific affinity interactions 164
spherical gold nanoparticles 154
square planar gold(III) complexes 44
– geometry 408
square planar pyramid geometry 79
stabilizers 131, 146, 159
– arsines 131
– boranes 131
– thiol-functionalized dendrons 159
– thiol ligands 146
stabilizing ligands 330
– phosphines 330
stereoselective tandem reaction 469
stilbazol ligands 426

Stone's tensor surface harmonic theory 6
sulfur-rich dithiolate ligands 262, 272, 323
supramolecular gold materials 296
supramolecular network 351–352
– binuclear 351
– mononuclear 351
– trinuclear 351
surface-catalyzed reactions 160
surface-initiated atom-transfer radical
  polymerization (SI-ATRP) 151
surface-nanoparticle stabilizers 157
surface plasmon bands (SPBs) 155
synthetic methods 9

t
target-oriented synthesis 437
TCNQ molecule 373
TD-DFT calculations 380, 389, 391, 393
tellurium atoms 326
transmission electron microscopy
  (TEM) 143, 151
– measurements 145
– micrographs 151
tetraazacyclotetradecane 112
– luminescent properties 112
– derivative 112
tetra-coordinate complexes 21
tetracyanoquinodimethane (TCNQ) 373
tetrahedral diphosphine complexes 23
tetranuclear gold clusters 132
thallium-gold interactions 386
thermo-responsive polymer 152
– PNIPAM 152
thermotropic liquid crystals 404, 407
– calamitic 404
– discotic 404
THF molecules 208, 216, 395
thin film transistor (TFT) 407
thioauracyl ligand 366
thioether ligands 36
thiolate-protected gold cluster 146, 172
– synthesis 146
thiolate-protected gold nanoparticles 144
thiolato benzoic acid groups 328
thiol-capped polystyrene
  macromolecules 156
thiol-functionalized gold nanoparticles 146
thiol-functionalized PAMAM
  dendrimers 158
thiol-stabilized gold nanoparticles 492
thiol-terminated PAMAM dendrimer 158
– synthesis 158
thiophene derivatives 444
three-dimensional arrays 65

three-dimensional network 300, 326
tiopronin water soluble Au clusters 145
transannular gold-mercury distance 187
transannular gold-mercury interaction 187
transannular metal-metal bond 300
trans-bromine atom 322
transformation reactions 429
transition metal catalysis 430, 466
transition-metal catalyst systems 442
transition metal-gold compounds 235–237
trialkoxyphenyl isocyanide 420
triangular trigold(I) complexes 24
triclinic polymorph 370–371
tricoordinate gold atoms 240
tri-cyclic ring systems 467
tridentate ligand 105
trifloroacetate derivatives 365
– crystal structures 365
trigonal bipyramidal geometry 26
trigonal bipyramidal/octahedral
  structures 28
trigonal-pyramidal units 81
trimetallic Hg-Au-Pt clusters 259
trinuclear derivatives 425
trinuclear heterobimetallic species 239
trinuclear pyrazolate coinage complexes
  374
trinuclear pyrazolate complex
  [Au(μ-C^2,N^3-bzim)]$_3$ 383, 397
trinuclear systems 367
– networks 367
– units 367
tri-phenylphosphine-based cluster 136
triphenylphosphine derivative 328
triphenylphosphine ligands 134
triphenylphosphine-stabilized gold
  nanoparticles 165
triphosphines complexes 111
triplet ligand-to-metal charge transfer
  (3LMCT) 360
TTF units 325
turnover frequencies (TOFs) 431
twisted nematic (TN) mode 407
two dimensional arrays 65
two-dimensional hydrogen-bonded
  framework 331
two-dimensional molecular interaction 324
two-dimensional networks 301
two- or three-step seeding process 162
two-phase liquid-liquid system 144

u
α,β-unsaturated aldehydes 478
– hydrogenation 478

α,β-unsaturated ketones   437
U-shaped polymers   20
UV-Vis spectroscopy   141, 372
– measurements   150

v

van der Waals interactions   297, 311, 341, 389, 403
van der Waals radius   182, 295, 298, 381, 384
vanadium-carbonyl fragments   237
vertex-sharing poly-icosahedral cluster series   8
vinyl chloride   446
– hydrochlorination   446
volatile organic compounds (VOCs)   202, 492
– vapors   99
VSEPR rules   78–79

w

Wacker oxidation   450
waste management   430
water gas shift reaction   481–482
water-soluble polymers   149
weak nucleophiles, see carboxylic acids
weak transannular gold-lead interactions   231
wingtip metal atoms   247

x

X-ray crystallography method   78–79
X-ray diffraction experiments   182, 366
– analysis   132
– measurements   423
– studies   417
– technique   491
X-ray photoelectron spectroscopy (XPS)   114
X-ray scattering method   140
XRD measurements   147
XRD spectra   145

y

yellow-colored metal, see gold
yellow potassium metallocryptate   185
ylide gold(III) complexes   43
ylide transfer reactions   43

z

zero-dimensional nanomaterials   141
zigzag infinite chains   213, 322
– two-dimensional array   322